INDEX OF APPLICATIONS

Mathematical Excursions

Second Edition

SPECIAL EDITION FOR RADFORD UNIVERSITY

Richard N. Aufmann
Palomar College, California

Joanne S. Lockwood
New Hampshire Community Technical College

Richard D. Nation
Palomar College, California

Daniel K. Clegg
Palomar College, California

Houghton Mifflin Company
Boston New York

MATHEMATICAL EXCURSIONS, SECOND EDITION
By Richard N. Aufmann, Joanne S. Lockwood, Richard D. Nation, and Daniel K. Clegg

Publisher: Jack Shira
Senior Sponsoring Editor: Lynn Cox
Development Editor: Lisa Collette
Assistant Editor: Noel Kamm
Senior Project Editor: Kerry Falvey
Manufacturing Manager: Florence Cadran
Senior Marketing Manager: Ben Rivera
Marketing Assistant: Lisa Lawler

Photo credits are found immediately after the Answer section in the back of the book.

Custom Publishing Editor: Brenda Hill
Custom Publishing Production Manager: Christina Battista
Project Coordinator: Georgia Young

Cover Design: Amy Files
Cover Image: Stock.xchng

This book contains select works from existing Houghton Mifflin Company resources and was produced by Houghton Mifflin Custom Publishing for collegiate use. As such, those adopting and/or contributing to this work are responsible for editorial content, accuracy, continuity and completeness.

Printed in the United States of America.

ISBN-13: 978-0-618-80924-0
ISBN-10: 0-618-80924-4
N-06651

1 2 3 4 5 6 7 8 9 – CM – 08 07 06

Houghton Mifflin
Custom Publishing
222 Berkeley Street • Boston, MA 02116

Address all correspondence and order information to the above address.

CONTENTS

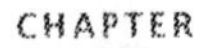

CHAPTER 4 Numeration Systems and Number Theory 177

APPLICATIONS

ASCII Code 207 **Binary Search/Sort** 204, 205 **Mathematics** 186, 194, 195, 206, 217, 218, 219, 229, 230, 238, 239, 240, 241 **Music CDs** 201 **Postnet Code** 207 **RSA Encryption Algorithm** 231

CHAPTER 6 Applications of Functions 327

APPLICATIONS

Aviation 343, 350, 364, 406 **Business** 350, 360, 361, 368, 372, 388, 404, 405, 406 **Compensation** 361, 372 **Construction** 352, 361, 370, 372, 373 **Education** 362, 387, 395 **Forestry** 338 **Geometry** 334, 337, 338, 339, 353, 404 **Health** 356, 405 **Money** 350, 351, 382, 384, 387, 388, 399, 404, 405 **Music** 387 **Populations** 387, 388 **Recreation** 343, 351, 363, 369, 388 **Science** 339, 347, 350, 352, 355, 359, 361, 362, 369, 372, 382, 383, 384, 387, 388, 396, 397, 398, 400, 401, 402, 404, 405, 406 **Sports** 338, 339, 351, 359, 372, 373, 395, 404, 405, 406 **Temperature** 346, 350, 382 **Vehicles** 349, 354, 355, 362, 373, 387 **World resources** 386, 401

CHAPTER

Geometry *449*

APPLICATIONS

PREFACE

Mathematical Excursions is about mathematics as a system of knowing or understanding our surroundings. It is similar to an English literature textbook, an Introduction to Philosophy textbook, or perhaps an Introductory Psychology textbook. Each of those books provide glimpses into the thoughts and perceptions of some of the world's greatest writers, philosophers, and psychologists. Reading and studying their thoughts enables us to better understand the world we inhabit.

In a similar way, *Mathematical Excursions* provides glimpses into the nature of mathematics and how it is used to understand our world. This understanding, in conjunction with other disciplines, contributes to a more complete portrait of our world. Our contention is that ancient Greek architecture is quite dramatic but even more so when the "Golden Ratio" is considered. That I. M. Pei's work becomes even more interesting with a knowledge of elliptical shapes. That the challenges of sending information across the Internet is better understood by examining prime numbers. That the perils of radioactive waste take on new meaning with a knowledge of exponential functions. That generally, a knowledge of mathematics strengthens the way we know, perceive, and understand our surroundings.

The central purpose of *Mathematical Excursions* is to explore those facets of mathematics that will strengthen your quantitative understandings of our environs. We hope you enjoy the journey.

New to This Edition

- Reading and interpreting graphs has been expanded to allow students more practice with this important topic.
- Chapter 8, Geometry, has been expanded to include the Pythagorean Theorem and congruent triangles.
- An introduction to right triangle trigonometry has been added to Chapter 8, Geometry.
- Chapter 10, Finance, has expanded coverage of stocks, bonds, and annuities.
- There is an Algebra Review Appendix that can be downloaded from the web as a PDF.
- All the exercise sets have been reviewed and new, contemporary problems have been added.

Interactive Method

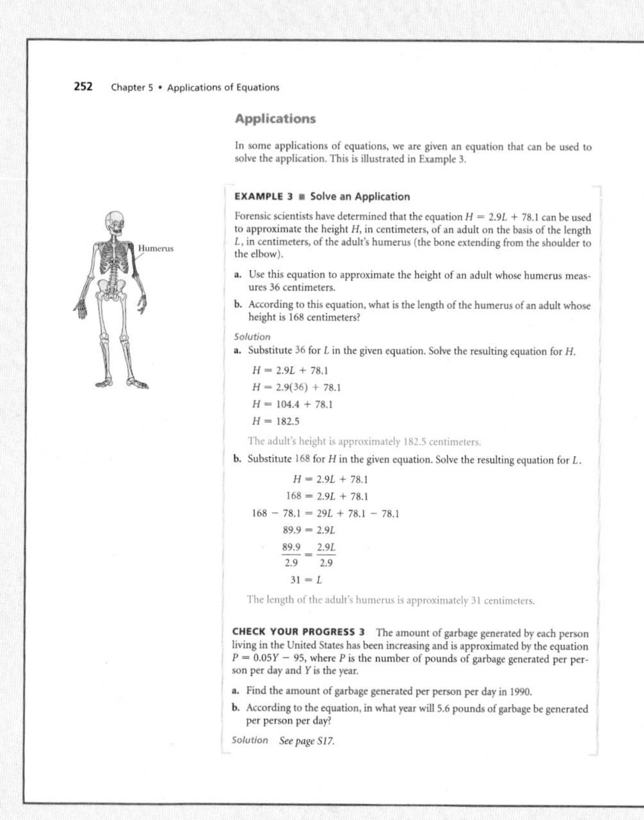

252 Chapter 5 • Applications of Equations

Applications

In some applications of equations, we are given an equation that can be used to solve the application. This is illustrated in Example 3.

EXAMPLE 3 ■ Solve an Application

Humerus

Forensic scientists have determined that the equation $H = 2.9L + 78.1$ can be used to approximate the height H, in centimeters, of an adult on the basis of the length L, in centimeters, of the adult's humerus (the bone extending from the shoulder to the elbow).

a. Use this equation to approximate the height of an adult whose humerus measures 36 centimeters.

b. According to this equation, what is the length of the humerus of an adult whose height is 168 centimeters?

Solution

a. Substitute 36 for L in the given equation. Solve the resulting equation for H.

$$H = 2.9L + 78.1$$
$$H = 2.9(36) + 78.1$$
$$H = 104.4 + 78.1$$
$$H = 182.5$$

The adult's height is approximately 182.5 centimeters.

b. Substitute 168 for H in the given equation. Solve the resulting equation for L.

$$H = 2.9L + 78.1$$
$$168 = 2.9L + 78.1$$
$$168 - 78.1 = 29L + 78.1 - 78.1$$
$$89.9 = 2.9L$$
$$\frac{89.9}{2.9} = \frac{2.9L}{2.9}$$
$$31 = L$$

The length of the adult's humerus is approximately 31 centimeters.

CHECK YOUR PROGRESS 3 The amount of garbage generated by each person living in the United States has been increasing and is approximated by the equation $P = 0.05Y - 95$, where P is the number of pounds of garbage generated per person per day and Y is the year.

a. Find the amount of garbage generated per person per day in 1990.

b. According to the equation, in what year will 5.6 pounds of garbage be generated per person per day?

Solution *See page S17.*

page 252

An Interactive Approach

Mathematical Excursions, Second Edition, is written in a style that encourages the student to interact with the textbook. Each section contains a variety of worked examples. Each example is given a title so that the student can see at a glance the type of problem that is being solved. Most examples include annotations that assist the student in moving from step to step, and the final answer is in color in order to be readily identifiable.

Check Your Progress Exercises

Following each worked example is a Check Your Progress exercise for the student to work. By solving this exercise, the student actively practices concepts as they are presented in the text. For each Check Your Progress exercise, there is a detailed solution in the Solutions appendix.

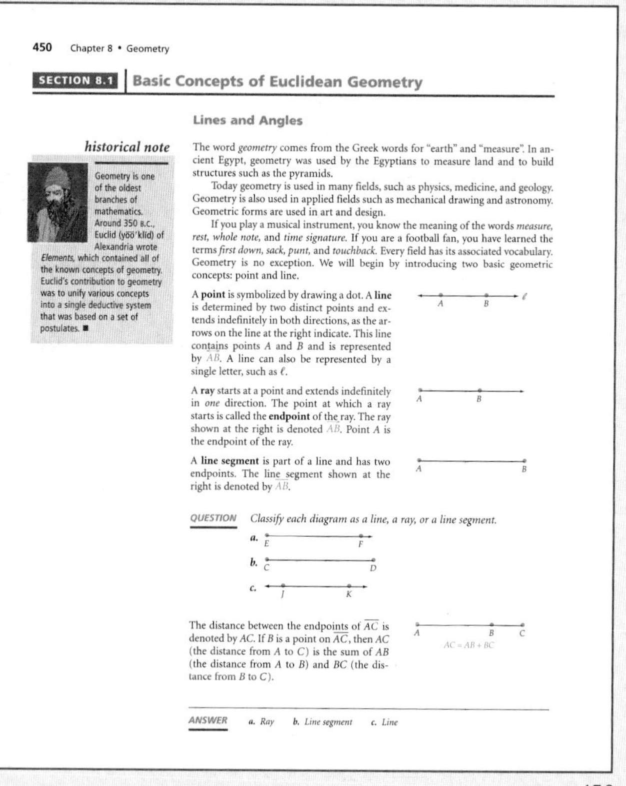

450 Chapter 8 • Geometry

SECTION 8.1 Basic Concepts of Euclidean Geometry

Lines and Angles

historical note

Geometry is one of the oldest branches of mathematics. Around 350 B.C., Euclid (yoo'klid) of Alexandria wrote *Elements*, which contained all of the known concepts of geometry. Euclid's contribution to geometry was to unify various concepts into a single deductive system that was based on a set of postulates. ■

The word *geometry* comes from the Greek words for "earth" and "measure". In ancient Egypt, geometry was used by the Egyptians to measure land and to build structures such as the pyramids.

Today geometry is used in many fields, such as physics, medicine, and geology. Geometry is also used in applied fields such as mechanical drawing and astronomy. Geometric forms are used in art and design.

If you play a musical instrument, you know the meaning of the words *measure, rest, whole note,* and *time signature.* If you are a football fan, you have learned the terms *first down, sack, punt,* and *touchback.* Every field has its associated vocabulary. Geometry is no exception. We will begin by introducing two basic geometric concepts: point and line.

A **point** is symbolized by drawing a dot. A **line** is determined by two distinct points and extends indefinitely in both directions, as the arrows on the line at the right indicate. This line contains points A and B and is represented by $\overleftrightarrow{AB}$. A line can also be represented by a single letter, such as ℓ.

A **ray** starts at a point and extends indefinitely in *one* direction. The point at which a ray starts is called the **endpoint** of the ray. The ray shown at the right is denoted $\overrightarrow{AB}$. Point A is the endpoint of the ray.

A **line segment** is part of a line and has two endpoints. The line segment shown at the right is denoted by $\overline{AB}$.

QUESTION *Classify each diagram as a line, a ray, or a line segment.*

The distance between the endpoints of $\overline{AC}$ is denoted by AC. If B is a point on $\overline{AC}$, then AC (the distance from A to C) is the sum of AB (the distance from A to B) and BC (the distance from B to C).

$AC = AB + BC$

ANSWER *a. Ray b. Line segment c. Line*

page 450

Question/Answer Feature

At various places throughout the text, a Question is posed about the topic that is being developed. This question encourages students to pause, think about the current discussion, and answer the question. Students can immediately check their understanding by referring to the Answer to the question provided in a footnote on the same page. This feature creates another opportunity for the student to interact with the textbook.

Interactive Method, *continued*

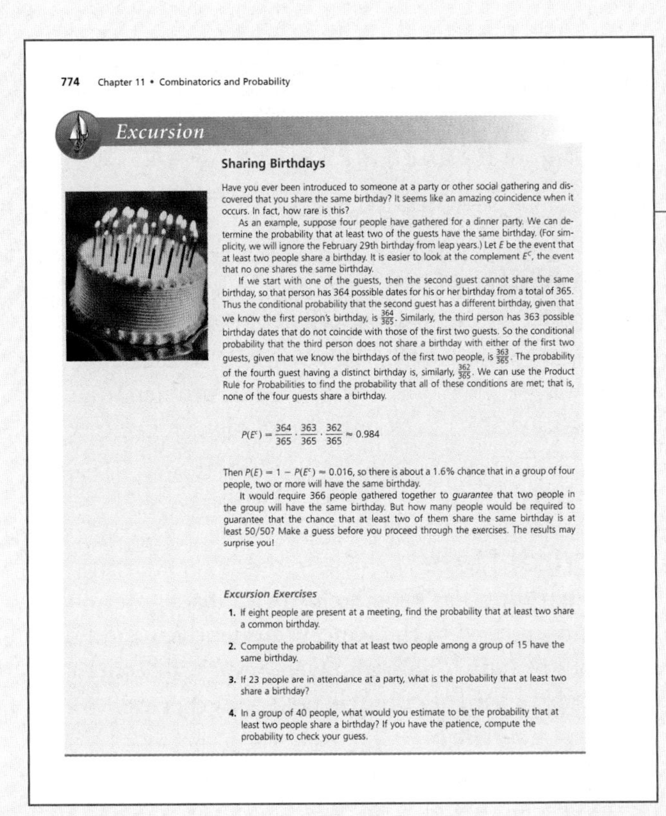

Excursion

Sharing Birthdays

Have you ever been introduced to someone at a party or other social gathering and discovered that you share the same birthday? It seems like an amazing coincidence when it occurs. In fact, how rare is this?

As an example, suppose four people have gathered for a dinner party. We can determine the probability that at least two of the guests have the same birthday. (For simplicity, we will ignore the February 29th birthday from leap years.) Let E be the event that at least two people share a birthday. It is easier to look at the complement E^C, the event that no one shares the same birthday.

If we start with one of the guests, then the second guest cannot share the same birthday, so that person has 364 possible dates for his or her birthday from a total of 365. Thus the conditional probability that the second guest has a different birthday, given that we know the first person's birthday, is $\frac{364}{365}$. Similarly, the third person has 363 possible birthday dates that do not coincide with those of the first two guests. So the conditional probability that the third person does not share a birthday with either of the first two guests, given that we know the birthdays of the first two people, is $\frac{363}{365}$. The probability of the fourth guest having a distinct birthday is, similarly, $\frac{362}{365}$. We can use the Product Rule for Probabilities to find the probability that all of these conditions are met; that is, none of the four guests share a birthday.

$$P(E^c) = \frac{364}{365} \cdot \frac{363}{365} \cdot \frac{362}{365} \approx 0.984$$

Then $P(E) = 1 - P(E^c) \approx 0.016$, so there is about a 1.6% chance that in a group of four people, two or more will have the same birthday.

It would require 366 people gathered together to *guarantee* that two people in the group will have the same birthday. But how many people would be required to guarantee that the chance that at least two of them share the same birthday is at least 50/50? Make a guess before you proceed through the exercises. The results may surprise you!

Excursion Exercises

1. If eight people are present at a meeting, find the probability that at least two share a common birthday.
2. Compute the probability that at least two people among a group of 15 have the same birthday.
3. If 23 people are in attendance at a party, what is the probability that at least two share a birthday?
4. In a group of 40 people, what would you estimate to be the probability that at least two people share a birthday? If you have the patience, compute the probability to check your guess.

page 774

Excursions

Each section ends with an Excursion along with corresponding Excursion Exercises. These activities engage students in the mathematics of the section. Some Excursions are designed as in-class cooperative learning activities that lend themselves to a hands-on approach. They can also be assigned as projects or extra credit assignments. The Excursions are a unique and important feature of this text. They provide opportunities for students to take an active role in the learning process.

AIM for Success Student Preface

This 'how to use this text' preface explains what is required of a student to be successful and how this text has been designed to foster student success. AIM for Success can be used as a lesson on the first day of class or as a project for students to complete to strengthen their study skills.

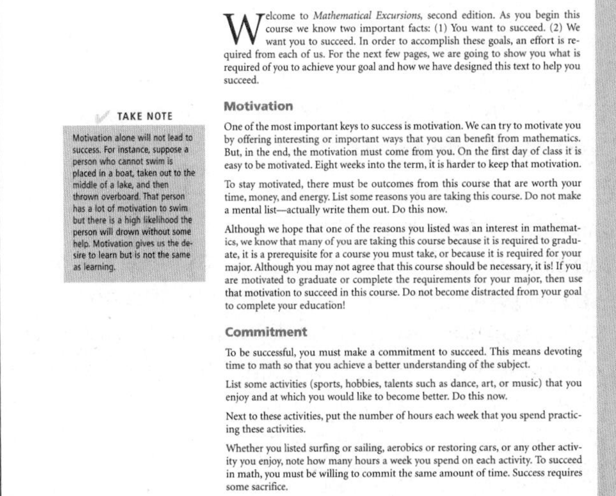

AIM FOR SUCCESS

Welcome to *Mathematical Excursions*, second edition. As you begin this course we know two important facts: (1) You want to succeed. (2) We want you to succeed. In order to accomplish these goals, an effort is required from each of us. For the next few pages, we are going to show you what is required of you to achieve your goal and how we have designed this text to help you succeed.

Motivation

One of the most important keys to success is motivation. We can try to motivate you by offering interesting or important ways that you can benefit from mathematics. But, in the end, the motivation must come from you. On the first day of class it is easy to be motivated. Eight weeks into the term, it is harder to keep that motivation.

To stay motivated, there must be outcomes from this course that are worth your time, money, and energy. List some reasons you are taking this course. Do not make a mental list—actually write them out. Do this now.

Although we hope that one of the reasons you listed was an interest in mathematics, we know that many of you are taking this course because it is required to graduate, it is a prerequisite for a course you must take, or because it is required for your major. Although you may not agree that this course should be necessary, it is! If you are motivated to graduate or complete the requirements for your major, then use that motivation to succeed in this course. Do not become distracted from your goal to complete your education!

> TAKE NOTE
> Motivation alone will not lead to success. For instance, suppose a person who cannot swim is placed in a boat, taken out to the middle of a lake, and then thrown overboard. That person has a lot of motivation to swim but there is a high likelihood the person will drown without some help. Motivation gives us the desire to learn but is not the same as learning.

Commitment

To be successful, you must make a commitment to succeed. This means devoting time to math so that you achieve a better understanding of the subject.

List some activities (sports, hobbies, talents such as dance, art, or music) that you enjoy and at which you would like to become better. Do this now.

Next to these activities, put the number of hours each week that you spend practicing these activities.

Whether you listed surfing or sailing, aerobics or restoring cars, or any other activity you enjoy, note how many hours a week you spend on each activity. To succeed in math, you must be willing to commit the same amount of time. Success requires some sacrifice.

The "I Can't Do Math" Syndrome

There may be things you cannot do, for instance, lift a two-ton boulder. You can, however, do math. It is much easier than lifting the two-ton boulder. When you first learned the activities you listed above, you probably could not do them well. With practice, you got better. With practice, you will be better at math. Stay focused, motivated, and committed to success.

page xix

Math Matters and Margin Notes

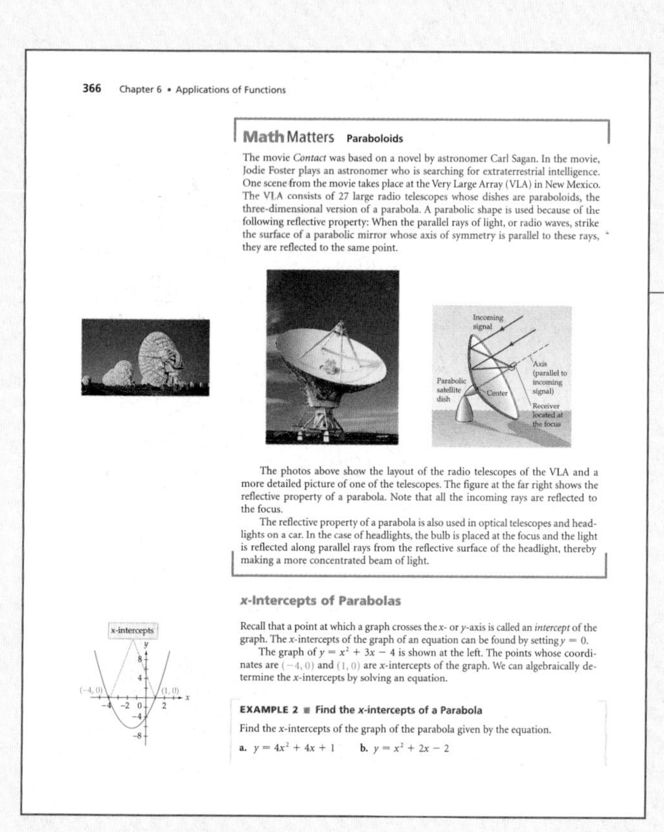
366 Chapter 6 • Applications of Functions

Math Matters **Paraboloids**

The movie *Contact* was based on a novel by astronomer Carl Sagan. In the movie, Jodie Foster plays an astronomer who is searching for extraterrestrial intelligence. One scene from the movie takes place at the Very Large Array (VLA) in New Mexico. The VLA consists of 27 large radio telescopes whose dishes are paraboloids, the three-dimensional version of a parabola. A parabolic shape is used because of the following reflective property: When the parallel rays of light, or radio waves, strike the surface of a parabolic mirror whose axis of symmetry is parallel to these rays, they are reflected to the same point.

The photos above show the layout of the radio telescopes of the VLA and a more detailed picture of one of the telescopes. The figure at the far right shows the reflective property of a parabola. Note that all the incoming rays are reflected to the focus.

The reflective property of a parabola is also used in optical telescopes and headlights on a car. In the case of headlights, the bulb is placed at the focus and the light is reflected along parallel rays from the reflective surface of the headlight, thereby making a more concentrated beam of light.

x-Intercepts of Parabolas

Recall that a point at which a graph crosses the x- or y-axis is called an *intercept* of the graph. The x-intercepts of the graph of an equation can be found by setting $y = 0$.

The graph of $y = x^2 + 3x - 4$ is shown at the left. The points whose coordinates are $(-4, 0)$ and $(1, 0)$ are x-intercepts of the graph. We can algebraically determine the x-intercepts by solving an equation.

EXAMPLE 2 ■ Find the x-intercepts of a Parabola

Find the x-intercepts of the graph of the parabola given by the equation.

a. $y = 4x^2 + 4x + 1$ **b.** $y = x^2 + 2x - 2$

page 366

Math Matters

This feature of the text typically contains an interesting sidelight about mathematics, its history, or its applications.

Historical Note

These margin notes provide historical background information related to the concept under discussion or vignettes of individuals who were responsible for major advancements in their fields of expertise.

Point of Interest

These notes provide interesting information related to the topics under discussion. Many of these are of a contemporary nature and, as such, they provide students with the needed motivation for studying concepts that may at first seem abstract and obscure without this information.

Take Note

These notes alert students to a point requiring special attention or are used to amplify the concepts that are currently being developed. Some Take Notes, identified by a CD icon, reference the student CD. A student who needs to review a prerequisite skill or concept can find the needed material on this CD.

Calculator Note

These notes provide information about how to use the various features of a calculator.

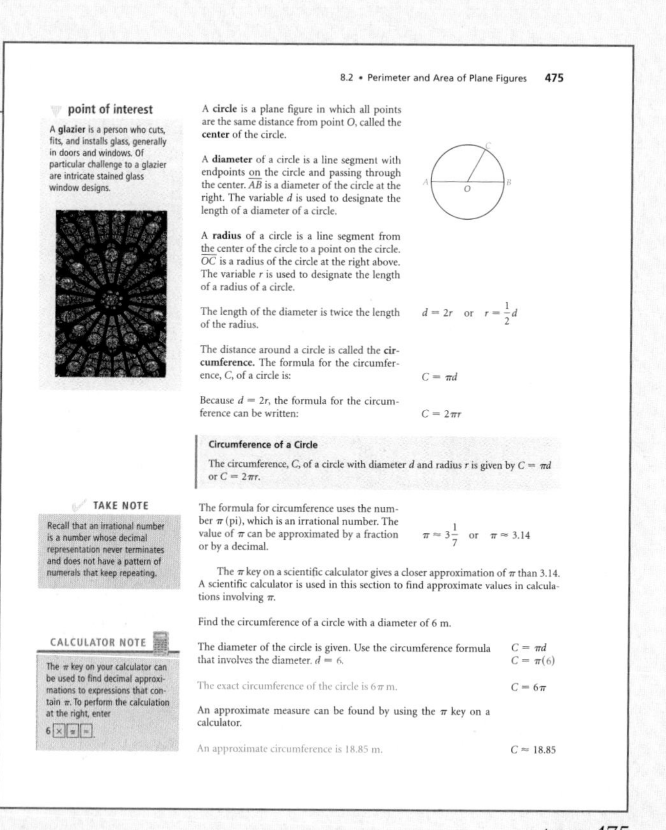
8.2 • Perimeter and Area of Plane Figures 475

point of interest

A **glazier** is a person who cuts, fits, and installs glass, generally in doors and windows. Of particular challenge to a glazier are intricate stained glass window designs.

A **circle** is a plane figure in which all points are the same distance from point O, called the **center** of the circle.

A **diameter** of a circle is a line segment with endpoints on the circle and passing through the center. $\overline{AB}$ is a diameter of the circle at the right. The variable d is used to designate the length of a diameter of a circle.

A **radius** of a circle is a line segment from the center of the circle to a point on the circle. $\overline{OC}$ is a radius of the circle at the right above. The variable r is used to designate the length of a radius of a circle.

The length of the diameter is twice the length of the radius. $d = 2r$ or $r = \frac{1}{2}d$

The distance around a circle is called the **circumference.** The formula for the circumference, C, of a circle is: $C = \pi d$

Because $d = 2r$, the formula for the circumference can be written: $C = 2\pi r$

Circumference of a Circle

The circumference, C, of a circle with diameter d and radius r is given by $C = \pi d$ or $C = 2\pi r$.

TAKE NOTE

Recall that an irrational number is a number whose decimal representation never terminates and does not have a pattern of numerals that keep repeating.

The formula for circumference uses the number π (pi), which is an irrational number. The value of π can be approximated by a fraction or by a decimal. $\pi \approx 3\frac{1}{7}$ or $\pi \approx 3.14$

The π key on a scientific calculator gives a closer approximation of π than 3.14. A scientific calculator is used in this section to find approximate values in calculations involving π.

CALCULATOR NOTE

The π key on your calculator can be used to find decimal approximations to expressions that contain π. To perform the calculation at the right, enter 6 [×] [π] [=].

Find the circumference of a circle with a diameter of 6 m.

The diameter of the circle is given. Use the circumference formula that involves the diameter. $d = 6$. $C = \pi d$, $C = \pi(6)$

The exact circumference of the circle is 6π m. $C = 6\pi$

An approximate measure can be found by using the π key on a calculator.

An approximate circumference is 18.85 m. $C \approx 18.85$

page 475

Exercises

Exercise Sets

The exercise sets were carefully written to provide a wide variety of exercises that range from drill and practice to interesting challenges. Exercise sets emphasize skill building, skill maintenance, concepts, and applications, when they are appropriate. Icons are used to identify various types of exercise.

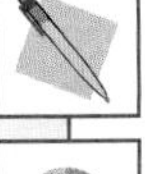

Writing exercises

Data analysis exercises

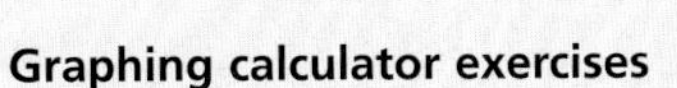

Graphing calculator exercises

Internet exercises

388 Chapter 6 • Applications of Functions

35. **Polonium** An initial amount of 100 micrograms of polonium decays to 75 micrograms in approximately 34.5 days. Find an exponential model for the amount of polonium in the sample after t days. Round to the nearest hundredth of a microgram.

36. **The Film Industry** The table below shows the number of multidisc DVDs, with three or more discs, released each year. (*Source:* DVD Release Report)

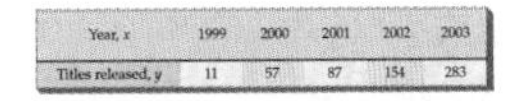

Year, x	1999	2000	2001	2002	2003
Titles released, y	11	57	87	154	283

a. Find an exponential regression equation for this data using $x = 0$ to represent 1995. Round to the nearest hundredth.

b. Use the equation to predict the number of multidisc DVDs released in 2008.

37. **Meteorology** The table below shows the saturation of water in air at various air temperatures.

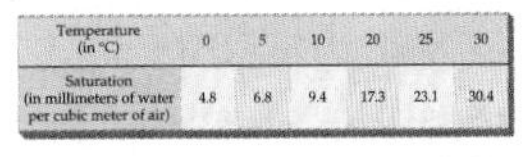

Temperature (in °C)	0	5	10	20	25	30
Saturation (in millimeters of water per cubic meter of air)	4.8	6.8	9.4	17.3	23.1	30.4

a. Find an exponential regression equation for these data. Round to the nearest thousandth.

b. Use the equation to predict the number of milliliters of water per cubic meter of air at a temperature of 15°C. Round to the nearest tenth.

38. **Snow Making** Artificial snow is made at a ski resort by combining air and water in a ratio that depends on the outside air temperature. The table below shows the rate of air flow needed for various temperatures.

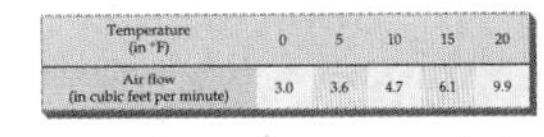

Temperature (in °F)	0	5	10	15	20
Air flow (in cubic feet per minute)	3.0	3.6	4.7	6.1	9.9

a. Find an exponential regression equation for these data. Round to the nearest hundredth.

b. Use the equation to predict the air flow needed when the temperature is 25°F. Round to the nearest tenth of a cubic foot per minute.

Extensions

CRITICAL THINKING

An exponential model for population growth or decay can be accurate over a short period of time. However, this model begins to fail because it does not account for the natural resources necessary to support growth, nor does it account for death within the population. Another model, called the *logistic model*, can account for some of these effects. The logistic model is given by $P(t) = \dfrac{mP_0}{P_0 + (m - P_0)e^{-kt}}$, where $P(t)$ is the population at time t, m is the maximum population that can be supported, P_0 is the population when $t = 0$, and k is a positive constant that is related to the growth of the population.

39. **Earth's Population** One model of Earth's population is given by $P(t) = \dfrac{280}{4 + 66e^{-0.021t}}$. In this equation, $P(t)$ is the population in billions and t is the number of years after 1980. Round answers to the nearest hundred million people.

a. According to this model, what was Earth's population in the year 2000?

b. According to this model, what will be Earth's population in the year 2010?

c. If t is very large, say greater than 500, then $e^{-0.021t} \approx 0$. What does this suggest about the maximum population that Earth can support?

40. **Wolf Population** Game wardens have determined that the maximum wolf population in a certain preserve is 1000 wolves. Suppose the population of wolves in the preserve in the year 2000 was 500, and that k is estimated to be 0.025.

a. Find a logistic function for the number of wolves in the preserve in year t, where t is the number of years after 2000.

b. Find the estimated wolf population in 2015.

EXPLORATIONS

41. **Car Payments** The formula used to calculate a monthly lease payment or a monthly car payment (for a purchase rather than a lease) is given by $P = \dfrac{Ar(1 + r)^n - Vr}{(1 + r)^n - 1}$, where P is the monthly payment, A is the amount of the loan, r is the *monthly* interest rate as a decimal, n is the number of months of the loan or lease, and V is the residual value of the car at the end of the lease. For a car purchase, $V = 0$.

page 388

Extensions

Extension exercises are placed at the end of each exercise set. These exercises are designed to extend concepts. In most cases these exercises are more challenging and require more time and effort than the preceding exercises. The Extension exercises always include at least two of the following types of exercises:

Critical Thinking

Cooperative Learning

Explorations

Some Critical Thinking exercises require the application of two or more procedures or concepts.

The Cooperative Learning exercises are designed for small groups of 2 to 4 students.

Many of the Exploration exercises require students to search on the Internet or through reference materials in a library.

756 Chapter 11 • Combinatorics and Probability

79. If a pair of fair dice are rolled once, what are the odds in favor of rolling a sum of 9?

80. If a single fair die is rolled, what are the odds in favor of rolling an even number?

81. If a card is randomly pulled from a standard deck of playing cards, what are the odds in favor of pulling a heart?

82. A coin is tossed four times. What are the odds against the coin showing heads all four times?

83. **Football** A bookmaker has placed 8 to 3 odds *against* a particular football team winning its next game. What is the probability, in the bookmaker's view, of the team winning?

84. **Contest Odds** A contest is advertising that the odds against winning first prize are 100 to 1. What is the probability of winning?

85. **Candy Colors** A snack-size bag of M&Ms candies contains 12 red candies, 12 blue, 7 green, 13 brown, 3 orange, and 10 yellow. If a candy is randomly picked from the bag, compute

a. the odds of getting a green M&M.

b. the probability of getting a green M&M.

86. **Candy Colors** A snack-size bag of Skittles candies contains 10 red candies, 15 blue, 9 green, 8 purple, 15 orange, and 13 yellow. If a candy is randomly picked from the bag, compute

a. the odds of picking a purple Skittle.

b. the probability of picking a purple Skittle.

Extensions

CRITICAL THINKING

87. If four cards labeled A, B, C, and D are randomly placed in four boxes also labeled A, B, C, and D, one to each box, find the probability that no card will be in a box with the same letter.

88. Determine the probability that if 10 coins are tossed, five heads and five tails will result.

89. In a family of three children, all of whom are girls, a family member new to probability reasons that the probability that each child would be a girl is 0.5. Therefore, the probability that the family would have three girls is $0.5 + 0.5 + 0.5 = 1.5$. Explain why this reasoning is not valid.

In Exercises 90 and 91, a hand of five cards is dealt from a standard deck of playing cards. You may want to review the material on combinations before doing these exercises.

90. Find the probability that the hand will contain all four aces.

91. Find the probability that the hand will contain three jacks and two queens.

EXPLORATIONS

Roulette Exercises 92 to 97 use the casino game roulette. Roulette is played by spinning a wheel with 38 numbered slots. The numbers 1 through 36 appear on the wheel, half of them colored black and half colored red. Two slots, numbered 0 and 00, are colored green. A ball is placed on the spinning wheel and allowed to come to rest in one of the slots. Bets are placed on where the ball will land.

92. You can place a bet that the ball will stop in a black slot. If you win, the casino will pay you \$1 for each dollar you bet. What is the probability of winning this bet?

93. You can bet that the ball will land on an odd number. If you win, the casino will pay you \$1 for each dollar you bet. What is the probability of winning this bet?

page 756

End of Chapter

Chapter Summary

At the end of each chapter there is a Chapter Summary that includes *Key Terms* and *Essential Concepts* that were covered in the chapter. These chapter summaries provide a single point of reference as the student prepares for an examination. Each key word references the page number where the word was first introduced.

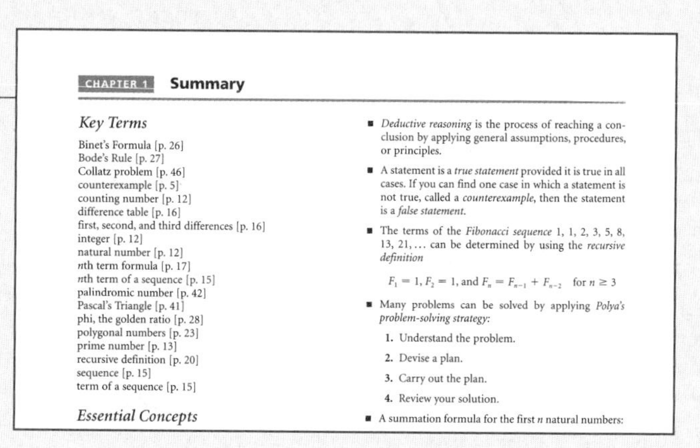

CHAPTER 1 **Summary**

Key Terms

Binet's Formula [p. 26]
Bode's Rule [p. 27]
Collatz problem [p. 46]
counterexample [p. 5]
counting number [p. 12]
difference table [p. 16]
first, second, and third differences [p. 16]
integer [p. 12]
natural number [p. 12]
*n*th term formula [p. 17]
*n*th term of a sequence [p. 15]
palindromic number [p. 42]
Pascal's Triangle [p. 41]
phi, the golden ratio [p. 28]
polygonal numbers [p. 23]
prime number [p. 13]
recursive definition [p. 20]
sequence [p. 15]
term of a sequence [p. 15]

Essential Concepts

- *Deductive reasoning* is the process of reaching a conclusion by applying general assumptions, procedures, or principles.
- A statement is a *true statement* provided it is true in all cases. If you can find one case in which a statement is not true, called a *counterexample*, then the statement is a *false statement*.
- The terms of the *Fibonacci sequence* 1, 1, 2, 3, 5, 8, 13, 21, ... can be determined by using the *recursive definition*

$$F_1 = 1, F_2 = 1, \text{ and } F_n = F_{n-1} + F_{n-2} \quad \text{for } n \geq 3$$

- Many problems can be solved by applying *Polya's problem-solving strategy:*
  1. Understand the problem.
  2. Devise a plan.
  3. Carry out the plan.
  4. Review your solution.
- A summation formula for the first *n* natural numbers:

page 46

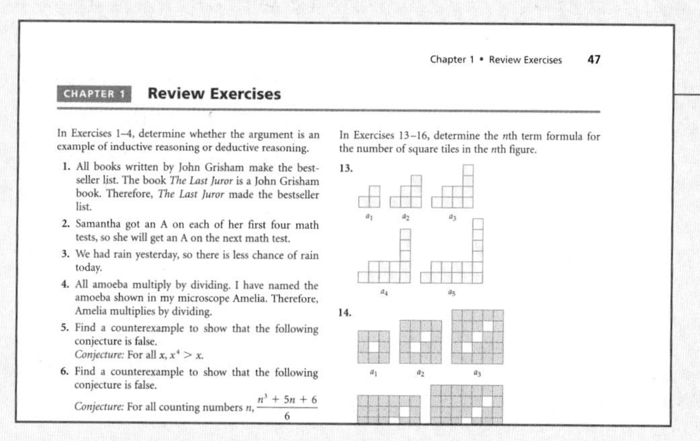

Chapter 1 • Review Exercises 47

CHAPTER 1 **Review Exercises**

In Exercises 1–4, determine whether the argument is an example of inductive reasoning or deductive reasoning.

1. All books written by John Grisham make the bestseller list. The book *The Last Juror* is a John Grisham book. Therefore, *The Last Juror* made the bestseller list.
2. Samantha got an A on each of her first four math tests, so she will get an A on the next math test.
3. We had rain yesterday, so there is less chance of rain today.
4. All amoeba multiply by dividing. I have named the amoeba shown in my microscope Amelia. Therefore, Amelia multiplies by dividing.
5. Find a counterexample to show that the following conjecture is false.
   Conjecture: For all x, $x^4 > x$.
6. Find a counterexample to show that the following conjecture is false.
   Conjecture: For all counting numbers n, $\frac{n^3 + 5n + 6}{6}$

In Exercises 13–16, determine the *n*th term formula for the number of square tiles in the *n*th figure.

13. a_1 a_2 a_3 a_4 a_5

14. a_1 a_2 a_3

page 47

Chapter Review Exercises

Review exercises are found near the end of each chapter. These exercises were selected to help the student integrate the major topics presented in the chapter. The answers to all the Chapter Review Exercises appear in the answer section along with a section reference for each exercise. These section references indicate the section or sections where a student can locate the concepts needed to solve each exercise.

Chapter Test

The Chapter Test exercises are designed to emulate a possible test of the material in the chapter. The answers to all the Chapter Test exercises appear in the answer section along with a section reference for each exercise. The section references indicate the section or sections where a student can locate the concepts needed to solve each exercise.

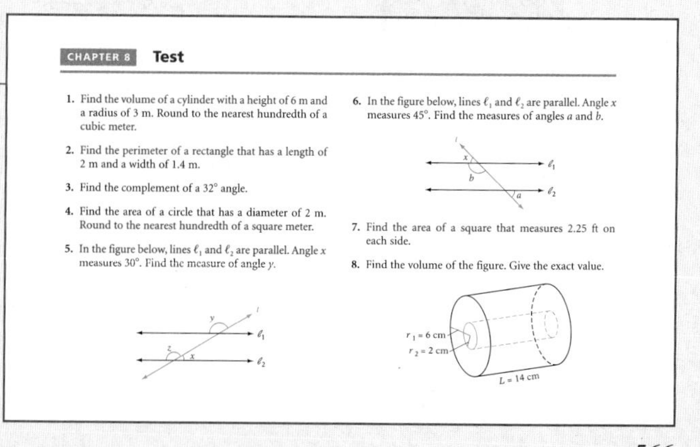

CHAPTER 8 **Test**

1. Find the volume of a cylinder with a height of 6 m and a radius of 3 m. Round to the nearest hundredth of a cubic meter.
2. Find the perimeter of a rectangle that has a length of 2 m and a width of 1.4 m.
3. Find the complement of a 32° angle.
4. Find the area of a circle that has a diameter of 2 m. Round to the nearest hundredth of a square meter.
5. In the figure below, lines ℓ_1 and ℓ_2 are parallel. Angle x measures 30°. Find the measure of angle y.
6. In the figure below, lines ℓ_1 and ℓ_2 are parallel. Angle x measures 45°. Find the measures of angles a and b.
7. Find the area of a square that measures 2.25 ft on each side.
8. Find the volume of the figure. Give the exact value.

page 566

Supplements for the Instructor

Mathematical Excursions, Second Edition, has an extensive support package for the instructor that includes:

Instructor's Annotated Edition (IAE): The *Instructor's Annotated Edition* is an exact replica of the student textbook with the following additional text-specific items for the instructor: answers to *all* of the end-of-section and end-of-chapter exercises, answers to *all* Excursion and Exploration exercises, Instructor Notes, Suggested Assignments, and P icons denoting tables and art that appear in PowerPoint® slides. (The files can be downloaded from our website at **math.college.hmco.com/instructors**).

Online Teaching Center: This free companion website contains an abundance of instructor resources such as solutions to all exercises in the text, digital art and tables, suggested course syllabi, Chapter Tests, Graphing Calculator Guide, and Microsoft® Excel spreadsheets. Visit **math.college.hmco.com/instructors** and choose *Mathematical Excursions,* 2/e, from the list provided on the site.

Online Instructor's Solutions Manual: The *Online Instructor's Solutions Manual* offers worked-out solutions to *all* of the exercises in each exercise set as well as solutions to the Excursion and Exploration exercises.

HM ClassPrep™ with HM Testing CD-ROM (powered by Diploma™): This CD-ROM is a combination of two course management tools.

- *HM Testing* (powered by *Diploma*) offers instructors a flexible and powerful tool for test generation and test management. Now supported by the Brownstone Research Group's market-leading *Diploma* software, this new version of *HM Testing* significantly improves on functionality and ease of use by offering all the tools needed to create, author, deliver, and customize multiple types of tests—including authoring and editing algorithmic questions. *Diploma* is currently in use at thousands of college and university campuses throughout the United States and Canada.
- HM ClassPrep also features the same text-specific resources for the instructor that are available on the Online Teaching Center.

Eduspace®: Eduspace, powered by Blackboard®, is Houghton Mifflin's customizable and interactive online learning tool.

Eduspace provides instructors with online courses and content. By pairing the widely recognized tools of Blackboard with quality, text-specific content from Houghton Mifflin Company, Eduspace makes it easy for instructors to create all or part of a course online. This online learning tool also contains ready-to-use homework exercises, quizzes, tests, tutorials, and supplemental study materials.

Visit **eduspace.com** for more information.

Supplements for the Student

Mathematical Excursions, Second Edition, has an extensive support package for the student that includes:

Student Solutions Manual: The *Student Solutions Manual* contains complete, worked-out solutions to *all* odd-numbered exercises and *all* of the solutions to the Chapter Reviews and Chapter Tests in the text.

Online Study Center: This free companion website contains an abundance of student resources such as binary cards, Graphing Calculator Guide, and Microsoft® Excel spreadsheets.

Online CLAST Preparation Guide: The CLAST Preparation Guide is a competency-based study guide that reviews and offers preparatory material for the CLAST (College Level Academic Skills Test) objectives required by the State of Florida for mathematics. The guide includes a correlation of the CLAST objectives to the *Mathematical Excursions,* Second Edition, text, worked-out examples, practice examples, cumulative reviews, and sample diagnostic tests with grading sheets.

HM mathSpace® Student Tutorial CD ROM: This tutorial provides opportunities for self-paced review and practice with algorithmically generated exercises and step-by-step solutions.

Houghton Mifflin Instructional DVDs: These text-specific DVDs, professionally produced by Dana Mosely, provide explanations of key concepts, examples, and exercises in a lecture-based format. They offer students a valuable resource for further instruction and review. They also provide support for students in online courses.

Eduspace®: Eduspace, powered by Blackboard®, is Houghton Mifflin's customizable and interactive online learning tool for instructors and students. Eduspace is a text-specific, web-based learning environment that your instructor can use to offer students a combination of practice exercises, multimedia tutorials, video explanations, online algorithmic homework and more. Specific content is available 24 hours a day to help you succeed in your course.

SMARTHINKING® Live, On-line Tutoring: Houghton Mifflin has partnered with SMARTHINKING to provide an easy-to-use, effective, online tutorial service. Through state-of-the-art tools and a two-way whiteboard, students communicate in real-time with qualified e-structors who can help the students understand difficult concepts and guide them through the problem-solving process while studying or completing homework.

Three levels of service are offered to the students.

- **Live Tutorial Help** provides real-time, one-on-one instruction.
- **Question submission** allows students to submit questions to the tutor outside the scheduled hours and receive a response within 24 hours.
- **Independent Study Resources** connects students around-the-clock to additional educational resources, ranging from interactive websites to Frequently Asked Questions.

Visit **smarthinking.com** for more information. *Limits apply; terms and hours of SMARTHINKING service are subject to change.*

Acknowledgments

The authors would like to thank the people who have reviewed this manuscript and provided many valuable suggestions.

Brenda Alberico, *College of DuPage*

Beverly R. Broomell, *Suffolk County Community College*

Henjin Chi, *Indiana State University*

Ivette Chuca, *El Paso Community College*

Marcella Cremer, *Richland Community College*

Kenny Fister, *Murray State University*

Luke Foster, *Northeastern State University*

Rita Fox, *Kalamazoo Valley Community College*

Sue Grapevine, *Northwest Iowa Community College*

Shane Griffith, *Lee University*

Dr. Nancy R. Johnson, *Manatee Community College*

Dr. Vernon Kays, *Richland Community College*

Dr. Suda Kunyosying, *Shepherd College*

Kathryn Lavelle, *Westchester Community College*

Roger Marty, *Cleveland State University*

Eric Matsuoka, *Leeward Community College*

Beverly Meyers, *Jefferson College*

Dr. Alec Mihailovs, *Shepherd University*

Bette Nelson, *Alvin Community College*

Kathleen Offenholley, *Brookdale Community College*

Kathy Pinchback, *University of Memphis*

Michael Polley, *Southeastern Community College*

Dr. Anne Quinn, *Edinboro University of Pennsylvania*

Brenda Reed, *Navarro College*

Marc Renault, *Shippensburg University*

Chistopher Rider, *North Greenville College*

Sharon M. Saxton, *Cascadia Community College*

Mary Lee Seitz, *Erie Community College—City Campus*

Dr. Sue Stokley, *Spartanburg Technical College*

Dr. Julie M. Theoret, *Lyndon State College*

Walter Jacob Theurer, *Fulton Montgomery Community College*

Jamie Thomas, *University of Wisconsin Colleges—Manitowoc*

William Twentyman, *ECPI College of Technology*

Denise A. Widup, *University of Wisconsin—Parkside*

Nancy Wilson, *Marshall University*

Jane-Marie Wright, *Suffolk Community College*

Welcome to *Mathematical Excursions,* Second Edition. As you begin this course, we know two important facts: (1) You want to succeed. (2) We want you to succeed. In order to accomplish these goals, an effort is required from each of us. For the next few pages, we are going to show you what is required of you to achieve your goal and how we have designed this text to help you succeed.

Motivation

TAKE NOTE

Motivation alone will not lead to success. For instance, suppose a person who cannot swim is placed in a boat, taken out to the middle of a lake, and then thrown overboard. That person has a lot of motivation to swim but there is a high likelihood the person will drown without some help. Motivation gives us the desire to learn but is not the same as learning.

One of the most important keys to success is motivation. We can try to motivate you by offering interesting or important ways that you can benefit from mathematics. But, in the end, the motivation must come from you. On the first day of class it is easy to be motivated. Eight weeks into the term, it is harder to keep that motivation.

To stay motivated, there must be outcomes from this course that are worth your time, money, and energy. List some reasons you are taking this course. Do not make a mental list—actually write them out. Do this now.

Although we hope that one of the reasons you listed was an interest in mathematics, we know that many of you are taking this course because it is required to graduate, it is a prerequisite for a course you must take, or because it is required for your major. Although you may not agree that this course should be necessary, it is! If you are motivated to graduate or complete the requirements for your major, then use that motivation to succeed in this course. Do not become distracted from your goal to complete your education!

Commitment

To be successful, you must make a commitment to succeed. This means devoting time to math so that you achieve a better understanding of the subject.

List some activities (sports, hobbies, talents such as dance, art, or music) that you enjoy and at which you would like to become better. Do this now.

Next to these activities, put the number of hours each week that you spend practicing these activities.

Whether you listed surfing or sailing, aerobics or restoring cars, or any other activity you enjoy, note how many hours a week you spend on each activity. To succeed in math, you must be willing to commit the same amount of time. Success requires some sacrifice.

The "I Can't Do Math" Syndrome

There may be things you cannot do, for instance, lift a two-ton boulder. You can, however, do math. It is much easier than lifting the two-ton boulder. When you first learned the activities you listed above, you probably could not do them well. With practice, you got better. With practice, you will be better at math. Stay focused, motivated, and committed to success.

It is difficult for us to emphasize how important it is to overcome the "I Can't Do Math Syndrome." If you listen to interviews of very successful athletes after a particularly bad performance, you will note that they focus on the positive aspect of what they did, not the negative. Sports psychologists encourage athletes to always be positive—to have a "Can Do" attitude. You need to develop this attitude toward math.

Strategies for Success

Know the Course Requirements To do your best in this course, you must know exactly what your instructor requires. Course requirements may be stated in a *syllabus,* which is a printed outline of the main topics of the course, or they may be presented orally. When they are listed in a syllabus or on other printed pages, keep them in a safe place. When they are presented orally, make sure to take complete notes. In either case, it is important that you understand them completely and follow them exactly. Be sure you know the answer to each of the following questions.

1. What is your instructor's name?
2. Where is your instructor's office?
3. At what times does your instructor hold office hours?
4. Besides the textbook, what other materials does your instructor require?
5. What is your instructor's attendance policy?
6. If you must be absent from a class meeting, what should you do before returning to class? What should you do when you return to class?
7. What is the instructor's policy regarding collection or grading of homework assignments?
8. What options are available if you are having difficulty with an assignment? Is there a math tutoring center?
9. If there is a math lab at your school, where is it located? What hours is it open?
10. What is the instructor's policy if you miss a quiz?
11. What is the instructor's policy if you miss an exam?
12. Where can you get help when studying for an exam?

Remember: Your instructor wants to see you succeed. If you need help, ask! Do not fall behind. If you were running a race and fell behind by 100 yards, you may be able to catch up but it will require more effort than had you not fallen behind.

TAKE NOTE

Besides time management, there must be realistic ideas of how much time is available. There are very few people who can *successfully* work full-time and go to school full-time. If you work 40 hours a week, take 15 units, spend the recommended study time given at the right, and sleep 8 hours a day, you will use over 80% of the available hours in a week. That leaves less than 20% of the hours in a week for family, friends, eating, recreation, and other activities.

Time Management We know that there are demands on your time. Family, work, friends, and entertainment all compete for your time. We do not want to see you receive poor job evaluations because you are studying math. However, it is also true that we do not want to see you receive poor math test scores because you devoted too much time to work. When several competing and important tasks require your time and energy, the only way to manage the stress of being successful at both is to manage your time efficiently.

Instructors often advise students to spend twice the amount of time outside of class studying as they spend in the classroom. Time management is important if you are to accomplish this goal and succeed in school. The following activity is intended to help you structure your time more efficiently.

Take out a sheet of paper and list the names of each course you are taking this term, the number of class hours each course meets, and the number of hours you should spend outside of class studying course materials. Now create a weekly calendar with the days of the week across the top and each hour of the day in a vertical column. Fill in the calendar with the hours you are in class, the hours you spend at work, and other commitments such as sports practice, music lessons, or committee meetings. Then fill in the hours that are more flexible, for example, study time, recreation, and meal times.

	Monday	**Tuesday**	**Wednesday**	**Thursday**	**Friday**	**Saturday**	**Sunday**
10–11 a.m.	History	Rev Spanish	History	Rev Span Vocab	History	Jazz Band	
11–12 p.m.	Rev History	Spanish	Study group	Spanish	Math tutor	Jazz Band	
12–1 p.m.	Math		Math		Math		Soccer

We know that many of you must work. If that is the case, realize that working 10 hours a week at a part-time job is equivalent to taking a three-unit class. If you must work, consider letting your education progress at a slower rate to allow you to be successful at both work and school. There is no rule that says you must finish school in a certain time frame.

Schedule Study Time As we encouraged you to do by filling out the time management form, schedule a certain time to study. You should think of this time like being at work or class. Reasons for "missing study time" should be as compelling as reasons for missing work or class. "I just didn't feel like it" is not a good reason to miss your scheduled study time. Although this may seem like an obvious exercise, list a few reasons you might want to study. Do this now.

Of course we have no way of knowing the reasons you listed, but from our experience one reason given quite frequently is "To pass the course." There is nothing wrong with that reason. If that is the most important reason for you to study, then use it to stay focused.

One method of keeping to a study schedule is to form a ***study group***. Look for people who are committed to learning, who pay attention in class, and who are punctual. Ask them to join your group. Choose people with similar educational goals but different methods of learning. You can gain from seeing the material from a new perspective. Limit groups to four or five people; larger groups are unwieldy.

There are many ways to conduct a study group. Begin with the following suggestions and see what works best for your group.

1. Test each other by asking questions. Each group member might bring two or three sample test questions to each meeting.
2. Practice teaching each other. Many of us who are teachers learned a lot about our subject when we had to explain it to someone else.
3. Compare class notes. You might ask other students about material in your notes that is difficult for you to understand.
4. Brainstorm test questions.
5. Set an agenda for each meeting. Set approximate time limits for each agenda item and determine a quitting time.

And now, probably the most important aspect of studying is that it should be done in relatively small chunks. If you can only study three hours a week for this course (probably not enough for most people), do it in blocks of one hour on three separate days, preferably after class. Three hours of studying on a Sunday is not as productive as three hours of paced study.

Features of This Text That Promote Success

TAKE NOTE

If you have difficulty with a particular algebra topic, there is a computer tutorial that accompanies this text that can be used to refresh your skills. You will see references to this tutorial as you go through this text.

Preparing for Class Before the class meeting in which your professor begins a new section, you should read the title of each section. Next, browse through the chapter material, being sure to note each word in bold type. These words indicate important concepts that you must know to learn the material. Do not worry about trying to understand all the material. Your professor is there to assist you with that endeavor. The purpose of browsing through the material is so that your brain will be prepared to accept and organize the new information when it is presented to you. Turn to page 794. Write down the title of Section 12.1.

Write down the words in the section that are in bold print. It is not necessary for you to understand the meaning of these words. You are in this class to learn their meaning.

Math is Not a Spectator Sport To learn mathematics you must be an active participant. Listening and watching your professor do mathematics is not enough. Mathematics requires that you interact with the lesson you are studying. If you have been writing down the things we have asked you to do, you were being interactive. There are other ways this textbook has been designed so that you can be an active learner.

Check Your Progress One of the key instructional features of this text is a completely worked-out example followed by a *Check Your Progress.*

11.2 • Permutations and Combinations 733

EXAMPLE 3 ■ Counting Permutations

In 2004, the Kentucky Derby had 18 horses entered in the race. How many different finishes of first, second, third, and fourth place were possible?

Solution

Because the order in which the horses finish the race is important, the number of possible finishes of first, second, third, and fourth place is $P(18, 4)$.

$$P(18, 4) = \frac{18!}{(18-4)!} = \frac{18!}{14!} = \frac{18 \cdot 17 \cdot 16 \cdot 15 \cdot 14!}{14!}$$
$$= 18 \cdot 17 \cdot 16 \cdot 15 = 73{,}440$$

There were 73,440 possible finishes of first, second, third, and fourth places.

CHECK YOUR PROGRESS 3 There were 42 cars entered in the 2004 Daytona 500 NASCAR race. How many different ways could first, second, and third place prizes be awarded?

Solution *See page S42.*

page 733

Note that each Example is completely worked out and the *Check Your Progress* following the example is not. Study the worked-out example carefully by working through each step. Your should do this with paper and pencil.

Now work the *Check Your Progress.* If you get stuck, refer to the page number following the word *Solution* which directs you to the page on which the *Check Your Progress* is solved—a complete worked-out solution is provided. Try to use the given solution to get a hint for the step you are stuck on. Then try to complete your solution.

When you have completed the solution, check your work against the solution we provide.

CHECK YOUR PROGRESS 3, *page 733*

The order in which the cars finish is important, so the number of ways to place first, second, and third is

$$P(42, 3) = \frac{42!}{(42 - 3)!} = \frac{42 \cdot 41 \cdot 40 \cdot \cancel{39!}}{\cancel{39!}} = 68{,}880$$

There are 68,880 different ways to award the first, second, and third place prizes.

page S42

Be aware that frequently there is more than one way to solve a problem. Your answer, however, should be the same as the given answer. If you have any question as to whether your method will "always work," check with your instructor or with someone in the math center.

Browse through the textbook and write down the page numbers where two other paired example features occur.

Remember: Be an active participant in your learning process. When you are sitting in class watching and listening to an explanation, you may think that you understand. However, until you actually try to do it, you will have no confirmation of the new knowledge or skill. Most of us have had the experience of sitting in class thinking we knew how to do something only to get home and realize we didn't.

Rule Boxes Pay special attention to rules placed in boxes. These rules give you the reasons certain types of problems are solved the way they are. When you see a rule, try to rewrite the rule in your own words.

Simple Interest Formula

The simple interest formula is

$$I = Prt$$

where I is the interest, P is the principal, r is the interest rate, and t is the time period.

page 642

Chapter Exercises When you have completed studying a section, do the section exercises. Math is a subject that needs to be learned in small sections and practiced continually in order to be mastered. Doing the exercises in each exercise set will help you master the problem-solving techniques necessary for success. As you work through the exercises, check your answers to the odd-numbered exercises with those in the back of the book.

Preparing for a Test There are important features of this text that can be used to prepare for a test.

- Chapter Summary
- Chapter Review Exercises
- Chapter Test

After completing a chapter, read the Chapter Summary. (See page 109 for the Chapter 2 Summary.) This summary highlights the important topics covered in the chapter. The page number following each topic refers you to the page in the text on which you can find more information about the concept.

Following the Chapter Summary are Chapter Review Exercises (see page 110). Doing the review exercises is an important way of testing your understanding of the chapter. The answer to each review exercise is given at the back of the book, along with, in brackets, the section reference from which the question was taken (see page A3). After checking your answers, restudy any section from which a question you missed was taken. It may be helpful to retry some of the exercises for that section to reinforce your problem-solving techniques.

Each chapter ends with a Chapter Test (see page 112). This test should be used to prepare for an exam. We suggest that you try the Chapter Test a few days before your actual exam. Take the test in a quiet place and try to complete the test in the same amount of time you will be allowed for your exam. When taking the Chapter Test, practice the strategies of successful test takers: 1) scan the entire test to get a feel for the questions; 2) read the directions carefully; 3) work the problems that are easiest for you first; and perhaps most importantly, 4) try to stay calm.

When you have completed the Chapter Test, check your answers (see page A7). Next to each answer is, in brackets, the reference to the section from which the question was taken. If you missed a question, review the material in that section and rework some of the exercises from that section. This will strengthen your ability to perform the skills in that section.

Your career goal goes here. →

Is it difficult to be successful? YES! Successful music groups, artists, professional athletes, teachers, sociologist, chefs, and ______________ have to work very hard to achieve their goals. They focus on their goals and ignore distractions. The things we ask you to do to achieve success take time and commitment. We are confident that if you follow our suggestions, you will succeed.

CHAPTER 2

Sets

In mathematics any group or collection of objects is called a *set.* A simple application of sets occurs when you use a search engine (such as Yahoo, AltaVista, Google, or Lycos) to find a topic on the Internet. You merely enter a few words describing what you are searching for and click the "Search" button. The search engine then creates a list (set) of websites that contain a match for the words you submitted.

For instance, suppose you wish to make a dessert. You decide to search the Internet for a chocolate cake recipe. You search for the words "chocolate cake" and you obtain a set containing 300,108 matches. This is a very large number, so you narrow your search. One method of narrowing your search is to use the AND option found in the Advanced Search link of some search engines. An AND search is an all-words search. That is, an AND search finds only those sites that contain all of the words submitted. An AND search for "chocolate cake" produces a set containing 74,400 matches. This is a more reasonable number, but it is still quite large.

Search for:

chocolate cake | Search

Advanced Search

You attempt to narrow the search even further by using an AND search for the words "chocolate cake recipe." This search returns 17,945 matches. An AND search for "flourless chocolate cake recipe" returns only 913 matches. The second of these sites provides you with a recipe and states that it is fabulous and foolproof.

Sometimes it is helpful to perform a search using the OR option. An OR search is an any-words search. That is, an OR search finds all those sites that contain any of the words you submitted.

Many additional applications of sets are given in this chapter.

For online student resources, visit this textbook's website at **math.college.hmco.com/students.**

SECTION 2.1 Basic Properties of Sets

Sets

The constellation Scorpius is a set of stars.

TAKE NOTE

Sets can also be designated by using *set-builder notation.* This method is described on page 56.

point of interest

Paper currency in denominations of \$500, \$1000, \$5000, and \$10,000 has been in circulation, but production of these bills ended in 1945. If you just happen to have some of these bills, you can still cash them for their face value.

In an attempt to better understand the universe, ancient astronomers classified certain groups of stars as constellations. Today we still find it extremely helpful to classify items into groups that enable us to find order and meaning in our complicated world.

Any group or collection of objects is called a **set.** The objects that belong in a set are the **elements,** or **members,** of the set. For example, the set consisting of the four seasons has spring, summer, fall and winter as its elements.

The following two methods are often used to designate a set.

- Describe the set using words.
- List the elements of the set inside a pair of braces, { }. This method is called the **roster method.** Commas are used to separate the elements.

For instance, let's use S to represent the set consisting of the four seasons. Using the roster method we would write

$S = \{\text{spring, summer, fall, winter}\}$

The order in which the elements of a set are listed is not important. Thus the set consisting of the four seasons can also be written as

$S = \{\text{winter, spring, fall, summer}\}$

The following table gives two examples of sets, where each set is designated by a word description and also by using the roster method.

Table 2.1 *Define Sets by Using a Word Description and the Roster Method*

Description	Roster Method
The set of denominations of U.S. paper currency in production at this time	{\$1, \$2, \$5, \$10, \$20, \$50, \$100}
The set of states in the United States that border the Pacific Ocean.	{California, Oregon, Washington, Alaska, Hawaii}

EXAMPLE 1 ■ Use The Roster Method to Represent a Set

Use the roster method to represent the set of the days in a week.

Solution {Sunday, Monday, Tuesday, Wednesday, Thursday, Friday, Saturday}

CHECK YOUR PROGRESS 1 Use the roster method to represent the set of months that start with the letter A.

Solution *See page S4.*

TAKE NOTE

Some sets can be described in more than one way. For instance, {Sunday, Saturday} can be described as the days of the week that begin with the letter *S*, as the days of the week that occur in a weekend, or as the first and last days of a week.

EXAMPLE 2 ■ Use a Word Description to Represent a Set

Write a word description for the set

$$A = \{a, b, c, d, e, f, g, h, i, j, k, l, m, n, o, p, q, r, s, t, u, v, w, x, y, z\}$$

Solution Set A is the set of letters of the English alphabet.

CHECK YOUR PROGRESS 2 Write a word description for the set {March, May}.

Solution *See page S4.*

The following sets of numbers are used extensively in many areas of mathematics.

TAKE NOTE

In this chapter the letters N, W, I, Q, $\mathcal{I}$, and R will often be used to represent the basic number sets defined at the right.

Basic Number Sets

Natural Numbers or Counting Numbers $N = \{1, 2, 3, 4, 5, \ldots\}$

Whole Numbers $W = \{0, 1, 2, 3, 4, 5, \ldots\}$

Integers $I = \{\ldots, -4, -3, -2, -1, 0, 1, 2, 3, 4, \ldots\}$

Rational Numbers $Q =$ the set of all terminating or repeating decimals

Irrational Numbers $\mathcal{I} =$ the set of all nonterminating, nonrepeating decimals

Real Numbers $R =$ the set of all rational or irrational numbers

The set of natural numbers is also called the set of counting numbers. The three dots ... are called an **ellipsis** and indicate that the elements of the set continue in a manner suggested by the elements that are listed.

The integers $\ldots, -4, -3, -2, -1$ are **negative integers.** The integers $1, 2, 3, 4, \ldots$ are **positive integers.** Note that the natural numbers and the positive integers are the same set of numbers. The integer zero is neither a positive nor a negative integer.

If a number in decimal form terminates or repeats a block of digits, then the number is a rational number. Rational numbers can also be written in the form $\frac{p}{q}$, where p and q are integers and $q \neq 0$. For example,

$$\frac{1}{4} = 0.25 \quad \text{and} \quad \frac{3}{11} = 0.\overline{27}$$

are rational numbers. The bar over the 27 means that the block of digits 27 repeats without end; that is, $0.\overline{27} = 0.27272727\ldots$.

A decimal that neither terminates nor repeats is an **irrational number.** For instance, 0.35335333533335... is a nonterminating, nonrepeating decimal and thus is an irrational number.

Every real number is either a rational number or an irrational number.

EXAMPLE 3 ■ Use The Roster Method to Represent a Set of Numbers

Use the roster method to write each of the given sets.

a. The set of natural numbers less than 5

b. The solution set of $x + 5 = -1$

c. The set of negative integers greater than -4

Solution

a. The set of natural numbers is given by $\{1, 2, 3, 4, 5, 6, 7, \ldots\}$. The natural numbers less than 5 are 1, 2, 3, and 4. Using the roster method, we write this set as $\{1, 2, 3, 4\}$.

b. Adding -5 to each side of the equation produces $x = -6$. The solution set of $x + 5 = -1$ is $\{-6\}$.

c. The set of negative integers greater than -4 is $\{-3, -2, -1\}$.

CHECK YOUR PROGRESS 3 Use the roster method to write each of the given sets.

a. The set of whole numbers less than 4

b. The set of counting numbers larger than 11 and less than or equal to 19

c. The set of negative integers between -5 and 7

Solution *See page S4.*

Definitions Regarding Sets

A set is **well defined** if it is possible to determine whether any given item is an element of the set. For instance, the set of letters of the English alphabet is well defined. The set of *great songs* is not a well-defined set. It is not possible to determine whether any given song is an element of the set or is not an element of the set because there is no standard method for making such a judgment.

The statement "4 is an element of the set of natural numbers" can be written using mathematical notation as $4 \in N$. The symbol $\in$ is read "is an element of." To state that "-3 is not an element of the set of natural numbers," we use the "is not an element of" symbol, $\notin$, and write $-3 \notin N$.

TAKE NOTE

Recall that N denotes the set of natural numbers, I denotes the set of integers, and W denotes the set of whole numbers.

EXAMPLE 4 ■ True or False

Determine whether each statement is true or false.

a. $4 \in \{2, 3, 4, 7\}$ **b.** $-5 \in N$ **c.** $\frac{1}{2} \notin I$

d. The set of nice cars is a well-defined set.

Solution

a. Since 4 is an element of the given set, the statement is true.

b. There are no negative natural numbers, so the statement is false.

c. Since $\frac{1}{2}$ is not an integer, the statement is true.

d. The word *nice* is not precise, so the statement is false.

CHECK YOUR PROGRESS 4 Determine whether each statement is true or false.

a. $5.2 \in \{1, 2, 3, 4, 5, 6\}$ **b.** $-101 \in I$ **c.** $2.5 \notin W$

d. The set of all integers larger than π is a well-defined set.

Solution *See page S4.*

TAKE NOTE

Neither the set $\{0\}$ nor the set $\{\emptyset\}$ represents the empty set because each set has one element.

The **empty set,** or **null set,** is the set that contains no elements. The symbol $\emptyset$ or $\{\ \}$ is used to represent the empty set. As an example of the empty set, consider the set of natural numbers that are negative integers.

Another method of representing a set is **set-builder notation.** Set-builder notation is especially useful when describing infinite sets. For instance, in set-builder notation, the set of natural numbers greater than 7 is written as follows:

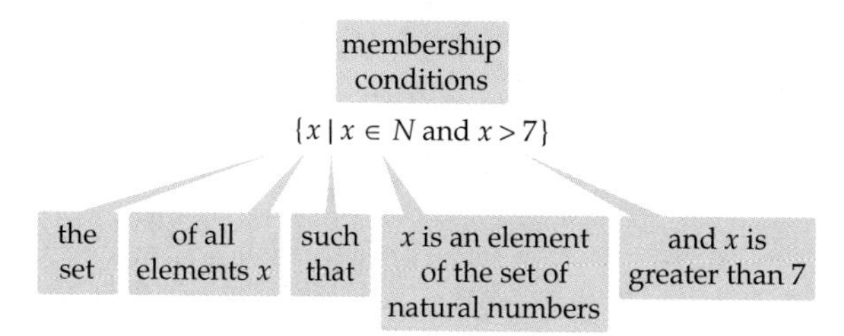

The above set-builder notation is read as "the set of all elements x such that x is an element of the set of natural numbers and x is greater than 7." It is impossible to list all the elements of the set, but set-builder notation defines the set by describing its elements.

EXAMPLE 5 ■ Set-Builder Notation

Use set-builder notation to write the following sets.

a. The set of integers greater than -3

b. The set of whole numbers less than 1000

Solution

a. $\{x \mid x \in I \text{ and } x > -3\}$ **b.** $\{x \mid x \in W \text{ and } x < 1000\}$

CHECK YOUR PROGRESS 5 Use set-builder notation to write the following sets.

a. The set of integers less than 9

b. The set of natural numbers greater than 4

Solution *See page S4.*

A set is **finite** if the number of elements in the set is a whole number. The **cardinal number** of a finite set is the number of elements in the set. The cardinal number of a finite set A is denoted by the notation $n(A)$. For instance, if $A = \{1, 4, 6, 9\}$, then $n(A) = 4$. In this case, A has a cardinal number of 4, which is sometimes stated as "A has a *cardinality* of 4."

EXAMPLE 6 ■ The Cardinality of a Set

Find the cardinality of each of the following sets.

a. $J = \{2, 5\}$ **b.** $S = \{3, 4, 5, 6, 7, \ldots, 31\}$ **c.** $T = \{3, 3, 7, 51\}$

Solution

a. Set J contains exactly two elements, so J has a cardinality of 2. Using mathematical notation we state this as $n(J) = 2$.

b. Only a few elements are actually listed. The number of natural numbers from 1 to 31 is 31. If we omit the numbers 1 and 2, then the number of natural numbers from 3 to 31 must be $31 - 2 = 29$. Thus $n(S) = 29$.

c. Elements that are listed more than once are counted only once. Thus $n(T) = 3$.

CHECK YOUR PROGRESS 6 Find the cardinality of the following sets.

a. $C = \{-1, 5, 4, 11, 13\}$ **b.** $D = \{0\}$ **c.** $E = \varnothing$

Solution See page S4.

The following definitions play an important role in our work with sets.

Definition of Equal Sets

Set A is **equal** to set B, denoted by $A = B$, if and only if A and B have exactly the same elements.

For instance $\{d, e, f\} = \{e, f, d\}$.

Definition of Equivalent Sets

Set A is **equivalent** to set B, denoted by $A \sim B$, if and only if A and B have the same number of elements.

QUESTION *If two sets are equal, must they also be equivalent?*

EXAMPLE 7 ■ Equal Sets and Equivalent Sets

State whether each of the following pairs of sets are equal, equivalent, both, or neither.

a. $\{a, e, i, o, u\}, \{3, 7, 11, 15, 19\}$ **b.** $\{4, -2, 7\}, \{3, 4, 7, 9\}$

Solution

a. The sets are not equal. However, each set has exactly five elements, so the sets are equivalent.

b. The first set has three elements and the second set has four elements, so the sets are not equal and are not equivalent.

CHECK YOUR PROGRESS 7 State whether each of the following pairs of sets are equal, equivalent, both, or neither.

a. $\{x \mid x \in W \text{ and } x \leq 5\}, \{\alpha, \beta, \Gamma, \Delta, \delta, \varepsilon\}$

b. $\{5, 10, 15, 20, 25, 30, \ldots, 80\}, \{x \mid x \in N \text{ and } x < 17\}$

Solution See page S4.

ANSWER *Yes. If the sets are equal, then they have exactly the same elements; therefore, they also have the same number of elements.*

Georg Cantor

Math Matters Georg Cantor

Georg Cantor (kăn′tər) (1845–1918) was a German mathematician who developed many new concepts regarding the theory of sets. Cantor studied under the famous mathematicians Karl Weirstrass and Leopold Kronecker at the University of Berlin. Although Cantor demonstrated a talent for mathematics, his professors were unaware that Cantor would produce extraordinary results that would cause a major stir in the mathematical community.

Cantor never achieved his lifelong goal of a professorship at the University of Berlin. Instead he spent his active career at the undistinguished University of Halle. It was during this period, when Cantor was between the ages of 29 and 39, that he produced his best work. Much of this work was of a controversial nature. One of the simplest of the controversial concepts concerned points on a line segment. For instance, consider the line segment $\overline{AB}$ and the line segment $\overline{CD}$ in the figure above. Which of these two line segments do you think contains the most points? Cantor was able to prove that they both contain the same number of points. In fact, he was able to prove that any line segment, no matter how short, contains the same number of points as a line, or a plane, or all of three-dimensional space. We will take a closer look at some of the mathematics developed by Cantor in the last section of this chapter.

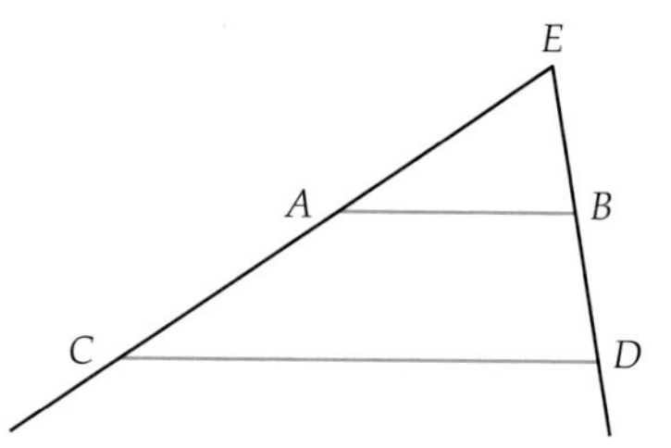

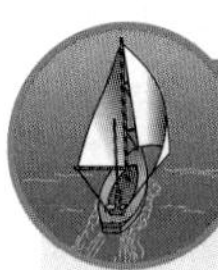

Excursion

Fuzzy Sets

In traditional set theory, an element either belongs to a set or does not belong to the set. For instance, let $A = \{x | x \text{ is an even integer}\}$. Given $x = 8$, we have $x \in A$. However, if $x = 11$, then $x \notin A$. For any given integer, we can decide whether x belongs to A.

Now consider the set $B = \{x | x \text{ is a number close to } 10\}$. Does 8 belong to this set? Does 9.9 belong to the set? Does 10.001 belong to the set? Does 10 belong to the set? Does -50 belong to the set? Given the imprecision of the words "close to," it is impossible to know which numbers belong to set B.

In 1965, Lotfi A. Zadeh of the University of California, Berkeley, published a paper titled *Fuzzy Sets* in which he described the mathematics of fuzzy set theory. This theory proposed that "to some degree," many of the numbers 8, 9.9, 10.001, 10 and -50 belong to set B defined in the previous paragraph. Zadeh proposed giving each element of a set a *membership grade* or *membership value*. This value is a number from 0 to 1. The closer the membership value is to 1, the greater the certainty that an element belongs to the set. The closer the membership value is to 0, the less the certainty that an element

(continued)

historical note

Lotfi Zadeh
Lotfi Zadeh's work in the area of fuzzy sets has led to a new area of mathematics called *soft computing.* On the topic of soft computing, Zadeh stated, "The essence of soft computing is that unlike the traditional, hard computing, soft computing is aimed at an accommodation with the pervasive imprecision of the real world. Thus the guiding principle of soft computing is: Exploit the tolerance for imprecision, uncertainty and partial truth to achieve tractability, robustness, low solution cost and better rapport with reality. In the final analysis, the role model for soft computing is the human mind." ■

belongs to the set. Elements of fuzzy sets are written in the form (element, membership value). Here is an example of a fuzzy set.

$$C = \{(8, 0.4), (9.9, 0.9), (10.001, 0.999), (10, 1), (-50, 0)\}$$

An examination of the membership values suggests that we are certain that 10 belongs to C (membership value is 1) and we are certain that -50 does not belong to C (membership value is 0). Every other element belongs to the set "to some degree."

The concept of a fuzzy set has been used in many real-world applications. Here are a few examples.

- Control of heating and air conditioning systems
- Compensation against vibrations in camcorders
- Voice recognition by computers
- Control of valves and dam gates at power plants
- Control of robots
- Control of subway trains
- Automatic camera focusing

A Fuzzy Heating System

Typical heating systems are controlled by a thermostat that turns a furnace on when the room temperature drops below a set point and turns the furnace off when the room temperature exceeds the set point. The furnace either runs at full force or it shuts down completely. This type of heating system is inefficient, and the frequent off and on changes can be annoying. A fuzzy heating system makes use of "fuzzy" definitions such as cold, warm, and hot to direct the furnace to run at low, medium, or full force. This results in a more efficient heating system and fewer temperature fluctuations.

Excursion Exercises

1. Mark, Erica, Larry, and Jennifer have each defined a fuzzy set to describe what they feel is a "good" grade. Each person paired the letter grades A, B, C, D, and F with a membership value. The results are as follows.

   Mark: $M = \{(A, 1), (B, 0.75), (C, 0.5), (D, 0.5), (F, 0)\}$

   Erica: $E = \{(A, 1), (B, 0), (C, 0), (D, 0), (F, 0)\}$

   Larry: $L = \{(A, 1), (B, 1), (C, 1), (D, 1), (F, 0)\}$

   Jennifer: $J = \{(A, 1), (B, 0.8), (C, 0.6), (D, 0.1), (F, 0)\}$

   a. Which of the four people considers an A grade to be the only good grade?
   b. Which of the four people is most likely to be satisfied with a grade of D or better?
   c. Write a fuzzy set that you would use to describe the set of good grades. Consider only the letter grades A, B, C, D, and F.

(continued)

2. In some fuzzy sets, membership values are given by a *membership graph* or by a *formula*. For instance, the following figure is a graph of the membership values of the fuzzy set *OLD*.

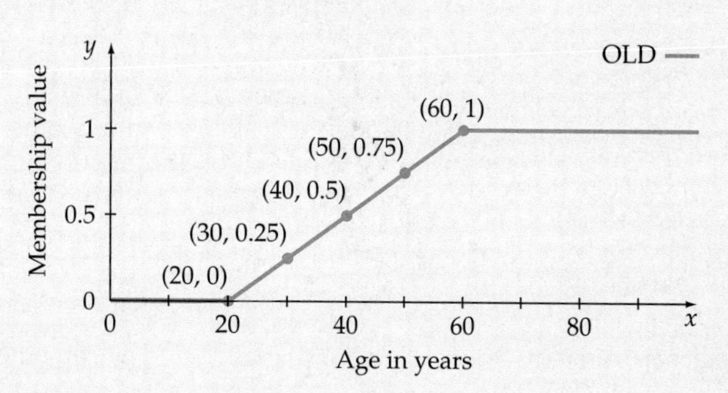

Use the membership graph of *OLD* to determine the membership value of each of the following.

a. $x = 15$ **b.** $x = 50$ **c.** $x = 65$

d. Use the graph of *OLD* to determine the age x with a membership value of 0.25.

An ordered pair (x, y) of a fuzzy set is a **crossover point** if its membership value is 0.5.

e. Find the crossover point for *OLD*.

3. The following membership graph provides a definition of real numbers x that are "about" 4.

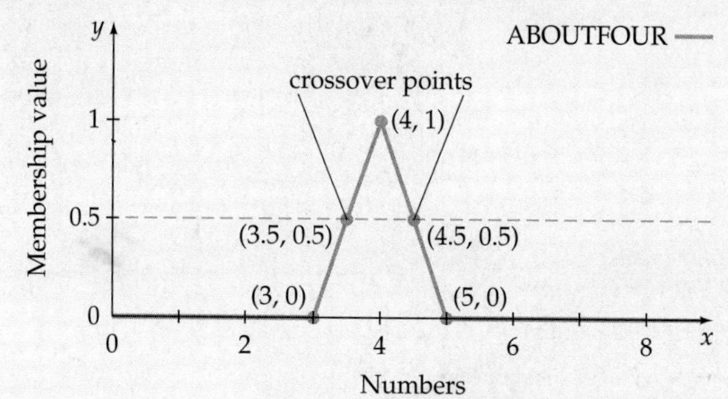

Use the graph of *ABOUTFOUR* to determine the membership value of:

a. $x = 2$ **b.** $x = 3.5$ **c.** $x = 7$

d. Use the graph of *ABOUTFOUR* to determine its crossover points.

4. The membership graphs in the following figure provide definitions of the fuzzy sets *COLD* and *WARM*.

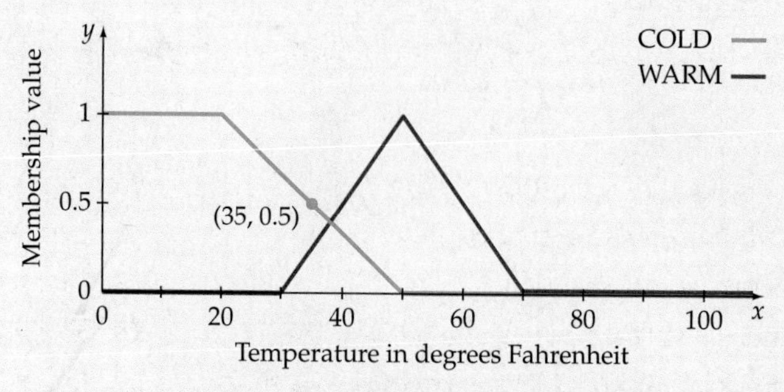

(continued)

The point (35, 0.5) on the membership graph of *COLD* indicates that the membership value for $x = 35$ is 0.5. Thus, by this definition, 35°F is 50% cold. Use the above graphs to estimate

a. the *WARM* membership value for $x = 40$.
b. the *WARM* membership value for $x = 50$.
c. the crossover points of *WARM*.

5. The membership graph in Excursion Exercise 2 shows one person's idea of what ages are "old." Use a grid similar to the following to draw a membership graph that you feel defines the concept of being "young" in terms of a person's age in years.

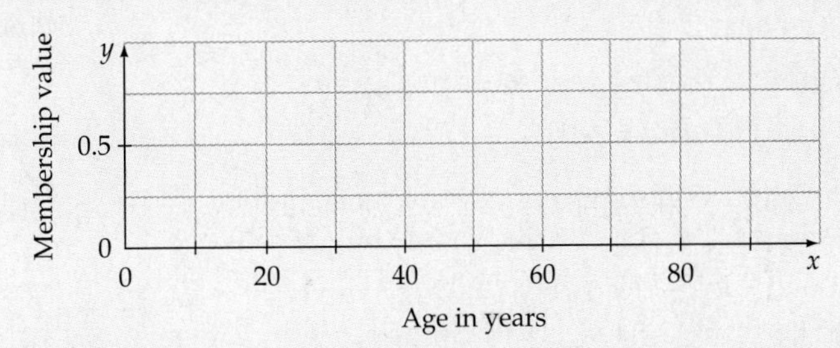

Show your membership graph to a few of your friends. Do they concur with your definition of "young?"

Exercise Set 2.1

In Exercises 1–12, use the roster method to write each of the given sets. For some exercises you may need to consult a reference, such as the Internet or an encyclopedia.

1. The set of U.S. coins with a value of less than 50¢
2. The set of months of the year with a name that ends with the letter y
3. The set of planets in our solar system with a name that starts with the letter M
4. The set of the seven dwarfs
5. The set of U.S. presidents who have served after Jimmy Carter
6. The set of months with exactly 30 days
7. The set of negative integers greater than −6
8. The set of whole numbers less than 8
9. The set of integers x that satisfy $x - 4 = 3$
10. The set of integers x that satisfy $2x - 1 = -11$
11. The set of natural numbers x that satisfy $x + 4 = 1$
12. The set of whole numbers x that satisfy $x - 1 < 4$

In Exercises 13–20, write a description of each set. There may be more than one correct description.

13. {Tuesday, Thursday}
14. {Libra, Leo}
15. {Mercury, Venus}
16. {penny, nickel, dime}
17. {1, 2, 3, 4, 5, 6, 7, 8, 9}
18. {2, 4, 6, 8}
19. $\{x \mid x \in N \text{ and } x \leq 7\}$
20. $\{x \mid x \in W \text{ and } x < 5\}$

In Exercises 21–30, determine whether each statement is true or false. If the statement is false, give the reason.

21. $b \in \{a, b, c\}$
22. $0 \notin N$
23. $\{b\} \in \{a, b, c\}$
24. $\{1, 5, 9\} \sim \{\Psi, \Pi, \Sigma\}$
25. $\{0\} \sim \varnothing$
26. The set of large numbers is a well-defined set.
27. The set of good teachers is a well-defined set.
28. The set $\{x \mid 2 \leq x \leq 3\}$ is a well-defined set.
29. $\{x^2 \mid x \in I\} = \{x^2 \mid x \in N\}$
30. $0 \in \varnothing$

In Exercises 31–40, use set-builder notation to write each of the following sets.

31. $\{1, 2, 3, 4, 5, 6, 7, 8, 9, 10, 11, 12\}$

32. $\{45, 55, 65, 75\}$

33. $\{5, 10, 15\}$

34. $\{1, 4, 9, 16, 25, 36, 49, 64, 81\}$

35. {January, March, May, July, August, October, December}

36. {Iowa, Ohio, Utah}

37. {Arizona, Alabama, Arkansas, Alaska}

38. {Mexico, Canada}

39. {spring, summer}

40. $\{1900, 1901, 1902, 1903, 1904, \ldots, 1999\}$

Charter Schools During recent years, the number of U.S. charter schools has increased dramatically. The following horizontal bar graph shows the eight states with the greatest percent of U.S. charter schools in the fall of 2004.

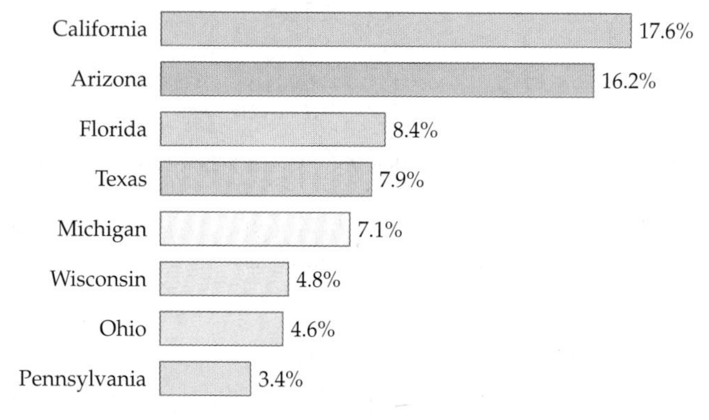

States with the Greatest Percent of Charter Schools

Source: U.S. Charter Schools webpage, **http://www.uscharterschools.org**

Use the data in the above graph and the roster method to represent each of the sets in Exercises 41–44.

41. The set of states in which more than 15% of the schools are charter schools

42. The set of states in which between 5% and 10% of the schools are charter schools

43. $\{x \mid x$ is one of the four states with the greatest percent of charter schools$\}$

44. $\{x \mid x$ is a state in which between 4% and 5% of the schools are charter schools$\}$

Affordability of Housing The following bar graph shows the monthly principal and interest payment needed to purchase an average-priced existing home in the United States for the years from 1997 to 2004.

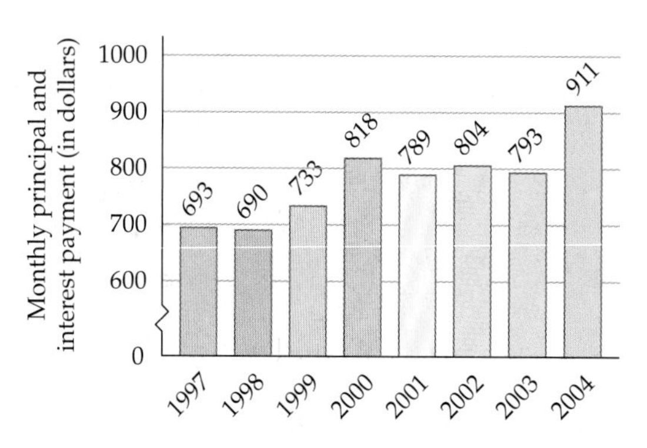

Monthly Principal and Interest Payment for an Average-Priced Existing Home

Source: National Association of REALTORS® as reported in the World Almanac, 2005, p. 483

Use the data in the above graph and the roster method to represent each of the sets in Exercises 45–48.

45. The set of years in which the monthly principal and interest payment, for an average-priced existing home, exceeded \$800

46. The set of years in which the monthly principal and interest payment, for an average-priced existing home, was between \$700 and \$800

47. $\{x \mid x$ is a year in which the monthly principal and interest payment, for an average-priced existing home, was between \$600 and \$700$\}$

48. $\{x \mid x$ is a year in which the monthly principal and interest payment, for an average-priced existing home, was more than \$750$\}$

Gasoline Prices The following graph shows the average cost for a gallon of regular unleaded gasoline in California and in the nation on the first day of each month in 2004.

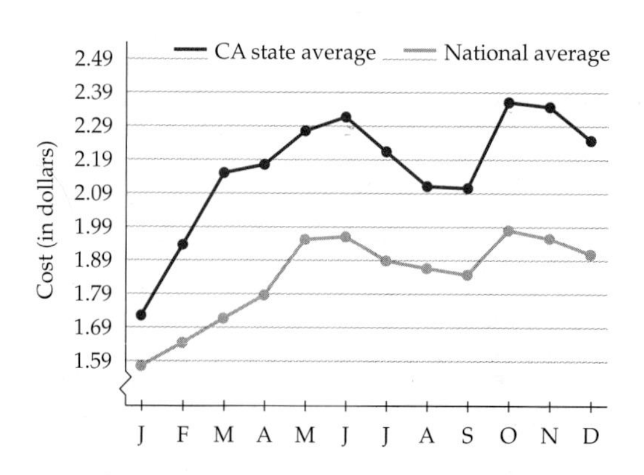

Average Cost of Gallon of Regular Unleaded Gasoline on the First Day of Each Month in 2004

Source: AAA's Media site for retail gasoline prices, **http://198.6.95.31/CAavg.asp**

Use the information in the previous graph and the roster method to represent each of the sets in Exercises 49 and 50.

49. The set of months for which the average cost for a gallon of regular unleaded gasoline was more than \$2.29 in the state of California on the first day of the month

50. The set of months for which the national average cost for a gallon of regular unleaded gasoline was less than \$1.69 on the first day of the month

Ticket Prices The following table shows the average U.S. movie theatre ticket prices for the years from 1985 to 2004.

Average U.S. Movie Theatre Ticket Prices

Year	Price	Year	Price
1985	\$3.55	1995	\$4.35
1986	3.71	1996	4.42
1987	3.91	1997	4.59
1988	4.11	1998	4.69
1989	3.99	1999	5.06
1990	4.22	2000	5.39
1991	4.21	2001	5.65
1992	4.15	2002	5.80
1993	4.14	2003	6.03
1994	4.08	2004	6.21

Source: National Association of Theatre Owners, **http://www.natoonline.org/statisticstickets.htm**

Use the information in the table and the roster method to represent each of the sets in Exercises 51–54.

51. $\{x \mid x$ is a year in the table for which the average ticket price was less than \$4.00$\}$

52. $\{x \mid x$ is a year in the table for which the average ticket price was greater than \$5.20$\}$

53. $\{x \mid x$ is a year in the table for which the average ticket price was greater than \$4.10 but less than \$4.50$\}$

54. $\{x \mid x$ is a year in the table for which the average ticket price was greater than \$4.25 but less than \$6.00$\}$

In Exercises 55–64, find the cardinality of each of the following sets. For some exercises you may need to consult a reference, such as the Internet or an encyclopedia.

55. $A = \{2, 4, 6, 8, 10, 12, 14, 16, 18, 20, 22\}$

56. $B = \{7, 14, 21, 28, 35, 42, 49, 56\}$

57. $D =$ the set of all dogs that can spell "elephant"

58. $S =$ the set of all states in the United States

59. $J =$ the set of all states of the United States that border Minnesota

60. $T =$ the set of all stripes on the U.S. flag

61. $N =$ the set of all baseball teams in the National League

62. $C =$ the set of all chess pieces on a chess board at the start of a chess game

63. $\{3, 6, 9, 12, 15, \ldots, 363\}$

64. $\{7, 11, 15, 19, 23, 27, \ldots, 407\}$

In Exercises 65–72, state whether each of the given pairs of sets are equal, equivalent, both, or neither.

65. The set of U.S. senators; the set of U.S. representatives

66. The set of single-digit natural numbers; the set of pins used in a regulation bowling game

67. The set of positive whole numbers; the set of counting numbers

68. The set of single-digit natural numbers; the set of single-digit integers

69. $\{1, 2, 3\}; \{\text{I}, \text{II}, \text{III}\}$

70. $\{6, 8, 10, 12\}; \{1, 2, 3, 4\}$

71. $\{2, 5\}; \{0, 1\}$

72. $\{\ \}; \{0\}$

In Exercises 73–84, determine whether each of the sets is a well-defined set.

73. The set of good foods

74. The set of the six most heavily populated cities in the United States

75. The set of tall buildings in the city of Chicago

76. The set of states that border Colorado

77. The set of even integers

78. The set of rational numbers of the form $\frac{1}{p}$, where p is a counting number

79. The set of former presidents of the United States who are alive at the present time

80. The set of real numbers larger than 89,000

81. The set of small countries

82. The set of great cities in which to live

83. The set consisting of the best soda drinks

84. The set of fine wines

Extensions

CRITICAL THINKING

In Exercises 85–88, determine whether the given sets are equal. Recall that W represents the set of whole numbers and N represents the set of natural numbers.

85. $A = \{2n + 1 \mid n \in W\}$
$B = \{2n - 1 \mid n \in N\}$

86. $A = \left\{16\left(\frac{1}{2}\right)^{n-1} \middle| \, n \in N\right\}$
$B = \left\{16\left(\frac{1}{2}\right)^{n} \middle| \, n \in W\right\}$

87. $A = \{2n - 1 \mid n \in N\}$
$B = \left\{\frac{n(n+1)}{2} \middle| \, n \in N\right\}$

88. $A = \{3n + 1 \mid n \in W\}$
$B = \{3n - 2 \mid n \in N\}$

89. Give an example of a set that cannot be written using the roster method.

EXPLORATIONS

90. In this section we have introduced the concept of *cardinal numbers.* Use the Internet or a mathematical textbook to find information about *ordinal numbers* and *nominal numbers.* Write a few sentences that explain the differences between these three types of numbers.

SECTION 2.2 Complements, Subsets, and Venn Diagrams

The Universal Set and the Complement of a Set

In complex problem-solving situations and even in routine daily activities, we need to understand the set of all elements that are under consideration. For instance, when an instructor assigns letter grades, the possible choices may include A, B, C, D, F, and I. In this case the letter H is not a consideration. When you place a telephone call, you know that the area code is given by a natural number with three digits. In this instance a rational number such as $\frac{2}{3}$ is not a consideration. The set of all elements that are being considered is called the **universal set.** We will use the letter U to denote the universal set.

The Complement of a Set

The **complement** of a set A, denoted by A', is the set of all elements of the universal set U that are not elements of A.

TAKE NOTE

To determine the complement of a set A you must know the elements of A as well as the elements of the universal set, U, that is being used.

EXAMPLE 1 ■ Find the Complement of a Set

Let $U = \{1, 2, 3, 4, 5, 6, 7, 8, 9, 10\}$, $S = \{2, 4, 6, 7\}$, and $T = \{x \mid x < 10 \text{ and } x \in \text{the odd counting numbers}\}$. Find

a. S' **b.** T'

Solution

a. The elements of the universal set are 1, 2, 3, 4, 5, 6, 7, 8, 9, and 10. From these elements we wish to exclude the elements of S, which are 2, 4, 6, and 7. Therefore $S' = \{1, 3, 5, 8, 9, 10\}$.

b. $T = \{1, 3, 5, 7, 9\}$. Excluding the elements of T from U gives us $T' = \{2, 4, 6, 8, 10\}$.

CHECK YOUR PROGRESS 1 Let $U = \{0, 2, 3, 4, 6, 7, 17\}$, $M = \{0, 4, 6, 17\}$, and $P = \{x \mid x < 7 \text{ and } x \in \text{the even natural numbers}\}$. Find

a. M' **b.** P'

Solution *See page S5.*

There are two fundamental results concerning the universal set and the empty set. Because the universal set contains all elements under consideration, the complement of the universal set is the empty set. Conversely, the complement of the empty set is the universal set, because the empty set has no elements and the universal set contains all the elements under consideration. Using mathematical notation, we state these fundamental results as follows:

The Complement of the Universal Set and the Complement of the Empty Set

$U' = \varnothing$ and $\varnothing' = U$

Subsets

Consider the set of letters in the alphabet and the set of vowels {a, e, i, o, u}. Every element of the set of vowels is an element of the set of letters in the alphabet. The set of vowels is said to be a *subset* of the set of letters in the alphabet. We will often find it useful to examine subsets of a given set.

A Subset of a Set

Set A is a **subset** of set B, denoted by $A \subseteq B$, if and only if every element of A is also an element of B.

Here are two fundamental subset relationships.

Subset Relationships

$A \subseteq A$, for any set A

$\varnothing \subseteq A$, for any set A

To convince yourself that the empty set is a subset of any set, consider the following. We know that a set is a subset of a second set provided every element of the first set is an element of the second set. Pick an arbitrary set A. Because every element of the empty set (*there are none*) is an element of A, we know that $\emptyset \subseteq A$.

The notation $A \not\subseteq B$ is used to denote that A is *not* a subset of B. To show that A is not a subset of B, it is necessary to find at least one element of A that is not an element of B.

TAKE NOTE

Recall that W represents the set of whole numbers and N represents the set of natural numbers.

EXAMPLE 2 ■ True or False

Determine whether each statement is true or false.

a. $\{5, 10, 15, 20\} \subseteq \{0, 5, 10, 15, 20, 25, 30\}$

b. $W \subseteq N$

c. $\{2, 4, 6\} \subseteq \{2, 4, 6\}$

d. $\emptyset \subseteq \{1, 2, 3\}$

Solution

a. True; every element of the first set is an element of the second set.

b. False; 0 is a whole number, but 0 is not a natural number.

c. True; every set is a subset of itself.

d. True; the empty set is a subset of every set.

CHECK YOUR PROGRESS 2 Determine whether each statement is true or false.

a. $\{1, 3, 5\} \subseteq \{1, 5, 9\}$

b. The set of counting numbers is a subset of the set of natural numbers.

c. $\emptyset \subseteq U$

d. $\{-6, 0, 11\} \subseteq I$

Solution *See page S5.*

U

A

A Venn diagram

The English logician John Venn (1834–1923) developed diagrams, which we now refer to as *Venn diagrams,* that can be used to illustrate sets and relationships between sets. In a **Venn diagram,** the universal set is represented by a rectangular region and subsets of the universal set are generally represented by oval or circular regions drawn inside the rectangle. The Venn diagram at the left shows a universal set and one of its subsets, labeled as set A. The size of the circle is not a concern. The region outside of the circle, but inside of the rectangle, represents the set A'.

QUESTION *What set is represented by $(A')'$?*

ANSWER *The set A' contains the elements of U that are not in A. By definition the set $(A')'$ contains only the elements of U that are elements of A. Thus $(A')' = A$.*

Proper Subsets of a Set

Proper Subset

Set A is a **proper subset** of set B, denoted by $A \subset B$, if every element of A is an element of B, and $A \neq B$.

To illustrate the difference between subsets and proper subsets, consider the following two examples.

1. Let $R = \{\text{Mars, Venus}\}$ and $S = \{\text{Mars, Venus, Mercury}\}$. The first set R is a subset of the second set S, because every element of R is an element of S. In addition, R is also a proper subset of S, because $R \neq S$.
2. Let $T = \{\text{Europe, Africa}\}$ and $V = \{\text{Africa, Europe}\}$. The first set T is a subset of the second set V; however, T is *not* a proper subset of V because $T = V$.

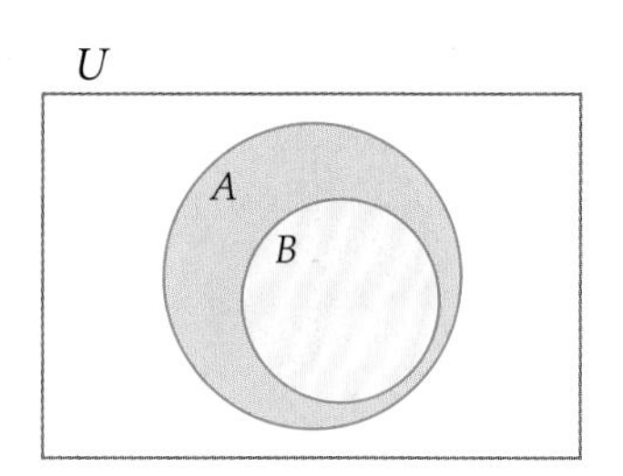

B is a proper subset of *A*.

Venn diagrams can be used to represent proper subset relationships. For instance, if a set B is a proper subset of a set A, then we illustrate this relationship in a Venn diagram by drawing a circle labeled B inside of a circle labeled A. See the Venn diagram at the left.

EXAMPLE 3 ■ Proper Subsets

For each of the following, determine whether the first set is a proper subset of the second set.

a. $\{a, e, i, o, u\}, \{e, i, o, u, a\}$ **b.** N, I

Solution

a. Because the sets are equal, the first set is not a proper subset of the second set.

b. Every natural number is an integer, so the set of natural numbers is a subset of the set of integers. The set of integers contains elements that are not natural numbers, such as -3. Thus the set of natural numbers is a proper subset of the set of integers.

CHECK YOUR PROGRESS 3 For each of the following, determine whether the first set is a proper subset of the second set.

a. N, W **b.** $\{1, 4, 5\}, \{5, 1, 4\}$

Solution *See page S5.*

Some counting problems in the study of probability require that we find all of the subsets of a given set. One way to find all the subsets of a given set is to use the method of making an organized list. First list the empty set, which has no elements. Next list all the sets that have exactly one element, followed by all the sets that contain exactly two elements, followed by all the sets that contain exactly three elements, and so on. This process is illustrated in the following example.

EXAMPLE 4 ■ List all the Subsets of a Set

List all the subsets of {1, 2, 3, 4}.

Solution

An organized list produces the following subsets.

{ } • Subsets with 0 elements

{1}, {2}, {3}, {4} • Subsets with 1 element

{1, 2}, {1, 3}, {1, 4}, {2, 3}, {2, 4}, {3, 4} • Subsets with 2 elements

{1, 2, 3}, {1, 2, 4}, {1, 3, 4}, {2, 3, 4} • Subsets with 3 elements

{1, 2, 3, 4} • Subsets with 4 elements

CHECK YOUR PROGRESS 4 List all of the subsets of {a, b, c, d, e}.

Solution *See page S5.*

Number of Subsets of a Set

The counting techniques developed in Section 1.3 can be used to produce the following result.

The Number of Subsets of a Set

A set with n elements has 2^n subsets.

CALCULATOR NOTE

A calculator can be used to compute powers of 2. The following results were produced on a TI-84 calculator.

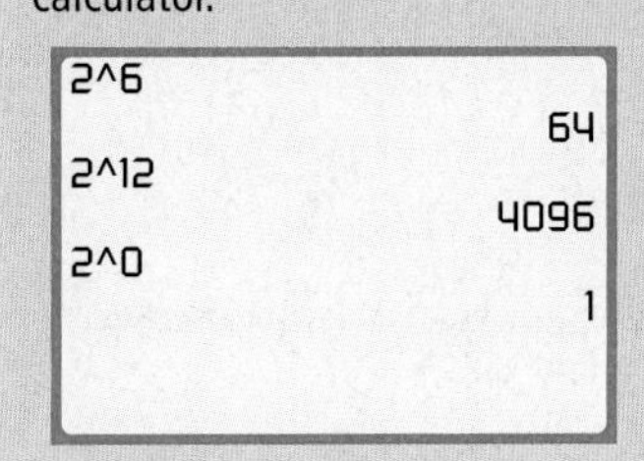

EXAMPLE 5 ■ The Number of Subsets of a Set

Find the number of subsets of each set.

a. {1, 2, 3, 4, 5, 6}

b. {4, 5, 6, 7, 8, . . . , 15}

c. $\varnothing$

Solution

a. {1, 2, 3, 4, 5, 6} has six elements. It has $2^6 = 64$ subsets.

b. {4, 5, 6, 7, 8, . . . , 15} has 12 elements. It has $2^{12} = 4096$ subsets.

c. The empty set has zero elements. It has only one subset ($2^0 = 1$), itself.

CHECK YOUR PROGRESS 5 Find the number of subsets of each set.

a. {Mars, Jupiter, Pluto}

b. $\{x \mid x \leq 15, x \in N\}$

c. {2, 3, 4, 5, . . . , 11}

Solution *See page S5.*

Math Matters **The Barber's Paradox**

Some problems that concern sets have led to paradoxes. For instance, in 1902, the mathematician Bertrand Russell developed the following paradox. "Is the set *A* of all sets that are not elements of themselves an element of itself?" Both the assumption that *A* is an element of *A* and the assumption that *A* is not an element of *A* lead to a contradiction. Russell's paradox has been popularized as follows.

> The town barber shaves all males who do not shave themselves, and he shaves only those males. The town barber is a male who shaves. Who shaves the barber?

The assumption that the barber shaves himself leads to a contradiction, and the assumption that the barber does not shave himself also leads to a contradiction.

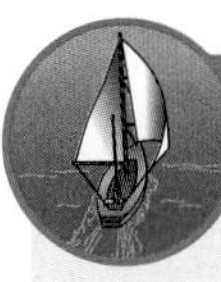

Excursion

Subsets and Complements of Fuzzy Sets

TAKE NOTE

A set such as {3, 5, 9} is called a *crisp set,* to distinguish it from a fuzzy set.

This excursion extends the concept of fuzzy sets that was developed in the Excursion in Section 2.1. Recall that the elements of a fuzzy set are ordered pairs. For any ordered pair (x, y) of a fuzzy set, the membership value y is a real number such that $0 \le y \le 1$.

The set of all x-values that are being considered is called the **universal set for the fuzzy set** and it is denoted by X.

Definition of a Fuzzy Subset

If the fuzzy sets $A = \{(x_1, a_1), (x_2, a_2), (x_3, a_3), \ldots\}$ and $B = \{(x_1, b_1), (x_2, b_2), (x_3, b_3), \ldots\}$ are both defined on the universal set $X = \{x_1, x_2, x_3, \ldots\}$, then $A \subseteq B$ if and only if $a_i \le b_i$ for all i.

A fuzzy set A is a **subset** of a fuzzy set B if and only if the membership value of each element of A *is less than or equal to* its corresponding membership value in set B. For instance, in Excursion Exercise 1 in Section 2.1, Mark and Erica used fuzzy sets to describe the set of good grades as follows:

Mark: $M = \{(\text{A}, 1), (\text{B}, 0.75), (\text{C}, 0.5), (\text{D}, 0.5), (\text{F}, 0)\}$

Erica: $E = \{(\text{A}, 1), (\text{B}, 0), (\text{C}, 0), (\text{D}, 0), (\text{F}, 0)\}$

In this case fuzzy set E is a subset of fuzzy set M because each membership value of set E *is less than or equal to* its corresponding membership value in set M.

Definition of the Complement of a Fuzzy Set

Let A be the fuzzy set $\{(x_1, a_1), (x_2, a_2), (x_3, a_3), \ldots\}$ defined on the universal set $X = \{x_1, x_2, x_3, \ldots\}$. Then the **complement** of A is the fuzzy set $A' = \{(x_1, b_1), (x_2, b_2), (x_3, b_3), \ldots\}$, where each $b_i = 1 - a_i$.

(continued)

Each element of the fuzzy set A' has a membership value that is 1 minus its membership value in the fuzzy set A. For example, the complement of

$$S = \{(\text{math}, 0.8), (\text{history}, 0.4), (\text{biology}, 0.3), (\text{art}, 0.1), (\text{music}, 0.7)\}$$

is the fuzzy set

$$S' = \{(\text{math}, 0.2), (\text{history}, 0.6), (\text{biology}, 0.7), (\text{art}, 0.9), (\text{music}, 0.3)\}.$$

The membership values in S' were calculated by subtracting the corresponding membership values in S from 1. For instance, the membership value of math in set S is 0.8. Thus the membership value of math in set S' is $1 - 0.8 = 0.2$.

Excursion Exercises

1. Let $K = \{(1, 0.4), (2, 0.6), (3, 0.8), (4, 1)\}$ and $J = \{(1, 0.3), (2, 0.6), (3, 0.5), (4, 0.1)\}$ be fuzzy sets defined on $X = \{1, 2, 3, 4\}$. Is $J \subseteq K$? Explain.
2. Consider the following membership graphs of *YOUNG* and *ADOLESCENT* defined on $X = \{x \mid 0 \le x \le 50\}$, where x is age in years.

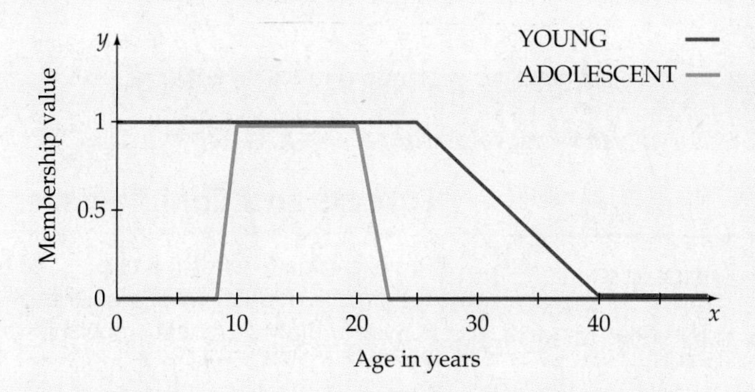

Is the fuzzy set *ADOLESCENT* a subset of the fuzzy set *YOUNG*? Explain.

3. Let the universal set be {A, B, C, D, F} and let

$$G = \{(A, 1), (B, 0.7), (C, 0.4), (D, 0.1), (F, 0)\}$$

be a fuzzy set defined by Greg to describe what he feels is a good grade. Determine G'.

4. Let $C = \{(\text{Ferrari}, 0.9), (\text{Ford Mustang}, 0.6), (\text{Dodge Neon}, 0.5), (\text{Hummer}, 0.7)\}$ be a fuzzy set defined on the universal set {Ferrari, Ford Mustang, Dodge Neon, Hummer}. Determine C'.

Consider the following membership graph.

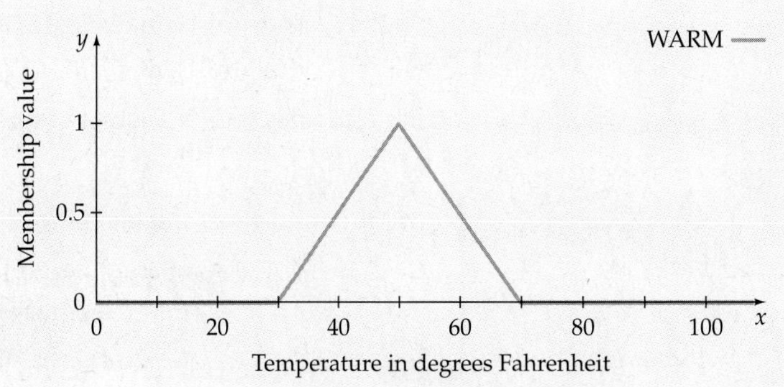

(continued)

The membership graph of *WARM′* can be drawn by *reflecting* the graph of *WARM* about the graph of the line $y = 0.5$, as shown in the following figure.

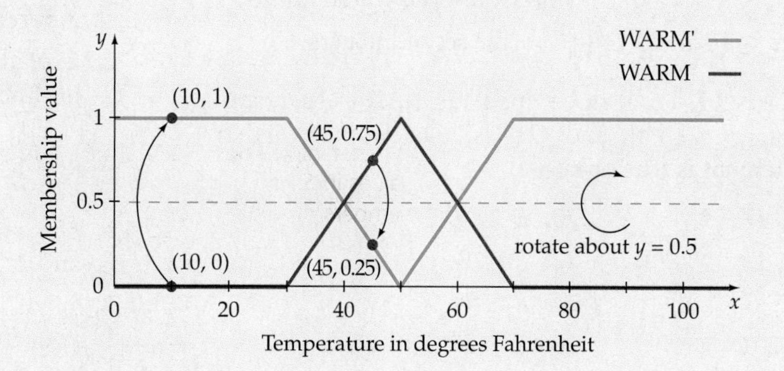

Note that when the membership graph of *WARM* is at a height of 0, the membership graph of *WARM′* is at a height of 1, and vice versa. In general, for any point (x, a) on the graph of *WARM*, there is a corresponding point $(x, 1 - a)$ on the graph of *WARM′*.

5. Use the following membership graph of *COLD* to draw the membership graph of *COLD′*.

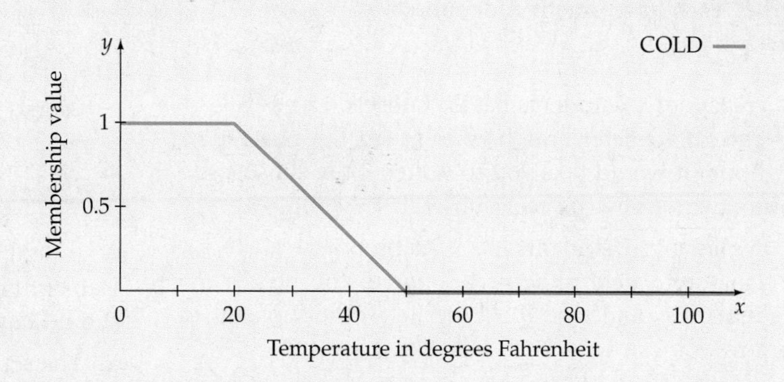

Exercise Set 2.2

In Exercises 1–8, find the complement of the set given that $U = \{0, 1, 2, 3, 4, 5, 6, 7, 8\}$.

1. $\{2, 4, 6, 7\}$
2. $\{3, 6\}$
3. $\emptyset$
4. $\{4, 5, 6, 7, 8\}$
5. $\{x \mid x < 7 \text{ and } x \in N\}$
6. $\{x \mid x < 6 \text{ and } x \in W\}$
7. The set of odd counting numbers less than 8
8. The set of even counting numbers less than 10

In Exercises 9–18, insert either $\subseteq$ or $\nsubseteq$ in the blank space between the sets to make a true statement.

9. {a, b, c, d} {a, b, c, d, e, f, g}
10. {3, 5, 7} {3, 4, 5, 6}
11. {big, small, little} {large, petite, short}
12. {red, white, blue} {the colors in the American flag}
13. the set of integers the set of rational numbers
14. the set of real numbers the set of integers

15. $\varnothing$ {a, e, i, o, u}
16. {all sandwiches} {all hamburgers}
17. {2, 4, 6, . . . , 5000} the set of even whole numbers
18. $\{x \mid x < 10 \text{ and } x \in Q\}$ the set of integers

In Exercises 19–36, let $U = \{p, q, r, s, t\}$, $D = \{p, r, s, t\}$, $E = \{q, s\}$, $F = \{p, t\}$, and $G = \{s\}$. Determine whether each statement is true or false.

19. $F \subseteq D$
20. $D \subseteq F$
21. $F \subset D$
22. $E \subset F$
23. $G \subset E$
24. $E \subset D$
25. $G' \subset D$
26. $E = F'$
27. $\varnothing \subset D$
28. $\varnothing \subset \varnothing$
29. $D' \subset E$
30. $G \in E$
31. $F \in D$
32. $G \not\subset F$
33. D has exactly eight subsets and seven proper subsets.
34. U has exactly 32 subsets.
35. F and F' each have exactly four subsets.
36. $\{0\} = \varnothing$
37. A class of 16 students has 2^{16} subsets. Use a calculator to determine how long (to the nearest hour) it would take you to write all the subsets, assuming you can write each subset in 1 second.
38. A class of 32 students has 2^{32} subsets. Use a calculator to determine how long (to the nearest year) it would take you to write all the subsets, assuming you can write each subset in 1 second.

In Exercises 39–42, list all subsets of the given set.

39. $\{\alpha, \beta\}$
40. $\{\alpha, \beta, \Gamma, \Delta\}$
41. {I, II, III}
42. $\varnothing$

In Exercises 43–50, find the number of subsets of the given set.

43. {2, 5}
44. {1, 7, 11}
45. $\{x \mid x \text{ is an even counting number between 7 and 21}\}$
46. $\{x \mid x \text{ is an odd integer between } -4 \text{ and } 8\}$
47. The set of eleven players on a football team
48. The set of all letters of our alphabet
49. The set of all negative whole numbers
50. The set of all single-digit natural numbers
51. Suppose you have a nickel, two dimes, and a quarter. One of the dimes was minted in 1976, and the other one was minted in 1992.
    a. Assuming you choose at least one coin, how many different sets of coins can you form?
    b. Assuming you choose at least one coin, how many different sums of money can you produce?
    c. Explain why the answers in part a and part b are not the same.
52. The number of subsets of a set with n elements is 2^n.
    a. Use a calculator to find the exact value of 2^{18}, 2^{19}, and 2^{20}.
    b. What is the largest integer power of 2 for which your calculator will display the exact value?
53. **Attribute Pieces** Elementary school teachers use plastic pieces called *attribute pieces* to illustrate subset concepts and to determine whether a student has learned to distinguish among different shapes, sizes, and colors. The following figure shows 12 attribute pieces.

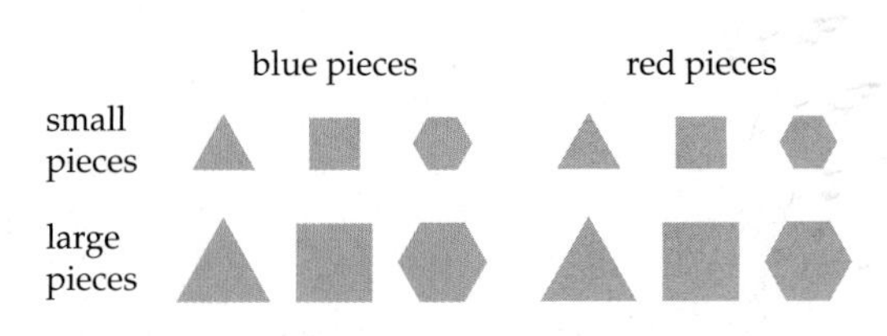

A set of 12 attribute pieces

A student has been asked to form the following sets. Determine the number of elements the student should have in each set.

    a. The set of red attribute pieces
    b. The set of red squares
    c. The set of hexagons
    d. The set of large blue triangles
54. **Sandwich Choices** A delicatessen makes a roast-beef-on-sour-dough sandwich for which you can choose from eight condiments.

a. How many different types of roast-beef-on-sour-dough sandwiches can the delicatessen prepare?

b. What is the minimum number of condiments the delicatessen must have available if it wished to offer at least 2000 different types of roast-beef-on-sour-dough sandwiches?

55. **Omelet Choices** A restaurant provides a brunch where the omelets are individually prepared. Each guest is allowed to choose from 10 different ingredients.

a. How many different types of omelets can the restaurant prepare?

b. What is the minimum number of ingredients that must be available if the restaurant wants to advertise that it offers over 4000 different omelets?

56. **Truck Options** A truck company makes a pickup truck with 12 upgrade options. Some of the options are air conditioning, chrome wheels, and a CD player.

a. How many different versions of this truck can the company produce?

b. What is the minimum number of upgrade options the company must be able to provide if it wishes to offer at least 14,000 different versions of this truck?

Extensions

CRITICAL THINKING

57.  a. Explain why $\{2\} \notin \{1, 2, 3\}$.

b. Explain why $1 \not\subseteq \{1, 2, 3\}$.

c. Consider the set $\{1, \{1\}\}$. Does this set have one or two elements? Explain.

58. a. A set has 1024 subsets. How many elements are in the set?

b. A set has 255 proper subsets. How many elements are in the set?

c. Is it possible for a set to have an odd number of subsets? Explain.

59. **Voting Coalitions** Five people, designated A, B, C, D, and E, serve on a committee. To pass a motion, at least three of the committee members must vote for the motion. In such a situation any set of three or more voters is called a **winning coalition** because if this set of people votes for a motion, the motion will pass. Any nonempty set of two or fewer voters is called a losing coalition.

a. List all the winning coalitions.

b. List all the losing coalitions.

EXPLORATIONS

60. **Subsets and Pascal's Triangle** Following is a list of all the subsets of {a, b, c, d}.
Subsets with

0 elements: { }
1 element: {a}, {b}, {c}, {d}
2 elements: {a, b}, {a, c}, {a, d}, {b, c}, {b, d}, {c, d}
3 elements: {a, b, c}, {a, b, d}, {a, c, d}, {b, c, d}
4 elements: {a, b, c, d}

There is 1 subset with zero elements, and there are 4 subsets with exactly one element, 6 subsets with exactly two elements, 4 subsets with exactly three elements, and 1 subset with exactly four elements. Note that the numbers 1, 4, 6, 4, 1 are the numbers in row 4 of Pascal's triangle, which is shown below. Recall that the numbers in Pascal's triangle are created in the following manner. Each row begins and ends with the number 1. Any other number in a row is the sum of the two closest numbers above it. For instance, the first 10 in row 5 is the sum of the first 4 and the 6 in row 4.

```
          1            row 0
        1   1          row 1
      1   2   1        row 2
    1   3   3   1      row 3
  1   4   6   4   1    row 4
1   5  10  10   5   1  row 5
```

Pascal's triangle

a. Use Pascal's triangle to make a conjecture about the numbers of subsets of {a, b, c, d, e} that have: zero elements, exactly one element, exactly two elements, exactly three elements, exactly four elements, and exactly five elements. Use your work from Check Your Progress 4 on page 68 to verify that your conjecture is correct.

b. Extend Pascal's triangle to show row 6. Use row 6 of Pascal's triangle to make a conjecture about the number of subsets of {a, b, c, d, e, f} that have exactly three elements. Make a list of all the subsets of {a, b, c, d, e, f} that have exactly three elements to verify that your conjecture is correct.

SECTION 2.3 Set Operations

Intersection and Union of Sets

In Section 2.2 we defined the operation of finding the complement of a set. In this section we define the set operations *intersection* and *union*. In everyday usage, the word "intersection" refers to the *common region* where two streets cross. See the figure at the left. The intersection of two sets is defined in a similar manner.

Intersection of Sets

The **intersection** of sets A and B, denoted by $A \cap B$, is the set of elements common to both A and B.

$$A \cap B = \{x \mid x \in A \quad \text{and} \quad x \in B\}$$

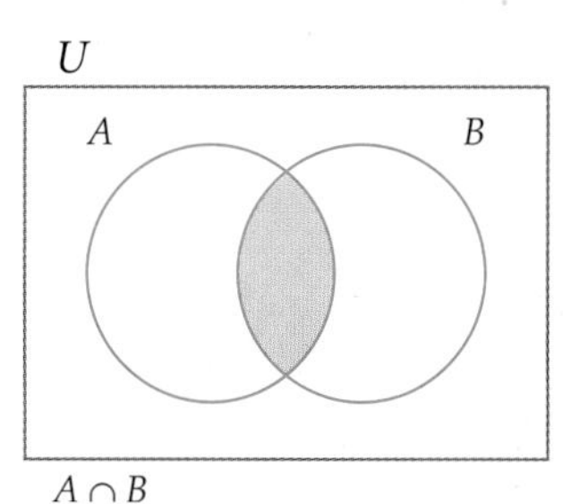

$A \cap B$

In the figure at the left, the region shown in blue represents the intersection of sets A and B.

EXAMPLE 1 ■ Find Intersections

Let $A = \{1, 4, 5, 7\}$, $B = \{2, 3, 4, 5, 6\}$, and $C = \{3, 6, 9\}$. Find

a. $A \cap B$ **b.** $A \cap C$

Solution

a. The elements common to both sets are 4 and 5.

$$A \cap B = \{1, 4, 5, 7\} \cap \{2, 3, 4, 5, 6\}$$
$$= \{4, 5\}$$

b. Sets A and C have no common elements. Thus $A \cap C = \emptyset$.

TAKE NOTE

It is a mistake to write

$$\{1, 5, 9\} \cap \{3, 5, 9\} = 5, 9$$

The intersection of two sets is a set. Thus

$$\{1, 5, 9\} \cap \{3, 5, 9\} = \{5, 9\}$$

CHECK YOUR PROGRESS 1 Let $D = \{0, 3, 8, 9\}$, $E = \{3, 4, 8, 9, 11\}$, and $F = \{0, 2, 6, 8\}$. Find

a. $D \cap E$ **b.** $D \cap F$

Solution *See page S5.*

U

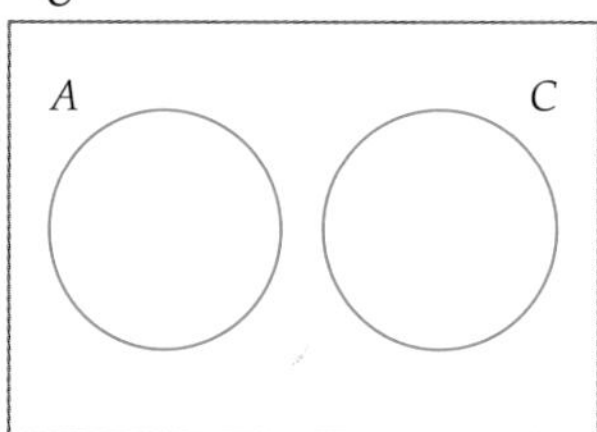

$A \cap C = \emptyset$

Two sets are **disjoint** if their intersection is the empty set. The sets A and C in Example 1b are disjoint. The Venn diagram at the left illustrates two disjoint sets.

In everyday usage, the word "union" refers to the act of uniting or joining together. The union of two sets has a similar meaning.

U

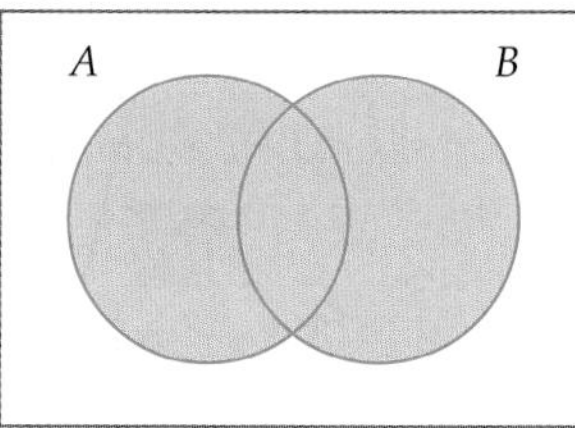

$A \cup B$

Union of Sets

The **union** of sets A and B, denoted by $A \cup B$, is the set that contains all the elements that belong to A or to B or to both.

$$A \cup B = \{x \mid x \in A \quad \text{or} \quad x \in B\}$$

In the figure at the left, the region shown in blue represents the union of sets A and B.

EXAMPLE 2 ■ Find Unions

Let $A = \{1, 4, 5, 7\}$, $B = \{2, 3, 4, 5, 6\}$, and $C = \{3, 6, 9\}$. Find

a. $A \cup B$ **b.** $A \cup C$

Solution

a. List all the elements of set A, which are 1, 4, 5, and 7. Then add to your list the elements of set B that have not already been listed—in this case 2, 3, and 6. Enclose all elements with a pair of braces. Thus

$$A \cup B = \{1, 4, 5, 7\} \cup \{2, 3, 4, 5, 6\}$$
$$= \{1, 2, 3, 4, 5, 6, 7\}$$

b. $A \cup C = \{1, 4, 5, 7\} \cup \{3, 6, 9\}$
$= \{1, 3, 4, 5, 6, 7, 9\}$

CHECK YOUR PROGRESS 2 Let $D = \{0, 4, 8, 9\}$, $E = \{1, 4, 5, 7\}$, and $F = \{2, 6, 8\}$. Find

a. $D \cup E$ **b.** $D \cup F$

Solution *See page S5.*

TAKE NOTE

Would you like soup or salad?

In a sentence, the word "or" can mean one or the other, but not both. For instance, if a menu states that you can have soup or salad with your meal, this generally means that you can have either soup or salad for the price of the meal, but not both. In this case the word "or" is said to be an *exclusive or.* In the mathematical statement "*A* or *B*," the "or" is an *inclusive or.* It means *A* or *B*, or both.

In mathematical problems that involve sets, the word "and" is interpreted to mean *intersection.* For instance, the phrase "the elements of A and B" means the elements of $A \cap B$. Similarly, the word "or" is interpreted to mean *union.* The phrase "the elements of A or B" means the elements of $A \cup B$.

EXAMPLE 3 ■ Describe Sets

Write a sentence that describes the set.

a. $A \cup (B \cap C)$ **b.** $J \cap K'$

Solution

a. The set $A \cup (B \cap C)$ can be described as "the set of all elements that are in A, or are in B and C."

b. The set $J \cap K'$ can be described as "the set of all elements that are in J and are not in K."

CHECK YOUR PROGRESS 3 Write a sentence that describes the set.

a. $D \cap (E' \cup F)$ **b.** $L' \cup M$

Solution *See page S5.*

Venn Diagrams and Equality of Sets

The equality $A \cup B = A \cap B$ is true for some sets A and B, but not for all sets A and B. For instance, if $A = \{1, 2\}$ and $B = \{1, 2\}$, then $A \cup B = A \cap B$. However, we can prove that, in general, $A \cup B \neq A \cap B$ by finding an example for which the expressions are not equal. One such example is $A = \{1, 2, 3\}$ and $B = \{2, 3\}$. In this case $A \cup B = \{1, 2, 3\}$, whereas $A \cap B = \{2, 3\}$. This example is called a *counterexample.* The point to remember is that if you wish to show that two set expressions are *not equal,* then you need to find just one counterexample.

In the next example, we present a technique that uses Venn diagrams to determine whether two set expressions are equal.

EXAMPLE 4 ■ Equality of Sets

Use Venn diagrams to determine whether $(A \cup B)' = A' \cap B'$ for all sets A and B.

Solution

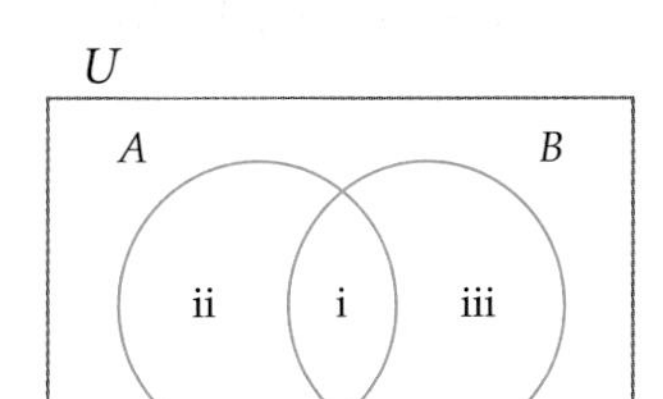

Draw a Venn diagram that shows the two sets A and B, as in the figure at the left. Label the four regions as shown. To determine what region(s) represents $(A \cup B)'$, first note that $A \cup B$ consists of regions i, ii, and iii. Thus $(A \cup B)'$ is represented by region iv. See Figure 2.1.

Draw a second Venn diagram. To determine what region(s) represents $A' \cap B'$, we shade A' (regions iii and iv) with a diagonal up pattern and we shade B' (regions ii and iv) with a diagonal down pattern. The intersection of these shaded regions, which is region iv, represents $A' \cap B'$. See Figure 2.2.

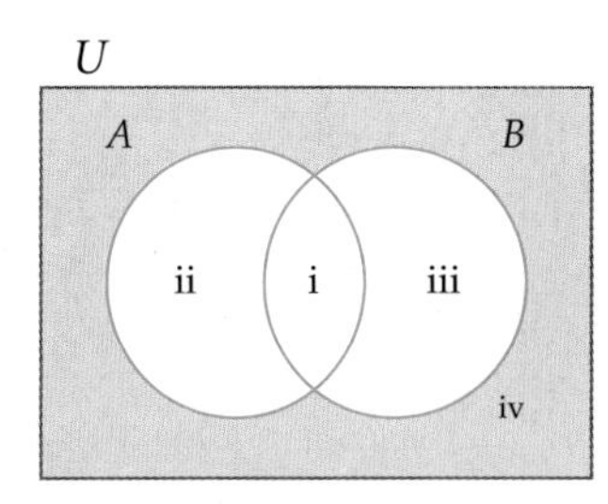

Figure 2.1

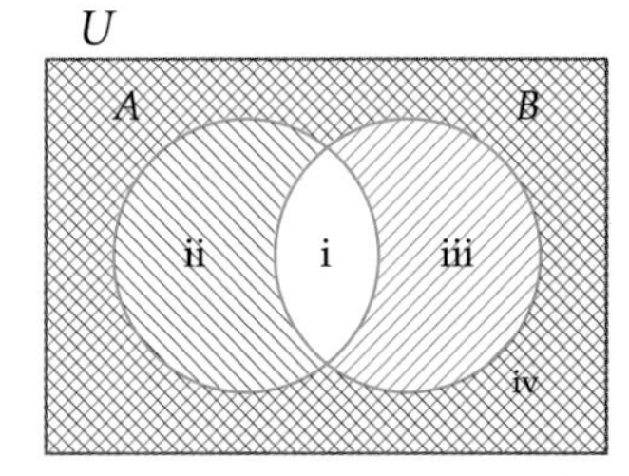

Figure 2.2

Because both $(A \cup B)'$ and $A' \cap B'$ are represented by the same region, we know that $(A \cup B)' = A' \cap B'$ for all sets A and B.

CHECK YOUR PROGRESS 4 Use Venn diagrams to determine whether $(A \cap B)' = A' \cup B'$ for all sets A and B.

Solution *See page S5.*

The properties that were verified in Example 4 and Check Your Progress 4 are known as **De Morgan's laws.**

De Morgan's Laws

For all sets A and B,

$$(A \cup B)' = A' \cap B' \qquad \text{and} \qquad (A \cap B)' = A' \cup B'$$

De Morgan's law $(A \cup B)' = A' \cap B'$ can be stated as "the complement of the union of two sets is the intersection of the complements of the sets. De Morgan's law $(A \cap B)' = A' \cup B'$ can be stated as "the complement of the intersection of two sets is the union of the complements of the sets."

Math Matters The Cantor Set

Consider the set of points formed by a line segment with a length of 1 unit. Remove the middle third of the line segment. Remove the middle third of each of the remaining two line segments. Remove the middle third of each of the remaining four line segments. Remove the middle third of each of the remaining eight line segments. Remove the middle third of each of the remaining sixteen line segments.

The first five steps in the formation of the Cantor set.

The **Cantor set,** also known as **Cantor's Dust,** is the set of points that *remain* after the above process is repeated infinitely many times. You might conjecture that there are no points in the Cantor set, but it can be shown that there are just as many points in the Cantor set as in the original line segment! This is remarkable because it can also be shown that the sum of the lengths of the *removed* line segments equals 1 unit, which is the length of the original line segment. You can find additional information about the remarkable properties of the Cantor set on the Internet.

Venn Diagrams Involving Three Sets

In the next example, we extend the Venn diagram procedure illustrated in the previous examples to expressions that involve three sets.

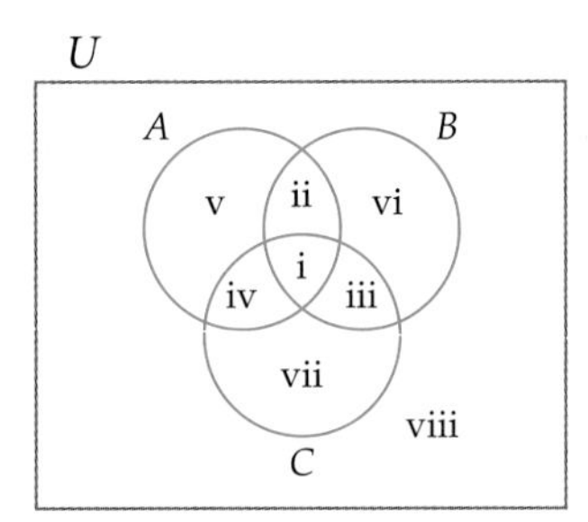

Figure 2.3

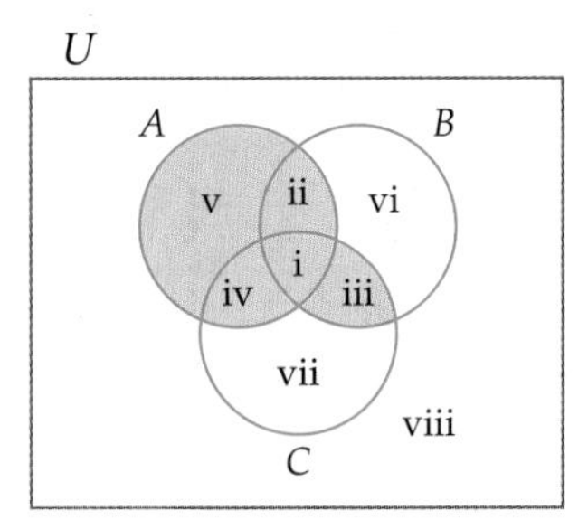

Figure 2.4

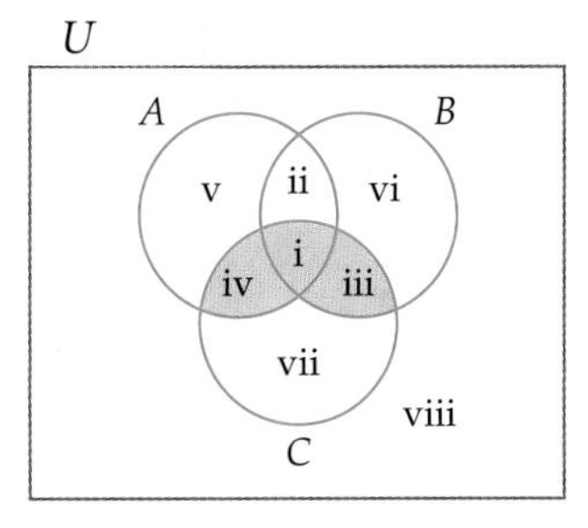

Figure 2.5

EXAMPLE 5 ■ Equality of Set Expressions

Use Venn diagrams to determine whether $A \cup (B \cap C) = (A \cup B) \cap C$ for all sets A, B, and C.

Solution

Draw a Venn diagram that shows the sets A, B, and C and the eight regions they form. See Figure 2.3. To determine what region(s) represents $A \cup (B \cap C)$, we first consider $(B \cap C)$, represented by regions i and iii, because it is in parentheses. Set A is represented by the regions i, ii, iv, and v. Thus $A \cup (B \cap C)$ is represented by all of the listed regions (namely i, ii, iii, iv, and v), as shown in Figure 2.4.

Draw a second Venn diagram showing the three sets A, B, and C, as in Figure 2.3. To determine what region(s) represents $(A \cup B) \cap C$, we first consider $(A \cup B)$ (regions i, ii, iii, iv, v, and vi) because it is in parentheses. Set C is represented by regions i, iii, iv, and vii. Therefore, the intersection of $(A \cup B)$ and C is represented by the overlap, or regions i, iii, and iv. See Figure 2.5.

Because the sets $A \cup (B \cap C)$ and $(A \cup B) \cap C$ are represented by different regions, we conclude that $A \cup (B \cap C) \neq (A \cup B) \cap C$.

CHECK YOUR PROGRESS 5 Use Venn diagrams to determine whether $A \cup (B \cap C) = (A \cup B) \cap (A \cup C)$ for all sets A, B, and C.

Solution *See page S5.*

Venn diagrams can be used to verify each of the following properties.

Properties of Sets

For all sets A and B,

Commutative Properties

$A \cap B = B \cap A$ — Commutative property of intersection

$A \cup B = B \cup A$ — Commutative property of union

For all sets A, B, and C,

Associative Properties

$(A \cap B) \cap C = A \cap (B \cap C)$ — Associative property of intersection

$(A \cup B) \cup C = A \cup (B \cup C)$ — Associative property of union

Distributive Properties

$A \cap (B \cup C) = (A \cap B) \cup (A \cap C)$ — Distributive property of intersection over union

$A \cup (B \cap C) = (A \cup B) \cap (A \cup C)$ — Distributive property of union over intersection

QUESTION *Does $(B \cup C) \cap A = (A \cap B) \cup (A \cap C)$?*

ANSWER *Yes. The commutative property of intersection allows us to write $(B \cup C) \cap A$ as $A \cap (B \cup C)$, and $A \cap (B \cup C) = (A \cap B) \cup (A \cap C)$ by the distributive property of intersection over union.*

Application: Blood Groups and Blood Types

historical note

The Nobel Prize is an award granted to people who have made significant contributions to society. Nobel Prizes are awarded annually for achievements in physics, chemistry, physiology or medicine, literature, peace, and economics. The prizes were first established in 1901 by the Swedish industrialist Alfred Nobel, who invented dynamite. ■

Karl Landsteiner won a Nobel Prize in 1930 for his discovery of the four different human blood groups. He discovered that the blood of each individual contains exactly one of the following combinations of antigens.

- Only A antigens (blood group A)
- Only B antigens (blood group B)
- Both A and B antigens (blood group AB)
- No A antigens and no B antigens (blood group O)

These four blood groups are represented by the Venn diagram at the left below.

In 1941, Landsteiner and Alexander Wiener discovered that human blood may or may not contain an Rh, or rhesus, factor. Blood with this factor is called Rh-positive and denoted by Rh+. Blood without this factor is called Rh-negative and is denoted by Rh−.

The Venn diagram in Figure 2.6 illustrates the eight blood types (A+, B+, AB+, O+, A−, B−, AB−, O−) that are possible if we consider antigens and the Rh factor.

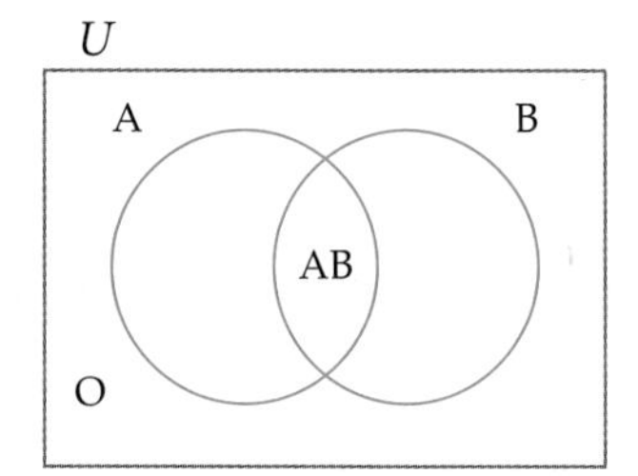

The four blood groups

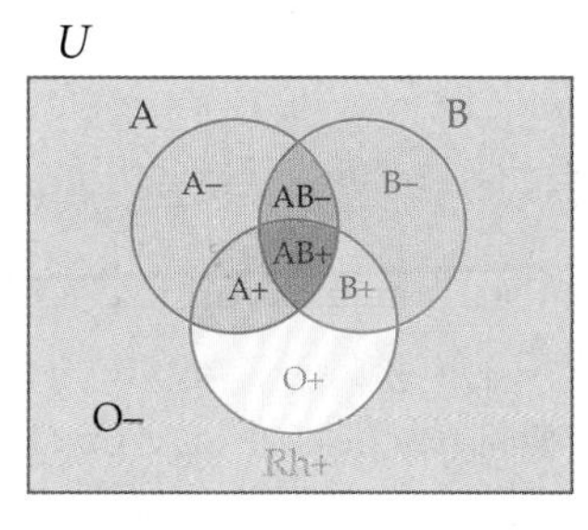

Figure 2.6 The eight blood types

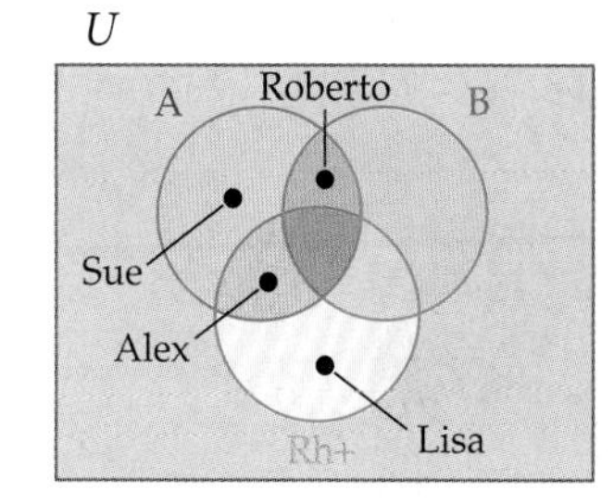

Figure 2.7

EXAMPLE 6 ■ Venn Diagrams and Blood Type

Use the Venn diagrams in Figures 2.6 and 2.7 to determine the blood type of each of the following people.

a. Sue **b.** Lisa

Solution

a. Because Sue is in blood group A, not in blood group B, and not Rh+, her blood type is A−.

b. Lisa is in blood group O and she is Rh+, so her blood type is O+.

CHECK YOUR PROGRESS 6 Use the Venn diagrams in Figures 2.6 and 2.7 to determine the blood type of each of the following people.

a. Alex **b.** Roberto

Solution *See page S6.*

The following table shows the blood types that can safely be given during a blood transfusion to persons of each of the eight blood types.

Blood Transfusion Table

Recipient Blood Type	Donor Blood Type
A+	A+, A−, O+, O−
B+	B+, B−, O+, O−
AB+	A+, A−, B+, B−, AB+, AB−, O+, O−
O+	O+, O−
A−	A−, O−
B−	B−, O−
AB−	A−, B−, AB−, O−
O−	O−

Source: American Red Cross

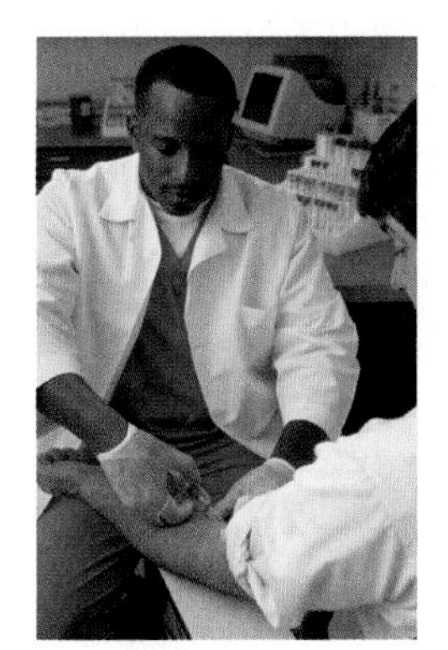

EXAMPLE 7 ■ Applications of the Blood Transfusion Table

Use the blood transfusion table and Figures 2.6 and 2.7 to answer the following questions.

a. Can Sue safely be given a type O+ blood transfusion?

b. Why is a person with type O− blood called a *universal donor?*

Solution

a. Sue's blood type is A−. The blood transfusion table shows that she can safely receive blood only if it is type A− or type O−. Thus it is not safe for Sue to receive type O+ blood in a blood transfusion.

b. The blood transfusion table shows that all eight blood types can safely receive type O− blood. Thus a person with type O− blood is said to be a universal donor.

CHECK YOUR PROGRESS 7 Use the blood transfusion table and Figures 2.6 and 2.7 to answer the following questions.

a. Is it safe for Alex to receive type A− blood in a blood transfusion?

b. What blood type do you have if you are classified as a *universal recipient?*

Solution *See page S6.*

Union and Intersection of Fuzzy Sets

This Excursion extends the concepts of fuzzy sets that were developed in Sections 2.1 and 2.2.

There are a number of ways in which the *union of two fuzzy sets* and the *intersection of two fuzzy sets* can be defined. The definitions we will use are called the **standard union operator** and the **standard intersection operator.** These standard operators preserve many of the set relations that exist in standard set theory.

> **Union and Intersection of Two Fuzzy Sets**
>
> Let $A = \{(x_1, a_1), (x_2, a_2), (x_3, a_3), \ldots\}$ and $B = \{(x_1, b_1), (x_2, b_2), (x_3, b_3), \ldots\}$ Then
>
> $$A \cup B = \{(x_1, c_1), (x_2, c_2), (x_3, c_3), \ldots\}$$
>
> where c_i is the *maximum* of the two numbers a_i and b_i and
>
> $$A \cap B = \{(x_1, c_1), (x_2, c_2), (x_3, c_3), \ldots\}$$
>
> where c_i is the *minimum* of the two numbers a_i and b_i.

Each element of the fuzzy set $A \cup B$ has a membership value that is the *maximum* of its membership value in the fuzzy set A and its membership value in the fuzzy set B. Each element of the fuzzy set $A \cap B$ has a membership value that is the *minimum* of its membership value in fuzzy set A and its membership value in the fuzzy set B. In the following example, we form the union and intersection of two fuzzy sets. Let P and S be defined as follows.

Paul: $P = \{(\text{math}, 0.2), (\text{history}, 0.5), (\text{biology}, 0.7), (\text{art}, 0.8), (\text{music}, 0.9)\}$

Sally: $S = \{(\text{math}, 0.8), (\text{history}, 0.4), (\text{biology}, 0.3), (\text{art}, 0.1), (\text{music}, 0.7)\}$

Then

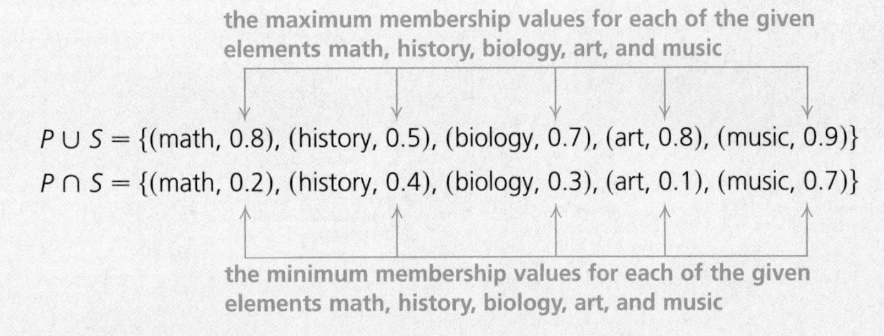

$$P \cup S = \{(\text{math}, 0.8), (\text{history}, 0.5), (\text{biology}, 0.7), (\text{art}, 0.8), (\text{music}, 0.9)\}$$

$$P \cap S = \{(\text{math}, 0.2), (\text{history}, 0.4), (\text{biology}, 0.3), (\text{art}, 0.1), (\text{music}, 0.7)\}$$

Excursion Exercises

In Excursion Exercise 1 of Section 2.1, we defined the following fuzzy sets.

Mark: $M = \{(A, 1), (B, 0.75), (C, 0.5), (D, 0.5), (F, 0)\}$

Erica: $E = \{(A, 1), (B, 0), (C, 0), (D, 0), (F, 0)\}$

Larry: $L = \{(A, 1), (B, 1), (C, 1), (D, 1), (F, 0)\}$

Jennifer: $J = \{(A, 1), (B, 0.8), (C, 0.6), (D, 0.1), (F, 0)\}$

(continued)

Use these fuzzy sets to find each of the following.

1. $M \cup J$ **2.** $M \cap J$ **3.** $E \cup J'$ **4.** $J \cap L'$ **5.** $J \cap (M' \cup L')$

Consider the following membership graphs.

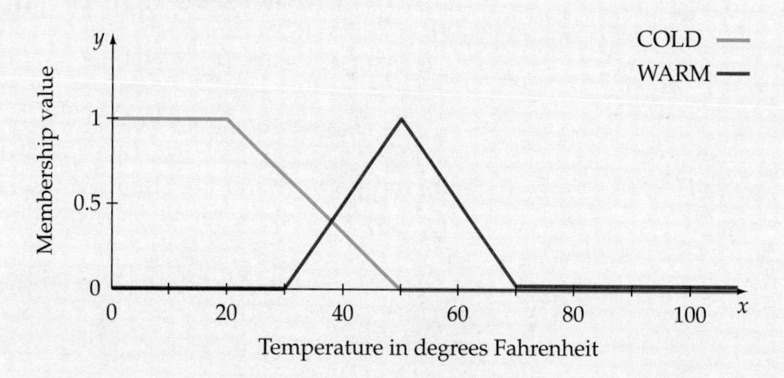

The membership graph of *COLD* ∪ *WARM* is shown in purple in the following figure. The membership graph of *COLD* ∪ *WARM* lies on either the membership graph of *COLD* or

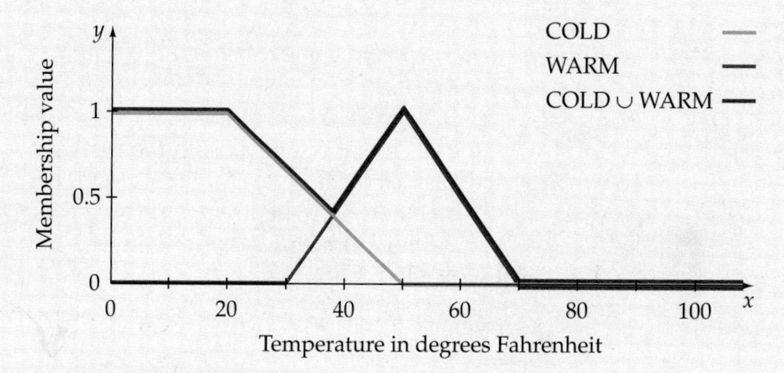

the membership graph of *WARM*, depending on which of these graphs is *higher* at any given temperature *x*.

The membership graph of *COLD* ∩ *WARM* is shown in green in the following figure. The membership graph of *COLD* ∩ *WARM* lies on either the membership graph of *COLD*

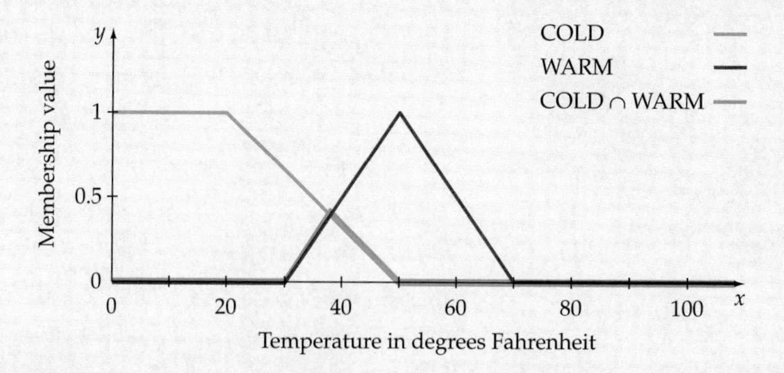

or the membership graph of *WARM*, depending on which of these graphs is *lower* at any given temperature *x*.

(continued)

TAKE NOTE

The following lyrics from the old Scottish song *Loch Lomond* provide an easy way to remember how to draw the graph of the union or intersection of two membership graphs.

Oh! ye'll take the high road and I'll take the low road, and I'll be in Scotland afore ye.

The graph of the union of two membership graphs takes the "high road" provided by the graphs, and the graph of the intersection of two membership graphs takes the "low road" provided by the graphs.

6. Use the following graphs to draw the membership graph of $WARM \cup HOT$.

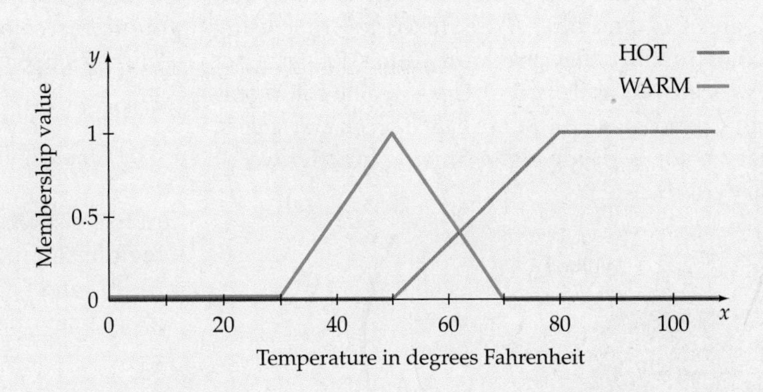

7. Let $X = \{a, b, c, d, e\}$ be the universal set. Determine whether De Morgan's Law $(A \cap B)' = A' \cup B'$ holds true for the fuzzy sets $A = \{(a, 0.3), (b, 0.8), (c, 1), (d, 0.2), (e, 0.75)\}$ and $B = \{(a, 0.5), (b, 0.4), (c, 0.9), (d, 0.7), (e, 0.45)\}$.

Exercise Set 2.3

In Exercises 1–20, let $U = \{1, 2, 3, 4, 5, 6, 7, 8\}$, $A = \{2, 4, 6\}$, $B = \{1, 2, 5, 8\}$, and $C = \{1, 3, 7\}$. Find each of the following.

1. $A \cup B$
2. $A \cap B$
3. $A \cap B'$
4. $B \cap C'$
5. $(A \cup B)'$
6. $(A' \cap B)'$
7. $A \cup (B \cup C)$
8. $A \cap (B \cup C)$
9. $A \cap (B \cap C)$
10. $A' \cup (B \cap C)$
11. $B \cap (B \cup C)$
12. $A \cap A'$
13. $B \cup B'$
14. $(A \cap (B \cup C))'$
15. $(A \cup C') \cap (B \cup A')$
16. $(A \cup C') \cup (B \cup A')$
17. $(C \cup B') \cup \varnothing$
18. $(A' \cup B) \cap \varnothing$
19. $(A \cup B) \cap (B \cap C')$
20. $(B \cap A') \cup (B' \cup C)$

In Exercises 21–28, write a sentence that describes the given mathematical expression.

21. $L' \cup T$
22. $J' \cap K$
23. $A \cup (B' \cap C)$
24. $(A \cup B) \cap C'$
25. $T \cap (J \cup K')$
26. $(A \cap B) \cup C$
27. $(W \cap V) \cup (W \cap Z)$
28. $D \cap (E \cup F)'$

In Exercises 29–36, draw a Venn diagram to show each of the following sets.

29. $A \cap B'$
30. $(A \cap B)'$
31. $(A \cup B)'$
32. $(A' \cap B) \cup B'$
33. $A \cap (B \cup C')$
34. $A \cap (B' \cap C)$
35. $(A \cup C) \cap (B \cup C')$
36. $(A' \cap B) \cup (A \cap C')$

In Exercises 37–40, draw two Venn diagrams to determine whether the following expressions are equal for all sets A and B.

37. $A \cap B'; A' \cup B$
38. $A' \cap B; A \cup B'$
39. $A \cup (A' \cap B); A \cup B$
40. $A' \cap (B \cup B'); A' \cup (B \cap B')$

In Exercises 41–46, draw two Venn diagrams to determine whether the following expressions are equal for all sets A, B, and C.

41. $(A \cup C) \cap B'; A' \cup (B \cup C)$
42. $A' \cap (B \cap C), (A \cup B') \cap C$
43. $(A' \cap B) \cup C, (A' \cap C) \cap (A' \cap B)$
44. $A' \cup (B' \cap C), (A' \cup B') \cap (A' \cup C)$
45. $((A \cup B) \cap C)', (A' \cap B') \cup C'$
46. $(A \cap B) \cap C, (A \cup B \cup C)'$

Computers and televisions make use of *additive color mixing*. The following figure shows that when the *primary colors* red R, green G, and blue B are mixed together using additive color mixing, they produce white, W. Using set notation, we state this as $R \cap B \cap G = W$. The colors yellow Y, cyan C, and magenta M are called *secondary colors*. A secondary color is produced by mixing exactly two of the primary colors.

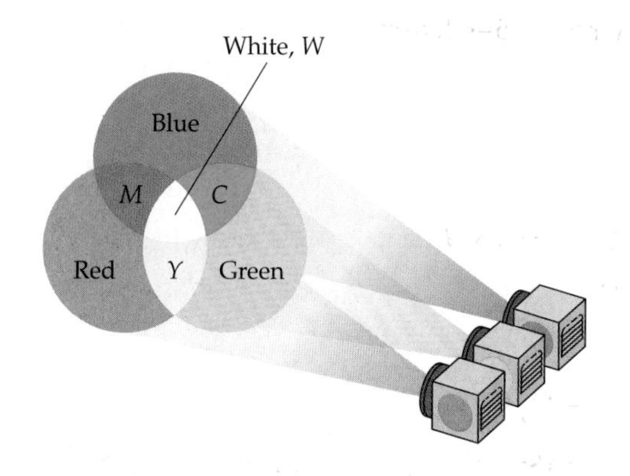

Additive Color Mixing In Exercises 47–49, determine which color is represented by each of the following. Assume the colors are being mixed using additive color mixing. (Use R for red, G for green, and B for blue.)

47. $R \cap G \cap B'$ **48.** $R \cap G' \cap B$

49. $R' \cap G \cap B$

Artists that paint with pigments use *subtractive color mixing* to produce different colors. In a subtractive color mixing system, the primary colors are cyan C, magenta M, and yellow Y. The following figure shows that when the three primary colors are mixed in equal amounts, using subtractive color mixing, they form black, K. Using set notation, we state this as $C \cap M \cap Y = K$. In subtractive color mixing the colors red R, blue B, and green G are the secondary colors. As mentioned previously, a secondary color is produced by mixing equal amounts of exactly two of the primary colors.

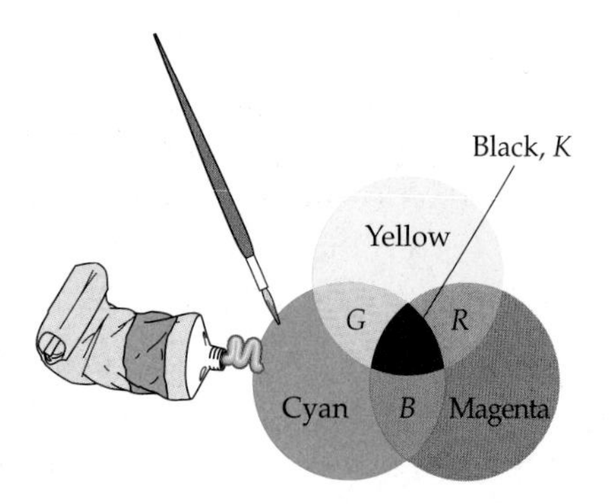

Subtractive Color Mixing In Exercises 50–52, determine which color is represented by each of the following. Assume the colors are being mixed using subtractive color mixing. (Use C for cyan, M for magenta, and Y for yellow.)

50. $C \cap M \cap Y'$ **51.** $C' \cap M \cap Y$

52. $C \cap M' \cap Y$

In Exercises 53–62, use set notation to describe the shaded region. You may use any of the following symbols: A, B, C, $\cap$, $\cup$, and $'$. Keep in mind that each shaded region has more than one set description.

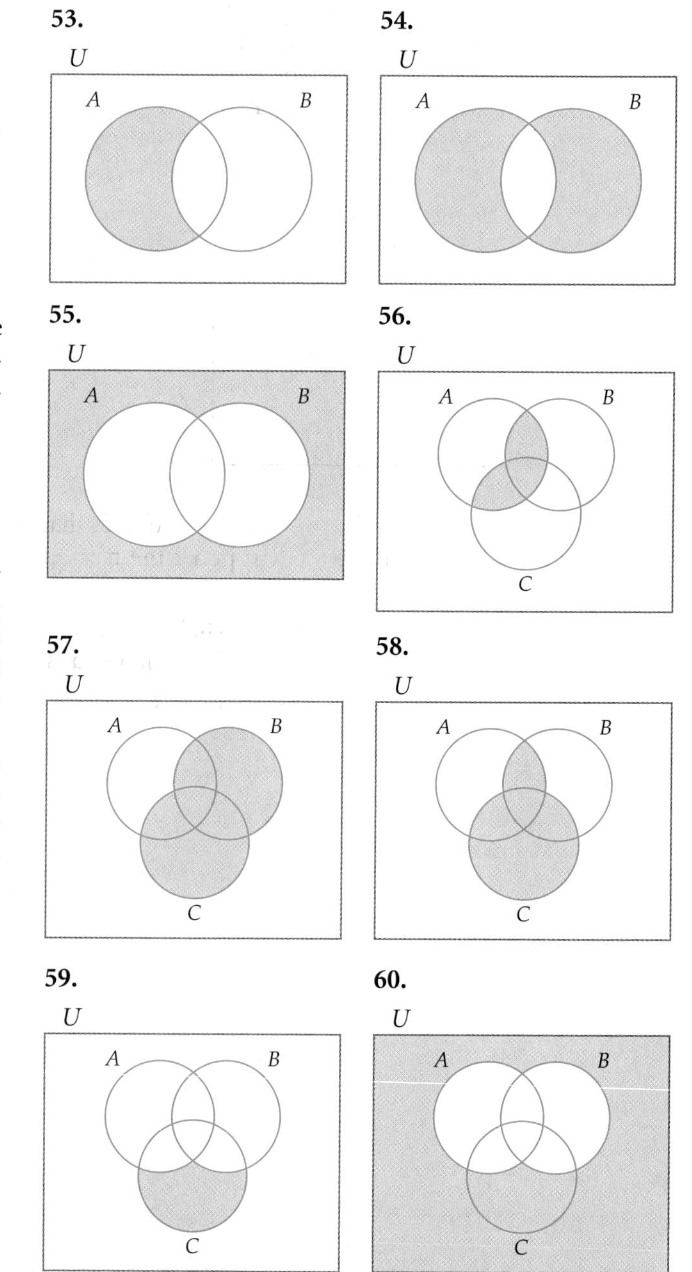

61.

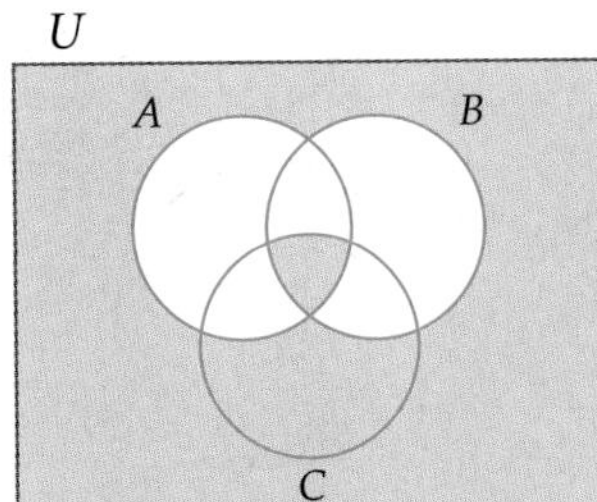

62.

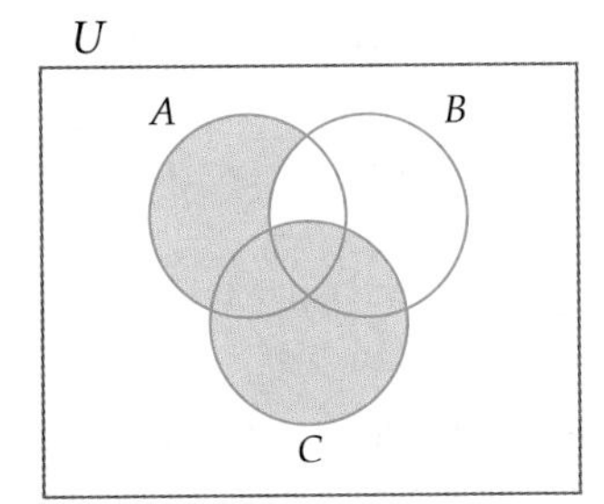

63. **A Survey** A special interest group plans to conduct a survey of households concerning a ban on hand guns. The special-interest group has decided to use the following Venn diagram to help illustrate the results of the survey. *Note:* A rifle is a gun, but it is not a hand gun.

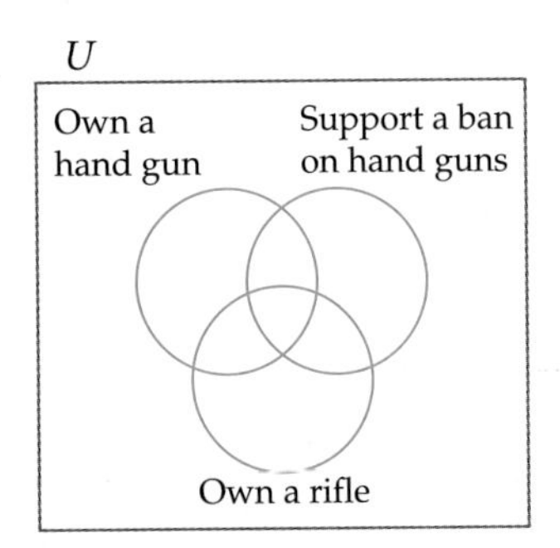

a. Shade in the regions that represent households that own a hand gun and do not support the ban on hand guns.

b. Shade in the region that represents households that own only a rifle and support the ban on hand guns.

c. Shade in the region that represents households that do not own a gun and do not support the ban on hand guns.

64. **A Music Survey** The administrators of an Internet music site plan to conduct a survey of college students to determine how the students acquire music. The administrators have decided to use the following Venn diagram to help tabulate the results of the survey.

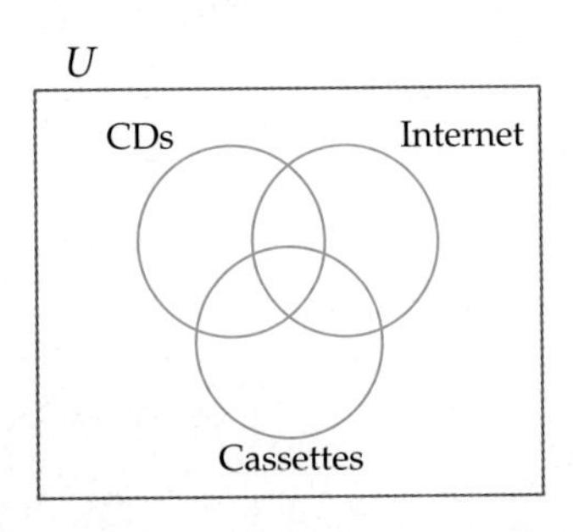

a. Shade in the region that represents students who acquire music from CDs and the Internet, but not from cassettes.

b. Shade in the regions that represent students who acquire music from CDs or the Internet.

c. Shade in the regions that represent students who acquire music from both CDs and cassettes.

In Exercises 65–68, draw a Venn diagram with each of the given elements placed in the correct region.

65. $U = \{-1, 0, 1, 2, 3, 4, 5, 6, 7, 8, 9\}$
$A = \{1, 3, 5\}$
$B = \{3, 5, 7, 8\}$
$C = \{-1, 8, 9\}$

66. $U = \{2, 4, 6, 8, 10, 12, 14\}$
$A = \{2, 10, 12\}$
$B = \{4, 8\}$
$C = \{6, 8, 10\}$

67. U = {Sue, Bob, Al, Jo, Ann, Herb, Eric, Mike, Sal}
A = {Sue, Herb}
B = {Sue, Eric, Jo, Ann}
$Rh+$ = {Eric, Sal, Al, Herb}

68. U = {Hal, Marie, Rob, Armando, Joel, Juan, Melody}
A = {Marie, Armando, Melody}
B = {Rob, Juan, Hal}
$Rh+$ = {Hal, Marie, Rob, Joel, Juan, Melody}

In Exercises 69 and 70, use two Venn diagrams to verify the following properties for all sets *A*, *B*, and *C*.

69. The associative property of intersection

$$(A \cap B) \cap C = A \cap (B \cap C)$$

70. The distributive property of intersection over union

$$A \cap (B \cup C) = (A \cap B) \cup (A \cap C)$$

Extensions

CRITICAL THINKING

Difference of Sets Another operation that can be defined on sets *A* and *B* is the **difference of the sets,** denoted by $A - B$. Here is a formal definition of the difference of sets *A* and *B*.

$$A - B = \{x \mid x \in A \quad \text{and} \quad x \notin B\}$$

Thus $A - B$ is the set of elements that belong to *A* but not to *B*. For instance, let $A = \{1, 2, 3, 7, 8\}$ and $B = \{2, 7, 11\}$. Then $A - B = \{1, 3, 8\}$.

In Exercises 71–76, determine each difference, given that $U = \{1, 2, 3, 4, 5, 6, 7, 8, 9\}$, $A = \{2, 4, 6, 8\}$, and $B = \{2, 3, 8, 9\}$.

71. $B - A$

72. $A - B$

73. $A - B'$

74. $B' - A$

75. $A' - B'$

76. $A' - B$

77. **John Venn** Write a few paragraphs about the life of John Venn and his work in the area of mathematics.

The following Venn diagram illustrates that four sets can partition the universal set into 16 different regions.

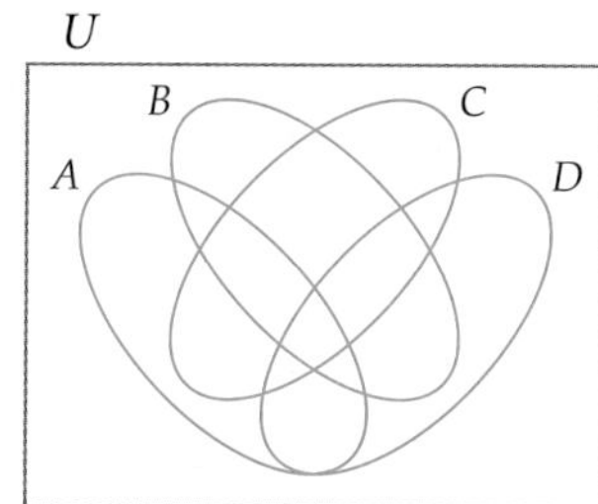

In Exercises 78 and 79, use a Venn diagram similar to the one at the left below to shade in the region represented by the given expression.

78. $(A \cap B) \cup (C' \cap D)$

79. $(A \cup B)' \cap (C \cap D)$

EXPLORATIONS

80. In an article in *New Scientist* magazine, Anthony W. F. Edwards illustrated how to construct Venn diagrams that involve many sets.[1] Search the Internet to find Edwards's method of constructing a Venn diagram for five sets and a Venn diagram for six sets. Use drawings to illustrate Edwards's method of constructing a Venn diagram for five sets and a Venn diagram for six sets. (*Source:* **http://www.combinatorics.org/Surveys/ds5/VennWhatEJC.html**)

SECTION 2.4 Applications of Sets

Surveys: An Application of Sets

Counting problems occur in many areas of applied mathematics. To solve these counting problems, we often make use of a Venn diagram and the inclusion-exclusion principle, which will be presented in this section.

EXAMPLE 1 ■ A Survey of Preferences

A movie company is making plans for future movies it wishes to produce. The company has done a random survey of 1000 people. The results of the survey are shown below.

695 people like action adventures.

340 people like comedies.

180 people like both action adventures and comedies.

Of the people surveyed, how many people:

a. like action adventures but not comedies?

b. like comedies but not action adventures?

c. do not like either of these types of movies?

---

1. Anthony W. F. Edwards, "Venn diagrams for many sets," *New Scientist*, 7 January 1989, pp. 51–56.

Solution
A Venn diagram can be used to illustrate the results of the survey. We use two overlapping circles (see Figure 2.8). One circle represents the set of people who like action adventures and the other represents the set of people who like comedies. The region i where the circles intersect represents the set of people who like both types of movies.

We start with the information that 180 people like both types of movies and write 180 in region i. See Figure 2.9.

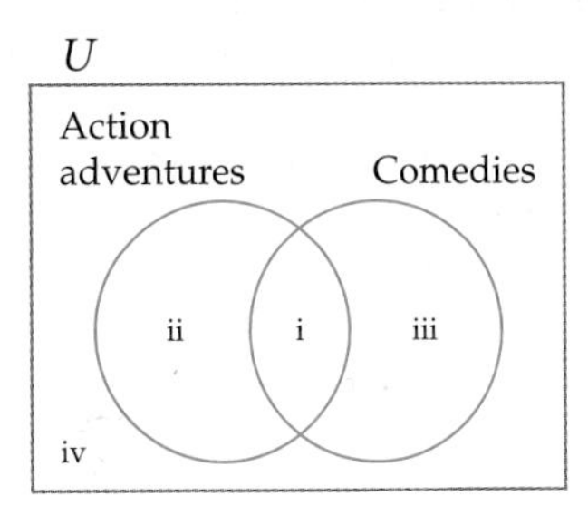

Figure 2.8

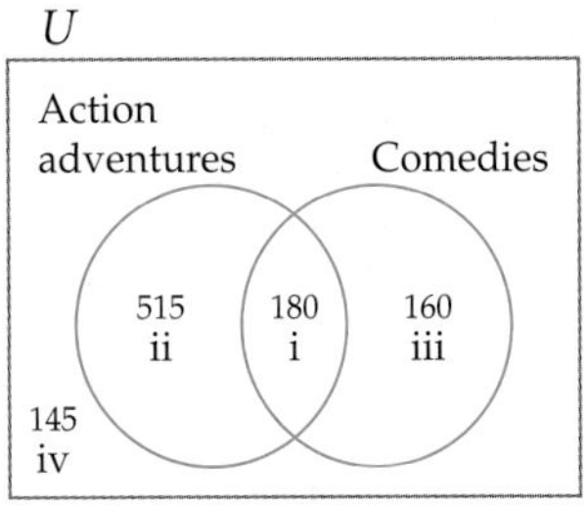

U
Action adventures
Comedies
515 ii
180 i
160 iii
145 iv

Figure 2.9

a. Regions i and ii have a total of 695 people. So far we have accounted for 180 of these people in region i. Thus the number of people in region ii, which is the set of people who like action adventures but do not like comedies, is $695 - 180 = 515$.

b. Regions i and iii have a total of 340 people. Thus the number of people in region iii, which is the set of people who like comedies but do not like action adventures, is $340 - 180 = 160$.

c. The number of people who do not like action adventure movies or comedies is represented by region iv. The number of people in region iv must be the total number of people, which is 1000, less the number of people accounted for in regions i, ii, and iii, which is 855. Thus the number of people who do not like either type of movie is $1000 - 855 = 145$.

CHECK YOUR PROGRESS 1 The athletic director of a school has surveyed 200 students. The survey results are shown below.

140 students like volleyball.
120 students like basketball.
85 students like both volleyball and basketball.

Of the students surveyed, how many students:

a. like volleyball but not basketball?
b. like basketball but not volleyball?
c. do not like either of these sports?

Solution *See page S6.*

In the next example we consider a more complicated survey that involves three types of music.

EXAMPLE 2 ■ A Music Survey

A music teacher has surveyed 495 students. The results of the survey are listed below.

320 students like rap music.
395 students like rock music.
295 students like heavy metal music.
280 students like both rap music and rock music.
190 students like both rap music and heavy metal music.
245 students like both rock music and heavy metal music.
160 students like all three.

How many students:

a. like exactly two of the three types of music?
b. like only rock music?
c. like only one of the three types of music?

Solution

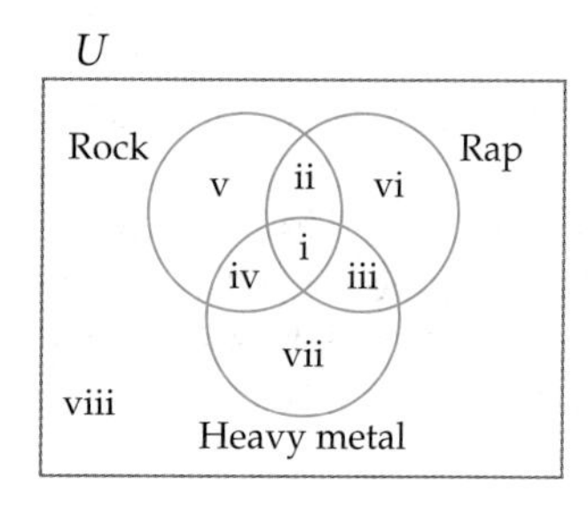

The Venn diagram at the left shows three overlapping circles. Region i represents the set of students who like all three types of music. Each of the regions v, vi, and vii represent the students who like only one type of music.

a. The survey shows that 245 students like rock and heavy metal music, so the numbers we place in regions i and iv must have a sum of 245. Since region i has 160 students, we see that region iv must have $245 - 160 = 85$ students. In a similar manner, we can determine that region ii has 120 students and region iii has 30 students. Thus $85 + 120 + 30 = 235$ students like exactly two of the three types of music.

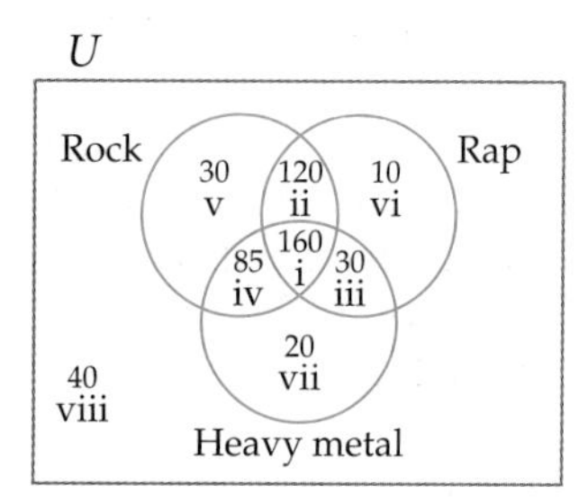

b. The sum of the students represented by regions i, ii, iv, and v must be 395. The number of students in region v must be the difference between this total and the sum of the numbers of students in region i, ii, and iv. Thus the number of students who like only rock music is $395 - (160 + 120 + 85) = 30$. See the Venn diagram at the left.

c. Using the same reasoning as in part b, we find that region vi has 10 students and region vii has 20 students. To find the number of students who like only one type of music, find the sum of the numbers of students in regions v, vi, and vii, which is $30 + 10 + 20 = 60$. See the Venn diagram at the left.

CHECK YOUR PROGRESS 2 An activities director for a cruise ship has surveyed 240 passengers. Of the 240 passengers,

135 like swimming.
150 like dancing.
65 like games.
80 like swimming and dancing.
40 like swimming and games.
25 like dancing and games.
15 like all three activities.

How many passengers:

a. like exactly two of the three types of activities?
b. like only swimming?
c. like none of these activities?

Solution *See page S6.*

Grace Chisholm Young

Math Matters **Grace Chisholm Young (1868–1944)**

Grace Chisholm Young studied mathematics at Girton College, which is part of Cambridge University. In England at that time, women were not allowed to earn a university degree, so she decided to continue her mathematical studies at the University of Göttingen in Germany, where her advisor was the renowned mathematician Felix Klein. She excelled while at Göttingen and at the age of 27 earned her doctorate in mathematics, magna cum laude. She was the first woman officially to earn a doctorate degree from a German university. Shortly after her graduation she married the mathematician William Young. Together they published several mathematical papers and books, one of which was the first textbook on set theory.

The Inclusion-Exclusion Principle

TAKE NOTE

Recall that $n(A)$ represents the number of elements in set A.

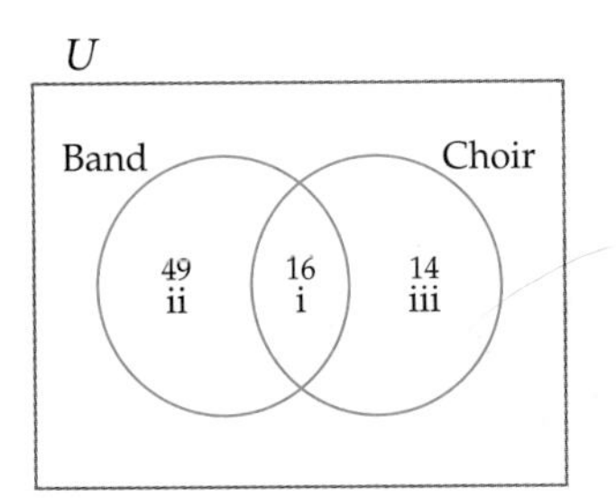

A music director wishes to take the band and the choir on a field trip. There are 65 students in the band and 30 students in the choir. The number of students in both the band and the choir is 16. How many students should the music director plan on taking on the field trip?

Using the process developed in the previous examples, we find that the number of students that are in only the band is $65 - 16 = 49$. The number of students that are in only the choir is $30 - 16 = 14$. See the Venn diagram at the left. Adding the numbers of students in regions i, ii, and iii gives us a total of $49 + 16 + 14 = 79$ students that might go on the field trip.

Although we can use Venn diagrams to solve counting problems, it is more convenient to make use of the following technique. First add the number of students in the band to the number of students in the choir. Then subtract the number of students who are in both the band and the choir. This technique gives us a total of $(65 + 30) - 16 = 79$ students, the same result as above. The reason we subtract the 16 students is that we have counted each of them twice. Note that first we include the students that are in both the band and the choir twice, and then we exclude them once. This procedure leads us to the following result.

The Inclusion-Exclusion Principle

For all finite sets A and B.

$$n(A \cup B) = n(A) + n(B) - n(A \cap B)$$

QUESTION *What must be true of the finite sets A and B if $n(A \cup B) = n(A) + n(B)$?*

ANSWER *A and B must be disjoint sets.*

EXAMPLE 3 ■ An Application of the Inclusion-Exclusion Principle

A school finds that 430 of its students are registered in chemistry, 560 are registered in mathematics, and 225 are registered in both chemistry and mathematics. How many students are registered in chemistry or mathematics?

Solution

Let $C = \{\text{students registered in chemistry}\}$ and let $M = \{\text{students registered in mathematics}\}$.

$$\begin{aligned} n(C \cup M) &= n(C) + n(M) - n(C \cap M) \\ &= 430 + 560 - 225 \\ &= 765 \end{aligned}$$

Using the inclusion-exclusion principle, we see that 765 students are registered in chemistry or mathematics.

CHECK YOUR PROGRESS 3 A high school has 80 athletes who play basketball, 60 athletes who play soccer, and 24 athletes who play both basketball and soccer. How many athletes play either basketball or soccer?

Solution *See page S6.*

The inclusion-exclusion principle can be used provided we know the number of elements in any three of the four sets in the formula.

EXAMPLE 4 ■ An Application of the Inclusion-Exclusion Principle

Given $n(A) = 15$, $n(B) = 32$, and $n(A \cup B) = 41$, find $n(A \cap B)$.

Solution

Substitute the given information in the inclusion-exclusion formula and solve for the unknown.

$$\begin{aligned} n(A \cup B) &= n(A) + n(B) - n(A \cap B) \\ 41 &= 15 + 32 - n(A \cap B) \\ 41 &= 47 - n(A \cap B) \end{aligned}$$

Thus

$$\begin{aligned} n(A \cap B) &= 47 - 41 \\ n(A \cap B) &= 6 \end{aligned}$$

CHECK YOUR PROGRESS 4 Given $n(A) = 785$, $n(B) = 162$, and $n(A \cup B) = 852$, find $n(A \cap B)$.

Solution *See page S6.*

The inclusion-exclusion formula can be adjusted and applied to problems that involve percents. In the following formula we denote "the percent in set A" by the notation $p(A)$.

The Percent Inclusion-Exclusion Formula

For all finite sets A and B,

$$p(A \cup B) = p(A) + p(B) - p(A \cap B).$$

EXAMPLE 5 ■ An Application of the Percent Inclusion-Exclusion Formula

The American Association of Blood Banks reports that about

44% of the U.S. population has the A antigen.

15% of the U.S. population has the B antigen.

4% of the U.S. population has both the A and the B antigen.

Use the percent inclusion-exclusion formula to estimate the percent of the U.S. population that has the A antigen or the B antigen.

Solution

We are given $p(\mathrm{A}) = 44\%$, $p(\mathrm{B}) = 15\%$, and $p(\mathrm{A} \cap \mathrm{B}) = 4\%$. Substituting in the percent inclusion-exclusion formula gives

$$\begin{aligned} p(\mathrm{A} \cup \mathrm{B}) &= p(\mathrm{A}) + p(\mathrm{B}) - p(\mathrm{A} \cap \mathrm{B}) \\ &= 44\% + 15\% - 4\% \\ &= 55\% \end{aligned}$$

Thus about 55% of the U.S. population has the A antigen or the B antigen. Notice that this result checks with the data given in the pie chart at the left (1% + 3% + 2% + 9% + 6% + 34% = 55%).

AB+ 3%
AB− 1%
B− 2%
B+ 9%
A− 6%
O+ 38%
A+ 34%
O− 7%

Percentage of U.S. Population with Each Blood Type

Source: American Association of Blood Banks, **http://www.aabb.org/All_About_Blood/FAQs/aabb_faqs.htm**

CHECK YOUR PROGRESS 5 The American Association of Blood Banks reports that about

44% of the U.S. population has the A antigen.

84% of the U.S. population is Rh+.

91% of the U.S. population either has the A antigen or is Rh+.

Use the percent inclusion-exclusion formula to estimate the percent of the U.S. population that has the A antigen *and* is Rh+.

Solution *See page S6.*

In the next example the data are provided in a table. The number in column G and row M represents the number of elements in $G \cap M$. The sum of all the numbers in column G and column L represents the number of elements in $G \cup L$.

EXAMPLE 6 ■ A Survey Presented in Tabular Form

A survey of men M, women W, and children C concerning the use of the Internet search engines Google G, Yahoo! Y, and Lycos L yielded the following results.

	Google (G)	Yahoo! (Y)	Lycos (L)
Men (M)	440	310	275
Women (W)	390	280	325
Children (C)	140	410	40

Use the data in the table to find each of the following.

a. $n(W \cap Y)$ **b.** $n(G \cap C')$ **c.** $n(M \cap (G \cup L))$

Solution

a. The table shows that 280 of the women surveyed use Yahoo! as a search engine. Thus, $n(W \cap Y) = 280$.

b. The set $G \cap C'$ is the set of surveyed Google users who are men or women. The number in this set is $440 + 390 = 830$.

c. The number of men in the survey that use either Google or Lycos is $440 + 275 = 715$.

CHECK YOUR PROGRESS 6 Use the table in Example 6 to find each of the following.

a. $n(Y \cap C)$ **b.** $n(L \cap M')$ **c.** $n((G \cap M) \cup (G \cap W))$

Solution *See page S6.*

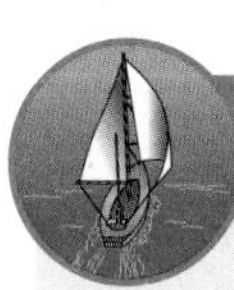

Excursion

Voting Systems

There are many types of voting systems. When people are asked to vote for or against a resolution, a one-person, one-vote *majority system* is often used to decide the outcome. In this type of voting, each voter receives one vote, and the resolution passes only if it receives *most* of the votes.

In any voting system, the number of votes that is required to pass a resolution is called the **quota.** A **coalition** is a set of voters each of whom votes the same way, either for or against a resolution. A **winning coalition** is a set of voters the sum of whose votes is greater than or equal to the quota. A **losing coalition** is a set of voters the sum of whose votes is less than the quota.

Sometimes you can find all the winning coalitions in a voting process by making an organized list. For instance, consider the committee consisting of Alice, Barry, Cheryl, and

(continued)

historical note

An ostrakon

In ancient Greece, the citizens of Athens adopted a procedure that allowed them to vote for the expulsion of any prominent person. The purpose of this procedure, known as an *ostracism*, was to limit the political power that any one person could attain.

In an ostracism, each voter turned in a *potsherd,* a piece of pottery fragment, on which was inscribed the name of the person the voter wished to ostracize. The pottery fragments used in the voting process became known as *ostrakon.*

The person who received the majority of the votes, above some set minimum, was exiled from Athens for a period of 10 years. ■

Dylan. To decide on any issues, they use a one-person, one-vote majority voting system. Since each of the four voters has a single vote, the quota for this majority voting system is 3. The winning coalitions consist of all subsets of the voters that have three or more people. We list these winning coalitions in the table at the left below, where A represents Alice, B represents Barry, C represents Cheryl, and D represents Dylan.

A **weighted voting system** is one in which some voters' votes carry more weight regarding the outcome of an election. As an example, consider a selection committee that consists of four people designated by A, B, C, and D. Voter A's vote has a weight of 2, and the vote of each other member of the committee has a weight of 1. The quota for this weighted voting system is 3. A winning coalition must have a weighted voting sum of at least 3. The winning coalitions are listed in the table at the right below.

Winning Coalition	Sum of the Votes
{A, B, C}	3
{A, B, D}	3
{A, C, D}	3
{B, C, D}	3
{A, B, C, D}	4

Winning Coalition	Sum of the Weighted Votes
{A, B}	3
{A, C}	3
{A, D}	3
{B, C, D}	3
{A, B, C}	4
{A, B, D}	4
{A, C, D}	4
{A, B, C, D}	5

A **minimal winning coalition** is a winning coalition that has no proper subset that is a winning coalition. In a minimal winning coalition each voter is said to be a **critical voter,** because if any of the voters leaves the coalition, the coalition will then become a losing coalition. In the table at the right above, the minimal winning coalitions are {A, B}, {A, C}, {A, D}, and {B, C, D}. If any single voter leaves one of these coalitions, then the coalition will become a losing coalition. The coalition {A, B, C, D} is not a minimal winning coalition, because it contains at least one proper subset, for instance {A, B, C}, that is a winning coalition.

Excursion Exercises

1. A selection committee consists of Ryan, Susan, and Trevor. To decide on issues, they use a one-person, one-vote majority voting system.
   a. Find all winning coalitions.
   b. Find all losing coalitions.
2. A selection committee consists of three people designated by M, N, and P. M's vote has a weight of 3, N's vote has a weight of 2, and P's vote has a weight of 1. The quota for this weighted voting system is 4. Find all winning coalitions.
3. Determine the minimal winning coalitions for the voting system in Excursion Exercise 2.

Additional information on the applications of mathematics to voting systems is given in Chapter 13.

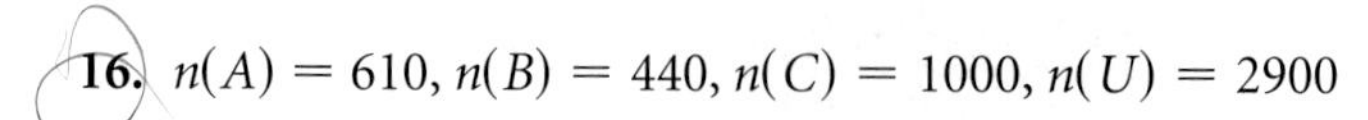

Exercise Set 2.4

In Exercises 1–8, let $U = \{\text{English, French, History, Math, Physics, Chemistry, Psychology, Drama}\}$,
$A = \{\text{English, History, Psychology, Drama}\}$,
$B = \{\text{Math, Physics, Chemistry, Psychology, Drama}\}$, and
$C = \{\text{French, History, Chemistry}\}$.
Find each of the following.

1. $n(B \cup C)$
2. $n(A \cup B)$
3. $n(B) + n(C)$
4. $n(A) + n(B)$
5. $n(A \cup B \cup C)$
6. $n(A \cap B)$
7. $n(A) + n(B) + n(C)$
8. $n(A \cap B \cap C)$
9. Verify that for A and B as defined in Exercises 1–8, $n(A \cup B) = n(A) + n(B) - n(A \cap B)$.
10. Verify that for A and C as defined in Exercises 1–8, $n(A \cup C) = n(A) + n(C) - n(A \cap C)$.
11. Given $n(J) = 245$, $n(K) = 178$, and $n(J \cup K) = 310$, find $n(J \cap K)$.
12. Given $n(L) = 780$, $n(M) = 240$, and $n(L \cap M) = 50$, find $n(L \cup M)$.
13. Given $n(A) = 1500$, $n(A \cup B) = 2250$, and $n(A \cap B) = 310$, find $n(B)$.
14. Given $n(A) = 640$, $n(B) = 280$, and $n(A \cup B) = 765$, find $n(A \cap B)$.

In Exercises 15 and 16, use the given information to find the number of elements in each of the regions labeled with a question mark.

15. $n(A) = 28$, $n(B) = 31$, $n(C) = 40$, $n(A \cap B) = 15$, $n(U) = 75$

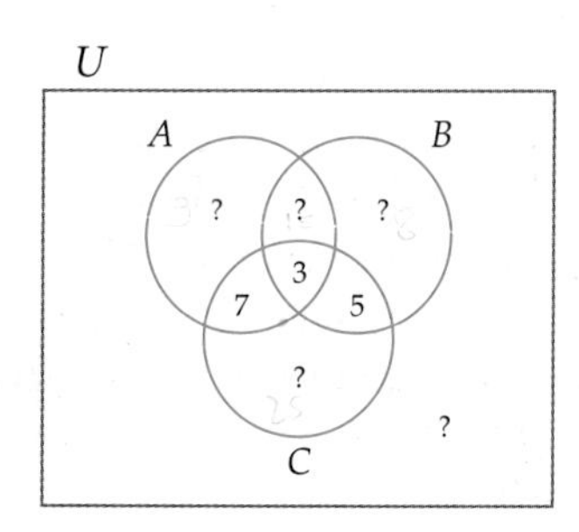

16. $n(A) = 610$, $n(B) = 440$, $n(C) = 1000$, $n(U) = 2900$

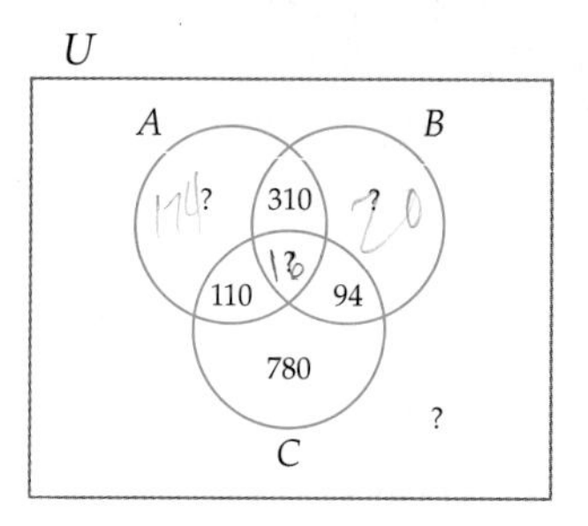

17. **A Survey** In a survey of 600 investors, it was reported that 380 had invested in stocks, 325 had invested in bonds, and 75 had not invested in either stocks or bonds.

a. How many investors had invested in both stocks and bonds?

b. How many investors had invested only in stocks?

18. **A Survey** A survey of 1500 commuters in New York City showed that 1140 take the subway, 680 take the bus, and 120 do not take either the bus or the subway.

a. How many commuters take both the bus and the subway?

b. How many commuters take only the subway?

19. **A Survey** A team physician has determined that of all the athletes who were treated for minor back pain, 72% responded to an analgesic, 59% responded to a muscle relaxant, and 44% responded to both forms of treatment.

a. What percent of the athletes who were treated responded to the muscle relaxant but not to the analgesic?

b. What percent of the athletes who were treated did not respond to either form of treatment?

20. A Survey The management of a hotel conducted a survey. It found that of the 2560 guests who were surveyed,

1785 tip the wait staff.
1219 tip the luggage handlers.
831 tip the maids.
275 tip the maids and the luggage handlers.
700 tip the wait staff and the maids.
755 tip the wait staff and the luggage handlers.
245 tip all three services.
210 do not tip these services.

How many of the surveyed guests tip:

a. exactly two of the three services?

b. only the wait staff?

c. only one of the three services?

21. A Survey A computer company advertises its computers in *PC World*, in *PC Magazine*, and on television. A survey of 770 customers finds that the numbers of customers who are familiar with the company's computers because of the different forms of advertising are as follows:

305, *PC World*
290, *PC Magazine*
390, television
110, *PC World* and *PC Magazine*
135, *PC Magazine* and television
150, *PC World* and television
85, all three sources

How many of the surveyed customers know about the computers because of:

a. exactly one of these forms of advertising?

b. exactly two of these forms of advertising?

c. *BYTE* magazine and neither of the other two forms of advertising?

22. Blood Types A report from the American Association of Blood Banks shows that in the United States:

44% of the population has the A antigen.
15% of the population has the B antigen.
84% of the population has the Rh+ factor.
34% of the population is blood type A+.
9% of the population is blood type B+.
4% of the population has the A antigen and the B antigen.
3% of the population is blood type AB+.

(*Source:* **http://www.aabb.org/All_About_Blood/FAQs/aabb_faqs.htm.**)

Find the percent of the U.S. population that is:

a. A− **b.** O+ **c.** O−

23. A Survey A special-interest group has conducted a survey concerning a ban on hand guns. *Note:* A rifle is a gun, but it is not a hand gun. The survey yielded the following results for the 1000 households that responded.

271 own a hand gun.
437 own a rifle.
497 supported the ban on hand guns.
140 own both a hand gun and a rifle.
202 own a rifle but no hand gun and do not support the ban on hand guns.
74 own a hand gun and support the ban on hand guns.
52 own both a hand gun and a rifle and also support the ban on hand guns.

How many of the surveyed households:

a. only own a hand gun and do not support the ban on hand guns?

b. do not own a gun and support the ban on hand guns?

c. do not own a gun and do not support the ban on hand guns?

24. A Survey A survey of college students was taken to determine how the students acquired music. The survey showed the following results.

365 students acquired music from CDs.
298 students acquired music from the Internet.
268 students acquired music from cassettes.
212 students acquired music from both CDs and cassettes.

155 students acquired music from both CDs and the Internet.
36 students acquired music from cassettes, but not from CDs or the Internet.
98 students acquired music from CDs, cassettes, and the Internet.

Of those surveyed,

a. how many acquired music from CDs, but not from the Internet or cassettes?

b. how many acquired music from the Internet, but not from CDs or cassettes?

c. how many acquired music from CDs or the Internet?

d. how many acquired music from the Internet and cassettes?

25. Diets A survey was completed by individuals who were currently on the Atkins diet (A), the South Beach diet (S), or the Weight Watchers diet (W). All persons surveyed were also asked whether they were currently in an exercise program (E), taking diet pills (P), or under medical supervision (M). The following table shows the results of the survey.

		Supplements			
		E	P	M	Totals
Diet	A	124	82	65	271
	S	101	66	51	218
	W	133	41	48	222
	Totals	358	189	164	711

Find the number of surveyed people in each of the following sets.

a. $S \cap E$

b. $A \cup M$

c. $S' \cap (E \cup P)$

d. $(A \cup S) \cap (M')$

e. $W' \cap (P \cup M)'$

f. $W' \cup P$

26. Financial Assistance A college study categorized its seniors (S), juniors (J), and sophomores (M) who are currently receiving financial assistance. The types of financial assistance consist of full scholarships (F), partial scholarships (P), and government loans (G). The following table shows the results of the survey.

		Financial Assistance			
		F	P	G	Totals
Year	S	210	175	190	575
	J	180	162	110	452
	M	114	126	86	326
	Totals	504	463	386	1353

Find the number of students who are currently receiving financial assistance in each of the following sets.

a. $S \cap P$

b. $J \cup G$

c. $M \cup F'$

d. $S \cap (F \cup P)$

e. $J \cap (F \cup P)'$

f. $(S \cup J) \cap (F \cup P)$

Extensions

CRITICAL THINKING

27. Given that set A has 47 elements and set B has 25 elements, determine each of the following.

a. The maximum possible number of elements in $A \cup B$.

b. The minimum possible number of elements in $A \cup B$.

c. The maximum possible number of elements in $A \cap B$.

d. The minimum possible number of elements in $A \cap B$.

28. Given that set A has 16 elements, set B has 12 elements, and set C has 7 elements, determine each of the following.

a. The maximum possible number of elements in $A \cup B \cup C$.

b. The minimum possible number of elements in $A \cup B \cup C$.

c. The maximum possible number of elements in $A \cap (B \cup C)$.

d. The minimum possible number of elements in $A \cap (B \cup C)$.

COOPERATIVE LEARNING

29. A Survey The following Venn diagram displays U parceled into 16 distinct regions by four sets.

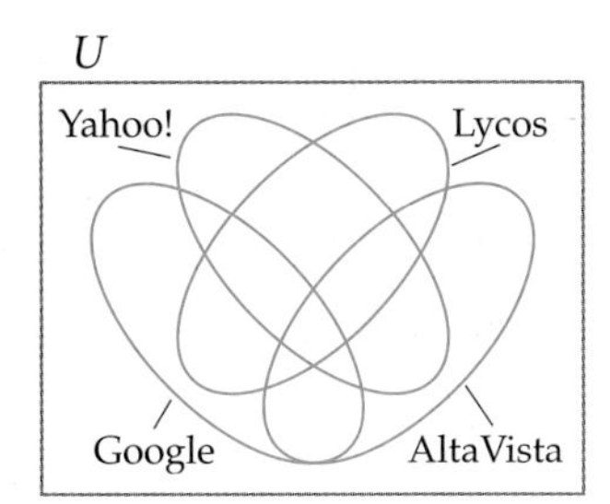

Use the above Venn diagram and the information in the column at right to answer the questions that follow.

A survey of 1250 Internet users shows the following results concerning the use of the search engines Google, AltaVista, Yahoo!, and Lycos.

585 use Google.
620 use Yahoo!.
560 use Lycos.
450 use AltaVista.
100 use only Google, Yahoo!, and Lycos.
41 use only Google, Yahoo!, and AltaVista.
50 use only Google, Lycos, and AltaVista.
80 use only Yahoo!, Lycos, and AltaVista.
55 use only Google and Yahoo!.
34 use only Google and Lycos.
45 use only Google and AltaVista.
50 use only Yahoo! and Lycos.
30 use only Yahoo! and AltaVista.
45 use only Lycos and AltaVista.
60 use all four.

How many of the Internet users:

a. use only Google?
b. use exactly three of the four search engines?
c. do not use any of the four search engines?

EXPLORATIONS

30. An Inclusion–Exclusion Formula for Three Sets Exactly one of the following equations is a valid inclusion-exclusion formula for the union of three finite sets. Which equation do you think is the valid formula? *Hint:* Use the data in Example 2 on page 88 to check your choice.

a. $n(A \cup B \cup C) = n(A) + n(B) + n(C)$

b. $n(A \cup B \cup C) = n(A) + n(B) + n(C) - n(A \cap B \cap C)$

c. $n(A \cup B \cup C) = n(A) + n(B) + n(C) - n(A \cap B) - n(A \cap C) - n(B \cap C)$

d. $n(A \cup B \cup C) = n(A) + n(B) + n(C) - n(A \cap B) - n(A \cap C) - n(B \cap C) + n(A \cap B \cap C)$

SECTION 2.5 Infinite Sets

One-To-One Correspondences

Much of Georg Cantor's work with sets concerned infinite sets. Some of Cantor's work with infinite sets was so revolutionary that it was not readily accepted by his contemporaries. Today, however, his work is generally accepted, and it provides unifying ideas in several diverse areas of mathematics.

Much of Cantor's set theory is based on the simple concept of a *one-to-one correspondence.*

One-to-One Correspondence

A **one-to-one correspondence** (or 1–1 correspondence) between two sets A and B is a rule or procedure that pairs each element of A with exactly one element of B and each element of B with exactly one element of A.

Many practical problems can be solved by applying the concept of a one-to-one correspondence. For instance, consider a concert hall that has 890 seats. During a performance the manager of the concert hall observes that every person occupies exactly one seat and that every seat is occupied. Thus, without doing any counting, the manager knows that there are 890 people in attendance. During a different performance the manager notes that all but six seats are filled, and thus there are $890 - 6 = 884$ people in attendance.

Recall that two sets are equivalent if and only if they have the same number of elements. One method of showing that two sets are equivalent is to establish a one-to-one correspondence between the elements of the sets.

One-to-One Correspondence and Equivalent Sets

Two sets A and B are equivalent, denoted by $A \sim B$, if and only if A and B can be placed in a one-to-one correspondence.

Set {a, b, c, d, e} is equivalent to set {1, 2, 3, 4, 5} because we can show that the elements of each set can be placed in a one-to-one correspondence. One method of establishing this one-to-one correspondence is shown in the following figure.

$$\begin{array}{ccccc} \{a, & b, & c, & d, & e\} \\ \updownarrow & \updownarrow & \updownarrow & \updownarrow & \updownarrow \\ \{1, & 2, & 3, & 4, & 5\} \end{array}$$

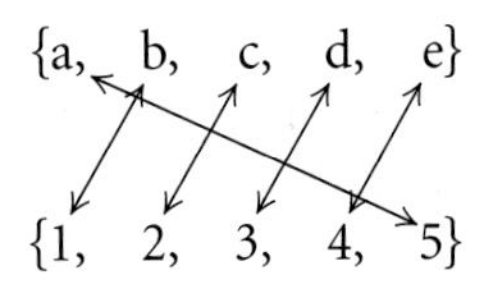

Each element of {a, b, c, d, e} has been paired with exactly one element of {1, 2, 3, 4, 5}, and each element of {1, 2, 3, 4, 5} has been paired with exactly one element of {a, b, c, d, e}. This is not the only one-to-one correspondence that we can establish. The figure at the left shows another one-to-one correspondence between the sets. In any case, we know that both sets have the same number of elements because we have established a one-to-one correspondence between the sets.

Sometimes a set is defined by including a general element. For instance, in the set $\{3, 6, 9, 12, 15, \ldots, 3n, \ldots\}$, the $3n$ (where n is a natural number) indicates that all the elements of the set are multiples of 3.

Some sets can be placed in a one-to-one correspondence with a proper subset of themselves. Example 1 illustrates this concept for the set of natural numbers.

TAKE NOTE

Many mathematicians and non-mathematicians have found the concept that $E \sim N$, as shown in Example 1, to be a bit surprising. After all, the set of natural numbers includes the even natural numbers as well as the odd natural numbers!

EXAMPLE 1 ■ Establish a One-to-One Correspondence

Establish a one-to-one correspondence between the set of natural numbers $N = \{1, 2, 3, 4, 5, \ldots, n, \ldots\}$ and the set of even natural numbers $E = \{2, 4, 6, 8, 10, \ldots, 2n, \ldots\}$.

Solution

Write the sets so that one is aligned below the other. Draw arrows to show how you wish to pair the elements of each set. One possible method is shown in the following figure.

$$\begin{array}{l} N = \{1, 2, 3, 4, \ldots, \; n, \ldots\} \\ \quad\quad\;\; \updownarrow \; \updownarrow \; \updownarrow \; \updownarrow \quad\quad\; \updownarrow \\ E = \{2, 4, 6, 8, \ldots, 2n, \ldots\} \end{array}$$

In the above correspondence, each natural number $n \in N$ is paired with the even number $(2n) \in E$. The *general correspondence* $n \leftrightarrow (2n)$ enables us to determine exactly which element of E will be paired with any given element of N, and vice versa. For instance, under this correspondence, $19 \in N$ is paired with the even number $2 \cdot 19 = 38 \in E$, and $100 \in E$ is paired with the natural number $\frac{1}{2} \cdot 100 = 50 \in N$. The general correspondence $n \leftrightarrow (2n)$ establishes a one-to-one correspondence between the sets.

CHECK YOUR PROGRESS 1 Establish a one-to-one correspondence between the set of natural numbers $N = \{1, 2, 3, 4, 5, \ldots, n, \ldots\}$ and the set of odd natural numbers $D = \{1, 3, 5, 7, 9, \ldots, 2n - 1, \ldots\}$.

Solution *See page S7.*

Infinite Sets

Definition of an Infinite Set

A set is an **infinite set** if it can be placed in a one-to-one correspondence with a proper subset of itself.

We know that the set of natural numbers N is an infinite set because in Example 1 we were able to establish a one-to-one correspondence between the elements of N and the elements of one of its proper subsets, E.

QUESTION *Can the set {1, 2, 3} be placed in a one-to-one correspondence with one of its proper subsets?*

ANSWER *No. The set {1, 2, 3} is a finite set with three elements. Every proper subset of {1, 2, 3} has two or fewer elements.*

TAKE NOTE

The solution shown in Example 2 is not the only way to establish that S is an infinite set. For instance, $R = \{10, 15, 20, \ldots 5n + 5, \ldots\}$ is also a proper set of S, and the sets S and R can be placed in a one-to-one correspondence as follows.

$$S = \{5, 10, 15, 20, \ldots, 5n, \ldots\}$$
$$\updownarrow \; \updownarrow \; \updownarrow \; \updownarrow \quad\quad \updownarrow$$
$$R = \{10, 15, 20, 25, \ldots, 5n + 5, \ldots\}$$

This one-to-one correspondence between S and one of its proper subsets R also establishes that S is an infinite set.

EXAMPLE 2 ■ Verify that a Set is an Infinite Set

Verify that $S = \{5, 10, 15, 20, \ldots, 5n, \ldots\}$ is an infinite set.

Solution

One proper subset of S is $T = \{10, 20, 30, 40, \ldots, 10n, \ldots\}$, which was produced by deleting the odd numbers in S. To establish a one-to-one correspondence between set S and set T, consider the following diagram.

$$S = \{5, 10, 15, 20, \ldots, 5n, \ldots\}$$
$$\updownarrow \; \updownarrow \; \updownarrow \; \updownarrow \quad\quad \updownarrow$$
$$T = \{10, 20, 30, 40, \ldots, 10n, \ldots\}$$

In the above correspondence, each $(5n) \in S$ is paired with $(10n) \in T$. The *general correspondence* $(5n) \leftrightarrow (10n)$ establishes a one-to-one correspondence between S and one of its proper subsets, namely T. Thus S is an infinite set.

CHECK YOUR PROGRESS 2 Verify that $V = \{40, 41, 42, 43, \ldots, 39 + n, \ldots\}$ is an infinite set.

Solution *See page S7.*

The Cardinality of Infinite Sets

The symbol $\aleph_0$ is used to represent the cardinal number for the set N of natural numbers. ($\aleph$ is the first letter of the Hebrew alphabet and is pronounced *aleph.* $\aleph_0$ is read as "aleph-null.") Using mathematical notation, we write this concept as $n(N) = \aleph_0$. Since $\aleph_0$ represents a cardinality larger than any finite number, it is called a **transfinite number.** Many infinite sets have a cardinality of $\aleph_0$. In Example 3, for instance, we show that the cardinality of the set of integers is $\aleph_0$ by establishing a one-to-one correspondence between the elements of the set of integers and the elements of the set of natural numbers.

EXAMPLE 3 ■ Establish the Cardinality of the Set of Integers

Show that the set of integers $I = \{\ldots, -5, -4, -3, -2, -1, 0, 1, 2, 3, 4, 5, \ldots\}$ has a cardinality of $\aleph_0$.

Solution

First we try to establish a one-to-one correspondence between I and N, with the elements in each set arranged as shown below. No general method of pairing the elements of N with the elements of I seems to emerge from this figure.

$$N = \{1, 2, 3, 4, 5, 6, 7, 8, 9, 10, 11, \ldots\}$$
$$?$$
$$I = \{\ldots, -5, -4, -3, -2, -1, 0, 1, 2, 3, 4, 5, \ldots\}$$

If we arrange the elements of I as shown in the figure below, then *two* general correspondences, shown by the blue arrows and the red arrows, can be identified.

$$N = \{1, 2, \ 3, \ 4, \ 5, \ 6, \ 7, \ 8, \ 9, \ 10, 11, \ldots, \ 2n - 1, 2n, \ldots\}$$

$$\updownarrow$$

$$I = \{0, 1, -1, 2, -2, 3, -3, 4, -4, \ 5, -5 \ldots, -n + 1, \ n, \ldots\}$$

- Each even natural number $2n$ of N is paired with the integer n of I. This correspondence is shown by the blue arrows.
- Each odd natural number $2n - 1$ of N is paired with the integer $-n + 1$ of I. This correspondence is shown by the red arrows.

Together the two general correspondences $(2n) \leftrightarrow n$ and $(2n - 1) \leftrightarrow (-n + 1)$ establish a one-to-one correspondence between the elements of I and the elements of N. Thus the cardinality of the set of integers must be the same as the cardinality of the set of natural numbers, which is $\aleph_0$.

CHECK YOUR PROGRESS 3 Show that $M = \left\{\frac{1}{2}, \frac{1}{3}, \frac{1}{4}, \frac{1}{5}, \ldots, \frac{1}{n+1}, \ldots\right\}$ has a cardinality of $\aleph_0$.

Solution *See page S7.*

Cantor was also able to show that the set of positive rational numbers is equivalent to the set of natural numbers. Recall that a rational number is a number that can be written as a fraction $\frac{p}{q}$ where p and q are integers and $q \neq 0$. Cantor's proof used an array of rational numbers similar to the array shown below.

Theorem The set $Q+$ of positive rational numbers is equivalent to the set N of natural numbers.

Proof Consider the following array of positive rational numbers.

TAKE NOTE

The rational number $\frac{2}{2}$ is not listed in the second row because $\frac{2}{2} = 1 = \frac{1}{1}$, which is already listed in the first row.

$$\begin{array}{ccccccccc}
\frac{1}{1} & \frac{2}{1} & \frac{3}{1} & \frac{4}{1} & \frac{5}{1} & \frac{6}{1} & \frac{7}{1} & \frac{8}{1} & \cdots \\
\frac{1}{2} & \frac{3}{2} & \frac{5}{2} & \frac{7}{2} & \frac{9}{2} & \frac{11}{2} & \frac{13}{2} & \frac{15}{2} & \cdots \\
\frac{1}{3} & \frac{2}{3} & \frac{4}{3} & \frac{5}{3} & \frac{7}{3} & \frac{8}{3} & \frac{10}{3} & \frac{11}{3} & \cdots \\
\frac{1}{4} & \frac{3}{4} & \frac{5}{4} & \frac{7}{4} & \frac{9}{4} & \frac{11}{4} & \frac{13}{4} & \frac{15}{4} & \cdots \\
\frac{1}{5} & \frac{2}{5} & \frac{3}{5} & \frac{4}{5} & \frac{6}{5} & \frac{7}{5} & \frac{8}{5} & \frac{9}{5} & \cdots \\
\vdots & \vdots & \vdots & \vdots & \vdots & \vdots & \vdots & \vdots & \ddots
\end{array}$$

An array of all the positive rational numbers

The first row of the above array contains, in order from smallest to largest, all the positive rational numbers *which when expressed in lowest terms* have a denominator of 1. The second row contains the positive rational numbers *which when expressed in lowest terms* have a denominator of 2. The third row contains the positive rational numbers *which when expressed in lowest terms* have a denominator of 3. This process continues indefinitely.

Cantor reasoned that every positive rational number appears once and only once in this array. Note that $\frac{3}{5}$ appears in the fifth row. In general, if $\frac{p}{q}$ is in lowest terms, then it appears in row q.

At this point Cantor used a numbering procedure that establishes a one-to-one correspondence between the natural numbers and the positive rational numbers in the array. The numbering procedure starts in the upper left corner with $\frac{1}{1}$. Cantor considered this to be the first number in the array, so he assigned the natural number 1 to this rational number. He then moved to the right and assigned the natural number 2 to the rational number $\frac{2}{1}$. From this point on, he followed the diagonal paths shown by the arrows and assigned each number he encountered to the next consecutive natural number. When he reached the bottom of a diagonal, he moved up to the top of the array and continued to number the rational numbers in the next diagonal. The following table shows the first 10 rational numbers Cantor numbered using this scheme.

Rational number in the array	$\frac{1}{1}$	$\frac{2}{1}$	$\frac{1}{2}$	$\frac{3}{1}$	$\frac{3}{2}$	$\frac{1}{3}$	$\frac{4}{1}$	$\frac{5}{2}$	$\frac{2}{3}$	$\frac{1}{4}$
Corresponding natural number	1	2	3	4	5	6	7	8	9	10

This numbering procedure shows that each element of $Q+$ can be paired with exactly one element of N, and each element of N can be paired with exactly one element of $Q+$. Thus $Q+$ and N are equivalent sets. ■

The negative rational numbers $Q-$ can also be placed in a one-to-one correspondence with the set of natural numbers in a similar manner.

QUESTION *Using Cantor's numbering scheme, which rational numbers in the array shown on page 101 would be assigned the natural numbers 11, 12, 13, 14, and 15?*

Definition of a Countable Set

A set is a **countable set** if and only if it is a finite set or an infinite set that is equivalent to the set of natural numbers.

Every infinite set that is countable has a cardinality of $\aleph_0$. Every infinite set that we have considered up to this point is countable. You might think that all infinite sets are countable; however, Cantor was able to show that this is not the case. Con-

---

ANSWER *The rational numbers in the next diagonal, namely $\frac{5}{1}, \frac{7}{2}, \frac{4}{3}, \frac{3}{4}$, and $\frac{1}{5}$, would be assigned to the natural numbers 11, 12, 13, 14, and 15, respectively.*

sider, for example, $A = \{x | x \in R \text{ and } 0 < x < 1\}$. To show that A is *not* a countable set, we use a *proof by contradiction,* where we assume that A is countable and then proceed until we arrive at a contradiction.

To better understand the concept of a proof by contradiction, consider the situation in which you are at a point where a road splits into two roads. See the figure at the left. Assume you know that only one of the two roads leads to your desired destination. If you can show that one of the roads cannot get you to your destination, then you know, without ever traveling down the other road, that it is the road that leads to your destination. In the following proof, we know that either set A is a countable set or set A is not a countable set. To establish that A is *not* countable, we show that the assumption that A is countable leads to a contradiction. In other words, our assumption that A is countable must be incorrect, and we are forced to conclude that A is not countable.

Theorem The set $A = \{x | x \in R \text{ and } 0 < x < 1\}$ is not a countable set.

Proof by contradiction Either A is countable or A is not countable. Assume A is countable. Then we can place the elements of A, which we will represent by a_1, a_2, a_3, $a_4, \ldots$, in a one-to-one correspondence with the elements of the natural numbers as shown below.

$$\begin{array}{l} N = \{1,\ 2,\ 3,\ 4,\ \ldots,\ n,\ \ldots\} \\ \qquad\ \ \updownarrow\ \ \updownarrow\ \ \updownarrow\ \ \updownarrow\ \qquad\ \ \updownarrow \\ A = \{a_1,\ a_2,\ a_3,\ a_4,\ \ldots,\ a_n,\ \ldots\} \end{array}$$

For example, the numbers $a_1, a_2, a_3, a_4, \ldots, a_n, \ldots$ could be as shown below.

$$\begin{array}{l} 1 \leftrightarrow a_1 = 0.\boxed{3}573485\ldots \\ 2 \leftrightarrow a_2 = 0.0\boxed{6}52891\ldots \\ 3 \leftrightarrow a_3 = 0.68\boxed{2}3514\ldots \\ 4 \leftrightarrow a_4 = 0.050\boxed{0}310\ldots \\ \quad\vdots \\ n \leftrightarrow a_n = 0.3155728\ldots\boxed{5}\ldots \\ \quad\vdots \end{array}$$

*n*th decimal digit of a_n

At this point we use a "diagonal technique" to construct a real number d that is greater than 0 and less than 1 and is not in the above list. We construct d by writing a decimal that *differs* from a_1 in the first decimal place, differs from a_2 in the second decimal place, differs from a_3 in the third decimal place, and, in general, differs from a_n in the nth decimal place. For instance, in the above list, a_1 has 3 as its first decimal digit. The first decimal digit of d can be any digit other than 3, say 4. The real number a_2 has 6 as its second decimal digit. The second decimal digit of d can be any digit other than 6, say 7. The real number a_3 has 2 as its third decimal digit. The third decimal digit of d can be any digit other than 2, say 3. Continue in this manner to determine the decimal digits of d. Now $d = 0.473\ldots$ must be in A because $0 < d < 1$. However, d is not in A, because d differs from each of the numbers in A in at least one decimal place.

We have reached a contradiction. Our assumption that the elements of A could be placed in a one-to-one correspondence with the elements of the natural numbers must be false. Thus A is not a countable set. ■

An infinite set that is not countable is said to be **uncountable.** Because the set $A = \{x | x \in R \text{ and } 0 < x < 1\}$ is uncountable, the cardinality of A is not $\aleph_0$. Cantor used the letter c, which is the first letter of the word *continuum,* to represent the cardinality of A. Cantor was also able to show that set A is equivalent to the set of all real numbers R. Thus the cardinality of R is also c. Cantor was able to prove that $c > \aleph_0$.

A Comparison of Transfinite Cardinal Numbers

$c > \aleph_0$

Up to this point, all of the infinite sets we have considered have a cardinality of either $\aleph_0$ or c. The following table lists several infinite sets and the transfinite cardinal number that is associated with each set.

The Cardinality of Some Infinite Sets

Set	Cardinal Number
Natural numbers, N	$\aleph_0$
Integers, I	$\aleph_0$
Rational numbers, Q	$\aleph_0$
Irrational Numbers, $\mathcal{J}$	c
Any set of the form $\{x \mid a \leq x \leq b\}$, where a and b are real numbers and $a \neq b$.	c
Real numbers, R	c

Your intuition may suggest that $\aleph_0$ and c are the only two cardinal numbers associated with infinite sets; however, this is not the case. In fact, Cantor was able to show that no matter how large the cardinal number of a set, we can find a set that has a larger cardinal number. Thus there are infinitely many transfinite numbers. Cantor's proof of this concept is now known as *Cantor's theorem.*

Cantor's Theorem

Let S be any set. The set of all subsets of S has a cardinal number that is larger than the cardinal number of S.

The set of all subsets of S is called the **power set** of S and is denoted by $P(S)$. We can see that Cantor's theorem is true for the finite set $S = \{a, b, c\}$ because the cardinality of S is 3 and S has $2^3 = 8$ subsets. The interesting part of Cantor's theorem is that it also applies to infinite sets.

Some of the following theorems can be established by using the techniques illustrated in the Excursion that follows.

Transfinite Arithmetic Theorems

- For any whole number a, $\aleph_0 + a = \aleph_0$ and $\aleph_0 - a = \aleph_0$
- $\aleph_0 + \aleph_0 = \aleph_0$ and, in general, $\underbrace{\aleph_0 + \aleph_0 + \aleph_0 + \cdots + \aleph_0}_{\text{a finite number of aleph nulls}} = \aleph_0$
- $c + c = c$ and, in general, $\underbrace{c + c + c + \cdots + c}_{\text{a finite number of } c\text{'s}} = c$
- $\aleph_0 + c = c$
- $\aleph_0 c = c$

Math Matters Criticism and Praise

Georg Cantor's work in the area of infinite sets was not well received by some of his colleagues. For instance, the mathematician Leopold Kronecker tried to stop the publication of some of Cantor's work. He felt many of Cantor's theorems were ridiculous and asked, "How can one infinity be greater than another?" The following quote illustrates that Cantor was aware that his work would attract harsh criticism.

> . . . I realize that in this undertaking I place myself in a certain opposition to views widely held concerning the mathematical infinite and to opinions frequently defended on the nature of numbers.[2]

A few mathematicians were willing to show support for Cantor's work. For instance, the famous mathematician David Hilbert stated that Cantor's work was

> . . . the finest product of mathematical genius and one of the supreme achievements of purely intellectual human activity.[3]

Excursion

Transfinite Arithmetic

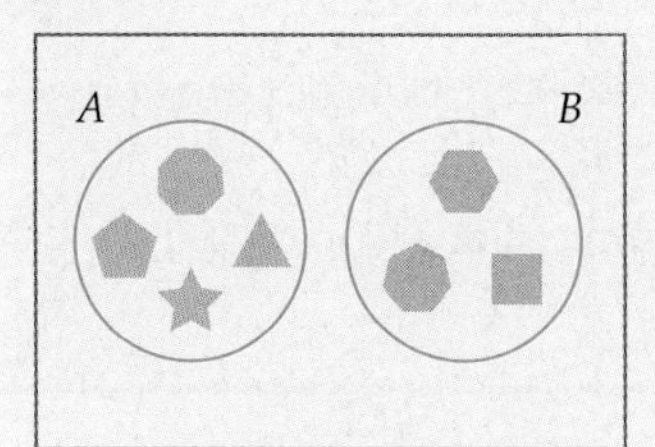

Disjoint sets are often used to explain addition. The sum 4 + 3, for example, can be deduced by selecting two disjoint sets, one with exactly four elements and one with exactly three elements. See the Venn diagram at the left. Now form the union of the two sets. The union of the two sets has exactly seven elements; thus, 4 + 3 = 7. In mathematical notation, we write

$$\begin{array}{ccccc} n(A) & + & n(B) & = & n(A \cup B) \\ 4 & + & 3 & = & 7 \end{array}$$

(continued)

2. *Source:* **http://www-groups.dcs.st-and.ac.uk/%7Ehistory/Mathematicians/Cantor.html**
3. See note 2 above.

Cantor extended this idea to infinite sets. He reasoned that the sum $\aleph_0 + 1$ could be determined by selecting two disjoint sets, one with cardinality of $\aleph_0$ and one with cardinality of 1. In this case the set N of natural numbers and the set $Z = \{0\}$ are appropriate choices. Thus

$$\begin{aligned} n(N) + n(Z) &= n(N \cup Z) \\ &= n(W) \\ \aleph_0 \;+\; 1 \;&=\; \aleph_0 \end{aligned}$$

and, in general, for any whole number a, $\aleph_0 + a = \aleph_0$.

To find the sum $\aleph_0 + \aleph_0$, use two disjoint sets, each with cardinality of $\aleph_0$. The set E of even natural numbers and the set D of odd natural numbers satisfy the necessary conditions. Since E and D are disjoint sets, we know

$$\begin{aligned} n(E) + n(D) &= n(E \cup D) \\ &= n(W) \\ \aleph_0 \;+\; \aleph_0 \;&=\; \aleph_0 \end{aligned}$$

Thus $\aleph_0 + \aleph_0 = \aleph_0$ and, in general,

$$\underbrace{\aleph_0 + \aleph_0 + \aleph_0 + \cdots + \aleph_0}_{\text{a finite number of aleph-nulls}} = \aleph_0$$

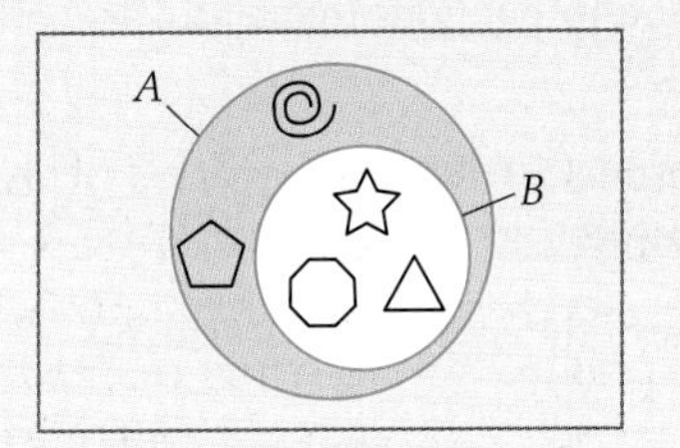

To determine a difference such as $5 - 3$ using sets, we first select a set A that has exactly five elements. We then find a subset B of this set that has exactly three elements. The difference $5 - 3$ is the cardinal number of the set $A \cap B'$, which is shown in blue in the figure at the left.

To determine $\aleph_0 - 3$, select a set with $\aleph_0$ elements, such as N, and then select a subset of this set that has exactly three elements. One such subset is $C = \{1, 2, 3\}$. The difference $\aleph_0 - 3$ is the cardinal number of the set $N \cap C' = \{4, 5, 6, 7, 8, \ldots\}$. Since $N \cap C'$ is a countably infinite set, we can conclude that $\aleph_0 - 3 = \aleph_0$. This procedure can be generalized to show that for any whole number a, $\aleph_0 - a = \aleph_0$.

Excursion Exercises

1. Use two disjoint sets to show that $\aleph_0 + 2 = \aleph_0$.
2. Use two disjoint sets other than the set of even natural numbers and the set of odd natural numbers to show that $\aleph_0 + \aleph_0 = \aleph_0$.
3. Use sets to show that $\aleph_0 - 6 = \aleph_0$.
4. **a.** Find two sets that can be used to show that $\aleph_0 - \aleph_0 = \aleph_0$. Now find another two sets that can be used to show that $\aleph_0 - \aleph_0 = 1$.

   b. Use the results of Excursion Exercise 4a to explain why subtraction of two transfinite numbers is an undefined operation.

Exercise Set 2.5

1. a. Use arrows to establish a one-to-one correspondence between $V = \{a, e, i\}$ and $M = \{3, 6, 9\}$.
   b. How many different one-to-one correspondences between V and M can be established?
2. Establish a one-to-one correspondence between the set of natural numbers $N = \{1, 2, 3, 4, 5, \ldots, n, \ldots\}$ and $F = \{5, 10, 15, 20, \ldots, 5n, \ldots\}$ by stating a general rule that can be used to pair the elements of the sets.
3. Establish a one-to-one correspondence between $D = \{1, 3, 5, \ldots, 2n - 1, \ldots\}$ and $M = \{3, 6, 9, \ldots, 3n, \ldots\}$ by stating a general rule that can be used to pair the elements of the sets.

In Exercises 4–10, state the cardinality of each set.

4. $\{2, 11, 19, 31\}$
5. $\{2, 9, 16, \ldots, 7n - 5, \ldots\}$, where n is a natural number
6. The set Q of rational numbers
7. The set R of real numbers
8. The set $\mathcal{I}$ of irrational numbers
9. $\{x \mid 5 \le x \le 9\}$
10. The set of subsets of $\{1, 5, 9, 11\}$

In Exercises 11–14, determine whether the given sets are equivalent.

11. The set of natural numbers and the set of integers
12. The set of whole numbers and the set of real numbers
13. The set of rational numbers and the set of integers
14. The set of rational numbers and the set of real numbers

In Exercises 15–18, show that the given set is an infinite set by placing it in a one-to-one correspondence with a proper subset of itself.

15. $A = \{5, 10, 15, 20, 25, 30, \ldots, 5n, \ldots\}$
16. $B = \{11, 15, 19, 23, 27, 31, \ldots, 4n + 7, \ldots\}$
17. $C = \left\{\frac{1}{2}, \frac{3}{4}, \frac{5}{6}, \frac{7}{8}, \frac{9}{10}, \ldots, \frac{2n - 1}{2n}, \ldots\right\}$
18. $D = \left\{\frac{1}{2}, \frac{1}{3}, \frac{1}{4}, \frac{1}{5}, \frac{1}{6}, \ldots, \frac{1}{n + 1}, \ldots\right\}$

In Exercises 19–26, show that the given set has a cardinality of $\aleph_0$ by establishing a one-to-one correspondence between the elements of the given set and the elements of N.

19. $\{50, 51, 52, 53, \ldots, n + 49, \ldots\}$
20. $\{10, 5, 0, -5, -10, -15, \ldots, -5n + 15, \ldots\}$
21. $\left\{1, \frac{1}{3}, \frac{1}{9}, \frac{1}{27}, \ldots, \frac{1}{3^{n-1}}, \ldots\right\}$
22. $\{-12, -18, -24, -30, \ldots, -6n - 6, \ldots\}$
23. $\{10, 100, 1000, \ldots, 10^n, \ldots\}$
24. $\left\{1, \frac{1}{2}, \frac{1}{4}, \frac{1}{8}, \ldots, \frac{1}{2^{n-1}}, \ldots\right\}$
25. $\{1, 8, 27, 64, \ldots, n^3, \ldots\}$
26. $\{0.1, 0.01, 0.001, 0.0001, \ldots, 10^{-n}, \ldots\}$

Extensions

CRITICAL THINKING

27. a. Place the set $M = \{3, 6, 9, 12, 15, \ldots\}$ of positive multiples of 3 in a one-to-one correspondence with the set K of all natural numbers that are not multiples of 3. Write a sentence or two that explains the general rule you used to establish the one-to-one correspondence.
    b. Use your rule to determine what number from K is paired with the number 606 from M.
    c. Use your rule to determine what number from M is paired with the number 899 from K.

In the figure below, every point on line segment AB corresponds to a real number from 0 to 1 and every real number from 0 to 1 corresponds to a point on line segment AB.

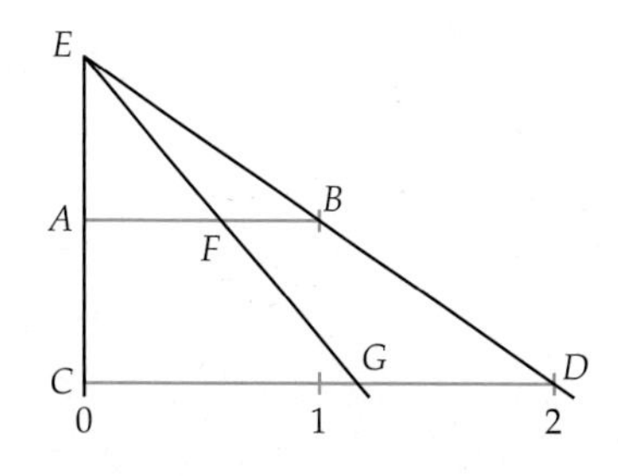

The line segment CD represents the real numbers from 0 to 2. Note that any point F on line segment AB can be paired with a unique point G on line segment CD by drawing a line from E through F. Also, any arbitrary point G on line segment CD can be paired with a unique point F on line segment AB by drawing the line EG. This geometric procedure establishes a one-to-one correspondence between the set $\{x \mid 0 \leq x \leq 1\}$ and the set $\{x \mid 0 \leq x \leq 2\}$. Thus $\{x \mid 0 \leq x \leq 1\} \sim \{x \mid 0 \leq x \leq 2\}$.

28. Draw a figure that can be used to verify each of the following.

a. $\{x \mid 0 \leq x \leq 1\} \sim \{x \mid 0 \leq x \leq 5\}$

b. $\{x \mid 2 \leq x \leq 5\} \sim \{x \mid 1 \leq x \leq 8\}$

29. Consider the semicircle with arc length 1 and center C and the line L_1 in the following figure. Each point on the semicircle, *other than the endpoints,* represents a unique real number between 0 and 1. Each point on line L_1 represents a unique real number.

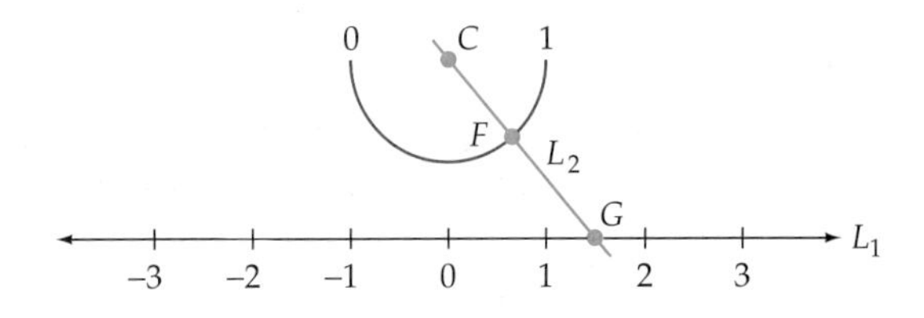

Any line through C that intersects the semicircle at a point other than one of its endpoints will intersect line L_1 at a unique point. Also, any line through C that intersects line L_1 will intersect the semicircle at a unique point that is not an endpoint of the semicircle. What can we conclude from this correspondence?

30. Explain how to use the following figure to verify that the set of all points on the circle is equivalent to the set of all points on the square.

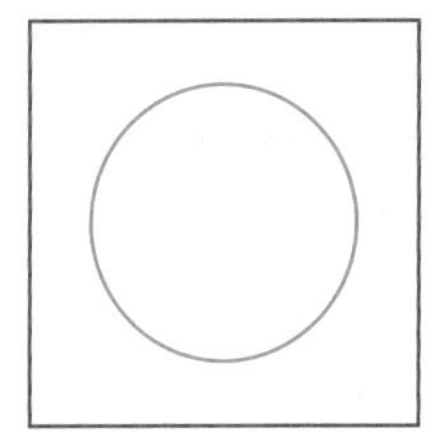

EXPLORATIONS

31. **The Hilbert Hotel** The Hilbert Hotel is an imaginary hotel created by the mathematician David Hilbert (1862–1943). The hotel has an infinite number of rooms. Each room is numbered with a natural number; room 1, room 2, room 3, and so on. Search the Internet for information on Hilbert's Hotel. Write a few paragraphs that explain some of the interesting questions that arise when guests arrive to stay at the hotel.

Mary Pat Campbell has written a song about a hotel with an infinite number of rooms. Her song is titled *Hotel Aleph Null—yeah.* Here are the lyrics for the chorus of her song, which is to be sung to the tune of *Hotel California* by the Eagles. (*Source:* **http://www.marypat.org/mathcamp/doc2001/hellrelays.html#hotel**)[4]

Hotel Aleph Null—yeah
Welcome to the Hotel Aleph Null—yeah
What a lovely place (what a lovely place)
Got a lot of space
Packin' em in at the Hotel Aleph Null—yeah
Any time of year
You can find space here

32. **The Continuum Hypothesis** Cantor conjectured that no set can have a cardinality larger than $\aleph_0$ but smaller than c. This conjecture has become known as the *Continuum Hypothesis.* Search the Internet for information on the Continuum Hypothesis and write a short report that explains how the Continuum Hypothesis was resolved.

4. Reprinted by permission of Mary Pat Campbell.

CHAPTER 2 Summary

Key Terms

aleph-null [p. 100]
cardinal number [p. 56]
complement of a set [p. 64]
countable set [p. 102]
counting number [p. 54]
difference of sets [p. 85]
disjoint sets [p. 75]
element (member) of a set [p. 53]
ellipsis [p. 54]
empty set or null set [p. 56]
equal sets [p. 57]
equivalent sets [p. 57]
finite set [p. 56]
infinite set [p. 99]
integer [p. 54]
intersection of sets [p. 74]
irrational number [p. 54]
natural number [p. 54]
one-to-one correspondence [p. 98]
power set [p. 104]
rational number [p. 54]
real number [p. 54]
roster method [p. 53]
set [p. 53]
set-builder notation [p. 56]
transfinite number [p. 100]
uncountable set [p. 104]
union of sets [p. 75]
universal set [p. 64]
Venn diagram [p. 66]
well-defined set [p. 55]
whole number [p. 54]

Essential Concepts

- **A Subset of a Set**
  Set A is a subset of set B, denoted by $A \subseteq B$, if and only if every element of A is also an element of B.
- **Proper Subset of a Set**
  Set A is a proper subset of set B, denoted by $A \subset B$, if every element of A is an element of B, and $A \neq B$.
- **Subset Relationships**
  $A \subseteq A$, for any set A
  $\varnothing \subseteq A$, for any set A
- **The Number of Subsets of a Set**
  A set with n elements has 2^n subsets.
- **Intersection of Sets**
  The *intersection* of sets A and B, denoted by $A \cap B$, is the set of elements common to both A and B.
  $$A \cap B = \{x \mid x \in A \quad \text{and} \quad x \in B\}$$
- **Union of Sets**
  The *union* of sets A and B, denoted by $A \cup B$, is the set that contains all the elements that belong to A or to B or to both.
  $$A \cup B = \{x \mid x \in A \quad \text{or} \quad x \in B\}$$
- **De Morgan's Laws**
  $$(A \cap B)' = A' \cup B' \qquad (A \cup B)' = A' \cap B'$$
- **Commutative Properties of Sets**
  $$A \cap B = B \cap A \qquad A \cup B = B \cup A$$
- **Associative Properties of Sets**
  $$(A \cap B) \cap C = A \cap (B \cap C)$$
  $$(A \cup B) \cup C = A \cup (B \cup C)$$
- **Distributive Properties of Sets**
  $$A \cap (B \cup C) = (A \cap B) \cup (A \cap C)$$
  $$A \cup (B \cap C) = (A \cup B) \cap (A \cup C)$$
- **The Inclusion-Exclusion Principle**
  For any finite sets A and B,
  $$n(A \cup B) = n(A) + n(B) - n(A \cap B)$$
- **Cantor's Theorem**
  Let S be any set. The set of all subsets of S has a larger cardinal number than the cardinal number of S.
- **Transfinite Arithmetic Theorems**
  For any whole number a,
  $$\aleph_0 + a = \aleph_0, \quad \aleph_0 - a = \aleph_0, \quad \aleph_0 + \aleph_0 = \aleph_0,$$
  $$\aleph_0 + c = c, \quad \aleph_0 c = c, \quad c + c = c$$

CHAPTER 2 Review Exercises

In Exercises 1–4, use the roster method to write each set.

1. The set of whole numbers less than 8
2. The set of integers that satisfy $x^2 = 64$
3. The set of natural numbers that satisfy $x + 3 \le 7$
4. The set of counting numbers larger than -3 and less than or equal to 6

In Exercises 5–8, use set-builder notation to write each set.

5. The set of integers greater than -6
6. {April, June, September, November}
7. {Kansas, Kentucky}
8. $\{1, 8, 27, 64, 125\}$

In Exercises 9–12, determine whether the statement is true or false.

9. $\{3\} \in \{1, 2, 3, 4\}$
10. $-11 \in I$
11. $\{a, b, c\} \sim \{1, 5, 9\}$
12. The set of small numbers is a well-defined set.

In Exercises 13–20, let $U = \{2, 6, 8, 10, 12, 14, 16, 18\}$, $A = \{2, 6, 10\}$, $B = \{6, 10, 16, 18\}$, and $C = \{14, 16\}$. Find each of the following.

13. $A \cap B$
14. $A \cup B$
15. $A' \cap C$
16. $B \cup C'$
17. $A \cup (B \cap C)$
18. $(A \cup C)' \cap B'$
19. $(A \cap B')'$
20. $(A \cup B \cup C)'$

In Exercises 21–24, determine whether the first set is a proper subset of the second set.

21. The set of natural numbers; the set of whole numbers
22. The set of integers; the set of real numbers
23. The set of counting numbers; the set of natural numbers
24. The set of real numbers; the set of rational numbers

In Exercises 25–28, list all the subsets of the given set.

25. {I, II}
26. {s, u, n}
27. {penny, nickel, dime, quarter}
28. {A, B, C, D, E}

In Exercises 29–32, find the number of subsets of the given set.

29. The set of the four musketeers
30. The set of the letters of the English alphabet
31. The set of the letters of "uncopyrightable," which is the longest English word with no repeated letters
32. The set of the seven dwarfs

In Exercises 33–36, draw a Venn diagram to represent the given set.

33. $A \cap B'$
34. $A' \cup B'$
35. $(A \cup B) \cup C'$
36. $A \cap (B' \cup C)$

In Exercises 37–40, draw Venn diagrams to determine whether the expressions are equal for all sets *A*, *B*, and *C*.

37. $A' \cup (B \cup C)$; $(A' \cup B) \cup (A' \cup C)$
38. $(A \cap B) \cap C'$; $(A' \cup B') \cup C$
39. $A \cap (B' \cap C)$; $(A \cup B') \cap (A \cup C)$
40. $A \cap (B \cup C)$; $A' \cap (B \cup C)$

In Exercises 41 and 42, use set notation to describe the shaded region.

41.

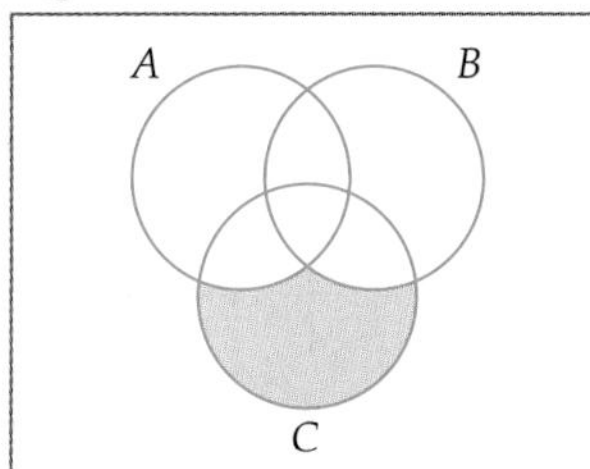

42.

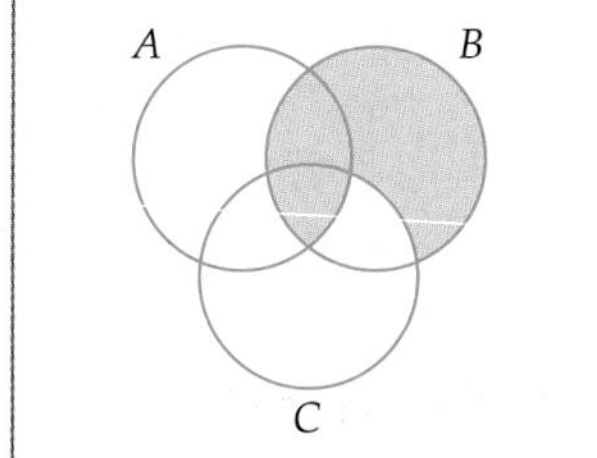

In Exercises 43 and 44, draw a Venn diagram with each of the given elements placed in the correct region.

43. $U = \{e, h, r, d, w, s, t\}$
$A = \{t, r, e\}$
$B = \{w, s, r, e\}$
$C' = \{s, r, d, h\}$

44. $U = \{\alpha, \beta, \Gamma, \gamma, \Delta, \delta, \varepsilon, \theta\}$
$A' = \{\beta, \Delta, \theta, \gamma\}$
$B = \{\delta, \varepsilon\}$
$C = \{\beta, \varepsilon, \Gamma\}$

45. A Survey In a survey at a health club, 208 members indicated that they enjoy aerobic exercises, 145 indicated that they enjoy weight training, 97 indicated that they enjoy both aerobics and weight training, and 135 indicated that they do not enjoy either of these types of exercise. How many members were surveyed?

46. A Survey A gourmet coffee bar conducted a survey to determine the preferences of its customers. Of the customers surveyed,

221 like espresso.
127 like cappuccino and chocolate-flavored coffee.
182 like cappuccino.
136 like espresso and chocolate-flavored coffee.
209 like chocolate-flavored coffee.
96 like all three types of coffee.
116 like espresso and cappuccino.
82 like none of these types of coffee.

How many of the customers in the survey:

a. like only chocolate-flavored coffee?

b. like cappuccino and chocolate-flavored coffee but not espresso?

c. like espresso and cappuccino but not chocolate-flavored coffee?

d. like exactly one of the three types of coffee?

In Exercises 47–50, establish a one-to-one correspondence between the sets.

47. $\{1, 3, 6, 10\}$; $\{1, 2, 3, 4\}$

48. $\{x \mid x > 10 \text{ and } x \in N\}$; $\{2, 4, 6, 8, \ldots, 2n, \ldots\}$

49. $\{3, 6, 9, 12, \ldots, 3n, \ldots\}$; $\{10, 100, 1000, \ldots, 10^n, \ldots\}$

50. $\{x \mid 0 \le x \le 1\}$; $\{x \mid 0 \le x \le 4\}$ (*Hint:* Use a drawing.)

In Exercises 51 and 52, show that the given set is an infinite set.

51. $A = \{6, 10, 14, 18, \ldots, 4n + 2, \ldots\}$

52. $B = \left\{1, \frac{1}{2}, \frac{1}{4}, \frac{1}{8}, \ldots, \frac{1}{2^{n-1}}, \ldots\right\}$

In Exercises 53–60, state the cardinality of each set.

53. $\{5, 6, 7, 8, 6\}$

54. $\{4, 6, 8, 10, 12, \ldots, 22\}$

55. $\{0, \varnothing\}$

56. The set of all states in the U.S. that border the Gulf of Mexico

57. The set of integers less than 1,000,000

58. The set of rational numbers between 0 and 1

59. The set of irrational numbers

60. The set of real numbers between 0 and 1

In Exercises 61–68, find each of the following, where $\aleph_0$ and c are transfinite cardinal numbers.

61. $\aleph_0 - 700$

62. $\aleph_0 + 4100$

63. $\aleph_0 + (\aleph_0 + \aleph_0)$

64. $\aleph_0 + c$

65. $c - 7$

66. $c + (c + c)$

67. $5\aleph_0$

68. $15c$

CHAPTER 2 Test

In Exercises 1–6, let $U = \{1, 2, 3, 4, 5, 6, 7, 8, 9, 10\}$ $A = \{3, 5, 7, 8\}$, $B = \{2, 3, 8, 9, 10\}$, and $C = \{1, 4, 7, 8\}$. Use the roster method to write each of the following sets.

1. $A \cup B$ **2.** $A' \cap B$

3. $(A \cap B)'$ **4.** $(A \cup B')'$

5. $A' \cup (B \cap C')$ **6.** $A \cap (B' \cup C)$

In Exercises 7 and 8, use set-builder notation to write each of the given sets.

7. $\{0, 1, 2, 3, 4, 5, 6\}$ **8.** $\{-3, -2, -1, 0, 1, 2\}$

In Exercise 9, state the cardinality of the given set.

9. a. The set of whole numbers less than 4

b. The set of rational numbers between 7 and 8

c. The set of natural numbers

d. The set of real numbers

In Exercises 10 and 11, state whether the given sets are equal, equivalent, both, or neither.

10. a. the set of natural numbers; the set of integers

b. the set of whole numbers; the set of positive integers

11. a. the set of rational numbers; the set of irrational numbers

b. the set of real numbers; the set of irrational numbers

12. List all of the subsets of {a, b, c, d}.

13. Determine the number of subsets of a set with 21 elements.

14. State whether each statement is true or false.

a. $\{4\} \in \{1, 2, 3, 4, 5, 6, 7\}$

b. The set of rational numbers is a well-defined set.

c. $A \subset A$

d. The set of positive even whole numbers is equivalent to the set of natural numbers.

15. Draw a Venn diagram to represent the given set.

a. $(A \cup B') \cap C$ **b.** $(A' \cap B) \cup (A \cap C')$

16. Use set notation to describe the shaded region.

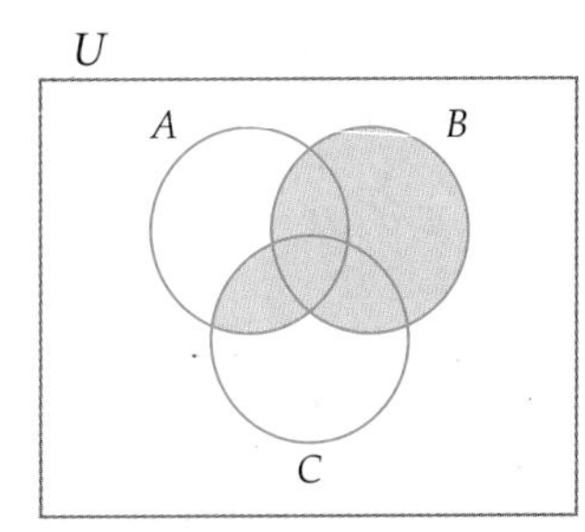

17. A Survey In the town of LeMars, 385 families have a CD player, 142 families have a DVD player, 41 families have both a CD player and a DVD player, and 55 families do not have a CD player or a DVD player. How many families live in LeMars?

18. A Survey A survey of 1000 households was taken to determine how they obtained news about current events. The survey considered only television, newspapers, and the Internet as sources for news. Of the households surveyed,

724 obtained news from television.
545 obtained news from newspapers.
280 obtained news from the Internet.
412 obtained news from both television and newspapers.
185 obtained news from both television and the Internet.
105 obtained news from television, newspapers, and the Internet.
64 obtained news from the Internet but not from television or newspapers.

Of those households that were surveyed,

a. how many obtained news from television but not from newspapers or the Internet?

b. how many obtained news from newspapers but not from television or the Internet?

c. how many obtained news from television or newspapers?

d. how many did not acquire news from television, newspapers, or the Internet?

19. Show a method that can be used to establish a one-to-one correspondence between the elements of the following sets.

$$\{5, 10, 15, 20, 25, \ldots, 5n, \ldots\}; \quad W$$

20. Prove that the following set is an infinite set by illustrating a one-to-one correspondence between the elements of the set and the elements of one of the set's proper subsets.

$$\{3, 6, 9, 12, \ldots, 3n \ldots\}$$

CHAPTER

3 Logic

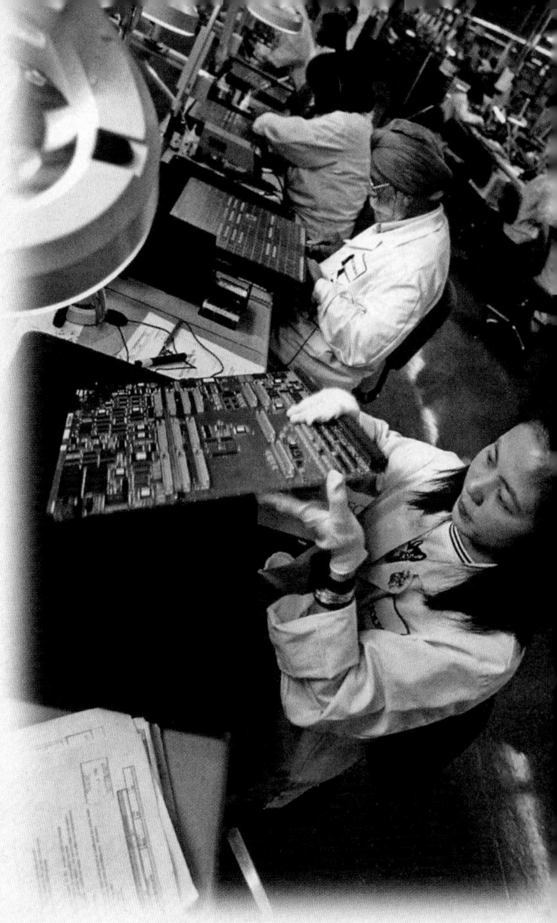

I know what you're thinking about," said Tweedledum (to Alice), "but it isn't so, nohow."

"Contrariwise," continued Tweedledee, "if it was so, it might be; and if it were so, it would be: but as it isn't, it ain't. That's logic."

The above excerpt, from Lewis Carroll's *Through the Looking-Glass,* sums up Tweedledee's understanding of logic. In today's complex world, it is not so easy to summarize the topic of logic. For lawyers and business people, logic is the science of correct reasoning. They often use logic to construct valid arguments, analyze legal contracts, and solve complicated problems. The principles of logic can also be used as a production tool. For example, programmers use logic to design computer software, engineers use logic to design the electronic circuits in computers, and mathematicians use logic to solve problems and construct mathematical proofs.

In this chapter, you will encounter several facets of logic. Specifically, you will use logic to

- analyze information and the relationship between statements,
- determine the validity of arguments,
- determine valid conclusions based on given assumptions, and
- analyze electronic circuits.

For online student resources, visit this textbook's website at **math.college.hmco.com/students.**

SECTION 3.1 Logic Statements and Quantifiers

Logic Statements

historical note

George Boole (bōōl) was born in 1815 in Lincoln, England. He was raised in poverty, but he was very industrious and had learned Latin and Greek by the age of 12. Later he mastered German, French, and Italian. His first profession, at the young age of 16, was that of an assistant school teacher. At the age of 20 he started his own school.

In 1849 Boole was appointed the chairperson of mathematics at Queens College in Cork, Ireland. He was known as a dedicated professor who gave detailed lectures. He continued to teach at Queens College until he died of pneumonia in 1864.

Many of Boole's mathematical ideas, such as Boolean algebra, have applications in the areas of computer programming and the design of telephone switching devices. ■

One of the first mathematicians to make a serious study of symbolic logic was Gottfried Wilhelm Leibniz (1646–1716). Leibniz tried to advance the study of logic from a merely philosophical subject to a formal mathematical subject. Leibniz never completely achieved this goal; however, several mathematicians, such as Augustus De Morgan (1806–1871) and George Boole (1815–1864), contributed to the advancement of symbolic logic as a mathematical discipline.

Boole published *The Mathematical Analysis of Logic* in 1848. In 1854 he published the more extensive work *An Investigation of the Laws of Thought.* Concerning this document, the mathematician Bertrand Russell stated, "Pure mathematics was discovered by Boole in a work which is called *The Laws of Thought.*" Although some mathematicians feel this is an exaggeration, the following paragraph, extracted from *An Investigation of the Laws of Thought,* gives some insight into the nature and scope of this document.

> The design of the following treatise is to investigate the fundamental laws of those operations of the mind by which reasoning is performed; to give expression to them in the language of a Calculus, and upon this foundation to establish the science of Logic and construct its method; to make the method itself the basis of a general method for the application of the mathematical doctrine of probabilities; and finally, to collect from the various elements of truth brought to view in the course of these inquiries some probable intimations concerning the nature and constitution of the human mind.[1]

Every language contains different types of sentences, such as statements, questions, and commands. For instance,

"Is the test today?" is a question.

"Go get the newspaper" is a command.

"This is a nice car" is an opinion.

"Denver is the capital of Colorado" is a statement of fact.

The symbolic logic that Boole was instrumental in creating applies only to sentences that are *statements* as defined below.

Definition of a Statement

A **statement** is a declarative sentence that is either true or false, but not both true and false.

1. Bell, E. T. *Men of Mathematics.* New York: Simon and Schuster, Inc., Touchstone Books, Reissue edition, 1986.

It may not be necessary to determine whether a sentence is true or false to determine whether it is a statement. For instance, the following sentence is either true or false:

> Every even number greater than 2 can be written as the sum of two prime numbers.

At this time mathematicians have not determined whether the sentence is true or false, but they do know that it is either true or false and that it is not both true and false. Thus the sentence is a statement.

TAKE NOTE

The following sentence is a famous paradox:

> This is a false sentence.

It is not a statement, because if we assume it to be a true sentence, then it is false, and if we assume it to be a false sentence, then it is true. Statements cannot be true and false at the same time.

EXAMPLE 1 ■ Identify Statements

Determine whether each sentence is a statement.

a. Florida is a state in the United States.
b. The word *dog* has four letters.
c. How are you?
d. $9^{(9^9)} + 2$ is a prime number.
e. $x + 1 = 5$

Solution

a. Florida is one of the 50 states in the United States, so this sentence is true and it is a statement.

b. The word *dog* consists of exactly three letters, so this sentence is false and it is a statement.

c. The sentence "How are you?" is a question; it is not a declarative sentence. Thus it is not a statement.

d. You may not know whether $9^{(9^9)} + 2$ is a prime number; however, you do know that it is a whole number larger than 1, so it is either a prime number or it is not a prime number. The sentence is either true or it is false, and it is not both true and false, so it is a statement.

e. $x + 1 = 5$ is a statement. It is known as an *open statement.* It is true for $x = 4$, and it is false for any other values of x. For any given value of x, it is true or false but not both.

CHECK YOUR PROGRESS 1 Determine whether each sentence is a statement.

a. Open the door.
b. 7055 is a large number.
c. $4 + 5 = 8$
d. In the year 2009, the president of the United States will be a woman.
e. $x > 3$

Solution *See page S7.*

Charles Dodgson (Lewis Carroll)

Math Matters Charles Dodgson

One of the best-known logicians is Charles Dodgson (1832–1898). Dodgson was educated at Rugby and Oxford, and in 1861 he became a lecturer in mathematics at Oxford. His mathematical works include *A Syllabus of Plane Algebraical Geometry, The Fifth Book of Euclid Treated Algebraically,* and *Symbolic Logic.* Although Dodgson was a distinguished mathematician in his time, he is best known by his pen name Lewis Carroll, which he used when he published *Alice's Adventures in Wonderland* and *Through the Looking-Glass.*

Queen Victoria of the United Kingdom enjoyed *Alice's Adventures in Wonderland* to the extent that she told Dodgson she was looking forward to reading another of his books. He promptly sent her his *Syllabus of Plane Algebraical Geometry,* and it was reported that she was less than enthusiastic about the latter book.

Compound Statements

Connecting statements with words and phrases such as *and, or, not, if ... then,* and *if and only if* creates a **compound statement.** For instance, "I will attend the meeting or I will go to school" is a compound statement. It is composed of the two **component statements** "I will attend the meeting" and "I will go to school." The word *or* is a **connective** for the two component statements.

George Boole used symbols such as p, q, r, and s to represent statements and the symbols $\wedge$, $\vee$, $\sim$, $\rightarrow$, and $\leftrightarrow$ to represent connectives. See Table 3.1.

Table 3.1 *Logic Symbols*

Original Statement	Connective	Statement in Symbolic Form	Type of Compound Statement
not p	not	$\sim p$	negation
p and q	and	$p \wedge q$	conjunction
p or q	or	$p \vee q$	disjunction
If p, then q	If ... then	$p \rightarrow q$	conditional
p if and only if q	if and only if	$p \leftrightarrow q$	biconditional

QUESTION *What connective is used in a conjunction?*

ANSWER *The connective and.*

Truth Value and Truth Tables

The **truth value** of a statement is true (T) if the statement is true and false (F) if the statement is false. A **truth table** is a table that shows the truth value of a statement for all possible truth values of its components.

The Truth Table for $\sim p$

p	$\sim p$
T	F
F	T

The *negation* of the statement "Today is Friday" is the statement "Today is not Friday." In symbolic logic, the tilde symbol $\sim$ is used to denote the negation of a statement. If a statement p is true, its negation $\sim p$ is false, and if a statement p is false, its negation $\sim p$ is true. See the table at the left. The negation of the negation of a statement is the original statement. Thus, $\sim(\sim p)$ can be replaced by p in any statement.

EXAMPLE 2 ■ Write the Negation of a Statement

Write the negation of each statement.

a. Bill Gates has a yacht.
b. The number 10 is a prime number.
c. The Dolphins lost the game.

Solution
a. Bill Gates does not have a yacht.
b. The number 10 is not a prime number.
c. The Dolphins did not lose the game.

CHECK YOUR PROGRESS 2 Write the negation of each statement.

a. 1001 is divisible by 7.
b. 5 is an even number.
c. The fire engine is not red.

Solution *See page S7.*

We will often find it useful to write compound statements in symbolic form.

EXAMPLE 3 ■ Write Compound Statements in Symbolic Form

Consider the following statements.

p: Today is Friday.
q: It is raining.
r: I am going to a movie.
s: I am not going to the basketball game.

Write the following compound statements in symbolic form.

a. Today is Friday and it is raining.
b. It is not raining and I am going to a movie.
c. I am going to the basketball game or I am going to a movie.
d. If it is raining, then I am not going to the basketball game.

Solution
a. $p \wedge q$ **b.** $\sim q \wedge r$ **c.** $\sim s \vee r$ **d.** $q \rightarrow s$

CHECK YOUR PROGRESS 3 Use p, q, r, and s as defined in Example 3 to write the following compound statements in symbolic form.

a. Today is not Friday and I am going to a movie.

b. I am going to the basketball game and I am not going to a movie.

c. I am going to the movie if and only if it is raining.

d. If today is Friday, then I am not going to a movie.

Solution *See page S7.*

In the next example, we translate symbolic logic statements into English sentences.

EXAMPLE 4 ■ Translate Symbolic Statements

Consider the following statements.

p: The game will be played in Atlanta.
q: The game will be shown on CBS.
r: The game will not be shown on ESPN.
s: The Dodgers are favored to win.

Write each of the following symbolic statements in words.

a. $q \wedge p$ **b.** $\sim r \wedge s$ **c.** $s \leftrightarrow \sim p$

Solution

a. The game will be shown on CBS and the game will be played in Atlanta.

b. The game will be shown on ESPN and the Dodgers are favored to win.

c. The Dodgers are favored to win if and only if the game will not be played in Atlanta.

CHECK YOUR PROGRESS 4 Consider the following statements.

e: All men are created equal.
t: I am trading places.
a: I get Abe's place.
g: I get George's place.

Use the above information to translate the dialogue in the following speech bubbles.

Solution *See page S7.*

The Truth Table for $p \wedge q$

p	q	$p \wedge q$
T	T	T
T	F	F
F	T	F
F	F	F

If you order cake *and* ice cream in a restaurant, the waiter will bring *both* cake and ice cream. In general, the **conjunction** $p \wedge q$ is true if both p and q are true, and the conjunction is false if either p or q is false. The truth table at the left shows the four possible cases that arise when we form a conjunction of two statements.

Truth Value of a Conjunction

The conjunction $p \wedge q$ is true if and only if both p and q are true.

Sometimes the word *but* is used in place of the connective *and* to form a conjunction. For instance, "My local phone company is *SBC*, but my long-distance carrier is Sprint" is equivalent to the conjunction "My local phone company is *SBC* and my long-distance carrier is Sprint."

The Truth Table for $p \vee q$

p	q	$p \vee q$
T	T	T
T	F	T
F	T	T
F	F	F

Any **disjunction** $p \vee q$ is true if p is true or q is true or both p and q are true. The truth table at the left shows that the disjunction p or q is false if both p and q are false; however, it is true in all other cases.

Truth Value of a Disjunction

The disjunction $p \vee q$ is true if p is true, if q is true, or if both p and q are true.

EXAMPLE 5 ■ Determine the Truth Value of a Statement

Determine whether each statement is true or false.

a. $7 \geq 5$

b. 5 is a whole number and 5 is an even number.

c. 2 is a prime number and 2 is an even number.

Solution

a. $7 \geq 5$ means $7 > 5$ or $7 = 5$. Because $7 > 5$ is true, the statement $7 \geq 5$ is a true statement.

b. This is a false statement because 5 is not an even number.

c. This is a true statement because each component statement is true.

CHECK YOUR PROGRESS 5 Determine whether each statement is true or false.

a. 21 is a rational number and 21 is a natural number.

b. $4 \leq 9$

c. $-7 \geq -3$

Solution *See page S7.*

Truth tables for the conditional and biconditional are given in Section 3.3.

Quantifiers and Negation

In a statement, the word *some* and the phrases *there exists* and *at least one* are called **existential quantifiers.** Existential quantifiers are used as prefixes to assert the existence of something.

In a statement, the words *none, no, all,* and *every* are called **universal quantifiers.** The universal quantifiers *none* and *no* deny the existence of something, whereas the universal quantifiers *all* and *every* are used to assert that every element of a given set satisfies some condition.

Recall that the negation of a false statement is a true statement and the negation of a true statement is a false statement. It is important to remember this fact when forming the negation of a quantified statement. For instance, what is the negation of the false statement, "All dogs are mean"? You may think that the negation is "No dogs are mean," but this is also a false statement. Thus the statement "No dogs are mean" is not the negation of "All dogs are mean." The negation of "All dogs are mean," which is a false statement, is in fact "Some dogs are not mean," which is a true statement. The statement "Some dogs are not mean" can also be stated as "At least one dog is not mean" or "There exists a dog that is not mean."

What is the negation of the false statement "No doctors write in a legible manner"? Whatever the negation is, we know it must be a true statement. The negation cannot be "All doctors write in a legible manner," because this is also a false statement. The negation is "Some doctors write in a legible manner." This can also be stated as "There exists at least one doctor who writes in a legible manner."

Table 3.2 summarizes the concepts needed to write the negations of statements that contain one of the quantifiers *all, none,* or *some.*

Table 3.2 *The Negation of a Statement that Contains a Quantifier*

Original Statement	Negation
All ______ are ______.	Some ______ are not ______.
No(ne) ______.	Some ______.
Some ______ are not ______.	All ______ are ______.
Some ______.	No(ne) ______.

EXAMPLE 6 ■ Write the Negation of a Quantified Statement

Write the negation of each of the following statements.

a. Some baseball players are worth a million dollars.

b. All movies are worth the price of admission.

c. No odd numbers are divisible by 2.

Solution

a. No baseball player is worth a million dollars.

b. Some movies are not worth the price of admission.

c. Some odd numbers are divisible by 2.

CHECK YOUR PROGRESS 6 Write the negation of the following statements.

a. All bears are brown.

b. No math class is fun.

c. Some vegetables are not green.

Solution *See page S7.*

Switching Networks

Claude E. Shannon

In 1939, Claude E. Shannon (1916–2001) wrote a thesis on an application of symbolic logic to *switching networks.* A switching network consists of wires and switches that can open or close. Switching networks are used in many electrical appliances, telephone equipment, and computers. Figure 3.1 shows a switching network that consists of a single switch P that connects two terminals. An electric current can flow from one terminal to the other terminal provided the switch P is in the closed position. If P is in the open position, then the current cannot flow from one terminal to the other. If a current can flow between the terminals we say that a network is closed, and if a current cannot flow between the terminals we say that the network is open. We designate this network by the letter P. There exists an analogy between a network P and a statement p in that a network is either open or it is closed, and a statement is either true or it is false.

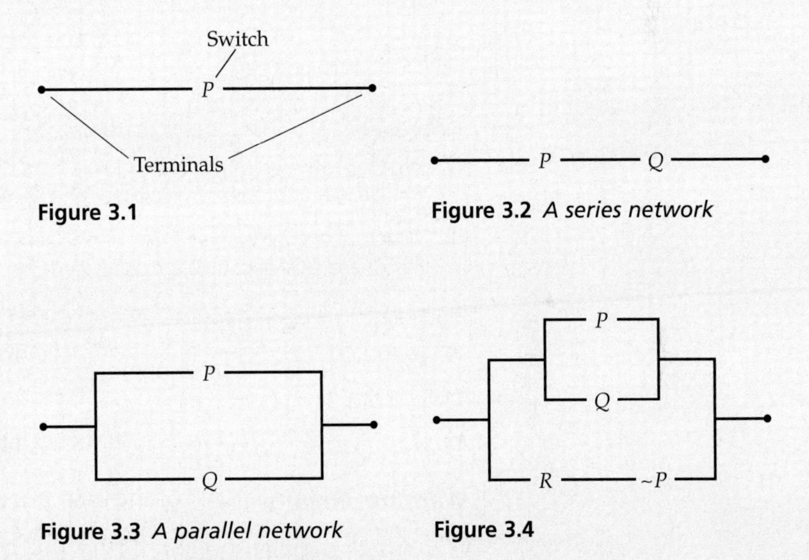

Figure 3.1

Figure 3.2 *A series network*

Figure 3.3 *A parallel network*

Figure 3.4

Figure 3.2 shows two switches P and Q connected in **series.** This series network is closed if and only if both switches are closed. We will use $P \wedge Q$ to denote this series network because it is analogous to the logic statement $p \wedge q$, which is true if and only if both p and q are true.

Figure 3.3 shows two switches P and Q connected in **parallel.** This parallel network is closed if either P or Q is closed. We will designate this parallel network by $P \vee Q$ because it is analogous to the logic statement $p \vee q$, which is true if p is true or if q is true.

Series and parallel networks can be combined to produce more complicated networks, as shown in Figure 3.4.

The network shown in Figure 3.4 is closed provided P or Q is closed or provided both R and $\sim P$ are closed. Note that the switch $\sim P$ is closed if P is open, and $\sim P$ is open if P is closed. We use the symbolic statement $(P \vee Q) \vee (R \wedge \sim P)$ to represent this network.

If two switches are always open at the same time and always closed at the same time, then we will use the same letter to designate both switches.

(continued)

Excursion Exercises

Write a symbolic statement to represent each of the networks in Excursion Exercises 1–6.

1.

2.

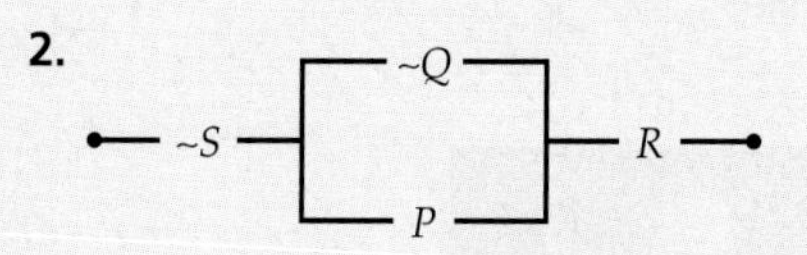

3.

4.

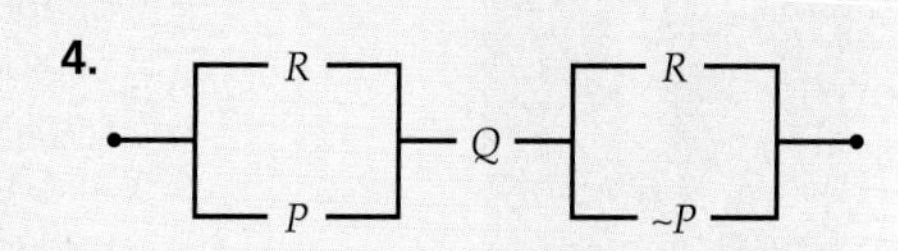

5.

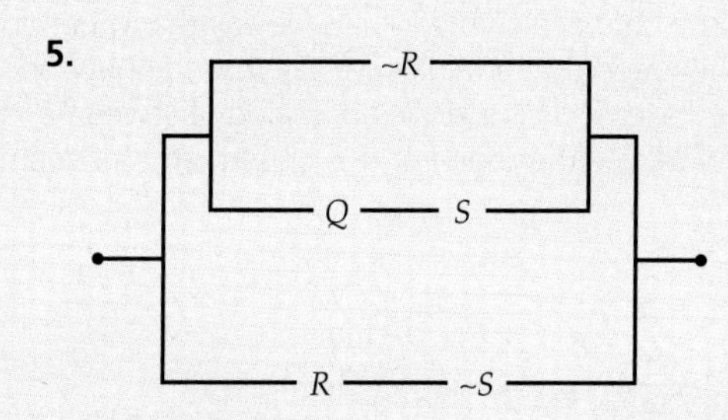

6.

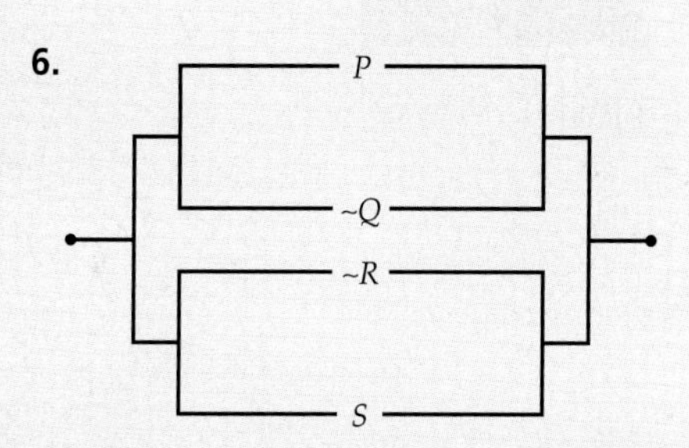

7. Which of the networks in Excursion Exercises 1–6 are closed networks, given that *P* is closed, *Q* is open, *R* is closed, and *S* is open?

8. Which of the networks in Excursion Exercises 1–6 are closed networks, given that *P* is open, *Q* is closed, *R* is closed, and *S* is closed?

In Excursion Exercises 9–14, draw a network to represent each statement.

9. $(\sim P \vee Q) \wedge (R \wedge P)$

10. $P \wedge [(Q \wedge \sim R) \vee R]$

11. $[\sim P \wedge Q \wedge R] \vee (P \wedge R)$

12. $(Q \vee R) \vee (S \vee \sim P)$

13. $[(\sim P \wedge R) \vee Q] \vee (\sim R)$

14. $(P \vee Q \vee R) \wedge S \wedge (\sim Q \vee R)$

Warning Circuits The circuits shown in Excursion Exercises 15 and 16 include a switching network, a warning light, and a battery. In each circuit the warning light will turn on only when the switching network is closed.

15. Consider the following circuit.

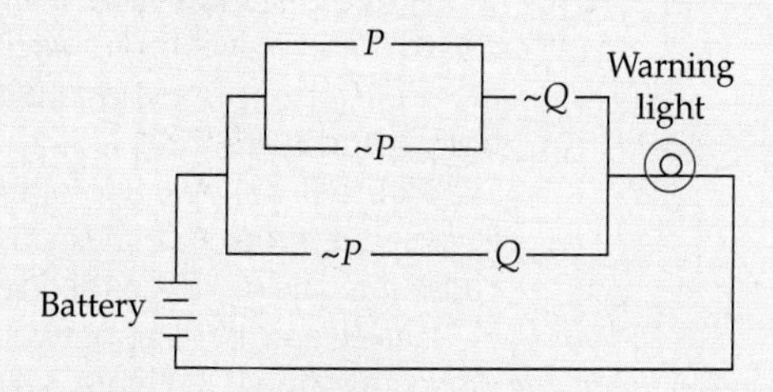

For each of the following conditions, determine whether the warning light will be on or off.

a. *P* is closed and *Q* is open.
b. *P* is closed and *Q* is closed.
c. *P* is open and *Q* is closed.
d. *P* is open and *Q* is open.

(continued)

16. An engineer thinks that the following circuit can be used in place of the circuit shown in Excursion Exercise 15. Do you agree? Explain.

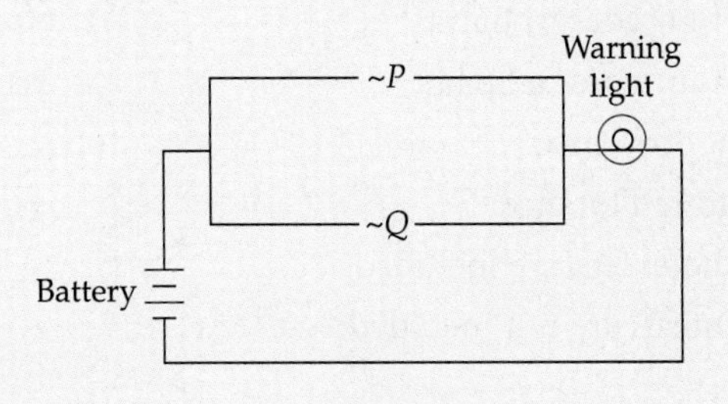

Exercise Set 3.1

In Exercises 1–10, determine whether each sentence is a statement.

1. West Virginia is west of the Mississippi River.
2. 1031 is a prime number.
3. The area code for Storm Lake, Iowa is 512.
4. Some negative numbers are rational numbers.
5. Have a fun trip.
6. Do you like to read?
7. All hexagons have exactly five sides.
8. If x is a negative number, then x^2 is a positive number.
9. Mathematics courses are better than history courses.
10. Every real number is a rational number.

In Exercises 11–18, determine the components of each compound statement.

11. The principal will attend the class on Tuesday or Wednesday.
12. 5 is an odd number and 6 is an even number.
13. A triangle is an acute triangle if and only if it has three acute angles.
14. Some birds can swim and some fish can fly.
15. I ordered a salad and a cola.
16. If this is Saturday, then tomorrow is Sunday.
17. $5 + 2 \geq 6$
18. $9 - 1 \leq 8$

In Exercises 19–22, write the negation of each statement.

19. The Giants lost the game.
20. The lunch was served at noon.
21. The game did not go into overtime.
22. The game was not shown on ABC.

In Exercises 23–32, write each sentence in symbolic form. Represent each component of the sentence with the letter indicated in parentheses. Also state whether the sentence is a conjunction, a disjunction, a negation, a conditional, or a biconditional.

23. If today is Wednesday (w), then tomorrow is Thursday (t).
24. It is not true that Sue took the tickets (t).
25. All squares (s) are rectangles (r).
26. I went to the post office (p) and the bookstore (s).
27. A triangle is an equilateral triangle (l) if and only if it is an equiangular triangle (a).
28. A number is an even number (e) if and only if it has a factor of 2 (t).
29. If it is a dog (d), it has fleas (f).
30. Polynomials that have exactly three terms (p) are called trinomials (t).
31. I will major in mathematics (m) or computer science (c).
32. All pentagons (p) have exactly five sides (s).

In Exercises 33–38, write each symbolic statement in words. Use p, q, r, s, t, and u as defined below.

p: The tour goes to Italy.
q: The tour goes to Spain.
r: We go to Venice.
s: We go to Florence.
t: The hotel fees are included.
u: The meals are not included.

33. $p \wedge \sim q$
34. $r \vee s$
35. $r \rightarrow \sim s$
36. $p \rightarrow r$
37. $s \leftrightarrow \sim r$
38. $\sim t \wedge u$

In Exercises 39–50, use the definitions presented in Table 3.2, page 120, to write the negation of each quantified statement.

39. Some cats do not have claws.
40. Some dogs are not friendly.
41. All classic movies were first produced in black and white.
42. Everybody enjoyed the dinner.
43. None of the numbers were even numbers.
44. At least one student received an A.
45. No irrational number can be written as a terminating decimal.
46. All cameras use film.
47. All cars run on gasoline.
48. None of the students took my advice.
49. Every item is on sale.
50. All of the telephone lines are not busy.

In Exercises 51–64, determine whether each statement is true or false.

51. $7 < 5$ or $3 > 1$.
52. $3 \leq 9$
53. $(-1)^{50} = 1$ and $(-1)^{99} = -1$.
54. $7 \neq 3$ or 9 is a prime number.
55. $-5 \geq -11$
56. $4.5 \leq 5.4$
57. 2 is an odd number or 2 is an even number.
58. 5 is a natural number and 5 is a rational number.
59. There exists an even prime number.
60. The square of any real number is a positive number.
61. Some real numbers are irrational.
62. All irrational numbers are real numbers.
63. Every integer is a rational number.
64. Every rational number is an integer.

Extensions

CRITICAL THINKING

Write Quotations in Symbolic Form In Exercises 65–70, translate each quotation into symbolic form. For each component, indicate what letter you used to represent the component.

65. If you can count your money, you don't have a billion dollars. *J. Paul Getty*
66. If you aren't fired with enthusiasm, then you will be fired with enthusiasm. *Vince Lombardi*
67. Those who do not learn from history are condemned to repeat it. *George Santayana*
68. We don't like their sound, and guitar music is on the way out. *Decca Recording Company,* rejecting the Beatles in 1962
69. If people concentrated on the really important things in life, there'd be a shortage of fishing poles. *Doug Larson*
70. If you're killed, you've lost a very important part of your life. *Brooke Shields*

Write Statements in Symbolic Form In Exercises 71–76, translate each mathematical statement into symbolic form. For each component, indicate what letter you used to represent the component.

71. An angle is a right angle if and only if its measure is 90°.
72. Any angle inscribed in a semicircle is a right angle.
73. If two sides of a triangle are equal in length, the angles opposite those sides are congruent.
74. The sum of the measures of the three angles of any triangle is 180°.
75. All squares are rectangles.
76. If the corresponding sides of two triangles are proportional, then the triangles are similar.

EXPLORATIONS

77. **Raymond Smullyan** is a logician, a philosopher, a professor, and an author of many books on logic and puzzles. Some of his fans rate his puzzle books as the best ever written. Search the Internet to find information on the life of Smullyan and his work in the area of logic. Write a few paragraphs that summarize your findings.

SECTION 3.2 Truth Tables, Equivalent Statements, and Tautologies

Truth Tables

In Section 3.1, we defined truth tables for the negation of a statement, the conjunction of two statements, and the disjunction of two statements. Each of these truth tables is shown below for review purposes.

Negation

p	$\sim p$
T	F
F	T

Conjunction

p	q	$p \wedge q$
T	T	T
T	F	F
F	T	F
F	F	F

Disjunction

p	q	$p \vee q$
T	T	T
T	F	T
F	T	T
F	F	F

p	q	Given Statement
T	T	
T	F	
F	T	
F	F	

Standard truth table form for a given statement that involves only the two simple statements p and q

In this section, we consider methods of constructing truth tables for a statement that involves a combination of conjunctions, disjunctions, and/or negations. If the given statement involves only the two simple statements, then start with a table with four rows (see the table at the left), called the **standard truth table form,** and proceed as shown in Example 1.

EXAMPLE 1 ■ Truth Tables

a. Construct a table for $\sim(\sim p \vee q) \vee q$.

b. Use the truth table from part a to determine the truth value of $\sim(\sim p \vee q) \vee q$, given that p is true and q is false.

Solution

a. Start with the standard truth table form and then include a $\sim p$ column.

p	q	$\sim p$
T	T	F
T	F	F
F	T	T
F	F	T

Now use the truth values from the $\sim p$ and q columns to produce the truth values for $\sim p \vee q$, as shown in the following table.

p	q	$\sim p$	$\sim p \vee q$
T	T	F	T
T	F	F	F
F	T	T	T
F	F	T	T

Negate the truth values in the $\sim p \vee q$ column to produce the following.

p	q	$\sim p$	$\sim p \vee q$	$\sim(\sim p \vee q)$
T	T	F	T	F
T	F	F	F	T
F	T	T	T	F
F	F	T	T	F

As our last step, we form the disjunction of $\sim(\sim p \vee q)$ with q and place the results in the rightmost column of the table. See the following table. The shaded column is the truth table for $\sim(\sim p \vee q) \vee q$.

p	q	$\sim p$	$\sim p \vee q$	$\sim(\sim p \vee q)$	$\sim(\sim p \vee q) \vee q$	
T	T	F	T	F	T	Row 1
T	F	F	F	T	T	Row 2
F	T	T	T	F	T	Row 3
F	F	T	T	F	F	Row 4

b. In row 2 of the above truth table, we see that when p is true, and q is false, the statement $\sim(\sim p \vee q) \vee q$ in the rightmost column is true.

CHECK YOUR PROGRESS 1

a. Construct a truth table for $(p \wedge \sim q) \vee (\sim p \vee q)$.

b. Use the truth table that you constructed in part a to determine the truth value of $(p \wedge \sim q) \vee (\sim p \vee q)$, given that p is true and q is false.

Solution *See page S8.*

p	q	r	Given Statement
T	T	T	
T	T	F	
T	F	T	
T	F	F	
F	T	T	
F	T	F	
F	F	T	
F	F	F	

Standard truth table form for a statement that involves the three simple statements p, q, and r

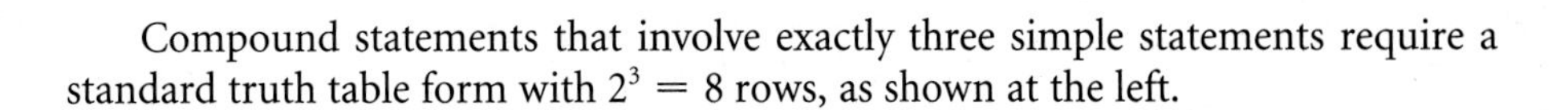

Compound statements that involve exactly three simple statements require a standard truth table form with $2^3 = 8$ rows, as shown at the left.

EXAMPLE 2 ■ Truth Tables

a. Construct a truth table for $(p \wedge q) \wedge (\sim r \vee q)$.

b. Use the truth table from part a to determine the truth value of $(p \wedge q) \wedge (\sim r \vee q)$, given that p is true, q is true, and r is false.

Solution

a. Using the procedures developed in Example 1, we can produce the following table. The shaded column is the truth table for $(p \wedge q) \wedge (\sim r \vee q)$. The numbers in the squares below the columns denote the order in which the columns were constructed. Each truth value in the column numbered 4 is the conjunction of the truth values to its left in the columns numbered 1 and 3.

p	q	r	$p \wedge q$	$\sim r$	$\sim r \vee q$	$(p \wedge q) \wedge (\sim r \vee q)$	
T	T	T	T	F	T	T	Row 1
T	T	F	T	T	T	T	Row 2
T	F	T	F	F	F	F	Row 3
T	F	F	F	T	T	F	Row 4
F	T	T	F	F	T	F	Row 5
F	T	F	F	T	T	F	Row 6
F	F	T	F	F	F	F	Row 7
F	F	F	F	T	T	F	Row 8
			1	2	3	4	

b. In row 2 of the above truth table we see that $(p \wedge q) \wedge (\sim r \vee q)$ is true when p is true, q is true, and r is false.

Plugger caller I.D.

Plugger Logic

CHECK YOUR PROGRESS 2

a. Construct a truth table for $(\sim p \wedge r) \vee (q \wedge \sim r)$.

b. Use the truth table that you constructed in part a to determine the truth value of $(\sim p \wedge r) \vee (q \wedge \sim r)$, given that p is false, q is true, and r is false.

Solution *See page S8.*

Alternative Method for the Construction of a Truth Table

In Example 3 we use an *alternative procedure* to construct a truth table. **This alternative procedure generally requires less writing, less time, and less effort than the procedure explained in Examples 1 and 2.**

Alternative Procedure for the Construction of a Truth Table

If the given statement has n simple statements, then start with a standard form that has 2^n rows.

1. In each row, enter the truth value for each *simple* statement and their negations.
2. Use the truth values from Step 1 to enter the truth value under each connective within a pair of grouping symbols (parentheses ( ), brackets [ ], braces { }). If some grouping symbols are nested inside other grouping symbols, then work from the inside out.
3. Use the truth values from Step 2 to determine the truth values under the remaining connectives.

TAKE NOTE

In a symbolic statement, grouping symbols are generally used to indicate the order in which logical connectives are applied. If grouping symbols are not used to specify the order in which logical connectives are applied, then we use the following **Order of Precedence Agreement:** First apply the negations from left to right, then apply the conjunctions from left to right, and finally apply the disjunctions from left to right.

EXAMPLE 3 ■ Use the Alternative Procedure to Construct a Truth Table

Construct a truth table for $p \vee [\sim(p \wedge \sim q)]$.

Solution

The given statement $p \vee [\sim(p \wedge \sim q)]$ has the two simple statements p and q. Thus we start with a standard form that has $2^2 = 4$ rows.

Step 1. In each column, enter the truth values for the statements p and $\sim q$, as shown in the columns numbered 1, 2, and 3 of the following table.

p	q	p	$\vee$	$[\sim$	$(p$	$\wedge$	$\sim q)]$
T	T	T			T		F
T	F	T			T		T
F	T	F			F		F
F	F	F			F		T
		1			2		3

Step 2. Use the truth values in columns 2 and 3 to determine the truth values to enter under the "and" connective. See the column numbered 4. Now negate the truth values in the column numbered 4 to produce the truth values in the column numbered 5.

p	q	p	$\vee$	$[\sim$	$(p$	$\wedge$	$\sim q)]$
T	T	T		T	T	F	F
T	F	T		F	T	T	T
F	T	F		T	F	F	F
F	F	F		T	F	F	T
		1		5	2	4	3

Step 3. Use the truth values in the columns numbered 1 and 5 to determine the truth values to enter under the "or" connective. See the column numbered 6, which is the truth table for $p \vee [\sim(p \wedge \sim q)]$.

p	q	p	$\vee$	$[\sim$	$(p$	$\wedge$	$\sim q)]$
T	T	T	T	T	T	F	F
T	F	T	T	F	T	T	T
F	T	F	T	T	F	F	F
F	F	F	T	T	F	F	T
		1	6	5	2	4	3

CHECK YOUR PROGRESS 3 Construct a truth table for $\sim p \vee (p \wedge q)$.

Solution *See page S8.*

Math Matters A Three-Valued Logic

In traditional logic either a statement is true or it is false. Many mathematicians have tried to extend traditional logic so that sentences that are *partially* true are assigned a truth value other than T or F. Jan Lukasiewicz was one of the first mathematicians to consider a three-valued logic in which a statement is true, false, or "somewhere between true and false." In his three-valued logic, Lukasiewicz classified the truth value of a statement as true (T), false (F), or maybe (M). The following table shows truth values for negation, conjunction, and disjunction in this three-valued logic.

p	q	Negation $\sim p$	Conjunction $p \wedge q$	Disjunction $p \vee q$
T	T	F	T	T
T	M	F	M	T
T	F	F	F	T
M	T	M	M	T
M	M	M	M	M
M	F	M	F	M
F	T	T	F	T
F	M	T	F	M
F	F	T	F	F

historical note

Jan Lukasiewicz (lo͞o-kä-shä-vēch) (1878–1956) was the Polish Minister of Education in 1919 and served as a professor of mathematics at Warsaw University from 1920–1939. Most of Lukasiewicz's work was in the area of logic. He is well known for developing *polish notation,* which was first used in logic to eliminate the need for parentheses in symbolic statements. Today *reverse polish notation* is used by many computers and calculators to perform computations without the need to enter parentheses. ■

TAKE NOTE

In the remaining sections of this chapter, the ≡ symbol will often be used to denote that two statements are equivalent.

Equivalent Statements

Two statements are **equivalent** if they both have the same truth value for all possible truth values of their component statements. Equivalent statements have identical truth values in the final columns of their truth tables. The notation $p \equiv q$ is used to indicate that the statements p and q are equivalent.

EXAMPLE 4 ■ Verify that Two Statements Are Equivalent

Show that $\sim(p \vee \sim q)$ and $\sim p \wedge q$ are equivalent statements.

Solution

Construct two truth tables and compare the results. The truth tables below show that $\sim(p \vee \sim q)$ and $\sim p \wedge q$ have the same truth values for all possible truth values of their component statements. Thus the statements are equivalent.

p	q	$\sim(p \vee \sim q)$
T	T	F
T	F	F
F	T	T
F	F	F

p	q	$\sim p \wedge q$
T	T	F
T	F	F
F	T	T
F	F	F

identical truth values

Thus $\sim(p \vee \sim q) \equiv \sim p \wedge q$.

CHECK YOUR PROGRESS 4 Show that $p \vee (p \wedge \sim q)$ and p are equivalent.

Solution *See page S9.*

The truth tables in Table 3.3 show that $\sim(p \vee q)$ and $\sim p \wedge \sim q$ are equivalent statements. The truth tables in Table 3.4 show that $\sim(p \wedge q)$ and $\sim p \vee \sim q$ are equivalent statements.

Table 3.3

p	q	$\sim(p \vee q)$	$\sim p \wedge \sim q$
T	T	F	F
T	F	F	F
F	T	F	F
F	F	T	T

Table 3.4

p	q	$\sim(p \wedge q)$	$\sim p \vee \sim q$
T	T	F	F
T	F	T	T
F	T	T	T
F	F	T	T

These equivalences are known as **De Morgan's laws for statements.**

De Morgan's Laws for Statements

For any statements p and q,

$$\sim(p \vee q) \equiv \sim p \wedge \sim q$$
$$\sim(p \wedge q) \equiv \sim p \vee \sim q$$

De Morgan's laws can be used to restate certain English sentences in an equivalent form.

EXAMPLE 5 ■ State an Equivalent Form

Use one of De Morgan's laws to restate the following sentence in an equivalent form.

It is not the case that I graduated or I got a job.

Solution
Let p represent the statement "I graduated." Let q represent the statement "I got a job." In symbolic form, the original sentence is $\sim(p \vee q)$. One of De Morgan's laws states that this is equivalent to $\sim p \wedge \sim q$. Thus a sentence that is equivalent to the original sentence is "I did not graduate and I did not get a job."

CHECK YOUR PROGRESS 5 Use one of De Morgan's laws to restate the following sentence in an equivalent form.

It is not true that I am going to the dance and I am going to the game.

Solution *See page S9.*

Tautologies and Self-Contradictions

A **tautology** is a statement that is always true. A **self-contradiction** is a statement that is always false.

EXAMPLE 6 ■ Verify Tautologies and Self-Contradictions

Show that $p \vee (\sim p \vee q)$ is a tautology.

Solution
Construct a truth table as shown below.

p	q	p	$\vee$	$(\sim p$	$\vee$	$q)$
T	T	T	T	F	T	T
T	F	T	T	F	F	F
F	T	F	T	T	T	T
F	F	F	T	T	T	F
		1	5	2	4	3

The table shows that $p \vee (\sim p \vee q)$ is always true. Thus $p \vee (\sim p \vee q)$ is a tautology.

CHECK YOUR PROGRESS 6 Show that $p \wedge (\sim p \wedge q)$ is a self-contradiction.

Solution *See page S9.*

QUESTION *Is the statement $x + 2 = 5$ a tautology or a self-contradiction?*

ANSWER *Neither. The statement is not true for all values of x, and it is not false for all values of x.*

Excursion

Switching Networks—Part II

The Excursion in Section 3.1 introduced the application of symbolic logic to switching networks. This Excursion makes use of *closure tables* to determine under what conditions a switching network is open or closed. **In a closure table, we use a 1 to designate that a switch or switching network is closed and a 0 to indicate that it is open.**

Figure 3.5 shows a switching network that consists of the single switch P and a second network that consists of the single switch $\sim P$. The table below shows that the switching network $\sim P$ is open when P is closed and is closed when P is open.

Negation Closure Table

P	$\sim P$
1	0
0	1

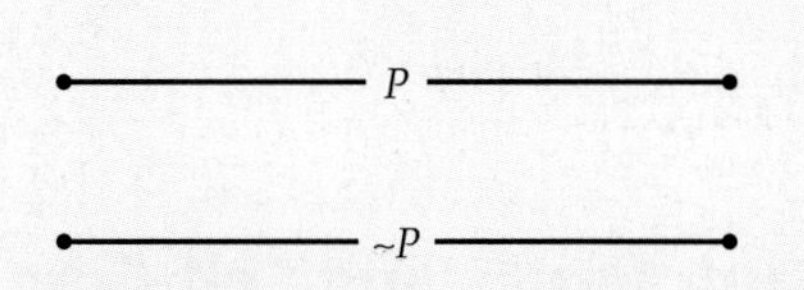

Figure 3.5

Figure 3.6 shows switches P and Q connected to form a series network. The table below shows that this series network is closed if and only if both P and Q are closed.

Series Network Closure Table

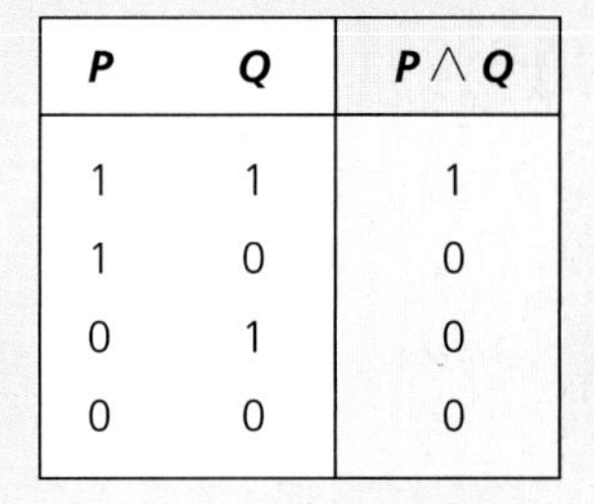

P	Q	$P \wedge Q$
1	1	1
1	0	0
0	1	0
0	0	0

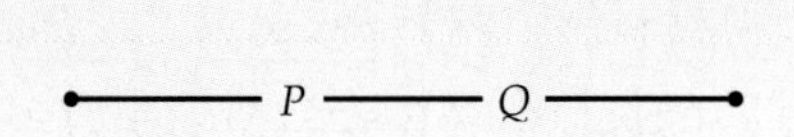

Figure 3.6 *A series network*

Figure 3.7 shows switches P and Q connected to form a parallel network. The table below shows that this parallel network is closed if P is closed or if Q is closed.

Parallel Network Closure Table

P	Q	$P \vee Q$
1	1	1
1	0	1
0	1	1
0	0	0

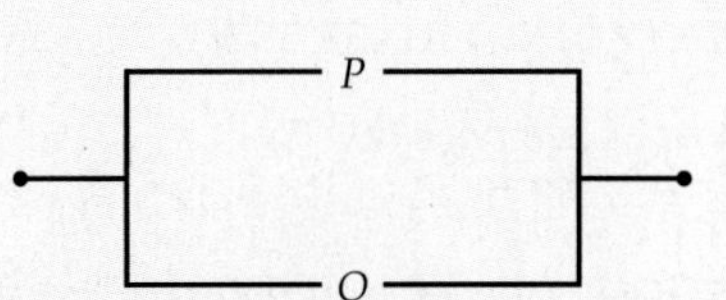

Figure 3.7 *A parallel network*

(continued)

Now consider the network shown in Figure 3.8. To determine the required conditions under which the network is closed, we first write a symbolic statement that represents the network, and then we construct a closure table.

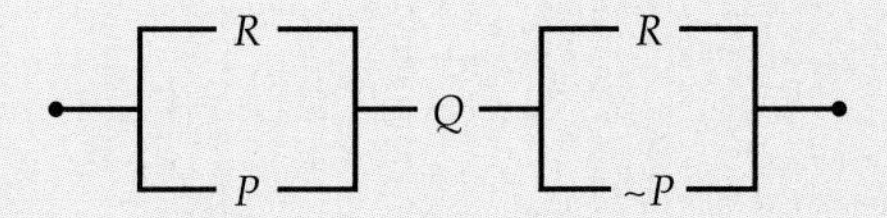

Figure 3.8

A symbolic statement that represents the network in Figure 3.8 is

$[(R \vee P) \wedge Q] \wedge (R \vee \sim P)$

The closure table for this network is shown below.

P	**Q**	**R**	[(**R**	∨	**P**)	∧	**Q**]	∧	(**R**	∨	~**P**)	
1	1	1	1	1	1	1	1	1	1	1	0	Row 1
1	1	0	0	1	1	1	1	0	0	0	0	Row 2
1	0	1	1	1	1	0	0	0	1	1	0	Row 3
1	0	0	0	1	1	0	0	0	0	0	0	Row 4
0	1	1	1	1	0	1	1	1	1	1	1	Row 5
0	1	0	0	0	0	0	1	0	0	1	1	Row 6
0	0	1	1	1	0	0	0	0	1	1	1	Row 7
0	0	0	0	0	0	0	0	0	0	1	1	Row 8
			1	6	2	7	3	9	4	8	5	

The rows numbered 1 and 5 of the above table show that the network is closed whenever

- P is closed, Q is closed, and R is closed, or
- P is open, Q is closed, and R is closed.

Thus the switching network in Figure 3.8 is closed provided Q is closed and R is closed. The switching network is open under all other conditions.

Excursion Exercises

Construct a closure table for each of the following switching networks. Use the closure table to determine the required conditions for the network to be closed.

1. P — [~P | Q]

2. ~P — [~Q | P] — Q

3. [~P | Q] — [~R | Q]

4. [R | P] — Q — [R | ~P]

(continued)

5.

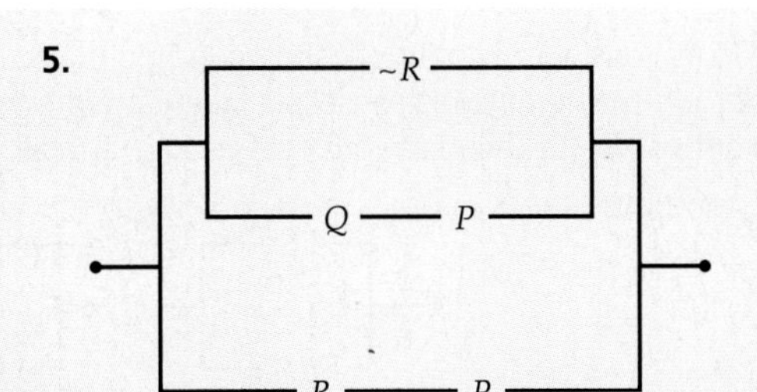

6.

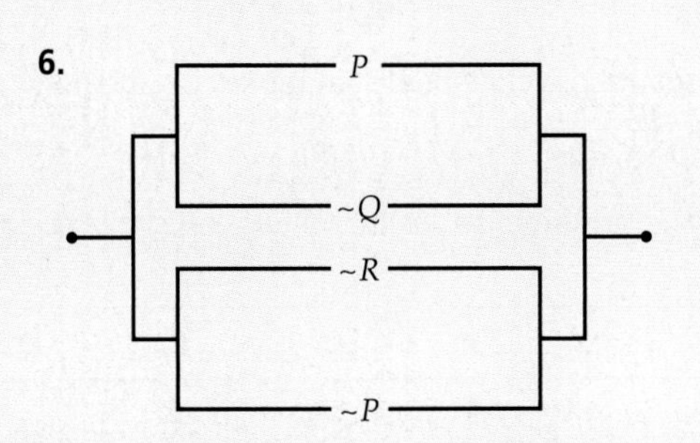

7. ***Warning Circuits***

a. The following circuit shows a switching network used in an automobile. The warning buzzer will buzz only when the switching network is closed. Construct a closure table for the switching network.

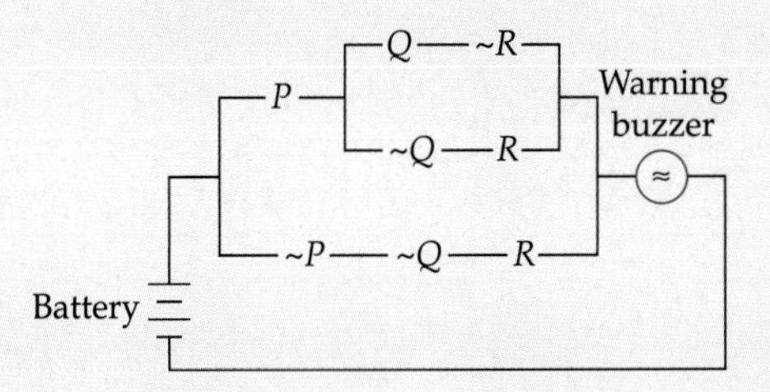

$$\{P \wedge [(Q \wedge \sim P) \vee (\sim Q \wedge R)]\} \vee [(\sim P \wedge \sim Q) \wedge R]$$

b. An engineer thinks that the following circuit can be used in place of the circuit in part a. Do you agree? *Hint:* Construct a closure table for the switching network and compare your closure table with the closure table in part a.

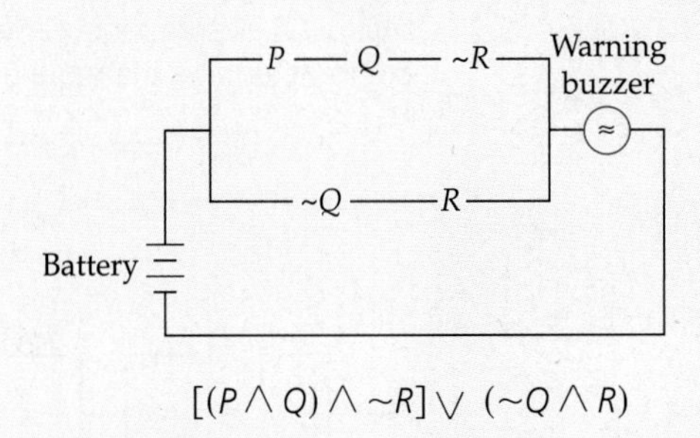

$$[(P \wedge Q) \wedge \sim R] \vee (\sim Q \wedge R)$$

Exercise Set 3.2

In Exercises 1–10, determine the truth value of the compound statement given that p is a false statement, q is a true statement, and r is a true statement.

1. $p \vee (\sim q \vee r)$
2. $r \wedge \sim(p \vee r)$
3. $(p \wedge q) \vee (\sim p \wedge \sim q)$
4. $(p \wedge q) \vee [(\sim p \wedge \sim q) \vee q]$
5. $[\sim(p \wedge \sim q) \vee r] \wedge (p \wedge \sim r)$
6. $(p \wedge \sim q) \vee [(p \wedge \sim q) \vee r]$
7. $[(p \wedge \sim q) \vee \sim r] \wedge (q \wedge r)$
8. $(\sim p \wedge q) \wedge [(p \wedge \sim q) \vee r]$
9. $[(p \wedge q) \wedge r] \vee [p \vee (q \wedge \sim r)]$
10. $\{[(\sim p \wedge q) \wedge r] \vee [(p \wedge q) \wedge \sim r]\} \vee [p \wedge (q \wedge r)]$
11. **a.** Given that p is a false statement, what can be said about $p \wedge (q \vee r)$?

    b. Explain why it is not necessary to know the truth values of q and r to determine the truth value of $p \wedge (q \vee r)$ in part a above.
12. **a.** Given that q is a true statement, what can be said about $q \vee \sim r$?

    b. Explain why it is not necessary to know the truth value of r to determine the truth value of $q \vee \sim r$ in part a above.

In Exercises 13–28, construct a truth table for each compound statement.

13. $\sim p \vee q$
14. $(q \wedge \sim p) \vee \sim q$
15. $p \wedge \sim q$
16. $p \vee [\sim(p \wedge \sim q)]$
17. $(p \wedge \sim q) \vee [\sim(p \wedge q)]$
18. $(p \vee q) \wedge [\sim(p \vee \sim q)]$
19. $\sim(p \vee q) \wedge (\sim r \vee q)$
20. $[\sim(r \wedge \sim q)] \vee (\sim p \vee q)$
21. $(p \wedge \sim r) \vee [\sim q \vee (p \wedge r)]$
22. $[r \wedge (\sim p \vee q)] \wedge (r \vee \sim q)$
23. $[(p \wedge q) \vee (r \wedge \sim p)] \wedge (r \vee \sim q)$
24. $(p \wedge q) \wedge \{[\sim(\sim p \vee r)] \wedge q\}$
25. $q \vee [\sim r \vee (p \wedge r)]$
26. $\{[\sim(p \vee \sim r)] \wedge \sim q\} \vee r$
27. $(\sim q \wedge r) \vee [p \wedge (q \wedge \sim r)]$
28. $\sim[\sim p \wedge (q \wedge r)]$

In Exercises 29–36, use two truth tables to show that each pair of compound statements are equivalent.

29. $p \vee (p \wedge r)$; p
30. $q \wedge (q \vee r)$; q
31. $p \wedge (q \vee r)$; $(p \wedge q) \vee (p \wedge r)$
32. $p \vee (q \wedge r)$; $(p \vee q) \wedge (p \vee r)$
33. $p \vee (q \wedge \sim p)$; $p \vee q$
34. $\sim[p \vee (q \wedge r)]$; $\sim p \wedge (\sim q \vee \sim r)$
35. $[(p \wedge q) \wedge r] \vee [p \wedge (q \wedge \sim r)]$; $p \wedge q$
36. $[(\sim p \wedge \sim q) \wedge r] \vee [(p \wedge q) \wedge \sim r] \vee [p \wedge (q \wedge r)]$; $(p \wedge q) \vee [(\sim p \wedge \sim q) \wedge r]$

In Exercises 37–42, make use of one of De Morgan's laws to write the given statement in an equivalent form.

37. It is not the case that it rained or it snowed.
38. I did not pass the test and I did not complete the course.
39. She did not visit France and she did not visit Italy.
40. It is not true that I bought a new car and I moved to Florida.
41. It is not true that she received a promotion or that she received a raise.
42. It is not the case that the students cut classes or took part in the demonstration.

In Exercises 43–48, use a truth table to determine whether the given statement is a tautology.

43. $p \vee \sim p$

44. $q \vee [\sim(q \wedge r) \wedge \sim q]$

45. $(p \vee q) \vee (\sim p \vee q)$

46. $(p \wedge q) \vee (\sim p \vee \sim q)$

47. $(\sim p \vee q) \vee (\sim q \vee r)$

48. $\sim[p \wedge (\sim p \vee q)] \vee q$

In Exercises 49–54, use a truth table to determine whether the given statement is a self-contradiction.

49. $\sim r \wedge r$

50. $\sim(p \vee \sim p)$

51. $p \wedge (\sim p \wedge q)$

52. $\sim[(p \vee q) \vee (\sim p \vee q)]$

53. $[p \wedge (\sim p \vee q)] \vee q$

54. $\sim[p \vee (\sim p \vee q)]$

55. Explain why the statement $7 \leq 8$ is a disjunction.

56. a. Why is the statement $5 \leq 7$ true?

 b. Why is the statement $7 \leq 7$ true?

Extensions

CRITICAL THINKING

57. How many rows are needed to construct a truth table for the statement $[p \wedge (q \vee \sim r)] \vee (s \wedge \sim t)$?

58. Explain why no truth table can have exactly 100 rows.

COOPERATIVE LEARNING

In Exercises 59 and 60, construct a truth table for the given compound statement. *Hint:* Use a table with 16 rows.

59. $[(p \wedge \sim q) \vee (q \wedge \sim r)] \wedge (r \vee \sim s)$

60. $s \wedge [\sim(\sim r \vee q) \vee \sim p]$

EXPLORATIONS

61. **Disjunctive Normal Form** Read about the *disjunctive normal form* of a statement in a logic text.

 a. What is the disjunctive normal form of a statement that has the following truth table?

p	q	r	given statement
T	T	T	T
T	T	F	F
T	F	T	T
T	F	F	F
F	T	T	F
F	T	F	F
F	F	T	T
F	F	F	F

 b. Explain why the disjunctive normal form is a valuable concept.

62. **Conjunctive Normal Form** Read about the *conjunctive normal form* of a statement in a logic text. What is the conjunctive normal form of the statement defined by the truth table in Exercise 61a?

SECTION 3.3 The Conditional and the Biconditional

Conditional Statements

If you don't get in that plane, you'll regret it. Maybe not today, maybe not tomorrow, but soon, and for the rest of your life.

The above quotation is from the movie *Casablanca.* Rick, played by Humphrey Bogart, is trying to convince Ilsa, played by Ingrid Bergman, to get on the plane with Laszlo. The sentence "If you don't get in that plane, you'll regret it" is a *conditional*

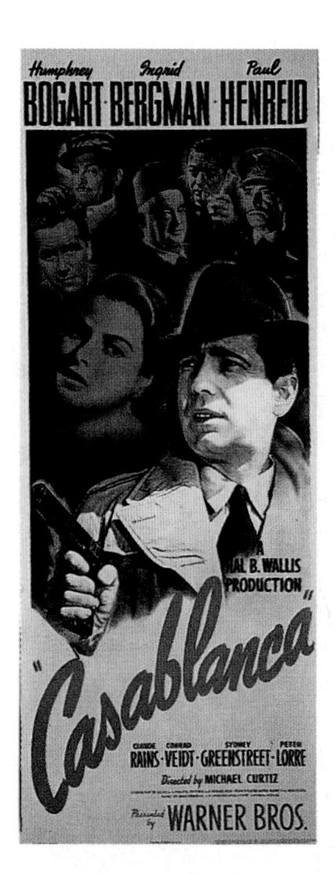

Humphrey Bogart and Ingrid Bergman star in *Casablanca* (1942).

statement. **Conditional statements** can be written in *if p, then q* form or in *if p, q* form. For instance, all of the following are conditional statements.

If we order pizza, then we can have it delivered.

If you go to the movie, you will not be able to meet us for dinner.

If n is a prime number greater than 2, then n is an odd number.

In any conditional statement represented by "If p, then q" or by "If p, q", the p statement is called the **antecedent** and the q statement is called the **consequent.**

EXAMPLE 1 ■ Identify the Antecedent and Consequent of a Conditional

Identify the antecedent and consequent in the following statements.

a. If our school was this nice, I would go there more than once a week.
— *The Basketball Diaries*

b. If you don't stop and look around once in a while, you could miss it.
—Ferris in *Ferris Bueller's Day Off*

c. If you strike me down, I shall become more powerful than you can possibly imagine.—Obi-Wan Kenobi, Star Wars, Episode IV, *A New Hope*

Solution

a. *Antecedent:* our school was this nice
Consequent: I would go there more than once a week

b. *Antecedent:* you don't stop and look around once in a while
Consequent: you could miss it

c. *Antecedent:* you strike me down
Consequent: I shall become more powerful than you can possibly imagine

CHECK YOUR PROGRESS 1 Identify the antecedent and consequent in each of the following conditional statements.

a. If I study for at least 6 hours, then I will get an A on the test.

b. If I get the job, I will buy a new car.

c. If you can dream it, you can do it.

Solution *See page S9.*

Arrow Notation

The conditional statement "If p, then q" can be written using the **arrow notation** $\boldsymbol{p \rightarrow q}$. The arrow notation $p \rightarrow q$ is read as "if p, then q" or as "p implies q."

The Truth Table for the Conditional $p \rightarrow q$

To determine the truth table for $p \rightarrow q$, consider the advertising slogan for a web authoring software product that states "If you can use a word processor, you can create a webpage." This slogan is a conditional statement. The antecedent is p, "you can use a word processor," and the consequent is q, "you can create a webpage." Now consider the truth value of $p \rightarrow q$ for each of the following four possibilities.

Table 3.5

p: you can use a word processor	q: you can create a webpage	$p \rightarrow q$	
T	T	?	Row 1
T	F	?	Row 2
F	T	?	Row 3
F	F	?	Row 4

Row 1: Antecedent T, consequent T You can use a word processor, and you can create a webpage. In this case the truth value of the advertisement is true. To complete Table 3.5, we place a T in place of the question mark in row 1.

Row 2: Antecedent T, consequent F You can use a word processor, but you cannot create a webpage. In this case the advertisement is false. We put an F in place of the question mark in row 2 of Table 3.5.

Row 3: Antecedent F, consequent T You cannot use a word processor, but you can create a webpage. Because the advertisement does not make any statement about what you might or might not be able to do if you cannot use a word processor, we cannot state that the advertisement is false, and we are compelled to place a T in place of the question mark in row 3 of Table 3.5.

Row 4: Antecedent F, consequent F You cannot use a word processor, and you cannot create a webpage. Once again we must consider the truth value in this case to be true because the advertisement does not make any statement about what you might or might not be able to do if you cannot use a word processor. We place a T in place of the question mark in row 4 of Table 3.5.

The truth table for the conditional $p \rightarrow q$ is given in Table 3.6.

Table 3.6 *The Truth Table for* $p \rightarrow q$

p	q	$p \rightarrow q$
T	T	T
T	F	F
F	T	T
F	F	T

Truth Value of the Conditional $p \rightarrow q$

The conditional $p \rightarrow q$ is false if p is true and q is false. It is true in all other cases.

EXAMPLE 2 ■ Find the Truth Value of a Conditional

Determine the truth value of each of the following.

a. If 2 is an integer, then 2 is a rational number.

b. If 3 is a negative number, then $5 > 7$.

c. If $5 > 3$, then $2 + 7 = 4$.

Solution

a. Because the consequent is true, this is a true statement.

b. Because the antecedent is false, this is a true statement.

c. Because the antecedent is true and the consequent is false, this is a false statement.

CHECK YOUR PROGRESS 2 Determine the truth value of each of the following.

a. If $4 \geq 3$, then $2 + 5 = 6$.

b. If $5 > 9$, then $4 > 9$.

c. If Tuesday follows Monday, then April follows March.

Solution *See page S9.*

CALCULATOR NOTE

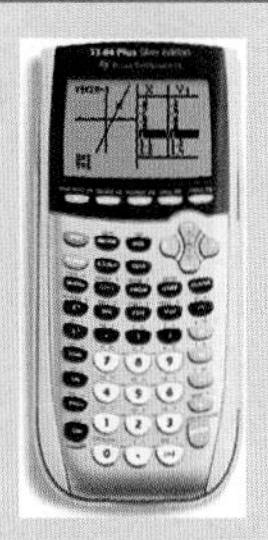

TI-84

Program FACTOR

```
0→dim (L1)
Prompt N
1→S: 2→F:0→E
√(N)→M
While F≤M
While fPart (N/F)=0
E+1→E:N/F→N
End
If E>0
Then
F→L1(S)
E→L1(S+1)
S+2→S:0→E
√(N)→M
End
If F=2
Then
3→F
Else
F+2→F
End: End
If N≠1
Then
N→L1(S)
1→L1(S+1)
End
If S=1
Then
Disp N, " IS PRIME"
Else
Disp L1
```

EXAMPLE 3 ■ Construct a Truth Table for a Statement Involving a Conditional

Construct a truth table for $[p \wedge (q \vee \sim p)] \rightarrow \sim p$.

Solution

Using the generalized procedure for truth table construction, we produce the following table.

p	q	$[p$	$\wedge$	$(q$	$\vee$	$\sim p)]$	$\rightarrow$	$\sim p$
T	T	T	T	T	T	F	F	F
T	F	T	F	F	F	F	T	F
F	T	F	F	T	T	T	T	T
F	F	F	F	F	T	T	T	T
		1	6	2	5	3	7	4

CHECK YOUR PROGRESS 3 Construct a truth table for $[p \wedge (p \rightarrow q)] \rightarrow q$.

Solution *See page S9.*

Math Matters Use Conditional Statements to Control a Calculator Program

Computer and calculator programs use conditional statements to control the flow of a program. For instance, the "If...Then" instruction in a TI-83 or TI-84 calculator program directs the calculator to execute a group of commands if a condition is true and to skip to the End statement if the condition is false. See the program steps below.

:If *condition*

:Then (skip to End if *condition* is false)

:*command* if *condition* is true

:*command* if *condition* is true

:End

:*command*

The TI-83/84 program FACTOR shown at the left factors a number N into its prime factors. Note the use of the "If...Then" instructions highlighted in red.

An Equivalent Form of the Conditional

Table 3.7 *The Truth Table for* $\sim p \vee q$

p	q	$\sim p \vee q$
T	T	T
T	F	F
F	T	T
F	F	T

The truth table for $\sim p \vee q$ is shown in Table 3.7. The truth values in this table are identical to the truth values in Table 3.6. Hence, the conditional $p \rightarrow q$ is equivalent to the disjunction $\sim p \vee q$.

An Equivalent Form of the Conditional $p \rightarrow q$

$$p \rightarrow q \equiv \sim p \vee q$$

EXAMPLE 4 ■ Write a Conditional in Its Equivalent Disjunctive Form

Write each of the following in its equivalent disjunctive form.

a. If I could play the guitar, I would join the band.

b. If Arnold cannot play, then the Dodgers will lose.

Solution

In each case we write the disjunction of the negation of the antecedent and the consequent.

a. I cannot play the guitar or I would join the band.

b. Arnold can play or the Dodgers will lose.

CHECK YOUR PROGRESS 4 Write each of the following in its equivalent disjunctive form.

a. If I don't move to Georgia, I will live in Houston.

b. If the number is divisible by 2, then the number is even.

Solution *See page S9.*

The Negation of the Conditional

Because $p \rightarrow q \equiv \sim p \vee q$, an equivalent form of $\sim(p \rightarrow q)$ is given by $\sim(\sim p \vee q)$, which, by one of De Morgan's laws, can be expressed as the conjunction $p \wedge \sim q$.

The Negation of $p \rightarrow q$

$$\sim(p \rightarrow q) \equiv p \wedge \sim q$$

EXAMPLE 5 ■ Write the Negation of a Conditional Statement

Write the negation of each conditional statement.

a. If they pay me the money, I will sign the contract.

b. If the lines are parallel, then they do not intersect.

Solution
In each case, we write the conjunction of the antecedent and the negation of the consequent.

a. They paid me the money and I did not sign the contract.

b. The lines are parallel and they intersect.

CHECK YOUR PROGRESS 5 Write the negation of each conditional statement.

a. If I finish the report, I will go to the concert.

b. If the square of n is 25, then n is 5 or -5.

Solution *See page S9.*

The Biconditional

The statement $(p \rightarrow q) \wedge (q \rightarrow p)$ is called a **biconditional** and is denoted by $p \leftrightarrow q$, which is read as "p if and only if q."

Definition of the Biconditional $p \leftrightarrow q$

$$p \leftrightarrow q \equiv [(p \rightarrow q) \wedge (q \rightarrow p)]$$

Table 3.8 *The Truth Table for $p \leftrightarrow q$*

p	q	$p \leftrightarrow q$
T	T	T
T	F	F
F	T	F
F	F	T

Table 3.8 shows that $p \leftrightarrow q$ is true only when the components p and q have the same truth value.

EXAMPLE 6 ■ Determine the Truth Value of a Biconditional

State whether each biconditional is true or false.

a. $x + 4 = 7$ if and only if $x = 3$.

b. $x^2 = 36$ if and only if $x = 6$.

Solution

a. Both components are true when $x = 3$ and both are false when $x \neq 3$. Both components have the same truth value for any value of x, so this is a true statement.

b. If $x = -6$, the first component is true and the second component is false. Thus this is a false statement.

CHECK YOUR PROGRESS 6 State whether each biconditional is true or false.

a. $x > 7$ if and only if $x > 6$.

b. $x + 5 > 7$ if and only if $x > 2$.

Solution *See page S9.*

Logic Gates

Modern digital computers use *gates* to process information. These gates are designed to receive two types of electronic impulses, which are generally represented as a 1 or a 0. Figure 3.9 shows a *NOT gate*. It is constructed so that a stream of impulses that enter the gate will exit the gate as a stream of impulses in which each 1 is converted to a 0 and each 0 is converted to a 1.

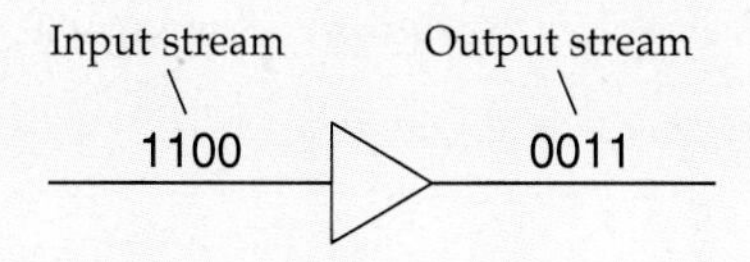

Figure 3.9 *NOT gate*

Note the similarity between the logical connective *not* and the logic gate NOT. The *not* connective converts the sequence of truth values T F to F T. The NOT gate converts the input stream 1 0 to 0 1. If the 1's are replaced with T's and the 0's with F's, then the NOT logic gate yields the same results as the *not* connective.

Many gates are designed so that two input streams are converted to one output stream. For instance, Figure 3.10 shows an *AND gate*. The AND gate is constructed so that a 1 is the output if and only if both input streams have a 1. In any other situation a 0 is produced as the output.

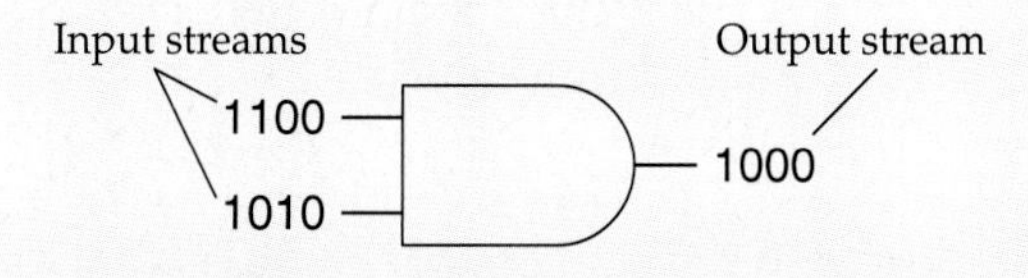

Figure 3.10 *AND gate*

Note the similarity between the logical connective *and* and the logic gate AND. The *and* connective combines the sequence of truth values T T F F with the truth values T F T F to produce T F F F. The AND gate combines the input stream 1 1 0 0 with the input stream 1 0 1 0 to produce 1 0 0 0. If the 1's are replaced with T's and the 0's with F's, then the AND logic gate yields the same result as the *and* connective.

The *OR gate* is constructed so that its output is a 0 if and only if both input streams have a 0. All other situations yield a 1 as the output. See Figure 3.11.

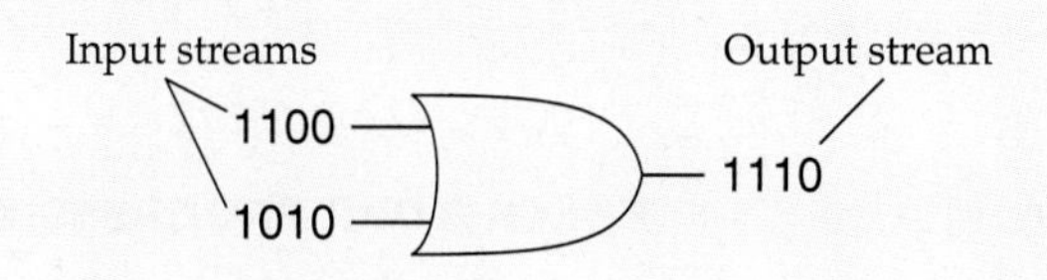

Figure 3.11 *OR gate*

(continued)

Figure 3.12 shows a network that consists of a NOT gate and an AND gate.

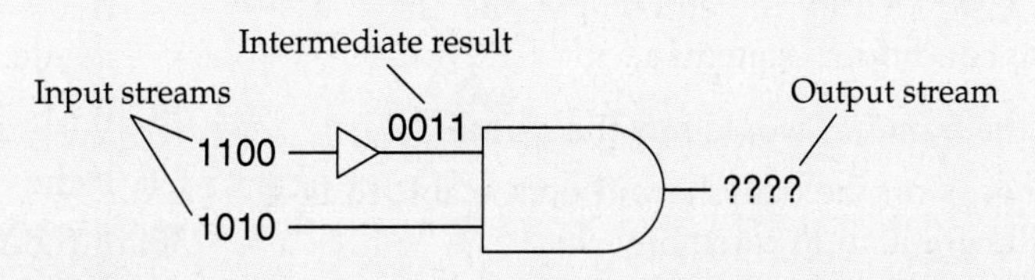

Figure 3.12

QUESTION *What is the output stream for the network in Figure 3.12?*

Excursion Exercises

1. For each of the following, determine the output stream for the given input streams.

a. 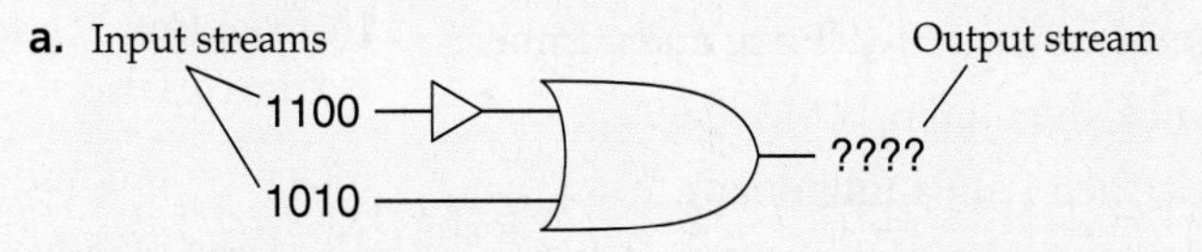

b. 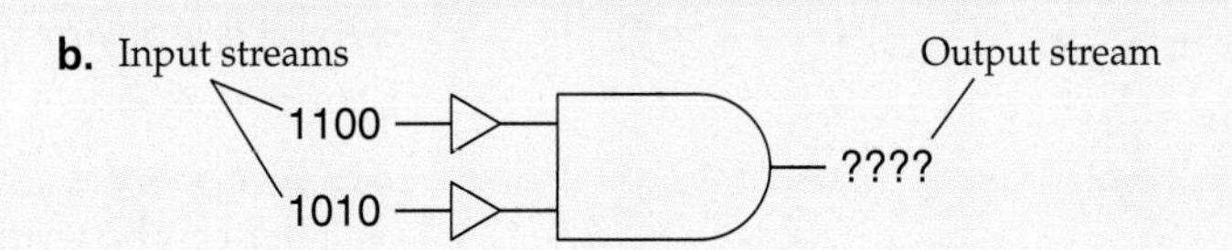

c. 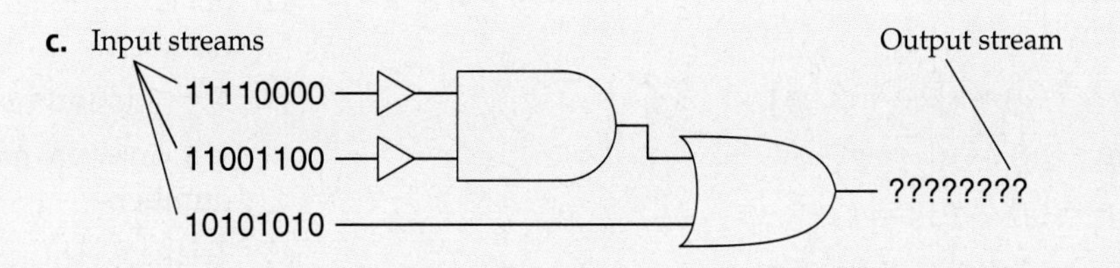

2. Construct a network using NOT, AND, and OR gates as needed that accepts the two input streams 1 1 0 0 and 1 0 1 0 and produces the output stream 0 1 1 1.

ANSWER *0010*

Exercise Set 3.3

In Exercises 1–6, identify the antecedent and the consequent of each conditional statement.

1. If I had the money, I would buy the painting.
2. If Shelly goes on the trip, she will not be able to take part in the graduation ceremony.
3. If they had a guard dog, then no one would trespass on their property.
4. If I don't get to school before 7:30, I won't be able to find a parking place.
5. If I change my major, I must reapply for admission.
6. If your blood type is type O−, then you are classified as a universal blood donor.

In Exercises 7–14, determine the truth value of the given statement.

7. If x is an even integer, then x^2 is an even integer.
8. If x is a prime number, then $x + 2$ is a prime number.
9. If all frogs can dance, then today is Monday.
10. If all cats are black, then I am a millionaire.
11. If $4 < 3$, then $7 = 8$.
12. If $x < 2$, then $x + 5 < 7$.
13. If $|x| = 6$, then $x = 6$.
14. If $\pi = 3$, then $2\pi = 6$.

In Exercises 15–24, construct a truth table for the given statement.

15. $(p \wedge \sim q) \rightarrow [\sim(p \wedge q)]$
16. $[(p \rightarrow q) \wedge p] \rightarrow p$
17. $[(p \rightarrow q) \wedge p] \rightarrow q$
18. $(\sim p \vee \sim q) \rightarrow \sim(p \wedge q)$
19. $[r \wedge (\sim p \vee q)] \rightarrow (r \vee \sim q)$
20. $[(p \rightarrow \sim r) \wedge q] \rightarrow \sim r$
21. $[(p \rightarrow q) \vee (r \wedge \sim p)] \rightarrow (r \vee \sim q)$
22. $\{p \wedge [(p \rightarrow q) \wedge (q \rightarrow r)]\} \rightarrow r$
23. $[\sim(p \rightarrow \sim r) \wedge \sim q] \rightarrow r$
24. $[p \wedge (r \rightarrow \sim q)] \rightarrow (r \vee q)$

In Exercises 25–30, write each conditional statement in its equivalent disjunctive form.

25. If she could sing, she would be perfect for the part.
26. If he does not get frustrated, he will be able to complete the job.
27. If x is an irrational number, then x is not a terminating decimal.
28. If Mr. Hyde had a brain, he would be dangerous.
29. If the fog does not lift, our flight will be cancelled.
30. If the Yankees win the pennant, Carol will be happy.

In Exercises 31–36, write the negation of each conditional statement in its equivalent conjunctive form.

31. If they offer me the contract, I will accept.
32. If I paint the house, I will get the money.
33. If pigs had wings, pigs could fly.
34. If we had a telescope, we could see that comet.
35. If she travels to Italy, she will visit her relatives.
36. If Paul could play better defense, he could be a professional basketball player.

In Exercises 37–46, state whether the given biconditional is true or false. Assume that x and y are real numbers.

37. $x^2 = 9$ if and only if $x = 3$.
38. x is a positive number if and only if $x > 0$.
39. $|x|$ is a positive number if and only if $x \neq 0$.
40. $|x + y| = x + y$ if and only if $x + y > 0$.
41. A number is a rational number if and only if the number can be written as a terminating decimal.
42. $0.\overline{3}$ is a rational number if and only if $\frac{1}{3}$ is a rational number.
43. $4 = 7$ if and only if $2 = 3$.
44. x is an even number if and only if x is not an odd number.
45. Triangle ABC is an equilateral triangle if and only if triangle ABC is an equiangular triangle.
46. Today is March 1 if and only if yesterday was February 28.

In Exercises 47–52, let v represent "I will take a vacation," let p represent "I get the promotion," and let t represent "I am transferred." Write each of the following statements in symbolic form.

47. If I get the promotion, I will take a vacation.
48. If I am not transferred, I will take a vacation.
49. If I am transferred, then I will not take a vacation.
50. If I will not take a vacation, then I will not be transferred and I get the promotion.

51. If I am not transferred and I get the promotion, then I will take a vacation.

52. If I get the promotion, then I am transferred and I will take a vacation.

In Exercises 53–58, construct a truth table for each statement to determine if the statements are equivalent.

53. $p \rightarrow \sim r$; $r \vee \sim p$

54. $p \rightarrow q$; $q \rightarrow p$

55. $\sim p \rightarrow (p \vee r)$; r

56. $p \rightarrow q$; $\sim q \rightarrow \sim p$

57. $p \rightarrow (q \vee r)$; $(p \rightarrow q) \vee (p \rightarrow r)$

58. $\sim q \rightarrow p$; $p \vee q$

Extensions

CRITICAL THINKING

The statement "All squares are rectangles" can be written as "If a figure is a square, then it is a rectangle." In Exercises 59–64, write each statement given in "All p are q" form in the form "If it is a p, then it is a q."

59. All rational numbers are real numbers.

60. All whole numbers are integers.

61. All repeating decimals are rational numbers.

62. All multiples of 5 end with a 0 or with a 5.

63. All Sauropods are herbivorous.

64. All paintings by Vincent van Gogh are valuable.

EXPLORATIONS

65. **A Factor Program** If you have access to a TI-83 or a TI-84 calculator, enter the program FACTOR on page 139 into the calculator and demonstrate the program to a classmate.

66. **Calculator Programs** Many calculator programs are available on the Internet. One source for Texas Instruments calculator programs is **ticalc.org.** Search the Internet and write a few paragraphs about the programs you found to be the most interesting.

SECTION 3.4 The Conditional and Related Statements

Equivalent Forms of the Conditional

Every conditional statement can be stated in many equivalent forms. It is not even necessary to state the antecedent before the consequent. For instance, the conditional "If I live in Boston, then I must live in Massachusetts" can also be stated as

I must live in Massachusetts, if I live in Boston.

Table 3.9 lists some of the various forms that may be used to write a conditional statement.

Table 3.9 *Common Forms of* $p \rightarrow q$

Every conditional statement $p \rightarrow q$ can be written in the following equivalent forms	
If p, then q.	Every p is a q.
If p, q.	q, if p.
p only if q.	q provided p.
p implies q.	q is a necessary condition for p.
Not p or q.	p is a sufficient condition for q.

EXAMPLE 1 ■ Write a Statement in an Equivalent Form

Write each of the following in "If p, then q" form.

a. The number is an even number provided it is divisible by 2.

b. Today is Friday, only if yesterday was Thursday.

Solution

a. The statement "The number is an even number provided it is divisible by 2" is in "q provided p" form. The antecedent is "it is divisible by 2," and the consequent is "the number is an even number." Thus its "If p, then q" form is

If it is divisible by 2, then the number is an even number.

b. The statement "Today is Friday, only if yesterday was Thursday" is in "p only if q" form. The antecedent is "today is Friday." The consequent is "yesterday was Thursday." Its "If p, then q" form is

If today is Friday, then yesterday was Thursday.

CHECK YOUR PROGRESS 1 Write each of the following in "If p, then q" form.

a. Every square is a rectangle.

b. Being older than 30 is sufficient to show I am at least 21.

Solution *See page S9.*

The Converse, the Inverse, and the Contrapositive

Every conditional statement has three related statements. They are called the *converse,* the *inverse,* and the *contrapositive.*

Statements Related to the Conditional Statement

The **converse** of $p \rightarrow q$ is $q \rightarrow p$.

The **inverse** of $p \rightarrow q$ is $\sim p \rightarrow \sim q$.

The **contrapositive** of $p \rightarrow q$ is $\sim q \rightarrow \sim p$.

The above definitions show the following:

- The converse of $p \rightarrow q$ is formed by interchanging the antecedent p with the consequent q.
- The inverse of $p \rightarrow q$ is formed by negating the antecedent p and negating the consequent q.
- The contrapositive of $p \rightarrow q$ is formed by negating both the antecedent p and the consequent q and interchanging these negated statements.

EXAMPLE 2 ■ Write the Converse, Inverse, and Contrapositive of a Conditional

Write the converse, inverse, and contrapositive of

If I get the job, then I will rent the apartment.

Solution
Converse: If I rent the apartment, then I get the job.
Inverse: If I do not get the job, then I will not rent the apartment.
Contrapositive: If I do not rent the apartment, then I did not get the job.

CHECK YOUR PROGRESS 2 Write the converse, inverse, and contrapositive of

If we have a quiz today, then we will not have a quiz tomorrow.

Solution *See page S9.*

Table 3.10 shows that any conditional statement is equivalent to its contrapositive, and that the converse of a conditional statement is equivalent to the inverse of the conditional statement.

Table 3.10 *Truth Tables for the Conditional and Related Statements*

p	q	Conditional $p \to q$	Converse $q \to p$	Inverse $\sim p \to \sim q$	Contrapositive $\sim q \to \sim p$
T	T	T	T	T	T
T	F	F	T	T	F
F	T	T	F	F	T
F	F	T	T	T	T

$q \to p \equiv \sim p \to \sim q$

$p \to q \equiv \sim q \to \sim p$

EXAMPLE 3 ■ Determine Whether Related Statements Are Equivalent

Determine whether the given statements are equivalent.

a. If a number ends with a 5, then the number is divisible by 5.
If a number is divisible by 5, then the number ends with a 5.

b. If two lines in a plane do not intersect, then the lines are parallel.
If two lines in a plane are not parallel, then the lines intersect.

Solution

a. The second statement is the converse of the first. The statements are not equivalent.

b. The second statement is the contrapositive of the first. The statements are equivalent.

CHECK YOUR PROGRESS 3 Determine whether the given statements are equivalent.

a. If $a = b$, then $a \cdot c = b \cdot c$.
If $a \neq b$, then $a \cdot c \neq b \cdot c$.

b. If I live in Nashville, then I live in Tennessee.
If I do not live in Tennessee, then I do not live in Nashville.

Solution *See page S9.*

In mathematics, it is often necessary to prove statements that are in "If p, then q" form. If a proof cannot be readily produced, mathematicians often try to prove the contrapositive "If $\sim q$, then $\sim p$." Because a conditional and its contrapositive are equivalent statements, a proof of either statement also establishes the proof of the other statement.

QUESTION *A mathematician wishes to prove the following statement about the integer x.*

If x^2 is an odd integer, then x is an odd integer. (I)

If the mathematician is able to prove the statement, "If x is an even integer, then x^2 is an even integer," does this also prove statement (I)?

EXAMPLE 4 ■ Use the Contrapositive to Determine a Truth Value

Write the contrapositive of each statement and use the contrapositive to determine whether the original statement is true or false.

a. If $a + b$ is not divisible by 5, then a and b are not both divisible by 5.

b. If x^3 is an odd integer, then x is an odd integer. (Assume x is an integer.)

c. If a geometric figure is not a rectangle, then it is not a square.

Solution

a. If a and b are both divisible by 5, then $a + b$ is divisible by 5. This is a true statement, so the original statement is also true.

b. If x is an even integer, then x^3 is an even integer. This is a true statement, so the original statement is also true.

c. If a geometric figure is a square, then it is a rectangle. This is a true statement, so the original statement is also true.

CHECK YOUR PROGRESS 4 Write the contrapositive of each statement and use the contrapositive to determine whether the original statement is true or false.

a. If $3 + x$ is an odd integer, then x is an even integer. (Assume x is an integer.)

b. If two triangles are not similar triangles, then they are not congruent triangles. *Note:* Similar triangles have the same shape. Congruent triangles have the same size and shape.

c. If today is not Wednesday, then tomorrow is not Thursday.

Solution *See page S10.*

ANSWER *Yes, because the second statement is the contrapositive of (I).*

Math Matters Grace Hopper

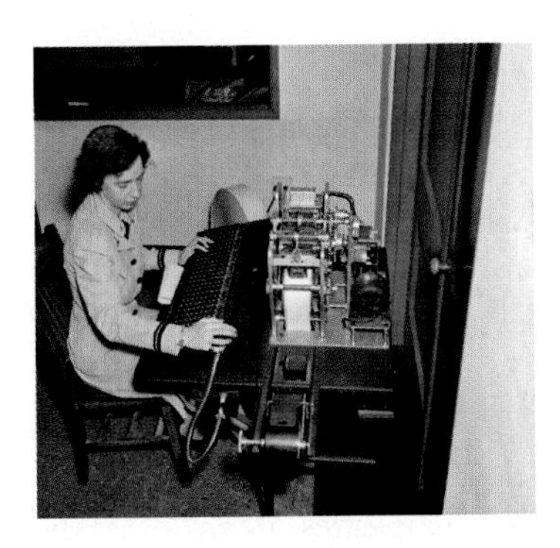

Rear Admiral Grace Hopper

Grace Hopper (1906–1992) was a visionary in the field of computer programming. She was a mathematics professor at Vassar from 1931 to 1943, but retired from teaching to start a career in the U.S. Navy at the age of 37.

The Navy assigned Hopper to the Bureau of Ordnance Computation at Harvard University. It was here that she was given the opportunity to program computers. It has often been reported that she was the third person to program the world's first large-scale digital computer. Grace Hopper had a passion for computers and computer programming. She wanted to develop a computer language that would be user-friendly and enable people to use computers in a more productive manner.

Grace Hopper had a long list of accomplishments. She designed some of the first computer compilers, she was one of the first to introduce English commands into computer languages, and she wrote the precursor to the computer language COBOL.

Grace Hopper retired from the Navy (for the first time) in 1966. In 1967 she was recalled to active duty and continued to serve in the Navy until 1986, at which time she was the nation's oldest active duty officer.

In 1951, the UNIVAC I computer that Grace Hopper was programming started to malfunction. The malfunction was caused by a moth that had become lodged in one of the computer's relays. Grace Hopper pasted the moth into the UNIVAC I logbook with a label that read, "computer bug." Since then computer programmers have used the word *bug* to indicate any problem associated with a computer program. Modern computers use logic gates instead of relays to process information, so actual bugs are not a problem; however, bugs such as the "Year 2000 bug" can cause serious problems.

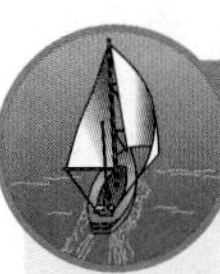

Excursion

Sheffer's Stroke and the NAND Gate

Table 3.11 *Sheffer's stroke*

p	q	$p\|q$
T	T	F
T	F	T
F	T	T
F	F	T

In 1913, the logician Henry M. Sheffer created a connective that we now refer to as *Sheffer's stroke* (or *NAND*). This connective is often denoted by the symbol $|$. Table 3.11 shows that $p|q$ is equivalent to $\sim(p \wedge q)$. Sheffer's stroke $p|q$ is false when both p and q are true and it is true in all other cases.

Any logic statement can be written using only Sheffer's stroke connectives. For instance, Table 3.12 shows that $p|p \equiv \sim p$ and $(p|p)|(q|q) \equiv p \vee q$.

Figure 3.13 shows a logic gate called a NAND gate. This gate models the Sheffer's stroke connective in that its output is 0 when both input streams are 1 and its output is 1 in all other cases.

(continued)

Table 3.12

p	q	$p\mid p$	$(p\mid p)\mid(q\mid q)$
T	T	F	T
T	F	F	T
F	T	T	T
F	F	T	F

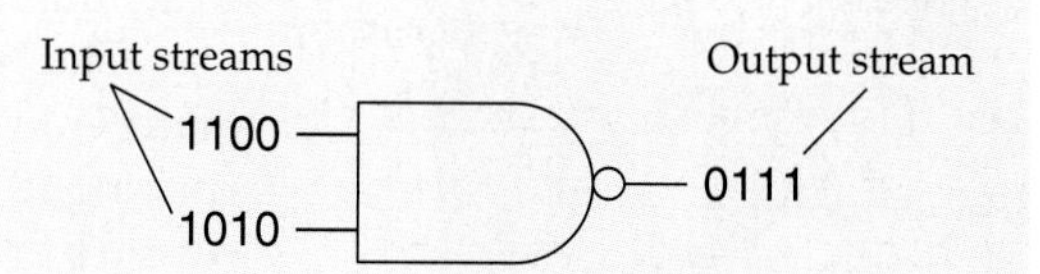

Figure 3.13 *NAND gate*

Excursion Exercises

1. **a.** Complete a truth table for $p\mid(q\mid q)$.
   b. Use the results of Excursion Exercise 1a to determine an equivalent statement for $p\mid(q\mid q)$.
2. **a.** Complete a truth table for $(p\mid q)\mid(p\mid q)$.
   b. Use the results of Excursion Exercise 2a to determine an equivalent statement for $(p\mid q)\mid(p\mid q)$.
3. **a.** Determine the output stream for the following network of NAND gates. *Note:* In a network of logic gates, a solid circle • is used to indicate a connection. A symbol such as ⤙ is used to indicate "no connection."

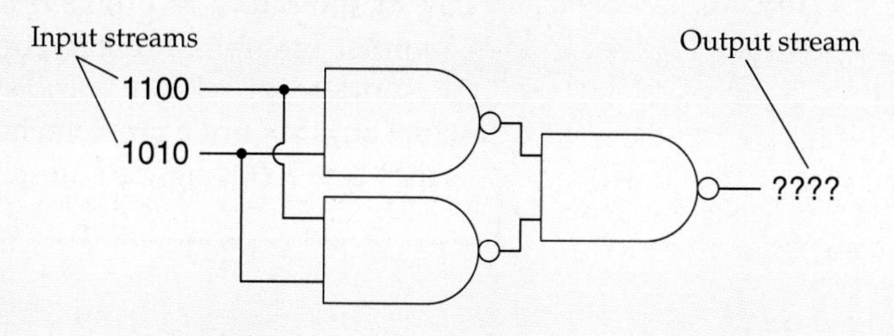

Figure 3.14

   b. What logic gate is modeled by the network in Figure 3.14?
4. NAND gates are functionally complete in that any logic gate can be constructed using only NAND gates. Construct a network of NAND gates that would produce the same output stream as an OR gate.

Exercise Set 3.4

In Exercises 1–10, write each statement in "If p, then q" form.

1. We will be in good shape for the ski trip provided we take the aerobics class.
2. We can get a dog only if we install a fence around the back yard.
3. Every odd prime number is greater than 2.
4. The triangle is a 30°-60°-90° triangle, if the length of the hypotenuse is twice the length of the shorter leg.
5. He can join the band, if he has the talent to play a keyboard.
6. Every theropod is carnivorous.

7. I will be able to prepare for the test only if I have the textbook.
8. I will be able to receive my credential provided Education 147 is offered in the spring semester.
9. Being in excellent shape is a necessary condition for running the Boston marathon.
10. If it is an ankylosaur, it is quadrupedal.

In Exercises 11–24, write the **a.** converse, **b.** inverse, and **c.** contrapositive of the given statement.

11. If I were rich, I would quit this job.
12. If we had a car, then we would be able to take the class.
13. If she does not return soon, we will not be able to attend the party.
14. I will be in the talent show only if I can do the same comedy routine I did for the banquet.
15. Every parallelogram is a quadrilateral.
16. If you get the promotion, you will need to move to Denver.
17. I would be able to get current information about astronomy provided I had access to the Internet.
18. You need four-wheel drive to make the trip to Death Valley.
19. We will not have enough money for dinner, if we take a taxi.
20. If you are the president of the United States, then your age is at least 35.
21. She will visit Kauai only if she can extend her vacation for at least two days.
22. In a right triangle, the acute angles are complementary.
23. Two lines perpendicular to a given line are parallel.
24. If $x + 5 = 12$, then $x = 7$.

In Exercises 25–30, determine whether the given statements are equivalent.

25. If Kevin wins, we will celebrate.
    If we celebrate, then Kevin will win.
26. If I save \$1000, I will go on the field trip.
    If I go on the field trip, then I saved \$1000.
27. If she attends the meeting, she will make the sale.
    If she does not make the sale, then she did not attend the meeting.
28. If you understand algebra, you can remember algebra.
    If you do not understand algebra, you cannot remember algebra.
29. If $a > b$, then $ac > bc$.
    If $a \leq b$, then $ac \leq bc$.
30. If $a < b$, then $\frac{1}{a} > \frac{1}{b}$.
    If $\frac{1}{a} \leq \frac{1}{b}$, then $a \geq b$.
    (Assume $a \neq 0$ and $b \neq 0$.)

In Exercises 31–36, write the contrapositive of the statement and use the contrapositive to determine whether the given statement is true or false.

31. If $3x - 7 = 11$, then $x \neq 7$.
32. If $x \neq 3$, then $5x + 7 \neq 22$.
33. If $a \neq 3$, then $|a| \neq 3$.
34. If $a + b$ is divisible by 3, then a is divisible by 3 and b is divisible by 3.
35. If $\sqrt{a + b} \neq 5$, then $a + b \neq 25$.
36. Assume x is an integer. If x^2 is an even integer, then x is an even integer.
37. What is the converse of the inverse of the contrapositive of $p \rightarrow q$?
38. What is the inverse of the converse of the contrapositive of $p \rightarrow q$?

Extensions

CRITICAL THINKING

39. Give an example of a true conditional statement whose
    a. converse is true. **b.** converse is false.
40. Give an example of a true conditional statement whose
    a. inverse is true. **b.** inverse is false.

In Exercises 41–44, determine the original statement if the given statement is related to the original in the manner indicated.

41. *Converse:* If you can do it, you can dream it.
42. *Inverse:* If I did not have a dime, I would not spend it.
43. *Contrapositive:* If I were a singer, I would not be a dancer.
44. *Negation:* Pigs have wings and pigs cannot fly.
45. Explain why it is not possible to find an example of a true conditional statement whose contrapositive is false.

46. If a conditional statement is false, must its converse be true? Explain.

47. **A Puzzle** Lewis Carroll (Charles Dodgson) wrote many puzzles, many of which he recorded in his diaries. Solve the following puzzle, which appears in one of his diaries.

The Dodo says that the Hatter tells lies.
The Hatter says that the March Hare tells lies.
The March Hare says that both the Dodo and the Hatter tell lies.
Who is telling the truth?[2]

Hint: Consider the three different cases in which only one of the characters is telling the truth. In only one of these cases can all three of the statements be true.

EXPLORATIONS

48. **Puzzles** Use a library or the Internet to find puzzles created by Lewis Carroll. For the puzzle that you think is most interesting, write an explanation of the puzzle and give its solution.

SECTION 3.5 Arguments

historical note

Aristotle (ăr'ĭ-stŏt'l) (384–322 B.C.) was an ancient Greek philosopher who studied under Plato. He wrote about many subjects, including logic, biology, politics, astronomy, metaphysics, and ethics. His ideas about logic and the reasoning process have had a major impact on mathematics and philosophy. ■

Arguments

In this section we consider methods of analyzing arguments to determine whether they are *valid* or *invalid.* For instance, consider the following argument.

If Aristotle was human, then Aristotle was mortal. Aristotle was human. Therefore, Aristotle was mortal.

To determine whether the above argument is a valid argument, we must first define the terms *argument* and *valid argument.*

Definition of an Argument and a Valid Argument

An **argument** consists of a set of statements called **premises** and another statement called the **conclusion.** An argument is **valid** if the conclusion is true whenever all the premises are assumed to be true. An argument is **invalid** if it is not a valid argument.

In the argument about Aristotle, the two premises and the conclusion are shown below. It is customary to place a horizontal line between the premises and the conclusion.

First Premise: If Aristotle was human, then Aristotle was mortal.
Second Premise: Aristotle was human.
Conclusion: Therefore, Aristotle was mortal.

2. The above puzzle is from *Lewis Carroll's Games and Puzzles,* newly compiled and edited by Edward Wakeling. New York: Dover Publications, Inc., copyright 1992, p. 11, puzzle 9, "Who's Telling the Truth?"

Arguments can be written in **symbolic form.** For instance, if we let h represent the statement "Aristotle was human" and m represent the statement "Aristotle was mortal," then the argument can be expressed as

$$\begin{array}{l} h \rightarrow m \\ \underline{h \quad\quad} \\ \therefore m \end{array}$$

The three dots $\therefore$ are a symbol for "therefore."

EXAMPLE 1 ■ Write an Argument in Symbolic Form

Write the following argument in symbolic form.

The fish is fresh or I will not order it. The fish is fresh. Therefore I will order it.

Solution

Let f represent the statement "The fish is fresh." Let o represent the statement "I will order it." The symbolic form of the argument is

$$\begin{array}{l} f \vee \sim o \\ \underline{f \quad\quad} \\ \therefore o \end{array}$$

CHECK YOUR PROGRESS 1 Write the following argument in symbolic form.

If she doesn't get on the plane, she will regret it. She does not regret it. Therefore, she got on the plane.

Solution *See page S10.*

Arguments and Truth Tables

The following truth table procedure can be used to determine whether an argument is valid or invalid.

Truth Table Procedure to Determine the Validity of an Argument

1. Write the argument in symbolic form.
2. Construct a truth table that shows the truth value of each premise and the truth value of the conclusion for all combinations of truth values of the component statements.
3. If the conclusion is true in every row of the truth table in which all the premises are true, the argument is valid. If the conclusion is false in any row in which all of the premises are true, the argument is invalid.

We will now use the above truth table procedure to determine the validity of the argument about Aristotle.

1. Once again we let h represent the statement "Aristotle was human" and m represent the statement "Aristotle was mortal." In symbolic form the argument is

$h \rightarrow m$ First premise

h Second premise

$\therefore m$ Conclusion

2. Construct a truth table as shown below.

h	m	First premise $h \rightarrow m$	Second premise h	Conclusion m	
T	T	T	T	T	Row 1
~~T~~	~~F~~	~~F~~	~~T~~	~~F~~	~~Row 2~~
~~F~~	~~T~~	~~T~~	~~F~~	~~T~~	~~Row 3~~
~~F~~	~~F~~	~~T~~	~~F~~	~~F~~	~~Row 4~~

3. Row 1 is the only row in which all the premises are true, so it is the only row that we examine. Because the conclusion is true in row 1, the argument is valid.

In Example 2, we use the truth table method to determine the validity of a more complicated argument.

EXAMPLE 2 ■ Determine the Validity of an Argument

Determine whether the following argument is valid or invalid.

If it rains, then the game will not be played. It is not raining. Therefore, the game will be played.

Solution

If we let r represent "it rains" and g represent "the game will be played," then the symbolic form is

$$\begin{array}{l} r \rightarrow \sim g \\ \underline{\sim r \quad\quad} \\ \therefore g \end{array}$$

The truth table for this argument is as follows:

r	g	First premise $r \rightarrow \sim g$	Second premise $\sim r$	Conclusion g	
~~T~~	~~T~~	~~F~~	~~F~~	~~T~~	~~Row 1~~
~~T~~	~~F~~	~~T~~	~~F~~	~~F~~	~~Row 2~~
F	T	T	T	T	Row 3
F	F	T	T	F	Row 4

QUESTION *Why do we need to examine only rows 3 and 4?*

Because the conclusion in row 4 is false and the premises are both true, we know the argument is invalid.

ANSWER *Rows 3 and 4 are the only rows in which all of the premises are true.*

CHECK YOUR PROGRESS 2 Determine the validity of the following argument.

If the stock market rises, then the bond market will fall.
The bond market did not fall.
∴The stock market did not rise.

Solution *See page S10.*

The argument in Example 3 involves three statements. Thus we use a truth table with $2^3 = 8$ rows to determine the validity of the argument.

EXAMPLE 3 ■ Determine the Validity of an Argument

Determine whether the following argument is valid or invalid.

If I am going to run the marathon, then I will buy new shoes.
If I buy new shoes, then I will not buy a television.
∴If I buy a television, I will not run the marathon.

Solution
Label the statements

m: I am going to run the marathon.
s: I will buy new shoes.
t: I will buy a television.

The symbolic form of the argument is

$$m \to s$$
$$s \to \sim t$$
$$\therefore t \to \sim m$$

The truth table for this argument is as follows:

m	s	t	First premise $m \to s$	Second premise $s \to \sim t$	Conclusion $t \to \sim m$	
T	T	T	T	F	F	Row 1
T	T	F	T	T	T	Row 2
T	F	T	F	T	F	Row 3
T	F	F	F	T	T	Row 4
F	T	T	T	F	T	Row 5
F	T	F	T	T	T	Row 6
F	F	T	T	T	T	Row 7
F	F	F	T	T	T	Row 8

The only rows in which both premises are true are rows 2, 6, 7, and 8. Because the conclusion is true in each of these rows, the argument is valid.

CHECK YOUR PROGRESS 3 Determine whether the following argument is valid or invalid.

If I arrive before 8 A.M., then I will make the flight.
If I make the flight, then I will give the presentation.
∴If I arrive before 8 A.M., then I will give the presentation.

Solution *See page S10.*

Standard Forms

Some arguments can be shown to be valid if they have the same symbolic form as an argument that is known to be valid. For instance, we have shown that the argument

$$\begin{array}{l} h \rightarrow m \\ \underline{h\quad\quad} \\ \therefore m \end{array}$$

is valid. This symbolic form is known as **modus ponens** or the **law of detachment.** All arguments that have this symbolic form are valid. Table 3.13 shows four symbolic forms and the name used to identify each form. Any argument that has a symbolic form identical to one of these symbolic forms is a valid argument.

TAKE NOTE

In logic, the ability to identify standard forms of arguments is an important skill. If an argument has one of the standard forms in Table 3.13, then it is a valid argument. If an argument has one of the standard forms in Table 3.14, then it is an invalid argument. The standard forms can be thought of as laws of logic. Concerning the laws of logic, the logician Gottlob Frege (frā′gə) (1848–1925) stated, "The laws of logic are not like the laws of nature. They...are laws of the laws of nature."

Table 3.13 *Standard Forms of Four Valid Arguments*

Modus ponens	Modus tollens	Law of syllogism	Disjunctive syllogism
$p \rightarrow q$	$p \rightarrow q$	$p \rightarrow q$	$p \vee q$
p	$\sim q$	$q \rightarrow r$	$\sim p$
$\therefore q$	$\therefore \sim p$	$\therefore p \rightarrow r$	$\therefore q$

The law of syllogism can be extended to include more than two conditional premises. For example, if the premises of an argument are $a \rightarrow b$, $b \rightarrow c$, $c \rightarrow d, \ldots,$ $y \rightarrow z$, then a valid conclusion for the argument is $a \rightarrow z$. We will refer to any argument of this form with more than two conditional premises as the **extended law of syllogism.**

Table 3.14 shows two symbolic forms associated with invalid arguments. Any argument that has one of these symbolic forms is invalid.

Table 3.14 *Standard Forms of Two Invalid Arguments*

Fallacy of the converse	Fallacy of the inverse
$p \rightarrow q$	$p \rightarrow q$
q	$\sim p$
$\therefore p$	$\therefore \sim q$

EXAMPLE 4 ■ Use a Standard Form to Determine the Validity of an Argument

Use a standard form to determine whether the following argument is valid or invalid.

The program is interesting or I will watch the basketball game.
The program is not interesting.
____
∴I will watch the basketball game.

Solution
Label the statements

i: The program is interesting.
w: I will watch the basketball game.

In symbolic form the argument is:

$$\begin{array}{l} i \vee w \\ \underline{\sim i} \\ \therefore w \end{array}$$

This symbolic form matches the standard form known as disjunctive syllogism. Thus the argument is valid.

CHECK YOUR PROGRESS 4 Use a standard form to determine whether the following argument is valid or invalid.

If I go to Florida for spring break, then I will not study.
I did not go to Florida for spring break.
____
∴I studied.

Solution *See page S11.*

Waterfall by M. C. Escher

M. C. Escher (1898–1972) created many works of art that defy logic. In this lithograph, the water completes a full cycle even though the water is always traveling downward.

Consider an argument with the following symbolic form.

$q \rightarrow r$	Premise 1
$r \rightarrow s$	Premise 2
$\sim t \rightarrow \sim s$	Premise 3
$\underline{q}$	Premise 4
$\therefore t$	

To determine whether the argument is valid or invalid using a truth table, we would require a table with $2^4 = 16$ rows. It would be time-consuming to construct such a table and, with the large number of truth values to be determined, we might make an error. Thus we consider a different approach that makes use of a sequence of valid arguments to arrive at a conclusion.

$q \rightarrow r$	Premise 1
$\underline{r \rightarrow s}$	Premise 2
$\therefore q \rightarrow s$	Law of syllogism

$q \rightarrow s$ The previous conclusion
$\underline{s \rightarrow t}$ Premise 3 expressed in an equivalent form
$\therefore q \rightarrow t$ Law of syllogism

$q \rightarrow t$ The previous conclusion
$\underline{q}$ Premise 4
$\therefore t$ Modus ponens

This sequence of valid arguments shows that t is a valid conclusion for the original argument.

EXAMPLE 5 ■ Determine the Validity of an Argument

Determine whether the following argument is valid.

> If the movie was directed by Steven Spielberg (s), then I want to see it (w). The movie's production costs must exceed 50 million dollars (c) or I do not want to see it. The movie's production costs were less than 50 million dollars. Therefore, the movie was not directed by Steven Spielberg.

Solution
In symbolic form the argument is

$s \rightarrow w$ Premise 1
$c \vee \sim w$ Premise 2
$\underline{\sim c}$ Premise 3
$\therefore \sim s$ Conclusion

Premise 2 can be written as $\sim w \vee c$, which is equivalent to $w \rightarrow c$. Applying the law of syllogism to Premise 1 and this equivalent form of Premise 2 produces

$s \rightarrow w$ Premise 1
$\underline{w \rightarrow c}$ Equivalent form of Premise 2
$\therefore s \rightarrow c$ Law of syllogism

Combining the above conclusion $s \rightarrow c$ with Premise 3 gives us

$s \rightarrow c$ Conclusion from above
$\underline{\sim c}$ Premise 3
$\therefore \sim s$ Modus tollens

This sequence of valid arguments has produced the desired conclusion, $\sim s$. Thus the original argument is valid.

CHECK YOUR PROGRESS 5 Determine whether the following argument is valid.

> I start to fall asleep if I read a math book. I drink soda whenever I start to fall asleep. If I drink a soda, then I must eat a candy bar. Therefore, I eat a candy bar whenever I read a math book.

Hint: p whenever q is equivalent to $q \rightarrow p$.

Solution *See page S11.*

In the next example, we use standard forms to determine a valid conclusion for an argument.

EXAMPLE 6 ■ Determine a Valid Conclusion for an Argument

Use all of the premises to determine a valid conclusion for the following argument.

> We will not go to Japan ($\sim j$) or we will go to Hong Kong (h). If we visit my uncle (u), then we will go to Singapore (s). If we go to Hong Kong, then we will not go to Singapore.

Solution

In symbolic form the argument is

$$\begin{array}{ll} \sim j \vee h & \text{Premise 1} \\ u \rightarrow s & \text{Premise 2} \\ \underline{h \rightarrow \sim s} & \text{Premise 3} \\ \therefore ? & \end{array}$$

TAKE NOTE

In Example 6 we are rewriting and reordering the statements so that the extended law of syllogism can be applied.

The first premise can be written as $j \rightarrow h$. The second premise can be written as $\sim s \rightarrow \sim u$. Therefore, the argument can be written as

$$\begin{array}{l} j \rightarrow h \\ \sim s \rightarrow \sim u \\ \underline{h \rightarrow \sim s} \\ \therefore ? \end{array}$$

Interchanging the second and third premises yields

$$\begin{array}{l} j \rightarrow h \\ h \rightarrow \sim s \\ \underline{\sim s \rightarrow \sim u} \\ \therefore ? \end{array}$$

An application of the extended law of syllogism produces

$$\begin{array}{l} j \rightarrow h \\ h \rightarrow \sim s \\ \underline{\sim s \rightarrow \sim u} \\ \therefore j \rightarrow \sim u \end{array}$$

Thus a valid conclusion for the original argument is "If we go to Japan (j), then we will not visit my uncle ($\sim u$)."

CHECK YOUR PROGRESS 6 Use all of the premises to determine a valid conclusion for the following argument.

$$\begin{array}{l} \sim m \vee t \\ t \rightarrow \sim d \\ e \vee g \\ \underline{e \rightarrow d} \\ \therefore ? \end{array}$$

Solution *See page S11.*

Math Matters The Paradox of the Unexpected Hanging

The following paradox, known as "the paradox of the unexpected hanging," has proved difficult to analyze.

> The man was sentenced on Saturday. "The hanging will take place at noon," said the judge to the prisoner, "on one of the seven days of next week. But you will not know which day it is until you are so informed on the morning of the day of the hanging."
>
> The judge was known to be a man who always kept his word. The prisoner, accompanied by his lawyer, went back to his cell. As soon as the two men were alone the lawyer broke into a grin. "Don't you see?" he exclaimed. "The judge's sentence cannot possibly be carried out."
>
> "I don't see," said the prisoner.
>
> "Let me explain. They obviously can't hang you next Saturday. Saturday is the last day of the week. On Friday afternoon you would still be alive and you would know with absolute certainty that the hanging would be on Saturday. You would know this *before* you were told so on Saturday morning. That would violate the judge's decree."
>
> "True," said the prisoner.
>
> "Saturday, then, is positively ruled out," continued the lawyer. "This leaves Friday as the last day they can hang you. But they can't hang you on Friday because by Thursday afternoon only two days would remain: Friday and Saturday. Since Saturday is not a possible day, the hanging would have to be on Friday. Your knowledge of that fact would violate the judge's decree again. So Friday is out. This leaves Thursday as the last possible day. But Thursday is out because if you're alive Wednesday afternoon, you'll know that Thursday is to be the day."
>
> "I get it," said the prisoner, who was beginning to feel much better. "In exactly the same way I can rule out Wednesday, Tuesday, and Monday. That leaves only tomorrow. But they can't hang me tomorrow because I know it today!"
>
> In brief, the judge's decree seems to be self-refuting. There is nothing logically contradictory in the two statements that make up his decree; nevertheless, it cannot be carried out in practice.
>
> He [the prisoner] is convinced, by what appears to be unimpeachable logic, that he cannot be hanged without contradicting the conditions specified in his sentence. Then, on Thursday morning, to his great surprise, the hangman arrives. Clearly he did not expect him. What is more surprising, the judge's decree is now seen to be perfectly correct. The sentence can be carried out exactly as stated.[3]

---

3. Reprinted with the permission of Simon & Schuster from *The Unexpected Hanging and Other Mathematical Diversions* by Martin Gardner. Copyright © 1969 by Martin Gardner.

Fallacies

Any argument that is not valid is called a **fallacy.** Ancient logicians enjoyed the study of fallacies and took pride in their ability to analyze and categorize different types of fallacies. In this Excursion we consider the four fallacies known as *circulus in probando,* the fallacy of experts, the fallacy of equivocation, and the fallacy of accident.

Circulus in Probando

A fallacy of *circulus in probando* is an argument that uses a premise as the conclusion. For instance, consider the following argument.

> The Chicago Bulls are the best basketball team because there is no basketball team that is better than the Chicago Bulls.

The fallacy of *circulus in probando* is also known as *circular reasoning* or *begging the question*.

Fallacy of Experts

A fallacy of experts is an argument that uses an expert (or a celebrity) to lend support to a product or an idea. Often the product or idea is outside the expert's area of expertise. The following endorsements may qualify as fallacy of experts arguments.

> Tiger Woods for Rolex watches
>
> Lindsey Wagner for Ford Motor Company

Fallacy of Equivocation

A fallacy of equivocation is an argument that uses a word with two interpretations in two different ways. The following argument is an example of a fallacy of equivocation.

> The highway sign read $268 fine for littering,
>
> so I decided fine, for $268, I will litter.

Fallacy of Accident

The following argument is an example of a fallacy of accident.

> Everyone should visit Europe.
>
> Therefore, prisoners on death row should be allowed to visit Europe.

Using more formal language, we can state the argument as follows.

> If you are a prisoner on death row (d), then you are a person (p).
>
> If you are a person (p), then you should be allowed to visit Europe (e).
>
> ∴If you are a prisoner on death row, then you should be allowed to visit Europe.

The symbolic form of the argument is

$$\begin{array}{l} d \rightarrow p \\ \underline{p \rightarrow e} \\ \therefore d \rightarrow e \end{array}$$

(continued)

This argument appears to be a valid argument because it has the standard form of the law of syllogism. Common sense tells us the argument is not valid, so where have we gone wrong in our analysis of the argument?

The problem occurs with the interpretation of the word "everyone." Often, when we say "everyone," we really mean "most everyone." A fallacy of accident may occur whenever we use a statement that is often true in place of a statement that is always true.

Excursion Exercises

1. Write an argument that is an example of *circulus in probando.*
2. Give an example of an argument that is a fallacy of experts.
3. Write an argument that is an example of a fallacy of equivocation.
4. Write an argument that is an example of a fallacy of accident.
5. Algebraic arguments often consist of a list of statements. In a valid algebraic argument, each statement (after the premises) can be deduced from the previous statements. The following argument that $1 = 2$ contains exactly one step that is not valid. Identify the step and explain why it is not valid.

Let	$a = b$	• Premise
	$a^2 = ab$	• Multiply each side by a.
	$a^2 - b^2 = ab - b^2$	• Subtract b^2 from each side.
	$(a + b)(a - b) = b(a - b)$	• Factor each side.
	$a + b = b$	• Divide each side by $(a - b)$.
	$b + b = b$	• Substitute b for a.
	$2b = b$	• Collect like terms.
	$2 = 1$	• Divide each side by b.

Exercise Set 3.5

In Exercises 1–8, use the indicated letters to write each argument in symbolic form.

1. If you can read this bumper sticker (r), you're too close (c). You can read the bumper sticker. Therefore, you're too close.
2. If Lois Lane marries Clark Kent (m), then Superman will get a new uniform (u). Superman does not get a new uniform. Therefore, Lois Lane did not marry Clark Kent.
3. If the price of gold rises (g), the stock market will fall (s). The price of gold did not rise. Therefore, the stock market did not fall.
4. I am going shopping (s) or I am going to the museum (m). I went to the museum. Therefore, I did not go shopping.
5. If we search the Internet (s), we will find information on logic (i). We searched the Internet. Therefore, we found information on logic.
6. If we check the sports results on the Excite channel (c), we will know who won the match (w). We know who won the match. Therefore, we checked the sports results on the Excite channel.
7. If the power goes off $(\sim p)$, then the air conditioner will not work $(\sim a)$. The air conditioner is working. Therefore, the power is not off.
8. If it snowed (s), then I did not go to my chemistry class $(\sim c)$. I went to my chemistry class. Therefore, it did not snow.

In Exercises 9–24, use a truth table to determine whether the argument is valid or invalid.

9. $p \vee \sim q$
   $\underline{\sim q}$
   $\therefore p$

10. $\sim p \wedge q$
    $\underline{\sim p}$
    $\therefore q$

11. $p \rightarrow \sim q$
    $\underline{\sim q}$
    $\therefore p$

12. $p \rightarrow \sim q$
    $\underline{p}$
    $\therefore \sim q$

13. $\sim p \rightarrow \sim q$
$\underline{\sim p}$
$\therefore \sim q$

14. $\sim p \rightarrow q$
$\underline{p}$
$\therefore \sim q$

15. $(p \rightarrow q) \wedge (\sim p \rightarrow q)$
$\underline{q}$
$\therefore p$

16. $(p \vee q) \wedge (p \wedge q)$
$\underline{p}$
$\therefore q$

17. $(p \wedge \sim q) \vee (p \rightarrow q)$
$\underline{q \vee p}$
$\therefore \sim p \wedge q$

18. $(p \wedge \sim q) \rightarrow (p \vee q)$
$\underline{q \rightarrow \sim p}$
$\therefore p \rightarrow q$

19. $(p \wedge \sim q) \vee (p \vee r)$
$\underline{r}$
$\therefore p \vee q$

20. $(p \rightarrow q) \rightarrow (r \rightarrow \sim q)$
$\underline{p}$
$\therefore \sim r$

21. $p \leftrightarrow q$
$\underline{p \rightarrow r}$
$\therefore \sim r \rightarrow \sim p$

22. $p \wedge r$
$\underline{p \rightarrow \sim q}$
$\therefore r \rightarrow q$

23. $p \wedge \sim q$
$\underline{p \leftrightarrow r}$
$\therefore q \vee r$

24. $p \rightarrow r$
$\underline{r \rightarrow q}$
$\therefore \sim p \rightarrow \sim q$

In Exercises 25–30, use the indicated letters to write the argument in symbolic form. Then use a truth table to determine whether the argument is valid or invalid.

25. If you finish your homework (h), you may attend the reception (r). You did not finish your homework. Therefore, you cannot go to the reception.

26. The X Games will be held in Oceanside (o) if and only if the city of Oceanside agrees to pay $100,000 in prize money (a). If San Diego agrees to pay $200,000 in prize money (s), then the city of Oceanside will not agree to pay $100,000 in prize money. Therefore, if the X Games were held in Oceanside, then San Diego did not agree to pay $200,000 in prize money.

27. If I can't buy the house $(\sim b)$, then at least I can dream about it (d). I can buy the house or at least I can dream about it. Therefore, I can buy the house.

28. If the winds are from the east (e), then we will not have a big surf $(\sim s)$. We do not have a big surf. Therefore, the winds are from the east.

29. If I master college algebra (c), then I will be prepared for trigonometry (t). I am prepared for trigonometry. Therefore, I mastered college algebra.

30. If it is a blot (b), then it is not a clot $(\sim c)$. If it is a zlot (z), then it is a clot. It is a blot. Therefore, it is not a zlot.

In Exercises 31–40, determine whether the argument is valid or invalid by comparing its symbolic form with the standard symbolic forms given in Tables 3.13 and 3.14. For each valid argument, state the name of its standard form.

31. If you take Art 151 in the fall, you will be eligible to take Art 152 in the spring. You were not eligible to take Art 152 in the spring. Therefore, you did not take Art 151 in the fall.

32. He will attend Stanford or Yale. He did not attend Yale. Therefore, he attended Stanford.

33. If I had a nickel for every logic problem I have solved, then I would be rich. I have not received a nickel for every logic problem I have solved. Therefore, I am not rich.

34. If it is a dog, then it has fleas. It has fleas. Therefore, it is a dog.

35. If we serve salmon, then Vicky will join us for lunch. If Vicky joins us for lunch, then Marilyn will not join us for lunch. Therefore, if we serve salmon, Marilyn will not join us for lunch.

36. If I go to college, then I will not be able to work for my Dad. I did not go to college. Therefore, I went to work for my Dad.

37. If my cat is left alone in the apartment, then she claws the sofa. Yesterday I left my cat alone in the apartment. Therefore, my cat clawed the sofa.

38. If I wish to use the new software, then I cannot continue to use this computer. I don't wish to use the new software. Therefore, I can continue to use this computer.

39. If Rita buys a new car, then she will not go on the cruise. Rita went on the cruise. Therefore, Rita did not buy a new car.

40. If Hideo Nomo pitches, then I will go to the game. I did not go to the game. Therefore, Hideo Nomo did not pitch.

In Exercises 41–46, use a sequence of valid arguments to show that each argument is valid.

41. $\sim p \rightarrow r$
$r \rightarrow t$
$\underline{\sim t}$
$\therefore p$

42. $r \rightarrow \sim s$
$s \vee \sim t$
$\underline{r}$
$\therefore \sim t$

43. If we sell the boat (s), then we will not go to the river $(\sim r)$. If we don't go to the river, then we will go camping (c). If we do not buy a tent $(\sim t)$, then we will not go camping. Therefore, if we sell the boat, then we will buy a tent.

44. If it is an ammonite (a), then it is from the Cretaceous period (c). If it is not from the Mesozoic era ($\sim m$), then it is not from the Cretaceous period. If it is from the Mesozoic era, then it is at least 65 million years old (s). Therefore, if it is an ammonite, then it is at least 65 million years old.

45. If the computer is not operating ($\sim o$), then I will not be able to finish my report ($\sim f$). If the office is closed (c), then the computer is not operating. Therefore, if I am able to finish my report, then the office is open.

46. If he reads the manuscript (r), he will like it (l). If he likes it, he will publish it (p). If he publishes it, then you will get royalties (m). You did not get royalties. Therefore, he did not read the manuscript.

Extensions

CRITICAL THINKING

In Exercises 47–50, use all of the premises to determine a valid conclusion for the given argument.

47.
$$\begin{array}{l} \sim(p \wedge \sim q) \\ \underline{p \qquad\quad} \\ \therefore ? \end{array}$$

48.
$$\begin{array}{l} \sim s \rightarrow q \\ \sim t \rightarrow \sim q \\ \underline{\sim t \qquad\quad} \\ \therefore ? \end{array}$$

49. If it is a theropod, then it is not herbivorous. If it is not herbivorous, then it is not a sauropod. It is a sauropod. Therefore, __________________.

50. If you buy the car, you will need a loan. You do not need a loan or you will make monthly payments. You buy the car. Therefore, __________________.

COOPERATIVE LEARNING

51. An Argument by Lewis Carroll The following argument is from *Symbolic Logic* by Lewis Carroll, written in 1896. Determine whether the argument is valid or invalid.

Babies are illogical.

Nobody is despised who can manage a crocodile.

Illogical persons are despised.

Hence, babies cannot manage crocodiles.

EXPLORATIONS

52. **Fallacies** Consult a logic text or search the Internet for information on fallacies. Write a report that includes examples of at least three of the following fallacies.

Ad hominem
Ad populum
Ad baculum
Ad vercundiam
Non sequitur
Fallacy of false cause
Pluriam interrogationem

SECTION 3.6 Euler Diagrams

Euler Diagrams

Many arguments involve sets whose elements are described using the quantifiers *all, some,* and *none.* The mathematician Leonhard Euler (laônhärt oi′lər) used diagrams to determine whether arguments that involved quantifiers were valid or invalid. The following figures show Euler diagrams that illustrate the four possible relationships that can exist between two sets.

All Ps are Qs.

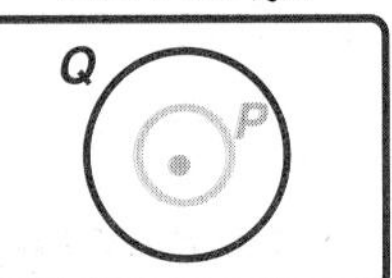

No Ps are Qs.

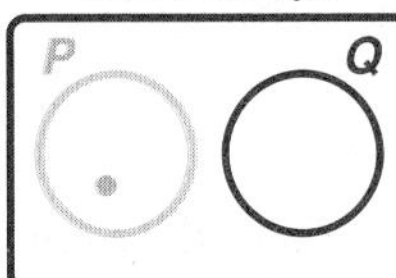

Some Ps are Qs.

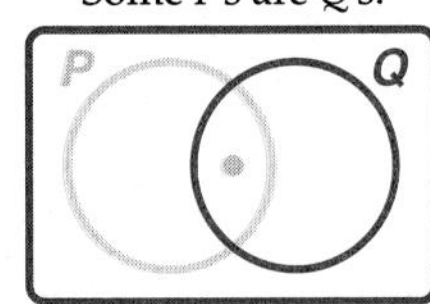

Some Ps are not Qs.

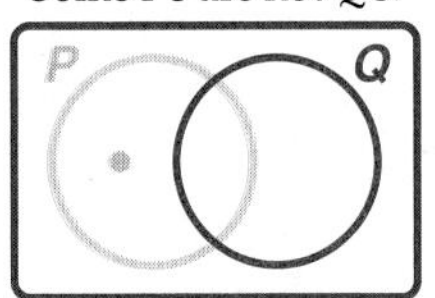

Euler diagrams

historical note

Leonhard Euler (1707–1783) Euler was an exceptionally talented Swiss mathematician. He worked in many different areas of mathematics and produced more written material about mathematics than any other mathematician. His mental computational abilities were remarkable. The French astronomer and statesman Dominque François Arago wrote,

> Euler calculated without apparent effort, as men breathe, or as eagles sustain themselves in the wind.

In 1776, Euler became blind; however, he continued to work in the disciplines of mathematics, physics, and astronomy. He even solved a problem that Newton had attempted concerning the motion of the moon. Euler performed all the necessary calculations in his head. ■

Euler used diagrams to illustrate logic concepts. Some 100 years later, John Venn extended the use of Euler's diagrams to illustrate many types of mathematics. In this section, we will construct diagrams to determine the validity of arguments. We will refer to these diagrams as Euler diagrams.

EXAMPLE 1 ■ Use an Euler Diagram to Determine the Validity of an Argument

Use an Euler diagram to determine whether the following argument is valid or invalid.

All college courses are fun.
<u>This course is a college course.</u>
∴This course is fun.

Solution
The first premise indicates that the set of college courses is a subset of the set of fun courses. We illustrate this subset relationship with an Euler diagram, as shown in Figure 3.15. The second premise tells us that "this course" is an element of the set of college courses. If we use c to represent "this course," then c must be placed inside the set of college courses, as shown in Figure 3.16.

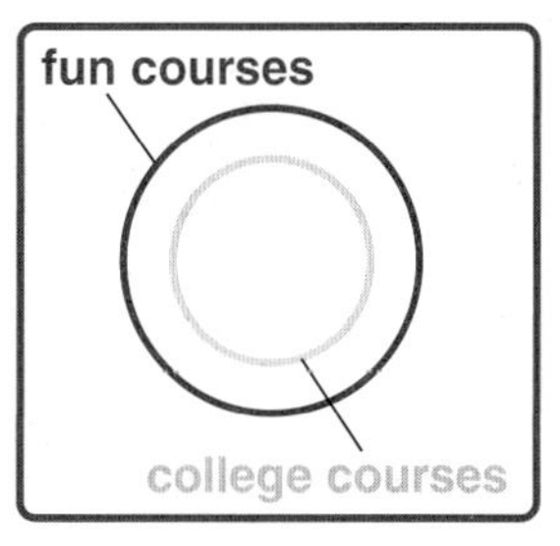

Figure 3.15

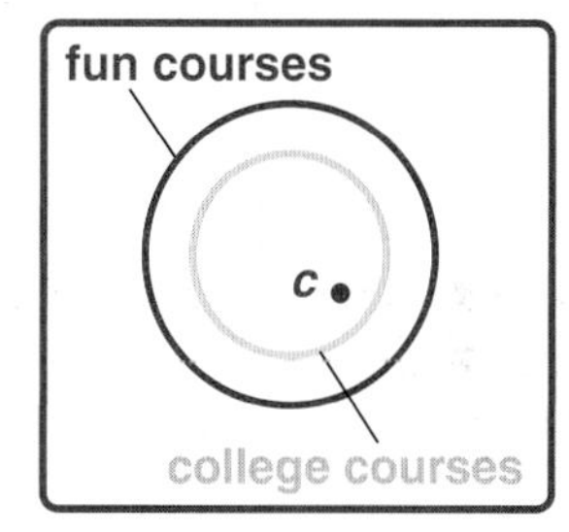

Figure 3.16

Figure 3.16 illustrates that c must also be an element of the set of fun courses. Thus the argument is valid.

CHECK YOUR PROGRESS 1 Use an Euler diagram to determine whether the following argument is valid or invalid.

All lawyers drive BMW's.
<u>Susan is a lawyer.</u>
∴Susan drives a BMW.

Solution *See page S11.*

If an Euler diagram can be drawn so that the conclusion does not necessarily follow from the premises, then the argument is invalid. This concept is illustrated in the next example.

This impressionist painting, *Dance at Bougival* by Renoir, is on display at the Museum of Fine Arts, Boston.

EXAMPLE 2 ■ Use an Euler Diagram to Determine the Validity of an Argument

Use an Euler diagram to determine whether the following argument is valid or invalid.

Some impressionists paintings are Renoirs.
Dance at Bougival is an impressionist painting.
∴*Dance at Bougival* is a Renoir.

Solution
The Euler diagram in Figure 3.17 illustrates the premise that some impressionist paintings are Renoirs. Let d represent the painting *Dance at Bougival.* Figures 3.18 and 3.19 show that d can be placed in one of two regions.

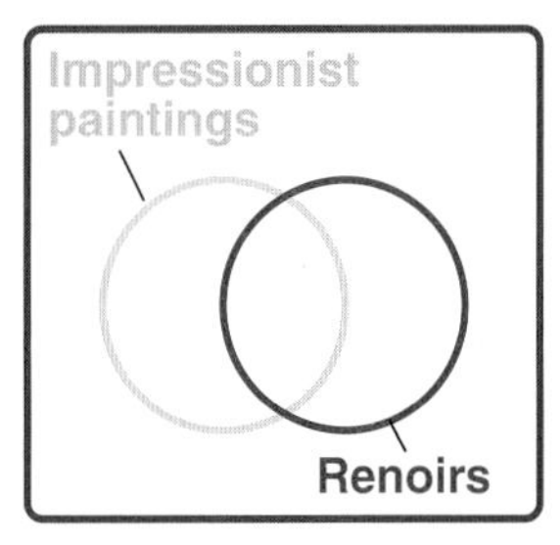

Figure 3.17

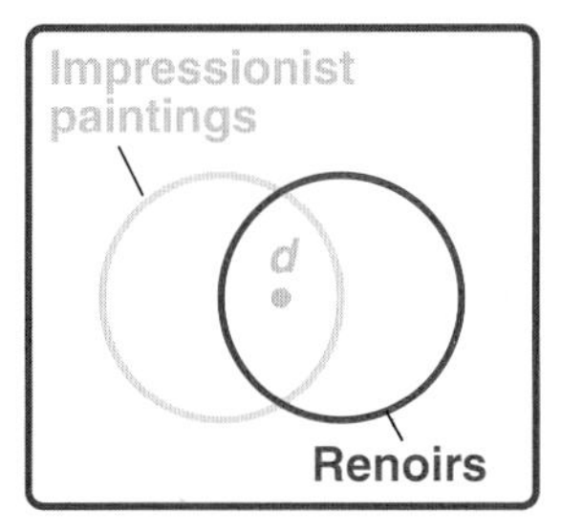

Figure 3.18

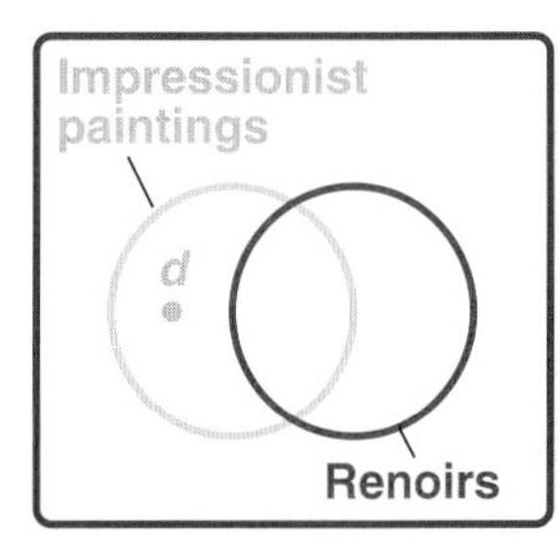

Figure 3.19

Although Figure 3.18 supports the argument, Figure 3.19 shows that the conclusion does not necessarily follow from the premises, and thus the argument is invalid.

TAKE NOTE

Even though the conclusion in Example 2 is true, the argument is invalid.

CHECK YOUR PROGRESS 2 Use an Euler diagram to determine whether the following argument is valid or invalid.

No prime numbers are negative.
The number 7 is not negative.
∴The number 7 is a prime number.

Solution *See page S12.*

QUESTION *If one particular example can be found for which the conclusion of an argument is true when its premises are true, must the argument be valid?*

Some arguments can be represented by an Euler diagram that involves three sets, as shown in Example 3.

ANSWER *No. To be a valid argument, the conclusion must be true whenever the premises are true. Just because the conclusion is true for one specific example, it does not mean the argument is a valid argument.*

EXAMPLE 3 ■ Use an Euler Diagram to Determine the Validity of an Argument

Use an Euler diagram to determine whether the following argument is valid or invalid.

No psychologist can juggle.
All clowns can juggle.
∴No psychologist is a clown.

Solution
The Euler diagram in Figure 3.20 shows that the set of psychologists and the set of jugglers are disjoint sets. Figure 3.21 shows that because the set of clowns is a subset of the set of jugglers, no psychologists *p* are elements of the set of clowns. Thus the argument is valid.

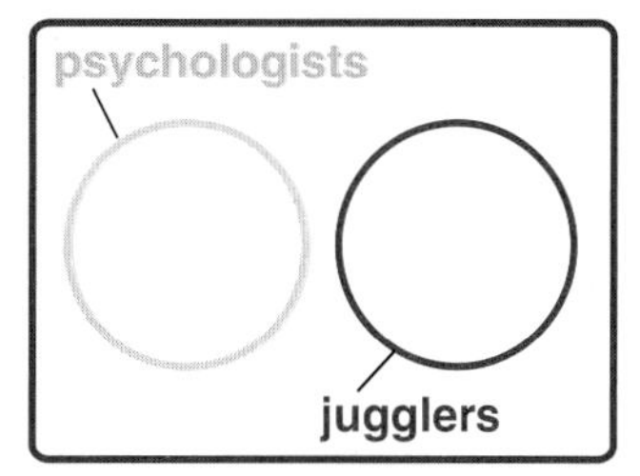

Figure 3.20

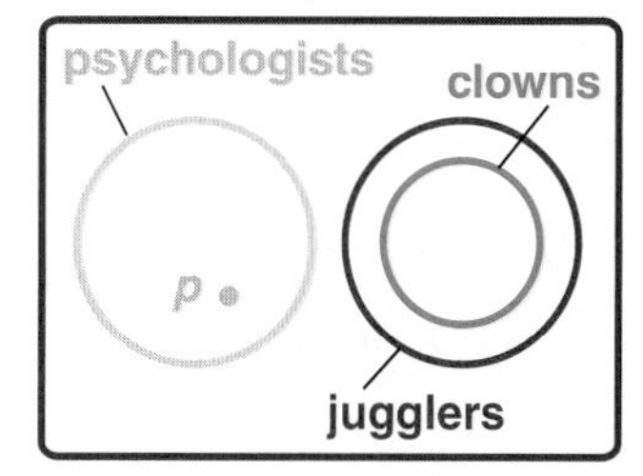

Figure 3.21

CHECK YOUR PROGRESS 3 Use an Euler diagram to determine whether the following argument is valid or invalid.

No mathematics professors are good-looking.
All good-looking people are models.
∴No mathematics professor is a model.

Solution *See page S12.*

Math Matters **A Famous Puzzle**

Three men decide to rent a room for one night. The regular room rate is $25; however, the desk clerk charges the men $30 because it will be easier for each man to pay one-third of $30 than it would be for each man to pay one-third of $25. Each man pays $10 and the porter shows them to their room.

After a short period of time, the desk clerk starts to feel guilty and gives the porter $5, along with instructions to return the $5 to the three men.

On the way to the room the porter decides to give each man $1 and pocket $2. After all, the men would find it difficult to split $5 evenly.

Thus each man has paid $10 and received a refund of $1. After the refund, the men have paid a total of $27. The porter has $2. The $27 added to the $2 equals $29.

QUESTION *Where is the missing dollar? (See Answer on the following page.)*

Euler Diagrams and the Extended Law of Syllogism

Example 4 uses Euler diagrams to visually illustrate the extended law of syllogism from Section 3.5.

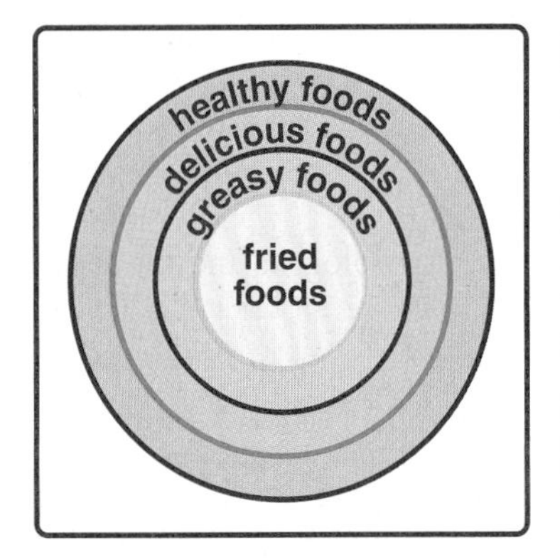

EXAMPLE 4 ■ Use an Euler Diagram to Determine the Validity of an Argument

Use an Euler diagram to determine whether the following argument is valid or invalid.

All fried foods are greasy.
All greasy foods are delicious.
All delicious foods are healthy.
∴All fried foods are healthy.

Solution
The figure at the left illustrates that every fried food is an element of the set of healthy foods, so the argument is valid.

TAKE NOTE

Although the conclusion in Example 4 is false, the argument in Example 4 is valid.

CHECK YOUR PROGRESS 4 Use an Euler diagram to determine whether the following argument is valid or invalid.

All squares are rhombi.
All rhombi are parallelograms.
All parallelograms are quadrilaterals.
∴All squares are quadrilaterals.

Solution *See page S12.*

Using Euler Diagrams to Form Conclusions

In Example 5, we make use of an Euler diagram to determine a valid conclusion for an argument.

ANSWER *The $2 the porter kept was added to the $27 the men spent to produce a total of $29. The fact that this amount just happens to be close to $30 is a coincidence. All the money can be located if we total the $2 the porter has, the $3 that was returned to the men, and the $25 the desk clerk has, to produce $30.*

EXAMPLE 5 ■ Use an Euler Diagram to Determine the Conclusion for an Argument

Use an Euler diagram and all of the premises in the following argument to determine a valid conclusion for the argument.

All *M*s are *N*s.
No *N*s are *P*s.
$\therefore$?

Solution
The first premise indicates that the set of *M*s is a subset of the set of *N*s. The second premise indicates that the set of *N*s and the set of *P*s are disjoint sets. The following Euler diagram illustrates these set relationships. An examination of the Euler diagram allows us to conclude that no *M*s are *P*s.

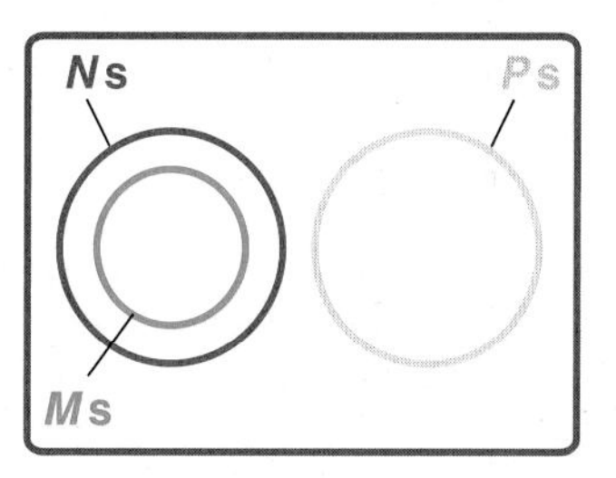

CHECK YOUR PROGRESS 5 Use an Euler diagram and all of the premises in the following argument to determine a valid conclusion for the argument.

Some rabbits are white.
All white animals like tomatoes.
$\therefore$?

Solution *See page S12.*

Excursion

Using Logic to Solve Puzzles

TAKE NOTE

When working with cryptarithms, we assume that the leading digit of each number is a nonzero digit.

Many puzzles can be solved by making an assumption and then checking to see if the assumption is supported by the conditions (premises) associated with the puzzle. For instance, consider the following addition problem in which each letter represents a digit from 0 through 9, and different letters represent different digits.

$$\begin{array}{r} \text{TA} \\ +\text{BT} \\ \hline \text{TEE} \end{array}$$

(continued)

Note that the T in T E E is a carry from the middle column. Because the sum of any two single digits plus a previous carry of at most 1 is 19 or less, the T in T E E must be a 1. Replacing all the T's with 1's produces:

```
  1A
 +B1
 ---
 1EE
```

Now B must be an 8 or a 9, because these are the only digits that would produce a carry into the leftmost column.

Case 1: Assume B is a 9. Then A must be an 8 or smaller, and A + 1 does not produce a carry into the middle column. The sum of the digits in the middle column is 10 thus E is a 0. This presents a dilemma because the units digit of A + 1 must also be a 0, which requires A to be a 9. The assumption that B is a 9 is not supported by the conditions of the problem; thus we reject the assumption that B is a 9.

Case 2: Assume B is an 8. To produce the required carry into the leftmost column, there must be a carry from the column on the right. Thus A must be a 9, and we have the result shown below.

$$\begin{array}{r} 19 \\ +81 \\ \hline 100 \end{array}$$

A check shows that this solution satisfies all the conditions of the problem.

Excursion Exercises

Solve the following cryptarithms. Assume that no leading digit is a 0. (*Source:* **http://www.geocities.com/Athens/Agora/2160/puzzles.html**)[4]

1.
```
  SO
 +SO
 ---
 TOO
```

2.
```
  US
 +AS
 ---
 ALL
```

3.
```
  COCA
 +COLA
 -----
 OASIS
```

4.
```
    AT
  EAST
 +WEST
 -----
 SOUTH
```

4. Copyright © 1998 by Jorge A C B Soares.

Exercise Set 3.6

In Exercises 1–20, use an Euler diagram to determine whether the argument is valid or invalid.

1. All frogs are poetical.
   Kermit is a frog.
   ∴Kermit is poetical.
2. All Oreo cookies have a filling.
   All Fig Newtons have a filling.
   ∴All Fig Newtons are Oreo cookies.
3. Some plants have flowers.
   All things that have flowers are beautiful.
   ∴Some plants are beautiful.
4. No squares are triangles.
   Some triangles are equilateral.
   ∴No squares are equilateral.
5. No rocker would do the Mariachi.
   All baseball fans do the Mariachi.
   ∴No rocker is a baseball fan.
6. Nuclear energy is not safe.
   Some electric energy is safe.
   ∴No electric energy is nuclear energy.
7. Some birds bite.
   All things that bite are dangerous.
   ∴Some birds are dangerous.
8. All fish can swim.
   That barracuda can swim.
   ∴That barracuda is a fish.
9. All men behave badly.
   Some hockey players behave badly.
   ∴Some hockey players are men.
10. All grass is green.
    That ground cover is not green.
    ∴That ground cover is not grass.
11. Most teenagers drink soda.
    No CEOs drink soda.
    ∴No CEO is a teenager.
12. Some students like history.
    Vern is a student.
    ∴Vern likes history.
13. No mathematics test is fun.
    All fun things are worth your time.
    ∴No mathematics test is worth your time.
14. All prudent people shun sharks.
    No accountant is imprudent.
    ∴No accountant fails to shun sharks.
15. All candidates without a master's degree will not be considered for the position of director.
    All candidates who are not considered for the position of director should apply for the position of assistant.
    ∴All candidates without a master's degree should apply for the position of assistant.
16. Some whales make good pets.
    Some good pets are cute.
    Some cute pets bite.
    ∴Some whales bite.
17. All prime numbers are odd.
    2 is a prime number.
    ∴2 is an odd number.
18. All Lewis Carroll arguments are valid.
    Some valid arguments are syllogisms.
    ∴Some Lewis Carroll arguments are syllogisms.
19. All aerobics classes are fun.
    Jan's class is fun.
    ∴Jan's class is an aerobics class.
20. No sane person takes a math class.
    Some students that take a math class can juggle.
    ∴No sane person can juggle.

In Exercises 21–26, use all of the premises in each argument to determine a valid conclusion for the argument.

21. All Reuben sandwiches are good.
    All good sandwiches have pastrami.
    All sandwiches with pastrami need mustard.
    ∴?
22. All cats are strange.
    Boomer is not strange.
    ∴?
23. All multiples of 11 end with a 5.
    1001 is a multiple of 11.
    ∴?
24. If it isn't broken, then I do not fix it.
    If I do not fix it, then I do not get paid.
    ∴?

25. Some horses are frisky.
All frisky horses are grey.

∴?

26. If we like to ski, then we will move to Vail.
If we move to Vail, then we will not buy a house.
If we do not buy a condo, then we will buy a house.

∴?

27. Examine the following three premises:

1. All people who have an Xbox play video games.
2. All people who play video games enjoy life.
3. Some mathematics professors enjoy life.

Now consider each of the following six conclusions. For each conclusion, determine whether the argument formed by the three premises and the conclusion is valid or invalid.

a. ∴Some mathematics professors have an Xbox.

b. ∴Some mathematics professors play video games.

c. ∴Some people who play video games are mathematics professors.

d. ∴Mathematics professors never play video games.

e. ∴All people who have an Xbox enjoy life.

f. ∴Some people who enjoy life are mathematics professors.

28. Examine the following three premises:

1. All people who drive pickup trucks like Willie Nelson.
2. All people who like Willie Nelson like country western music.
3. Some people who like heavy metal music like Willie Nelson.

Now consider each of the following five conclusions. For each conclusion, determine whether the argument formed by the three premises and the conclusion is valid or invalid.

a. ∴ Some people who like heavy metal music drive a pickup truck.

b. ∴Some people who like heavy metal music like country western music.

c. ∴Some people who like Willie Nelson like heavy metal music.

d. ∴All people who drive a pickup truck like country western music.

e. ∴People who like heavy metal music never drive a pickup truck.

Extensions

CRITICAL THINKING

29. A Crossnumber Puzzle In the following *crossnumber puzzle,* each square holds a single digit from 0 through 9. Use the clues under the *Across* and *Down* headings to solve the puzzle.

Across

1. One-fourth of 3 across

3. Two more than 1 down with its digits reversed

Down

1. Larger than 20 and less than 30

2. Half of 1 down

1	2
3	4

EXPLORATIONS

30. **Bilateral Diagrams** Lewis Carroll (Charles Dodgson) devised a *bilateral diagram* (two-part board) to analyze syllogisms. His method has some advantages over Euler diagrams and Venn diagrams. Use a library or the Internet to find information on Carroll's method of analyzing syllogisms. Write a few paragraphs that explain his method and its advantages.

CHAPTER 3 Summary

Key Terms

antecedent [p. 137]
argument [p. 152]
arrow notation [p. 137]
biconditional [p. 141]
component statement [p. 116]
compound statement [p. 116]
conclusion [p. 152]
conditional [p. 136]
conjunction [p. 118]
connective [p. 116]
consequent [p. 137]
contrapositive [p. 146]
converse [p. 146]
disjunction [p. 119]
disjunctive form of the conditional [p. 140]
disjunctive syllogism [p. 156]
equivalent statements [p. 129]
Euler diagram [p. 164]
existential quantifier [p. 119]
extended law of syllogism [p. 156]
fallacy of the converse [p. 156]
fallacy of the inverse [p. 156]
invalid argument [p. 152]
inverse [p. 146]
law of syllogism [p. 156]
modus ponens [p. 156]
modus tollens [p. 156]
negation [p. 117]
premise [p. 152]
quantifier [p. 119]
self-contradiction [p. 131]
standard forms of arguments [p. 156]
standard truth table form [p. 125]
statement [p. 114]
symbolic form [p. 116]
tautology [p. 131]
truth table [p. 117]
truth value [p. 117]
universal quantifier [p. 119]
valid argument [p. 152]

Essential Concepts

- **Truth Values**
  $\sim p$ is true if and only if p is false.
  $p \wedge q$ is true if and only if both p and q are true.
  $p \vee q$ is true if and only if p is true, q is true, or both p and q are true.
  The *conditional* $p \rightarrow q$ is false if p is true and q is false. It is true in all other cases.

- **De Morgan's Laws for Statements**
  $\sim(p \wedge q) \equiv \sim p \vee \sim q$ and $\sim(p \vee q) \equiv \sim p \wedge \sim q$

- **Equivalent Forms**
  $p \rightarrow q \equiv \sim p \vee q$
  $\sim(p \rightarrow q) \equiv p \wedge \sim q$
  $p \leftrightarrow q \equiv [(p \rightarrow q) \wedge (q \rightarrow p)]$

- **Statements Related to the Conditional Statement**
  The *converse* of $p \rightarrow q$ is $q \rightarrow p$.
  The *inverse* of $p \rightarrow q$ is $\sim p \rightarrow \sim q$.
  The *contrapositive* of $p \rightarrow q$ is $\sim q \rightarrow \sim p$.

- **Valid Arguments**

Modus ponens	Modus tollens	Law of syllogism	Disjunctive syllogism
$p \rightarrow q$	$p \rightarrow q$	$p \rightarrow q$	$p \vee q$
p	$\sim q$	$q \rightarrow r$	$\sim p$
$\therefore q$	$\therefore \sim p$	$\therefore p \rightarrow r$	$\therefore q$

- **Invalid Arguments**

Fallacy of the converse	Fallacy of the inverse
$p \rightarrow q$	$p \rightarrow q$
q	$\sim p$
$\therefore p$	$\therefore \sim q$

- **Order of Precedence Agreement**
  If grouping symbols are not used to specify the order in which logical connectives are applied, then we use the following Order of Precedence Agreement.

  First apply the negations from left to right, then apply the conjunctions from left to right, and finally apply the disjunctions from left to right.

CHAPTER 3 Review Exercises

In Exercises 1–6, determine whether each sentence is a statement. Assume that a and b are real numbers.

1. How much is a ticket to London?
2. 91 is a prime number.
3. $a > b$
4. $a^2 \geq 0$
5. Lock the car.
6. Clark Kent is Superman.

In Exercises 7–10, write each sentence in symbolic form. Represent each component of the sentence with the letter indicated in parentheses. Also state whether the sentence is a conjunction, a disjunction, a negation, a conditional, or a biconditional.

7. Today is Monday (m) and it is my birthday (b).
8. If x is divisible by 2 (d), then x is an even number (e).
9. I am going to the dance (g) if and only if I have a date (d).
10. All triangles (t) have exactly three sides (s).

In Exercises 11–16, write the negation of each quantified statement.

11. Some dogs bite.
12. Every dessert at the Cove restaurant is good.
13. All winners receive a prize.
14. Some cameras do not use film.
15. None of the students received an A.
16. At least one person enjoyed the story.

In Exercises 17–22, determine whether each statement is true or false.

17. $5 > 2$ or $5 = 2$.
18. $3 \neq 5$ and 7 is a prime number.
19. $4 \leq 7$
20. $-3 < -1$
21. Every repeating decimal is a rational number.
22. There exists a real number that is not positive and not negative.

In Exercises 23–28, determine the truth value of the statement given that p is true, q is false, and r is false.

23. $(p \wedge q) \vee (\sim p \vee q)$
24. $(p \rightarrow \sim q) \leftrightarrow \sim(p \vee q)$
25. $(p \wedge \sim q) \wedge (\sim r \vee q)$
26. $(r \wedge \sim p) \vee [(p \vee \sim q) \leftrightarrow (q \rightarrow r)]$
27. $[p \wedge (r \rightarrow q)] \rightarrow (q \vee \sim r)$
28. $(\sim q \vee \sim r) \rightarrow [(p \leftrightarrow \sim r) \wedge q]$

In Exercises 29–36, construct a truth table for the given statement.

29. $(\sim p \rightarrow q) \vee (\sim q \wedge p)$
30. $\sim p \leftrightarrow (q \vee p)$
31. $\sim(p \vee \sim q) \wedge (q \rightarrow p)$
32. $(p \leftrightarrow q) \vee (\sim q \wedge p)$
33. $(r \leftrightarrow \sim q) \vee (p \rightarrow q)$
34. $(\sim r \vee \sim q) \wedge (q \rightarrow p)$
35. $[p \leftrightarrow (q \rightarrow \sim r)] \wedge \sim q$
36. $\sim(p \wedge q) \rightarrow (\sim q \vee \sim r)$

In Exercises 37–40, make use of De Morgan's laws to write the given statement in an equivalent form.

37. It is not true that Bob failed the English proficiency test and he registered for a speech course.
38. Ellen did not go to work this morning and she did not take her medication.
39. Wendy will go to the store this afternoon or she will not be able to prepare her fettuccine al pesto recipe.
40. Gina enjoyed the movie, but she did not enjoy the party.

In Exercises 41–44, use a truth table to show that the given pairs of statements are equivalent.

41. $\sim p \rightarrow \sim q$; $p \vee \sim q$
42. $\sim p \vee q$; $\sim(p \wedge \sim q)$
43. $p \vee (q \wedge \sim p)$; $p \vee q$
44. $p \leftrightarrow q$; $(p \wedge q) \vee (\sim p \wedge \sim q)$

In Exercises 45–48, use a truth table to determine whether the given statement is a tautology or a self-contradiction.

45. $p \wedge (q \wedge \sim p)$
46. $(p \wedge q) \vee (p \rightarrow \sim q)$
47. $[\sim(p \rightarrow q)] \leftrightarrow (p \wedge \sim q)$
48. $p \vee (p \rightarrow q)$

In Exercises 49–52, identify the antecedent and the consequent of each conditional statement.

49. If he has talent, he will succeed.
50. If I had a credential, I could get the job.
51. I will follow the exercise program provided I join the fitness club.
52. I will attend only if it is free.

In Exercises 53–56, write each conditional statement in its equivalent disjunctive form.

53. If she were tall, she would be on the volleyball team.
54. If he can stay awake, he can finish the report.
55. Rob will start provided he is not ill.
56. Sharon will be promoted only if she closes the deal.

In Exercises 57–60, write the negation of each conditional statement in its equivalent conjunctive form.

57. If I get my paycheck, I will purchase a ticket.
58. The tomatoes will get big only if you provide them with plenty of water.
59. If you entered Cleggmore University, then you had a high score on the SAT exam.
60. If Ryan enrolls at a university, then he will enroll at Yale.

In Exercises 61–66, determine whether the given statement is true or false. Assume that x and y are real numbers.

61. $x = y$ if and only if $|x| = |y|$.
62. $x > y$ if and only if $x - y > 0$.
63. If $x + y = 2x$, then $y = x$.
64. If $x > y$, then $\frac{1}{x} > \frac{1}{y}$.
65. If $x^2 > 0$, then $x > 0$.
66. If $x^2 = y^2$, then $x = y$.

In Exercises 67–70, write each statement in "If p, then q" form.

67. Every nonrepeating, nonterminating decimal is an irrational number.
68. Being well known is a necessary condition for a politician.
69. I could buy the house provided I could sell my condominium.
70. Being divisible by 9 is a sufficient condition for being divisible by 3.

In Exercises 71–76, write the **a.** converse, **b.** inverse, and **c.** contrapositive of the given statement.

71. If $x + 4 > 7$, then $x > 3$.
72. All recipes in this book can be prepared in less than 20 minutes.
73. If a and b are both divisible by 3, then $(a + b)$ is divisible by 3.
74. If you build it, they will come.
75. Every trapezoid has exactly two parallel sides.
76. If they like it, they will return.
77. What is the inverse of the contrapositive of $p \rightarrow q$?
78. What is the converse of the contrapositive of the inverse of $p \rightarrow q$?

In Exercises 79–82, determine the original statement if the given statement is related to the original statement in the manner indicated.

79. *Converse:* If $x > 2$, then x is an odd prime number.
80. *Negation:* The senator will attend the meeting and she will not vote on the motion.
81. *Inverse:* If their manager will not contact me, then I will not purchase any of their products.
82. *Contrapositive:* If Ginny can't rollerblade, then I can't rollerblade.

In Exercises 83–86, use a truth table to determine whether the argument is valid or invalid.

83.
$$\begin{array}{l} (p \wedge \sim q) \wedge (\sim p \rightarrow q) \\ \underline{p \qquad\qquad\qquad\qquad} \\ \therefore \sim q \end{array}$$

84.
$$\begin{array}{l} p \rightarrow \sim q \\ \underline{q \qquad\quad} \\ \therefore \sim p \end{array}$$

85.
$$\begin{array}{l} r \\ p \rightarrow \sim r \\ \underline{\sim p \rightarrow q} \\ \therefore p \wedge q \end{array}$$

86.
$$\begin{array}{l} (p \vee \sim r) \rightarrow (q \wedge r) \\ \underline{r \wedge p \qquad\qquad\qquad} \\ \therefore p \vee q \end{array}$$

In Exercises 87–92, determine whether the argument is valid or invalid by comparing its symbolic form with the symbolic forms in Tables 3.13 and 3.14, page 156.

87. We will serve either fish or chicken for lunch. We did not serve fish for lunch. Therefore, we served chicken for lunch.
88. If Mike is a CEO, then he will be able to afford to make a donation. If Mike can afford to make a donation, then he loves to ski. Therefore, if Mike does not love to ski, he is not a CEO.
89. If we wish to win the lottery, we must buy a lottery ticket. We did not win the lottery. Therefore, we did not buy a lottery ticket.
90. Robert can charge it on his MasterCard or his Visa. Robert does not use his MasterCard. Therefore, Robert charged it to his Visa.
91. If we are going to have a caesar salad, then we need to buy some eggs. We did not buy eggs. Therefore, we are not going to have a caesar salad.
92. If we serve lasagna, then Eva will not come to our dinner party. We did not serve lasagna. Therefore, Eva came to our dinner party.

In Exercises 93–96, use an Euler diagram to determine whether the argument is valid or invalid.

93. No wizard can yodel.
All lizards can yodel.
∴No wizard is a lizard.

94. Some dogs have tails.
Some dogs are big.
∴Some big dogs have tails.

95. All Italian villas are wonderful. It is not wise to invest in expensive villas. Some wonderful villas are expensive. Therefore, it is not wise to invest in Italian villas.

96. All logicians like to sing "It's a small world after all." Some logicians have been presidential candidates. Therefore, some presidential candidates like to sing "It's a small world after all."

CHAPTER 3 Test

1. Determine whether each sentence is a statement.

a. Look for the cat.

b. Clark Kent is afraid of the dark.

2. Write the negation of each statement.

a. Some trees are not green.

b. None of the kids had seen the movie.

3. Determine whether each statement is true or false.

a. $5 \leq 4$

b. $-2 \geq -2$

4. Determine the truth value of each statement given that p is true, q is false, and r is true.

a. $(p \vee \sim q) \wedge (\sim r \wedge q)$

b. $(r \vee \sim p) \vee [(p \vee \sim q) \leftrightarrow (q \rightarrow r)]$

In Exercises 5 and 6, construct a truth table for the given statement.

5. $\sim(p \wedge \sim q) \vee (q \rightarrow p)$ **6.** $(r \leftrightarrow \sim q) \wedge (p \rightarrow q)$

7. Use one of De Morgan's laws to write the following in an equivalent form.

Elle did not eat breakfast and she did not take a lunch break.

8. What is a tautology?

9. Write $p \rightarrow q$ in its equivalent disjunctive form.

10. Determine whether the given statement is true or false. Assume that x, y, and z are real numbers.

a. $x = y$ if $|x| = |y|$. **b.** If $x > y$, then $xz > yz$.

11. Write the **a.** converse, **b.** inverse, and **c.** contrapositive of the following statement.

If $x + 7 > 11$, then $x > 4$.

12. Write the standard form known as modus ponens.

13. Write the standard form known as the law of syllogism.

In Exercises 14 and 15, use a truth table to determine whether the argument is valid or invalid.

14.
$(p \wedge \sim q) \wedge (\sim p \rightarrow q)$
p
$\therefore \sim q$

15.
r
$p \rightarrow \sim r$
$\sim p \rightarrow q$
$\therefore p \wedge q$

In Exercises 16–20, determine whether the argument is valid or invalid. Explain how you made your decision.

16. If we wish to win the talent contest, we must practice. We did not win the contest. Therefore, we did not practice.

17. Gina will take a job in Atlanta or she will take a job in Kansas City. Gina did not take a job in Atlanta. Therefore, Gina took a job in Kansas City.

18. No wizard can glow in the dark.
Some lizards can glow in the dark.
∴No wizard is a lizard.

19. Some novels are worth reading.
War and Peace is a novel.
∴*War and Peace* is worth reading.

20. If I cut my night class, then I will go to the party. I went to the party. Therefore, I cut my night class.

CHAPTER 4

Numeration Systems and Number Theory

We start this chapter with an examination of several numeration systems. A working knowledge of these numeration systems will enable you better to understand and appreciate the advantages of our current Hindu-Arabic numeration system.

The last two sections of this chapter cover prime numbers and topics from the field of number theory. Many of the concepts in number theory are easy to comprehend but difficult, or impossible, to prove. The mathematician Karl Friedrich Gauss (1777–1855) remarked that "it is just this which gives the higher arithmetic (number theory) that magical charm which has made it the favorite science of the greatest mathematicians, not to mention its inexhaustible wealth, wherein it so greatly surpasses other parts of mathematics." Gauss referred to mathematics as "the queen of the sciences," and he considered the field of number theory "the queen of mathematics."

There are many unsolved problems in the field of number theory. One unsolved problem, dating from the year 1742, is *Goldbach's Conjecture,* which states that every even number greater than 2 can be written as the sum of two prime numbers. This conjecture has yet to be proved or disproved, despite the efforts of the world's best mathematicians. A British publishing company has recently offered a $1 million prize to the first person who proves or disproves Goldbach's Conjecture. The company hopes that the prize money will entice young, mathematically talented people to work on the problem. This scenario is similar to the story line in the movie *Good Will Hunting,* in which a mathematics problem posted on a bulletin board attracts the attention of a yet-to-be-discovered math genius, played by Matt Damon.

For online student resources, visit this textbook's website at **math.college.hmco.com/students.**

SECTION 4.1 Early Numeration Systems

The Egyptian Numeration System

In mathematics, symbols that are used to represent numbers are called **numerals.** A number can be represented by many different numerals. For instance, the concept of "eightness" is represented by each of the following.

Hindu-Arabic: 8 Tally: 卌 ||| Roman: VIII

Chinese: 八 Egyptian: |||||||| Babylonian: 𒐜

A **numeration system** consists of a set of numerals and a method of arranging the numerals to represent numbers. The numeration system that most people use today is known as the *Hindu-Arabic numeration system.* It makes use of the 10 numerals 0, 1, 2, 3, 4, 5, 6, 7, 8, and 9. Before we examine the Hindu-Arabic numeration system in detail, it will be helpful to study some of the earliest numeration systems that were developed by the Egyptians, the Romans, and the Chinese.

The Egyptian numeration system uses pictorial symbols called **hieroglyphics** as numerals. The Egyptian hieroglyphic system is an **additive system** because any given number is written by using numerals whose sum equals the number. Table 4.1 gives the Egyptian hieroglyphics for powers of 10 from one to one million.

Table 4.1 *Egyptian Hieroglyphics for Powers of 10*

Hindu-Arabic Numeral	Egyptian Hieroglyphic	Description of Hieroglyphic
1	|	stroke
10	∩	heel bone
100	9	scroll
1000	𓆼	lotus flower
10,000	𓂭	pointing finger
100,000	𓆛	fish
1,000,000	𓁨	astonished person

To write the number 300, the Egyptians wrote the scroll hieroglyphic three times: 999. In the Egyptian hieroglyphic system, the order of the hieroglyphics is of no importance. Each of the following Egyptian numerals represents 321.

999∩∩|, ∩∩9|99, 9|∩9∩9, |∩999

EXAMPLE 1 ■ Write a Numeral Using Egyptian Hieroglyphics

Write 3452 using Egyptian hieroglyphics.

Solution

3452 = 3000 + 400 + 50 + 2. Thus the Egyptian numeral for 3452 is

𓆼𓆼𓆼𓍢𓍢𓍢𓍢𓎆𓎆𓎆𓎆𓎆𓏺𓏺

CHECK YOUR PROGRESS 1 Write 201,473 using Egyptian hieroglyphics.

Solution *See page S12.*

QUESTION *Do the Egyptian hieroglyphics* 𓍢𓍢𓎆𓏺 *and* 𓎆𓏺𓍢𓍢 *represent the same number?*

EXAMPLE 2 ■ Evaluate a Numeral Written Using Egyptian Hieroglyphics

Write 𓆐𓆐𓂭𓂭𓂭𓆼𓆼𓍢𓍢𓍢𓍢𓎆𓏺𓏺𓏺 as a Hindu-Arabic numeral.

Solution

(2 × 100,000) + (3 × 10,000) + (2 × 1000) + (4 × 100) + (1 × 10) + (3 × 1) = 232,413

CHECK YOUR PROGRESS 2 Write 𓁨𓆐𓆐𓆐𓂭𓆼𓆼𓆼𓆼𓍢𓍢𓍢𓎆𓎆𓏺𓏺𓏺𓏺𓏺𓏺 as a Hindu-Arabic numeral.

Solution *See page S12.*

historical note

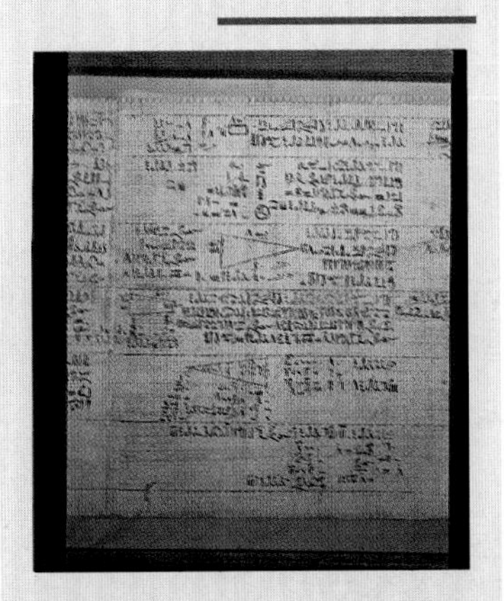

A portion of the Rhind papyrus

The Rhind papyrus is named after Alexander Henry Rhind, who purchased the papyrus in Egypt in A.D. 1858. Today the Rhind papyrus is preserved in the British Museum in London. ■

One of the earliest written documents of mathematics is the Rhind papyrus (see the figure at the left). This tablet was found in Egypt in A.D. 1858, but it is estimated that the writings date back to 1650 B.C. The Rhind papyrus contains 85 mathematical problems. Studying these problems has enabled mathematicians and historians to understand some of the mathematical procedures used in the early Egyptian numeration system.

The operation of addition with Egyptian hieroglyphics is a simple grouping process. In some cases the final sum can be simplified by replacing a group of hieroglyphics by a single hieroglyphic with a larger numeric value. This technique is illustrated in Example 3.

ANSWER *Yes, they both represent 211.*

EXAMPLE 3 ■ Use Egyptian Hieroglyphics to Find a Sum

Use Egyptian hieroglyphics to find 2452 + 1263.

Solution

The sum is found by combining the hieroglyphics.

$$\begin{array}{r} 2452 \\ +\ 1263 \\ \hline \end{array}$$

Replacing 10 heel bones with one scroll produces

or 3715

The sum is 3715.

CHECK YOUR PROGRESS 3 Use Egyptian hieroglyphics to find 23,341 + 10,562.

Solution *See page S12.*

In the Egyptian numeration system, subtraction is performed by removing some of the hieroglyphics from the larger numeral. In some cases it is necessary to "borrow," as shown in the next example.

TAKE NOTE

Five scrolls cannot be removed from two scrolls, so one lotus flower is replaced by ten scrolls, resulting in a total of twelve scrolls. Now five scrolls can be removed from twelve scrolls.

EXAMPLE 4 ■ Use Egyptian Hieroglyphics to Find a Difference

Use Egyptian hieroglyphics to find 332,246 − 101,512.

Solution

The numerical value of one lotus flower is equivalent to the numerical value of 10 scrolls. Thus

$$\begin{array}{r} 332{,}246 \\ -\ 101{,}512 \\ \hline \end{array}$$

The difference is 230,734.

CHECK YOUR PROGRESS 4 Use Egyptian hieroglyphics to find 61,432 − 45,121.

Solution *See page S12.*

Math Matters Early Egyptian Fractions

Evidence gained from the Rhind papyrus shows that the Egyptian method of calculating with fractions was much different from the methods we use today. All Egyptian fractions (except for $\frac{2}{3}$) were represented in terms of unit fractions, which are fractions of the form $\frac{1}{n}$, for some natural number $n > 1$. The Egyptians wrote these unit fractions by placing an oval over the numeral that represented the denominator. For example,

$$\overset{\bigcirc}{|||} = \frac{1}{3} \qquad \overset{\bigcirc}{\cap|||||} = \frac{1}{15}$$

If a fraction was not a unit fraction, then the Egyptians wrote the fraction as the sum of *distinct* unit fractions. For instance,

$\frac{2}{5}$ was written as the sum of $\frac{1}{3}$ and $\frac{1}{15}$.

Of course, $\frac{2}{5} = \frac{1}{5} + \frac{1}{5}$, but (for some mysterious reason) the early Egyptian numeration system didn't allow repetitions. The Rhind papyrus includes a table that shows how to write fractions of the form $\frac{2}{k}$, where k is an odd number from 5 to 101, in terms of unit fractions. Some of these are listed below.

$$\frac{2}{7} = \frac{1}{4} + \frac{1}{28} \qquad \frac{2}{11} = \frac{1}{6} + \frac{1}{66} \qquad \frac{2}{19} = \frac{1}{12} + \frac{1}{76} + \frac{1}{114}$$

Table 4.2 *Roman Numerals*

Hindu-Arabic Numeral	Roman Numeral
1	I
5	V
10	X
50	L
100	C
500	D
1000	M

The Roman Numeration System

The Roman numeration system was used in Europe during the reign of the Roman Empire. Today we still make limited use of Roman numerals on clock faces, on the cornerstones of buildings, and in numbering the volumes of periodicals and books. Table 4.2 shows the numerals used in the Roman numeration system. If the Roman numerals are listed so that each numeral has a larger value than the numeral to its right, then the value of the Roman numeral is found by adding the values of each numeral. For example,

$$\text{CLX} = 100 + 50 + 10 = 160$$

If a Roman numeral is repeated two or three times in succession, we add to determine its numerical value. For instance, XX = 10 + 10 = 20 and CCC = 100 + 100 + 100 = 300. Each of the numerals I, X, C, and M may be repeated up to three times. The numerals V, L, and D are not repeated.

Although the Roman numeration system is an additive system, it also incorporates a subtraction property. In the Roman numeration system, the value of a numeral is determined by adding the values of the numerals from left to right. However, if the value of a numeral is less than the value of the numeral to its right, the smaller value is subtracted from the next larger value. For instance, VI = 5 + 1 = 6; however, IV = 5 − 1 = 4. In the Roman numeration system the only numerals whose values can be subtracted from the value of the numeral to

historical note

The Roman numeration system evolved over a period of several years, and thus some Roman numerals displayed on ancient structures do not adhere to the basic rules given at the right. For instance, in the Colosseum in Rome (c. A.D. 80), the numeral XXVIIII appears above archway 29 instead of the numeral XXIX. ■

the right are I, X, and C. Also, the subtraction of these values is allowed only if the value of the numeral to the right is within two rows as shown in Table 4.2. That is, the value of the numeral to be subtracted must be *no less than* one-tenth of the value of the numeral it is to be subtracted from. For instance, XL = 40 and XC = 90, but XD does not represent 490 because the value of X is less than one-tenth the value of D. To write 490 using Roman numerals, we write CDXC.

A Summary of the Basic Rules Employed in the Roman Numeration System

I = 1, V = 5, X = 10, L = 50, C = 100, D = 500, M = 1000

1. If the numerals are listed so that each numeral has a larger value than the numeral to the right, then the value of the Roman numeral is found by adding the values of the numerals.
2. Each of the numerals, I, X, C, and M may be repeated up to three times. The numerals V, L, and D are not repeated. If a numeral is repeated two or three times in succession, we add to determine its numerical value.
3. The only numerals whose values can be subtracted from the value of the numeral to the right are I, X, and C. The value of the numeral to be subtracted must be no less than one-tenth of the value of the numeral to its right.

EXAMPLE 5 ■ Evaluate a Roman Numeral

Write DCIV as a Hindu-Arabic numeral.

Solution

Because the value of D is larger than the value of C, we add their numerical values. The value of I is less than the value of V, so we subtract the smaller value from the larger value. Thus

DCIV = (DC) + (IV) = (500 + 100) + (5 − 1) = 600 + 4 = 604

CHECK YOUR PROGRESS 5 Write MCDXLV as a Hindu-Arabic numeral.

Solution *See page S12.*

✔ TAKE NOTE

The spreadsheet program Excel has a function that converts Hindu-Arabic numerals to Roman numerals. In the Edit Formula dialogue box, type

= ROMAN(n),

where n is the number you wish to convert to a Roman numeral.

EXAMPLE 6 ■ Write a Hindu-Arabic Numeral as a Roman Numeral

Write 579 as a Roman numeral.

Solution

579 = 500 + 50 + 10 + 10 + 9

In Roman numerals 9 is written as IX. Thus 579 = DLXXIX.

CHECK YOUR PROGRESS 6 Write 473 as a Roman numeral.

Solution *See page S12.*

In the Roman numeration system, a bar over a numeral is used to denote a value 1000 times the value of the numeral. For instance,

$\overline{\text{V}} = 5 \times 1000 = 5000 \qquad \overline{\text{IV}}\text{LXX} = (4 \times 1000) + 70 = 4070$

TAKE NOTE

The method of writing a bar over a numeral should be used only to write Roman numerals that cannot be written using the basic rules. For instance, the Roman numeral for 2003 is MMIII, not $\overline{\text{I}}\text{III}$.

EXAMPLE 7 ■ Convert Between Roman Numerals and Hindu-Arabic Numerals

a. Write $\overline{\text{IV}}\text{DLXXII}$ as a Hindu-Arabic numeral.

b. Write 6125 as a Roman numeral.

Solution

a.
$$\begin{aligned}\overline{\text{IV}}\text{DLXXII} &= (\overline{\text{IV}}) + (\text{DLXXII}) \\ &= (4 \times 1000) + (572) \\ &= 4572\end{aligned}$$

b. The Roman numeral 6 is written VI and 125 is written as CXXV. Thus in Roman numerals 6125 is $\overline{\text{VI}}\text{CXXV}$.

CHECK YOUR PROGRESS 7

a. Write $\overline{\text{VII}}\text{CCLIV}$ as a Hindu-Arabic numeral.

b. Write 8070 as a Roman numeral.

Solution *See page S12.*

Excursion

A Rosetta Tablet for the Traditional Chinese Numeration System

The Rosetta Stone

Most of the knowledge we have gained about early numeration systems has been obtained from inscriptions found on ancient tablets or stones. The information provided by these inscriptions has often been difficult to interpret. For several centuries archeologists had little success in interpreting the Egyptian hieroglyphics they had discovered. Then, in 1799, a group of French military engineers discovered a basalt stone near Rosetta in the Nile delta. This stone, which we now call the Rosetta Stone, has an inscription in three scripts: Greek, Egyptian Demotic, and Egyptian hieroglyphic. It was soon discovered that all three scripts contained the same message. The Greek script was easy to translate, and from its translation, clues were uncovered that enabled scholars to translate many of the documents that up to that time had been unreadable.

Pretend that you are an archeologist. Your team has just discovered an old tablet that displays Roman numerals and traditional Chinese numerals. It also provides hints in the form of a crossword puzzle about the traditional Chinese numeration system. Study the inscriptions on the following tablet and then complete the Excursion Exercises that follow.

(continued)

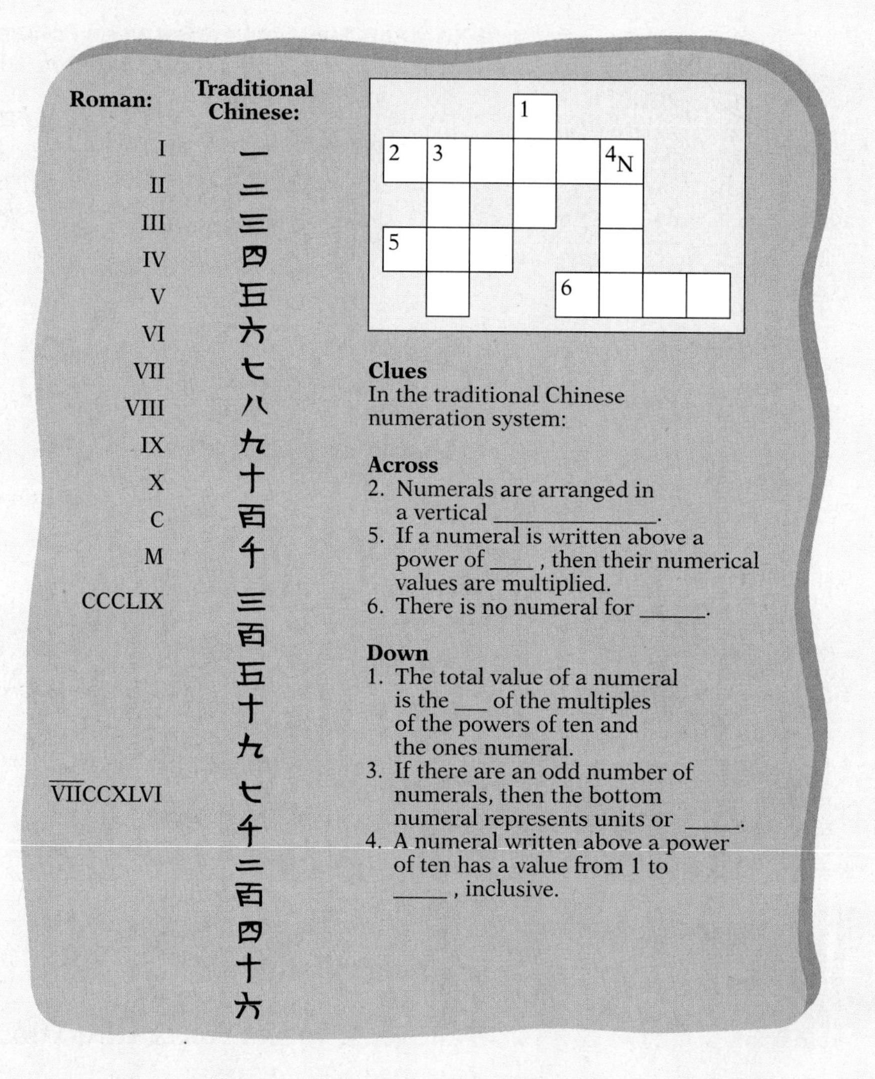

Excursion Exercises

1. Complete the crossword puzzle shown on the above tablet.

2. Write 26 as a traditional Chinese numeral.

3. Write 357 as a traditional Chinese numeral.

4. Write the Hindu-Arabic numeral given by each of the following traditional Chinese numerals.

a. 八
百
九
十
六

b. 二
千
四
百
六
十
五

(continued)

5. **a.** How many Hindu-Arabic numerals are required to write four thousand five hundred twenty-eight?
   b. How many traditional Chinese numerals are required to write four thousand five hundred twenty-eight?

6. The traditional Chinese numeration system is no longer in use. Give a reason that may have contributed to its demise.

Exercise Set 4.1

In Exercises 1–12, write each Hindu-Arabic numeral using Egyptian hieroglyphics.

1. 46
2. 82
3. 103
4. 157
5. 2568
6. 3152
7. 23,402
8. 15,303
9. 65,800
10. 43,217
11. 1,405,203
12. 653,271

In Exercises 13–24, write each Egyptian numeral as a Hindu-Arabic numeral.

13. 𓆼𓆼𓍢𓎆𓎆𓎆𓏺𓏺𓏺𓏺
14. 𓆼𓍢𓍢𓍢𓎆𓏺𓏺
15. 𓍢𓍢𓍢𓍢𓎆𓎆𓏺𓏺𓏺
𓍢𓍢𓍢𓍢𓎆𓎆𓏺𓏺
16. 𓍢𓎆𓎆𓎆𓎆𓏺𓏺𓏺𓏺
𓍢𓎆𓎆𓎆𓎆𓏺𓏺𓏺
17. 𓆼𓍢𓍢𓎆𓎆𓎆𓏺𓏺
18. 𓂭𓂭 𓍢𓍢𓎆𓏺
𓆐𓆐𓂭𓆼𓆼𓍢𓎆𓎆𓏺𓏺
19. 𓆐𓆐𓂭𓂭𓆼𓎆𓏺
20. 𓂭𓂭𓂭𓆼𓆼𓍢𓍢𓍢𓎆𓎆𓏺𓏺𓏺
21. 𓂭𓂭𓂭𓆼𓆼𓆼𓍢𓍢𓍢𓎆𓎆𓎆𓏺𓏺𓏺𓏺𓏺
𓂭𓂭𓂭𓆼𓆼𓍢𓍢𓍢𓍢𓎆𓎆𓎆𓏺𓏺𓏺𓏺
22. 𓁨𓆐𓆐𓂭𓂭𓆼𓆼𓆼𓍢𓍢𓍢𓍢𓎆
𓁨𓆐𓆐𓂭 𓆼𓆼𓆼𓍢𓍢𓍢𓎆𓏺𓏺𓏺
23. 𓁨𓁨𓁨𓆐𓆼𓍢𓍢𓏺𓏺𓏺
𓁨𓁨𓂭𓂭 𓆼𓍢𓍢𓏺𓏺𓏺
24. 𓁨𓆐𓆐𓆼𓍢𓍢
𓆐𓆐
𓁨𓆐𓆐𓍢𓍢

In Exercises 25–32, use Egyptian hieroglyphics to find each sum or difference.

25. 51 + 43
26. 67 + 58
27. 231 + 435
28. 623 + 124
29. 83 − 51
30. 94 − 23
31. 254 − 198
32. 640 − 278

In Exercises 33–44, write each Roman numeral as a Hindu-Arabic numeral.

33. DCL
34. MCX
35. MCDIX
36. MDCCII
37. MCCXL
38. MMDCIV
39. DCCCXL
40. CDLV
41. $\overline{\text{IX}}$XLIV
42. $\overline{\text{VII}}$DXVII
43. $\overline{\text{XI}}$CDLXI
44. $\overline{\text{IV}}$CCXXI

In Exercises 45–56, write each Hindu-Arabic numeral as a Roman numeral.

45. 157
46. 231
47. 542
48. 783
49. 1197
50. 1039
51. 787
52. 1343
53. 683
54. 959
55. 6898
56. 4357

Egyptian Multiplication The Rhind papyrus contains problems that show a *doubling procedure* used by the Egyptians to find the product of two whole numbers. The following examples illustrate this doubling procedure. In the examples we have used Hindu-Arabic numerals so that you can concentrate on the doubling procedure and not be distracted by the Egyptian hieroglyphics. The first example determines the product 5×27 by computing two successive doublings of 27 and then forming the sum of the blue numbers in the rows marked with a check. Note that the rows marked with a check show that one 27 is 27 and four 27's is 108. Thus five 27's is the sum of 27 and 108, or 135.

√ 1	27	double
2	54	double
√ 4	108	
5	135	← This sum is the product of 5 and 27.

In the next example, we use the Egyptian doubling procedure to find the product of 35 and 94. Because the sum of 1, 2, and 32 is 35, we add only the blue numbers in the rows marked with a check to find that $35 \times 94 = 94 + 188 + 3008 = 3290$.

√ 1	94	double
√ 2	188	double
4	376	double
8	752	double
16	1504	double
√ 32	3008	
35	3290	← This sum is the product of 35 and 94.

In Exercises 57–64, use the Egyptian doubling procedure to find each product.

57. 8×63

58. 4×57

59. 7×29

60. 9×33

61. 17×35

62. 26×43

63. 23×108

64. 72×215

Extensions

CRITICAL THINKING

65. a. State a reason why you might prefer to use the Egyptian hieroglyphic numeration system rather than the Roman numeration system.

b. State a reason why you might prefer to use the Roman numeration system rather than the Egyptian hieroglyphic numeration system.

66. What is the largest number that can be written using Roman numerals without using the bar over a numeral or the subtraction property?

EXPLORATIONS

67. **The Ionic Greek Numeration System** The Ionic Greek numeration system assigned numerical values to the letters of the Greek alphabet. Research the Ionic Greek numeration system and write a report that explains this numeration system. Include information about some of the advantages and disadvantages of this system compared with our present Hindu-Arabic numeration system.

68. Some clock faces display the Roman numeral IV as IIII. Research this topic and write a few paragraphs that explain at least three possible reasons for this variation.

69. **The Method of False Position** The Rhind papyrus (see page 179) contained solutions to several mathematical problems. Some of these solutions made use of a procedure called the *method of false position.* Research the method of false position and write a report that explains this method. In your report, include a specific mathematical problem and its solution by the method of false position.

SECTION 4.2 Place-Value Systems

Expanded Form

historical note

Abu Ja'far Muhammad ibn Musa al'Khwarizmi (ca. A.D. 790–850) Al'Khwarizmi (ăl'khwăhr'ĭz mee) produced two important texts. One of these texts advocated the use of the Hindu-Arabic numeration system. The twelfth century Latin translation of this book is called *Liber Algoritmi de Numero Indorum,* or *Al-Khwarizmi on the Hindu Art of Reckoning.* In Europe, the people who favored the adoption of the Hindu-Arabic numeration system became known as *algorists* because of Al'Khwarizmi's *Liber Algoritmi de Numero Indorum* text. The Europeans who opposed the Hindu-Arabic system were called *abacists.* They advocated the use of Roman numerals and often performed computations with the aid of an abacus. ■

The most common numeration system used by people today is the Hindu-Arabic numeration system. It is called the Hindu-Arabic system because it was first developed in India (around A.D. 800) and then refined by the Arabs. It makes use of the 10 symbols 0, 1, 2, 3, 4, 5, 6, 7, 8, and 9. The reason for the 10 symbols, called *digits,* is related to the fact that we have 10 fingers. The Hindu-Arabic numeration system is also called the *decimal system,* where the word *decimal* is a derivation of the Latin word *decem,* which means "ten."

One important feature of the Hindu-Arabic numeration system is that it is a *place-value* or *positional-value system.* This means that the numerical value of each digit in a Hindu-Arabic numeral depends on its *place* or *position* in the numeral. For instance, the 3 in 31 represents 3 tens, whereas the 3 in 53 represents 3 ones. The Hindu-Arabic numeration system is a **base ten numeration system** because the place values are the powers of 10:

$$\ldots, 10^5, 10^4, 10^3, 10^2, 10^1, 10^0$$

The place value associated with the nth digit of a numeral (counting from right to left) is 10^{n-1}. For instance, in the numeral 7532, the 7 is the fourth digit from the right and is in the $10^{4-1} = 10^3$, or thousands', place. The numeral 2 is the first digit from the right and is in the $10^{1-1} = 10^0$, or ones', place. The *indicated sum* of each digit of a numeral multiplied by its respective place value is called the **expanded form** of the numeral.

EXAMPLE 1 ■ Write a Numeral in its Expanded Form

Write 4672 in expanded form.

Solution

$$\begin{aligned} 4672 &= 4000 + 600 + 70 + 2 \\ &= (4 \times 1000) + (6 \times 100) + (7 \times 10) + (2 \times 1) \end{aligned}$$

The above expanded form can also be written as

$$(4 \times 10^3) + (6 \times 10^2) + (7 \times 10^1) + (2 \times 10^0)$$

CHECK YOUR PROGRESS 1 Write 17,325 in expanded form.

Solution *See page S13.*

If a number is written in expanded form, it can be simplified to its ordinary decimal form by performing the indicated operations. The *Order of Operations Agreement* states that we should first perform the exponentiations, then perform the multiplications, and finish by performing the additions.

EXAMPLE 2 ■ Simplify a Number Written in Expanded Form

Simplify: $(2 \times 10^3) + (7 \times 10^2) + (6 \times 10^1) + (3 \times 10^0)$

Solution

$$\begin{aligned}&(2 \times 10^3) + (7 \times 10^2) + (6 \times 10^1) + (3 \times 10^0)\\ &\quad = (2 \times 1000) + (7 \times 100) + (6 \times 10) + (3 \times 1)\\ &\quad = 2000 + 700 + 60 + 3\\ &\quad = 2763\end{aligned}$$

CHECK YOUR PROGRESS 2 Simplify:

$$(5 \times 10^4) + (9 \times 10^3) + (2 \times 10^2) + (7 \times 10^1) + (4 \times 10^0)$$

Solution *See page S13.*

In the next few examples, we make use of the expanded form of a numeral to compute sums and differences. An examination of these examples will help you better understand the computational algorithms used in the Hindu-Arabic numeration system.

EXAMPLE 3 ■ Use Expanded Forms to Find a Sum

Use expanded forms of 26 and 31 to find their sum.

Solution

$$\begin{aligned}26 &= (2 \times 10) + 6\\ +\ 31 &= (3 \times 10) + 1\\ \hline &\quad (5 \times 10) + 7 = 50 + 7 = 57\end{aligned}$$

CHECK YOUR PROGRESS 3 Use expanded forms to find the sum of 152 and 234.

Solution *See page S13.*

TAKE NOTE

From the expanded forms in Example 4, note that 12 is 1 ten and 2 ones. The 1 ten is added to the 13 tens, resulting in a total of 14 tens. When we add columns of numbers, this is shown as "carry a 1." Because the 1 is placed in the tens column, we are actually adding 10.

$$\begin{array}{r} 1 \\ 85 \\ +\ 57 \\ \hline 142 \end{array}$$

If the expanded form of a sum contains one or more powers of 10 that have multipliers larger than 9, then we simplify by rewriting the sum with multipliers that are less than or equal to 9. This process is known as *carrying.*

EXAMPLE 4 ■ Use Expanded Forms to Find a Sum

Use expanded forms of 85 and 57 to find their sum.

Solution

$$\begin{aligned}85 &= (8 \times 10) + 5\\ +\ 57 &= (5 \times 10) + 7\\ \hline &\quad (13 \times 10) + 12\\ &\quad (10 + 3) \times 10 + 10 + 2\\ &\quad 100 + 30 + 10 + 2 = 100 + 40 + 2 = 142\end{aligned}$$

CHECK YOUR PROGRESS 4 Use expanded forms to find the sum of 147 and 329.

Solution *See page S13.*

In the next example, we use the expanded forms of numerals to analyze the concept of "borrowing" in a subtraction problem.

TAKE NOTE

From the expanded forms in Example 5, note that we "borrowed" 1 hundred as 10 tens. This explains how we show borrowing when numbers are subtracted using place value form.

$$\begin{array}{r} \overset{3}{\cancel{4}}{}^{1}57 \\ -\ 2\ 83 \\ \hline 1\ 74 \end{array}$$

EXAMPLE 5 ■ Use Expanded Forms to Find a Difference

Use the expanded forms of 457 and 283 to find 457 − 283.

Solution

$$\begin{aligned} 457 &= (4 \times 100) + (5 \times 10) + 7 \\ -\ 283 &= (2 \times 100) + (8 \times 10) + 3 \end{aligned}$$

At this point, this example is similar to Example 4 in Section 4.1. We cannot remove 8 tens from 5 tens, so 1 hundred is replaced by 10 tens.

$$\begin{aligned} 457 &= (4 \times 100) + (5 \times 10) + 7 \\ &= (3 \times 100) + (10 \times 10) + (5 \times 10) + 7 \\ &= (3 \times 100) + (15 \times 10) + 7 \end{aligned}$$

• $4 \times 100 = 3 \times 100 + 100$
$= 3 \times 100 + 10 \times 10$

We can now remove 8 tens from 15 tens.

$$\begin{aligned} 457 &= (3 \times 100) + (15 \times 10) + 7 \\ -\ 283 &= (2 \times 100) + (8 \times 10) + 3 \\ \hline &= (1 \times 100) + (7 \times 10) + 4 = 100 + 70 + 4 = 174 \end{aligned}$$

CHECK YOUR PROGRESS 5 Use expanded forms to find the difference 382 − 157.

Solution *See page S13.*

The Babylonian Numeration System

The Babylonian numeration system uses a base of 60. The place values in the Babylonian system are given in the following table.

Table 4.3 *Place Values in the Babylonian Numeration System*

. . .	60^3 = 216,000	60^2 = 3600	60^1 = 60	60^0 = 1

The Babylonians recorded their numerals on damp clay using a wedge-shaped stylus. A vertical wedge shape represented one unit and a sideways "vee" shape represented 10 units.

𒁹 1

𒌋 10

To represent a number smaller than 60, the Babylonians used an *additive* feature similar to that used by the Egyptians. For example, the Babylonian numeral for 32 is

<<<▼▼

For the number 60 and larger numbers, the Babylonians left a small space between groups of symbols to indicate a different place value. This procedure is illustrated in the following example.

EXAMPLE 6 ■ Write a Babylonian Numeral as a Hindu-Arabic Numeral

Write ▼ <<<▼ <<▼▼▼▼▼ as a Hindu-Arabic numeral.

Solution

▼ — 1 group of 60^2; <<<▼ — 31 groups of 60; <<▼▼▼▼▼ — 25 ones

$$= (1 \times 60^2) + (31 \times 60) + (25 \times 1)$$
$$= 3600 + 1860 + 25$$
$$= 5485$$

CHECK YOUR PROGRESS 6 Write <<▼ ▼▼▼▼▼ <<<▼▼▼▼ as a Hindu-Arabic numeral.

Solution *See page S13.*

QUESTION *In the Babylonian numeration system, does* ▼▼ = ▼ ▼?

In the next example we illustrate a division process that can be used to convert Hindu-Arabic numerals to Babylonian numerals.

EXAMPLE 7 ■ Write a Hindu-Arabic Numeral as a Babylonian Numeral

Write 8503 as a Babylonian numeral.

Solution
The Babylonian numeration system uses place values of

$60^0, 60^1, 60^2, 60^3, \ldots.$

Evaluating the powers produces

$1, 60, 3600, 216{,}000, \ldots$

The largest of these powers that is contained in 8503 is 3600. One method of finding how many groups of 3600 are in 8503 is to divide 3600 into 8503. Refer to the

---

ANSWER *No.* ▼▼ $= 2$, *whereas* ▼ ▼ $= (1 \times 60) + (1 \times 1) = 61$.

first division shown below. Now divide to determine how many groups of 60 are contained in the remainder 1303.

```
         2             21
3600)8503         60)1303
      7200            120
      ----            ---
      1303            103
                       60
                      ---
                       43
```

The above computations show that 8503 consists of 2 groups of 3600 and 21 groups of 60, with 43 left over. Thus

$$8503 = (2 \times 60^2) + (21 \times 60) + (43 \times 1)$$

As a Babylonian numeral, 8503 is written

𒁹𒁹 𒌋𒌋𒁹 𒌋𒌋𒌋𒌋𒁹𒁹𒁹

CHECK YOUR PROGRESS 7 Write 12,578 as a Babylonian numeral.

Solution *See page S13.*

In Example 8 we find the sum of two Babylonian numerals. If a numeral for any power of 60 is larger than 59, then simplify by decreasing that numeral by 60 and increasing the place value to its left by 1.

EXAMPLE 8 ■ Find the Sum of Babylonian Numerals

Find the sum:

𒌋𒌋𒁹 𒌋𒌋𒌋𒌋𒁹𒁹
\+ 𒌋𒌋𒌋𒌋𒁹𒁹𒁹 𒌋𒌋𒌋𒁹𒁹𒁹

Solution

		𒌋𒌋𒁹	𒌋𒌋𒌋𒌋𒁹𒁹	
+		𒌋𒌋𒌋𒌋𒁹𒁹𒁹	𒌋𒌋𒌋𒁹𒁹𒁹	
=		𒌋𒌋𒌋𒌋𒌋𒌋𒁹𒁹𒁹𒁹	𒌋𒌋𒌋𒌋𒌋𒌋𒌋𒁹𒁹𒁹𒁹𒁹	• Combine the symbols for each place value.
=		𒌋𒌋𒌋𒌋𒌋𒌋𒁹𒁹𒁹𒁹𒁹	𒌋𒁹𒁹𒁹𒁹𒁹	• Take away 60 from the ones' place and add 1 to the 60s' place.
=	𒁹	𒁹𒁹𒁹𒁹𒁹	𒌋𒁹𒁹𒁹𒁹𒁹	• Take away 60 from the 60s' place and add 1 to the 60^2 place.

𒌋𒌋𒁹 𒌋𒌋𒌋𒌋𒁹𒁹 + 𒌋𒌋𒌋𒌋𒁹𒁹𒁹 𒌋𒌋𒌋𒁹𒁹𒁹 = 𒁹 𒁹𒁹𒁹𒁹𒁹 𒌋𒁹𒁹𒁹𒁹𒁹

CHECK YOUR PROGRESS 8 Find the sum:

𒌋𒌋𒌋𒁹𒁹 𒌋𒌋𒌋𒌋𒌋𒁹𒁹𒁹𒁹𒁹𒁹
\+ 𒌋𒌋𒌋𒌋𒁹𒁹𒁹𒁹 𒌋𒌋𒌋𒌋𒌋𒁹𒁹𒁹𒁹𒁹𒁹𒁹

Solution *See page S13.*

Math Matters **Zero as a Placeholder and as a Number**

When the Babylonian numeration system first began to develop around 1700 B.C., it did not make use of a symbol for zero. The Babylonians merely used an empty space to indicate that a place value was missing. This procedure of "leaving a space" can be confusing. How big is an empty space? Is that one empty space or two empty spaces? Around 300 B.C., the Babylonians started to use the symbol to indicate that a particular place value was missing. For instance, represented $(2 \times 60^2) + (11 \times 1) = 7211$. In this case the zero placeholder indicates that there are no 60's. There is evidence that although the Babylonians used the zero placeholder, they did not use the number zero.

The Mayan Numeration System

The Mayan civilization existed in the Yucatan area of southern Mexico and in Guatemala, Belize, and parts of El Salvador and Honduras. It started as far back as 9000 B.C. and reached its zenith during the period from A.D. 200 to A.D. 900. Among their many accomplishments, the Maya are best known for their complex hieroglyphic writing system, their sophisticated calendars, and their remarkable numeration system.

The Maya used three calendars—the solar calendar, the ceremonial calendar, and the Venus calendar. The solar calendar consisted of about 365.24 days. Of these, 360 days were divided into 18 months, each with 20 days. The Mayan numeration system was strongly influenced by this solar calendar, as evidenced by the use of the numbers 18 and 20 in determining place values. See Table 4.4.

TAKE NOTE

Observe that the place values used in the Mayan numeration system are not all powers of 20.

Table 4.4 *Place Values in the Mayan Numeration System*

. . .	$18 \times 20^3 = 144{,}000$	$18 \times 20^2 = 7200$	$18 \times 20^1 = 360$	$20^1 = 20$	$20^0 = 1$

The Mayan numeration system was one of the first systems to use a symbol for zero as a placeholder. The Mayan numeration system used only three symbols. A dot was used to represent 1, a horizontal bar represented 5, and a conch shell represented 0. The following table shows how the Maya used a combination of these three symbols to write the whole numbers from 0 to 19. Note that each numeral contains at most four dots and at most three horizontal bars.

Table 4.5 *Mayan Numerals*

0	1	2	3	4	5	6	7	8	9
10	11	12	13	14	15	16	17	18	19

To write numbers larger than 19, the Maya used a vertical arrangement with the largest place value at the top. The following example illustrates the process of converting a Mayan numeral to a Hindu-Arabic numeral.

TAKE NOTE

In the Mayan numeration system,

does not represent 3 because the three dots are not all in the same row. The Mayan numeral

represents one group of 20 and two ones, or 22.

EXAMPLE 9 ■ Write a Mayan Numeral as a Hindu-Arabic Numeral

Write each of the following as a Hindu-Arabic numeral.

a.

b.

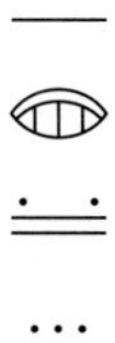

Solution

a.

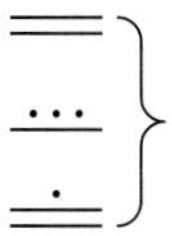

$10 \times 360 = 3600$

$8 \times 20 = 160$

$11 \times 1 = +11$

3771

b.

$5 \times 7200 = 36{,}000$

$0 \times 360 = 0$

$12 \times 20 = 240$

$3 \times 1 = +3$

36,243

CHECK YOUR PROGRESS 9 Write each of the following as a Hindu-Arabic numeral.

a.

b.

Solution *See page S13.*

In the next example, we illustrate how the concept of place value is used to convert Hindu-Arabic numerals to Mayan numerals.

EXAMPLE 10 ■ Write a Hindu-Arabic Numeral as a Mayan Numeral

Write 7495 as a Mayan numeral.

Solution

The place values used in the Mayan numeration system are

$$20^0, 20^1, 18 \times 20^1, 18 \times 20^2, 18 \times 20^3, \ldots$$

or

$$1, 20, 360, 7200, 144{,}000, \ldots$$

Removing 1 group of 7200 from 7495 leaves 295. No groups of 360 can be obtained from 295, so we divide 295 by the next smaller place value of 20 to find that 295 equals 14 groups of 20 with 15 left over.

$$\begin{array}{r} 1 \\ 7200\overline{)7495} \\ \underline{7200} \\ 295 \end{array} \qquad \begin{array}{r} 14 \\ 20\overline{)295} \\ \underline{20} \\ 95 \\ \underline{80} \\ 15 \end{array}$$

Thus

$$7495 = (1 \times 7200) + (0 \times 360) + (14 \times 20) + (15 \times 1)$$

In Mayan numerals, 7495 is written as

CHECK YOUR PROGRESS 10 Write 11,480 as a Mayan numeral.

Solution *See page S13.*

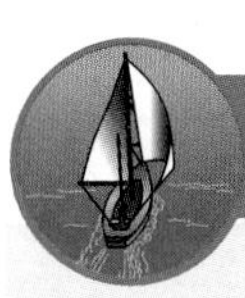

Excursion

Subtraction via the Nines Complement and the End-Around Carry

In the subtraction 5627 − 2564 = 3063, the number 5627 is called the *minuend,* 2564 is called the *subtrahend,* and 3063 is called the *difference.* In the Hindu-Arabic base ten system, subtraction can be performed by a process that involves addition and the *nines complement* of the subtrahend. The **nines complement** of a single digit n is the number $9 - n$. For instance, the nines complement of 3 is 6, the nines complement of 1 is 8, and the nines complement of 0 is 9. The nines complement of a number with more than one digit is the number that is formed by taking the nines complement of each digit. The nines complement of 25 is 74 and the nines complement of 867 is 132.

Subtraction by Using the Nines Complement and the End-Around Carry

To subtract by using the nines complement:

1. Add the nines complement of the subtrahend to the minuend.
2. Take away 1 from the leftmost digit and add 1 to the units digit. This is referred to as the end-around carry procedure.

(continued)

The following example illustrates the process of subtracting 2564 from 5627 by using the nines complement.

$$\begin{array}{r} 5627 \\ -\ 2564 \\ \hline \end{array}$$ **Minuend**, **Subtrahend**

$$\begin{array}{r} 5627 \\ +\ 7435 \\ \hline 13062 \end{array}$$

Minuend

Replace the subtrahend with the nines complement of the subtrahend and add.

$$\begin{array}{r} 13062 \\ +\ \quad 1 \\ \hline 3063 \end{array}$$

Take away 1 from the leftmost digit and add 1 to the units digit. This is the end-around carry procedure.

Thus

$$\begin{array}{r} 5627 \\ -\ 2564 \\ \hline 3063 \end{array}$$

If the subtrahend has fewer digits than the minuend, leading zeros should be inserted in the subtrahend so that it has the same number of digits as the minuend. This process is illustrated below for 2547 − 358.

$$\begin{array}{r} 2547 \\ -\ 358 \\ \hline \end{array}$$ **Minuend**, **Subtrahend**

$$\begin{array}{r} 2547 \\ -\ 0358 \\ \hline \end{array}$$ **Insert a leading zero.**

$$\begin{array}{r} 2547 \\ +\ 9641 \\ \hline 12188 \end{array}$$

Minuend

Nines complement of subtrahend

$$\begin{array}{r} 12188 \\ +\ \quad 1 \\ \hline 2189 \end{array}$$

Take away 1 from the leftmost digit and add 1 to the units digit.

Verify that 2189 is the correct difference.

Excursion Exercises

For Exercises 1–6, use the nines complement of the subtrahend to find the indicated difference.

1. 724 − 351

2. 2405 − 1608

3. 91,572 − 7824

4. 214,577 − 48,231

5. 3,156,782 − 875,236

6. 54,327,105 − 7,678,235

7.  Explain why the nines complement and the end-around carry procedure produce the correct answer to a subtraction problem.

Exercise Set 4.2

In Exercises 1–8, write each numeral in its expanded form.

1. 48
2. 93
3. 420
4. 501
5. 6803
6. 9045
7. 10,208
8. 67,482

In Exercises 9–16, simplify each expansion.

9. $(4 \times 10^2) + (5 \times 10^1) + (6 \times 10^0)$
10. $(7 \times 10^2) + (6 \times 10^1) + (3 \times 10^0)$
11. $(5 \times 10^3) + (0 \times 10^2) + (7 \times 10^1) + (6 \times 10^0)$
12. $(3 \times 10^3) + (1 \times 10^2) + (2 \times 10^1) + (8 \times 10^0)$
13. $(3 \times 10^4) + (5 \times 10^3) + (4 \times 10^2) + (0 \times 10^1) + (7 \times 10^0)$
14. $(2 \times 10^5) + (3 \times 10^4) + (0 \times 10^3) + (6 \times 10^2) + (7 \times 10^1) + (5 \times 10^0)$
15. $(6 \times 10^5) + (8 \times 10^4) + (3 \times 10^3) + (0 \times 10^2) + (4 \times 10^1) + (0 \times 10^0)$
16. $(5 \times 10^7) + (3 \times 10^6) + (0 \times 10^5) + (0 \times 10^4) + (7 \times 10^3) + (9 \times 10^2) + (0 \times 10^1) + (2 \times 10^0)$

In Exercises 17–22, use expanded forms to find each sum.

17. 35 + 41
18. 42 + 56
19. 257 + 138
20. 352 + 461
21. 1023 + 1458
22. 3567 + 2651

In Exercises 23–28, use expanded forms to find each difference.

23. 62 − 35
24. 193 − 157
25. 4725 − 1362
26. 85,381 − 64,156
27. 23,168 − 12,857
28. 59,163 − 47,956

In Exercises 29–36, write each Babylonian numeral as a Hindu-Arabic numeral.

29. 𒌋𒌋𒁹𒁹𒁹
30. 𒌋𒌋𒌋𒌋𒁹𒁹𒁹𒁹𒁹
31. 𒁹 𒌋𒌋𒌋𒁹𒁹𒁹𒁹𒁹𒁹𒁹
32. 𒌋𒁹𒁹 𒌋𒌋𒁹𒁹𒁹𒁹𒁹𒁹
33. 𒌋𒌋 𒁹𒁹 𒌋𒁹𒁹𒁹
34. 𒌋𒌋𒁹 𒌋𒁹 𒌋𒁹𒁹
35. 𒌋 𒁹𒁹𒁹 𒌋𒁹 𒁹𒁹𒁹𒁹𒁹𒁹
36. 𒌋𒌋𒁹 𒌋𒁹 𒁹 𒌋𒌋𒌋𒁹𒁹𒁹𒁹

In Exercises 37–46, write each Hindu-Arabic numeral as a Babylonian numeral.

37. 42
38. 57
39. 128
40. 540
41. 5678
42. 7821
43. 10,584
44. 12,687
45. 21,345
46. 24,567

In Exercises 47–52, find the sum of the Babylonian numerals.

47. 𒌋𒌋𒌋𒌋𒁹𒁹𒁹𒁹𒁹
+ 𒌋𒌋𒁹𒁹𒁹

48. 𒌋𒌋𒌋𒌋𒁹𒁹𒁹𒁹𒁹𒁹𒁹𒁹𒁹
+ 𒌋𒌋𒌋𒁹𒁹𒁹

49. 𒌋𒌋𒌋𒁹𒁹𒁹 𒌋𒌋𒌋𒌋𒁹𒁹
+ 𒌋𒌋𒌋𒁹𒁹 𒌋𒌋𒁹

50. 𒌋𒌋𒌋𒌋𒁹𒁹 𒌋𒌋𒌋𒌋𒌋𒁹𒁹
+ 𒌋𒌋𒌋𒁹 𒌋𒌋𒌋𒁹𒁹𒁹

51. 𒌋 𒌋𒌋𒌋𒁹 𒌋𒌋𒌋𒌋𒁹𒁹𒁹
+ 𒁹 𒌋𒌋𒁹 𒌋𒌋𒌋𒁹𒁹

52. 𒌋𒌋 𒌋𒌋𒌋𒁹𒁹𒁹 𒌋𒌋𒌋𒁹𒁹𒁹𒁹𒁹𒁹
+ 𒌋𒁹 𒌋𒌋𒌋𒁹𒁹 𒌋𒌋𒌋𒌋𒁹𒁹𒁹𒁹

In Exercises 53–60, write each Mayan numeral as a Hindu-Arabic numeral.

53.

54.

55.

56.

57.

58.

59.

60.

In Exercises 61–68, write each Hindu-Arabic numeral as a Mayan numeral.

61. 137

62. 253

63. 948

64. 1265

65. 1693

66. 2728

67. 7432

68. 8654

Extensions

CRITICAL THINKING

69. a. State a reason why you might prefer to use the Babylonian numeration system instead of the Mayan numeration system.

b. State a reason why you might prefer to use the Mayan numeration system instead of the Babylonian numeration system.

70. Explain why it might be easy to mistake the number 122 for the number 4 when 122 is written as a Babylonian numeral.

EXPLORATIONS

71. **A Base Three Numeration System** A student has created a *base three* numeration system. The student has named this numeration system ZUT because Z, U, and T are the symbols used in this system: Z represents 0, U represents 1, and T represents 2. The place values in this system are: ..., $3^3 = 27$, $3^2 = 9$, $3^1 = 3$, $3^0 = 1$.

Write each ZUT numeral as a Hindu-Arabic numeral.

a. TU **b.** TZT **c.** UZTT

Write each Hindu-Arabic numeral as a ZUT numeral.

d. 37 **e.** 87 **f.** 144

SECTION 4.3 Different Base Systems

Converting Non-Base-Ten Numerals to Base Ten

> **TAKE NOTE**
>
> Recall that in the expression
>
> b^n
>
> *b* is the *base,* and *n* is the *exponent.*

Recall that the Hindu-Arabic numeration system is a base ten system because its place values

$$\ldots, 10^5, 10^4, 10^3, 10^2, 10^1, 10^0$$

all have 10 as their base. The Babylonian numeration system is a base sixty system because its place values

$$\ldots, 60^5, 60^4, 60^3, 60^2, 60^1, 60^0$$

all have 60 as their base. In general, a base *b* (where *b* is a natural number greater than 1) numeration system has place values of

$$\ldots, b^5, b^4, b^3, b^2, b^1, b^0$$

Many people think that our base ten numeration system was chosen because it is the easiest to use, but this is not the case. In reality most people find it easier to use our base ten system only because they have had a great deal of experience with the base ten system and have not had much experience with non-base-ten systems. In this section, we examine some non-base-ten numeration systems. To reduce the amount of memorization that would be required to learn new symbols for each of these new systems, we will (as far as possible) make use of our familiar Hindu-Arabic symbols. For instance, if we discuss a base four numeration system that requires four basic symbols, then we will use the four Hindu-Arabic symbols 0, 1, 2, and 3 and the place values

$$\ldots, 4^5, 4^4, 4^3, 4^2, 4^1, 4^0$$

The base eight, or **octal,** numeration system uses the Hindu-Arabic symbols 0, 1, 2, 3, 4, 5, 6, and 7 and the place values

$$\ldots, 8^5, 8^4, 8^3, 8^2, 8^1, 8^0$$

To differentiate between bases, we will label each non-base-ten numeral with a subscript that indicates the base. For instance, 23_{four} represents a base four numeral. If a numeral is written without a subscript, then it is understood that the base is ten. Thus 23 written without a subscript is understood to be the base ten numeral 23.

To convert a non-base-ten numeral to base ten, we write the numeral in its expanded form, as shown in the following example.

TAKE NOTE

Because 23_{four} is *not* equal to the base ten number 23, it is important *not* to read 23_{four} as "twenty-three." To avoid confusion, read 23_{four} as "two three base four."

EXAMPLE 1 ■ Convert to Base Ten

Convert 2314_{five} to base ten.

Solution

In the base five numeration system, the place values are

$$\ldots, 5^4, 5^3, 5^2, 5^1, 5^0$$

The expanded form of 2314_{five} is

$$\begin{aligned} 2314_{five} &= (2 \times 5^3) + (3 \times 5^2) + (1 \times 5^1) + (4 \times 5^0) \\ &= (2 \times 125) + (3 \times 25) + (1 \times 5) + (4 \times 1) \\ &= 250 + 75 + 5 + 4 \\ &= 334 \end{aligned}$$

Thus $2314_{five} = 334$.

CHECK YOUR PROGRESS 1 Convert 3156_{seven} to base ten.

Solution *See page S13.*

QUESTION *Does the notation 26_{five} make sense?*

ANSWER *No. The expression 26_{five} is a meaningless expression because there is no 6 in base five.*

In base two, which is called the **binary numeration system,** the place values are the powers of two.

$$\ldots, 2^7, 2^6, 2^5, 2^4, 2^3, 2^2, 2^1, 2^0$$

The binary numeration system uses only the two digits 0 and 1. These *bi*nary digi*ts* are often called **bits.** To convert a base two numeral to base ten, write the numeral in its expanded form and then evaluate the expanded form.

EXAMPLE 2 ■ Convert to Base Ten

Convert 10110111_{two} to base ten.

Solution

$$\begin{aligned} 10110111_{\text{two}} &= (1 \times 2^7) + (0 \times 2^6) + (1 \times 2^5) + (1 \times 2^4) + (0 \times 2^3) \\ &\quad + (1 \times 2^2) + (1 \times 2^1) + (1 \times 2^0) \\ &= (1 \times 128) + (0 \times 64) + (1 \times 32) + (1 \times 16) + (0 \times 8) \\ &\quad + (1 \times 4) + (1 \times 2) + (1 \times 1) \\ &= 128 + 0 + 32 + 16 + 0 + 4 + 2 + 1 \\ &= 183 \end{aligned}$$

CHECK YOUR PROGRESS 2 Convert 111000101_{two} to base ten.

Solution *See page S13.*

TAKE NOTE

The base twelve numeration system is called the **duodecimal system.** A group called the Dozenal Society of America advocates the replacement of our base ten decimal system with the duodecimal system. If you wish to find out more about this organization, you can contact them at: The Dozenal Society of America, Nassau Community College, Garden City, New York. Not surprisingly, the dues are \$12 per year and \$144 ($12^2 = 144$) for a lifetime membership.

The base twelve numeration system requires 12 distinct symbols. We will use the symbols 0, 1, 2, 3, 4, 5, 6, 7, 8, 9, A, and B as our base twelve numeration system symbols. The symbols 0 through 9 have their usual meaning; however, A is used to represent 10 and B to represent 11.

EXAMPLE 3 ■ Convert to Base Ten

Convert $\text{B}37_{\text{twelve}}$ to base ten.

Solution

In the base twelve numeration system, the place values are

$$\ldots, 12^4, 12^3, 12^2, 12^1, 12^0$$

Thus

$$\begin{aligned} \text{B}37_{\text{twelve}} &= (11 \times 12^2) + (3 \times 12^1) + (7 \times 12^0) \\ &= 1584 + 36 + 7 \\ &= 1627 \end{aligned}$$

CHECK YOUR PROGRESS 3 Convert $\text{A5B}_{\text{twelve}}$ to base ten.

Solution *See page S14.*

Computer programmers often write programs that use the base sixteen numeration system, which is also called the **hexadecimal system.** This system uses the symbols 0, 1, 2, 3, 4, 5, 6, 7, 8, 9, A, B, C, D, E, and F. Table 4.6 shows that A represents 10, B represents 11, C represents 12, D represents 13, E represents 14, and F represents 15.

Table 4.6 *Decimal and Hexadecimal Equivalents*

Base Ten Decimal	Base Sixteen Hexadecimal
0	0
1	1
2	2
3	3
4	4
5	5
6	6
7	7
8	8
9	9
10	A
11	B
12	C
13	D
14	E
15	F

EXAMPLE 4 ■ Convert to Base Ten

Convert $3E8_{\text{sixteen}}$ to base ten.

Solution

In the base sixteen numeration system the place values are

$$\ldots, 16^4, 16^3, 16^2, 16^1, 16^0$$

Thus

$$\begin{aligned} 3E8_{\text{sixteen}} &= (3 \times 16^2) + (14 \times 16^1) + (8 \times 16^0) \\ &= 768 + 224 + 8 \\ &= 1000 \end{aligned}$$

CHECK YOUR PROGRESS 4 Convert $C24F_{\text{sixteen}}$ to base ten.

Solution *See page S14.*

Converting from Base Ten to Another Base

The most efficient method of converting a number written in base ten to another base makes use of a *successive division process.* For example, to convert 219 to base four, divide 219 by 4 and write the quotient 54 and the remainder 3, as shown below. Now divide the quotient 54 by the base to get a new quotient of 13 and a new remainder of 2. Continuing the process, divide the quotient 13 by 4 to get a new quotient of 3 and a remainder of 1. Because our last quotient, 3, is less than the base, 4, we stop the division process. The answer is given by the last quotient, 3, and the remainders, shown in red in the following diagram. That is, $219 = 3123_{\text{four}}$.

4 | 219
4 | 54   3
4 | 13   2
3   1

You can understand how the successive division process converts a base ten numeral to another base by analyzing the process. The first division shows there are 54 fours in 219, with **3 ones** left over. The second division shows that there are 13 sixteens (two successive divisions by 4 is the same as dividing by 16) in 219, and the remainder 2 indicates that there are **2 fours** left over. The last division shows that there are 3 sixty-fours (three successive divisions by 4 is the same as dividing by 64) in 219, and the remainder 1 indicates that there is **1 sixteen** left over. In mathematical notation these results are written as follows.

$$\begin{aligned} 219 &= (3 \times 64) + (1 \times 16) + (2 \times 4) + (3 \times 1) \\ &= (3 \times 4^3) + (1 \times 4^2) + (2 \times 4^1) + (3 \times 4^0) \\ &= 3123_{\text{four}} \end{aligned}$$

EXAMPLE 5 ■ Convert a Base Ten Numeral to Another Base

Convert 5821 to **a.** base three and **b.** base sixteen.

Solution

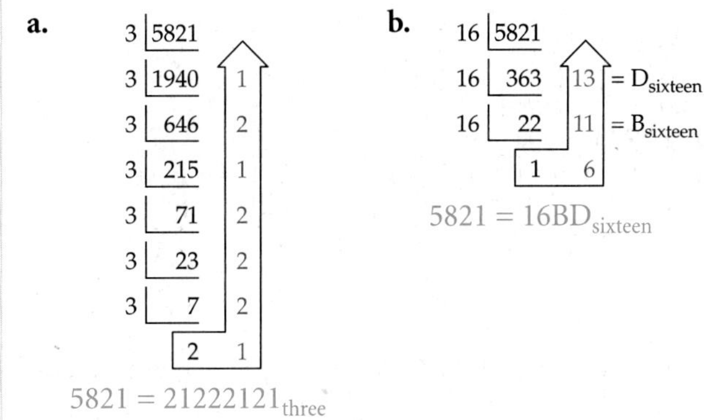

CHECK YOUR PROGRESS 5 Convert 1952 to **a.** base five and **b.** base twelve.

Solution *See page S14.*

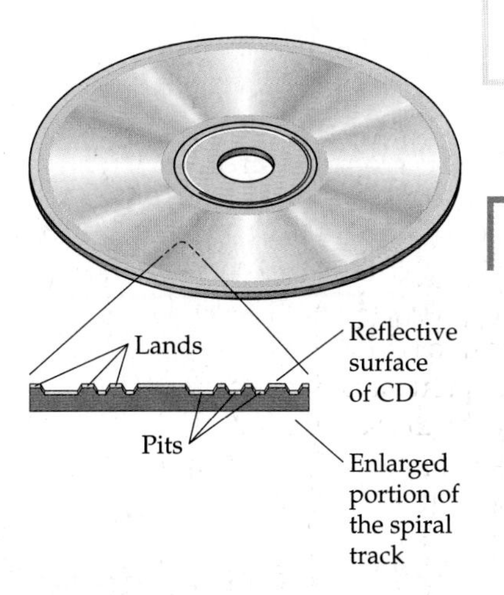

Math Matters Music by the Numbers

The binary numeration system is used to encode music on a CD (compact disc). The figure at the left shows the surface of a CD, which consists of flat regions called *lands* and small indentations called *pits.* As a laser beam tracks along a spiral path, the beam is reflected to a sensor when it shines on a land, but it is not reflected to the sensor when it shines on a pit. The sensor interprets a reflection as a 1 and no reflection as a 0. As the CD is playing, the sensor receives a series of 1's and 0's, which the CD player converts to music. On a typical CD, the spiral path that the laser follows loops around the disc over 20,000 times and contains about 650 megabytes of data. A **byte** is eight bits, so this amounts to 5,200,000,000 bits, each of which is represented by a pit or a land.

Table 4.7 *Octal and Binary Equivalents*

Octal	Binary
0	000
1	001
2	010
3	011
4	100
5	101
6	110
7	111

Converting Directly Between Computer Bases

Although computers compute internally by using base two (binary system), humans generally find it easier to compute with a larger base. Fortunately, there are easy conversion techniques that can be used to convert a base two numeral directly to a base eight (octal) numeral or a base sixteen (hexadecimal) numeral. Before we explain the techniques, it will help to become familiar with the information in Table 4.7, which shows the eight octal symbols and their binary equivalents.

To convert from octal to binary, just replace each octal symbol with its three-bit binary equivalent.

EXAMPLE 6 ■ Convert Directly from Base Eight to Base Two

Convert 5724_{eight} directly to binary form.

Solution

$$\begin{array}{cccc} 5 & 7 & 2 & 4_{eight} \\ \parallel & \parallel & \parallel & \parallel \\ 101 & 111 & 010 & 100_{two} \end{array}$$

$5724_{eight} = 101111010100_{two}$

CHECK YOUR PROGRESS 6 Convert 63210_{eight} directly to binary form.

Solution *See page S14.*

Because every group of three binary bits is equivalent to an octal symbol, we can convert from binary directly to octal by breaking a binary numeral into groups of three (from right to left) and replacing each group with its octal equivalent.

EXAMPLE 7 ■ Convert Directly from Base Two to Base Eight

Convert 11100101_{two} directly to octal form.

Solution

Starting from the right, break the binary numeral into groups of three. Then replace each group with its octal equivalent.

This zero was inserted to make a group of three.

$$\begin{array}{ccc} 011 & 100 & 101_{two} \\ \parallel & \parallel & \parallel \\ 3 & 4 & 5_{eight} \end{array}$$

$11100101_{two} = 345_{eight}$

CHECK YOUR PROGRESS 7 Convert 111010011100_{two} directly to octal form.

Solution *See page S14.*

Table 4.8 shows the hexadecimal symbols and their binary equivalents. To convert from hexadecimal to binary, replace each hexadecimal symbol with its four-bit binary equivalent.

Table 4.8 *Hexadecimal and Binary Equivalents*

Hexadecimal	Binary
0	0000
1	0001
2	0010
3	0011
4	0100
5	0101
6	0110
7	0111
8	1000
9	1001
A	1010
B	1011
C	1100
D	1101
E	1110
F	1111

EXAMPLE 8 ■ Convert Directly from Base Sixteen to Base Two

Convert $BAD_{sixteen}$ directly to binary form.

Solution

$$\begin{array}{ccc} B & A & D_{sixteen} \\ \parallel & \parallel & \parallel \\ 1011 & 1010 & 1101_{two} \end{array}$$

$BAD_{sixteen} = 101110101101_{two}$

CHECK YOUR PROGRESS 8 Convert $C5A_{sixteen}$ directly to binary form.

Solution *See page S14.*

Because every group of four binary bits is equivalent to a hexadecimal symbol, we can convert from binary to hexadecimal by breaking the binary numeral into groups of four (from right to left) and replacing each group with its hexadecimal equivalent.

EXAMPLE 9 ■ Convert Directly from Base Two to Base Sixteen

Convert 10110010100011_{two} directly to hexadecimal form.

Solution

Starting from the right, break the binary numeral into groups of four. Replace each group with its hexadecimal equivalent.

Insert two zeros to make a group of four.

0010	1100	1010	0011_{two}
‖	‖	‖	‖
2	C	A	3

$10110010100011_{two} = 2CA3_{sixteen}$

CHECK YOUR PROGRESS 9 Convert 101000111010010_{two} directly to hexadecimal form.

Solution *See page S14.*

The Double-Dabble Method

There is a short cut that can be used to convert a base two numeral to base ten. The advantage of this short cut, called the *double-dabble method,* is that you can start at the left of the numeral and work your way to the right without first determining the place value of each bit in the base two numeral.

EXAMPLE 10 ■ Apply the Double-Dabble Method

Use the double-dabble method to convert 1011001_{two} to base ten.

Solution

Start at the left with the first 1 and move to the right. Every time you pass by a 0, double your current number. Every time you pass by a 1, dabble. Dabbling is accomplished by doubling your current number and adding 1.

	$2 \cdot 1$	$2 \cdot 2 + 1$	$2 \cdot 5 + 1$	$2 \cdot 11$	$2 \cdot 22$	$2 \cdot 44 + 1$
	double	dabble	dabble	double	double	dabble
	2	5	11	22	44	89
1	0	1	1	0	0	1_{two}

As we pass by the final 1 in the units place, we dabble 44 to get 89. Thus $1011001_{two} = 89$.

CHECK YOUR PROGRESS 10 Use the double-dabble method to convert 1110010_{two} to base ten.

Solution *See page S14.*

Excursion

Information Retrieval via a Binary Search

To complete this Excursion, you must first construct a set of 31 cards that we refer to as a deck of *binary cards.* Templates for constructing the cards are available at our website, **math.hmco.com/students,** under the file name Binary Cards. Use a computer to print the templates onto a medium-weight card stock similar to that used for playing cards. Specific directions are provided with the templates.

We are living in the information age, but information is not useful if it cannot be retrieved when you need it. The binary numeration system is vital to the retrieval of information. To illustrate the connection between retrieval of information and the binary system, examine the card in the following figure. The card is labeled with the base ten numeral 20, and the holes and notches at the top of the card represent 20 in binary notation. A hole is used to indicate a 1 and a notch is used to indicate a 0. In the figure, the card has holes in the third and fifth binary-place-value positions (counting from right to left) and notches cut out of the first, second, and fourth positions.

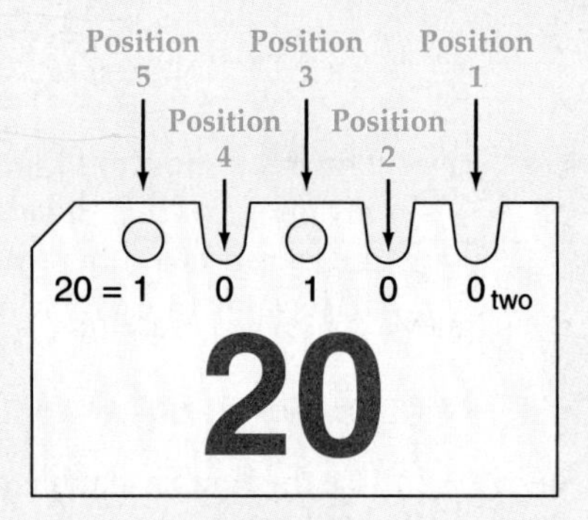

After you have constructed your deck of binary cards, take a few seconds to shuffle the deck. To find the card labeled with the numeral 20, complete the following process.

1. Use a thin dowel (or the tip of a sharp pencil) to lift out the cards that have a hole in the fifth position. *Keep* these cards and set the other cards off to the side.
2. From the cards that are *kept,* use the dowel to lift out the cards with a hole in the fourth position. Set these cards off to the side.
3. From the cards that are *kept,* use the dowel to lift out the cards that have a hole in the third position. *Keep* these cards and place the others off to the side.

(continued)

4. From the cards that are *kept,* use the dowel to lift out the cards with a hole in the second position. Set these cards off to the side.

5. From the cards that are *kept,* use the dowel to lift out the card that has a hole in the first position. Set this card off to the side.

The card that remains is the card labeled with the numeral 20. You have just completed a binary search.

Excursion Exercises

The binary numeration system can also be used to implement a *sorting* procedure. To illustrate, shuffle your deck of cards. Use the dowel to lift out the cards that have a hole in the rightmost position. Place these cards, *face up,* in the back of the other cards. Now use the dowel to lift out the cards that have a hole in the next position to the left. Place these cards, face up, in back of the other cards. Continue this process of lifting out the cards in the next position to the left and placing them in back of the other cards until you have completed the process for all five positions.

1. Examine the numerals on the cards. What do you notice about the order of the numerals? Explain why they are in this order.

2. If you wanted to sort 1000 cards from smallest to largest value by using the binary sort procedure, how many positions (where each position is either a hole or a notch) would be required at the top of each card? How many positions are needed to sort 10,000 cards?

3. Explain why the above sorting procedure cannot be implemented with base three cards.

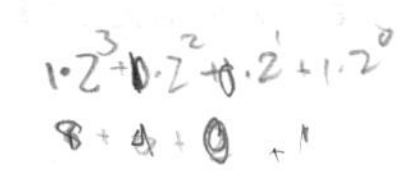

Exercise Set 4.3

In Exercises 1–10, convert the given numeral to base ten.

1. 243_{five}
2. 145_{seven}
3. 67_{nine}
4. 573_{eight}
5. 3154_{six}
6. 735_{eight}
7. 13211_{four}
8. 102022_{three}
9. $B5_{\text{sixteen}}$
10. $4A_{\text{twelve}}$

In Exercises 11–20, convert the given base ten numeral to the indicated base.

11. 267 to base five
12. 362 to base eight
13. 1932 to base six
14. 2024 to base four
15. 15,306 to base nine
16. 18,640 to base seven
17. 4060 to base two
18. 5673 to base three
19. 283 to base twelve
20. 394 to base sixteen

In Exercises 21–28, use expanded forms to convert the given base two numeral to base ten.

21. 1101_{two}
22. 10101_{two}
23. 11011_{two}
24. 101101_{two}
25. 1100100_{two}
26. 11110101000_{two}
27. 10001011_{two}
28. 110110101_{two}

In Exercises 29–34, use the double-dabble method to convert the given base two numeral to base ten.

29. 101001_{two}
30. 1110100_{two}
31. 1011010_{two}
32. 10001010_{two}
33. 10100111010_{two}
34. 10000000100_{two}

In Exercises 35–46, convert the given numeral to the indicated base.

35. 34_{six} to base eight

36. 71_{eight} to base five

37. 878_{nine} to base four

38. 546_{seven} to base six

39. 1110_{two} to base five

40. 21200_{three} to base six

41. 3440_{eight} to base nine

42. 1453_{six} to base eight

43. $56_{sixteen}$ to base eight

44. 43_{twelve} to base six

45. $A4_{twelve}$ to base sixteen

46. $C9_{sixteen}$ to base twelve

In Exercises 47–56, convert the given numeral *directly* (without first converting to base ten) to the indicated base.

47. 352_{eight} to base two

48. $A4_{sixteen}$ to base two

49. 11001010_{two} to base eight

50. 111011100101_{two} to base sixteen

51. 101010001_{two} to base sixteen

52. 56721_{eight} to base two

53. $BEF3_{sixteen}$ to base two

54. $6A7B8_{sixteen}$ to base two

55. $BA5CF_{sixteen}$ to base two

56. 47134_{eight} to base two

57. **An Extension** There is a procedure that can be used to convert a base three numeral directly to base ten without using the expanded form of the numeral. Write an explanation of this procedure, which we will call the *triple-whipple-zipple* method. *Hint:* The method is an extension of the double-dabble method.

58. Determine whether the following statements are true or false.

a. A number written in base two is divisible by 2 if and only if the number ends with a 0.

b. In base six, the next counting number after 55_{six} is 100_{six}.

c. In base sixteen, the next counting number after $3BF_{sixteen}$ is $3C0_{sixteen}$.

Extensions

CRITICAL THINKING

The D'ni Numeration System In the computer game *Riven*, a D'ni numeration system is used. Although the D'ni numeration system is a base twenty-five numeration system with 25 distinct numerals, you really need to memorize only the first five numerals, which are shown below.

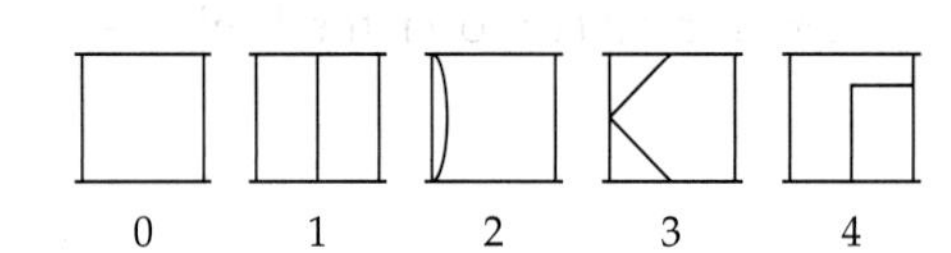

The basic D'ni numerals

If two D'ni numerals are placed side-by-side, then the numeral on the left is in the twenty-fives' place and the numeral on the right is in the ones' place. Thus

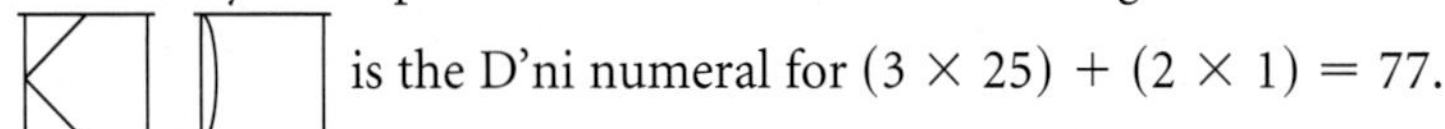

is the D'ni numeral for $(3 \times 25) + (2 \times 1) = 77$.

59. Convert the following D'ni numeral to base ten.

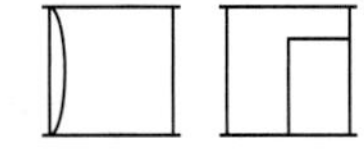

60. Convert the following D'ni numeral to base ten.

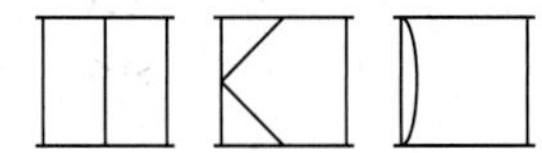

Rotating any of the D'ni numerals for 1, 2, 3, and 4 by a 90° counterclockwise rotation produces a numeral with a value five times its original value. For instance, rotating the numeral for 1 produces 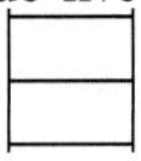 , which is the D'ni numeral for 5, and rotating the numeral for 2 produces 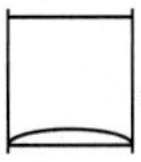 , which is the D'ni numeral for 10.

61. Write the D'ni numeral for 15.

62. Write the D'ni numeral for 20.

In the D'ni numeration system, explained above, many numerals are obtained by rotating a basic numeral and then overlaying it on one of the basic numerals. For instance, if you rotate the D'ni numeral for 1, you get the numeral for 5. If you then overlay the numeral for 5 on the numeral for 1, you get the numeral for $5 + 1 = 6$.

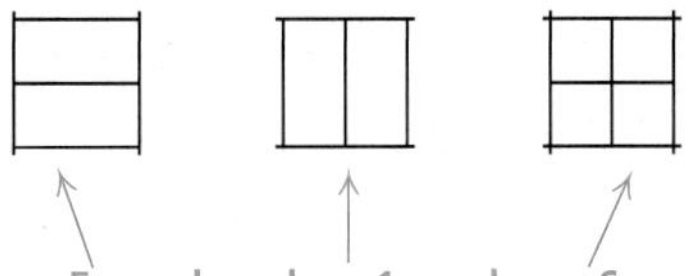

5 overlayed on 1 produces 6.

63. Write the D'ni numeral for 8.

64. Write the D'ni numeral for 22.

65. Convert the following D'ni numeral to base ten.

66. Convert the following D'ni numeral to base ten.

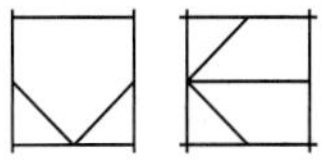

67. a. State one advantage of the hexadecimal numeration system over the decimal numeration system.

b. State one advantage of the decimal numeration system over the hexadecimal numeration system.

68. a. State one advantage of the D'ni numeration system over the decimal numeration system.

b. State one advantage of the decimal numeration system over the D'ni numeration system.

EXPLORATIONS

69. **The ASCII Code** ASCII, pronounced *ask-key,* is an acronym for the American Standard Code for Information Interchange. In this code, each of the characters that can be typed on a computer keyboard is represented by a number. For instance, the letter A is assigned the number 65, which when written as an 8-bit binary numeral is 01000001. Research the topic of ASCII. Write a report about ASCII and its applications.

70. **The Postnet Code** The U.S. Postal Service uses a *Postnet code* to write zip codes + 4 on envelopes. The Postnet code is a bar code that is based on the binary numeration system. Postnet code is very useful because it can be read by a machine. Write a few paragraphs that explain how to convert a zip code + 4 to its Postnet code. What is the Postnet code for your zip code + 4?

SECTION 4.4 Arithmetic in Different Bases

Addition in Different Bases

Most computers and calculators make use of the base two (binary) numeration system to perform arithmetic computations. For instance, if you use a calculator to find the sum of 9 and 5, the calculator first converts the 9 to 1001_{two} and the 5 to 101_{two}. The calculator uses electronic circuitry called *binary adders* to find the sum of 1001_{two} and 101_{two} as 1110_{two}. The calculator then converts 1110_{two} to base ten and displays the sum 14. All of the conversions and the base two addition are done internally in a fraction of a second, which gives the user the impression that the calculator performed the addition in base ten.

The following examples illustrate how to perform arithmetic in different bases. We first consider the operation of addition in the binary numeration system. Table 4.9 is an addition table for base two. It is similar to the base ten addition table that you memorized in elementary school, except that it is much smaller because base two involves only the bits 0 and 1. The numerals shown in red in Table 4.9 illustrate that $1_{two} + 1_{two} = 10_{two}$.

Table 4.9 *A Binary Addition Table*

Second addend

+	0	1
0	0	1
1	1	10

First addend (rows); Sums (body)

EXAMPLE 1 ■ Add Base Two Numerals

Find the sum of 11110_{two} and 1011_{two}.

Solution

Arrange the numerals vertically, keeping the bits of the same place value in the same column.

	THIRTY-TWOS	SIXTEENS	EIGHTS	FOURS	TWOS	ONES
		1	1	1	1	0_{two}
+			1	0	1	1_{two}
						1_{two}

Start by adding the bits in the ones' column: $0_{two} + 1_{two} = 1_{two}$. Then move left and add the bits in the twos' column. When the sum of the bits in a column exceeds 1, the addition will involve carrying, as shown below.

	THIRTY-TWOS	SIXTEENS	EIGHTS	FOURS	TWOS	ONES
				1		
		1	1	1	1	0_{two}
+			1	0	1	1_{two}
					0	1_{two}

Add the bits in the twos' column.
$1_{two} + 1_{two} = 10_{two}$
Write the 0 in the twos' column and carry the 1 to the fours' column.

TAKE NOTE

In this section assume that the small numerals, used to indicate a carry, are written in the same base as the numerals in the given addition (multiplication) problem.

$$\begin{array}{rccccc} & & 1 & 1 & & \\ & 1 & 1 & 1 & 1 & 0_{two} \\ + & & 1 & 0 & 1 & 1_{two} \\ \hline & & & 0 & 0 & 1_{two} \end{array}$$

Add the bits in the fours' column.
$(1_{two} + 1_{two}) + 0_{two} = 10_{two} + 0_{two} = 10_{two}$
Write the 0 in the fours' column and carry the 1 to the eights' column.

$$\begin{array}{rcccccc} & 1 & 1 & 1 & 1 & & \\ & & 1 & 1 & 1 & 1 & 0_{two} \\ + & & & 1 & 0 & 1 & 1_{two} \\ \hline & 1 & 0 & 1 & 0 & 0 & 1_{two} \end{array}$$

Add the bits in the eights' column.
$(1_{two} + 1_{two}) + 1_{two} = 10_{two} + 1_{two} = 11_{two}$
Write a 1 in the eights' column and carry a 1 to the sixteens' column. Continue to add the bits in each column to the left of the eights' column.

The sum of 11110_{two} and 1011_{two} is 101001_{two}.

CHECK YOUR PROGRESS 1 Find the sum of 11001_{two} and 1101_{two}.

Solution *See page S14.*

Table 4.10 *A Base Four Addition Table*

+	0	1	2	3
0	0	1	2	3
1	1	2	3	10
2	2	3	10	11
3	3	10	11	12

There are four symbols in base four, namely 0, 1, 2, and 3. Table 4.10 shows a base four addition table that lists all the sums that can be produced by adding two base four digits. The numerals shown in red in Table 4.10 illustrate that $2_{four} + 3_{four} = 11_{four}$.

In the next example we compute the sum of two numbers written in base four.

EXAMPLE 2 ■ Add Base Four Numerals

Find the sum of 23_{four} and 13_{four}.

Solution

Arrange the numerals vertically, keeping the digits of the same place value in the same column.

SIXTEENS FOURS ONES

$$\begin{array}{rccc} & & 1 & \\ & & 2 & 3_{four} \\ + & & 1 & 3_{four} \\ \hline & & & 2_{four} \end{array}$$

Add the digits in the ones' column.
Table 4.10 shows that $3_{four} + 3_{four} = 12_{four}$.
Write the 2 in the ones' column and carry the 1 to the fours' column.

$$\begin{array}{rccc} & 1 & 1 & \\ & & 2 & 3_{four} \\ + & & 1 & 3_{four} \\ \hline & 1 & 0 & 2_{four} \end{array}$$

Add the digits in the fours' column:
$(1_{four} + 2_{four}) + 1_{four} = 3_{four} + 1_{four} = 10_{four}$.
Write the 0 in the fours' column and carry the 1 to the sixteens' column. Bring down the 1 that was carried to the sixteens' column to form the sum 102_{four}.

The sum of 23_{four} and 13_{four} is 102_{four}.

CHECK YOUR PROGRESS 2 Find $32_{\text{four}} + 12_{\text{four}}$.

Solution *See page S14.*

In the previous examples we used a table to determine the necessary sums. However, it is generally quicker to find a sum by computing the base ten sum of the digits in each column and then converting each base ten sum back to its equivalent in the given base. The next two examples illustrate this summation technique.

EXAMPLE 3 ■ Add Base Six Numerals

Find $25_{\text{six}} + 32_{\text{six}} + 42_{\text{six}}$.

Solution

Arrange the numerals vertically, keeping the digits of the same place value in the same column.

	THIRTY-SIXES	SIXES	ONES
		1	
		2	5_{six}
		3	2_{six}
+		4	2_{six}
			3_{six}

Add the digits in the ones' column:
$5_{\text{six}} + 2_{\text{six}} + 2_{\text{six}} = 5 + 2 + 2 = 9$.
Convert 9 to base six. ($9 = 13_{\text{six}}$)
Write the 3 in the ones' column and carry the 1 to the sixes' column.

	THIRTY-SIXES	SIXES	ONES
	1	1	
		2	5_{six}
		3	2_{six}
+		4	2_{six}
	1	4	3_{six}

Add the digits in the sixes' column and convert the sum to base six.
$1_{\text{six}} + 2_{\text{six}} + 3_{\text{six}} + 4_{\text{six}} = 1 + 2 + 3 + 4 = 10 = 14_{\text{six}}$
Write the 4 in the sixes' column and carry the 1 to the thirty-sixes' column. Bring down the 1 that was carried to the thirty-sixes' column to form the sum 143_{six}.

$$25_{\text{six}} + 32_{\text{six}} + 42_{\text{six}} = 143_{\text{six}}$$

CHECK YOUR PROGRESS 3 Find $35_{\text{seven}} + 46_{\text{seven}} + 24_{\text{seven}}$.

Solution *See page S14.*

In the next example, we solve an addition problem that involves a base greater than ten.

EXAMPLE 4 ■ Add Base Twelve Numerals

Find $A97_{twelve} + 8BA_{twelve}$.

Solution

	ONE HUNDRED FORTY-FOURS	TWELVES	ONES
		1	
	A	9	7_{twelve}
+	8	B	A_{twelve}
			5_{twelve}

Add the digits in the ones' column.
$7_{twelve} + A_{twelve} = 7 + 10 = 17$
Convert 17 to base twelve. ($17 = 15_{twelve}$)
Write the 5 in the ones' column and carry the 1 to the twelves' column.

	1	1	
	A	9	7_{twelve}
+	8	B	A_{twelve}
		9	5_{twelve}

Add the digits in the twelves' column.
$1_{twelve} + 9_{twelve} + B_{twelve} = 1 + 9 + 11 = 21 = 19_{twelve}$
Write the 9 in the twelves' column and carry the 1 to the one hundred forty-fours' column.

	1	1	1	
		A	9	7_{twelve}
+		8	B	A_{twelve}
	1	7	9	5_{twelve}

Add the digits in the one hundred forty-fours' column.
$1_{twelve} + A_{twelve} + 8_{twelve} = 1 + 10 + 8 = 19 = 17_{twelve}$
Write the 7 in the one hundred forty-fours' column and carry the 1 to the one thousand seven hundred twenty-eights' column. Bring down the 1 that was carried to the one thousand seven hundred twenty-eights' column to form the sum 1795_{twelve}.

$A97_{twelve} + 8BA_{twelve} = 1795_{twelve}$

CHECK YOUR PROGRESS 4 Find $AC4_{sixteen} + 6E8_{sixteen}$.

Solution *See page S14.*

Subtraction in Different Bases

TAKE NOTE

In the following subtraction, 7 is the *minuend* and 4 is the *subtrahend.*

$7 - 4 = 3$

To subtract two numbers written in the same base, begin by arranging the numbers vertically, keeping digits that have the same place value in the same column. It will be necessary to borrow whenever a digit in the subtrahend is greater than its corresponding digit in the minuend. Every number that is borrowed will be a power of the base.

EXAMPLE 5 ■ Subtract Base Seven Numerals

Find $463_{seven} - 124_{seven}$.

Solution

Arrange the numerals vertically, keeping the digits of the same place value in the same column.

TAKE NOTE

In this section assume that the small numerals, used to illustrate the borrowing process, are written in the same base as the numerals in the given subtraction problem.

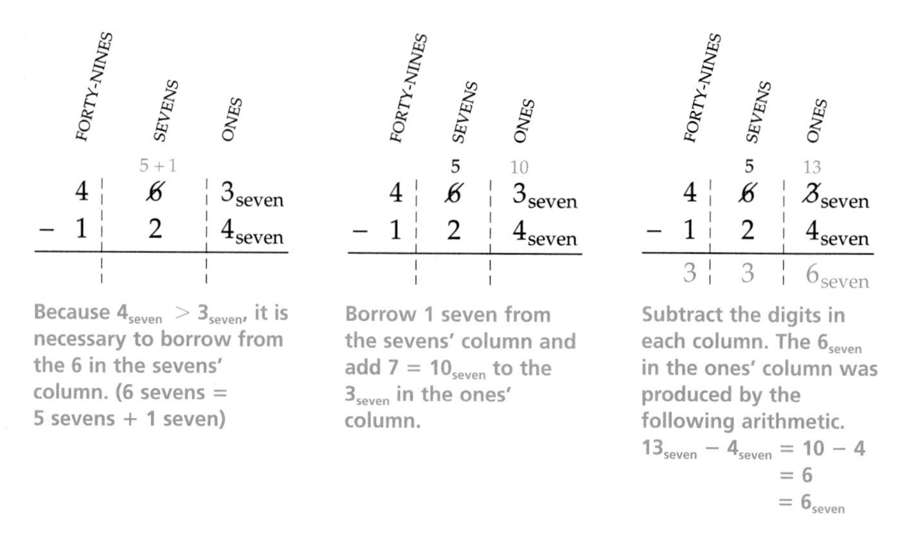

$463_{seven} - 124_{seven} = 336_{seven}$

CHECK YOUR PROGRESS 5 Find $365_{nine} - 183_{nine}$.

Solution *See page S14.*

Table 4.11 *Decimal and Hexadecimal Equivalents*

Base Ten Decimal	Base Sixteen Hexadecimal
0	0
1	1
2	2
3	3
4	4
5	5
6	6
7	7
8	8
9	9
10	A
11	B
12	C
13	D
14	E
15	F

EXAMPLE 6 ■ Subtract Base Sixteen Numerals

Find $7AB_{sixteen} - 3E4_{sixteen}$.

Solution

Table 4.11 shows the hexadecimal digits and their decimal equivalents. Because $B_{sixteen}$ is greater than $4_{sixteen}$, there is no need to borrow to find the difference in the ones' column. However, $A_{sixteen}$ is less than $E_{sixteen}$, so it is necessary to borrow to find the difference in the sixteens' column.

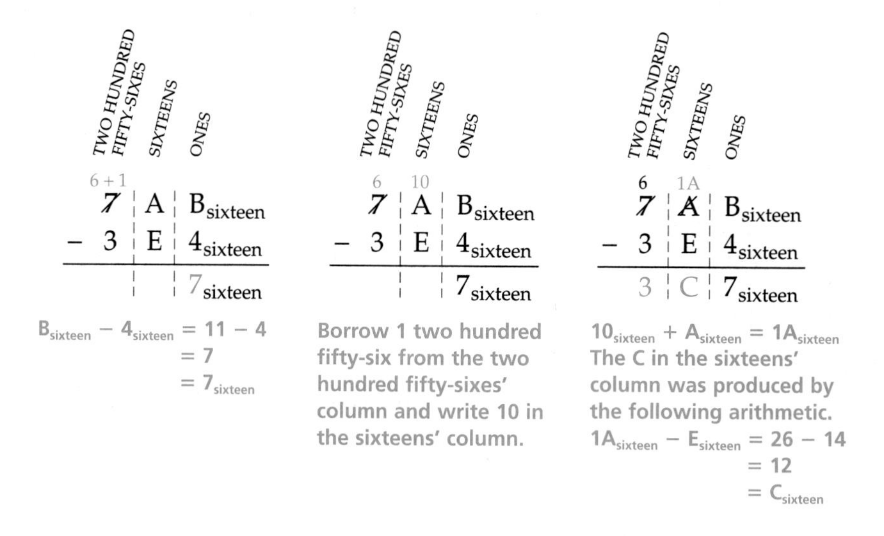

$7AB_{sixteen} - 3E4_{sixteen} = 3C7_{sixteen}$

CHECK YOUR PROGRESS 6 Find $83A_{twelve} - 467_{twelve}$.

Solution *See page S14.*

Multiplication in Different Bases

Table 4.12 *A Base Four Multiplication Table*

×	0	1	2	3
0	0	0	0	0
1	0	1	2	3
2	0	2	10	12
3	0	3	12	21

To perform multiplication in bases other than base ten, it is often helpful to first write a multiplication table for the given base. Table 4.12 shows a multiplication table for base four. The numbers shown in red in the table illustrate that $2_{four} \times 3_{four} = 12_{four}$. You can verify this result by converting the numbers to base ten, multiplying in base ten, and then converting back to base four. Here is the actual arithmetic.

$$2_{four} \times 3_{four} = 2 \times 3 = 6 = 12_{four}$$

QUESTION *What is $5_{six} \times 4_{six}$?*

EXAMPLE 7 ■ Multiply Base Four Numerals

Use the base four multiplication table to find $3_{four} \times 123_{four}$.

Solution

Arrange the numerals vertically, keeping the digits of the same place value in the same column. Use Table 4.12 to multiply 3_{four} times each digit in 123_{four}. If any of these multiplications produces a two-digit product, then write down the digit on the right and carry the digit on the left.

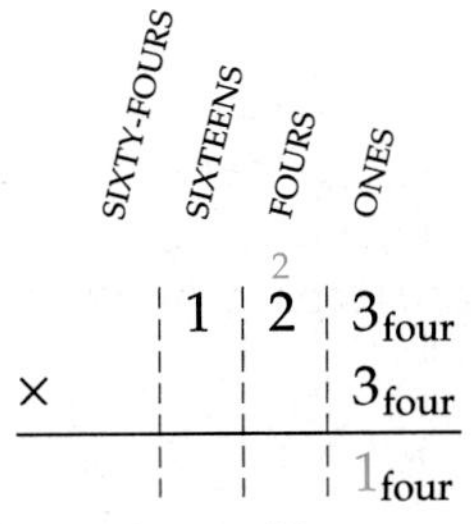

$3_{four} \times 3_{four} = 21_{four}$
Write the 1 in the ones' column and carry the 2.

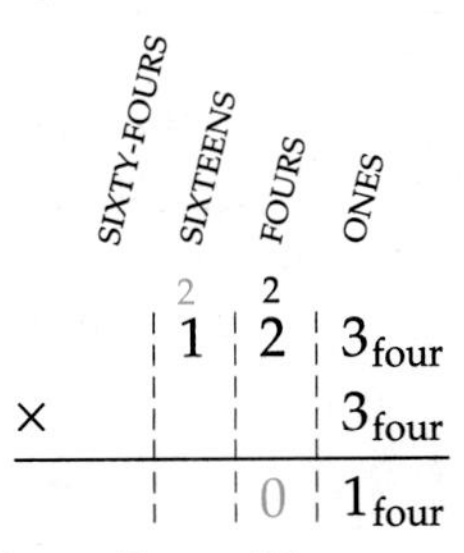

$3_{four} \times 2_{four} = 12_{four}$
$12_{four} + 2_{four}$(the carry) $= 20_{four}$
Write the 0 in the fours' column and carry the 2.

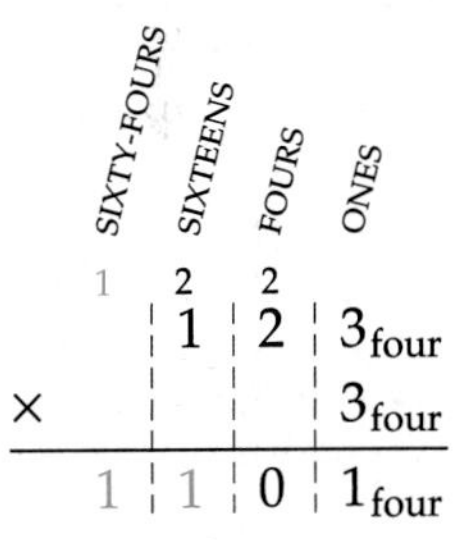

$3_{four} \times 1_{four} = 3_{four}$
$3_{four} + 2_{four}$(the carry) $= 11_{four}$
Write a 1 in the sixteens' column and carry a 1 to the sixty-fours' column. Bring down the 1 that was carried to the sixty-fours' column to form the product 1101_{four}.

$3_{four} \times 123_{four} = 1101_{four}$

CHECK YOUR PROGRESS 7 Find $2_{four} \times 213_{four}$.

Solution *See page S14.*

ANSWER $5_{six} \times 4_{six} = 20 = 32_{six}$

Writing all of the entries in a multiplication table for a large base such as base twelve can be time-consuming. In such cases you may prefer to multiply in base ten and then convert each product back to the given base. The next example illustrates this multiplication method.

EXAMPLE 8 ■ Multiply Base Twelve Numerals

Find $53_{\text{twelve}} \times 27_{\text{twelve}}$.

Solution

Arrange the numerals vertically, keeping the digits of the same place value in the same column. Start by multiplying each digit of the multiplicand (53_{twelve}) by the ones' digit of the multiplier (27_{twelve}).

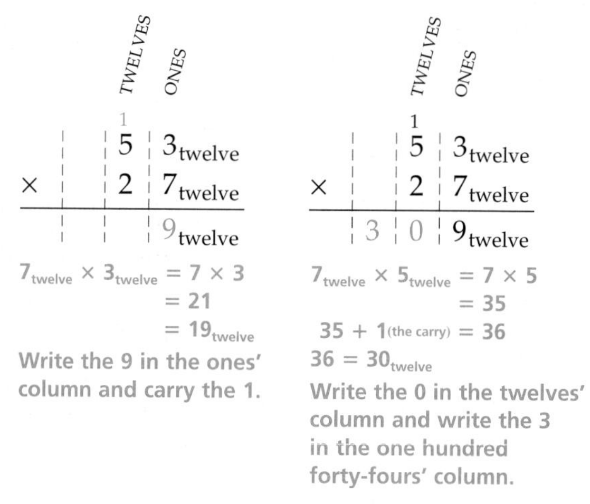

Now multiply each digit of the multiplicand by the twelves' digit of the multiplier.

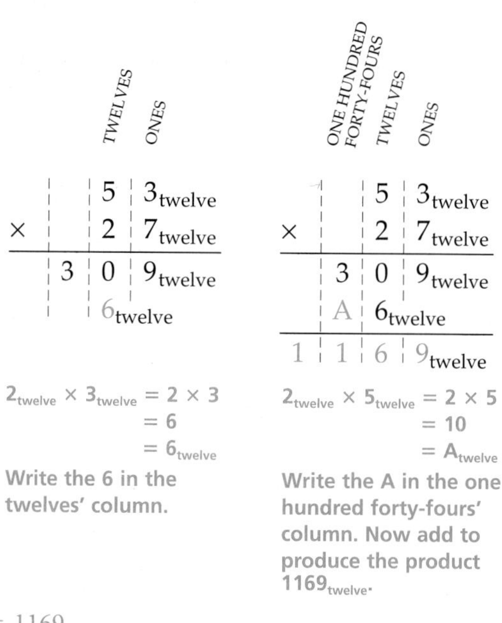

$53_{\text{twelve}} \times 27_{\text{twelve}} = 1169_{\text{twelve}}$

CHECK YOUR PROGRESS 8 Find $25_{eight} \times 34_{eight}$.

Solution *See page S15.*

MathMatters **The Fields Medal**

The Fields Medal

A Nobel Prize is awarded each year in the categories of chemistry, physics, physiology, medicine, literature, and peace. However, no award is given in mathematics. Why Alfred Nobel chose not to provide an award in the category of mathematics is unclear. There has been some speculation that Nobel had a personal conflict with the mathematician Gosta Mittag-Leffler.

The Canadian mathematician John Charles Fields (1863–1932) felt that a prestigious award should be given in the area of mathematics. Fields helped establish the Fields Medal, which was first given to Lars Valerian Ahlfors and Jesse Douglas in 1936. The International Congress of Mathematicians had planned to give two Fields Medals every four years after 1936, but because of World War II, the next Fields Medals were not given until 1950.

It was Fields's wish that the Fields Medal recognize both existing work and the promise of future achievement. Because of this concern for future achievement, the International Congress of Mathematicians decided to restrict those eligible for the Fields Medal to mathematicians under the age of 40.

Division in Different Bases

To perform a division in a base other than base ten, it is helpful to first make a list of a few multiples of the divisor. This procedure is illustrated in the following example.

EXAMPLE 9 ■ Divide Base Seven Numerals

Find $253_{seven} \div 3_{seven}$.

Solution

First list a few multiples of the divisor 3_{seven}.

$$3_{seven} \times 0_{seven} = 3 \times 0 = 0 = 0_{seven} \qquad 3_{seven} \times 4_{seven} = 3 \times 4 = 12 = 15_{seven}$$
$$3_{seven} \times 1_{seven} = 3 \times 1 = 3 = 3_{seven} \qquad 3_{seven} \times 5_{seven} = 3 \times 5 = 15 = 21_{seven}$$
$$3_{seven} \times 2_{seven} = 3 \times 2 = 6 = 6_{seven} \qquad 3_{seven} \times 6_{seven} = 3 \times 6 = 18 = 24_{seven}$$
$$3_{seven} \times 3_{seven} = 3 \times 3 = 9 = 12_{seven}$$

Because $3_{seven} \times 6_{seven} = 24_{seven}$ is slightly less than 25_{seven}, we pick 6 as our first numeral in the quotient when dividing 25_{seven} by 3_{seven}.

$$\begin{array}{r} 6\phantom{\,5\,3_{seven}} \\ 3_{seven}\overline{)2\,5\,3_{seven}} \\ \underline{2\,4}\phantom{\,3_{seven}} \\ 1\phantom{\,3_{seven}} \end{array}$$

$3_{seven} \times 6_{seven} = 24_{seven}$
Subtract 24_{seven} from 25_{seven}.

quotient

$$\begin{array}{r} 6\ 3_{seven} \\ 3_{seven}\overline{)2\ 5\ 3_{seven}} \\ \underline{2\ 4} \\ 1\ 3 \\ \underline{1\ 2} \\ 1 \end{array}$$

remainder

Bring down the 3.

$3_{seven} \times 3_{seven} = 12_{seven}$

Subtract 12_{seven} from 13_{seven}.

Thus $253_{seven} \div 3_{seven} = 63_{seven}$ with a remainder of 1_{seven}.

CHECK YOUR PROGRESS 9 Find $324_{five} \div 3_{five}$.

Solution *See page S15.*

In a base two division problem, the only multiples of the divisor that are used are zero times the divisor and one times the divisor.

EXAMPLE 10 ■ Divide Base Two Numerals

Find $101011_{two} \div 11_{two}$.

Solution

The divisor is 11_{two}. The multiples of the divisor that may be needed are $11_{two} \times 0_{two} = 0_{two}$ and $11_{two} \times 1_{two} = 11_{two}$. Also note that because $10_{two} - 1_{two} = 2 - 1 = 1 = 1_{two}$, we know that

$$\begin{array}{r} 10_{two} \\ -\ \underline{1_{two}} \\ 1_{two} \end{array}$$

$$\begin{array}{r} 1\ 1\ 1\ 0_{two} \\ 11_{two}\overline{)1\ 0\ 1\ 0\ 1\ 1_{two}} \\ \underline{1\ 1} \\ 1\ 0\ 0 \\ \underline{1\ 1} \\ 1\ 1 \\ \underline{1\ 1} \\ 0\ 1 \\ \underline{0} \\ 1 \end{array}$$

Therefore, $101011_{two} \div 11_{two} = 1110_{two}$ with a remainder of 1_{two}.

CHECK YOUR PROGRESS 10 Find $1110011_{two} \div 10_{two}$.

Solution *See page S15.*

Subtraction in Base Two via the Ones Complement and the End-Around Carry

Computers and calculators are often designed so that the number of required circuits is minimized. Instead of using separate circuits to perform addition and subtraction, engineers make use of an *end-around carry procedure* that uses addition to perform subtraction. The end-around carry procedure also makes use of the ones complement of a number. In base two, the ones complement of 0 is 1 and the ones complement of 1 is 0. Thus the ones complement of any base two number can be found by changing each 1 to a 0 and each 0 to a 1.

Subtraction Using the Ones Complement and the End-Around Carry

To subtract a base two number from a larger base two number:

1. Add the ones complement of the subtrahend to the minuend.
2. Take away 1 from the leftmost bit and add 1 to the units bit.

The following example illustrates the process of subtracting 1001_{two} from 1101_{two} using the ones complement and the end-around carry procedure.

$$\begin{array}{r} 1101_{two} \\ -\ 1001_{two} \\ \hline \end{array}$$ Minuend / Subtrahend

$$\begin{array}{r} 1101_{two} \\ +\ 0110_{two} \\ \hline 10011_{two} \end{array}$$ Replace the subtrahend with the ones complement of the subtrahend and add.

$$\begin{array}{r} 10011_{two} \\ +\ 1_{two} \\ \hline 100_{two} \end{array}$$ Take away 1 from the leftmost bit and add 1 to the ones bit. This is the end-around carry procedure.

$1101_{two} - 1001_{two} = 100_{two}$

If the subtrahend has fewer bits than the minuend, leading zeros should be inserted in the subtrahend so that it has the same number of bits as the minuend. This process is illustrated below for the subtraction $1010110_{two} - 11001_{two}$.

$$\begin{array}{r} 1010110_{two} \\ -\ 11001_{two} \\ \hline \end{array}$$ Minuend / Subtrahend

$$\begin{array}{r} 1010110_{two} \\ -\ 0011001_{two} \\ \hline \end{array}$$ Insert two leading zeros.

$$\begin{array}{r} 1010110_{two} \\ +\ 1100110_{two} \\ \hline 10111100_{two} \end{array}$$ Ones complement of subtrahend

(continued)

$$\begin{array}{r} 10111100_{two} \\ + \quad 1_{two} \\ \hline 111101_{two} \end{array}$$

Take away 1 from the leftmost bit and add 1 to the ones bit.

$1010110_{two} - 11001_{two} = 111101_{two}$

Excursion Exercises

Use the ones complement of the subtrahend and the end-around carry method to find each difference.

1. $1110_{two} - 1001_{two}$

2. $101011_{two} - 100010_{two}$

3. $101001010_{two} - 1011101_{two}$

4. $111011100110_{two} - 101010100_{two}$

5. $1111101011_{two} - 1001111_{two}$

6. $1110010101100_{two} - 100011110_{two}$

Exercise Set 4.4

In Exercises 1–12, find each sum in the same base as the addends.

1. $204_{five} + 123_{five}$

2. $323_{four} + 212_{four}$

3. $5625_{seven} + 634_{seven}$

4. $1011_{two} + 101_{two}$

5. $110101_{two} + 10011_{two}$

6. $11001010_{two} + 1100111_{two}$

7. $8B5_{twelve} + 578_{twelve}$

8. $379_{sixteen} + 856_{sixteen}$

9. $C489_{sixteen} + BAD_{sixteen}$

10. $221_{three} + 122_{three}$

11. $435_{six} + 245_{six}$

12. $5374_{eight} + 615_{eight}$

In Exercises 13–24, find each difference.

13. $434_{five} - 143_{five}$

14. $534_{six} - 241_{six}$

15. $7325_{eight} - 563_{eight}$

16. $6148_{nine} - 782_{nine}$

17. $11010_{two} - 1011_{two}$

18. $111001_{two} - 10101_{two}$

19. $11010100_{two} - 1011011_{two}$

20. $9C5_{sixteen} - 687_{sixteen}$

21. $43A7_{twelve} - 289_{twelve}$

22. $BAB2_{twelve} - 475_{twelve}$

23. $762_{nine} - 367_{nine}$

24. $3223_{four} - 133_{four}$

In Exercises 25–38, find each product.

25. $3_{six} \times 145_{six}$

26. $5_{seven} \times 542_{seven}$

27. $2_{three} \times 212_{three}$

28. $4_{five} \times 4132_{five}$

29. $5_{eight} \times 7354_{eight}$

30. $11_{two} \times 11011_{two}$

31. $10_{two} \times 101010_{two}$

32. $101_{two} \times 110100_{two}$

33. $25_{eight} \times 453_{eight}$

34. $43_{six} \times 1254_{six}$

35. $132_{four} \times 1323_{four}$

36. $43_{twelve} \times 895_{twelve}$

37. $5_{sixteen} \times BAD_{sixteen}$

38. $23_{sixteen} \times 798_{sixteen}$

In Exercises 39–49, find each quotient and remainder.

39. $132_{four} \div 2_{four}$

40. $124_{five} \div 2_{five}$

41. $231_{four} \div 3_{four}$

42. $672_{eight} \div 5_{eight}$

43. $5341_{six} \div 4_{six}$

44. $11011_{two} \div 10_{two}$

45. $101010_{two} \div 11_{two}$

46. $1011011_{two} \div 100_{two}$

47. $457_{twelve} \div 5_{twelve}$

48. $832_{sixteen} \div 7_{sixteen}$

49. $234_{five} \div 12_{five}$

50. If $232_x = 92$, find the base x.

51. If $143_x = 10200_{\text{three}}$, find the base x.

52. If $46_x = 101010_{\text{two}}$, find the base x.

53. Consider the addition $384 + 245$.

a. Use base ten addition to find the sum.

b. Convert 384 and 245 to base two.

c. Find the base two sum of the base two numbers you found in part b.

d. Convert the base two sum from part c to base ten.

e. How does the answer to part a compare with the answer to part d?

54. Consider the subtraction $457 - 318$.

a. Use base ten subtraction to find the difference.

b. Convert 457 and 318 to base two.

c. Find the base two difference of the base two numbers you found in part b.

d. Convert the base two difference from part c to base ten.

e. How does the answer to part a compare with the answer to part d?

55. Consider the multiplication 247×26.

a. Use base ten multiplication to find the product.

b. Convert 247 and 26 to base two.

c. Find the base two product of the base two numbers you found in part b.

d. Convert the base two product from part c to base ten.

e. How does the answer to part a compare with the answer to part d?

Extensions

CRITICAL THINKING

56. Explain the error in the following base eight subtraction.

$$\begin{array}{r} 751_{\text{eight}} \\ -\ 126_{\text{eight}} \\ \hline 625_{\text{eight}} \end{array}$$

57. Determine the base used in the following multiplication.

$$314_{\text{base } x} \times 24_{\text{base } x} = 11202_{\text{base } x}$$

58. The base ten number 12 is an even number. In base seven, 12 is written as 15_{seven}. Is 12 an odd number in base seven?

59. Explain why there is no numeration system with a base of 1.

60. A Cryptarithm In the following base four addition problem, each letter represents one of the numerals 0, 1, 2, or 3. No two different letters represent the same numeral. Determine which digit is represented by each letter.

$$\begin{array}{r} \text{NO}_{\text{four}} \\ +\ \text{AT}_{\text{four}} \\ \hline \text{NOT}_{\text{four}} \end{array}$$

61. A Cryptarithm In the following base six addition problem, each letter represents one of the numerals 0, 1, 2, 3, 4, or 5. No two different letters represent the same numeral. Determine which digit is represented by each letter.

$$\begin{array}{r} \text{MA}_{\text{six}} \\ +\ \text{AS}_{\text{six}} \\ \hline \text{MOM}_{\text{six}} \end{array}$$

EXPLORATIONS

62. Negative Base Numerals It is possible to use a negative number as the base of a numeration system. For instance, the negative base four numeral $32_{\text{negative four}}$ represents the number $3 \times (-4)^1 + 2 \times (-4)^0 = -12 + 2 = -10$.

a. Convert each of the following negative base numerals to base 10:

$143_{\text{negative five}}$

$74_{\text{negative nine}}$

$10110_{\text{negative two}}$

b. Write -27 as a negative base five numeral.

c. Write 64 as a negative base three numeral.

d. Write 112 as a negative base ten numeral.

SECTION 4.5 | Prime Numbers

Prime Numbers

Number theory is a mathematical discipline that is primarily concerned with the properties that are exhibited by the natural numbers. The mathematician Carl Friedrich Gauss established many theorems in number theory. As we noted in the chapter opener, Gauss called mathematics the queen of the sciences and number theory the queen of mathematics. Many topics in number theory involve the concept of a *divisor* or *factor.*

Definition of Divisor

The natural number a is a **divisor** or **factor** of the natural number b provided there exists a natural number j such that $aj = b$.

In less formal terms, a natural number a is a divisor of the natural number b provided $b \div a$ has a remainder of 0. For instance, 10 has divisors of 1, 2, 5, and 10 because each of these numbers divides into 10 with a remainder of 0.

EXAMPLE 1 ■ Find Divisors

Determine all of the natural number divisors of each number.

a. 6 **b.** 42 **c.** 17

Solution

a. Divide 6 by 1, 2, 3, 4, 5, and 6. The division of 6 by 1, 2, 3, and 6 each produces a natural number quotient and a remainder of 0. Thus 1, 2, 3, and 6 are divisors of 6. Dividing 6 by 4 and 6 by 5 does not produce a remainder of 0. Therefore 4 and 5 are not divisors of 6.

b. The only natural numbers from 1 to 42 that divide into 42 with a remainder of 0 are 1, 2, 3, 6, 7, 14, 21, and 42. Thus the divisors of 42 are 1, 2, 3, 6, 7, 14, 21, and 42.

c. The only divisors of 17 are 1 and 17.

CHECK YOUR PROGRESS 1 Determine all of the natural number divisors of each number.

a. 9 **b.** 11 **c.** 24

Solution *See page S15.*

It is worth noting that every natural number greater than 1 has itself as a factor and 1 as a factor. If a natural number greater than 1 has only 1 and itself as factors, then it is a very special number known as a *prime number.*

Definition of a Prime Number and a Composite Number

A **prime number** is a natural number greater than 1 that has exactly two factors (divisors): itself and 1.
A **composite number** is a natural number greater than 1 that is not a prime number.

TAKE NOTE

The natural number 1 is neither a prime number nor a composite number.

The ten smallest prime numbers are 2, 3, 5, 7, 11, 13, 17, 19, 23, and 29. Each of these numbers has only itself and 1 as factors. The ten smallest composite numbers are 4, 6, 8, 9, 10, 12, 14, 15, 16, and 18.

EXAMPLE 2 ■ Classify a Number as a Prime Number or a Composite Number

Determine whether each number is a prime number or a composite number.

a. 41 **b.** 51 **c.** 119

Solution

a. The only divisors of 41 and 1 are 41. Thus 41 is a prime number.

b. The divisors of 51 are 1, 3, 17, and 51. Thus 51 is a composite number.

c. The divisors of 119 are 1, 7, 17, and 119. Thus 119 is a composite number.

CHECK YOUR PROGRESS 2 Determine whether each number is a prime number or a composite number.

a. 47 **b.** 171 **c.** 91

Solution *See page S15.*

QUESTION *Are all prime numbers odd numbers?*

Divisibility Tests

To determine whether one number is divisible by a smaller number, we often apply a **divisibility test,** which is a procedure that enables one to determine whether the smaller number is a divisor of the larger number without actually dividing the smaller number into the larger number. Table 4.13 provides divisibility tests for the numbers 2, 3, 4, 5, 6, 8, 9, 10, and 11.

ANSWER *No. The even number 2 is a prime number.*

Table 4.13 *Base Ten Divisibility Tests*

A number is divisible by the following divisor if:	Divisibility Test	Example
2	The number is an even number.	846 is divisible by 2 because 846 is an even number.
3	The sum of the digits of the number is divisible by 3.	531 is divisible by 3 because $5 + 3 + 1 = 9$ is divisible by 3.
4	The last two digits of the number form a number that is divisible by 4.	1924 is divisible by 4 because the last two digits form the number 24, which is divisible by 4.
5	The number ends with a 0 or a 5.	8785 is divisible by 5 because it ends with 5.
6	The number is divisible by 2 and by 3.	972 is divisible by 6 because it is divisible by 2 and also by 3.
8	The last three digits of the number form a number that is divisible by 8.	19,168 is divisible by 8 because the last three digits form the number 168, which is divisible by 8.
9	The sum of the digits of the number is divisible by 9.	621,513 is divisible by 9 because the sum of the digits is 18, which is divisible by 9.
10	The last digit is 0.	970 is divisible by 10 because it ends with 0.
11	Start at one end of the number and compute the sum of every other digit. Next compute the sum of the remaining digits. If the difference of these sums is divisible by 11, then the original number is divisible by 11.	4807 is divisible by 11 because the difference of the sum of the digits shown in blue ($8 + 7 = 15$) and the sum of the remaining digits shown in red ($4 + 0 = 4$) is $15 - 4 = 11$, which is divisible by 11.

EXAMPLE 3 ■ Apply Divisibility Tests

Use divisibility tests to determine whether 16,278 is divisible by the following numbers.

a. 2 **b.** 3 **c.** 5 **d.** 8 **e.** 11

Solution

a. Because 16,278 is an even number, it is divisible by 2.

b. The sum of the digits of 16,278 is 24, which is divisible by 3. Therefore, 16,278 is divisible by 3.

c. The number 16,278 does not end with a 0 or a 5. Therefore, 16,278 is not divisible by 5.

d. The last three digits of 16,278 form the number 278, which is not divisible by 8. Thus 16,278 is not divisible by 8.

e. The sum of the digits with even place-value powers is $1 + 2 + 8 = 11$. The sum of the digits with odd place-value powers is $6 + 7 = 13$. The difference of these sums is $13 - 11 = 2$. This difference is not divisible by 11, so 16,278 is not divisible by 11.

point of interest

Frank Nelson Cole (1861–1926) concentrated his mathematical work in the areas of number theory and group theory. He is well known for his 1903 presentation to the American Mathematical Society. Without speaking a word, he went to a chalkboard and wrote

$2^{67} - 1 =$
147573952589676412927

Many mathematicians considered this large number to be a prime number. Then Cole moved to a second chalkboard and computed the product

$761838257287 \cdot 193707721$

When the audience saw that this product was identical to the result on the first chalkboard, they gave Cole the only standing ovation ever given during an American Mathematical Society presentation. They realized that Cole had factored $2^{67} - 1$, which was a most remarkable feat, considering that no computers existed at that time.

CHECK YOUR PROGRESS 3 Use divisibility tests to determine whether 341,565 is divisible by each of the following numbers.

a. 3 **b.** 4 **c.** 10 **d.** 11

Solution *See page S15.*

Prime Factorization

The **prime factorization** of a composite number is a factorization that contains only prime numbers. Many proofs in number theory make use of the following important theorem.

The Fundamental Theorem of Arithmetic

Every composite number can be written as a unique product of prime numbers (disregarding the order of the factors).

To find the prime factorization of a composite number, rewrite the number as a product of two smaller natural numbers. If these smaller numbers are both prime numbers, then you are finished. If either of the smaller numbers is not a prime number, then rewrite it as a product of smaller natural numbers. Continue this procedure until all factors are primes. In Example 4 we make use of a *tree diagram* to organize the factorization process.

EXAMPLE 4 ■ Find the Prime Factorization of a Number

Determine the prime factorization of the following numbers.

a. 84 **b.** 495 **c.** 4004

Solution

a. The following tree diagrams show two different ways of finding the prime factorization of 84, which is $2 \cdot 2 \cdot 3 \cdot 7 = 2^2 \cdot 3 \cdot 7$. Each number in the tree is equal to the product of the two smaller numbers below it. The numbers (in red) at the extreme ends of the branches are the prime factors.

84 → 2, 42; 42 → 2, 21; 21 → 3, 7 or 84 → 4, 21; 4 → 2, 2; 21 → 3, 7

$84 = 2^2 \cdot 3 \cdot 7$

b.

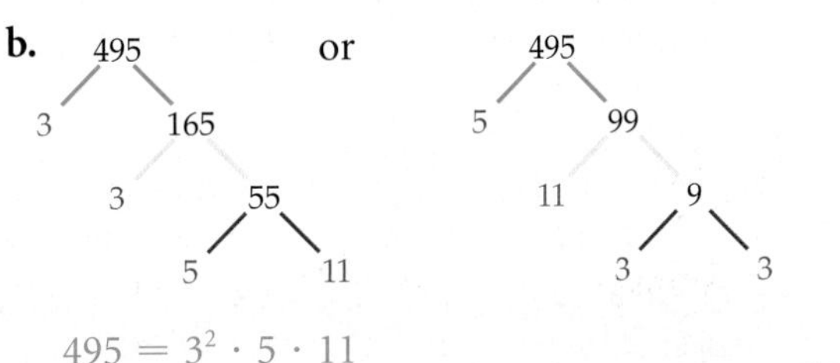

$495 = 3^2 \cdot 5 \cdot 11$

TAKE NOTE

The TI-83/84 program listed on page 139 can be used to find the prime factorization of a given natural number less than ten billion.

TAKE NOTE

The following compact division procedure can also be used to determine the prime factorization of a number.

2	4004
2	2002
7	1001
11	143
	13

In this procedure we use only prime number divisors, and we continue until the last quotient is a prime number. The prime factorization is the product of all the prime numbers, which are shown in red.

c.

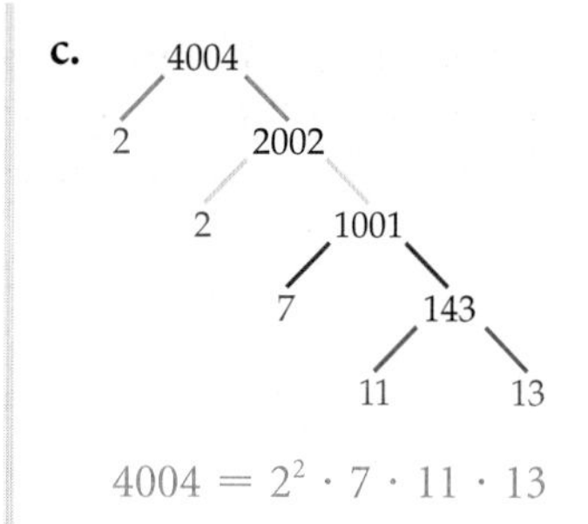

$4004 = 2^2 \cdot 7 \cdot 11 \cdot 13$

CHECK YOUR PROGRESS 4 Determine the prime factorization of the following.

a. 315 **b.** 273 **c.** 1309

Solution *See page S15.*

Math Matters Srinivasa Ramanujan

Srinivasa Ramanujan (1887–1920)

On January 16, 1913, the young 26-year-old Srinivasa Ramanujan (Rä-mä′noo-jûn) sent a letter from Madras, India, to the illustrious English mathematician G. H. Hardy. The letter requested that Hardy give his opinion about several mathematical ideas that Ramanujan had developed. In the letter Ramanujan explained, "I have not trodden through the conventional regular course which is followed in a University course, but I am striking out a new path for myself." Much of the mathematics was written using unconventional terms and notation; however, Hardy recognized (after many detailed readings and with the help of other mathematicians at Cambridge University) that Ramanujan was "a mathematician of the highest quality, a man of altogether exceptional originality and power."

On March 17, 1914, Ramanujan set sail for England, where he joined Hardy in a most unusual collaboration that lasted until Ramanujan returned to India in 1919. The following famous story is often told to illustrate the remarkable mathematical genius of Ramanujan.

> After Hardy had taken a taxicab to visit Ramanujan, he made the remark that the license plate number for the taxi was "1729, a rather dull number." Ramanujan immediately responded by saying that 1729 was a most interesting number, because it is the smallest natural number that can be expressed in two different ways as the sum of two cubes.

$$1^3 + 12^3 = 1729 \quad \text{and} \quad 9^3 + 10^3 = 1729$$

An interesting biography of the life of Srinivasa Ramanujan is given in *The Man Who Knew Infinity: A Life of the Genius Ramanujan* by Robert Kanigel.[1]

1. Kanigel, Robert. *The Man Who Knew Infinity: A Life of the Genius Ramanujan.* New York: Simon & Schuster, 1991.

TAKE NOTE

To determine whether a natural number is a prime number, it is necessary to consider only divisors from 2 up to the square root of the number, because every composite number n has at least one divisor less than or equal to $\sqrt{n}$. The proof of this statement is outlined in Exercise 77 of this section.

point of interest

In article 329 of *Disquisitiones Arithmeticae,* Gauss wrote:

"The problem of distinguishing prime numbers from composite numbers and of resolving the latter into their prime factors is known to be one of the most important and useful in arithmetic... The dignity of the science itself seems to require that every possible means be explored for the solution of a problem so elegant and so celebrated." (*Source: The Little Book of Big Primes* by Paulo Ribenboim. New York: Springer-Verlag, 1991.)

It is possible to determine whether a natural number n is a prime number by checking each natural number from 2 up to the largest integer not greater than $\sqrt{n}$ to see whether each is a divisor of n. If none of these numbers is a divisor of n, then n is a prime number. For large values of n, this division method is generally time-consuming and tedious. The Greek astronomer and mathematician Eratosthenes (about 276–192 B.C.) recognized that multiplication is generally easier than division, and he devised a method that makes use of multiples to determine every prime number in a list of natural numbers. Today we call this method the *Sieve of Eratosthenes.*

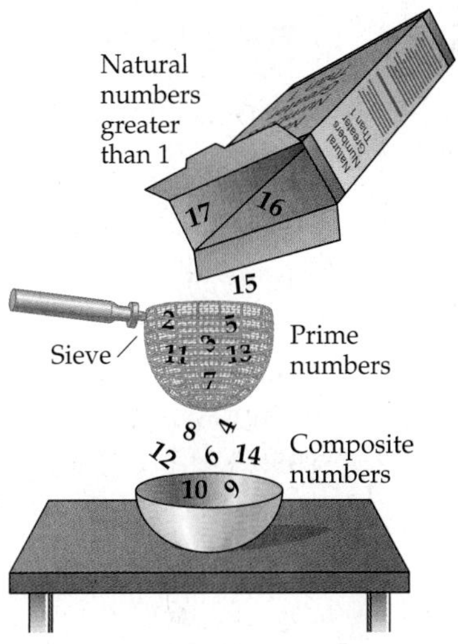

To sift prime numbers, first make a list of consecutive natural numbers. In Table 4.14 we have listed the consecutive counting numbers from 2 to 100.

- Cross out every multiple of 2 larger than 2. The next smallest remaining number in the list is 3. Cross out every multiple of 3 larger than 3.
- Call the next smallest remaining number in the list k. Cross out every multiple of k larger than k. Repeat this step for all $k < \sqrt{100}$.

Table 4.14 *The Sieve Method of Finding Primes*

	2	3	~~4~~	5	~~6~~	7	~~8~~	~~9~~	~~10~~
11	~~12~~	13	~~14~~	~~15~~	~~16~~	17	~~18~~	19	~~20~~
~~21~~	~~22~~	23	~~24~~	~~25~~	~~26~~	~~27~~	~~28~~	29	~~30~~
31	~~32~~	~~33~~	~~34~~	~~35~~	~~36~~	37	~~38~~	~~39~~	~~40~~
41	~~42~~	43	~~44~~	~~45~~	~~46~~	47	~~48~~	~~49~~	~~50~~
~~51~~	~~52~~	53	~~54~~	~~55~~	~~56~~	~~57~~	~~58~~	59	~~60~~
61	~~62~~	~~63~~	~~64~~	~~65~~	~~66~~	67	~~68~~	~~69~~	~~70~~
71	~~72~~	73	~~74~~	~~75~~	~~76~~	~~77~~	~~78~~	79	~~80~~
~~81~~	~~82~~	83	~~84~~	~~85~~	~~86~~	~~87~~	~~88~~	89	~~90~~
~~91~~	~~92~~	~~93~~	~~94~~	~~95~~	~~96~~	97	~~98~~	~~99~~	~~100~~

The numbers in blue that are not crossed out are prime numbers. Table 4.14 shows that there are 25 prime numbers less than 100.

Over 2000 years ago, Euclid proved that the set of prime numbers is an infinite set. Euclid's proof is an *indirect proof* or a *proof by contradiction.* Essentially his proof shows that for any finite list of prime numbers, we can create a number T, as described below, such that any prime factor of T can be shown to be a prime number that is not in the list. Thus there must be an infinite number of primes because it is not possible for all of the primes to be in any finite list.

Euclid's Proof Assume *all* of the prime numbers are contained in the list $p_1, p_2, p_3, \ldots, p_r$. Let $T = (p_1 \cdot p_2 \cdot p_3 \cdots p_r) + 1$. Either T is a prime number or T has a prime divisor. If T is a prime then it is a prime that is not in our list and we have reached a contradiction. If T is not prime, then one of the primes $p_1, p_2, p_3, \ldots, p_r$ must be a divisor of T. However, the number T is not divisible by any of the primes $p_1, p_2, p_3, \ldots, p_r$ because each p_i divides $p_1 \cdot p_2 \cdot p_3 \cdots p_r$ but does not divide 1. Hence any prime divisor of T, say p, is a prime number that is not in the list $p_1, p_2, p_3, \ldots, p_r$. So p is yet another prime number, and $p_1, p_2, p_3, \ldots, p_r$ is not a complete list of all the prime numbers.

Paul Erdös

We conclude this section with two quotations about prime numbers. The first is by the illustrious mathematician Paul Erdös (1913–1996); the second by the mathematics professor Don B. Zagier of the Max-Planck Institute, Bonn, Germany.

> It will be millions of years before we'll have any understanding, and even then it won't be a complete understanding, because we're up against the infinite.—P. Erdös, about prime numbers in *Atlantic Monthly,* November 1987, p. 74. *Source:* **http://www.mlahanas.de/Greeks/Primes.htm.**

In a 1975 lecture, D. Zagier commented,

Don B. Zagier

> There are two facts about the distribution of prime numbers of which I hope to convince you so overwhelmingly that they will be permanently engraved in your hearts. The first is that, despite their simple definition and role as the building blocks of the natural numbers, the prime numbers grow like weeds among the natural numbers, seeming to obey no other law than that of chance, and nobody can predict where the next one will sprout. The second fact is even more astonishing, for it states just the opposite: that the prime numbers exhibit stunning regularity, that there are laws governing their behavior, and that they obey these laws with almost military precision.—(Havil 2003, p. 171). *Source:* **http://mathworld.wolfram.com/PrimeNumber.html.**

Excursion

The Distribution of the Primes

Many mathematicians have searched without success for a mathematical formula that can be used to generate the sequence of prime numbers. We know that the prime numbers form an infinite sequence. However, the distribution of the prime numbers within the sequence of natural numbers is very complicated. The ratio of prime numbers to composite numbers appears to become smaller and smaller as larger and larger numbers are considered. In general, the number of consecutive composite numbers that come

(continued)

between two prime numbers tends to increase as the size of the numbers becomes larger; however, this increase is erratic and appears to be unpredictable.

In this Excursion we refer to a list of two or more consecutive composite numbers as a **prime desert.** For instance, 8, 9, 10 is a prime desert because it consists of three consecutive composite numbers. The longest prime desert shown in Table 4.14 is the seven consecutive composite numbers 90, 91, 92, 93, 94, 95, and 96. A formula that involves *factorials* can be used to form prime deserts of any finite length.

Definition of *n* factorial

If n is a natural number, then $n!$, which is read "n factorial," is defined as

$$n! = n \cdot (n-1) \cdot \cdots \cdot 3 \cdot 2 \cdot 1$$

As an example of a factorial, consider $4! = 4 \cdot 3 \cdot 2 \cdot 1 = 24$.

The sequence

$$4! + 2, 4! + 3, 4! + 4$$

is a prime desert of the three composite numbers 26, 27, and 28. Figure 4.1 below shows a prime desert of 10 consecutive composite numbers. Figure 4.2 shows a procedure that can be used to produce a prime desert of n composite numbers, where n is any natural number greater than 2.

$$\left.\begin{array}{l} 11! + 2 \\ 11! + 3 \\ 11! + 4 \\ 11! + 5 \\ 11! + 6 \\ \vdots \\ 11! + 10 \\ 11! + 11 \end{array}\right\} \text{A prime desert of 10 consecutive composite numbers}$$

Figure 4.1

$$\left.\begin{array}{l} (n+1)! + 2 \\ (n+1)! + 3 \\ (n+1)! + 4 \\ (n+1)! + 5 \\ (n+1)! + 6 \\ \vdots \\ (n+1)! + n \\ (n+1)! + (n+1) \end{array}\right\} \text{A prime desert of } n \text{ consecutive composite numbers}$$

Figure 4.2

A prime desert of length one million is shown by the sequence

$$1{,}000{,}001! + 2; 1{,}000{,}001! + 3; 1{,}000{,}001! + 4; \ldots; 1{,}000{,}001! + 1{,}000{,}001$$

It appears that the distribution of prime numbers is similar to the situation wherein a mathematical gardener plants an infinite number of grass seeds on a windy day. Many of the grass seeds fall close to the gardener, but some are blown down the street and into the next neighborhood. There are gaps where no grass seeds are within 1 mile of each other. Farther down the road there are gaps where no grass seeds are within 10 miles of each other. No matter how far the gardener travels and how long it has been since the last grass seed was spotted, the gardener knows that more grass seeds will appear.

Excursion Exercises

1. Explain how you know that each of the numbers

$$1{,}000{,}001! + 2; 1{,}000{,}001! + 3; 1{,}000{,}001! + 4; \ldots; 1{,}000{,}001! + 1{,}000{,}001$$

is a composite number.

(continued)

2. Use factorials to generate the numbers in a prime desert of 12 consecutive composite numbers. Now use a calculator to evaluate each number in this prime desert.
3. Use factorials and "..." notation to represent a prime desert of
   a. 20 consecutive composite numbers.
   b. 500,000 consecutive composite numbers.
   c. 7 billion consecutive composite numbers.

Exercise Set 4.5

In Exercises 1–10, determine all natural number divisors of the given number.

1. 20 **2.** 32
3. 65 **4.** 75
5. 41 **6.** 79
7. 110 **8.** 150
9. 385 **10.** 455

In Exercises 11–20, determine whether each number is a prime number or a composite number.

11. 21 **12.** 31
13. 37 **14.** 39
15. 101 **16.** 81
17. 79 **18.** 161
19. 203 **20.** 211

In Exercises 21–28, use the divisibility tests in Table 4.13 to determine whether the given number is divisible by each of the following: 2, 3, 4, 5, 6, 8, 9, and 10.

21. 210 **22.** 314
23. 51 **24.** 168
25. 2568 **26.** 3525
27. 4190 **28.** 6123

In Exercises 29–40, write the prime factorization of the number.

29. 18 **30.** 48
31. 120 **32.** 380
33. 425 **34.** 625
35. 1024 **36.** 1410
37. 6312 **38.** 3155
39. 18,234 **40.** 19,345

41. Use the Sieve of Eratosthenes procedure to find all prime numbers from 2 to 200. *Hint:* Because $\sqrt{200} \approx 14.1$, you need to continue the sieve procedure up to $k = 13$. Note: You do not need to consider $k = 14$ because 14 is not a prime number.

42. Use your list of prime numbers from Exercise 41 to find the number of prime numbers from:
   a. 2 to 50 **b.** 51 to 100
   c. 101 to 150 **d.** 151 to 200

43. **Twin Primes** If the natural numbers n and $n + 2$ are both prime numbers, then they are said to be **twin primes.** For example, 11 and 13 are twin primes. It is not known whether the set of twin primes is an infinite set or a finite set. Use the list of primes from Exercise 41 to write all twin primes less than 200.

44. **Twin Primes** Find a pair of twin primes between 200 and 300. See Exercise 43.

45. **Twin Primes** Find a pair of twin primes between 300 and 400. See Exercise 43.

46. **A Prime Triplet** If the natural numbers n, $n + 2$, and $n + 4$ are all prime numbers, then they are said to be **prime triplets.** Write a few sentences that explain why the prime triplets 3, 5, and 7 are the only prime triplets.

47. **Goldbach's Conjecture** In 1742, Christian Goldbach conjectured that every even number greater than 2 can be written as the sum of two prime numbers. Many mathematicians have tried to prove or disprove this conjecture without succeeding. Show that *Goldbach's conjecture* is true for each of the following even numbers.

a. 24 **b.** 50
c. 86 **d.** 144
e. 210 **f.** 264

48. **Perfect Squares** The square of a natural number is called a **perfect square.** Pick six perfect squares. For each perfect square, determine the number of distinct natural-number factors of the perfect square. Make a conjecture about the number of distinct natural-number factors of any perfect square.

Every prime number has a divisibility test. Many of these divisibility tests are slight variations of the following divisibility test for 7.

A divisibility test for 7 To determine whether a given base ten number is divisible by 7, double the ones digit of the given number. Find the difference between this number and the number formed by omitting the ones digit from the given number. If necessary, repeat this procedure until you obtain a small final difference.

If the final difference is divisible by 7, then the given number is also divisible by 7.

If the final difference is not divisible by 7, then the given number is not divisible by 7.

Example Use the above divisibility test to determine whether 301 is divisible by 7.

Solution The double of the ones digit is 2. Subtracting 2 from 30, which is the number formed by omitting the ones digit from the original number, yields 28. Because 28 is divisible by 7, the original number 301 is divisible by 7.

In Exercises 49–56, use the above divisibility test for 7 to determine whether each number is divisible by 7.

49. 182 **50.** 203
51. 1001 **52.** 2403
53. 11,561 **54.** 13,842
55. 204,316 **56.** 789,327

A divisibility test for 13 To determine whether a given base ten number is divisible by 13, multiply the ones digit of the given number by 4. Find the sum of this multiple of 4 and the number formed by omitting the ones digit from the given number. If necessary, repeat this procedure until you obtain a small final sum.

If the final sum is divisible by 13, then the given number is divisible by 13.

If the final sum is not divisible by 13, then the given number is not divisible by 13.

Example Use the above divisibility test to determine whether 1079 is divisible by 13.

Solution Four times the ones digit is 36. The number formed by omitting the ones digit is 107. The sum of 36 and 107 is 143. Now repeat the procedure on 143. Four times the ones digit is 12. The sum of 12 and 14, which is the number formed by omitting the ones digit, is 26. Because 26 is divisible by 13, the original number 1079 is divisible by 13.

In Exercises 57–64, use the above divisibility test for 13 to determine whether each number is divisible by 13.

57. 91 **58.** 273
59. 1885 **60.** 8931
61. 14,507 **62.** 22,184
63. 13,351 **64.** 85,657

Extensions

CRITICAL THINKING

65. **Factorial Primes** A prime number of the form $n! \pm 1$ is called a **factorial prime.** Recall that the notation $n!$ is called n factorial and represents the product of all natural numbers from 1 to n. For example, $4! = 4 \cdot 3 \cdot 2 \cdot 1 = 24$. Factorial primes are of interest to mathematicians because they often signal the end or the beginning of a lengthy string of consecutive composite numbers. See the Excursion on page 226.

a. Find the smallest value of n such that $n! + 1$ and $n! - 1$ are twin primes.

b. Find the smallest value of n for which $n! + 1$ is a composite number and $n! - 1$ is a prime number.

66. Primorial Primes The notation $p\#$ represents the product of all the prime numbers less than or equal to the prime number p. For instance,

$$3\# = 2 \cdot 3 = 6$$

$$5\# = 2 \cdot 3 \cdot 5 = 30$$

$$11\# = 2 \cdot 3 \cdot 5 \cdot 7 \cdot 11 = 2310$$

A **primorial prime** is a prime number of the form $p\# \pm 1$. For instance, $3\# + 1 = 2 \cdot 3 + 1 = 7$ and $3\# - 1 = 2 \cdot 3 - 1 = 5$ are both primorial primes. Large primorial primes are often examined in the search for a pair of large twin primes.

a. Find the smallest prime number p, where $p \geq 7$, such that $p\# + 1$ and $p\# - 1$ are twin primes.

b. Find the smallest prime number p, such that $p\# + 1$ is a prime number but $p\# - 1$ is a composite number.

67. A Divisibility Test for 17 Determine a divisibility test for 17. *Hint*: One divisibility test for 17 is similar to the divisibility test for 7 on the previous page in that it involves the last digit of the given number and the operation of subtraction.

68. A Divisibility Test for 19 Determine a divisibility test for 19. *Hint*: One divisibility test for 19 is similar to the divisibility test for 13 on the previous page in that it involves the last digit of the given number and the operation of addition.

Number of Divisors of a Composite Number The following method can be used to determine the number of divisors of a composite number. First find the prime factorization (in exponential form) of the composite number. Add 1 to each exponent in the prime factorization and then compute the product of these exponents. This product is equal to the number of divisors of the composite number. To illustrate that this procedure yields the correct result, consider the composite number 12, which has the six divisors 1, 2, 3, 4, 6, and 12. The prime factorization of 12 is $2^2 \cdot 3^1$. Adding 1 to each of the exponents produces the numbers 3 and 2. The product of 3 and 2 is 6, which agrees with the result obtained by listing all of the divisors.

In Exercises 69–74, determine the number of divisors of each composite number.

69. 60

70. 84

71. 297

72. 288

73. 360

74. 875

EXPLORATIONS

75. Kummer's Proof In the 1870s, the mathematician Eduard Kummer used a proof similar to the following to show that there exist an infinite number of prime numbers. Supply the missing reasons in parts a and b.

a. ***Proof*** Assume there exist only a finite number of prime numbers, say $p_1, p_2, p_3, \ldots, p_r$. Let $N = p_1 p_2 p_3 \cdots p_r > 2$. The natural number $N - 1$ has at least one common prime factor with N. Why?

b. Call the common prime factor from part a p_i. Now p_i divides N and p_i divides $N - 1$. Thus p_i divides their difference: $N - (N - 1) = 1$. Why?

This leads to a contradiction, because no prime number is a divisor of 1. Hence Kummer concluded that the original assumption was incorrect, and there must exist an infinite number of prime numbers.

76. *Theorem* If a number of the form 111...1 is a prime number, then the number of 1's in the number is a prime number. For instance,

11 and 1,111,111,111,111,111,111

are prime numbers, and the number of 1's in each number (two in the first number and 19 in the second number) is a prime number.

a. What is the converse of the above theorem? Is the converse of a theorem always true?

b. The number $111 = 3 \cdot 37$, so 111 is not a prime number. Explain why this does not contradict the above theorem.

77. State the missing reasons in parts a, b, and c of the following proof.

Theorem Every composite number n has at least one divisor less than or equal to $\sqrt{n}$.

a. ***Proof*** Assume a is a divisor of n. Then there exists a natural number j such that $aj = n$. Why?

b. Now a and j cannot both be greater than $\sqrt{n}$, because this would imply that $aj > \sqrt{n}\sqrt{n}$. However, $\sqrt{n}\sqrt{n}$ simplifies to n, which equals aj. What contradiction does this lead to?

c. Thus either a or j must be less than or equal to $\sqrt{n}$. Because j is also a divisor of n, the proof is complete. How do we know that j is a divisor of n?

78. **The RSA Algorithm** In 1977, Ron Rivest, Adi Shamir, and Leonard Adleman invented a method for encrypting information. Their method is known as the RSA algorithm. Today the RSA algorithm is used by both the Microsoft Internet Explorer and Netscape Navigator Web browsers, as well as by VISA and MasterCard to ensure secure electronic credit card transactions. The RSA algorithm involves large prime numbers. Research the RSA algorithm and write a report about some of the reasons why this algorithm has become one of the most popular of all the encryption algorithms.

79. **Theorems and Conjectures** Use the Internet or a text on prime numbers to determine whether each of the following statements is an established theorem or a conjecture. (*Note:* n represents a natural number).

a. There are infinitely many twin primes.

b. There are infinitely many primes of the form $n^2 + 1$.

c. There is always a prime number between n and $2n$ for $n \geq 2$.

d. There is always a prime number between n^2 and $(n + 1)^2$.

e. Every odd number greater than 5 can be written as the sum of three primes.

f. Every positive even number can be written as the difference of two primes.

SECTION 4.6 Topics from Number Theory

Perfect, Deficient, and Abundant Numbers

The ancient Greek mathematicians personified the natural numbers. For instance, they considered the odd natural numbers as male and the even natural numbers as female. They also used the concept of a *proper factor* to classify a natural number as *perfect, deficient,* or *abundant.* The **proper factors** of a natural number n include all natural number factors of n except for the number n itself. For instance, the proper factors of 10 are 1, 2, and 5. The proper factors of 16 are 1, 2, 4, and 8.

Perfect, Deficient, and Abundant Numbers

A natural number is

- **perfect** if it is equal to the sum of its proper factors.
- **deficient** if it is greater than the sum of its proper factors.
- **abundant** if it is less than the sum of its proper factors.

point of interest

Six is a number perfect in itself, and not because God created the world in six days; rather the contrary is true. God created the world in six days because this number is perfect, and it would remain perfect, even if the work of the six days did not exist.
—*St. Augustine* (354–430)

EXAMPLE 1 ■ Classify a Number as Perfect, Deficient, or Abundant
Determine whether the following numbers are perfect, deficient, or abundant.

a. 6 **b.** 20 **c.** 25

Solution

a. The proper factors of 6 are 1, 2, and 3. The sum of these proper factors is $1 + 2 + 3 = 6$. Because 6 is equal to the sum of its proper divisors, 6 is a perfect number.

b. The proper factors of 20 are 1, 2, 4, 5, and 10. The sum of these proper factors is $1 + 2 + 4 + 5 + 10 = 22$. Because 20 is less than the sum of its proper factors, 20 is an abundant number.

c. The proper factors of 25 are 1 and 5. The sum of these proper factors is $1 + 5 = 6$. Because 25 is greater than the sum of its proper factors, 25 is a deficient number.

CHECK YOUR PROGRESS 1 Determine whether the following numbers are perfect, deficient, or abundant.

a. 24 **b.** 28 **c.** 35

Solution *See page S15.*

Mersenne Numbers and Perfect Numbers

Marin Mersenne (1588–1648)

As a French monk in the religious order known as the Minims, Marin Mersenne (mər-sĕn′) devoted himself to prayer and his studies, which included topics from number theory. Mersenne took it upon himself to collect and disseminate mathematical information to scientists and mathematicians through Europe. Mersenne was particularly interested in prime numbers of the form $2^n - 1$, where n is a prime number. Today numbers of the form $2^n - 1$, where n is a prime number, are known as **Mersenne numbers.**

Some Mersenne numbers are prime and some are composite. For instance, the Mersenne numbers $2^2 - 1$ and $2^3 - 1$ are prime numbers, but the Mersenne number $2^{11} - 1 = 2047$ is not prime because $2047 = 23 \cdot 89$.

EXAMPLE 2 ■ Determine Whether a Mersenne Number is a Prime Number

Determine whether the Mersenne number $2^5 - 1$ is a prime number.

Solution
$2^5 - 1 = 31$ and 31 is a prime number. Thus $2^5 - 1$ is a Mersenne prime.

CHECK YOUR PROGRESS 2 Determine whether the Mersenne number $2^7 - 1$ is a prime number.

Solution *See page S16.*

The ancient Greeks knew that the first four perfect numbers were 6, 28, 496, and 8128. In fact, proposition 36 from Volume IX of Euclid's *Elements* states a procedure that uses Mersenne primes to produce a perfect number.

Euclid's Procedure for Generating a Perfect Number

If n and $2^n - 1$ are both prime numbers, then $2^{n-1}(2^n - 1)$ is a perfect number.

Euclid's procedure shows how every Mersenne prime can be used to produce a perfect number. For instance,

If $n = 2$, then $2^{2-1}(2^2 - 1) = 2(3) = 6$.
If $n = 3$, then $2^{3-1}(2^3 - 1) = 4(7) = 28$.
If $n = 5$, then $2^{5-1}(2^5 - 1) = 16(31) = 496$.
If $n = 7$, then $2^{7-1}(2^7 - 1) = 64(127) = 8128$.

The fifth perfect number was not discovered until the year 1461. It is $2^{12}(2^{13} - 1) = 33{,}550{,}336$. The sixth and seventh perfect numbers were discovered in 1588 by P. A. Cataldi. In exponential form, they are $2^{16}(2^{17} - 1)$ and $2^{18}(2^{19} - 1)$. It is interesting to observe that the ones' digits of the first five perfect numbers alternate: 6, 8, 6, 8, 6. Evaluate $2^{16}(2^{17} - 1)$ and $2^{18}(2^{19} - 1)$ to determine if this alternating pattern continues for the first seven perfect numbers.

EXAMPLE 3 ■ Use a Given Mersenne Prime Number to Write a Perfect Number

In 1750, Leonhard Euler proved that $2^{31} - 1$ is a Mersenne prime. Use Euclid's theorem to write the perfect number associated with this prime.

Solution
Euler's theorem states that if n and $2^n - 1$ are both prime numbers, then $2^{n-1}(2^n - 1)$ is a perfect number. In this example $n = 31$, which is a prime number. We are given that $2^{31} - 1$ is a prime number, so the perfect number we seek is $2^{30}(2^{31} - 1)$.

CHECK YOUR PROGRESS 3 In 1883, I. M. Pervushin proved that $2^{61} - 1$ is a Mersenne prime. Use Euclid's theorem to write the perfect number associated with this prime.

Solution *See page S16.*

QUESTION *Must a perfect number produced by Euclid's perfect-number-generating procedure be an even number?*

The search for Mersenne primes and their associated perfect numbers still continues. As of July 2005, only 42 Mersenne primes (and 42 perfect numbers) had been discovered. The largest of these 42 Mersenne primes is $(2^{25{,}964{,}951} - 1)$. See Table 4.15 on page 234. This gigantic Mersenne prime number is also the largest known prime number as of July 2005. It has 7,816,230 digits and it was discovered on February 18, 2005 by Dr. Martin Nowak, who used his 2.4-gigahertz Pentium 4 personal computer and a program that is available on the Internet. (*Source:* **http://www.mersenne.org**) More information about this program is given in the *Math Matters* on page 235.

ANSWER *Yes. The 2^{n-1} factor of $2^{n-1}(2^n - 1)$ ensures that this product will be an even number.*

Table 4.15 *Some Mersenne Primes and Their Associated Perfect Numbers*

	Mersenne Prime	Perfect Number	Date	Discoverer
1	$2^2 - 1$	$2^1(2^2 - 1) = 6$	B.C.	
2	$2^3 - 1$	$2^2(2^3 - 1) = 28$	B.C.	
3	$2^5 - 1$	$2^4(2^5 - 1) = 496$	B.C.	
4	$2^7 - 1$	$2^6(2^7 - 1) = 8128$	B.C.	
5	$2^{13} - 1$	$2^{12}(2^{13} - 1) = 33{,}550{,}336$	1461	Unknown
6	$2^{17} - 1$	$2^{16}(2^{17} - 1) = 8{,}589{,}869{,}056$	1588	Cataldi
7	$2^{19} - 1$	$2^{18}(2^{19} - 1) = 137{,}438{,}691{,}328$	1588	Cataldi
	⋮	⋮	⋮	⋮
32	$2^{756839} - 1$	$2^{756838}(2^{756839} - 1)$	1992	Slowinski and Gage
33	$2^{859433} - 1$	$2^{859432}(2^{859433} - 1)$	1994	Slowinski and Gage
34	$2^{1257787} - 1$	$2^{1257786}(2^{1257787} - 1)$	1996	Slowinski and Gage
35	$2^{1398269} - 1$	$2^{1398268}(2^{1398269} - 1)$	1996	Armengaud and Woltman, et al. (GIMPS)
36	$2^{2976221} - 1$	$2^{2976220}(2^{2976221} - 1)$	1997	Spence and Woltman, et al. (GIMPS)
37	$2^{3021377} - 1$	$2^{3021376}(2^{3021377} - 1)$	1998	Clarkson, Woltman, and Kurowski, et al. (GIMPS, PrimeNet)
38	$2^{6972593} - 1$	$2^{6972592}(2^{6972593} - 1)$	1999	Hajratwala, Woltman, and Kurowski, et al. (GIMPS, PrimeNet)
?	$2^{13466917} - 1$	$2^{13466916}(2^{13466917} - 1)$	2001	Cameron, Woltman, and Kurowski, et al. (GIMPS, PrimeNet)
?	$2^{20996011} - 1$	$2^{20996010}(2^{20996011} - 1)$	2003	Shafer, Woltman, Kurowski, et al. (GIMPS, PrimeNet)
?	$2^{24036583} - 1$	$2^{24036582}(2^{24036583} - 1)$	2004	Findley, Woltman, Kurowski, et al. (GIMPS, PrimeNet)
?	$2^{25964951} - 1$	$2^{25964950}(2^{25964951} - 1)$	2005	Nowak, Woltman, Kurowski, et al. (GIMPS, PrimeNet)

Source: The Little Book of Big Primes and the GIMPS homepage (**http://www.mersenne.org**)
Note: A complete listing of Mersenne primes is given at **http://www.utm.edu/research/primes/mersenne/**

The numbers to the right of the question marks in the bottom four rows of Table 4.15 are the 39th through the 42nd Mersenne primes that have been discovered. They may not be the 39th, 40th, 41st, and 42nd Mersenne primes because there are many smaller numbers that have yet to be tested.

Math Matters The Great Internet Mersenne Prime Search

From 1952 to 1996, large-scale computers were used to find Mersenne primes. However, during the past few years, small personal computers working in parallel have joined in the search. This search using personal computers has been organized by a fast-growing Internet organization known as the Great Internet Mersenne Prime Search (GIMPS). Members of this group use the Internet to download a Mersenne prime program that runs on their personal computers. Each member is assigned a range of numbers to check for Mersenne primes. A search over a specified range of numbers can take several weeks, but because the program is designed to run in the background, you can still use your computer to perform its regular duties. As of July 2005, eight Mersenne primes had been discovered by the members of GIMPS (see the bottom eight rows of Table 4.15). One of the current goals of GIMPS is to find a prime number that has more than 10 million digits. In fact, the Electronic Frontier Foundation is offering a $100,000 reward to the person or group to discover a 10-million-digit prime number. If you are using your computer just to run a screen saver, why not join in the search? You can get the needed program and additional information at

http://www.mersenne.org

Who knows, maybe you will discover a new Mersenne prime and share in the reward money. Of course, we know that you really just want to have your name added to Table 4.15.

The Number of Digits in b^x

To determine the number of digits in a Mersenne number, mathematicians make use of the following formula, which involves finding the *greatest integer* of a number. **The greatest integer** of a number k is the greatest integer less than or equal to k. For instance, the greatest integer of 5 is 5 and the greatest integer of 7.8 is 7.

TAKE NOTE

If 2^{19} were a power of 10, such as 100,000, then $2^{19} - 1$ would equal 99,999 and it would have one less digit than 2^{19}. However, we know 2^{19} is not a power of 10 and thus $2^{19} - 1$ has the same number of digits as 2^{19}.

The Number of Digits in b^x

The number of digits in the number b^x, where b is a natural number less than 10 and x is a natural number, is the greatest integer of $(x \log b) + 1$.

CALCULATOR NOTE

To evaluate 19 log 2 on a TI-83/84 calculator press

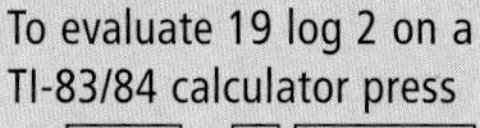

19 LOG 2) ENTER

On a scientific calculator press

19 × 2 LOG =

EXAMPLE 4 ■ Determine the Number of Digits in a Mersenne Number

Find the number of digits in the Mersenne prime number $2^{19} - 1$.

Solution

First consider just 2^{19}. The base b is 2. The exponent x is 19.

$$\begin{aligned}(x \log b) + 1 &= (19 \log 2) + 1 \\ &\approx 5.72 + 1 \\ &= 6.72\end{aligned}$$

The greatest integer of 6.72 is 6. Thus 2^{19} has six digits. The Mersenne prime number $2^{19} - 1$ also has six digits. See the Take Note in the margin.

CHECK YOUR PROGRESS 4 Find the number of digits in the Mersenne prime number $2^{2976221} - 1$.

Solution See page S16.

Euclid's perfect-number-generating formula produces only *even* perfect numbers. Do odd perfect numbers exist? As of July 2005, no odd perfect number had been discovered and no one had been able to prove either that odd perfect numbers exist or that they do not exist. The question of the existence of an odd perfect number is one of the major unanswered questions of number theory.

Fermat's Last Theorem

historical note

Pierre de Fermat (1601–1665) In the mathematical community, Pierre de Fermat (fĕhr′məh) is known as the Prince of Amateurs because although he spent his professional life as a councilor and judge, he spent his leisure time working on mathematics. As an amateur mathematician, Fermat did important work in analytic geometry and calculus, but he is remembered today for his work in the area of number theory. Fermat stated theorems he had developed, but seldom did he provide the actual proofs. He did not want to waste his time showing the details required of a mathematical proof, and then spend additional time defending his work once it was scrutinized by other mathematicians.

By the twentieth century all but one of Fermat's proposed theorems had been proved by other mathematicians. The remaining unproved theorem became known as Fermat's Last Theorem. ■

In 1637, the French mathematician Pierre de Fermat wrote in the margin of a book:

> It is impossible to divide a cube into two cubes, or a fourth power into two fourth powers, or in general any power greater than the second into two like powers, and I have a truly marvelous demonstration of it. But this margin will not contain it.

This problem, which became known as **Fermat's Last Theorem,** can also be stated in the following manner.

Fermat's Last Theorem

There are no natural numbers x, y, z, and n that satisfy $x^n + y^n = z^n$, where n is greater than 2.

Fermat's Last Theorem has attracted a great deal of attention over the last three centuries. The theorem has become so well known that it is simply called "FLT." Here are some of the reasons for its popularity:

- Very little mathematical knowledge is required to understand the statement of FLT.
- FLT is an extension of the well-known Pythagorean theorem $x^2 + y^2 = z^2$, which has several natural number solutions. Two such solutions are $3^2 + 4^2 = 5^2$ and $5^2 + 12^2 = 13^2$.
- It seems so simple. After all, while reading the text *Arithmetica* by Diophantus, Fermat wrote that he had discovered a *truly marvelous proof* of FLT. The only reason that Fermat gave for not providing his proof was that the margin of *Arithmetica* was too narrow to contain it.

Many famous mathematicians have worked on FLT. Some of these mathematicians tried to disprove FLT by searching for natural numbers x, y, z, and n that satisfied $x^n + y^n = z^n$, $n > 2$.

EXAMPLE 5 ■ Check a Possible Solution to Fermat's Last Theorem

Determine whether $x = 6$, $y = 8$, and $n = 3$ satisfies the equation $x^n + y^n = z^n$, where z is a natural number.

Solution

Substituting 6 for x, 8 for y, and 3 for n in $x^n + y^n = z^n$ yields

$$6^3 + 8^3 = z^3$$
$$216 + 512 = z^3$$
$$728 = z^3$$

The real solution of $z^3 = 728$ is $\sqrt[3]{728} \approx 8.99588289$, which is not a natural number. Thus $x = 6$, $y = 8$, and $n = 3$ does not satisfy the equation $x^n + y^n = z^n$, where z is a natural number.

CHECK YOUR PROGRESS 5 Determine whether $x = 9$, $y = 11$, and $n = 4$ satisfies the equation $x^n + y^n = z^n$, where z is a natural number.

Solution *See page S16.*

point of interest

Andrew Wiles

Fermat's Last Theorem had been labeled by some mathematicians as the world's hardest mathematical problem. After solving Fermat's Last Theorem, Andrew Wiles made the following remarks: "Having solved this problem there's certainly a sense of loss, but at the same time there is this tremendous sense of freedom. I was so obsessed by this problem that for eight years I was thinking about it all the time—when I woke up in the morning to when I went to sleep at night. That's a long time to think about one thing. That particular odyssey is now over. My mind is at rest."

In the eighteenth century, the great mathematician Leonhard Euler was able to make some progress on a proof of FLT. He adapted a technique that he found in Fermat's notes about another problem. In this problem, Fermat gave an outline of how to prove that the special case $x^4 + y^4 = z^4$ has no natural number solutions. Using a similar procedure, Euler was able to show that $x^3 + y^3 = z^3$ also has no natural number solutions. Thus all that was left was to show that $x^n + y^n = z^n$ has no solutions with n greater than 4.

In the nineteenth century, additional work on FLT was done by Sophie Germain, Augustin Louis Cauchy, and Gabriel Lame. Each of these mathematicians produced some interesting results, but FLT still remained unsolved.

In 1983, the German mathematician Gerd Faltings used concepts from differential geometry to prove that the number of solutions to FLT must be finite. Then, in 1988, the Japanese mathematician Yoichi Miyaoka claimed he could show that the number of solutions to FLT was not only finite, but that the number of solutions was zero, and thus he had proved FLT. At first Miyaoka's work appeared to be a valid proof, but after a few weeks of examination a flaw was discovered. Several mathematicians looked for a way to repair the flaw, but eventually they came to the conclusion that although Miyaoka had developed some interesting mathematics, he had failed to establish the validity of FLT.

In 1993, Andrew Wiles of Princeton University made a major advance toward a proof of FLT. Wiles first became familiar with FLT when he was only 10 years old. At his local public library, Wiles first learned about FLT in the book *The Last Problem* by Eric Temple Bell. In reflecting on his first thoughts about FLT, Wiles recalled

> It looked so simple, and yet all the great mathematicians in history couldn't solve it. Here was a problem that I, a ten-year-old, could understand and I knew from that moment that I would never let it go. I had to solve it.[2]

---

2. Singh, Simon. *Fermat's Enigma: The Quest to Solve the World's Greatest Mathematical Problem.* New York: Walker Publishing Company, Inc., 1997, p. 6.

Wiles took a most unusual approach to solving FLT. Whereas most contemporary mathematicians share their ideas and coordinate their efforts, Wiles decided to work alone. After 7 years of working in the attic of his home, Wiles was ready to present his work. In June of 1993 Wiles gave a series of three lectures at the Isaac Newton Institute in Cambridge, England. After showing that FLT was a corollary of his major theorem, Wiles's concluding remark was "I think I'll stop here." Many mathematicians in the audience felt that Wiles had produced a valid proof of FLT, but a formal verification by several mathematical referees was required before Wiles's work could be classified as an official proof. The verification process was lengthy and complex. Wiles's written work was about 200 pages in length, it covered several different areas of mathematics, and it used hundreds of sophisticated logical arguments and a great many mathematical calculations. Any mistake could result in an invalid proof. Thus it was not too surprising when a flaw was discovered in late 1993. The flaw did not necessarily imply that Wiles's proof could not be repaired, but it did indicate that it was not a valid proof in its present form.

It appeared that once again the proof of FLT had eluded a great effort by a well-known mathematician. Several months passed and Wiles was still unable to fix the flaw. It was a most depressing period for Wiles. He felt that he was close to solving one of the world's hardest mathematical problems, yet all of his creative efforts failed to turn the flawed proof into a valid proof. Several mathematicians felt that it was not possible to repair the flaw, but Wiles did not give up. Finally, in late 1994, Wiles had an insight that eventually led to a valid proof. The insight required Wiles to seek additional help from Richard Taylor, who had been a student of Wiles. On October 15, 1994, Wiles and Taylor presented to the world a proof that has now been judged to be a valid proof of FLT. Their proof is certainly not the "truly marvelous proof" that Fermat said he had discovered. But it has been deemed a wonderful proof that makes use of several new mathematical procedures and concepts.

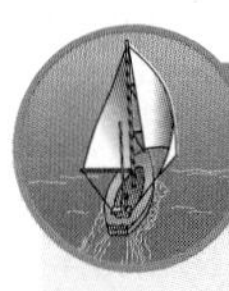

Excursion

A Sum of the Divisors Formula

Consider the numbers 10, 12, and 28 and the sums of the proper factors of these numbers.

$$10: 1 + 2 + 5 = 8 \qquad 12: 1 + 2 + 3 + 4 + 6 = 16$$

$$28: 1 + 2 + 4 + 7 + 14 = 28$$

From the above sums we see that 10 is a deficient number, 12 is an abundant number, and 28 is a perfect number. The goal of this Excursion is to find a method that will enable us to determine whether a number is deficient, abundant, or perfect without having to first find all of its proper factors and then compute their sum.

(continued)

In the following example, we use the number 108 and its prime factorization $2^2 \cdot 3^3$ to illustrate that every factor of 108 can be written as a product of powers of its prime factors. Table 4.16 includes all the proper factors of 108 (the numbers in blue) plus the factor $2^2 \cdot 3^3$, which is 108 itself. The sum of each column is shown at the bottom (the numbers in red).

Table 4.16 *Every Factor of 108 Expressed as a Product of Powers of its Prime Factors*

	1	$1 \cdot 3$	$1 \cdot 3^2$	$1 \cdot 3^3$
	2	$2 \cdot 3$	$2 \cdot 3^2$	$2 \cdot 3^3$
	2^2	$2^2 \cdot 3$	$2^2 \cdot 3^2$	$2^2 \cdot 3^3$
Sum	7	$7 \cdot 3$	$7 \cdot 3^2$	$7 \cdot 3^3$

The sum of *all* the factors of 108 is the sum of the numbers in the bottom row.

$$\text{Sum of all factors of } 108 = 7 + 7 \cdot 3 + 7 \cdot 3^2 + 7 \cdot 3^3$$
$$= 7(1 + 3 + 3^2 + 3^3) = 7(40) = 280$$

To find the sum of just the *proper* factors, we must subtract 108 from 280, which gives us 172. Thus 108 is an abundant number.

We now look for a pattern for the sum of all the factors. Note that

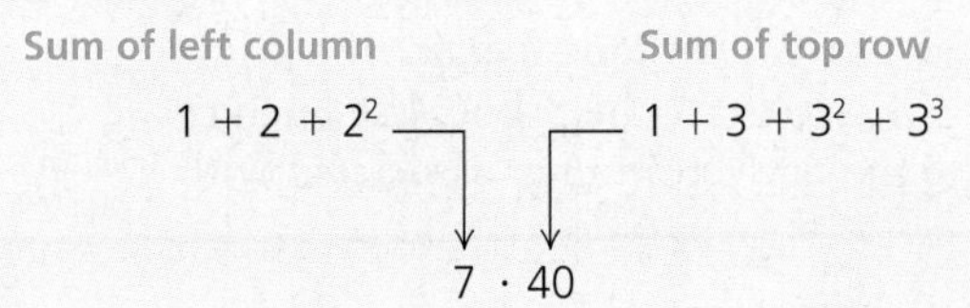

This result suggests that the sum of the factors of a number can be found by finding the sum of all the prime power factors of each prime factor and then computing the product of those sums. Because we are interested only in the sum of the proper factors, we subtract the original number. Although we have not proved this result, it is a true statement and can save much time and effort. For instance, the sum of the proper factors of 3240 can be found as follows:

$$3240 = 2^3 \cdot 3^4 \cdot 5$$

Compute the sum of *all* the prime power factors of each prime factor.

$$1 + 2 + 2^2 + 2^3 = 15 \qquad 1 + 3 + 3^2 + 3^3 + 3^4 = 121 \qquad 1 + 5 = 6$$

The sum of the proper factors of 3240 is $(15)(121)(6) - 3240 = 7650$. Thus 3240 is an abundant number.

Excursion Exercises

Use the above technique to find the sum of the proper factors of each number and then state whether the number is deficient, abundant, or perfect.

1. 200 **2.** 262 **3.** 325 **4.** 496

5. Use deductive reasoning to prove that every prime number is deficient.

6. Use inductive reasoning to decide whether every multiple of 6 greater than 6 is abundant.

Exercise Set 4.6

In Exercises 1–16, determine whether each number is perfect, deficient, or abundant.

1. 18

2. 32

3. 91

4. 51

5. 19

6. 144

7. 204

8. 128

9. 610

10. 508

11. 291

12. 1001

13. 176

14. 122

15. 260

16. 258

In Exercises 17–20, determine whether each Mersenne number is a prime number.

17. $2^3 - 1$

18. $2^5 - 1$

19. $2^7 - 1$

20. $2^{13} - 1$

21. In 1876, E. Lucas proved, without the aid of a computer, that $2^{127} - 1$ is a Mersenne prime. Use Euclid's theorem to write the perfect number associated with this prime.

22. In 1952, R. M. Robinson proved, with the aid of a computer, that $2^{521} - 1$ is a Mersenne prime. Use Euclid's theorem to write the perfect number associated with this prime.

In Exercises 23–28, determine the number of digits in the given Mersenne prime.

23. $2^{17} - 1$

24. $2^{132049} - 1$

25. $2^{1398269} - 1$

26. $2^{3021377} - 1$

27. $2^{6972593} - 1$

28. $2^{20996011} - 1$

29. Verify that $x = 9$, $y = 15$, and $n = 5$ do not yield a solution to the equation $x^n + y^n = z^n$ where z is a natural number.

30. Verify that $x = 7$, $y = 19$, and $n = 6$ do not yield a solution to the equation $x^n + y^n = z^n$ where z is a natural number.

31. Determine whether each of the following statements is a true statement, a false statement, or a conjecture.

a. If n is a prime number, then $2^n - 1$ is also a prime number.

b. Fermat's Last Theorem is called his last theorem because we believe that it was the last theorem he proved.

c. All perfect numbers of the form $2^{n-1}(2^n - 1)$ are even numbers.

d. Every perfect number is an even number.

32. Prove that $4078^n + 3433^n = 12{,}046^n$ cannot be a solution to the equation $x^n + y^n = z^n$ where n is a natural number. *Hint:* Examine the ones' digits of the powers.

33. Fermat's Little Theorem A theorem known as *Fermat's Little Theorem* states, "If n is a prime number and a is any natural number, then $a^n - a$ is divisible by n." Verify Fermat's Little Theorem for

a. $n = 7$ and $a = 12$.

b. $n = 11$ and $a = 8$.

34. Amicable Numbers The Greeks considered the pair of numbers 220 and 284 to be *amicable* or *friendly* numbers because the sum of the proper divisors of one of the numbers is the other number.

The sum of the proper factors of 220 is

$$1 + 2 + 4 + 5 + 10 + 11 + 20 + 22 + 44 + 55 + 110 = 284$$

The sum of the proper factors of 284 is

$$1 + 2 + 4 + 71 + 142 = 220$$

Determine whether

a. 60 and 84 are amicable numbers.

b. 1184 and 1210 are amicable numbers.

35. A Sum of Cubes Property The perfect number 28 can be written as $1^3 + 3^3$. The perfect number 496 can be written as $1^3 + 3^3 + 5^3 + 7^3$. Verify that the next perfect number, 8128, can also be written as the sum of the cubes of consecutive odd natural numbers, starting with 1^3.

36. A Sum of the Digits Theorem If you sum the digits of any even perfect number (except 6), then sum the digits of the resulting number, and repeat this process until you get a single digit, that digit will be 1.

As an example, consider the perfect number 28. The sum of its digits is 10. The sum of the digits of 10 is 1.

Verify the previous theorem for each of the following perfect numbers.

a. 496 b. 8128

c. 33,550,336 d. 8,589,869,056

37. **A Sum of Reciprocals Theorem** The sum of the reciprocals of all the positive divisors of a perfect number is always 2.

Verify the above theorem for each of the following perfect numbers.

a. 6 b. 28

Extensions

CRITICAL THINKING

38. **The Smallest Odd Abundant Number** Determine the smallest odd abundant number. *Hint:* It is greater than 900 but less than 1000.

39. **Fermat Numbers** Numbers of the form $2^{2^m} + 1$, where m is a whole number, are called *Fermat numbers.* Fermat believed that all Fermat numbers were prime. Prove that Fermat was wrong.

EXPLORATIONS

40. **Semiperfect Numbers** Any number that is the sum of *some or all* of its proper divisors is called a **semiperfect number.** For instance, 12 is a semiperfect number because it has 1, 2, 3, 4, and 6 as proper factors, and $12 = 1 + 2 + 3 + 6$.

The first twenty-five semiperfect numbers are 6, 12, 18, 20, 24, 28, 30, 36, 40, 42, 48, 54, 56, 60, 66, 72, 78, 80, 84, 88, 90, 96, 100, 102, and 104. It has been established that every natural number multiple of a semiperfect number is semiperfect and that a semiperfect number cannot be a deficient number.

a. Use the definition of a semiperfect number to verify that 20 is a semiperfect number.

b. Explain how to verify that 200 is a semiperfect number without examining its proper factors.

41. **Weird Numbers** Any number that is an abundant number but not a semiperfect number (see Exercise 40) is called a **weird number.** Find the only weird number less than 100. *Hint:* The abundant numbers less than 100 are 12, 18, 20, 24, 30, 36, 40, 42, 48, 54, 56, 60, 66, 70, 72, 78, 80, 84, 88, 90, and 96.

42. **A False Prediction** In 1811 Peter Barlow wrote, in his text *Theory of Numbers,* that the eighth perfect number, $2^{30}(2^{31} - 1) =$ 2,305,843,008,139,952,128, which was discovered by Leonhard Euler in 1772, "is the greatest [perfect number] that will be discovered; for as they are merely curious, without being useful, it is not likely that any person will attempt to find one beyond it." (*Source:* **http://www.utm-edu/research/primes/mersenne/index.html**)

The current search for larger and larger perfect numbers shows that Barlow's prediction did not come true. Search the Internet for answers to the question "Why do people continue the search for large perfect numbers (or large prime numbers)?" Write a brief summary of your findings.

CHAPTER 4 Summary

Key Terms

additive system [p. 178]
amicable numbers [p. 240]
Babylonian numeration system [p. 189]
base ten [p. 187]
binary adders [p. 208]
binary numeration system [p. 199]
bit [p. 199]
decimal system [p. 187]
digits [p. 187]
divisibility test [p. 221]
double-dabble method [p. 203]
duodecimal system [p. 199]
Egyptian multiplication procedure [p. 186]
expanded form [p. 187]
factorial prime [p. 230]
Fermat numbers [p. 241]
Fermat's Little Theorem [p. 240]
Goldbach's conjecture [p. 228]
hexadecimal system [p. 199]
hieroglyphics [p. 178]
Hindu-Arabic numeration system [p. 178]
indirect proof [p. 226]
Mayan numeration system [p. 192]
Mersenne number [p. 232]
Mersenne prime [p. 232]
number theory [p. 220]
numeral [p. 178]
numeration system [p. 178]
octal numeration system [p. 201]
perfect square [p. 229]
place-value or positional-value system [p. 187]
prime factorization [p. 223]
prime triplets [p. 228]
primorial prime [p. 230]
proof by contradiction [p. 226]
proper factor [p. 231]
Roman numeration system [p. 181]
Sieve of Eratosthenes [p. 225]
successive division process [p. 200]
twin primes [p. 228]
zero as a place holder [p. 192]

Essential Concepts

- A *base b numeration system* (where b is a natural number greater than 1) has place values of $\ldots, b^5, b^4, b^3, b^2, b^1, b^0$.
- The natural number a is a *divisor,* or *factor,* of the natural number b provided there exists a natural number j such that $aj = b$.
- A *prime number* is a natural number greater than 1 that has exactly two factors (divisors), itself and 1. A *composite number* is a natural number greater than 1 that is not a prime number.
- **The Fundamental Theorem of Arithmetic**
  Every composite number can be written as a unique product of prime numbers (disregarding the order of the factors).
- A natural number is *perfect* if it is equal to the sum of its proper factors. It is *deficient* if it is greater than the sum of its proper factors. It is *abundant* if it is less than the sum of its proper factors.
- **Euclid's Perfect-Number-Generating Procedure**
  If n and $2^n - 1$ are both prime numbers, then $2^{n-1}(2^n - 1)$ is a perfect number.
- **Fermat's Last Theorem**
  There are no natural numbers x, y, z, and n that satisfy $x^n + y^n = z^n$ with n greater than 2.

CHAPTER 4 Review Exercises

1. Write 4,506,325 using Egyptian hieroglyphics.
2. Write 3,124,043 using Egyptian hieroglyphics.
3. Write the Egyptian hieroglyphic

   as a Hindu-Arabic numeral.
4. Write the Egyptian hieroglyphic

   as a Hindu-Arabic numeral.

In Exercises 5–8, write each Roman numeral as a Hindu-Arabic numeral.

5. CCCXLIX

6. DCCLXXIV

7. $\overline{\text{IX}}$DCXL

8. $\overline{\text{XCII}}$CDXLIV

In Exercises 9–12, write each Hindu-Arabic numeral as a Roman numeral.

9. 567

10. 823

11. 2489

12. 1335

In Exercises 13 and 14, write each Hindu-Arabic numeral in expanded form.

13. 432

14. 456,327

In Exercises 15 and 16, simplify each expanded form.

15. $(5 \times 10^6) + (3 \times 10^4) + (8 \times 10^3) + (2 \times 10^2) + (4 \times 10^0)$

16. $(3 \times 10^5) + (8 \times 10^4) + (7 \times 10^3) + (9 \times 10^2) + (6 \times 10^1)$

In Exercises 17–20, write each Babylonian numeral as a Hindu-Arabic numeral.

17. 𒌋𒁹𒁹𒁹 𒌋𒌋𒁹

18. 𒌋𒌋𒁹𒁹𒁹𒁹𒁹𒁹 𒌋𒌋𒌋𒌋𒁹𒁹𒁹

19. 𒌋𒌋𒁹 𒌋𒁹𒁹𒁹𒁹 𒁹

20. 𒌋𒌋𒁹𒁹𒁹𒁹 𒌋𒁹𒁹𒁹𒁹𒁹𒁹 𒌋𒌋𒌋𒁹𒁹𒁹

In Exercises 21–24, write each Hindu-Arabic numeral as a Babylonian numeral.

21. 721

22. 1080

23. 12,543

24. 19,281

In Exercises 25–28, write each Mayan numeral as a Hindu-Arabic numeral.

25. 𝋩
𝋳

26. 𝋭
𝋧

27. 𝋦
𝋠
𝋲

28. 𝋲
𝋥
𝋠

In Exercises 29–32, write each Hindu-Arabic numeral as a Mayan numeral.

29. 522

30. 346

31. 1862

32. 1987

In Exercises 33–36, convert each numeral to base ten.

33. 45_{six}

34. 172_{nine}

35. $\text{E3}_{\text{sixteen}}$

36. $\text{1BA}_{\text{twelve}}$

In Exercises 37–40, convert each numeral to the indicated base.

37. 346_{nine} to base six

38. 1532_{six} to base eight

39. 275_{twelve} to base nine

40. $\text{67A}_{\text{sixteen}}$ to base twelve

In Exercises 41–48, convert each numeral directly (without first converting to base ten) to the indicated base.

41. 11100_{two} to base eight

42. 1010100_{two} to base eight

43. 1110001101_{two} to base sixteen

44. 11101010100_{two} to base sixteen

45. 25_{eight} to base two

46. 1472_{eight} to base two

47. $\text{4A}_{\text{sixteen}}$ to base two

48. $\text{C72}_{\text{sixteen}}$ to base two

In Exercises 49–56, perform the indicated operation.

49. $235_{\text{six}} + 144_{\text{six}}$

50. $673_{\text{eight}} + 345_{\text{eight}}$

51. $672_{\text{nine}} - 135_{\text{nine}}$

52. $1332_{\text{four}} - 213_{\text{four}}$

53. $25_{\text{eight}} \times 542_{\text{eight}}$

54. $43_{\text{five}} \times 3421_{\text{five}}$

55. $1010101_{\text{two}} \div 11_{\text{two}}$

56. $321_{\text{four}} \div 12_{\text{four}}$

In Exercises 57–60, determine the prime factorization of the given number.

57. 45

58. 54

59. 153

60. 285

In Exercises 61–64, determine whether the given number is a prime number or a composite number.

61. 501

62. 781

63. 689

64. 1003

In Exercises 65–68, determine whether the given number is perfect, deficient, or abundant.

65. 28 **66.** 81

67. 144 **68.** 200

In Exercises 69 and 70, use Euclid's perfect-number-generating procedure to write the perfect number associated with the given Mersenne prime.

69. $2^{61} - 1$ **70.** $2^{1279} - 1$

In Exercises 71–74, use the Egyptian doubling procedure to find the given product.

71. 8×46 **72.** 9×57

73. 14×83 **74.** 21×143

In Exercises 75–78, use the double-dabble method to convert each base two numeral to base ten.

75. 110011010_{two} **76.** 100010101_{two}

77. 10000010001_{two} **78.** 11001010000_{two}

79. State the (base ten) divisibility test for 3.

80. State the (base ten) divisibility test for 6.

81. State the Fundamental Theorem of Arithmetic.

82. How many odd perfect numbers had been discovered as of July 2005?

83. Find the number of digits in the Mersenne number $2^{132049} - 1$.

84. Find the number of digits in the Mersenne number $2^{2976221} - 1$.

CHAPTER 4 Test

1. Write 3124 using Egyptian hieroglyphics.

2. Write the Egyptian hieroglyphic

as a Hindu-Arabic numeral.

3. Write the Roman numeral MCDXLVII as a Hindu-Arabic numeral.

4. Write 2609 as a Roman numeral.

5. Write 67,485 in expanded form.

6. Simplify:

$$(5 \times 10^5) + (3 \times 10^4) + (2 \times 10^2) + (8 \times 10^1) + (4 \times 10^0)$$

7. Write the Babylonian numeral

as a Hindu-Arabic numeral.

8. Write 9675 as a Babylonian numeral.

9. Write the Mayan numeral

as a Hindu-Arabic numeral.

10. Write 502 as a Mayan numeral.

11. Convert 3542_{six} to base ten.

12. Convert 2148 to **a.** base eight and **b.** base twelve.

13. Convert 4567_{eight} to binary form.

14. Convert $101010110111_{\text{two}}$ to hexadecimal form.

In Exercises 15–18, perform the indicated operation.

15. $34_{\text{five}} + 23_{\text{five}}$

16. $462_{\text{eight}} - 147_{\text{eight}}$

17. $101_{\text{two}} \times 101110_{\text{two}}$

18. $431_{\text{seven}} \div 5_{\text{seven}}$

19. Determine the prime factorization of 230.

20. Determine whether 1001 is a prime number or a composite number.

21. Use divisibility tests to determine whether 1,737,285,147 is divisible by **a.** 2, **b.** 3, or **c.** 5.

22. Use divisibility tests to determine whether 19,531,333,276 is divisible by **a.** 4, **b.** 6, or **c.** 11.

23. Determine whether 96 is perfect, deficient, or abundant.

24. Use Euclid's perfect-number-generating procedure to write the perfect number associated with the Mersenne prime $2^{17} - 1$.

CHAPTER

6 Applications of Functions

The Clarence Buckingham Memorial Fountain, better known as Buckingham Fountain, is one of the largest fountains in the world. A major landmark of Chicago, it is located at Columbus Drive in Grant Park and is made of Georgia pink marble.

Kate Buckingham, who dedicated the fountain to the people of Chicago in memory of her brother Clarence, funded the project. Edward H. Bennett designed the fountain to represent Lake Michigan, with four sea horses, built by sculptor Marcel Loyau, to symbolize the four states that touch the lake: Wisconsin, Illinois, Indiana, and Michigan. Buckingham Fountain opened on May 26, 1927.

The fountain operates from April 1 to November 1 each year and runs from 10:00 A.M. to 11:00 P.M. each day. Every hour on the hour for 20 minutes, the fountain produces a major water display. Beginning at dusk, the water display is accompanied by lights and music. The fountain has 133 nozzles that spray 14,000 gallons of water per minute. The water that shoots upward from the center nozzle of the fountain can reach heights of 135 feet in the air. A quadratic function, one of the topics of this chapter, can be used to approximate the height of a given volume of water t seconds after it shoots upward from the center nozzle. For more information on this application, see Exercise 43, page 373.

For online student resources, visit this textbook's website at **math.college.hmco.com/students.**

SECTION 6.1 Rectangular Coordinates and Functions

Introduction to Rectangular Coordinate Systems

When archeologists excavate a site, a *coordinate grid* is laid over the site so that records can be kept not only of what was found but also of *where* it was found. The grid below is from an archeological dig at Poggio Colla, a site in the Mugello about 20 miles northeast of Florence, Italy.

historical note

The concept of a coordinate system developed over time, culminating in 1637 with the publication of *Discourse on the Method for Rightly Directing One's Reason and Searching for Truth in the Sciences* by René Descartes (1596–1650) and *Introduction to Plane and Solid Loci* by Pierre de Fermat (1601–1665). Of the two mathematicians, Descartes is usually given more credit for developing the concept of a coordinate system. In fact, he became so famous in Le Haye, the town in which he was born, that the town was renamed Le Haye–Descartes. ■

In mathematics we encounter a similar problem, that of locating a point in a plane. One way to solve the problem is to use a *rectangular coordinate system.*

A **rectangular coordinate system** is formed by two number lines, one horizontal and one vertical, that intersect at the zero point of each line. The point of intersection is called the **origin.** The two number lines are called the **coordinate axes,** or simply the **axes.** Frequently, the horizontal axis is labeled the x-axis and the vertical axis is labeled the y-axis. In this case, the axes form what is called the ***xy*-plane.**

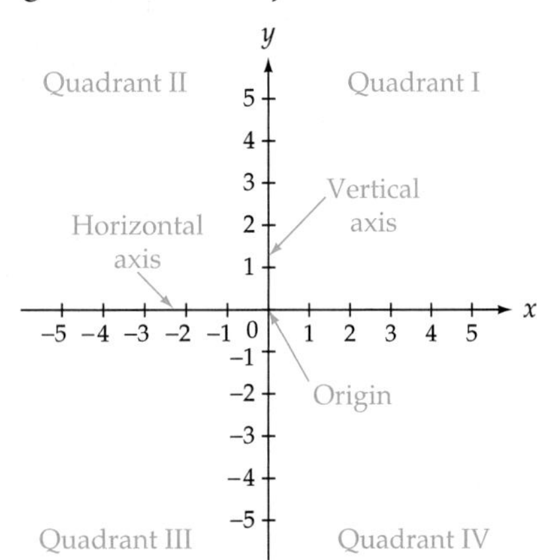

The two axes divide the plane into four regions called **quadrants,** which are numbered counterclockwise, using Roman numerals, from I to IV, starting at the upper right.

point of interest

The word *abscissa* has the same root as the word *scissors.* When open, a pair of scissors looks like an *x*.

Each point in the plane can be identified by a pair of numbers called an **ordered pair.** The first number of the ordered pair measures a horizontal change from the y-axis and is called the **abscissa,** or **x-coordinate.** The second number of the ordered pair measures a vertical change from the x-axis and is called the **ordinate,** or **y-coordinate.** The ordered pair (x, y) associated with a point is also called the **coordinates** of the point.

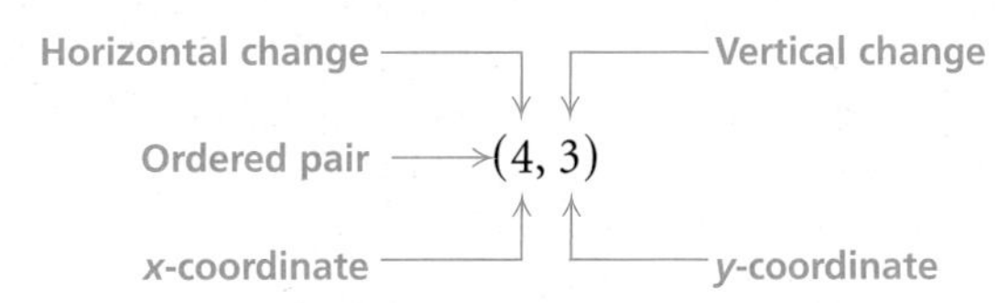

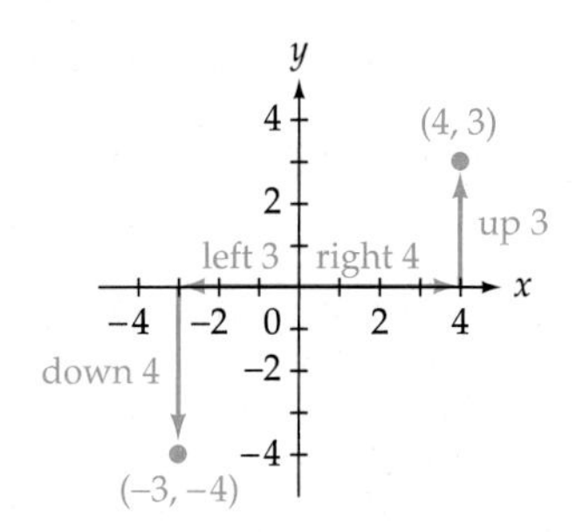

To **graph,** or **plot,** a point means to place a dot at the coordinates of the point. For example, to graph the ordered pair $(4, 3)$, start at the origin. Move 4 units to the right and then 3 units up. Draw a dot. To graph $(-3, -4)$, start at the origin. Move 3 units left and then 4 units down. Draw a dot.

TAKE NOTE

This is *very* important. An *ordered pair* is a pair of coordinates, and the *order* in which the coordinates are listed matters.

The **graph of an ordered pair** is the dot drawn at the coordinates of the point in the plane. The graphs of the ordered pairs $(4, 3)$ and $(-3, -4)$ are shown above.

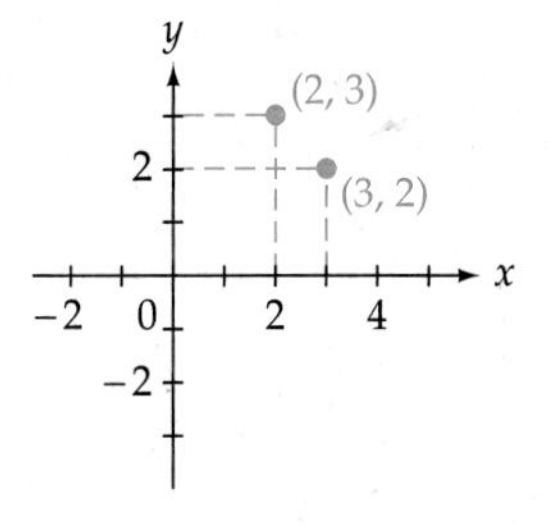

The graphs of the points whose coordinates are $(2, 3)$ and $(3, 2)$ are shown at the right. Note that they are different points. The order in which the numbers in an ordered pair are listed is important.

If the axes are labeled with letters other than x or y, then we refer to the ordered pair using the given labels. For instance, if the horizontal axis is labeled t and the vertical axis is labeled d, then the ordered pairs are written as (t, d). We sometimes refer to the first number in an ordered pair as the **first coordinate** of the ordered pair and to the second number as the **second coordinate** of the ordered pair.

One purpose of a coordinate system is to draw a picture of the solutions of an **equation in two variables.** Examples of equations in two variables are shown at the right.

$$y = 3x - 2$$
$$x^2 + y^2 = 25$$
$$s = t^2 - 4t + 1$$

A **solution of an equation in two variables** is an ordered pair that makes the equation a true statement. For instance, as shown below, $(2, 4)$ is a solution of $y = 3x - 2$ but $(3, -1)$ is not a solution of the equation.

$y = 3x - 2$		
4	$3(2) - 2$	• $x = 2, y = 4$
4	$6 - 2$	
$4 = 4$		• Checks.

$y = 3x - 2$		
-1	$3(3) - 2$	• $x = 3, y = -1$
-1	$9 - 2$	
$-1 \neq 7$		• Does not check.

QUESTION *Is $(-2, 1)$ a solution of $y = 3x + 7$?*

ANSWER *Yes, because $1 = 3(-2) + 7$.*

historical note

Maria Graëtana Agnesi (an'yayzee) (1718–1799) was probably the first woman to write a calculus text. A report on the text made by a committee of the Académie des Sciences in Paris stated: "It took much skill and sagacity to reduce, as the author has done, to almost uniform methods these discoveries scattered among the works of modern mathematicians and often presented by methods very different from each other. Order, clarity and precision reign in all parts of this work.... We regard it as the most complete and best made treatise."

There is a graph named after Agnesi called the Witch of Agnesi. This graph came by its name because of an incorrect translation from Italian to English of a work by Agnesi. There is also a crater on Venus named after Agnesi called the Crater of Agnesi. ■

The **graph of an equation in two variables** is a drawing of all the ordered-pair solutions of the equation. To create a graph of an equation, find some ordered-pair solutions of the equation, plot the corresponding points, and then connect the points with a smooth curve.

EXAMPLE 1 ■ Graph an Equation in Two Variables

Graph $y = 3x - 2$.

Solution

To find ordered-pair solutions, select various values of x and calculate the corresponding values of y. Plot the ordered pairs. After the ordered pairs have been graphed, draw a smooth curve through the points. It is convenient to keep track of the solutions in a table.

When choosing values of x, we often choose integer values because the resulting ordered pairs are easier to graph.

x	$3x - 2 = y$	(x, y)
−2	3(−2) − 2 = −8	(−2, −8)
−1	3(−1) − 2 = −5	(−1, −5)
0	3(0) − 2 = −2	(0, −2)
1	3(1) − 2 = 1	(1, 1)
2	3(2) − 2 = 4	(2, 4)
3	3(3) − 2 = 7	(3, 7)

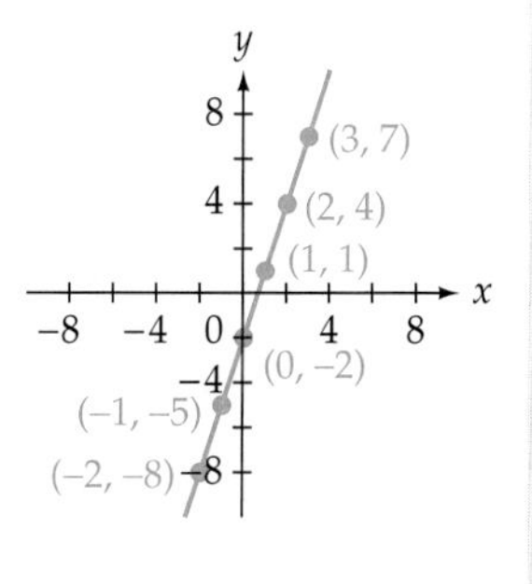

CHECK YOUR PROGRESS 1 Graph $y = -2x + 3$.

Solution *See page S21.*

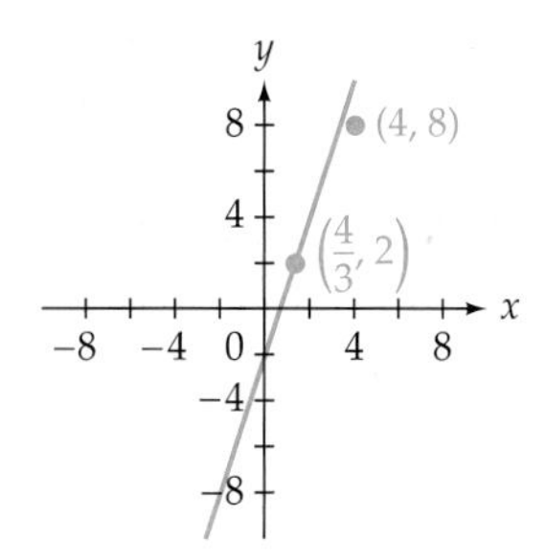

The graph of $y = 3x - 2$ is shown again at the left. Note that the ordered pair $\left(\frac{4}{3}, 2\right)$ is a solution of the equation and is a point on the graph. The ordered pair $(4, 8)$ is *not* a solution of the equation and is *not* a point on the graph. Every ordered-pair solution of the equation is a point on the graph, and every point on the graph is an ordered-pair solution of the equation.

EXAMPLE 2 ■ Graph an Equation in Two Variables

Graph $y = x^2 + 4x$.

Solution

Select various values of x and calculate the corresponding values of y. Plot the ordered pairs. After the ordered pairs have been graphed, draw a smooth curve through the points. Here is a table showing some possible ordered pairs.

TAKE NOTE

As this example shows, it may be necessary to graph quite a number of points before a reasonably accurate graph can be drawn.

x	$x^2 + 4x = y$	(x, y)
−5	$(-5)^2 + 4(-5) = 5$	(−5, 5)
−4	$(-4)^2 + 4(-4) = 0$	(−4, 0)
−3	$(-3)^2 + 4(-3) = -3$	(−3, −3)
−2	$(-2)^2 + 4(-2) = -4$	(−2, −4)
−1	$(-1)^2 + 4(-1) = -3$	(−1, −3)
0	$(0)^2 + 4(0) = 0$	(0, 0)
1	$(1)^2 + 4(1) = 5$	(1, 5)

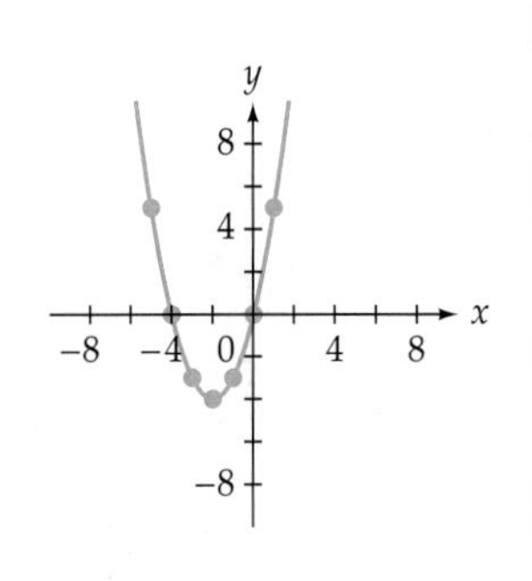

CHECK YOUR PROGRESS 2 Graph $y = -x^2 + 1$.

Solution *See page S21.*

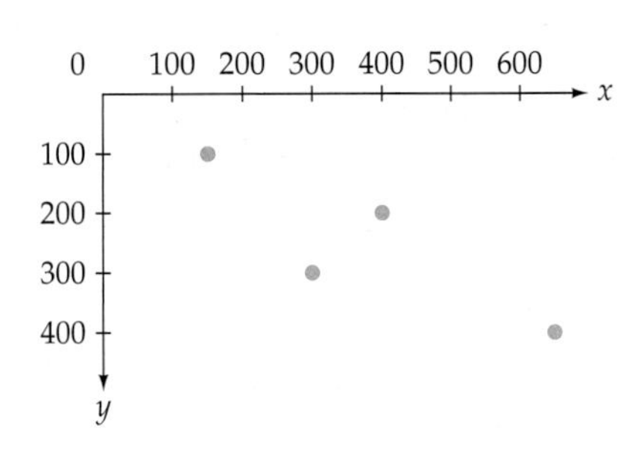

Math Matters Computer Software Program Coordinate Systems

Some computer software programs use a coordinate system that is different from the xy-coordinate system we have discussed. For instance, in one particular software program, the origin (0, 0) represents the top left point of a computer screen, as shown at the left. The points (150, 100), (300, 300), (400, 200), and (650, 400) are shown on the graph.

Introduction to Functions

TAKE NOTE

A car that uses 1 gallon of gas to travel 25 miles uses $\frac{1}{25} = 0.04$ gallon to travel 1 mile.

An important part of mathematics is the study of the relationship between known quantities. Exploring relationships between known quantities frequently results in equations in two variables. For instance, as a car is driven, the fuel in the gas tank is burned. There is a correspondence between the number of gallons of fuel used and the number of miles traveled. If a car gets 25 miles per gallon, then the car consumes 0.04 gallon of fuel for each mile driven. For the sake of simplicity, we will assume that the car always consumes 0.04 gallon of gasoline for each mile driven. The equation $g = 0.04d$ defines how the number of gallons used, g, depends on the number of miles driven, d.

Distance traveled (in miles), d	25	50	100	250	300
Fuel used (in gallons), g	1	2	4	10	12

The ordered pairs in this table are only some of the possible ordered pairs. Other possibilities are (90, 3.6), (125, 5), and (235, 9.4). If all of the ordered pairs of the equation were drawn, the graph would appear as a portion of a line. The graph of the equation and the ordered pairs we have calculated are shown on the following page. Note that the graphs of all the ordered pairs are on the same line.

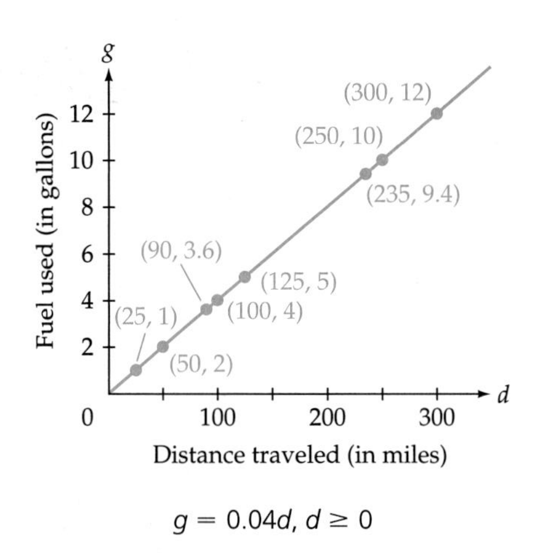

$g = 0.04d, d \geq 0$

QUESTION *What is the meaning of the ordered pair (125, 5)?*

The ordered pairs, the graph, and the equation are all different ways of expressing the correspondence, or relationship, between the two variables. This correspondence, which pairs the number of miles driven with the number of gallons of fuel used, is called a *function.*

Here are some additional examples of functions, along with a specific example of each correspondence.

To each real number 5	there corresponds ⟶	its square 25
To each score on an exam 87	there corresponds ⟶	a grade B
To each student Alexander Sterling	there corresponds ⟶	a student identification number S18723519

TAKE NOTE

Because the square of any real number is a positive number or zero ($0^2 = 0$), the range of the function that pairs a number with its square contains the positive numbers and zero. Therefore, the range is the nonnegative real numbers. The difference between *nonnegative* numbers and *positive* numbers is that nonnegative numbers include zero; positive numbers do not include zero.

An important fact about each of these correspondences is that each result is *unique.* For instance, for the real number 5, there is *exactly one* square, 25. With this in mind, we now state the definition of a function.

Definition of a Function

A **function** is a correspondence, or relationship, between two sets called the **domain** and **range** such that for each element of the domain there corresponds *exactly one* element of the range.

Test score	Grade
90–100	A
80–89	B
70–79	C
60–69	D
0–59	F

As an example of domain and range, consider the function that pairs a test score with a letter grade. The domain is the real numbers from 0 to 100. The range is the letters A, B, C, D, and F.

ANSWER *A car that gets 25 miles per gallon can travel 125 miles on 5 gallons of fuel.*

As shown above, a function can be described in terms of ordered pairs or by a graph. Functions can also be defined by equations in two variables. For instance, when gravity is the only force acting on a falling body, a function that describes the distance s, in feet, an object will fall in t seconds is given by $s = 16t^2$.

Given a value of t (time), the value of s (the distance the object falls) can be found. For instance, given $t = 3$,

$$s = 16t^2$$

$$s = 16(3)^2 \quad \text{• Replace } t \text{ by } 3.$$

$$s = 16(9) \quad \text{• Simplify.}$$

$$s = 144$$

The object falls 144 feet in 3 seconds.

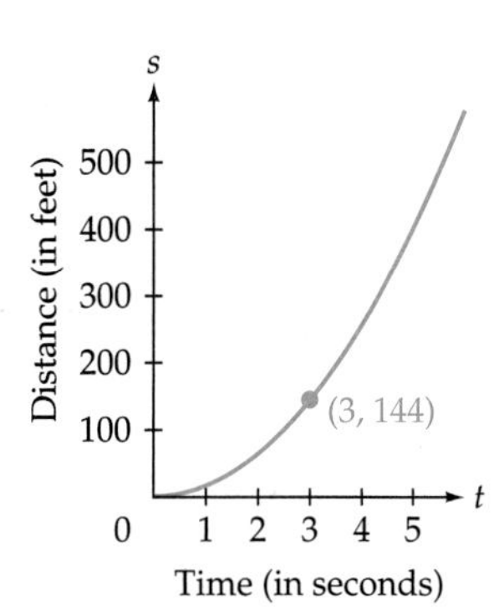

Because the distance the object falls *depends on* how long it has been falling, s is called the **dependent variable** and t is called the **independent variable.** Some of the ordered pairs of this function are (3, 144), (1, 16), (0, 0), and $\left(\frac{1}{4}, 1\right)$. The ordered pairs can be written as (t, s), where $s = 16t^2$. By substituting $16t^2$ for s, we can also write the ordered pairs as $(t, 16t^2)$. For the equation $s = 16t^2$, we say that "distance is a function of time." A graph of the function is shown at the left.

Not all equations in two variables define a function. For instance,

$$y^2 = x^2 + 9$$

is not an equation that defines a function because

$$5^2 = 4^2 + 9 \qquad \text{and} \qquad (-5)^2 = 4^2 + 9$$

The ordered pairs (4, 5) and (4, −5) both belong to the equation. Consequently, there are two ordered pairs with the same first coordinate, 4, but *different* second coordinates, 5 and −5. By definition, the equation does not define a function. The phrase "y is a function of x," or a similar phrase with different variables, is used to describe those equations in two variables that define functions.

Functional notation is frequently used for equations that define functions. Just as the letter x is commonly used as a variable, the letter f is commonly used to name a function.

TAKE NOTE

The notation $f(x)$ does *not* mean "f times x." The letter f stands for the name of the function, and $f(x)$ is the value of the function at x.

To describe the relationship between a number and its square using functional notation, we can write $f(x) = x^2$. The symbol $f(x)$ is read "the value of f at x" or "f of x." The symbol $f(x)$ is the **value of the function** and represents the value of the dependent variable for a given value of the independent variable. We will often write $y = f(x)$ to emphasize the relationship between the independent variable, x, and the dependent variable, y. Remember: y and $f(x)$ are different symbols for the same number. Also, the *name* of the function is f; the *value* of the function at x is $f(x)$.

The letters used to represent a function are somewhat arbitrary. All of the following equations represent the same function.

$$\left.\begin{aligned} f(x) &= x^2 \\ g(t) &= t^2 \\ P(v) &= v^2 \end{aligned}\right\} \quad \text{Each of these equations represents the square function.}$$

The process of finding $f(x)$ for a given value of x is called **evaluating the function.** For instance, to evaluate $f(x) = x^2$ when $x = 4$, replace x by 4 and simplify.

$$f(x) = x^2$$

$$f(4) = 4^2 = 16 \quad \text{• Replace } x \text{ by } 4. \text{ Then simplify.}$$

The *value* of the function is 16 when $x = 4$. This means that an ordered pair of the function is $(4, 16)$.

TAKE NOTE

To evaluate a function, you can use open parentheses in place of the variable. For instance,

$$s(t) = 2t^2 - 3t + 1$$
$$s(\ ) = 2(\ )^2 - 3(\ ) + 1$$

To evaluate the function, fill in each set of parentheses with the same number and then use the Order of Operations Agreement to evaluate the numerical expression on the right side of the equation.

EXAMPLE 3 ■ Evaluate a Function

Evaluate $s(t) = 2t^2 - 3t + 1$ when $t = -2$.

Solution

$$s(t) = 2t^2 - 3t + 1$$
$$s(-2) = 2(-2)^2 - 3(-2) + 1 \quad \text{• Replace } t \text{ by } -2\text{. Then simplify.}$$
$$= 15$$

The value of the function is 15 when $t = -2$.

CHECK YOUR PROGRESS 3 Evaluate $f(z) = z^2 - z$ when $z = -3$.

Solution *See page S22.*

Any letter or combination of letters can be used to name a function. In the next example, the letters *SA* are used to name a *S*urface *A*rea function.

EXAMPLE 4 ■ Application of Evaluating a Function

The surface area of a cube (the sum of the areas of each of the six faces) is given by $SA(s) = 6s^2$, where $SA(s)$ is the surface area of the cube and s is the length of one side of the cube. Find the surface area of a cube that has a side of length 10 centimeters.

Solution

$$SA(s) = 6s^2$$
$$SA(10) = 6(10)^2 \quad \text{• Replace } s \text{ by } 10\text{.}$$
$$= 6(100) \quad \text{• Simplify.}$$
$$= 600$$

The surface area of the cube is 600 square centimeters.

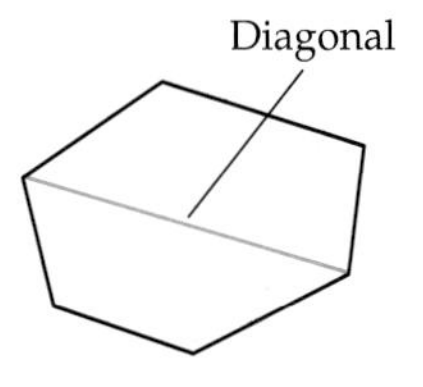

CHECK YOUR PROGRESS 4 A **diagonal** of a polygon is a line segment from one vertex to a nonadjacent vertex, as shown at the left. The total number of diagonals of a polygon is given by $N(s) = \dfrac{s^2 - 3s}{2}$, where $N(s)$ is the total number of diagonals and s is the number of sides of the polygon. Find the total number of diagonals of a polygon with 12 sides.

Solution *See page S22.*

Math Matters The Special Theory of Relativity

In 1905, Albert Einstein published a paper that set the framework for relativity theory. This theory, now called the Special Theory of Relativity, explains, among other things, how mass changes for a body in motion. Essentially, the theory states that the mass of a body is a function of its velocity. That is, the mass of a body changes as its speed increases.

The function can be given by $M(v) = \dfrac{m_0}{\sqrt{1 - \dfrac{v^2}{c^2}}}$, where m_0 is the mass of the body at rest (its mass when its speed is zero) and c is the velocity of light, which Einstein showed was the same for all observers. The table below shows how a 5-kilogram mass increases as its speed becomes closer and closer to the speed of light.

Speed	Mass (kilograms)
30 meters/second—speed of a car on an expressway	5
240 meters/second—speed of a commercial jet	5
3.0×10^7 meters/second—10% of the speed of light	5.025
1.5×10^8 meters/second—50% of the speed of light	5.774
2.7×10^8 meters/second—90% of the speed of light	11.471

Note that for speeds of everyday objects, such as a car or plane, the increase in mass is negligible. Physicists have verified these increases using particle accelerators that can accelerate a particle such as an electron to more than 99.9% of the speed of light.

historical note

Albert Einstein (īn′stīn) (1879–1955) was honored by *Time* magazine as Person of the Century. According to the magazine, "He was the pre-eminent scientist in a century dominated by science. The touchstones of the era—the Bomb, the Big Bang, quantum physics and electronics—all bear his imprint." ■

Graphs of Functions

Often the graph of a function can be drawn by finding ordered pairs of the function, plotting the points corresponding to the ordered pairs, and then connecting the points with a smooth curve.

For example, to graph $f(x) = x^3 + 1$, select several values of x and evaluate the function at each value. Recall that $f(x)$ and y are different symbols for the same quantity.

TAKE NOTE

We are basically creating the graph of an equation in two variables, as we did earlier. The only difference is the use of functional notation.

x	$f(x) = x^3 + 1$	(x, y)
-2	$f(-2) = (-2)^3 + 1 = -7$	$(-2, -7)$
-1	$f(-1) = (-1)^3 + 1 = 0$	$(-1, 0)$
0	$f(0) = (0)^3 + 1 = 1$	$(0, 1)$
1	$f(1) = (1)^3 + 1 = 2$	$(1, 2)$
2	$f(2) = (2)^3 + 1 = 9$	$(2, 9)$

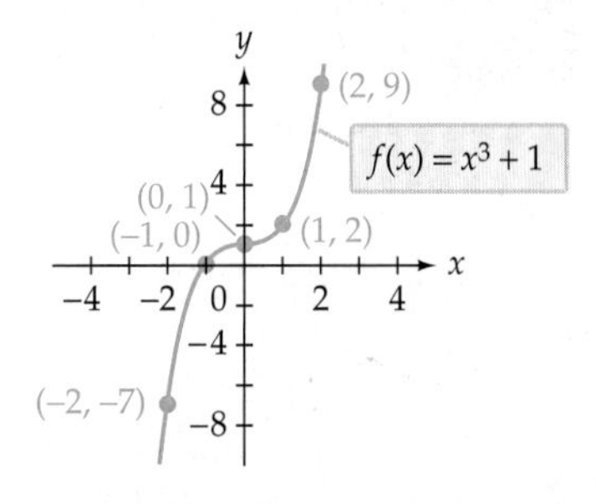

Plot the ordered pairs and draw a smooth curve through the points.

EXAMPLE 5 ■ Graph a Function

Graph $h(x) = x^2 - 3$.

Solution

x	$h(x) = x^2 - 3$	(x, y)
−3	$h(-3) = (-3)^2 - 3 = 6$	(−3, 6)
−2	$h(-2) = (-2)^2 - 3 = 1$	(−2, 1)
−1	$h(-1) = (-1)^2 - 3 = -2$	(−1, −2)
0	$h(0) = (0)^2 - 3 = -3$	(0, −3)
1	$h(1) = (1)^2 - 3 = -2$	(1, −2)
2	$h(2) = (2)^2 - 3 = 1$	(2, 1)
3	$h(3) = (3)^2 - 3 = 6$	(3, 6)

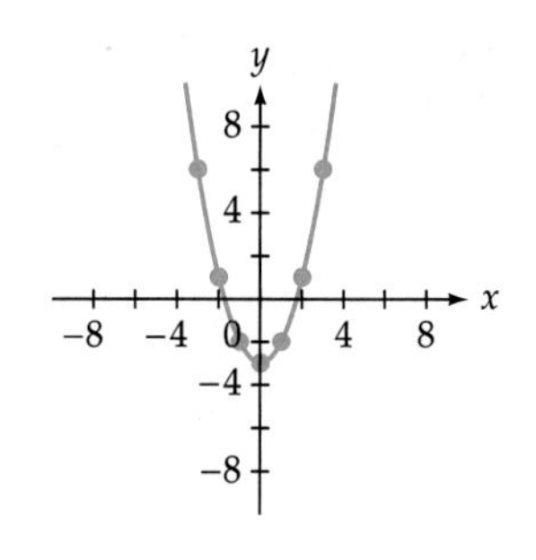

Plot the ordered pairs and draw a smooth curve through the points.

CHECK YOUR PROGRESS 5 Graph $f(x) = 2 - \frac{3}{4}x$.

Solution *See page S22.*

Excursion

Dilations of a Geometric Figure

A **dilation** of a geometric figure changes the size of the figure by either enlarging it or reducing it. This is accomplished by multiplying the coordinates of the figure by a positive number called the **dilation constant.** Examples of enlarging (multiplying the coordinates by a number greater than 1) and reducing (multiplying the coordinates by a number between 0 and 1) a geometric figure are shown at the top of the following page.

(continued)

point of interest

Photocopy machines have reduction and enlargement features that function essentially as constants of dilation. The numbers are usually expressed as a percent. A copier selection of 50% reduces the size of the object being copied by 50%. A copier selection of 125% increases the size of the object being copied by 25%.

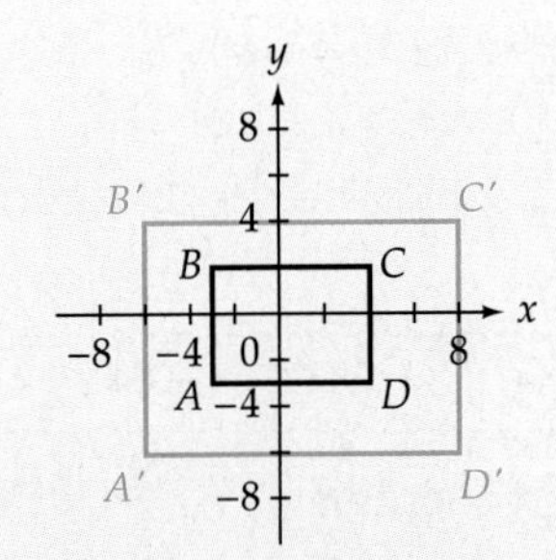

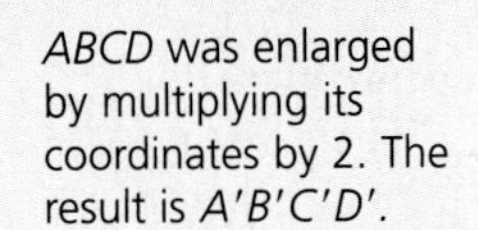

ABCD was enlarged by multiplying its coordinates by 2. The result is $A'B'C'D'$.

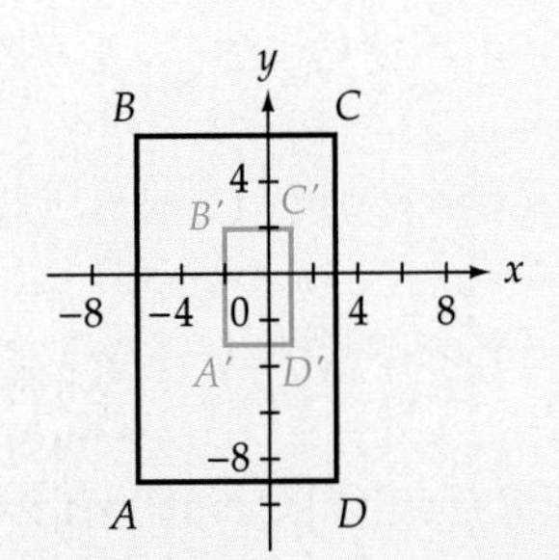

ABCD was reduced by multiplying its coordinates by $\frac{1}{3}$. The result is $A'B'C'D'$.

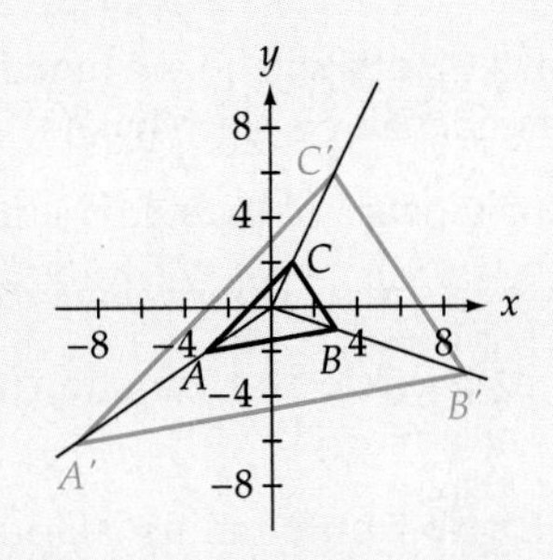

When each of the coordinates of a figure is multiplied by the same number in order to produce a dilation, the **center of dilation** will be the origin of the coordinate system. For triangle *ABC* at the left, a constant of dilation of 3 was used to produce triangle $A'B'C'$. Note that lines through the corresponding vertices of the two triangles intersect at the origin, the center of dilation. The center of dilation, however, can be any point in the plane.

Excursion Exercises

1. A dilation is performed on the figure with vertices $A(-2, 0)$, $B(2, 0)$, $C(4, -2)$, $D(2, -4)$, and $E(-2, -4)$.

   a. Draw the original figure and a new figure using 2 as the dilation constant.

   b. Draw the original figure and a new figure using $\frac{1}{2}$ as the dilation constant.

2. Because each of the coordinates of a geometric figure is multiplied by a number, the lengths of the sides of the figure will change. It is possible to show that the lengths change by a factor equal to the constant of dilation. In this exercise, you will examine the effect of a dilation on the angles of a geometric figure. Draw some figures and then draw a dilation of each figure using the origin as the center of dilation. Using a protractor, determine whether the measures of the angles of the dilated figure are different from the measures of the corresponding angles of the original figure.

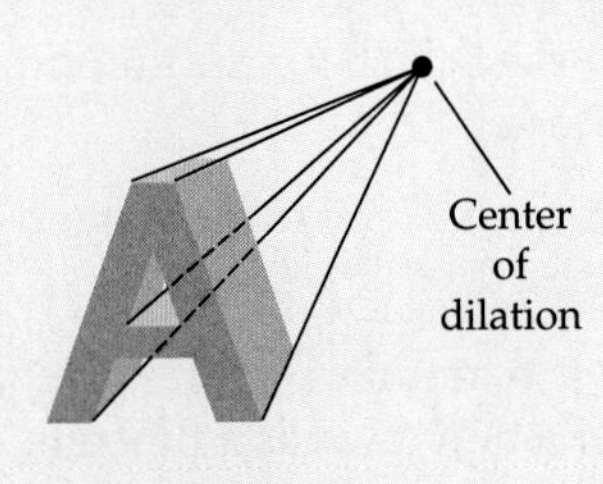

3. Graphic artists use centers of dilation to create three-dimensional effects. Consider the block letter A shown at the left. Draw another block letter A by changing the center of dilation to see how it affects the 3-D look of the letter. Programs such as PowerPoint use these methods to create various shading options for design elements in a presentation.

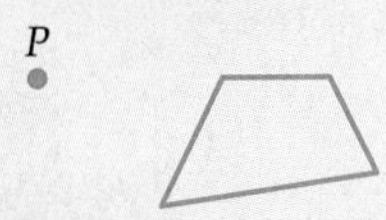

4. Draw an enlargement and a reduction of the figure at the left for the given center of dilation *P*.

5. On a blank piece of paper, draw a rectangle 4 inches by 6 inches with the center of the rectangle in the center of the paper. Make various photocopies of the rectangle using the reduction and enlargement settings on a copy machine. Where is the center of dilation for the copy machine?

Exercise Set 6.1

1. Graph the ordered pairs $(0, -1)$, $(2, 0)$, $(3, 2)$, and $(-1, 4)$.

2. Graph the ordered pairs $(-1, -3)$, $(0, -4)$, $(0, 4)$, and $(3, -2)$.

3. Draw a line through all points with an x-coordinate of 2.

4. Draw a line through all points with an x-coordinate of -3.

5. Draw a line through all points with a y-coordinate of -3.

6. Draw a line through all points with a y-coordinate of 4.

7. Graph the ordered-pair solutions of $y = x^2$ when $x = -2, -1, 0, 1$, and 2.

8. Graph the ordered-pair solutions of $y = -x^2 + 1$ when $x = -2, -1, 0, 1$, and 2.

9. Graph the ordered-pair solutions of $y = |x + 1|$ when $x = -5, -3, 0, 3$, and 5.

10. Graph the ordered-pair solutions of $y = -2|x|$ when $x = -3, -1, 0, 1$, and 3.

11. Graph the ordered-pair solutions of $y = -x^2 + 2$ when $x = -2, -1, 0$, and 1.

12. Graph the ordered-pair solutions of $y = -x^2 + 4$ when $x = -3, -1, 0, 1$, and 3.

13. Graph the ordered-pair solutions of $y = x^3 - 2$ when $x = -1, 0$, and 1.

14. Graph the ordered-pair solutions of $y = -x^3 + 1$ when $x = -1, 0, 1$, and 2.

In Exercises 15–24, graph each equation.

15. $y = 2x - 1$

16. $y = -3x + 2$

17. $y = \frac{2}{3}x + 1$

18. $y = -\frac{x}{2} - 3$

19. $y = \frac{1}{2}x^2$

20. $y = \frac{1}{3}x^2$

21. $y = 2x^2 - 1$

22. $y = -3x^2 + 2$

23. $y = |x - 1|$

24. $y = |x - 3|$

In Exercises 25–32, evaluate the function for the given value.

25. $f(x) = 2x + 7; x = -2$

26. $y(x) = 1 - 3x; x = -4$

27. $f(t) = t^2 - t - 3; t = 3$

28. $P(n) = n^2 - 4n - 7; n = -3$

29. $v(s) = s^3 + 3s^2 - 4s - 2; s = -2$

30. $f(x) = 3x^3 - 4x^2 + 7; x = 2$

31. $T(p) = \dfrac{p^2}{p - 2}; p = 0$

32. $s(t) = \dfrac{4t}{t^2 + 2}; t = 2$

33. **Geometry** The perimeter P of a square is a function of the length s of one of its sides and is given by $P(s) = 4s$.
    a. Find the perimeter of a square whose side is 4 meters.
    b. Find the perimeter of a square whose side is 5 feet.

34. **Geometry** The area of a circle is a function of its radius and is given by $A(r) = \pi r^2$.
    a. Find the area of a circle whose radius is 3 inches. Round to the nearest tenth of a square inch.
    b. Find the area of a circle whose radius is 12 centimeters. Round to the nearest tenth of a square centimeter.

35. **Sports** The height h, in feet, of a ball that is released 4 feet above the ground with an initial velocity of 80 feet per second is a function of the time t, in seconds, the ball is in the air and is given by

$$h(t) = -16t^2 + 80t + 4, \quad 0 \le t \le 5.04$$

    a. Find the height of the ball above the ground 2 seconds after it is released.
    b. Find the height of the ball above the ground 4 seconds after it is released.

36. **Forestry** The distance d, in miles, a forest fire ranger can see from an observation tower is a function of the height h, in feet, of the tower above level ground and is given by $d(h) = 1.5\sqrt{h}$.
    a. Find the distance a ranger can see whose eye level is 20 feet above level ground. Round to the nearest tenth of a mile.
    b. Find the distance a ranger can see whose eye level is 35 feet above level ground. Round to the nearest tenth of a mile.

37. **Sound** The speed s, in feet per second, of sound in air depends on the temperature t of the air in degrees Celsius and is given by $s(t) = \dfrac{1087\sqrt{t + 273}}{16.52}$.
    a. What is the speed of sound in air when the temperature is 0°C (the temperature at which water freezes)? Round to the nearest foot per second.
    b. What is the speed of sound in air when the temperature is 25°C? Round to the nearest foot per second.

38. **Softball** In a softball league in which each team plays every other team three times, the number of games N that must be scheduled depends on the number of teams n in the league and is given by $N(n) = \frac{3}{2}n^2 - \frac{3}{2}n$.
    a. How many games must be scheduled for a league that has five teams?
    b. How many games must be scheduled for a league that has six teams?

39. **Mixtures** The percent concentration P of salt in a particular salt water solution depends on the number of grams x of salt that are added to the solution and is given by $P(x) = \dfrac{100x + 100}{x + 10}$.
    a. What is the original percent concentration of salt?
    b. What is the percent concentration of salt after 5 more grams of salt are added?

40. **Pendulums** The time T, in seconds, it takes a pendulum to make one swing depends on the length of the pendulum and is given by $T(L) = 2\pi\sqrt{\frac{L}{32}}$, where L is the length of the pendulum in feet.
    a. Find the time it takes the pendulum to make one swing if the length of the pendulum is 3 feet. Round to the nearest hundredth of a second.
    b. Find the time it takes the pendulum to make one swing if the length of the pendulum is 9 inches. Round to the nearest tenth of a second.

In Exercises 41–56, graph the function.

41. $f(x) = 2x - 5$
42. $f(x) = -2x + 4$
43. $f(x) = -x + 4$
44. $f(x) = 3x - 1$
45. $g(x) = \dfrac{2}{3}x + 2$
46. $h(x) = \dfrac{5}{2}x - 1$
47. $F(x) = -\dfrac{1}{2}x + 3$
48. $F(x) = -\dfrac{3}{4}x - 1$
49. $f(x) = x^2 - 1$
50. $f(x) = x^2 + 2$
51. $f(x) = -x^2 + 4$
52. $f(x) = -2x^2 + 5$
53. $g(x) = x^2 - 4x$
54. $h(x) = x^2 + 4x$
55. $P(x) = x^2 - x - 6$
56. $P(x) = x^2 - 2x - 3$

Extensions

CRITICAL THINKING

57. **Geometry** Find the area of the rectangle.

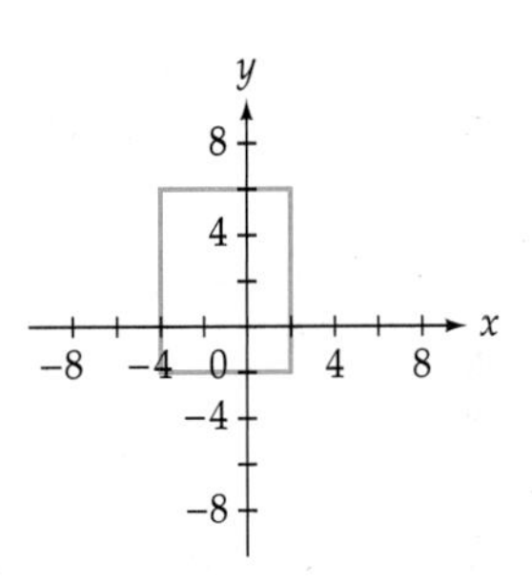

58. **Geometry** Find the area of the triangle.

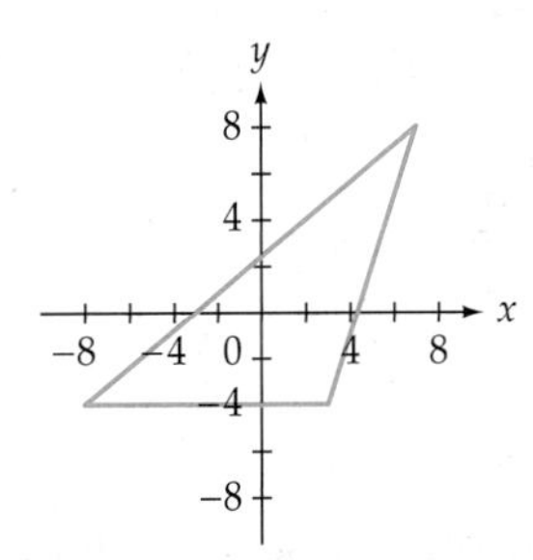

59. Suppose f is a function. Is it possible to have $f(2) = 4$ and $f(2) = 7$? Explain your answer.

60. Suppose f is a function and $f(a) = 4$ and $f(b) = 4$. Does this mean that $a = b$?

61. If $f(x) = 2x + 5$ and $f(a) = 9$, find a.

62. If $f(x) = x^2$ and $f(a) = 9$, find a.

63. Let $f(a, b) =$ the sum of a and b.
    Let $g(a, b) =$ the product of a and b.
    Find $f(2, 5) + g(2, 5)$.

64. Let $f(a, b) =$ the greatest common factor of a and b and let $g(a, b) =$ the least common multiple of a and b. Find $f(14, 35) + g(14, 35)$.

65. Given $f(x) = x^2 + 3$, for what value of x is $f(x)$ least?

66. Given $f(x) = -x^2 + 4x$, for what value of x is $f(x)$ greatest?

EXPLORATIONS

67. Consider the function given by

$$M(x, y) = \frac{x + y}{2} + \frac{|x - y|}{2}.$$

a. Complete the following table.

x	y	$M(x, y) = \frac{x+y}{2} + \frac{\lvert x-y \rvert}{2}$
−5	11	$M(-5, 11) = \frac{-5 + 11}{2} + \frac{\lvert -5 - 11 \rvert}{2} = 11$
10	8	
−3	−1	
12	−13	
−11	15	

b. Extend the table by choosing some additional values of x and y.

c. How is the value of the function related to the values of x and y? *Hint:* For $x = -5$ and $y = 11$, the value of the function was 11, the value of y.

d. The function $M(x, y)$ is sometimes referred to as the *maximum function.* Why is this a good name for this function?

e. Create a *minimum function*—that is, a function that yields the minimum of two numbers x and y. *Hint:* The function is similar in form to the maximum function.

SECTION 6.2 Properties of Linear Functions

Intercepts

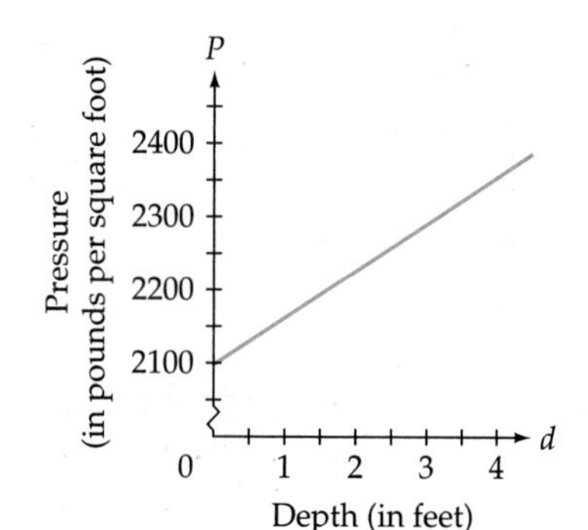

The graph at the left shows the pressure on a diver as the diver descends in the ocean. The equation of this graph can be represented by $P(d) = 64d + 2100$, where $P(d)$ is the pressure, in pounds per square foot, on a diver d feet below the surface of the ocean. By evaluating the function for various values of d, we can determine the pressure on the diver at those depths. For instance, when $d = 2$, we have

$$\begin{aligned} P(d) &= 64d + 2100 \\ P(2) &= 64(2) + 2100 \\ &= 128 + 2100 \\ &= 2228 \end{aligned}$$

The pressure on a diver 2 feet below the ocean's surface is 2228 pounds per square foot.

The function $P(d) = 64d + 2100$ is an example of a *linear function.*

Linear Function

A **linear function** is one that can be written in the form $f(x) = mx + b$, where m is the coefficient of x and b is a constant.

For the linear function $P(d) = 64d + 2100$, $m = 64$ and $b = 2100$.

Here are some other examples of linear functions.

$f(x) = 2x + 5$ • $m = 2, b = 5$

$g(t) = \frac{2}{3}t - 1$ • $m = \frac{2}{3}, b = -1$

$v(s) = -2s$ • $m = -2, b = 0$

$h(x) = 3$ • $m = 0, b = 3$

$f(x) = 2 - 4x$ • $m = -4, b = 2$

Note that different variables can be used to designate a linear function.

QUESTION *Which of the following are linear functions?*

a. $f(x) = 2x^2 + 5$ ***b.*** $g(x) = 1 - 3x$

Consider the linear function $f(x) = 2x + 4$. The graph of the function is shown below, along with a table listing some of its ordered pairs.

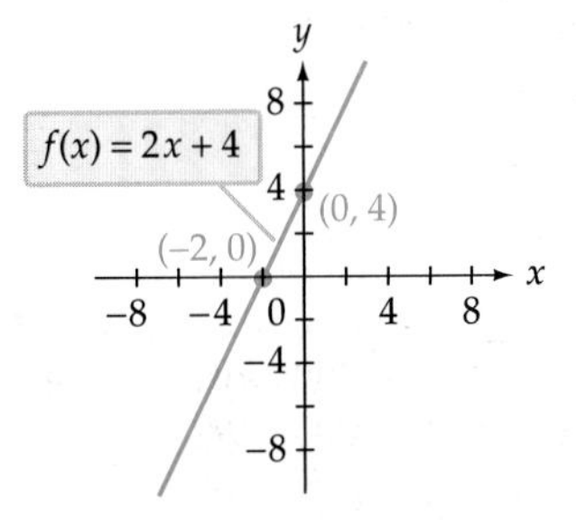

x	$f(x) = 2x + 4$	(x, y)
−3	$f(-3) = 2(-3) + 4 = -2$	$(-3, -2)$
−2	$f(-2) = 2(-2) + 4 = 0$	$(-2, 0)$
−1	$f(-1) = 2(-1) + 4 = 2$	$(-1, 2)$
0	$f(0) = 2(0) + 4 = 4$	$(0, 4)$
1	$f(1) = 2(1) + 4 = 6$	$(1, 6)$

TAKE NOTE

Note that the graph of a *lin*ear function is a straight *line.* Observe that when the graph crosses the *x*-axis, the *y*-coordinate is 0. When the graph crosses the *y*-axis, the *x*-coordinate is 0. The table confirms these observations.

From the table and the graph, we can see that when $x = -2$, $y = 0$, and the graph crosses the *x*-axis at $(-2, 0)$. The point $(-2, 0)$ is called the ***x*-intercept** of the graph. When $x = 0$, $y = 4$, and the graph crosses the *y*-axis at $(0, 4)$. The point $(0, 4)$ is called the ***y*-intercept** of the graph.

ANSWER ***a.*** *Because* $f(x) = 2x^2 + 5$ *has an* x^2 *term, f is not a linear function.* ***b.*** *Because* $g(x) = 1 - 3x$ *can be written in the form* $f(x) = mx + b$ *as* $g(x) = -3x + 1$ $(m = -3$ *and* $b = 1)$*, g is a linear function.*

EXAMPLE 1 ■ Find the x- and y-intercepts of a Graph

Find the x- and y-intercepts of the graph of $g(x) = -3x + 2$.

Solution

When a graph crosses the x-axis, the y-coordinate of the point is 0. Therefore, to find the x-intercept, replace $g(x)$ by 0 and solve the equation for x. [Recall that $g(x)$ is another name for y.]

$$\begin{aligned} g(x) &= -3x + 2 \\ 0 &= -3x + 2 \qquad \text{• Replace } g(x) \text{ by } 0. \\ -2 &= -3x \\ \frac{2}{3} &= x \end{aligned}$$

The x-intercept is $\left(\frac{2}{3}, 0\right)$.

When a graph crosses the y-axis, the x-coordinate of the point is 0. Therefore, to find the y-intercept, evaluate the function when x is 0.

$$\begin{aligned} g(x) &= -3x + 2 \\ g(0) &= -3(0) + 2 \qquad \text{• Evaluate } g(x) \text{ when } x = 0. \text{ Then simplify.} \\ &= 2 \end{aligned}$$

The y-intercept is $(0, 2)$.

CHECK YOUR PROGRESS 1 Find the x- and y-intercepts of the graph of $f(x) = \frac{1}{2}x + 3$.

Solution See page S22.

TAKE NOTE

To find the y-intercept of $y = mx + b$ [we have replaced $f(x)$ by y], let $x = 0$. Then

$$\begin{aligned} y &= mx + b \\ y &= m(0) + b \\ &= b \end{aligned}$$

The y-intercept is $(0, b)$. This result is shown at the right.

In Example 1, note that the y-coordinate of the y-intercept of $g(x) = -3x + 2$ has the same value as b in the equation $f(x) = mx + b$. This is always true.

y-intercept

The y-intercept of the graph of $f(x) = mx + b$ is $(0, b)$.

If we evaluate the linear function that models pressure on a diver, $P(d) = 64d + 2100$, at 0, we have

$$\begin{aligned} P(d) &= 64d + 2100 \\ P(0) &= 64(0) + 2100 = 2100 \end{aligned}$$

TAKE NOTE

We are working with the function $P(d) = 64d + 2100$. Therefore, the intercepts on the horizontal axis of a graph of the function are d-intercepts rather than x-intercepts, and the intercept on the vertical axis is a P-intercept rather than a y-intercept.

In this case, the P-intercept (the intercept on the vertical axis) is $(0, 2100)$. In the context of the application, this means that the pressure on a diver 0 feet below the ocean's surface is 2100 pounds per square foot. Another way of saying "zero feet below the ocean's surface" is "at sea level." Thus the pressure on the diver, or anyone else for that matter, at sea level is 2100 pounds per square foot.

Both the x- and y-intercept can have meaning in an application problem. This is demonstrated in the next example.

EXAMPLE 2 ■ Application of the Intercepts of a Linear Function

After a parachute is deployed, a function that models the height of the parachutist above the ground is $f(t) = -10t + 2800$, where $f(t)$ is the height, in feet, of the parachutist t seconds after the parachute is deployed. Find the intercepts on the vertical and horizontal axes and explain what they mean in the context of the problem.

Solution

To find the intercept on the vertical axis, evaluate the function when t is 0.

$$f(t) = -10t + 2800$$
$$f(0) = -10(0) + 2800 = 2800$$

The intercept on the vertical axis is (0, 2800). This means that the parachutist is 2800 feet above the ground when the parachute is deployed.

To find the intercept on the horizontal axis, set $f(t) = 0$ and solve for t.

$$f(t) = -10t + 2800$$
$$0 = -10t + 2800$$
$$-2800 = -10t$$
$$280 = t$$

The intercept on the horizontal axis is (280, 0). This means that the parachutist reaches the ground 280 seconds after the parachute is deployed. Note that the parachutist reaches the ground when $f(t) = 0$.

CHECK YOUR PROGRESS 2 A function that models the descent of a certain small airplane is given by $g(t) = -20t + 8000$, where $g(t)$ is the height, in feet, of the airplane t seconds after it begins its descent. Find the intercepts on the vertical and horizontal axes, and explain what they mean in the context of the problem.

Solution *See page S22.*

Slope of a Line

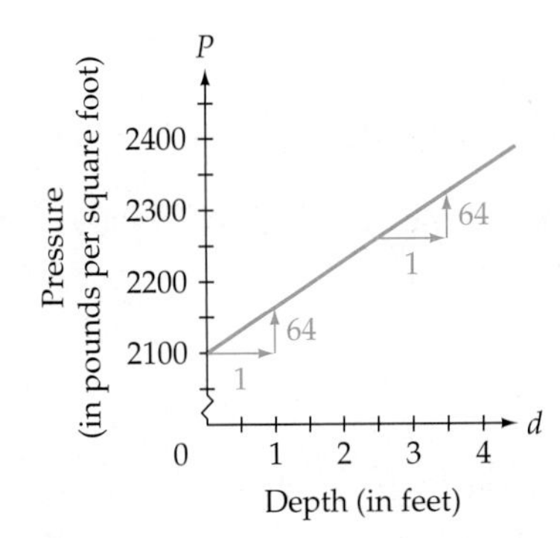

Consider again the linear function $P(d) = 64d + 2100$, which models the pressure on a diver as the diver descends below the ocean's surface. From the graph at the left, you can see that when the depth of the diver increases by 1 foot, the pressure on the diver increases by 64 pounds per square foot. This can be verified algebraically.

$P(0) = 64(0) + 2100 = 2100$ • Pressure at sea level

$P(1) = 64(1) + 2100 = 2164$ • Pressure after descending **1** foot

$2164 - 2100 = 64$ • Change in pressure

If we choose two other depths that differ by 1 foot, such as 2.5 feet and 3.5 feet (see the graph at the left), the change in pressure is the same.

$P(2.5) = 64(2.5) + 2100 = 2260$ • Pressure at **2.5** feet below the surface

$P(3.5) = 64(3.5) + 2100 = 2324$ • Pressure at **3.5** feet below the surface

$2324 - 2260 = 64$ • Change in pressure

The **slope** of a line is the change in the vertical direction caused by one unit of change in the horizontal direction. For $P(d) = 64d + 2100$, the slope is 64. In the context of the problem, the slope means that the pressure on a diver increases by 64 pounds per square foot for each additional foot the diver descends. Note that the slope (64) has the same value as the coefficient of d in $P(d) = 64d + 2100$. This connection between the slope and the coefficient of the variable in a linear function always holds.

Slope

For a linear function given by $f(x) = mx + b$, the slope of the graph of the function is m, the coefficient of the variable.

QUESTION *What is the slope of each of the following?*

a. $y = -2x + 3$ **b.** $f(x) = x + 4$ **c.** $g(x) = 3 - 4x$

d. $y = \frac{1}{2}x - 5$

The slope of a line can be calculated by using the coordinates of any two distinct points on the line and the following formula.

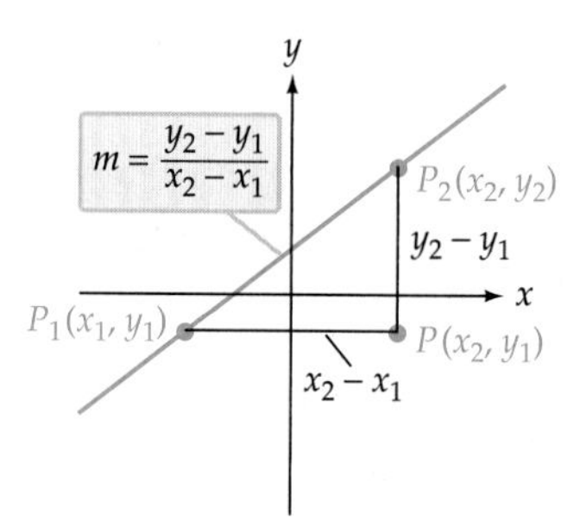

Slope of a Line

Let (x_1, y_1) and (x_2, y_2) be two points on a nonvertical line. Then the **slope** of the line through the two points is the ratio of the change in the y-coordinates to the change in the x-coordinates.

$$m = \frac{\text{change in } y}{\text{change in } x} = \frac{y_2 - y_1}{x_2 - x_1}, x_1 \neq x_2$$

QUESTION *Why is the restriction $x_1 \neq x_2$ required in the definition of slope?*

ANSWER **a.** -2 **b.** 1 **c.** -4 **d.** $\frac{1}{2}$

ANSWER *If $x_1 = x_2$, then the difference $x_2 - x_1 = 0$. This would make the denominator 0, and division by 0 is not defined.*

EXAMPLE 3 ■ Find the Slope of a Line Between Two Points

Find the slope of the line between the two points.

a. $(-4, -3)$ and $(-1, 1)$ **b.** $(-2, 3)$ and $(1, -3)$

c. $(-1, -3)$ and $(4, -3)$ **d.** $(4, 3)$ and $(4, -1)$

Solution

a. $(x_1, y_1) = (-4, -3), (x_2, y_2) = (-1, 1)$

$$m = \frac{y_2 - y_1}{x_2 - x_1} = \frac{1 - (-3)}{-1 - (-4)} = \frac{4}{3}$$

The slope is $\frac{4}{3}$. A *positive* slope indicates that the line slopes *upward* to the right. For this particular line, the value of y *increases* by $\frac{4}{3}$ when x increases by 1.

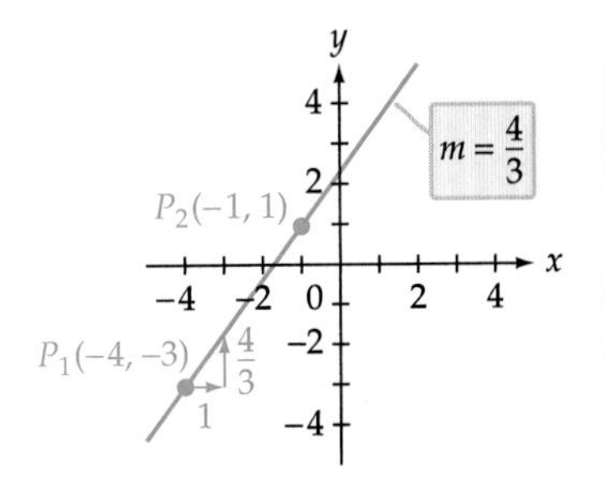

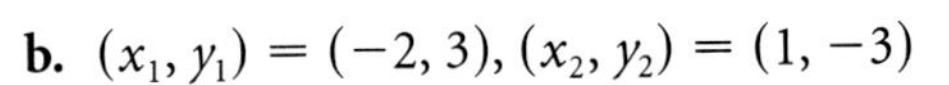

b. $(x_1, y_1) = (-2, 3), (x_2, y_2) = (1, -3)$

$$m = \frac{y_2 - y_1}{x_2 - x_1} = \frac{-3 - 3}{1 - (-2)} = \frac{-6}{3} = -2$$

The slope is -2. A *negative* slope indicates that the line slopes *downward* to the right. For this particular line, the value of y *decreases* by 2 when x increases by 1.

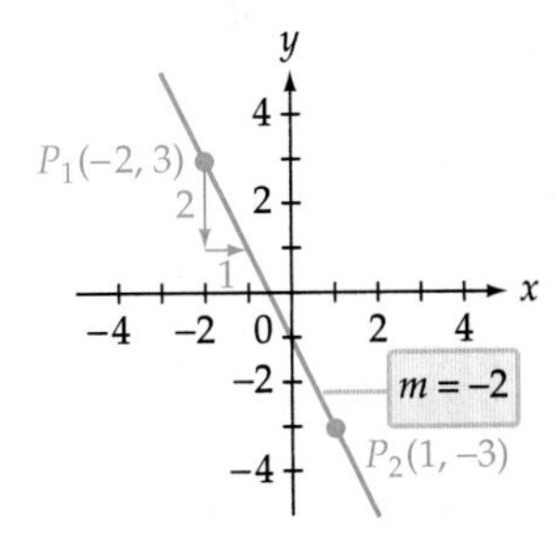

✓ **TAKE NOTE**

When we talk about y values increasing (as in part a) or decreasing (as in part b), we always mean as we move from left to right.

c. $(x_1, y_1) = (-1, -3), (x_2, y_2) = (4, -3)$

$$m = \frac{y_2 - y_1}{x_2 - x_1} = \frac{-3 - (-3)}{4 - (-1)} = \frac{0}{5} = 0$$

The slope is 0. A *zero* slope indicates that the line is *horizontal*. For this particular line, the value of y *stays the same* when x increases by any amount.

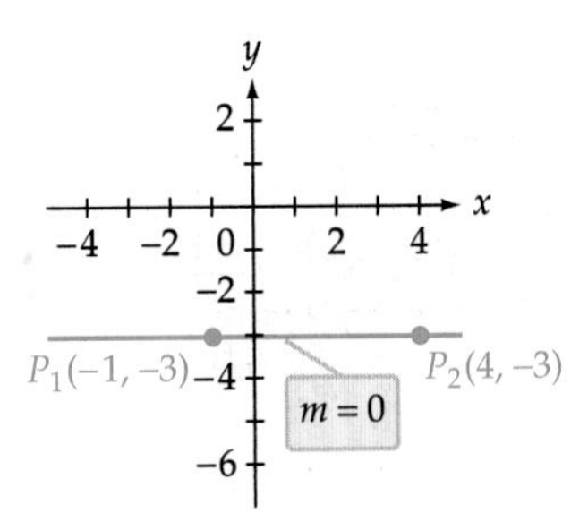

d. $(x_1, y_1) = (4, 3), (x_2, y_2) = (4, -1)$

$$m = \frac{y_2 - y_1}{x_2 - x_1} = \frac{-1 - 3}{4 - 4} = \frac{-4}{0}$$ Division by 0 is undefined.

If the denominator of the slope formula is zero, the line has *no slope*. Sometimes we say that the slope of a vertical line is *undefined*.

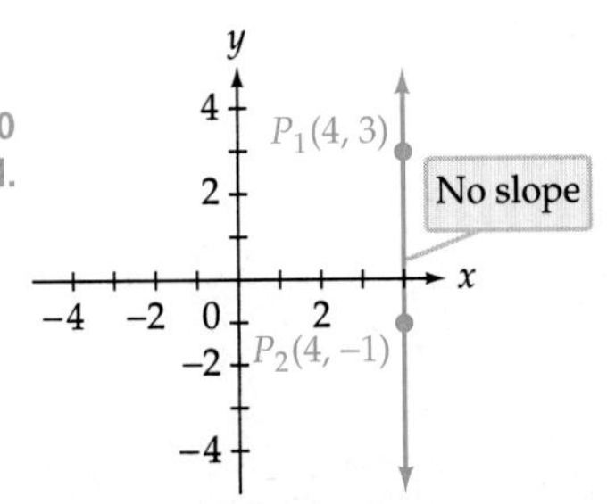

✓ **TAKE NOTE**

A horizontal line has zero slope. A line that has no slope, or whose slope is undefined, is a vertical line. Note that when $y_1 = y_2$ in the formula for slope, the slope of the line through the two points is zero. When $x_1 = x_2$, the slope is undefined.

CHECK YOUR PROGRESS 3 Find the slope of the line between the two points.

a. $(-6, 5)$ and $(4, -5)$ **b.** $(-5, 0)$ and $(-5, 7)$
c. $(-7, -2)$ and $(8, 8)$ **d.** $(-6, 7)$ and $(1, 7)$

Solution *See page S22.*

Suppose a jogger is running at a constant velocity of 6 miles per hour. Then the linear function $d = 6t$ relates the time t, in hours, spent running to the distance traveled d, in miles. A table of values is shown below.

Time, *t*, in hours	0	0.5	1	1.5	2	2.5
Distance, *d*, in miles	0	3	6	9	12	15

TAKE NOTE

Whether we write $f(t) = 6t$ or $d = 6t$, the equation represents a linear function. $f(t)$ and d are different symbols for the same quantity.

Because the equation $d = 6t$ represents a linear function, the slope of the graph of the equation is 6. This can be confirmed by choosing any two points on the graph shown below and finding the slope of the line between the two points. The points $(0.5, 3)$ and $(2, 12)$ are used here.

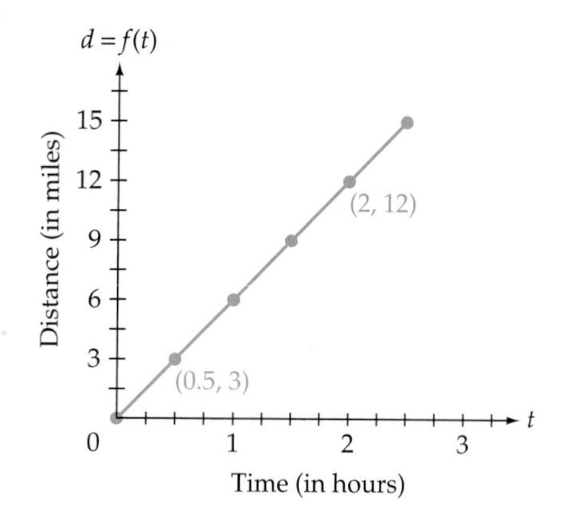

$$m = \frac{\text{change in } d}{\text{change in } t} = \frac{12 \text{ miles} - 3 \text{ miles}}{2 \text{ hours} - 0.5 \text{ hours}} = \frac{9 \text{ miles}}{1.5 \text{ hours}} = 6 \text{ miles per hour}$$

This example demonstrates that the slope of the graph of an object in uniform motion is the same as the velocity of the object. In a more general way, any time we discuss the velocity of an object, we are discussing the slope of the graph that describes the relationship between the distance the object travels and the time it travels.

EXAMPLE 4 ■ Application of the Slope of a Linear Function

The function $T(x) = -6.5x + 20$ approximates the temperature $T(x)$, in degrees Celsius, at x kilometers above sea level. What is the slope of this function? Write a sentence that explains the meaning of the slope in the context of this application.

Solution

For the linear function $T(x) = -6.5x + 20$, the slope is the coefficient of x. Therefore, the slope is -6.5. The slope means that the temperature is decreasing (because the slope is negative) 6.5°C for each 1-kilometer increase in height above sea level.

CHECK YOUR PROGRESS 4 The distance that a homing pigeon can fly can be approximated by $d(t) = 50t$, where $d(t)$ is the distance, in miles, flown by the pigeon in t hours. Find the slope of this function. What is the meaning of the slope in the context of the problem?

Solution *See page S23.*

Math Matters

Galileo Galilei (găl-ĭ′lā-ē) (1564–1642) was one of the most influential scientists of his time. In addition to inventing the telescope, with which he discovered the moons of Jupiter, Galileo successfully argued that Aristotle's assertion that heavy objects drop at a greater velocity than lighter ones was incorrect. According to legend, Galileo went to the top of the Leaning Tower of Pisa and dropped two balls at the same time, one weighing twice the other. His assistant, standing on the ground, observed that both balls reached the ground at the same time.

There is no historical evidence that Galileo actually performed this experiment, but he did do something similar. Galileo correctly reasoned that if Aristotle's assertion were true, then balls of different weights should roll down a ramp at different speeds. Galileo did carry out this experiment and was able to show that, in fact, balls of different weights reached the end of the ramp at the same time. Galileo was not able to determine why this happened, and it took Issac Newton, born the same year that Galileo died, to formulate the first theory of gravity.

Slope–Intercept Form of a Straight Line

The value of the slope of a line gives the change in y for a *1-unit* change in x. For instance, a slope of -3 means that y changes by -3 as x changes by 1; a slope of $\frac{4}{3}$ means that y changes by $\frac{4}{3}$ as x changes by 1. Because it is difficult to graph a change of $\frac{4}{3}$, it is easier to think of a fractional slope in terms of integer changes in x and y. As shown at the right, for a slope of $\frac{4}{3}$ we have

$$m = \frac{\text{change in } y}{\text{change in } x} = \frac{4}{3}$$

That is, for a slope of $\frac{4}{3}$, y changes by 4 as x changes by 3.

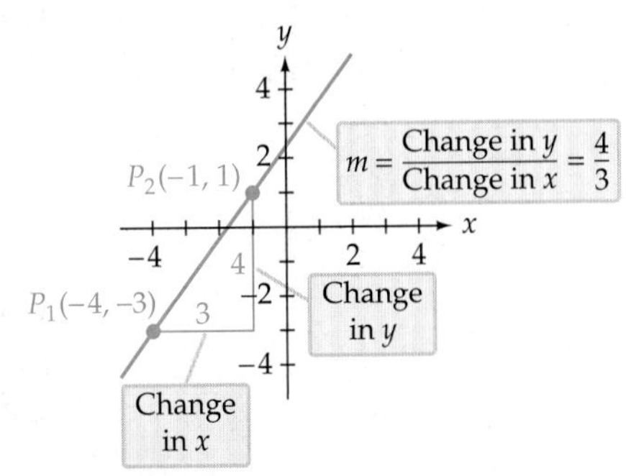

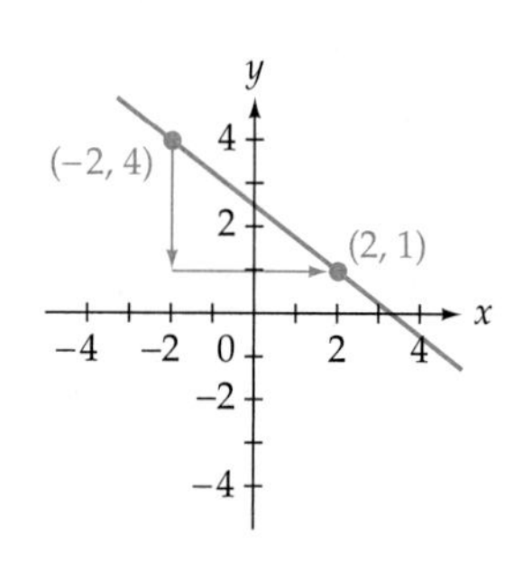

EXAMPLE 5 ■ Graph a Line Given a Point on the Line and the Slope

Draw the line that passes through $(-2, 4)$ and has slope $-\frac{3}{4}$.

Solution

Place a dot at $(-2, 4)$ and then rewrite $-\frac{3}{4}$ as $\frac{-3}{4}$. Starting from $(-2, 4)$, move 3 units down (the change in y) and then 4 units to the right (the change in x). Place a dot at that location and then draw a line through the two points.

CHECK YOUR PROGRESS 5 Draw the line that passes through $(2, 4)$ and has slope -1.

Solution *See page S23.*

Because the slope and y-intercept can be determined directly from the equation $f(x) = mx + b$, this equation is called the *slope–intercept form* of a straight line.

Slope–Intercept Form of the Equation of a Line

The graph of $f(x) = mx + b$ is a straight line with slope m and y-intercept $(0, b)$.

When a function is written in this form, it is possible to create a quick graph of the function.

EXAMPLE 6 ■ Graph a Linear Function Using the Slope and y-intercept

Graph $f(x) = -\frac{2}{3}x + 4$ by using the slope and y-intercept.

Solution

From the equation, the slope is $-\frac{2}{3}$ and the y-intercept is $(0, 4)$. Place a dot at the y-intercept. We can write the slope as $m = -\frac{2}{3} = \frac{-2}{3}$. Starting from the y-intercept, move 2 units down and 3 units to the right and place another dot. Now draw a line through the two points.

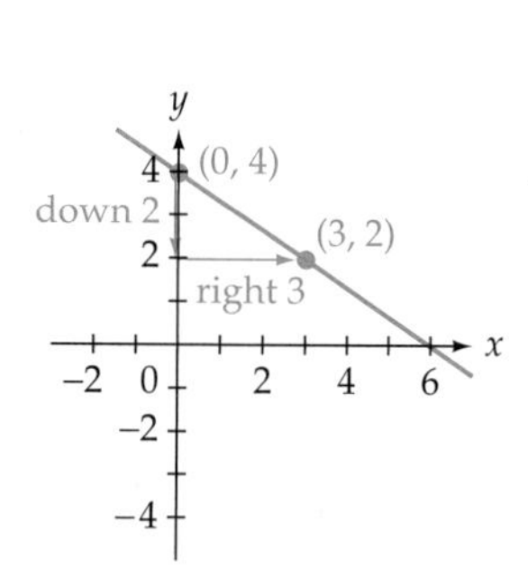

CHECK YOUR PROGRESS 6 Graph $y = \frac{3}{4}x - 5$ by using the slope and y-intercept.

Solution *See page S23.*

Excursion

Negative Velocity

We can expand the concept of velocity to include negative velocity. Suppose a car travels in a straight line starting at a given point. If the car is moving to the right, then we say that its velocity is positive. If the car is moving to the left, then we say that its velocity is negative. For instance, a velocity of −45 miles per hour means the car is moving to the left at 45 miles per hour.

If we were to graph the motion of an object on a distance–time graph, a positive velocity would be indicated by a positive slope; a negative velocity would be indicated by a negative slope.

The graph at the left represents a car traveling on a straight road. Answer the following questions on the basis of this graph.

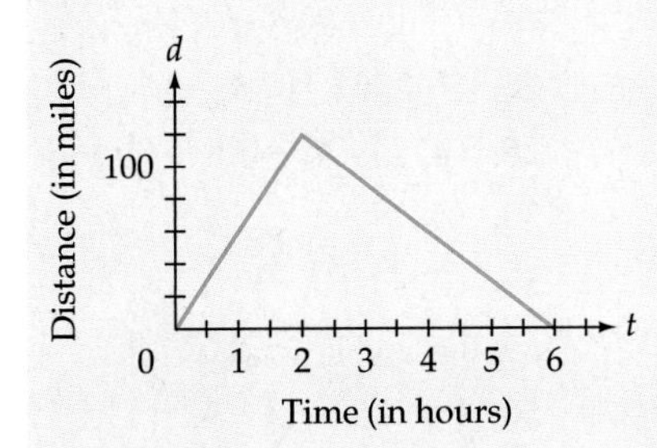

1. Between what two times is the car moving to the right?
2. Between what two times does the car have a positive velocity?
3. Between what two times is the car moving to the left?
4. Between what two times does the car have a negative velocity?
5. After 2 hours, how far is the car from its starting position?
6. How long after the car leaves its starting position does it return to its starting position?
7. What is the velocity of the car during its first 2 hours of travel?
8. What is the velocity of the car during its last 4 hours of travel?

The graph below represents another car traveling on a straight road, but this car's motion is a little more complicated. Use this graph for the questions below.

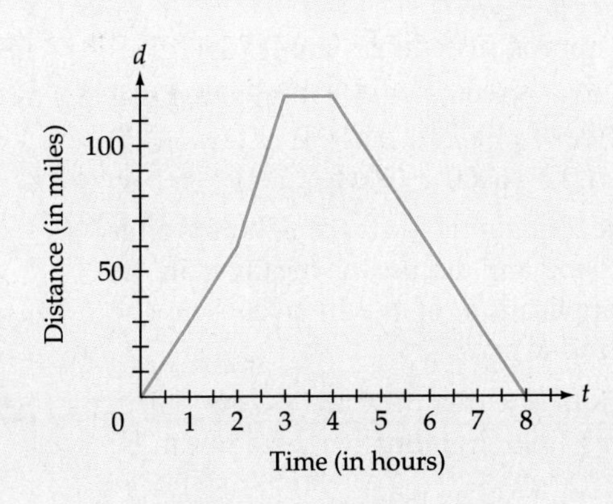

9. What is the slope of the line between hours 3 and 4?
10. What is the velocity of the car between hours 3 and 4? Is the car moving?
11. During which of the following intervals of time is the absolute value of the velocity greatest: 0 to 2 hours, 2 to 3 hours, 3 to 4 hours, or 4 to 8 hours? (Recall that the absolute value of a real number a is the distance between a and 0 on the number line.)

Exercise Set 6.2

In Exercises 1–14, find the x- and y-intercepts of the graph of the equation.

1. $f(x) = 3x - 6$
2. $f(x) = 2x + 8$
3. $y = \frac{2}{3}x - 4$
4. $y = -\frac{3}{4}x + 6$
5. $y = -x - 4$
6. $y = -\frac{x}{2} + 1$
7. $3x + 4y = 12$
8. $5x - 2y = 10$
9. $2x - 3y = 9$
10. $4x + 3y = 8$
11. $\frac{x}{2} + \frac{y}{3} = 1$
12. $\frac{x}{3} - \frac{y}{2} = 1$
13. $x - \frac{y}{2} = 1$
14. $-\frac{x}{4} + \frac{y}{3} = 1$

15. **Crickets** There is a relationship between the number of times a cricket chirps per minute and the air temperature. A linear model of this relationship is given by

$$f(x) = 7x - 30$$

where x is the temperature in degrees Celsius and $f(x)$ is the number of chirps per minute. Find and discuss the meaning of the x-intercept in the context of this application.

16. **Travel** An approximate linear model that gives the remaining distance, in miles, a plane must travel from Los Angeles to Paris is given by

$$s(t) = 6000 - 500t$$

where $s(t)$ is the remaining distance t hours after the flight begins. Find and discuss the meaning, in the context of this application, of the intercepts on the vertical and horizontal axes.

17. **Refrigeration** The temperature of an object taken from a freezer gradually rises and can be modeled by

$$T(x) = 3x - 15$$

where $T(x)$ is the Fahrenheit temperature of the object x minutes after being removed from the freezer. Find and discuss the meaning, in the context of this application, of the intercepts on the vertical and horizontal axes.

18. **Retirement Account** A retired biologist begins withdrawing money from a retirement account according to the linear model

$$A(t) = 100{,}000 - 2500t$$

where $A(t)$ is the amount, in dollars, remaining in the account t months after withdrawals begin. Find and discuss the meaning, in the context of this application, of the intercepts on the vertical and horizontal axes.

In Exercises 19–36, find the slope of the line containing the two points.

19. $(1, 3), (3, 1)$
20. $(2, 3), (5, 1)$
21. $(-1, 4), (2, 5)$
22. $(3, -2), (1, 4)$
23. $(-1, 3), (-4, 5)$
24. $(-1, -2), (-3, 2)$
25. $(0, 3), (4, 0)$
26. $(-2, 0), (0, 3)$
27. $(2, 4), (2, -2)$
28. $(4, 1), (4, -3)$
29. $(2, 5), (-3, -2)$
30. $(4, 1), (-1, -2)$
31. $(2, 3), (-1, 3)$
32. $(3, 4), (0, 4)$
33. $(0, 4), (-2, 5)$
34. $(-2, 3), (-2, 5)$
35. $(-3, -1), (-3, 4)$
36. $(-2, -5), (-4, -1)$

37. **Travel** The graph below shows the relationship between the distance traveled by a motorist and the time of travel. Find the slope of the line between the two points shown on the graph. Write a sentence that states the meaning of the slope in the context of this application.

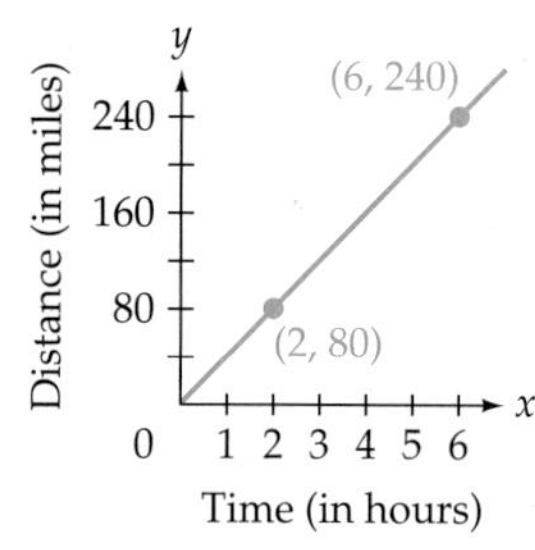

38. **Depreciation** The graph below shows the relationship between the value of a building and the depreciation allowed for income tax purposes. Find the slope of the line between the two points shown on the graph. Write a sentence that states the meaning of the slope in the context of this application.

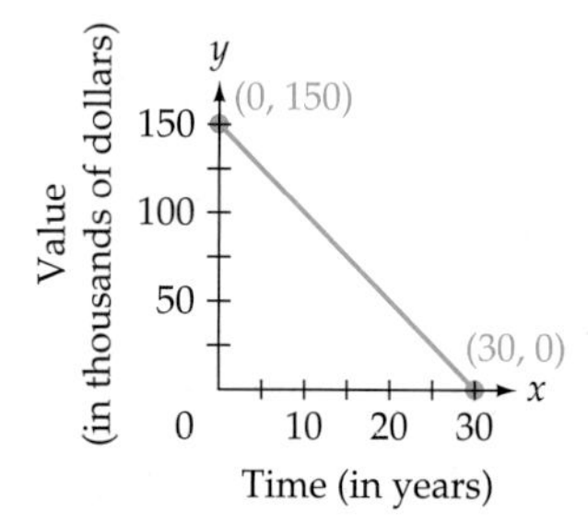

39. **Income tax** The graph below shows the relationship between the amount of tax and the amount of taxable income between \$29,050 and \$70,350. Find the slope of the line between the two points shown on the graph. Write a sentence that states the meaning of the slope in the context of this application.

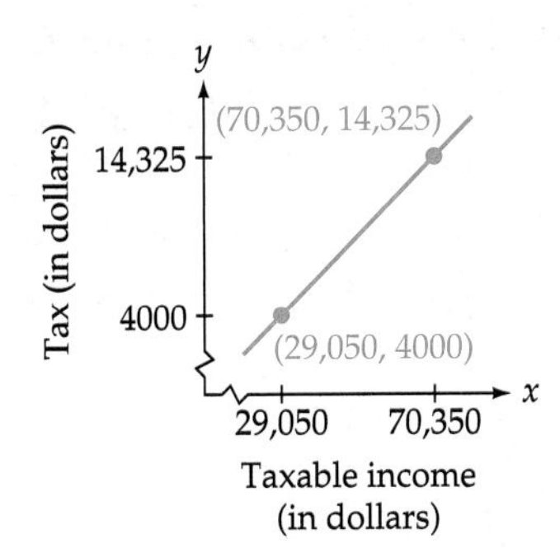

40. **Mortgages** The graph below shows the relationship between the monthly payment on a mortgage and the amount of the mortgage. Find the slope of the line between the two points shown on the graph. Write a sentence that states the meaning of the slope in the context of this application.

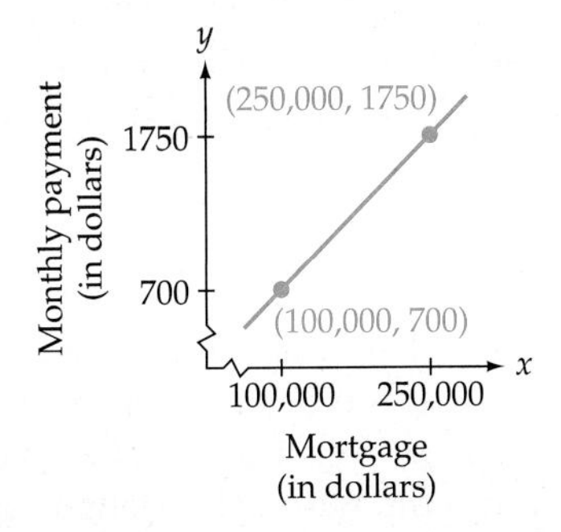

41. **Foot Races** The graph below shows the relationship between distance and time for the 5000-meter run for the world record by Deena Drossin in 2002. (Assume Drossin ran the race at a constant rate.) Find the slope of the line between the two points shown on the graph. Round to the nearest tenth. Write a sentence that states the meaning of the slope in the context of this application.

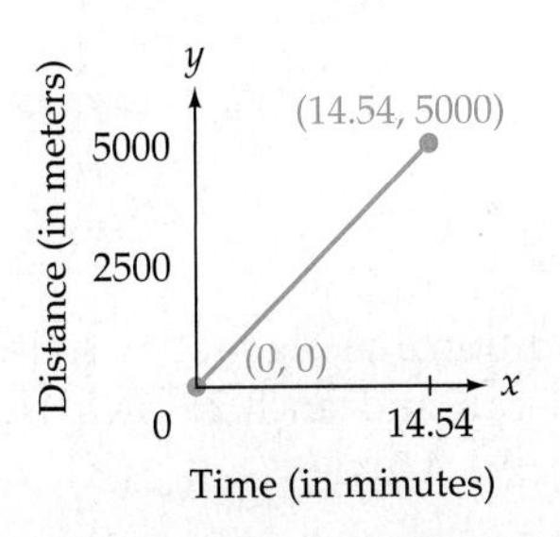

42. **Foot Races** The graph below shows the relationship between distance and time for the 10,000-meter run for the world record by Sammy Kipketer in 2002. (Assume Kipketer ran the race at a constant rate.) Find the slope of the line between the two points shown on the graph. Round to the nearest tenth. Write a sentence that states the meaning of the slope in the context of this application.

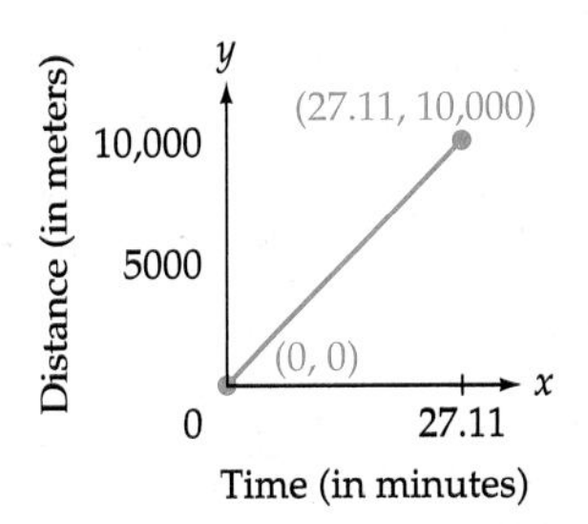

43. Graph the line that passes through the point $(-1, -3)$ and has slope $\frac{4}{3}$.

44. Graph the line that passes through the point $(-2, -3)$ and has slope $\frac{5}{4}$.

45. Graph the line that passes through the point $(-3, 0)$ and has slope -3.

46. Graph the line that passes through the point $(2, 0)$ and has slope -1.

In Exercises 47–52, graph using the slope and y-intercept.

47. $f(x) = \frac{1}{2}x + 2$

48. $f(x) = \frac{2}{3}x - 3$

49. $f(x) = -\frac{3}{2}x$

50. $f(x) = \frac{3}{4}x$

51. $f(x) = \frac{1}{3}x - 1$

52. $f(x) = -\frac{3}{2}x + 6$

Extensions

CRITICAL THINKING

53. **Jogging** Lois and Tanya start from the same place on a straight jogging course, at the same time, and jog in the same direction. Lois is jogging at 9 kilometers per hour, and Tanya is jogging at 6 kilometers per hour. The graphs on the following page show the distance each jogger has traveled in x hours and the distance between the joggers in x hours. Which graph represents the distance Lois has traveled in x hours? Which graph represents the distance Tanya has traveled in x

hours? Which graph represents the distance between Lois and Tanya in x hours?

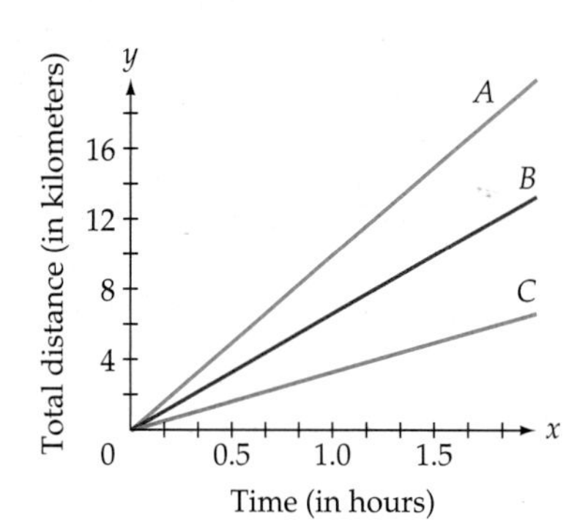

54. **Chemistry** A chemist is filling two cans from a faucet that releases water at a constant rate. Can 1 has a diameter of 20 millimeters and can 2 has a diameter of 30 millimeters.

 a. In the following graph, which line represents the depth of the water in can 1 after x seconds?

 b. Use the graph to estimate the difference in the depths of the water in the two cans after 15 seconds.

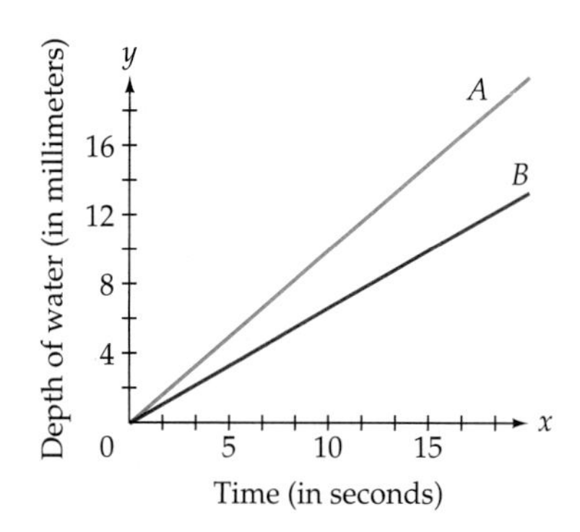

55. **ANSI** The American National Standards Institute (ANSI) states that the slope of a wheelchair ramp must not exceed $\frac{1}{12}$.

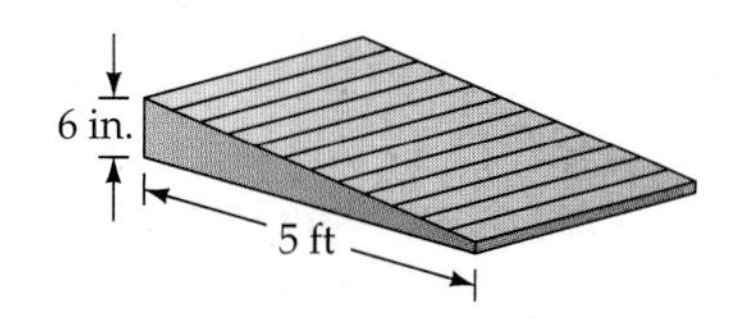

Does the ramp pictured above meet the requirements of ANSI?

56. **ANSI** A ramp for a wheelchair must be 14 inches high. What is the minimum length of this ramp so that it meets the ANSI requirements stated in Exercise 55?

57. If $(2, 3)$ are the coordinates of a point on a line that has slope 2, what is the y-coordinate of the point on the line at which $x = 4$?

58. If $(-1, 2)$ are the coordinates of a point on a line that has slope -3, what is the y-coordinate of the point on the line at which $x = 1$?

59. If $(1, 4)$ are the coordinates of a point on a line that has slope $\frac{2}{3}$, what is the y-coordinate of the point on the line at which $x = -2$?

60. If $(-2, -1)$ are the coordinates of a point on a line that has slope $\frac{3}{2}$, what is the y-coordinate of the point on the line at which $x = -6$?

61. What effect does increasing the coefficient of x have on the graph of $y = mx + b$?

62. What effect does decreasing the coefficient of x have on the graph of $y = mx + b$?

63. What effect does increasing the constant term have on the graph of $y = mx + b$?

64. What effect does decreasing the constant term have on the graph of $y = mx + b$?

65. Do the graphs of all straight lines have a y-intercept? If not, give an example of one that does not.

66. If two lines have the same slope and the same y-intercept, must the graphs of the lines be the same? If not, give an example.

EXPLORATIONS

67. **Construction** When you climb a staircase, the flat part of a stair that you step on is called the *tread* of the stair. The *riser* is the vertical part of the stair. The slope of a staircase is the ratio of the length of the riser to the length of the tread. Because the design of a staircase may affect safety, most cities have building codes that give rules for the design of a staircase.

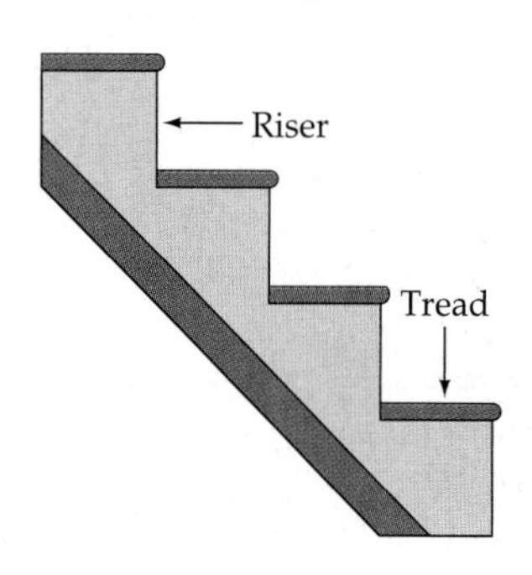

 a. The traditional design of a staircase calls for a 9-inch tread and an 8.25-inch riser. What is the slope of this staircase?

b. A newer design for a staircase uses an 11-inch tread and a 7-inch riser. What is the slope of this staircase?

c. An architect is designing a house with a staircase that is 8 feet high and 12 feet long. Is the architect using the traditional design in part a or the newer design in part b? Explain your answer.

d. Staircases that have a slope between 0.5 and 0.7 are usually considered safer than those with a slope greater than 0.7. Design a safe staircase that goes from the first floor of a house to the second floor, which is 9 feet above the first floor.

e. Measure the tread and riser for three staircases you encounter. Do these staircases match the traditional design in part a or the newer design in part b?

68. Geometry In the diagram at the right, lines l_1 and l_2 are perpendicular with slopes m_1 and m_2, respectively, and $m_1 > 0$ and $m_2 < 0$. Line segment $\overline{AC}$ has length 1.

a. Show that the length of $\overline{BC}$ is m_1.

b. Show that the length of $\overline{CD}$ is $-m_2$. Note that because m_2 is a negative number, $-m_2$ is a positive number.

c. Show that right triangles ACB and DCA are similar triangles. **Similar triangles** have the same shape; the corresponding angles are equal, and corresponding sides are in proportion. (*Suggestion:* Show that the measure of angle ADC equals the measure of angle BAC.)

d. Show that $\frac{m_1}{1} = \frac{1}{-m_2}$. Use the fact that the ratios of corresponding sides of similar triangles are equal.

e. Use the equation in part d to show that $m_1 m_2 = -1$.

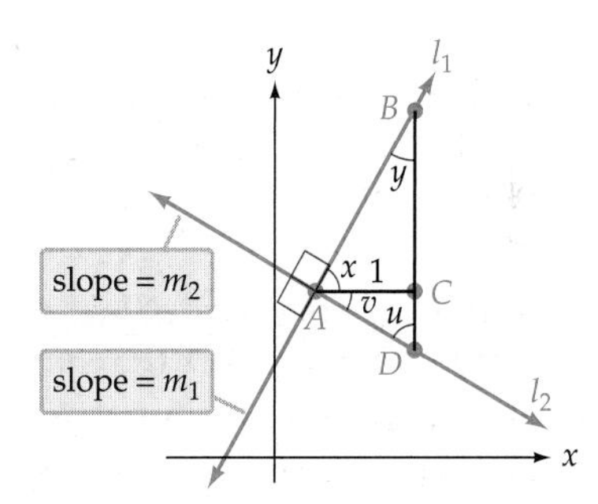

SECTION 6.3 Finding Linear Models

Finding Linear Models

TAKE NOTE

When creating a linear model, the slope will be the quantity that is expressed using the word *per*. The car discussed at the right uses 0.04 gallon *per* mile. The slope is negative because the amount of fuel in the tank is decreasing.

Suppose that a car uses 0.04 gallon of gas per mile driven and that the fuel tank, which holds 18 gallons of gas, is full. Using this information, we can determine a linear model for the amount of fuel remaining in the gas tank after driving x miles.

Recall that a linear function is one that can be written in the form $f(x) = mx + b$, where m is the slope of the line and b is the y-intercept. The slope is the rate at which the car is using fuel, 0.04 gallon per mile. Because the car is consuming the fuel, the amount of fuel in the tank is decreasing. Therefore, the slope is negative and we have $m = -0.04$.

The amount of fuel in the tank depends on the number of miles, x, the car has been driven. Before the car starts (that is, when $x = 0$), there are 18 gallons of gas in the tank. The y-intercept is $(0, 18)$.

Using this information, we can create the linear function.

$$f(x) = mx + b$$
$$f(x) = -0.04x + 18$$
• Replace m by -0.04 and b by 18.

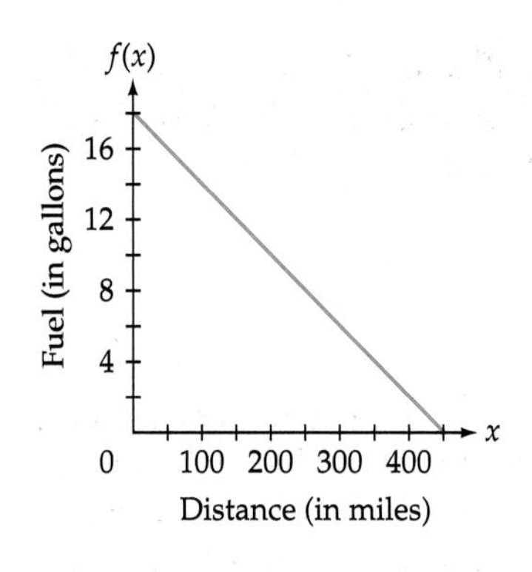

The linear function that models the amount of fuel remaining in the tank is given by $f(x) = -0.04x + 18$, where $f(x)$ is the amount of fuel, in gallons, remaining after driving x miles. The graph of the function is shown at the left.

The x-intercept of a graph is the point at which $f(x) = 0$. For this application, $f(x) = 0$ when there are 0 gallons of fuel remaining in the tank. Thus, replacing $f(x)$ by 0 in $f(x) = -0.04x + 18$ and solving for x will give the number of miles the car can be driven before running out of gas.

$$f(x) = -0.04x + 18$$
$$0 = -0.04x + 18 \quad \text{• Replace } f(x) \text{ by } 0.$$
$$-18 = -0.04x$$
$$450 = x$$

The car can travel 450 miles before running out of gas.

Recall that the domain of a function is all possible values of x, and the range of a function is all possible values of $f(x)$. For the function $f(x) = -0.04x + 18$, which was used above to model the fuel remaining in the gas tank of the car, the domain is $\{x \mid 0 \le x \le 450\}$ because the fuel tank is empty when the car has traveled 450 miles. The range is $\{y \mid 0 \le y \le 18\}$ because the tank can hold up to 18 gallons of fuel. Sometimes it is convenient to write the domain $\{x \mid 0 \le x \le 450\}$ as $[0, 450]$ and the range $\{y \mid 0 \le y \le 18\}$ as $[0, 18]$. The notation $[0, 450]$ and $[0, 18]$ is called **interval notation.** Using interval notation, a domain of $\{x \mid a \le x \le b\}$ is written $[a, b]$ and a range of $\{y \mid c \le y \le d\}$ is written $[c, d]$.

QUESTION *Why does it not make sense for the domain of $f(x) = -0.04x + 18$ to exceed 450?*

EXAMPLE 1 ■ Application of Finding a Linear Model Given the Slope and y-intercept

Suppose a 20-gallon gas tank contains 2 gallons when a motorist decides to fill up the tank. If the gas pump fills the tank at a rate of 0.1 gallon per second, find a linear function that models the amount of fuel in the tank t seconds after fueling begins.

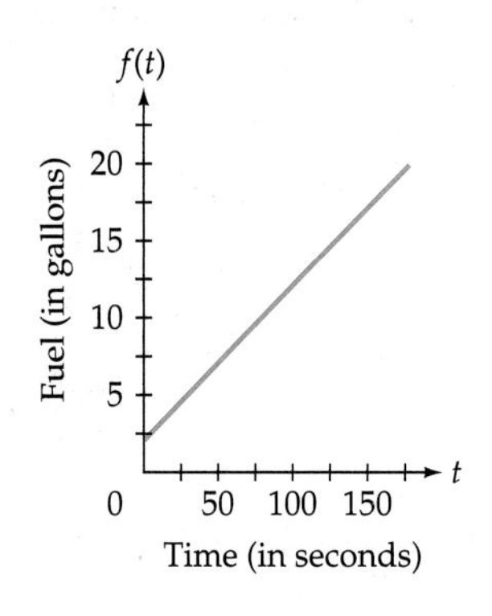

Solution

When fueling begins, at $t = 0$, there are 2 gallons of gas in the tank. Therefore, the y-intercept is $(0, 2)$. The slope is the rate at which fuel is being added to the tank. Because the amount of fuel in the tank is increasing, the slope is positive and we have $m = 0.1$. To find the linear function, replace m and b by their respective values.

$$f(t) = mt + b$$
$$f(t) = 0.1t + 2 \quad \text{• Replace } m \text{ by } \mathbf{0.1} \text{ and } b \text{ by } \mathbf{2}.$$

The linear function is $f(t) = 0.1t + 2$, where $f(t)$ is the number of gallons of fuel in the tank t seconds after fueling begins.

ANSWER *If $x > 450$, then $f(x) < 0$. This would mean that the tank has negative gallons of gas. For instance, $f(500) = -2$.*

CHECK YOUR PROGRESS 1 The boiling point of water at sea level is 100°C. The boiling point decreases 3.5°C per 1 kilometer increase in altitude. Find a linear function that gives the boiling point of water as a function of altitude.

Solution *See page S23.*

For each of the previous examples, the known point on the graph of the linear function was the y-intercept. This information enabled us to determine b for the linear function $f(x) = mx + b$. In some cases, a point other than the y-intercept is given. In such a case, the *point–slope formula* is used to find the equation of the line.

TAKE NOTE

Using parentheses may help when substituting into the point–slope formula.

$$y - y_1 = m(x - x_1)$$
$$\downarrow \quad \downarrow \quad \downarrow$$
$$y - (\ ) = (\ )[x - (\ )]$$

Point–Slope Formula of a Straight Line

Let (x_1, y_1) be a point on a line and let m be the slope of the line. Then the equation of the line can be found using the point–slope formula

$$y - y_1 = m(x - x_1)$$

EXAMPLE 2 ■ Find the Equation of a Line Given the Slope and a Point on the Line

Find the equation of the line that passes through $(1, -3)$ and has slope -2.

Solution

$$y - y_1 = m(x - x_1) \qquad \text{• Use the point–slope formula.}$$
$$y - (-3) = -2(x - 1) \qquad \text{• } m = -2, (x_1, y_1) = (1, -3)$$
$$y + 3 = -2x + 2$$
$$y = -2x - 1$$

Note that we wrote the equation of the line as $y = -2x - 1$. We could also write the equation in functional notation as $f(x) = -2x - 1$.

TAKE NOTE

Recall that $f(x)$ and y are different symbols for the same quantity, the value of the function at x.

CHECK YOUR PROGRESS 2 Find the equation of the line that passes through $(-2, 2)$ and has slope $-\frac{1}{2}$.

Solution *See page S23.*

EXAMPLE 3 ■ Application of Finding a Linear Model Given a Point and the Slope

Based on data from the *Kelley Blue Book*, the value of a certain car decreases approximately \$250 per month. If the value of the car 2 years after it was purchased was \$14,000, find a linear function that models the value of the car after x months of ownership. Use this function to find the value of the car after 3 years of ownership.

Solution

Let V represent the value of the car after x months. Then $V = 14{,}000$ when $x = 24$ (2 years is 24 months). A solution of the equation is (24, 14,000). The car is decreasing in value at a rate of \$250 per month. Therefore, the slope is -250. Now use the point–slope formula to find the linear equation that models the function.

$$V - V_1 = m(x - x_1)$$
$$V - 14{,}000 = -250(x - 24) \quad \bullet\ x_1 = 24,\ V_1 = 14{,}000,\ m = -250$$
$$V - 14{,}000 = -250x + 6000$$
$$V = -250x + 20{,}000$$

A linear function that models the value of the car after x months of ownership is $V(x) = -250x + 20{,}000$.

To find the value of the car after 3 years (36 months), evaluate the function when $x = 36$.

$$V(x) = -250x + 20{,}000$$
$$V(36) = -250(36) + 20{,}000 = 11{,}000$$

The value of the car is \$11,000 after 3 years of ownership.

CHECK YOUR PROGRESS 3 During a brisk walk, a person burns about 3.8 calories per minute. If a person has burned 191 calories in 50 minutes, determine a linear function that models the number of calories burned after t minutes.

Solution *See page S23.*

TAKE NOTE

There are many ways to find the equation of a line. However, in every case, there must be enough information to determine a point on the line and to find the slope of the line. When you are doing problems of this type, look for different ways that information may be presented. For instance, in Example 4, even though the slope of the line is not given, knowing two points enables us to find the slope.

The next example shows how to find the equation of a line given two points on the line.

EXAMPLE 4 ■ Find the Equation of a Line Given Two Points on the Line

Find the equation of the line that passes through $P_1(6, -4)$ and $P_2(3, 2)$.

Solution

Find the slope of the line between the two points.

$$m = \frac{y_2 - y_1}{x_2 - x_1} = \frac{2 - (-4)}{3 - 6} = \frac{6}{-3} = -2$$

Use the point–slope formula to find the equation of the line.

$$y - y_1 = m(x - x_1)$$
$$y - (-4) = -2(x - 6) \quad \bullet\ m = -2,\ x_1 = 6,\ y_1 = -4$$
$$y + 4 = -2x + 12$$
$$y = -2x + 8$$

CHECK YOUR PROGRESS 4 Find the equation of the line that passes through $P_1(-2, 3)$ and $P_2(4, 1)$.

Solution *See page S23.*

Math Matters Perspective: Using Straight Lines in Art

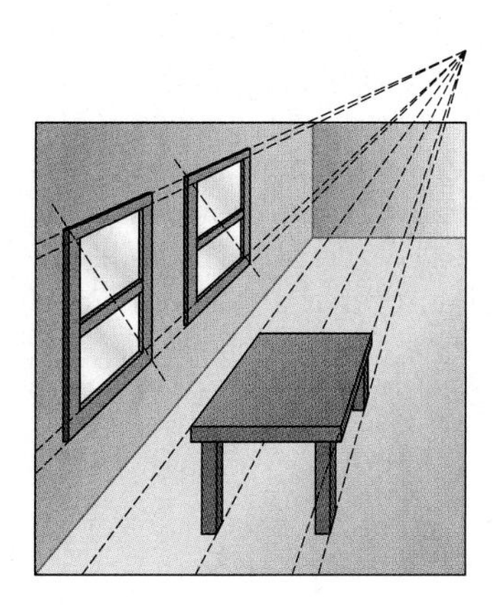

Many paintings we see today have a three-dimensional quality to them, even though they are painted on a flat surface. This was not always the case. It wasn't until the Renaissance that artists started to paint "in perspective." Using lines is one way to create this perspective. Here is a simple example.

Draw a dot, called the *vanishing point,* and a rectangle on a piece of paper. Draw windows as shown. To keep the perspective accurate, the lines through opposite corners of the windows should be parallel. A table in proper perspective is created in the same way.

This method of creating perspective was employed by Leonardo da Vinci. Use the Internet to find and print a copy of his painting *The Last Supper.* Using a ruler, see whether you can find the vanishing point by drawing two lines along the top edges of the windows on the sides of the painting.

Regression Lines

There are many situations in which a linear function can be used to approximate collected data. For instance, the table below shows the maximum exercise heart rates for specific individuals of various ages who exercise regularly.

Age, x, in years	20	25	30	32	43	55	28	42	50	55	62
Heart rate, y, in maximum beats per minute	160	150	148	145	140	130	155	140	132	125	125

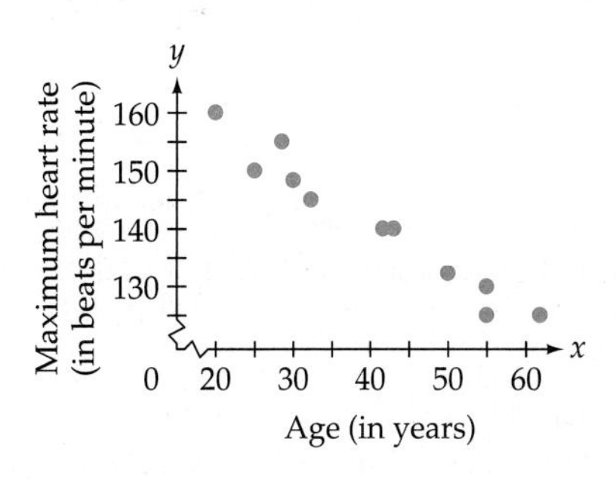

The graph at the left, called a **scatter diagram,** is a graph of the ordered pairs of the table. These ordered pairs suggest that the maximum exercise heart rate for an individual decreases as the person's age increases.

Although these points do not lie on one line, it is possible to find a line that *approximately fits* the data. One way to do this is to select two data points and then find the equation of the line that passes through the two points. To do this, we first find the slope of the line between the two points and then use the point–slope formula to find the equation of the line. Suppose we choose (20, 160) as P_1 and (62, 125) as P_2. Then the slope of the line between P_1 and P_2 is

$$m = \frac{y_2 - y_1}{x_2 - x_1} = \frac{125 - 160}{62 - 20} = -\frac{35}{42} = -\frac{5}{6}$$

Now use the point–slope formula.

$$y - y_1 = m(x - x_1)$$

$$y - 160 = -\frac{5}{6}(x - 20) \quad \bullet\ m = -\frac{5}{6},\ x_1 = 20,\ y_1 = 160$$

$$y - 160 = -\frac{5}{6}x + \frac{50}{3} \quad \bullet\ \text{Multiply by } -\frac{5}{6}.$$

$$y = -\frac{5}{6}x + \frac{530}{3} \quad \bullet\ \text{Add 160 to each side of the equation.}$$

The graph of $y = -\frac{5}{6}x + \frac{530}{3}$

The graph of $y = -\frac{5}{6}x + \frac{530}{3}$ is shown at the left. This line *approximates* the data.

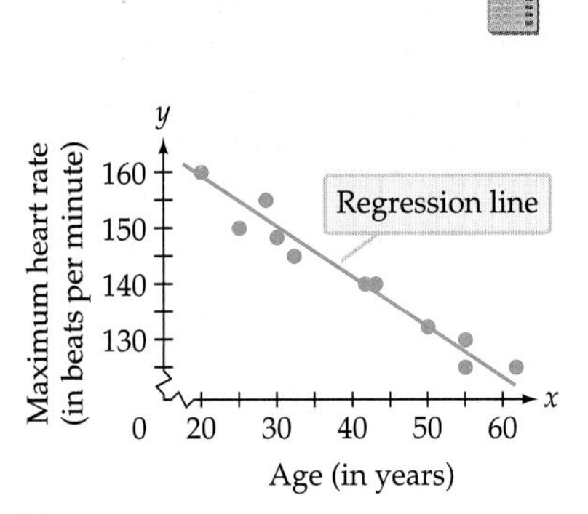

The equation of the line we found by choosing two data points gives an approximate linear model for the data. If we had chosen different points, the result would have been a different equation. Among all the lines that can be chosen, statisticians generally pick the *least-squares line,* which is also called the *regression line.* The **regression line** is the line for which the sum of the squares of the vertical distances between the data points and the line is a minimum.

A graphing calculator can be used to find the equation of the regression line for a given set of data. For instance, the equation of the regression line for the maximum heart rate data is given by $y = -0.827x + 174$. The graph of this line is shown at the left. Using this model, an exercise physiologist can determine the recommended maximum exercise heart rate for an individual of any particular age.

For example, suppose an individual is 28 years old. The physiologist would replace x by 28 and determine the value of y.

$$y = -0.827x + 174$$
$$y = -0.827(28) + 174 \quad \text{• Replace } x \text{ by } 28.$$
$$= 150.844$$

The maximum exercise heart rate recommended for a 28-year-old person is approximately 151 beats per minute.

The calculation of the equation of the regression line can be accomplished with a graphing calculator by using the STAT key.

The table below shows the data collected by a chemistry student who is trying to determine a relationship between the temperature, in degrees Celsius, and volume, in liters, of 1 gram of oxygen at a constant pressure. Chemists refer to this relationship as Charles's Law.

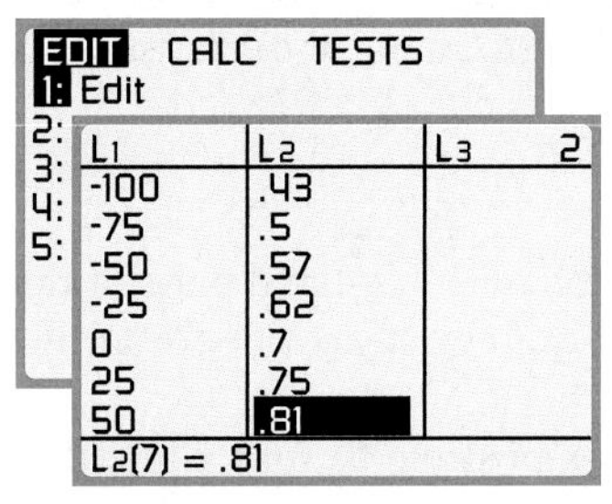

Temperature, *T*, in degrees Celsius	−100	−75	−50	−25	0	25	50
Volume, *V*, in liters	0.43	0.5	0.57	0.62	0.7	0.75	0.81

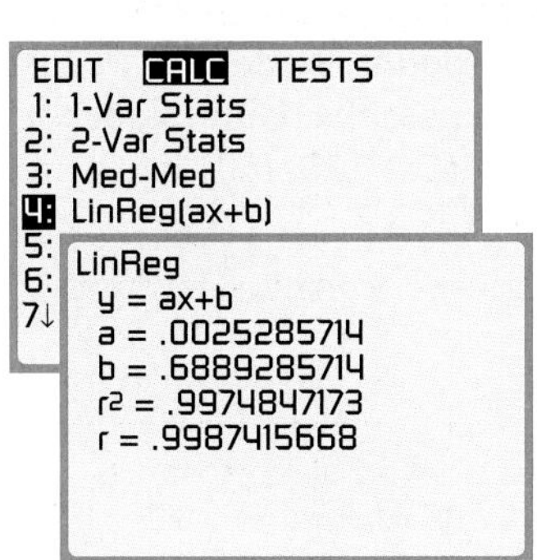

To find the equation of the regression line for these data, press the STAT key and then select EDIT from the menu. This will bring up a table into which you can enter data. Let L1 be the independent variable (temperature) and L2 be the dependent variable (volume). Screens from a TI-83/84 are shown at the left.

Once the data have been entered, select the STAT key again, highlight CALC, and arrow down to LinReg(ax+b) to see the linear regression equation. Selecting this option will paste LinReg(ax+b) onto the home screen. Now you can just press ENTER and the values for a and b will appear on the screen. Entering LinReg(ax+b)L1,L2,Y1 not only shows the results on the home screen, but pastes the regression equation into Y1 in the Y= editor. This will enable you to easily graph or evaluate the regression equation.

For this set of data, the regression equation, with the coefficient and constant rounded to the nearest ten-millionth, is $V = 0.0025286T + 0.6889286$. To determine the volume of 1 gram of oxygen when the temperature is -30°C, replace T by -30 and evaluate the expression. This can be done using your calculator. Use the Y-VARS menu to place Y1 on the screen (VARS ▷ ENTER ENTER). Then enter the value -30, within parentheses, as shown at the left. After hitting ENTER, the volume will be displayed as approximately 0.61 liter.

CALCULATOR NOTE 

If r and r^2 do not appear on the screen of your TI-83/84, press 2nd CATALOG D (above the 0 key) and then scroll down to DiagnosticsOn. Press ENTER twice. Now repeat the procedure, described on the previous page, for finding the linear regression equation. This time the r and r^2 values will also be displayed.

You may have noticed some additional results on the screen when the regression equation was calculated. The variable r is called the **correlation coefficient,** and r^2 is called the **coefficient of determination.** Statisticians use these numbers to determine how well the regression equation approximates the data. If $r = 1$, the data exactly fit a line of positive slope. If $r = -1$, the data exactly fit a line of negative slope. In general, the closer r^2 is to 1, the closer the data fit a linear model. For our purposes, we will assume that the given data are approximated by a linear function.

EXAMPLE 5 ■ Find a Linear Regression Equation

Sodium thiosulfate is used by photographers to develop some types of film. The amount of this chemical that will dissolve in water depends on the temperature of the water. The table below gives the number of grams of sodium thiosulfate that will dissolve in 100 milliliters of water at various temperatures.

Temperature, x, in degrees Celsius	20	35	50	60	75	90	100
Sodium thiosulfate dissolved, y, in grams	50	80	120	145	175	205	230

a. Find the linear regression equation for these data.

b. How many grams of sodium thiosulfate does the model predict will dissolve in 100 milliliters of water when the temperature is 70°C? Round to the nearest tenth of a gram.

Solution

a. Using a calculator, the regression equation is $y = 2.2517731x + 5.2482270$.

b. Evaluate the regression equation when $x = 70$.

$$y = 2.2517731x + 5.2482270$$
$$= 2.2517731(70) + 5.2482270 \quad \text{• Replace } x \text{ by 70.}$$
$$= 162.872344$$

Approximately 162.9 grams of sodium thiosulfate will dissolve when the temperature is 70°C.

CHECK YOUR PROGRESS 5 The heights and weights of women swimmers on a college swim team are given in the table below.

Height, x, in inches	68	64	65	67	62	67	65
Weight, y, in pounds	132	108	108	125	102	130	105

a. Find the linear regression equation for these data.

b. Use your regression equation to estimate the weight of a woman swimmer who is 63 inches tall. Round to the nearest pound.

Solution *See page S23.*

Excursion

A Linear Business Model

Two people decide to open a business reconditioning toner cartridges for copy machines. They rent a building for \$7000 per year and estimate that building maintenance, taxes, and insurance will cost \$6500 per year. Each person wants to make \$12 per hour in the first year and will work 10 hours per day for 260 days of the year. Assume that it costs \$28 to restore a cartridge and that the restored cartridge can be sold for \$45.

1. Write a linear function for the total cost C to operate the business and restore n cartridges during the first year, not including the hourly wage the owners wish to earn.
2. Write a linear function for the total revenue R the business will earn during the first year by selling n cartridges.
3. How many cartridges must the business restore and sell annually to break even, not including the hourly wage the owners wish to earn?
4. How many cartridges must the business restore and sell annually for the owners to pay all expenses and earn the hourly wage they desire?
5. Suppose the entrepreneurs are successful in their business and are restoring and selling 25 cartridges each day of the 260 days they are open. What will be their hourly wage for the year if all the profit is shared equally?
6. As the company becomes successful and is selling and restoring 25 cartridges each day of the 260 days it is open, the entrepreneurs decide to hire a part-time employee 4 hours per day and to pay the employee \$8 per hour. How many additional cartridges must be restored and sold each year just to cover the cost of the new employee? You can neglect employee costs such as social security, worker's compensation, and other benefits.
7. Suppose the company decides that it could increase its business by advertising. Answer Exercises 1, 2, 3, and 5 if the owners decide to spend \$400 per month on advertising.

Exercise Set 6.3

In Exercises 1–8, find the equation of the line that passes through the given point and has the given slope.

1. $(0, 5), m = 2$
2. $(2, 3), m = \dfrac{1}{2}$
3. $(-1, 7), m = -3$
4. $(0, 0), m = \dfrac{1}{2}$
5. $(3, 5), m = -\dfrac{2}{3}$
6. $(0, -3), m = -1$
7. $(-2, -3), m = 0$
8. $(4, -5), m = -2$

In Exercises 9–16, find the equation of the line that passes through the given points.

9. $(0, 2), (3, 5)$
10. $(0, -3), (-4, 5)$
11. $(0, 3), (2, 0)$
12. $(-2, -3), (-1, -2)$
13. $(2, 0), (0, -1)$
14. $(3, -4), (-2, -4)$
15. $(-2, 5), (2, -5)$
16. $(2, 1), (-2, -3)$
17. **Hotel Industry** The operator of a hotel estimates that 500 rooms per night will be rented if the room rate per night is \$75. For each \$10 increase in the price of a room, six fewer rooms per night will be rented. Determine a linear function that will predict the number of rooms that will be rented per night for a given price per room. Use this model to predict the number of rooms that will be rented if the room rate is \$100 per night.

18. **Construction** A general building contractor estimates that the cost to build a new home is $30,000 plus $85 for each square foot of floor space in the house. Determine a linear function that gives the cost of building a house that contains x square feet of floor space. Use this model to determine the cost to build a house that contains 1800 square feet of floor space.

19. **Travel** A plane travels 830 miles in 2 hours. Determine a linear model that will predict the number of miles the plane can travel in a given interval of time. Use this model to predict the distance the plane will travel in $4\frac{1}{2}$ hours.

20. **Compensation** An account executive receives a base salary plus a commission. On $20,000 in monthly sales, an account executive would receive compensation of $1800. On $50,000 in monthly sales, an account executive would receive compensation of $3000. Determine a linear function that yields the compensation of a sales executive for x dollars in monthly sales. Use this model to determine the compensation of an account executive who has $85,000 in monthly sales.

21. **Car Sales** A manufacturer of economy cars has determined that 50,000 cars per month can be sold at a price of $9000 per car. At a price of $8750, the number of cars sold per month would increase to 55,000. Determine a linear function that predicts the number of cars that will be sold at a price of x dollars. Use this model to predict the number of cars that will be sold at a price of $8500.

22. **Calculator Sales** A manufacturer of graphing calculators has determined that 10,000 calculators per week will be sold at a price of $95 per calculator. At a price of $90, it is estimated that 12,000 calculators will be sold. Determine a linear function that predicts the number of calculators that will be sold per week at a price of x dollars. Use this model to predict the number of calculators that will be sold at a price of $75.

23. **Stress** A research hospital did a study on the relationship between stress and diastolic blood pressure. The results from eight patients in the study are given in the table below. The units for blood pressure values are measured in milliliters of mercury.

Stress test score, x	55	62	58	78	92	88	75	80
Blood pressure, y	70	85	72	85	96	90	82	85

a. Find the linear regression equation for these data. Round to the nearest hundredth.

b. Use the regression equation to estimate the diastolic blood pressure of a person whose stress test score was 85. Round to the nearest whole number.

24. **Hourly wages** The average hourly earnings, in dollars, of nonfarm workers in the United States for the years 1996 to 2003 are given in the table below. (*Source:* Bureau of Labor Statistics)

Year, x	1996	1997	1998	1999	2000	2001	2002	2003
Hourly wage, y	12.03	12.49	13.00	13.47	14.00	14.53	14.95	15.35

a. Using $x = 0$ to correspond to 1990, find the linear regression equation for these data. Round to the nearest hundredth.

b. Use the regression equation to estimate the expected average hourly wage, to the nearest cent, in 2015.

25. **High School Graduates** The table below shows the numbers of students, in thousands, who have graduated from or are projected to graduate from high school for the years 1996 to 2005. (*Source:* U.S. Bureau of Labor Statistics)

Year, x	1996	1997	1998	1999	2000	2001	2002	2003	2004	2005
Number of students (in thousands), y	2540	2633	2740	2786	2820	2837	2886	2929	2935	2944

a. Using $x = 0$ to correspond to 1996, find the linear regression equation for these data. Round to the nearest hundredth.

b. Use the regression equation to estimate the expected number of high school graduates in 2010. Round to the nearest thousand students.

26. **Fuel Efficiency** An automotive engineer studied the relationship between the speed of a car and the number of miles traveled per gallon of fuel consumed at that speed. The results of the study are shown in the table below.

Speed (in miles per hour), x	40	25	30	50	60	80	55	35	45
Consumption (in miles per gallon), y	26	27	28	24	22	21	23	27	25

a. Find the linear regression equation for these data. Round to the nearest hundredth.

b. Use the regression equation to estimate the expected number of miles traveled per gallon of fuel consumed for a car traveling at 65 miles per hour. Round to the nearest mile per hour.

27. **Meteorology** A meteorologist studied the maximum temperatures at various latitudes for January of a certain year. The results of the study are shown in the table below.

Latitude (in °N), x	22	30	36	42	56	51	48
Maximum temperature (in °F), y	80	65	47	54	21	44	52

a. Find the linear regression equation for these data. Round to the nearest hundredth.

b. Use the regression equation to estimate the expected maximum temperature in January at a latitude of 45°N. Round to the nearest degree.

28. **Zoology** A zoologist studied the running speeds of animals in terms of the animals' body lengths. The results of the study are shown in the table below.

Body length (in centimeters), x	1	9	15	16	24	25	60
Running speed (in meters per second), y	1	2.5	7.5	5	7.4	7.6	20

a. Find the linear regression equation for these data. Round to the nearest hundredth.

b. Use the regression equation to estimate the expected running speed of a deer mouse, whose body length is 10 centimeters. Round to the nearest tenth of a centimeter.

Extensions

CRITICAL THINKING

29. A line contains the points $(4, -1)$ and $(2, 1)$. Find the coordinates of three other points on the line.

30. If f is a linear function for which $f(1) = 3$ and $f(-1) = 5$, find $f(4)$.

31. The ordered pairs $(0, 1)$, $(4, 9)$ and $(3, n)$ are solutions of the same linear equation. Find n.

32. The ordered pairs $(2, 2)$, $(-1, 5)$ and $(3, n)$ are solutions of the same linear equation. Find n.

33. Is there a linear function that contains the ordered pairs $(2, 4)$, $(-1, -5)$, and $(0, 2)$? If so, find the function and explain why there is such a function. If not, explain why there is no such function.

34. Is there a linear function that contains the ordered pairs $(5, 1)$, $(4, 2)$, and $(0, 6)$? If so, find the function and explain why there is such a function. If not, explain why there is no such function.

35. Travel Assume that the maximum speed your car will travel varies linearly with the steepness of the hill it is climbing or descending. If the hill is 5° up, your car can travel 77 kilometers per hour. If the hill is 2° down (−2°), your car can travel 154 kilometers per hour. When your car's top speed is 99 kilometers per hour, how steep is the hill? State your answer in degrees, and note whether the car is climbing or descending.

EXPLORATIONS

36. Boating A person who can row at a rate of 3 miles per hour in calm water is trying to cross a river in which a current of 4 miles per hour runs perpendicular to the direction of rowing. See the figure at the right.

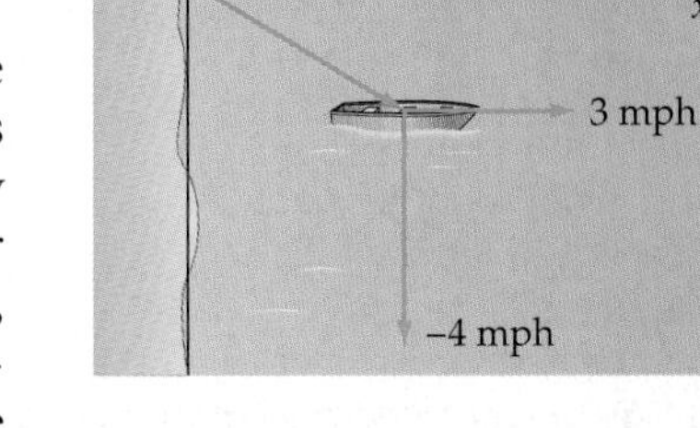

Because of the current, the boat is being pushed downstream at the same time that it is moving across the river. Because the boat is traveling at 3 miles per hour in the x direction, its horizontal position after t hours is given by $x = 3t$. The current is pushing the boat in the negative y direction at 4 miles per hour. Therefore, the boat's vertical position after t hours is given by $y = -4t$, where -4 indicates that the boat is moving downstream. The set of equations $x = 3t$ and $y = -4t$ are called **parametric equations,** and t is called the **parameter.**

a. What is the location of the boat after 15 minutes (0.25 hour)?

b. If the river is 1 mile wide, how far down the river will the boat be when it reaches the other shore? *Hint:* Find the time it takes the boat to cross the river by solving $x = 3t$ for t when $x = 1$. Then replace t by this value in $y = -4t$ and simplify.

c. For the parametric equations $x = 3t$ and $y = -4t$, write y in terms of x by solving $x = 3t$ for t and then substituting this expression into $y = -4t$.

37. Aviation In the diagram below, a plane flying at 5000 feet above sea level begins a gradual ascent.

a. Determine parametric equations for the path of the plane. *Hint:* See Exercise 36.

b. What is the altitude of the plane 5 minutes after it begins its ascent?

c. What is the altitude of the plane after it has traveled 12,000 feet in the positive x direction?

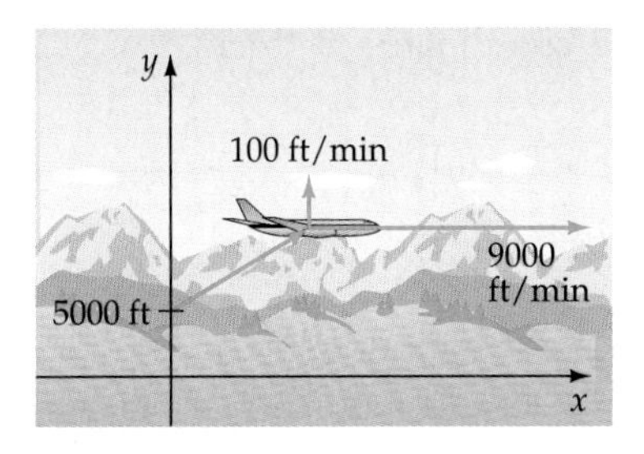

SECTION 6.4 Quadratic Functions

Properties of Quadratic Functions

point of interest

The suspension cables of some bridges, such as the Golden Gate Bridge, have the shape of a parabola.

The photo above shows the roadway of the bridge being assembled in sections and attached to suspender ropes. The bridge was opened to vehicles on May 28, 1937.

Recall that a linear function is a function of the form $f(x) = mx + b$. The graph of a linear function has certain characteristics. It is a straight line with slope m and y-intercept $(0, b)$.

A **quadratic function** in a single variable x is a function of the form $f(x) = ax^2 + bx + c$, $a \neq 0$. Examples of quadratic functions are given below.

$f(x) = x^2 - 3x + 1$ • $a = 1, b = -3, c = 1$

$g(t) = -2t^2 - 4$ • $a = -2, b = 0, c = -4$

$h(p) = 4 - 2p - p^2$ • $a = -1, b = -2, c = 4$

$f(x) = 2x^2 + 6x$ • $a = 2, b = 6, c = 0$

The graph of a quadratic function in a single variable x is a **parabola.** The graphs of two of these quadratic functions are shown below.

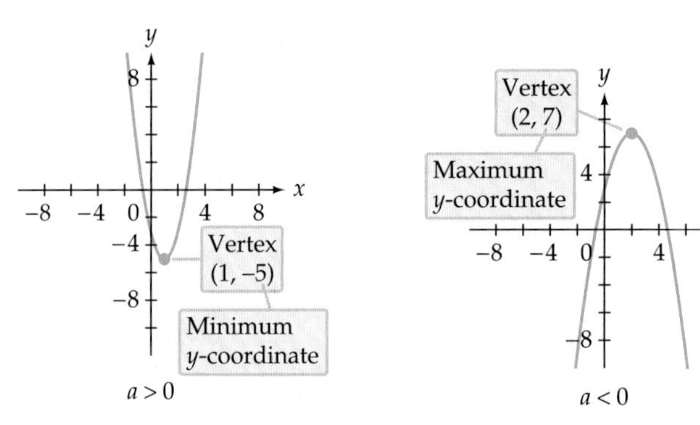

For the figure on the left above, $f(x) = 2x^2 - 4x - 3$. The value of a is *positive* $(a = 2)$ and the graph opens up. For the figure on the right above,

$f(x) = -x^2 + 4x + 3$. The value of a is *negative* $(a = -1)$ and the graph opens down. The point at which the graph of a parabola has a minimum or a maximum is called the *vertex* of the parabola. The **vertex** of a parabola is the point with the smallest y-coordinate when $a > 0$ and the point with largest y-coordinate when $a < 0$.

TAKE NOTE

The axis of symmetry is a vertical line. The vertex of the parabola lies on the axis of symmetry.

The **axis of symmetry** of the graph of a quadratic function is a vertical line that passes through the vertex of the parabola. To understand the concept of the axis of symmetry of a graph, think of folding the graph along that line. The two portions of the graph will match up.

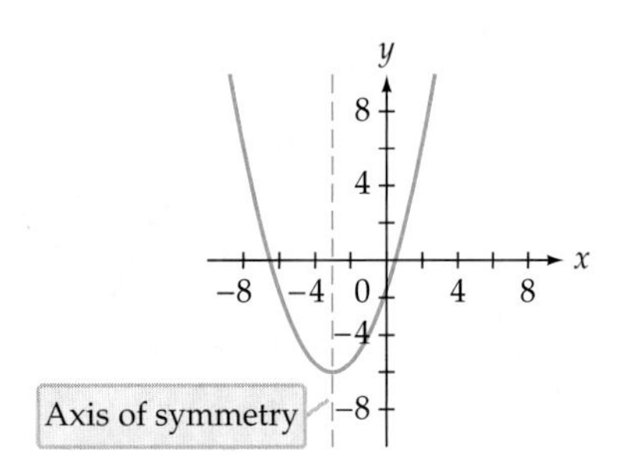

The following formula enables us to determine the vertex of a parabola.

Vertex of a Parabola

Let $f(x) = ax^2 + bx + c$ be the equation of a parabola. The coordinates of the vertex are

$$\left(-\frac{b}{2a}, f\left(-\frac{b}{2a}\right)\right).$$

EXAMPLE 1 ■ Find the Vertex of a Parabola

Find the vertex of the parabola whose equation is $y = -3x^2 + 6x + 1$.

Solution

$$x = -\frac{b}{2a} = -\frac{6}{2(-3)} = 1$$

• Find the x-coordinate of the vertex. $a = -3$, $b = 6$

$$y = -3x^2 + 6x + 1$$

• Find the y-coordinate of the vertex by replacing x by **1** and solving for y.

$$y = -3(1)^2 + 6(1) + 1$$
$$y = 4$$

The vertex is $(1, 4)$.

CHECK YOUR PROGRESS 1 Find the vertex of the parabola whose equation is $y = x^2 - 2$.

Solution *See page S23.*

Math Matters Paraboloids

The movie *Contact* was based on a novel by astronomer Carl Sagan. In the movie, Jodie Foster plays an astronomer who is searching for extraterrestrial intelligence. One scene from the movie takes place at the Very Large Array (VLA) in New Mexico. The VLA consists of 27 large radio telescopes whose dishes are paraboloids, the three-dimensional version of a parabola. A parabolic shape is used because of the following reflective property: When the parallel rays of light, or radio waves, strike the surface of a parabolic mirror whose axis of symmetry is parallel to these rays, they are reflected to the same point.

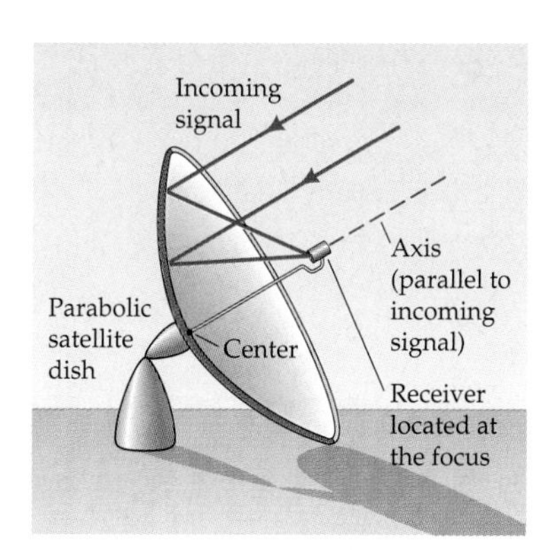

The photos above show the layout of the radio telescopes of the VLA and a more detailed picture of one of the telescopes. The figure at the far right shows the reflective property of a parabola. Note that all the incoming rays are reflected to the focus.

The reflective property of a parabola is also used in optical telescopes and headlights on a car. In the case of headlights, the bulb is placed at the focus and the light is reflected along parallel rays from the reflective surface of the headlight, thereby making a more concentrated beam of light.

x-Intercepts of Parabolas

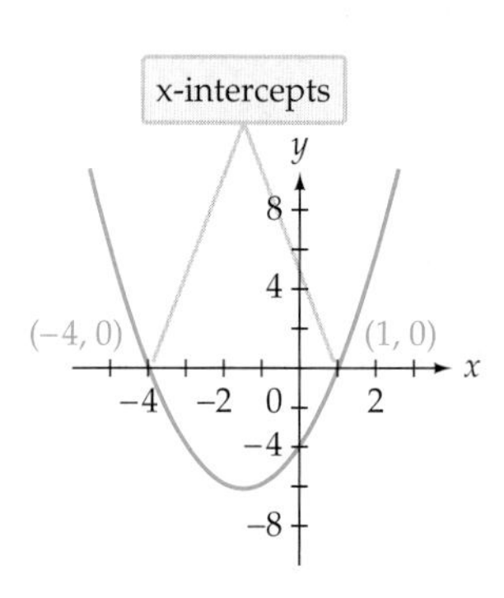

Recall that a point at which a graph crosses the *x*- or *y*-axis is called an *intercept* of the graph. The *x*-intercepts of the graph of an equation can be found by setting $y = 0$.

The graph of $y = x^2 + 3x - 4$ is shown at the left. The points whose coordinates are $(-4, 0)$ and $(1, 0)$ are *x*-intercepts of the graph. We can algebraically determine the *x*-intercepts by solving an equation.

EXAMPLE 2 ■ Find the *x*-intercepts of a Parabola

Find the *x*-intercepts of the graph of the parabola given by the equation.

a. $y = 4x^2 + 4x + 1$ **b.** $y = x^2 + 2x - 2$

TAKE NOTE

When a quadratic equation has a double root, as in part a, the graph of the quadratic equation just touches the x-axis and is said to be *tangent* to the x-axis. See the graph at the right. Some graphs of parabolas do not have any x-intercepts.

TAKE NOTE

Try to solve the quadratic equation by factoring (as we did in part a). If that cannot be done easily, then use the quadratic formula, as we did in part b.

Solution

a. $y = 4x^2 + 4x + 1$

$0 = 4x^2 + 4x + 1$ • Let $y = 0$.

$0 = (2x + 1)(2x + 1)$ • Solve for x by factoring.

$2x + 1 = 0 \qquad 2x + 1 = 0$

$x = -\frac{1}{2} \qquad x = -\frac{1}{2}$

The x-intercept is $\left(-\frac{1}{2}, 0\right)$.

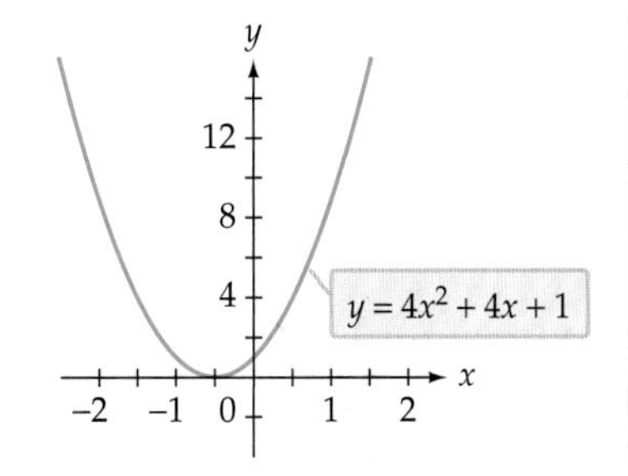

b. $y = x^2 + 2x - 2$

$0 = x^2 + 2x - 2$ • Let $y = 0$. The trinomial $x^2 + 2x - 2$ is nonfactorable over the integers.

$x = \frac{-b \pm \sqrt{b^2 - 4ac}}{2a}$ • Use the quadratic formula to solve for x. $a = 1$, $b = 2$, $c = -2$

$= \frac{-(2) \pm \sqrt{(2)^2 - 4(1)(-2)}}{2(1)} = \frac{-2 \pm \sqrt{4 + 8}}{2}$

$= \frac{-2 \pm \sqrt{12}}{2} = \frac{-2 \pm 2\sqrt{3}}{2} = -1 \pm \sqrt{3}$

The x-intercepts are $\left(-1 + \sqrt{3}, 0\right)$ and $\left(-1 - \sqrt{3}, 0\right)$.

CHECK YOUR PROGRESS 2 Find the x-intercepts of the graph of the parabola given by the equation.

a. $y = 2x^2 - 5x + 2$ **b.** $y = x^2 + 4x + 4$

Solution *See page S23.*

point of interest

Calculus is a branch of mathematics that demonstrates, among other things, how to find the maximum or minimum of functions other than quadratic functions. These are very important problems in applied mathematics. For instance, an automotive engineer wants to design a car whose shape will *minimize* the effect of air flow. Another engineer tries to *maximize* the efficiency of a car's engine. Similarly, an economist may try to determine what business practices will *minimize* cost and *maximize* profit.

Minimum and Maximum of a Quadratic Function

Note that for the graphs below, when $a > 0$, the vertex is the point with the minimum y-coordinate. When $a < 0$, the vertex is the point with the maximum y-coordinate.

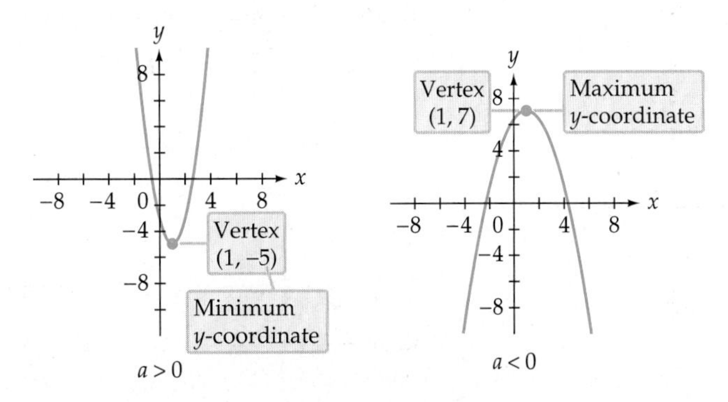

Finding the minimum or maximum value of a quadratic function is a matter of finding the vertex of the graph of the function.

EXAMPLE 3 ■ Find the Minimum or Maximum Value of a Quadratic Function

Find the maximum value of $f(x) = -2x^2 + 4x + 3$.

Solution

$$x = -\frac{b}{2a} = -\frac{4}{2(-2)} = 1$$

• Find the x-coordinate of the vertex. $a = -2$, $b = 4$

$$f(x) = -2x^2 + 4x + 3$$

• Find the y-coordinate of the vertex by replacing x by **1** and solving for y.

$$f(1) = -2(1)^2 + 4(1) + 3$$
$$f(1) = 5$$

The vertex is $(1, 5)$.

The maximum value of the function is 5, the y-coordinate of the vertex.

CHECK YOUR PROGRESS 3 Find the minimum value of $f(x) = 2x^2 - 3x + 1$.

Solution *See page S24.*

QUESTION *The vertex of a parabola that opens up is $(-4, 7)$. What is the minimum value of the function?*

Applications of Quadratic Functions

EXAMPLE 4 ■ Application of Finding the Minimum of a Quadratic Function

A mining company has determined that the cost c, in dollars per ton, of mining a mineral is given by $c(x) = 0.2x^2 - 2x + 12$, where x is the number of tons of the mineral that are mined. Find the number of tons of the mineral that should be mined to minimize the cost. What is the minimum cost?

Solution

To find the number of tons of the mineral that should be mined to minimize the cost and to find the minimum cost, find the x- and y-coordinates of the vertex of the graph of $c(x) = 0.2x^2 - 2x + 12$.

$$x = -\frac{b}{2a} = -\frac{-2}{2(0.2)} = 5$$

• Find the x-coordinate of the vertex. $a = 0.2$, $b = -2$

ANSWER *The minimum value of the function is 7, the y-coordinate of the vertex.*

To minimize the cost, 5 tons of the mineral should be mined.

$c(x) = 0.2x^2 - 2x + 12$ • Find the y-coordinate of the vertex by replacing x by **5** and solving for y.

$c(5) = 0.2(5)^2 - 2(5) + 12$

$c(5) = 7$

The minimum cost per ton is \$7.

CHECK YOUR PROGRESS 4 The height s, in feet, of a ball thrown straight up is given by $s(t) = -16t^2 + 64t + 4$, where t is the time in seconds after the ball is released. Find the time it takes the ball to reach its maximum height. What is the maximum height?

Solution *See page S24.*

EXAMPLE 5 ■ Application of Finding the Maximum of a Quadratic Function

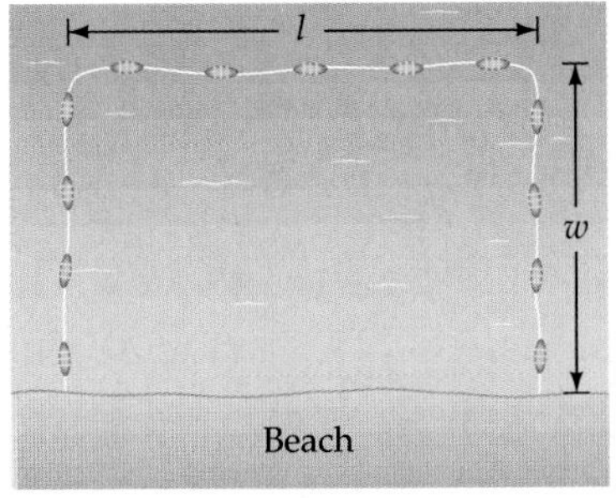

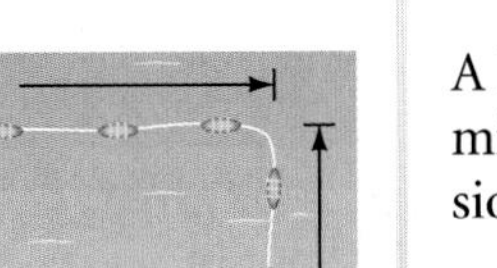

A lifeguard has 600 feet of rope with buoys attached to lay out a rectangular swimming area on a lake. If the beach forms one side of the rectangle, find the dimensions of the rectangle that will enclose the greatest swimming area.

Solution

Let l represent the length of the rectangle, let w represent the width of the rectangle, and let A (which is unknown) represent the area of the rectangle. See the figure at the left. Use these variables to write expressions for the perimeter and area of the rectangle.

Perimeter: $w + l + w = 600$ • There are **600** feet of rope.

$2w + l = 600$

Area: $A = lw$

The goal is to maximize A. To do this, first write A in terms of a single variable. This can be accomplished by solving $2w + l = 600$ for l and then substituting into $A = lw$.

$2w + l = 600$ • Solve for l.

$l = -2w + 600$

$A = lw$

$= (-2w + 600)w$ • Substitute $-2w + 600$ for l.

$A = -2w^2 + 600w$ • Multiply. This is now a quadratic equation. $a = -2$, $b = 600$

Find the w-coordinate of the vertex.

$$w = -\frac{b}{2a} = -\frac{600}{2(-2)} = 150$$ • Find the w-coordinate of the vertex. $a = -2$, $b = 600$

The width is 150 feet. To find l, replace w by 150 in $l = -2w + 600$ and solve for l.

$l = -2w + 600$

$l = -2(150) + 600 = -300 + 600 = 300$

The length is 300 feet. The dimensions of the rectangle with maximum area are 150 feet by 300 feet.

CHECK YOUR PROGRESS 5 A mason is forming a rectangular floor for a storage shed. The perimeter of the rectangle is 44 feet. What dimensions will give the floor a maximum area?

Solution *See page S24.*

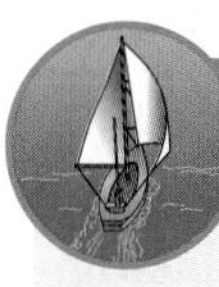

Excursion

Reflective Properties of a Parabola

The fact that the graph of $y = ax^2 + bx + c$ is a parabola is based on the following geometric definition of a parabola.

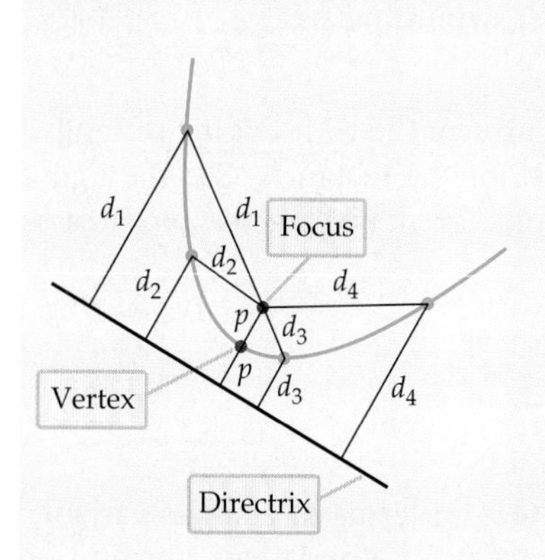

Definition of a Parabola

A parabola is the set of points in the plane that are equidistant from a fixed line (the **directrix**) and a fixed point (the **focus**) not on the line.

This geometric definition of a parabola is illustrated in the figure at the left. Basically, for a point to be on a parabola, the distance from the point to the focus must equal the distance from the point to the directrix. Note also that the vertex is halfway between the focus and the directrix. This distance is traditionally labeled p.

The general form of the equation of a parabola that opens up with vertex at the origin can be written in terms of the distance p between the vertex and focus as $\boldsymbol{y = \frac{1}{4p}x^2}$. For this equation, the coordinates of the focus are $(0, p)$. For instance, to find the coordinates of the focus for $y = \frac{1}{4}x^2$, let $\frac{1}{4p} = \frac{1}{4}$ and solve for p.

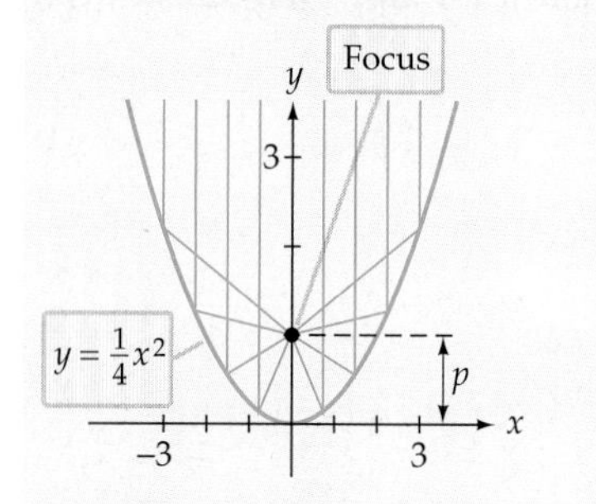

$$\frac{1}{4p} = \frac{1}{4}$$

$$1 = \frac{4p}{4}$$ • Multiply each side of the equation by $4p$.

$$1 = p$$ • Simplify the right side of the equation.

The coordinates of the focus are (0, 1).

1. Find the coordinates of the focus for the parabola whose equation is $y = 0.4x^2$.

Optical telescopes work on the same principle as radio telescopes (see Math Matters, p. 366) except that light hits a mirror that has been shaped into a paraboloid. The light is reflected to the focus, where another mirror reflects it through a lens to the observer. See the diagram at the top of the following page.

(continued)

Palomar Observatory with the shutters open.

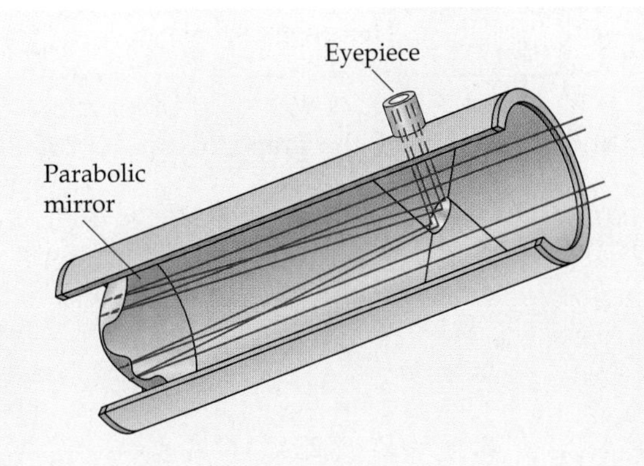

2. The telescope at the Palomar Observatory in California has a parabolic mirror. The circle at the top of the parabolic mirror has a 200-inch diameter. An equation that approximates the parabolic cross-section of the surface of the mirror is $y = \frac{1}{2639}x^2$, where x and y are measured in inches. How far is the focus from the vertex of the mirror?

If a point on a parabola whose vertex is at the origin is known, then the equation of the parabola can be found. For instance, if (4, 1) is a point on a parabola with vertex at the origin, then we can find the equation as follows:

$y = \frac{1}{4p}x^2$ • **Begin with the general form of the equation of a parabola.**

$1 = \frac{1}{4p}(4)^2$ • **The known point is (4, 1). Replace x by 4 and y by 1.**

$1 = \frac{4}{p}$ • **Solve for p.**

$p = 4$ • **$p = 4$ in the equation $y = \frac{1}{4p}x^2$.**

The equation of the parabola is $y = \frac{1}{16}x^2$.

Find a flashlight and measure the diameter of its lens cover and the depth of the reflecting parabolic surface. See the diagram below.

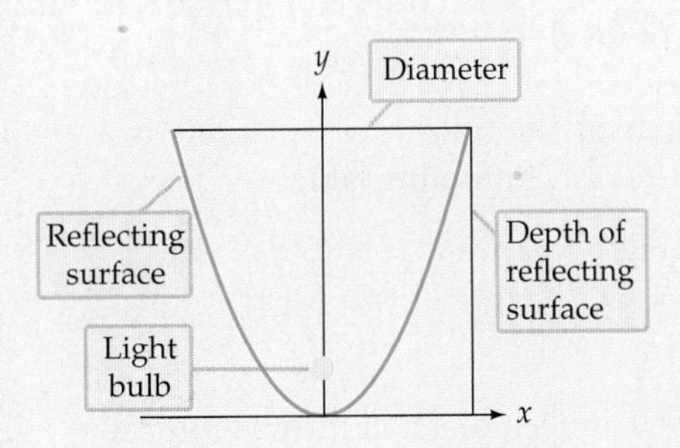

3. If coordinate axes are set up as shown, find the equation of the parabola.

4. Find the location of the focus. Explain why the light bulb should be placed at this point.

Exercise Set 6.4

In Exercises 1–12, find the vertex of the graph of the equation.

1. $y = x^2 - 2$
2. $y = x^2 + 2$
3. $y = -x^2 - 1$
4. $y = -x^2 + 3$
5. $y = -\frac{1}{2}x^2 + 2$
6. $y = \frac{1}{2}x^2$
7. $y = 2x^2 - 1$
8. $y = x^2 - 2x$
9. $y = x^2 - x - 2$
10. $y = x^2 - 3x + 2$
11. $y = 2x^2 - x - 5$
12. $y = 2x^2 - x - 3$

In Exercises 13–24, find the x-intercepts of the parabola given by the equation.

13. $y = 2x^2 - 4x$
14. $y = 3x^2 + 6x$
15. $y = 4x^2 + 11x + 6$
16. $y = x^2 - 9$
17. $y = x^2 + 2x - 1$
18. $y = x^2 + 4x - 3$
19. $y = -x^2 - 4x - 5$
20. $y = -x^2 - 2x + 1$
21. $y = -x^2 + 4x + 1$
22. $y = x^2 + 6x + 10$
23. $y = 2x^2 - 5x - 3$
24. $y = x^2 - 2$

In Exercises 25–32, find the minimum or maximum value of the quadratic function. State whether the value is a minimum or a maximum.

25. $f(x) = x^2 - 2x + 3$
26. $f(x) = 2x^2 + 4x$
27. $f(x) = -2x^2 + 4x - 5$
28. $f(x) = 3x^2 + 3x - 2$
29. $f(x) = x^2 - 5x + 3$
30. $f(x) = 2x^2 - 3x$
31. $f(x) = -x^2 - x + 2$
32. $f(x) = -3x^2 + 4x - 2$

33. The graph of which of the following equations is a parabola with the largest minimum value?

    a. $y = x^2 - 2x - 3$

    b. $y = x^2 - 10x + 20$

    c. $y = 3x^2 - 1$

34. **Sports** The height s, in feet, of a ball thrown upward at an initial speed of 64 feet per second from a cliff 50 feet above an ocean beach is given by the function

$$s(t) = -16t^2 + 64t + 50$$

where t is the time in seconds after the ball is released. Find the maximum height above the beach that the ball will attain.

35. **Sports** The height s, in feet, of a ball thrown upward at an initial speed of 80 feet per second from a platform 50 feet high is given by

$$s(t) = -16t^2 + 80t + 50$$

where t is the time in seconds after the ball is released. Find the maximum height above the ground that the ball will attain.

36. **Revenue** A manufacturer of microwave ovens estimates that the revenue R, in dollars, the company receives is related to the price p, in dollars, of an oven by the function

$$R(p) = 125p - 0.25p^2$$

What price per oven will give the maximum revenue?

37. **Manufacturing** A manufacturer of camera lenses estimates that the average monthly cost C of producing camera lenses is given by the function

$$C(x) = 0.1x^2 - 20x + 2000$$

where x is the number of lenses produced each month. Find the number of lenses the company should produce to minimize the average cost.

38. **Water Treatment** A pool is treated with a chemical to reduce the number of algae. The number of algae in the pool t days after the treatment can be approximated by the function

$$A(t) = 40t^2 - 400t + 500$$

How many days after treatment will the pool have the least number of algae?

39. **Civil Engineering** The suspension cable that supports a footbridge hangs in the shape of a parabola. The height h, in feet, of the cable above the bridge is given by the function

$$h(x) = 0.25x^2 - 0.8x + 25,\ 0 < x < 3.2$$

where x is the distance in feet measured from the left tower toward the right tower. What is the minimum height of the cable above the bridge?

40. **Annual Income** The net annual income I, in dollars, of a family physician can be modeled by the equation

$$I(x) = -290(x - 48)^2 + 148{,}000$$

where x is the age of the physician and $27 \le x \le 70$. Find (a) the age at which the physician's income will be a maximum and (b) the maximum income.

41. **Pitching** Karen is throwing an orange to her brother Saul, who is standing on the balcony of their home. The height h, in feet, of the orange above the ground t seconds after it is thrown is given by

$$h(t) = -16t^2 + 32t + 4, 0 \leq t \leq 2.118$$

If Saul's outstretched arms are 18 feet above the ground, will the orange ever be high enough so that he can catch it?

42. **Football** Some football fields are built in a parabolic-mound shape so that water will drain off the field. A model for the shape of such a field is given by

$$h(x) = -0.00023475x^2 + 0.0375x$$

where h is the height of the field in feet at a distance of x feet from the sideline and $0 \leq x \leq 159.744$. What is the maximum height of the field? Round to the nearest tenth of a foot.

43. **Fountains** The Buckingham Memorial Fountain in Chicago shoots water from a nozzle within the fountain. The height h, in feet, of the water above the ground t seconds after it leaves the nozzle is given by

$$h(t) = -16t^2 + 90t + 15, h \geq 15$$

What is the maximum height of the water spout? Round to the nearest tenth of a foot.

44. **Stopping Distance** On wet concrete, the stopping distance s, in feet, of a car traveling v miles per hour is given by

$$s(v) = 0.055v^2 + 1.1v, s \geq 0$$

At what maximum speed could a car be traveling on wet concrete and still stop at a stop sign 44 feet away?

45. **Fuel Efficiency** The fuel efficiency of an average car is given by the equation

$$E(v) = -0.018v^2 + 1.476v + 3.4, E \geq 0$$

where E is the fuel efficiency in miles per gallon and v is the speed of the car in miles per hour.

a. What speed will yield the maximum fuel efficiency?

b. What is the maximum fuel efficiency?

46. **Ranching** A rancher has 200 feet of fencing to build a rectangular corral alongside an existing fence. Determine the dimensions of the corral that will maximize the enclosed area.

Extensions

CRITICAL THINKING

In Exercises 47 and 48, find the value of k such that the graph of the function contains the given point.

47. $f(x) = x^2 - 3x + k$; $(2, 5)$

48. $f(x) = 2x^2 + kx - 3$; $(4, -3)$

49. For $f(x) = 2x^2 - 5x + k$, we have $f\left(-\frac{3}{2}\right) = 0$. Find another value of x for which $f(x) = 0$.

50. For what values of k does the graph of $f(x) = 2x^2 - kx + 8$ just touch the x-axis without crossing it?

EXPLORATIONS

A real number x is called a **zero of a function** if the function evaluated at x is 0. That is, if $f(x) = 0$, then x is called a zero of the function. For instance, evaluating $f(x) = x^2 + x - 6$ when $x = -3$, we have

$$f(x) = x^2 + x - 6$$
$$f(-3) = (-3)^2 + (-3) - 6 \quad \text{• Replace } x \text{ by } -3.$$
$$f(-3) = 9 - 3 - 6 = 0$$

For this function, $f(-3) = 0$, so -3 is a zero of the function.

51. Verify that 2 is a zero of $f(x) = x^2 + x - 6$ by showing that $f(2) = 0$.

The graph of $f(x) = x^2 + x - 6$ is shown below. Note that the graph crosses the x-axis at -3 and 2, the two zeros of the function. The points $(-3, 0)$ and $(2, 0)$ are x-intercepts of the graph.

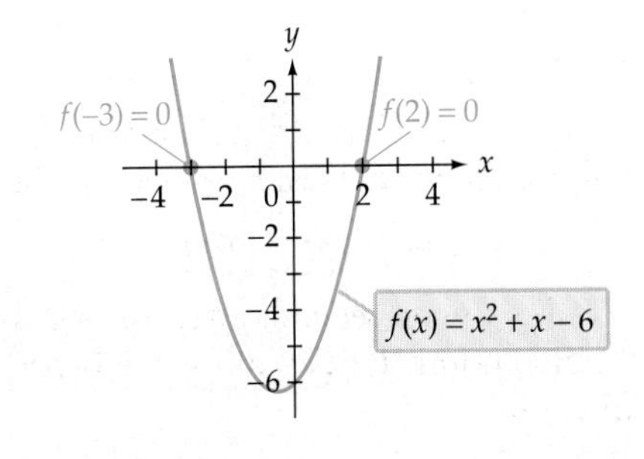

Consider the equation $0 = x^2 + x - 6$, which is $f(x) = x^2 + x - 6$ with $f(x)$ replaced by 0. By solving $0 = x^2 + x - 6$, we get

$$0 = x^2 + x - 6$$

$$0 = (x + 3)(x - 2) \quad \text{• Solve by factoring.}$$

$$x + 3 = 0 \qquad x - 2 = 0$$

$$x = -3 \qquad x = 2$$

Observe that the solutions of the equation are the zeros of the function. This important connection among the real zeros of a function, the x-intercepts of the graph, and the real solutions of an equation is the basis of using a graphing calculator to solve an equation. The method discussed below of solving a quadratic equation using a graphing calculator is based on a TI-83/84 calculator. Other calculators will necessitate a slightly different approach.

Approximate the solutions of $x^2 + 4x = 6$ by using a graphing calculator.

i. Write the equation in standard form.

$$x^2 + 4x = 6$$

$$x^2 + 4x - 6 = 0$$

Press [Y=] and enter $x^2 + 4x - 6$ for Y1.

Plot1 Plot2 Plot3
\Y1 = X²+4X−6
\Y2 =
\Y3 =
\Y4 =
\Y5 =
\Y6 =
\Y7 =

ii. Press [GRAPH]. If the graph does not appear on the screen, press [ZOOM] 6.

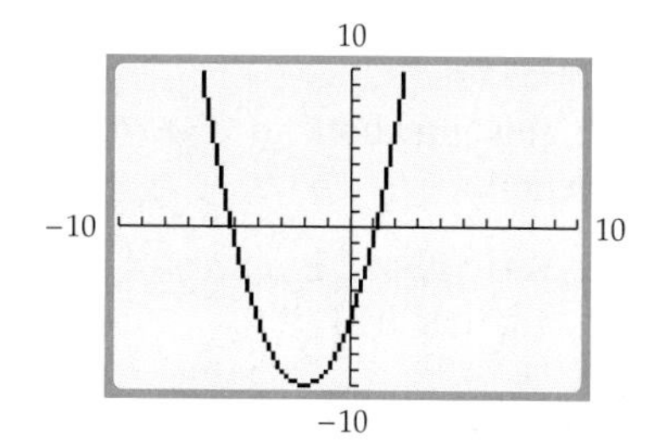

iii. Press [2nd] CALC 2. Note that the second menu item is `zero`. This will begin the calculation of the zeros of the function, which are the solutions of the equation.

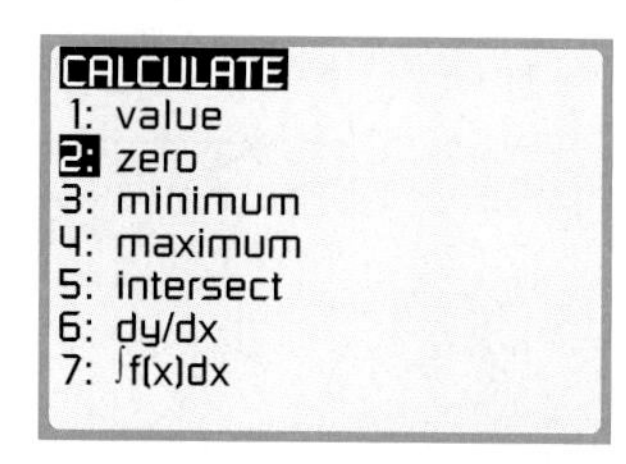

iv. At the bottom of the screen you will see `LeftBound?` This is asking you to move the blinking cursor so that it is to the *left* of the first x-intercept. Use the left arrow key to move the cursor to the left of the first x-intercept. The values of x and y that appear on your calculator may be different from the ones shown here. Just be sure you are to the left of the x-intercept. When you are done, press ENTER.

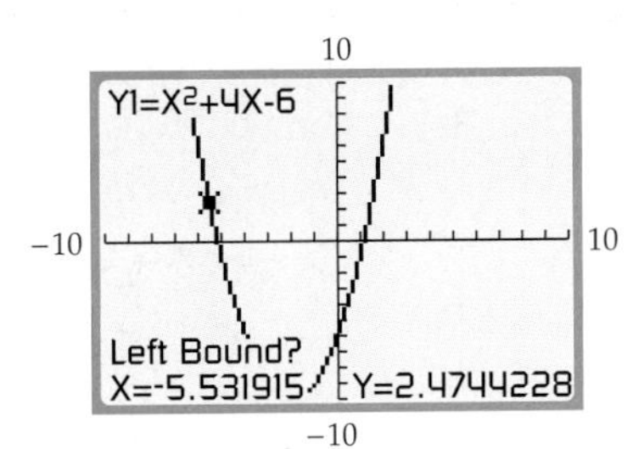

v. At the bottom of the screen you will see `RightBound?` This is asking you to move the blinking cursor so that it is to the *right* of the x-intercept. Use the right arrow key to move the cursor to the right of the x-intercept. The values of x and y that appear on your calculator may be different from the ones shown here. Just be sure you are to the right of the x-intercept. When you are done, press ENTER.

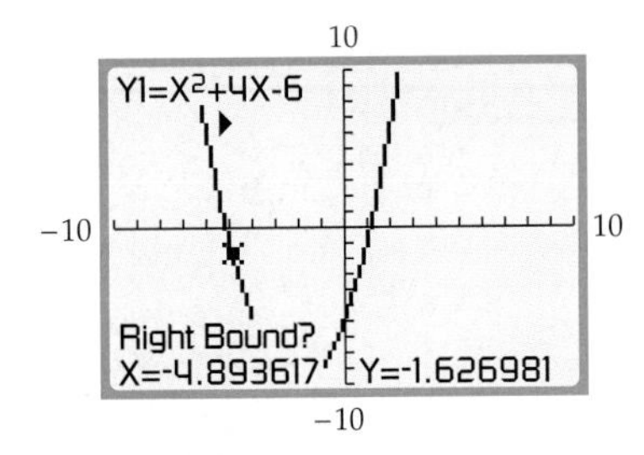

vi. At the bottom of the screen you will see `Guess?` This is asking you to move the blinking cursor so that it is close to the x-intercept. Use the arrow keys to move the cursor to the approximate x-intercept. The values of x and y that appear on your calculator may be different from the ones shown here. When you are done, press ENTER.

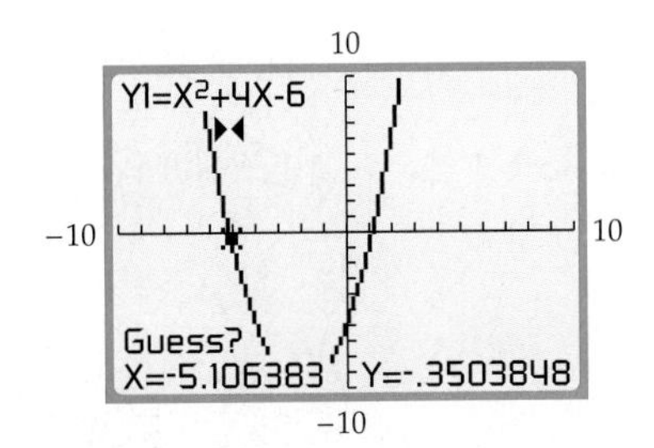

vii. The zero of the function is approximately -5.162278. Thus one approximate solution of $x^2 + 4x = 6$ is -5.162278. Note that the value of y is given as Y1 = ⁻1E⁻12. This is how the calculator writes a number in scientific notation. We would normally write $\text{Y1} = -1.0 \times 10^{-12}$. This number is very close to zero.

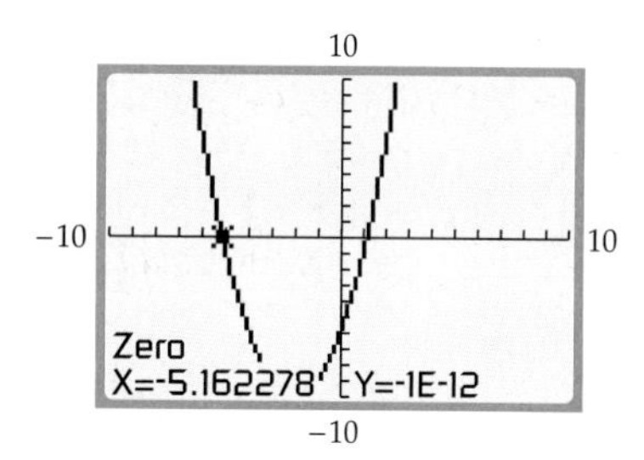

To find the other solution, repeat steps **iii** through **vi.** The screens are shown below.

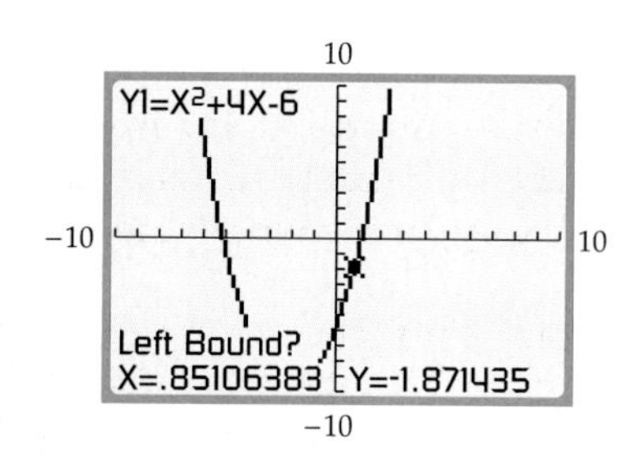

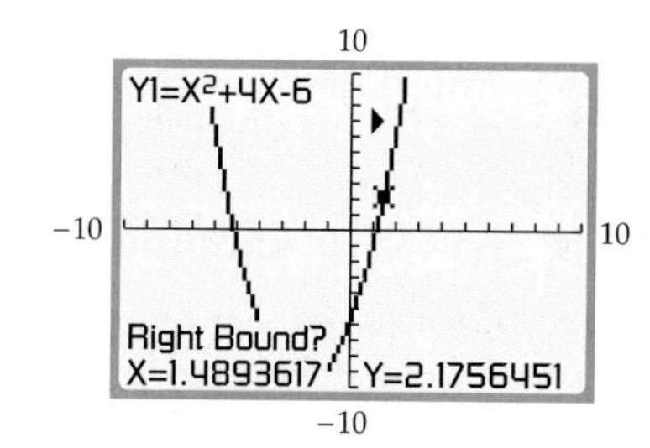

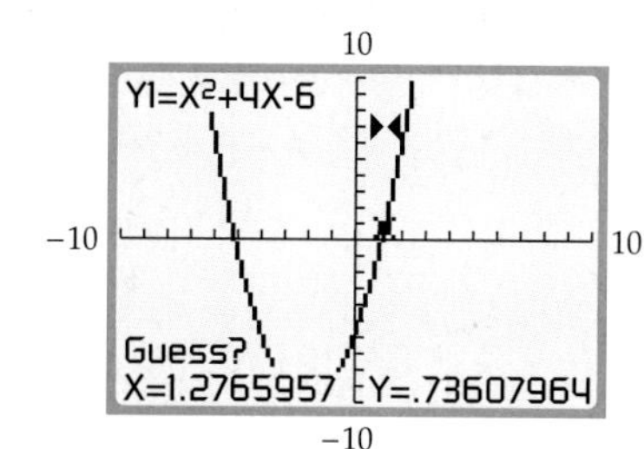

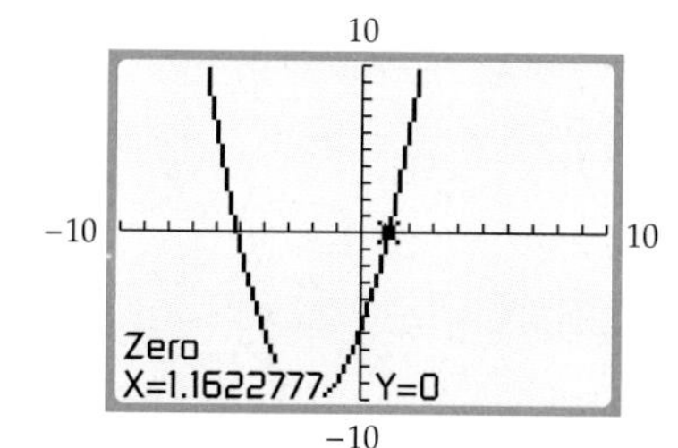

A second zero of the function is approximately 1.1622777. Thus the two solutions of $x^2 + 4x = 6$ are approximately -5.162278 and 1.1622777.

In Exercises 52–57, find the solutions of the equation. Round to the nearest hundredth.

52. $x^2 + 3x - 4 = 0$

53. $x^2 - 4x - 5 = 0$

54. $x^2 + 3.4x = 4.15$

55. $2x^2 - \frac{5}{9}x = \frac{3}{8}$

56. $\pi x^2 - \sqrt{17}x - 2 = 0$

57. $\sqrt{2}x^2 + x - \sqrt{7} = 0$

This method of finding solutions can be extended to find the real zeros, and therefore the real solutions, of any equation that can be graphed.

In Exercises 58–60, find the real solutions of the equation.

58. $x^3 - 2x^2 - 5x + 6 = 0$

59. $x^3 + 2x^2 - 11x - 12 = 0$

60. $x^4 + 2x^3 - 13x^2 - 14x + 24 = 0$

SECTION 6.5 Exponential Functions

Introduction to Exponential Functions

In 1965, Gordon Moore, one of the cofounders of Intel Corporation, observed that the maximum number of transistors that could be placed on a microprocessor seemed to be doubling every 18 to 24 months. The table below shows how the maximum number of transistors on various Intel processors has changed over time. (*Source:* **www.intel.com/technology/mooreslaw/index.htm**)

Year, x	1971	1978	1982	1985	1989	1993	1997	1999	2000	2002	2003
Number of transistors per microprocessor (in thousands), y	2.3	29	120	275	1180	3100	7500	24,000	42,000	220,000	410,000

The curve in Figure 6.1 that approximately passes through the points associated with the data for the years 1971 to 2003 is the graph of a mathematical model of the data. The model is an *exponential function.* Because y is increasing (growing), it is an *exponential growth function.*

When light enters water, the intensity of the light decreases with the depth of the water. The graph in Figure 6.2 shows a model, for Lake Michigan, of the decrease in the percent of light available as the depth of the water increases. This model is also an exponential function. In this case y is decreasing (decaying), and the model is an *exponential decay function.*

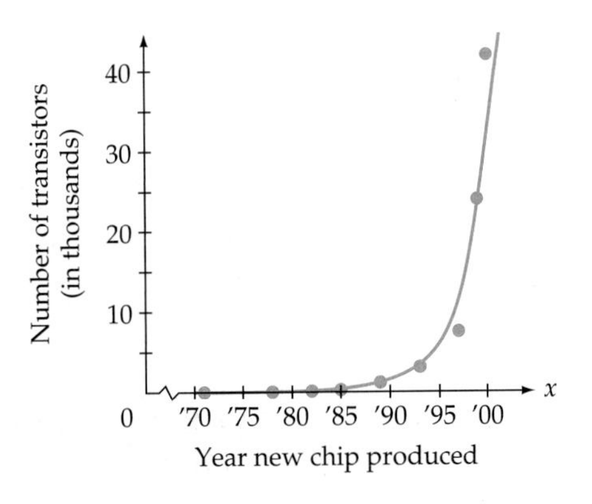

Figure 6.1

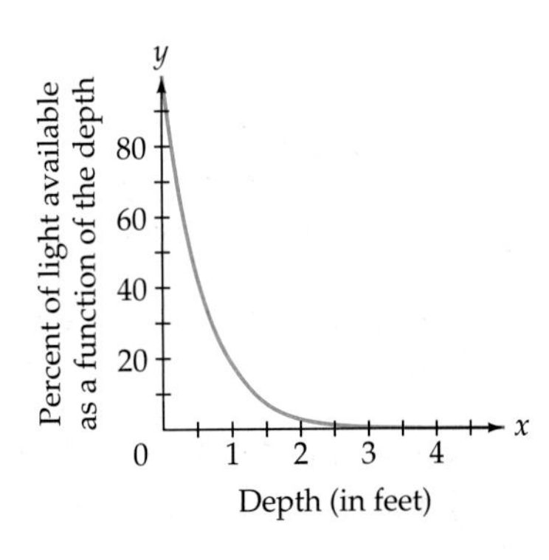

Figure 6.2

> The **exponential function** is defined by $f(x) = b^x$, where b is called the base, $b > 0$, $b \neq 1$, and x is any real number.

The base b of $f(x) = b^x$ must be positive. If the base were a negative number, the value of the function would not be a real number for some values of x. For instance, if $b = -4$ and $x = \frac{1}{2}$, then $f\left(\frac{1}{2}\right) = (-4)^{1/2} = \sqrt{-4}$, which is not a real number. Also, $b \neq 1$ because when $b = 1$, $b^x = 1^x = 1$, a constant function.

TAKE NOTE

If you need to review material on fractional exponents or the square root of a negative number, see Section 9.1 on the CD that you received with this book.

At the right we evaluate $f(x) = 2^x$ for $x = 3$ and $x = -2$.

$$f(x) = 2^x$$
$$f(3) = 2^3 = 8$$
$$f(-2) = 2^{-2} = \frac{1}{2^2} = \frac{1}{4}$$

To evaluate the exponential function $f(x) = 2^x$ for an irrational number such as $x = \sqrt{2}$, we use a rational approximation of $\sqrt{2}$ (for instance, 1.4142) and a calculator to obtain an approximation of the function.

$$f(\sqrt{2}) = 2^{\sqrt{2}} \approx 2^{1.4142} \approx 2.6651$$

CALCULATOR NOTE

To evaluate $f(\pi)$ we used a calculator. For a scientific calculator, enter

3 [yˣ] π [=]

For the TI-83/84 graphing calculator, enter

3 [^] [2nd] π [ENTER]

EXAMPLE 1 ■ Evaluate an Exponential Function

Evaluate $f(x) = 3^x$ at $x = 2$, $x = -4$, and $x = \pi$. Round approximate results to the nearest hundred thousandth.

Solution

$$f(2) = 3^2 = 9$$
$$f(-4) = 3^{-4} = \frac{1}{3^4} = \frac{1}{81}$$
$$f(\pi) = 3^{\pi} \approx 3^{3.1415927} \approx 31.54428$$

CHECK YOUR PROGRESS 1 Evaluate $g(x) = \left(\frac{1}{2}\right)^x$ when $x = 3$, $x = -1$, and $x = \sqrt{3}$. Round approximate results to the nearest thousandth.

Solution *See page S24.*

Graphs of Exponential Functions

The graph of $f(x) = 2^x$ is shown below. The coordinates of some of the points on the graph are given in the table.

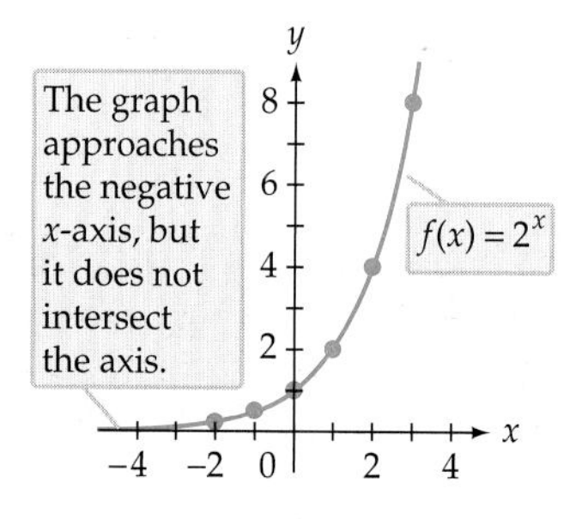

Figure 6.3

x	$f(x) = 2^x$	(x, y)
−2	$f(-2) = 2^{-2} = \frac{1}{4}$	$\left(-2, \frac{1}{4}\right)$
−1	$f(-1) = 2^{-1} = \frac{1}{2}$	$\left(-1, \frac{1}{2}\right)$
0	$f(0) = 2^0 = 1$	(0, 1)
1	$f(1) = 2^1 = 2$	(1, 2)
2	$f(2) = 2^2 = 4$	(2, 4)
3	$f(3) = 2^3 = 8$	(3, 8)

Observe that the values of y *increase* as x increases. This is an exponential growth function. This is typical of the graph of all exponential functions for which the base is *greater than* 1. For the function $f(x) = 2^x$, $b = 2$, which is greater than 1.

Now consider the graph of an exponential function for which the base is between 0 and 1. The graph of $f(x) = \left(\frac{1}{2}\right)^x$ is shown below. The coordinates of some of the points on the graph are given in the table.

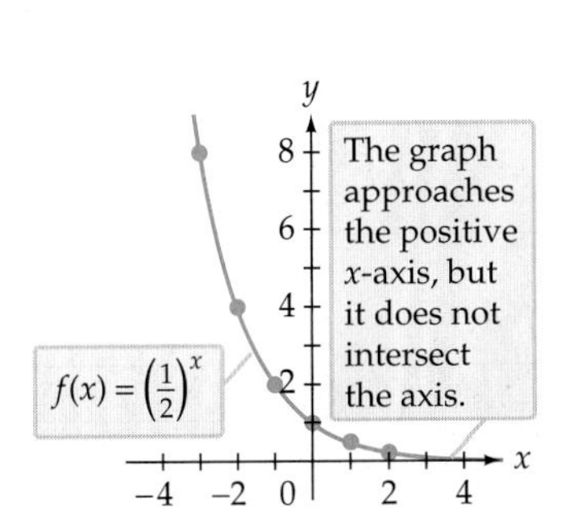

Figure 6.4

x	$f(x) = \left(\frac{1}{2}\right)^x$	(x, y)
-3	$f(-3) = \left(\frac{1}{2}\right)^{-3} = 8$	$(-3, 8)$
-2	$f(-2) = \left(\frac{1}{2}\right)^{-2} = 4$	$(-2, 4)$
-1	$f(-1) = \left(\frac{1}{2}\right)^{-1} = 2$	$(-1, 2)$
0	$f(0) = \left(\frac{1}{2}\right)^{0} = 1$	$(0, 1)$
1	$f(1) = \left(\frac{1}{2}\right)^{1} = \frac{1}{2}$	$\left(1, \frac{1}{2}\right)$
2	$f(2) = \left(\frac{1}{2}\right)^{2} = \frac{1}{4}$	$\left(2, \frac{1}{4}\right)$

Observe that the values of y *decrease* as x increases. This is an exponential decay function. This is typical of the graph of all exponential functions for which the positive base is *less than* 1. For the function $f(x) = \left(\frac{1}{2}\right)^x$, $b = \frac{1}{2}$, which is less than 1.

QUESTION *Is $f(x) = 0.25^x$ an exponential growth or exponential decay function?*

EXAMPLE 2 ■ Graph an Exponential Function

State whether $g(x) = \left(\frac{3}{4}\right)^x$ is an exponential growth function or an exponential decay function. Then graph the function.

Solution

Because the base $\frac{3}{4}$ is less than 1, g is an exponential decay function. Because it is an exponential decay function, the y-values will decrease as x increases. The y-intercept of the graph is the point $(0, 1)$, and the graph also passes through $\left(1, \frac{3}{4}\right)$. Plot a few additional points. (See the table on the following page.) Then draw a smooth curve through the points, as shown on the following page.

ANSWER *The base is 0.25, which is less than 1. It is an exponential decay function.*

x	$g(x) = \left(\frac{3}{4}\right)^x$	(x, y)
−3	$g(-3) = \left(\frac{3}{4}\right)^{-3} = \frac{64}{27}$	$\left(-3, \frac{64}{27}\right)$
−2	$g(-2) = \left(\frac{3}{4}\right)^{-2} = \frac{16}{9}$	$\left(-2, \frac{16}{9}\right)$
−1	$g(-1) = \left(\frac{3}{4}\right)^{-1} = \frac{4}{3}$	$\left(-1, \frac{4}{3}\right)$
0	$g(0) = \left(\frac{3}{4}\right)^{0} = 1$	(0, 1)
1	$g(1) = \left(\frac{3}{4}\right)^{1} = \frac{3}{4}$	$\left(1, \frac{3}{4}\right)$
2	$g(2) = \left(\frac{3}{4}\right)^{2} = \frac{9}{16}$	$\left(2, \frac{9}{16}\right)$
3	$g(3) = \left(\frac{3}{4}\right)^{3} = \frac{27}{64}$	$\left(3, \frac{27}{64}\right)$

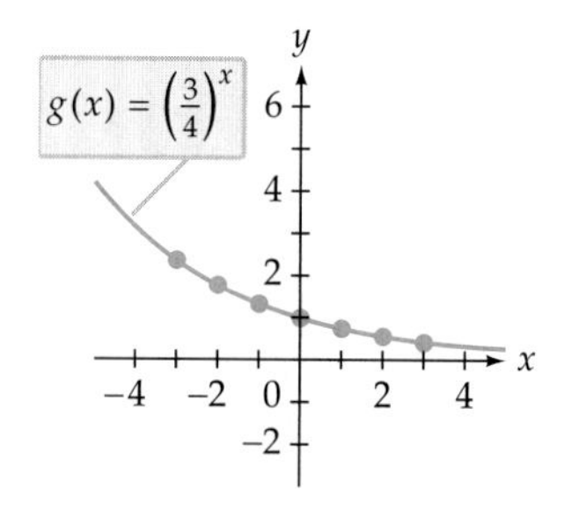

CHECK YOUR PROGRESS 2 State whether $f(x) = \left(\frac{3}{2}\right)^x$ is an exponential growth function or an exponential decay function. Then graph the function.

Solution *See page S24.*

The Natural Exponential Function

The irrational number π is often used in applications that involve circles. Another irrational number, denoted by the letter e, is useful in applications that involve growth or decay.

n	$\left(1 + \frac{1}{n}\right)^n$
10	2.59374246
100	2.70481383
1000	2.71692393
10,000	2.71814593
100,000	2.71826824
1,000,000	2.71828047

Definition of e

The number e is defined as the number that

$$\left(1 + \frac{1}{n}\right)^n$$

approaches as n increases without bound.

The letter e was chosen in honor of the Swiss mathematician Leonhard Euler. He was able to compute the value of e to several decimal places by evaluating $\left(1 + \frac{1}{n}\right)^n$ for large values of n, as shown at the left. The value of e accurate to eight decimal places is 2.71828183.

The Natural Exponential Function

For all real numbers x, the function defined by $f(x) = e^x$ is called the **natural exponential function.**

historical note

Leonhard Euler (oi'lər) (1707–1783) Some mathematicians consider Euler to be the greatest mathematician of all time. He certainly was the most prolific writer of mathematics of all time. He was the first to introduce many of the mathematical notations that we use today. For instance, he introduced the symbol π for pi, the functional notation $f(x)$, and the letter e as the base of the natural exponential function. ■

A calculator with an e^x key can be used to evaluate e^x for specific values of x. For instance,

$$e^2 \approx 7.389056, \qquad e^{3.5} \approx 33.115452, \qquad \text{and} \qquad e^{-1.4} \approx 0.246597$$

The graph of the natural exponential function can be constructed by plotting a few points or by using a graphing utility.

EXAMPLE 3 ■ Graph a Natural Exponential Function

Graph $f(x) = e^x$.

Solution

Use a calculator to find range values for a few domain values. The range values in the table below have been rounded to the nearest tenth.

x	−2	−1	0	1	2
$f(x) = e^x$	0.1	0.4	1.0	2.7	7.4

Plot the points given in the table and then connect the points with a smooth curve. Because $e > 1$, as x increases, e^x increases. Thus the values of y increase as x increases. As x decreases, e^x becomes closer to zero. For instance, when $x = -5$, $e^{-5} \approx 0.0067$. Thus as x decreases, the graph gets closer and closer to the x-axis. The y-intercept is $(0, 1)$.

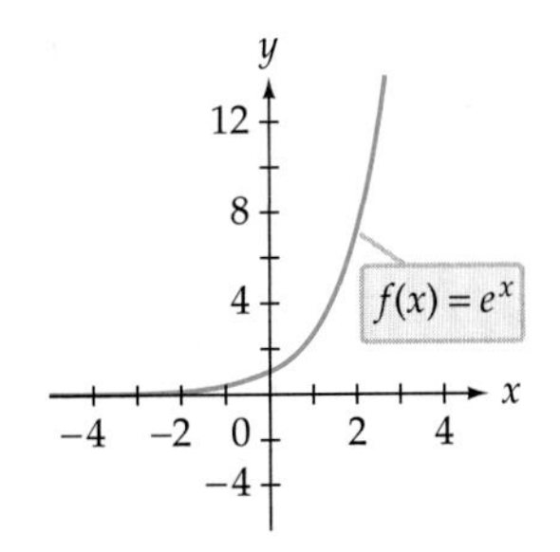

In the figure at the right, compare the graph of $f(x) = e^x$ with the graphs of $g(x) = 2^x$ and $h(x) = 3^x$. Because $2 < e < 3$, the graph of $f(x) = e^x$ is between the graphs of g and h.

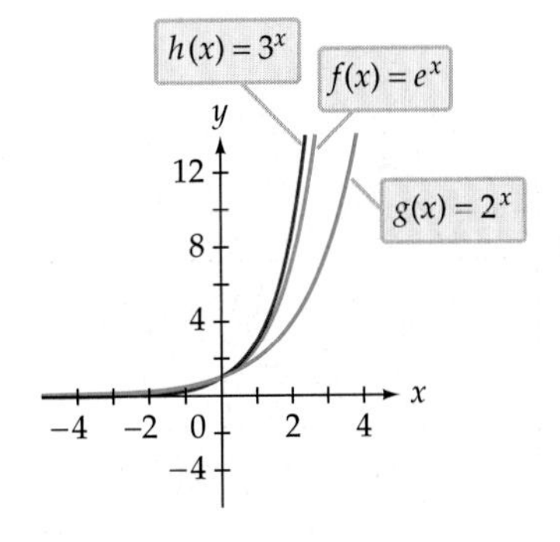

CALCULATOR NOTE

The graph below was produced on a TI-83/84 graphing calculator by entering e^x in the Y= menu.

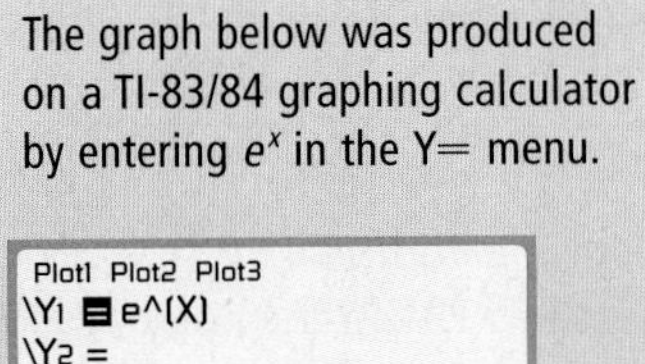

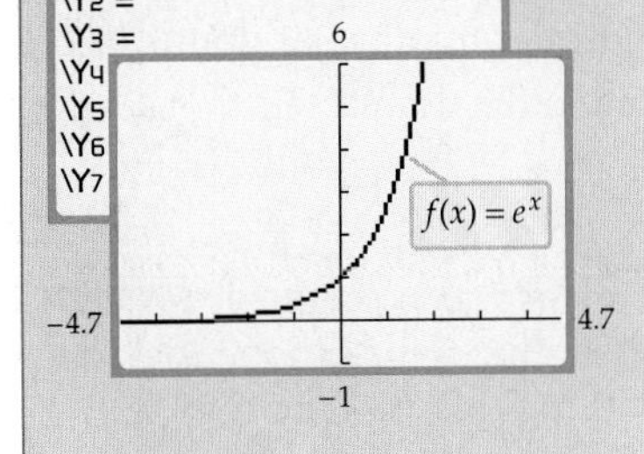

CHECK YOUR PROGRESS 3 Graph $f(x) = e^{-x} + 2$.

Solution *See page S24.*

Applications of Exponential Functions

TAKE NOTE

Here is an application of exponential functions. Assume that you have a long strip of paper that is $\frac{1}{16}$ inch thick and can be folded in half infinitely many times. After how many folds would the paper be as high as the Empire State Building (1250 feet)?

Many applications can be effectively modeled by an exponential function. For instance, when money is deposited into a compound interest account, the value of the money can be represented by an exponential growth function. When physicians use a test that involves a radioactive element, the amount of radioactivity remaining in the patient's body can be modeled by an exponential decay function.

EXAMPLE 4 ■ Application of an Exponential Function

When an amount of money P is placed in an account that earns compound interest, the value A of the money after t years is given by the compound interest formula

$$A = P\left(1 + \frac{r}{n}\right)^{nt}$$

where r is the annual interest rate as a decimal and n is the number of compounding periods per year. Suppose \$500 is placed in an account that earns 8% interest compounded daily. Find the value of the investment after 5 years.

Solution

Use the compound interest formula. Because interest is compounded daily, $n = 365$.

$$A = P\left(1 + \frac{r}{n}\right)^{nt}$$

$$= 500\left(1 + \frac{0.08}{365}\right)^{365(5)} \qquad \bullet\ P = 500,\ r = 0.08,\ n = 365,\ t = 5$$

$$\approx 500(1.491759) \approx 745.88$$

After 5 years, there is \$745.88 in the account.

CHECK YOUR PROGRESS 4 The radioactive isotope iodine-131 is used to monitor thyroid activity. The number of grams N of iodine-131 in the body t hours after an injection is given by $N(t) = 1.5\left(\frac{1}{2}\right)^{t/193.7}$. Find the number of grams of the isotope in the body 24 hours after an injection. Round to the nearest ten-thousandth.

Solution *See page S25.*

The next example is based on Newton's Law of Cooling. This exponential function can be used to model the temperature of something that is being cooled.

EXAMPLE 5 ■ Application of an Exponential Function

A cup of coffee is heated to 160°F and placed in a room that maintains a temperature of 70°F. The temperature of the coffee after t minutes is given by $T(t) = 70 + 90e^{-0.0485t}$. Find the temperature of the coffee 20 minutes after it is placed in the room. Round to the nearest degree.

Solution

Evaluate the function $T(t) = 70 + 90e^{-0.0485t}$ for $t = 20$.

$$T(t) = 70 + 90e^{-0.0485t}$$
$$T(20) = 70 + 90e^{-0.0485(20)} \quad \text{• Substitute 20 for } t.$$
$$\approx 70 + 34.1$$
$$\approx 104.1$$

After 20 minutes the temperature of the coffee is about 104°F.

CHECK YOUR PROGRESS 5 The function $A(t) = 200e^{-0.014t}$ gives the amount of aspirin, in milligrams, in a patient's bloodstream t minutes after the aspirin has been administered. Find the amount of aspirin in the patient's bloodstream after 45 minutes. Round to the nearest milligram.

Solution *See page S25.*

Math Matters Marie Curie

In 1903, Marie Sklodowska-Curie (1867–1934) became the first woman to receive a Nobel prize in physics. She was awarded the prize along with her husband, Pierre Curie, and Henri Becquerel for their discovery of radioactivity. In fact, Marie Curie coined the word *radioactivity*.

In 1911, Marie Curie became the first person to win a second Nobel prize, this time alone and in chemistry, for the isolation of radium. She also discovered the element polonium, which she named after Poland, her birthplace. The radioactive phenomena that she studied are modeled by exponential decay functions.

The stamp at the left was printed in 1938 to commemorate Pierre and Marie Curie. When the stamp was issued, there was a surcharge added to the regular price of the stamp. The additional revenue was donated to cancer research. Marie Curie died in 1934 from leukemia, which was caused by her constant exposure to radiation from the radioactive elements she studied.

In 1935, one of Marie Curie's daughters, Irene Joloit-Curie, won a Nobel prize in chemistry.

Exponential Regression

Earlier in this chapter we examined linear regression models. In some cases, an exponential function may more closely model a set of data.

EXAMPLE 6 ■ Find an Exponential Regression Equation

A diamond merchant has determined the values of several white diamonds that have different weights, measured in carats, but are similar in quality. See the table below.

4.00 ct	3.00 ct	2.00 ct	1.75 ct	1.50 ct	1.25 ct	1.00 ct	0.75 ct	0.50 ct
$14,500	$10,700	$7900	$7300	$6700	$6200	$5800	$5000	$4600

Find an exponential growth function that models the values of the diamonds as a function of their weights and use the model to predict the value of a 3.5-carat diamond of similar quality.

Solution

Use a graphing calculator to find the regression equation. The calculator display below shows that the exponential regression equation is $y \approx 4067.6(1.3816)^x$, where x is the carat weight of the diamond and y is the value of the diamond.

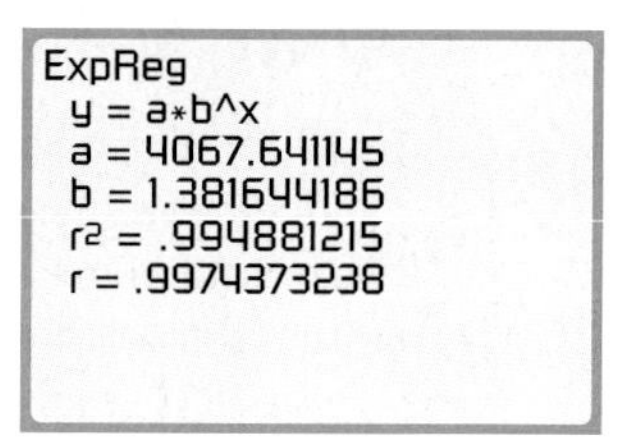

To use the regression equation to predict the value of a 3.5-carat diamond of similar quality, substitute 3.5 for x and evaluate.

$$y \approx 4067.6(1.3816)^x$$
$$y \approx 4067.6(1.3816)^{3.5}$$
$$y \approx 12{,}609$$

According to the modeling function, the value of a 3.5-carat diamond of similar quality is $12,609.

Notice on the calculator screen above that r^2, the coefficient of determination, is about 0.9949, which is very close to 1. This indicates that the equation $y \approx 4067.6(1.3816)^x$ provides a good fit for the data.

CHECK YOUR PROGRESS 6 The following table shows Earth's atmospheric pressure P at an altitude of a kilometers. Find an exponential function that models the atmospheric pressure as a function of altitude. Use the function to estimate, to the nearest tenth, the atmospheric pressure at an altitude of 24 kilometers.

TAKE NOTE

The value of a diamond is generally determined by its color, cut, clarity, and carat weight. These characteristics of a diamond are known as the four c's. In Example 6 we have assumed that the color, cut, and clarity of all of the diamonds are similar. This assumption enables us to model the value of each diamond as a function of just its carat weight.

point of interest

The Hope Diamond, shown below, is the world's largest deep-blue diamond. It weighs 45.52 carats. We should not expect the function $y \approx 4067.6(1.3816)^x$ in Example 6 to yield an accurate value of the Hope Diamond because the Hope Diamond is not the same type of diamond as the diamonds in Example 6, and its weight is much larger.

The Hope Diamond is on display at the Smithsonian Museum of Natural History in Washington, D.C.

Altitude, a, above sea level, in kilometers	Atmospheric pressure, P, in newtons per square centimeter
0	10.3
2	8.0
4	6.4
6	5.1
8	4.0
10	3.2
12	2.5
14	2.0
16	1.6
18	1.3

Solution *See page S25.*

Chess and Exponential Functions

According to legend, when Sissa Ben Dahir of India invented the game of chess, King Shirham was so impressed with the game that he summoned the game's inventor and offered him the reward of his choosing. The inventor pointed to the chessboard and requested, for his reward, one grain of wheat on the first square, two grains of wheat on the second square, four grains on the third square, eight grains on the fourth square, and so on for all 64 squares on the chessboard. The king considered this a very modest reward and said he would grant the inventor's wish.

1. This portion of this Excursion will enable you to find a formula for the total amount of wheat on the first n squares of the chessboard. You may want to use the chart below as you answer the questions. It may help you to see a pattern.

Square number, n	1	2	3	4	5	6	7
Number of grains of wheat on square n	1	2	4				
Total number of grains of wheat on squares 1 through n	1	$1 + 2 =$	$3 + 4 =$				

a. How many grains of wheat are on each of the first seven squares?

(continued)

What is the total number (the sum) of grains of wheat on the first

b. two squares?

c. three squares?

d. four squares?

e. five squares?

f. six squares?

g. seven squares?

2. Use inductive reasoning to find a function that gives the total number (the sum) of grains of wheat on the first *n* squares of the chessboard. Test your function to ensure that it works for parts b through g.

3. If all 64 squares of the chessboard are piled with wheat as requested by Sissa Ben Dahir, how many grains of wheat are on the board?

4. A grain of wheat weighs approximately 0.000008 kilogram. Find the total weight of the wheat requested by Sissa Ben Dahir.

5. In a recent year, a total of 6.5×10^8 metric tons of wheat were produced in the world. At this level, how many years of wheat production would be required to fill the request of Sissa Ben Dahir? *Hint:* One metric ton equals 1000 kilograms.

Exercise Set 6.5

1. Given $f(x) = 3^x$, evaluate:
a. $f(2)$ **b.** $f(0)$ **c.** $f(-2)$

2. Given $H(x) = 2^x$, evaluate:
a. $H(-3)$ **b.** $H(0)$ **c.** $H(2)$

3. Given $g(x) = 2^{x+1}$, evaluate:
a. $g(3)$ **b.** $g(1)$ **c.** $g(-3)$

4. Given $F(x) = 3^{x-2}$, evaluate:
a. $F(-4)$ **b.** $F(-1)$ **c.** $F(0)$

5. Given $G(r) = \left(\frac{1}{2}\right)^{2r}$, evaluate:
a. $G(0)$ **b.** $G\left(\frac{3}{2}\right)$ **c.** $G(-2)$

6. Given $R(t) = \left(\frac{1}{3}\right)^{3t}$, evaluate:
a. $R\left(-\frac{1}{3}\right)$ **b.** $R(1)$ **c.** $R(-2)$

7. Given $h(x) = e^{x/2}$, evaluate the following. Round to the nearest ten-thousandth.
a. $h(4)$ **b.** $h(-2)$ **c.** $h\left(\frac{1}{2}\right)$

8. Given $f(x) = e^{2x}$, evaluate the following. Round to the nearest ten-thousandth.
a. $f(-2)$ **b.** $f\left(-\frac{2}{3}\right)$ **c.** $f(2)$

9. Given $H(x) = e^{-x+3}$, evaluate the following. Round to the nearest ten-thousandth.
a. $H(-1)$ **b.** $H(3)$ **c.** $H(5)$

10. Given $g(x) = e^{-x/2}$, evaluate the following. Round to the nearest ten-thousandth.
a. $g(-3)$ **b.** $g(4)$ **c.** $g\left(\frac{1}{2}\right)$

11. Given $F(x) = 2^{x^2}$, evaluate the following. Round to the nearest ten-thousandth.
a. $F(2)$ **b.** $F(-2)$ **c.** $F\left(\frac{3}{4}\right)$

12. Given $Q(x) = 2^{-x^2}$, evaluate:
a. $Q(3)$ **b.** $Q(-1)$ **c.** $Q(-2)$

13. Given $f(x) = e^{-x^2/2}$, evaluate the following. Round to the nearest ten-thousandth.
a. $f(-2)$ **b.** $f(2)$ **c.** $f(-3)$

14. Given $h(x) = e^{-2x} + 1$, evaluate the following. Round to the nearest ten-thousandth.
a. $h(-1)$ **b.** $h(3)$ **c.** $h(-2)$

In Exercises 15–24, graph the function.

15. $f(x) = 2^x + 1$
16. $f(x) = 3^x - 2$
17. $g(x) = 3^{x/2}$
18. $h(x) = 2^{-x/2}$
19. $f(x) = 2^{x+3}$
20. $g(x) = 4^{-x} + 1$
21. $H(x) = 2^{2x}$
22. $F(x) = 2^{-x}$
23. $f(x) = e^{-x}$
24. $y(x) = e^{2x}$

Investments In Exercises 25 and 26, use the compound interest formula $A = P\left(1 + \frac{r}{n}\right)^{nt}$, where P is the amount deposited, A is the value of the money after t years, r is the annual interest rate as a decimal, and n is the number of compounding periods per year.

25. A computer network specialist deposits \$2500 into a retirement account that earns 7.5% annual interest, compounded daily. What is the value of the investment after 20 years?

26. A \$10,000 certificate of deposit (CD) earns 5% annual interest, compounded daily. What is the value of the investment after 20 years?

27. **Investments** Some banks now use continuous compounding of an amount invested. In this case, the equation that models the value of an initial investment of P dollars in t years at an annual interest rate of r is given by $A = Pe^{rt}$. Using this equation, find the value in 5 years of an investment of \$2500 that earns 5% annual interest.

28. **Isotopes** An isotope of technetium is used to prepare images of internal body organs. This isotope has a half-life (time required for half the material to erode) of approximately 6 hours. If a patient is injected with 30 milligrams of this isotope, what will be the technetium level in the patient after 3 hours? Use the function $A(t) = 30\left(\frac{1}{2}\right)^{t/6}$, where A is the technetium level, in milligrams, in the patient after t hours. Round to the nearest tenth.

29. **Isotopes** Iodine-131 is an isotope that is used to study the functioning of the thyroid gland. This isotope has a half-life (time required for half the material to erode) of approximately 8 days. If a patient is given an injection that contains 8 micrograms of iodine-131, what will be the iodine level in the patient after 5 days? Use the function $A(t) = 8\left(\frac{1}{2}\right)^{t/8}$, where A is the amount of the isotope, in micrograms, in the patient after t days. Round to the nearest tenth.

30. **Welding** The percent of correct welds that a student can make will increase with practice and can be approximated by the function $P(t) = 100[1 - (0.75)^t]$, where P is the percent of correct welds and t is the number of weeks of practice. Find the percent of correct welds that a student will make after 4 weeks of practice. Round to the nearest percent.

31. **Music** The "concert A" note on a piano is the first A below middle C. When that key is struck, the string associated with the key vibrates 440 times per second. The next A above concert A vibrates twice as fast. An exponential function with a base of two is used to determine the frequency of the 11 notes between the two A's. Find this function. *Hint:* The function is of the form $f(x) = k \cdot 2^{(cx)}$, where k and c are constants. Also $f(0) = 440$ and $f(12) = 880$.

32. **Atmospheric Pressure** Atmospheric pressure changes as you rise above Earth's surface. At an altitude of h kilometers, where $0 < h < 80$, the pressure P in newtons per square centimeter is approximately modeled by the function

$$P(h) = 10.13e^{-0.116h}.$$

a. What is the approximate pressure at 40 kilometers above Earth?

b. What is the approximate pressure on Earth's surface?

c. Does atmospheric pressure increase or decrease as you rise above Earth's surface?

33. **Automobiles** The number of automobiles in the United States in 1900 was around 8000. In the year 2000, the number of automobiles in the United States reached 200 million. Find an exponential model for the data and use the model to predict the number of automobiles in the United States in 2010. Use $t = 0$ to represent the year 1900. Round to the nearest hundred thousand automobiles.

34. **Panda Population** One estimate gives the world panda population as 3200 in 1980 and 590 in 2000. Find an exponential model for the data and use the model to predict the panda population in 2040. Use $t = 0$ to represent the year 1980. Round to the nearest whole number.

35. **Polonium** An initial amount of 100 micrograms of polonium decays to 75 micrograms in approximately 34.5 days. Find an exponential model for the amount of polonium in the sample after t days. Round to the nearest hundredth of a microgram.

36. **The Film Industry** The table below shows the number of multidisc DVDs, with three or more discs, released each year. (*Source:* DVD Release Report)

Year, x	1999	2000	2001	2002	2003
Titles released, y	11	57	87	154	283

a. Find an exponential regression equation for this data using $x = 0$ to represent 1995. Round to the nearest hundredth.

b. Use the equation to predict the number of multidisc DVDs released in 2008.

37. **Meteorology** The table below shows the saturation of water in air at various air temperatures.

Temperature (in °C)	0	5	10	20	25	30
Saturation (in millimeters of water per cubic meter of air)	4.8	6.8	9.4	17.3	23.1	30.4

a. Find an exponential regression equation for these data. Round to the nearest thousandth.

b. Use the equation to predict the number of milliliters of water per cubic meter of air at a temperature of 15°C. Round to the nearest tenth.

38. **Snow Making** Artificial snow is made at a ski resort by combining air and water in a ratio that depends on the outside air temperature. The table below shows the rate of air flow needed for various temperatures.

Temperature (in °F)	0	5	10	15	20
Air flow (in cubic feet per minute)	3.0	3.6	4.7	6.1	9.9

a. Find an exponential regression equation for these data. Round to the nearest hundredth.

b. Use the equation to predict the air flow needed when the temperature is 25°F. Round to the nearest tenth of a cubic foot per minute.

Extensions

CRITICAL THINKING

An exponential model for population growth or decay can be accurate over a short period of time. However, this model begins to fail because it does not account for the natural resources necessary to support growth, nor does it account for death within the population. Another model, called the *logistic model*, can account for some of these effects. The logistic model is given by $P(t) = \dfrac{mP_0}{P_0 + (m - P_0)e^{-kt}}$, where $P(t)$ is the population at time t, m is the maximum population that can be supported, P_0 is the population when $t = 0$, and k is a positive constant that is related to the growth of the population.

39. **Earth's Population** One model of Earth's population is given by $P(t) = \dfrac{280}{4 + 66e^{-0.021t}}$. In this equation, $P(t)$ is the population in billions and t is the number of years after 1980. Round answers to the nearest hundred million people.

a. According to this model, what was Earth's population in the year 2000?

b. According to this model, what will be Earth's population in the year 2010?

c. If t is very large, say greater than 500, then $e^{-0.021t} \approx 0$. What does this suggest about the maximum population that Earth can support?

40. **Wolf Population** Game wardens have determined that the maximum wolf population in a certain preserve is 1000 wolves. Suppose the population of wolves in the preserve in the year 2000 was 500, and that k is estimated to be 0.025.

a. Find a logistic function for the number of wolves in the preserve in year t, where t is the number of years after 2000.

b. Find the estimated wolf population in 2015.

EXPLORATIONS

41. **Car Payments** The formula used to calculate a monthly lease payment or a monthly car payment (for a purchase rather than a lease) is given by $P = \dfrac{Ar(1 + r)^n - Vr}{(1 + r)^n - 1}$, where P is the monthly payment, A is the amount of the loan, r is the *monthly* interest rate as a decimal, n is the number of months of the loan or lease, and V is the residual value of the car at the end of the lease. For a car purchase, $V = 0$.

a. If the annual interest rate for a loan is 9%, what is the monthly interest rate as a decimal?

b. Write the formula for a monthly car payment when the car is purchased rather than leased.

c. Suppose you lease a car for 5 years. Find the monthly lease payment if the lease amount is \$10,000, the residual value is \$6000, and the annual interest rate is 6%.

d. Suppose you purchase a car and secure a 5-year loan for \$10,000 at an annual interest rate of 6%. Find the monthly payment.

e. Why are the answers to parts c and d different?

The total amount C that has been repaid on a loan or lease is given by $C = \dfrac{(P - Ar)[(1 + r)^n - 1]}{r}$.

f. Using the lease payment in part c, find the total amount that will be repaid in 5 years. How much remains to be paid?

g. Using the monthly payment in part d, find the total amount that will be repaid in 5 years. How much remains to be paid?

h. Explain why the answers to parts f and g make sense.

SECTION 6.6 Logarithmic Functions

Introduction to Logarithmic Functions

Time (in hours)	Number of Bacteria
0	1000
1	2000
2	4000
3	8000

Suppose a bacteria colony that originally contained 1000 bacteria doubled in size every hour. Then the table at the left would show the number of bacteria in that colony after 1, 2, and 3 hours.

The exponential function $A = 1000(2^t)$, where A is the number of bacteria in the colony at time t, is a model of the growth of the colony. For instance, when $t = 3$ hours, we have

$$A = 1000(2^t)$$
$$A = 1000(2^3) \quad \text{• Replace } t \text{ by } 3.$$
$$A = 1000(8) = 8000$$

After 3 hours there are 8000 bacteria in the colony.

Now we ask, "How long will it take for there to be 32,000 bacteria in the colony?" To answer the question, we must solve the *exponential equation* $32{,}000 = 1000(2^t)$. By trial and error, we find that when $t = 5$,

$$A = 1000(2^t)$$
$$A = 1000(2^5) \quad \text{• Replace } t \text{ by } 5.$$
$$A = 1000(32) = 32{,}000$$

After 5 hours there will be 32,000 bacteria in the colony.

Now suppose we want to know how long it will be before the colony reaches 50,000 bacteria. To answer that question, we must find t such that $50{,}000 = 1000(2^t)$. Using trial and error again, we find that

$$1000(2^5) = 32{,}000 \qquad \text{and} \qquad 1000(2^6) = 64{,}000$$

Because 50,000 is between 32,000 and 64,000, we conclude that t is between 5 and 6 hours. If we try $t = 5.5$ (halfway between 5 and 6), we get

$$A = 1000(2^t)$$
$$A = 1000(2^{5.5}) \quad \text{• Replace } t \text{ by } 5.5.$$
$$A \approx 1000(45.25) = 45{,}250$$

In 5.5 hours, there are approximately 45,250 bacteria in the colony. Because this is less than 50,000, the value of t must be a little greater than 5.5.

We could continue to use trial and error to find the correct value of t, but it would be more efficient if we could just solve the exponential equation $50{,}000 = 1000(2^t)$ for t. If we follow the procedures for solving equations that were discussed earlier in the text, we have

$$50{,}000 = 1000(2^t)$$

$$50 = 2^t \quad \text{• Divide each side of the equation by 1000.}$$

To proceed to the next step, it would be helpful to have a function that would find the power of 2 that produces 50.

Around the mid-sixteenth century, mathematicians created such a function, which we now call a *logarithmic function*. We write the solution of $50 = 2^t$ as $t = \log_2 50$. This is read "t equals the logarithm base 2 of 50." and it means "t equals the power of 2 that produces 50." When logarithms were first introduced, tables were used to find a numerical value of t. Today, a calculator is used. Using a calculator, we can approximate the value of t as 5.644. This means that $2^{5.644} \approx 50$.

The equivalence of the expressions $50 = 2^t$ and $t = \log_2 50$ are described in the following definition of logarithm.

CALCULATOR NOTE 

Using a calculator, we can verify that $2^{5.644} \approx 50$. On a graphing calculator, press 2 ^ 5.644 ENTER.

TAKE NOTE

Note that on page 389 when we tried $t = 5.5$, we stated that the actual value of t must be greater than 5.5 and that 5.644 is a little greater than 5.5.

Definition of Logarithm

For $b > 0$, $b \neq 1$, $y = \log_b x$ is equivalent to $x = b^y$.

TAKE NOTE

Read $\log_b x$ as "the logarithm of x, base b" or "log base b of x."

For every exponential equation there is a corresponding logarithmic equation, and for every logarithmic equation there is a corresponding exponential equation. Here are some examples.

Exponential Equation	Logarithmic Equation
$2^5 = 32$	$\log_2 32 = 5$
$3^2 = 9$	$\log_3 9 = 2$
$5^{-2} = \frac{1}{25}$	$\log_5 \frac{1}{25} = -2$
$7^0 = 1$	$\log_7 1 = 0$

TAKE NOTE

The idea of a function that performs the opposite of a given function occurs frequently. For instance, the opposite of a "doubling" function, one that doubles a given number, is a "halving" function, one that takes one-half of a given number. We could write $f(x) = 2x$ for the doubling function and $g(x) = \frac{1}{2}x$ for the halving function. We call these functions *inverse functions* of one another. In a similar manner, exponential and logarithmic functions are inverses of one another.

QUESTION *Which of the following is the logarithmic form of $4^3 = 64$?*

a. $\log_4 3 = 64$ **b.** $\log_3 4 = 64$ **c.** $\log_4 64 = 3$

The equation $y = \log_b x$ is the logarithmic form of $b^y = x$, and the equation $b^y = x$ is the *exponential form* of $y = \log_b x$. These two forms state exactly the same relationship between x and y.

ANSWER c. *$\log_4 64 = 3$ is equivalent to $4^3 = 64$.*

historical note

Logarithms were developed independently by Jobst Burgi (1552–1632) and John Napier (1550–1617) as a means of simplifying the calculations of astronomers. The idea was to devise a method by which two numbers could be multiplied by performing additions. Napier is usually given credit for logarithms because he published his results first.

In Napier's original work, the logarithm of 10,000,000 was 0. After this work was published, Napier, in discussions with Henry Briggs (1561–1631), decided that tables of logarithms would be easier to use if the logarithm of 1 were 0. Napier died before new tables could be prepared, and Briggs took on the task. His table consisted of logarithms accurate to 30 decimal places, all accomplished without a calculator! ■

EXAMPLE 1 ■ Write a Logarithmic Equation in Exponential Form and an Exponential Equation in Logarithmic Form

a. Write $2 = \log_{10}(x + 5)$ in exponential form.

b. Write $2^{3x} = 64$ in logarithmic form.

Solution

Use the definition of logarithm: $y = \log_b x$ if and only if $b^y = x$.

a. $2 = \log_{10}(x + 5)$ if and only if $10^2 = x + 5$.

b. $2^{3x} = 64$ if and only if $\log_2 64 = 3x$.

CHECK YOUR PROGRESS 1

a. Write $\log_2(4x) = 10$ in exponential form.

b. Write $10^3 = 2x$ in logarithmic form.

Solution *See page S25.*

The relationship between the exponential and logarithmic forms can be used to evaluate some logarithms. The solutions to these types of problems are based on the Equality of Exponents Property.

Equality of Exponents Property

If $b > 0$ and $b^x = b^y$, then $x = y$.

EXAMPLE 2 ■ Evaluate Logarithmic Expressions

Evaluate the logarithms.

a. $\log_8 64$ **b.** $\log_2\left(\frac{1}{8}\right)$

Solution

a.

$\log_8 64 = x$ • Write an equation.

$8^x = 64$ • Write the equation in its equivalent exponential form.

$8^x = 8^2$ • Write 64 in exponential form using 8 as the base.

$x = 2$ • Solve for x using the Equality of Exponents Property.

$\log_8 64 = 2$

b.

$\log_2\left(\frac{1}{8}\right) = x$ • Write an equation.

$2^x = \frac{1}{8}$ • Write the equation in its equivalent exponential form.

$2^x = 2^{-3}$ • Write $\frac{1}{8}$ in exponential form using 2 as the base.

$x = -3$ • Solve for x using the Equality of Exponents Property.

$\log_2\left(\frac{1}{8}\right) = -3$

TAKE NOTE

If you need to review material on negative exponents, see Lesson 6.1B on the CD that you received with this book.

CHECK YOUR PROGRESS 2 Evaluate the logarithms.

a. $\log_{10} 0.001$ **b.** $\log_5 125$

Solution *See page S25.*

EXAMPLE 3 ■ Solve a Logarithmic Equation

Solve: $\log_3 x = 2$

Solution

$$\log_3 x = 2$$

$$3^2 = x$$ • Write the equation in its equivalent exponential form.

$$9 = x$$ • Simplify the exponential expression.

CHECK YOUR PROGRESS 3 Solve: $\log_2 x = 6$

Solution *See page S25.*

Not all logarithms can be evaluated by rewriting the logarithm in its equivalent exponential form and using the Equality of Exponents Property. For instance, if we tried to evaluate $\log_{10} 18$, it would be necessary to solve the equivalent exponential equation $10^x = 18$. The difficulty here is trying to rewrite 18 in exponential form with 10 as a base.

Common and Natural Logarithms

Two of the most frequently used logarithmic functions are *common logarithms,* which have base 10, and *natural logarithms,* which have base e (the base of the natural exponential function).

Common and Natural Logarithms

The function defined by $f(x) = \log_{10} x$ is called the **common logarithmic function.** It is customarily written without the base as $f(x) = \log x$.

The function defined by $f(x) = \log_e x$ is called the **natural logarithmic function.** It is customarily written as $f(x) = \ln x$.

Most scientific or graphing calculators have a `log` key for evaluating common logarithms and a `ln` key to evaluate natural logarithms. For instance,

$$\log 24 \approx 1.3802112 \qquad \text{and} \qquad \ln 81 \approx 4.3944492$$

EXAMPLE 4 ■ Solve Common and Natural Logarithmic Equations

Solve each of the following equations. Round to the nearest thousandth.

a. $\log x = -1.5$ **b.** $\ln x = 3$

Solution

a. $\log x = -1.5$

$10^{-1.5} = x$ • Write the equation in its equivalent exponential form.

$0.032 \approx x$ • Simplify the exponential expression.

b. $\ln x = 3$

$e^3 = x$ • Write the equation in its equivalent exponential form.

$20.086 \approx x$ • Simplify the exponential expression.

CHECK YOUR PROGRESS 4 Solve each of the following equations. Round to the nearest thousandth.

a. $\log x = -2.1$ **b.** $\ln x = 2$

Solution *See page S25.*

Math Matters Zipf's Law

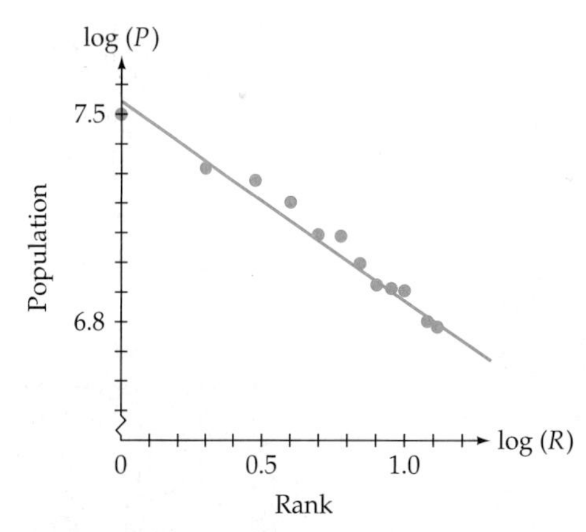

George Zipf (1902–1950) was a lecturer in German and philology (the study of the change in a language over time) at Harvard University. Zipf's Law, as originally formulated, referred to the frequency of occurrence of words in a book, magazine article, or other written material, and its rank. The word used most frequently had rank 1, the next most frequent word had rank 2, and so on. Zipf hypothesized that if the x-axis were the logarithm of a word's rank and the y-axis were the logarithm of the frequency of the word, then the graph of rank versus frequency would lie on approximately a straight line.

Zipf's Law has been extended to demographics. For instance, let the population of a U.S. state be P and its rank in population be R. The graph of the points $(\log R, \log P)$ lie on approximately a straight line. The graph above shows Zipf's Law as applied to the populations and ranks of the 12 most populated states according to the 2000 census.

Recently, Zipf's Law has been applied to web site traffic, where web sites are ranked according to the number of hits they receive. Assumptions about web traffic are used by engineers and programmers who study ways to make the web more efficient.

Graphs of Logarithmic Functions

The graph of a logarithmic function can be drawn by first rewriting the function in its exponential form. This procedure is illustrated in Example 5.

EXAMPLE 5 ■ Graph a Logarithmic Function

Graph $f(x) = \log_3 x$.

Solution

To graph $f(x) = \log_3 x$, first write the equation in the form $y = \log_3 x$. Then write the equivalent exponential equation $x = 3^y$. Because this equation is solved for x, choose values of y and calculate the corresponding values of x, as shown in the table below.

$x = 3^y$	$\frac{1}{9}$	$\frac{1}{3}$	1	3	9
y	−2	−1	0	1	2

Plot the ordered pairs and connect the points with a smooth curve, as shown at the left.

CHECK YOUR PROGRESS 5 Graph $f(x) = \log_5 x$.

Solution *See page S25.*

The graphs of $y = \log x$ and $y = \ln x$ can be drawn on a graphing calculator by using the `log` and `ln` keys. The graphs are shown below for a TI-83/84 calculator.

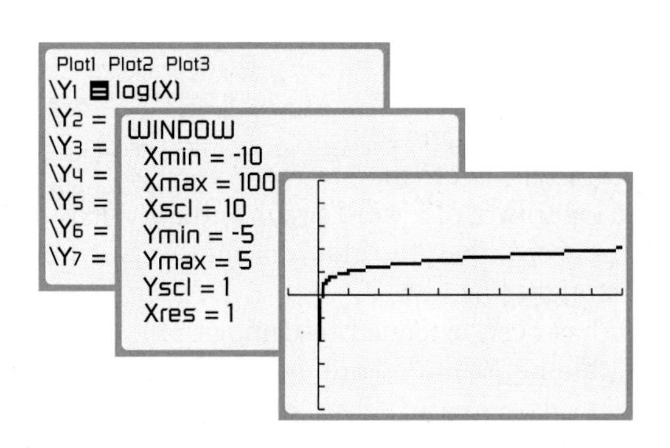

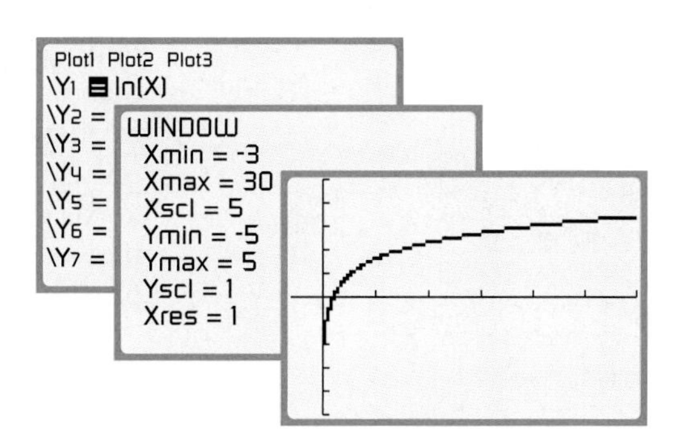

Applications of Logarithmic Functions

Many applications can be modeled by logarithmic functions.

EXAMPLE 6 ■ Application of a Logarithmic Function

During the 1980s and 1990s, the average time T of a major league baseball game tended to increase each year. If the year 1981 is represented by $x = 1$, then the function

$$T(x) = 149.57 + 7.63 \ln x$$

approximates the average time T, in minutes, of a major league baseball game for the years $x = 1$ to $x = 19$.

a. Use the function to determine the average time of a major league baseball game during the 1981 season and during the 1999 season. Round to the nearest hundredth of a minute.

b. By how much did the average time of a major league baseball game increase from 1981 to 1999?

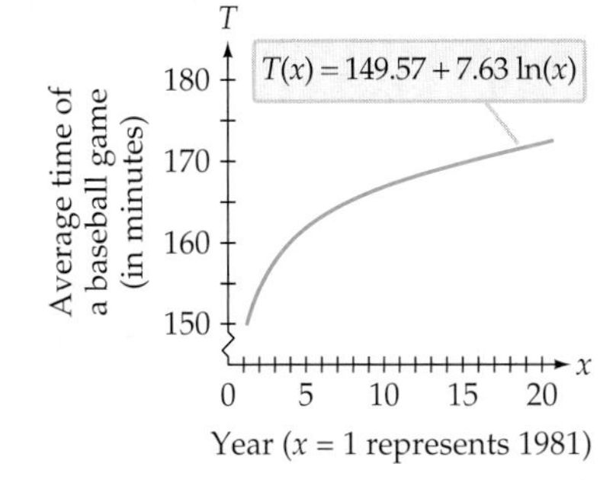

Solution

a. The year 1981 is represented by $x = 1$ and the year 1999 by $x = 19$.

$$T(1) = 149.57 + 7.63 \ln(1) = 149.57$$

In 1981 the average time of a major league baseball game was about 149.57 minutes.

$$T(19) = 149.57 + 7.63 \ln(19) \approx 172.04$$

In 1999 the average time of a major league baseball game was about 172.04 minutes.

b. $T(19) - T(1) \approx 172.04 - 149.57 \approx 22.47$

From 1981 to 1999, the average time of a major league baseball game increased by about 22.47 minutes.

CHECK YOUR PROGRESS 6 The following function models the average typing speed S, in words per minute, of a student who has been typing for t months.

$$S(t) = 5 + 29 \ln(t + 1), \quad 0 \le t \le 9$$

a. Use the function to determine the student's average typing speed when the student first started to type and the student's average typing speed after 3 months. Round to the nearest whole word per minute.

b. By how much did the typing speed increase during the 3 months?

Solution *See page S25.*

historical note

The Richter scale was created by the seismologist Charles Francis Richter (rĭk'tər) (1900–1985) in 1935.

Richter was born in Ohio. At the age of 16, he and his mother moved to Los Angeles, where he enrolled at the University of Southern California. He went on to study physics at Stanford University. Richter was a professor of seismology at the Seismological Laboratory at California Institute of Technology (Caltech) from 1936 until he retired in 1970. ■

Logarithmic functions are often used to convert very large or very small numbers into numbers that are easier to comprehend. For instance, the *Richter scale*, which measures the magnitude of an earthquake, uses a logarithmic function to scale the intensity of an earthquake's shock waves I into a number M, which for most earthquakes is in the range of 0 to 10. The intensity I of an earthquake is often given in terms of the constant I_0, where I_0 is the intensity of the smallest earthquake (called a **zero-level earthquake**) that can be measured on a seismograph near the earthquake's epicenter. The following formula is used to compute the Richter scale magnitude of an earthquake.

The Richter Scale Magnitude of an Earthquake

An earthquake with an intensity of I has a Richter scale magnitude of

$$M = \log\left(\frac{I}{I_0}\right)$$

where I_0 is the measure of the intensity of a zero-level earthquake.

EXAMPLE 7 ■ Find the Magnitude of an Earthquake

Find the Richter scale magnitude of the 2003 Amazonas, Brazil, earthquake, which had an intensity of $I = 12{,}589{,}254I_0$. Round to the nearest tenth.

TAKE NOTE

Notice in Example 7 that we did not need to know the value of I_0 to determine the Richter scale magnitude of the quake.

Solution

$$M = \log\left(\frac{I}{I_0}\right) = \log\left(\frac{12{,}589{,}254I_0}{I_0}\right) = \log(12{,}589{,}254) \approx 7.1$$

The 2003 Amazonas, Brazil, earthquake had a Richter scale magnitude of 7.1.

CHECK YOUR PROGRESS 7 What is the Richter scale magnitude of an earthquake whose intensity is twice that of the Amazonas, Brazil, earthquake in Example 7?

Solution *See page S25.*

If you know the Richter scale magnitude of an earthquake, then you can determine the intensity of the earthquake.

EXAMPLE 8 ■ Find the Intensity of an Earthquake

Find the intensity of the 2003 Colina, Mexico, earthquake, which measured 7.6 on the Richter scale. Round to the nearest thousand.

Solution

$$\log\left(\frac{I}{I_0}\right) = 7.6$$

$$\frac{I}{I_0} = 10^{7.6} \quad \text{• Write in exponential form.}$$

$$I = 10^{7.6}I_0 \quad \text{• Solve for } I.$$

$$I \approx 39{,}810{,}717I_0$$

The 2003 Colina, Mexico, earthquake had an intensity that was approximately 39,811,000 times the intensity of a zero-level earthquake.

CHECK YOUR PROGRESS 8 On April 29, 2003, an earthquake measuring 4.6 on the Richter scale struck Fort Payne, Alabama. Find the intensity of the quake. Round to the nearest thousand.

Solution *See page S25.*

point of interest

The pH scale was created by the Danish biochemist Søren Sørensen (sû'rn-sn) in 1909 to measure the acidity of water used in the brewing of beer. pH is an abbreviation for *pondus hydrogenii,* which translates as "potential hydrogen."

Logarithmic scales are also used in chemistry. Chemists use logarithms to determine the pH of a liquid, which is a measure of the liquid's **acidity** or **alkalinity.** (You may have tested the pH of a swimming pool or an aquarium.) Pure water, which is considered neutral, has a pH of 7.0. The pH scale ranges from 0 to 14, with 0 corresponding to the most acidic solutions and 14 to the most alkaline. Lemon juice has a pH of about 2, whereas household ammonia measures about 11.

Specifically, the acidity of a solution is a function of the hydronium-ion concentration of the solution. Because the hydronium-ion concentration of a solution can be very small (with values as low as 0.00000001), pH measures the acidity or alkalinity of a solution using a logarithmic scale.

TAKE NOTE

One mole is equivalent to 6.022×10^{23} ions.

The pH of a Solution

The pH of a solution with a hydronium-ion concentration of H^+ moles per liter is given by

$$pH = -\log[H^+]$$

EXAMPLE 9 ■ Calculate the pH of a Liquid

Find the pH of each liquid. Round to the nearest tenth.

a. Orange juice containing an H^+ concentration of 2.8×10^{-4} mole per liter

b. Milk containing an H^+ concentration of 3.97×10^{-7} mole per liter

c. A baking soda solution containing an H^+ concentration of 3.98×10^{-9} mole per liter

Solution

a. $pH = -\log[H^+]$

$pH = -\log(2.8 \times 10^{-4}) \approx 3.6$

The orange juice has a pH of about 3.6.

b. $pH = -\log[H^+]$

$pH = -\log(3.97 \times 10^{-7}) \approx 6.4$

The milk has a pH of about 6.4.

c. $pH = -\log[H^+]$

$pH = -\log(3.98 \times 10^{-9}) \approx 8.4$

The baking soda solution has a pH of about 8.4.

CHECK YOUR PROGRESS 9 Find the pH of each liquid. Round to the nearest tenth.

a. A cleaning solution containing an H^+ concentration of 2.41×10^{-13} mole per liter

b. A cola soft drink containing an H^+ concentration of 5.07×10^{-4} mole per liter

c. Rainwater containing an H^+ concentration of 6.31×10^{-5} mole per liter

Solution *See pages S25–S26.*

In Example 9, the hydronium-ion concentrations of orange juice and milk were given as 2.8×10^{-4} and 3.97×10^{-7}, respectively.

The following figure illustrates the pH scale, along with the corresponding hydronium-ion concentrations. A solution with a pH less than 7 is an **acid,** and a solution with a pH greater than 7 is an **alkaline solution,** or a **base.** Because the scale is logarithmic, a solution with a pH of 5 is 10 times more acidic than a solution with a pH of 6. From Example 9 we see that the orange juice and milk are acids, whereas the baking soda solution is a base.

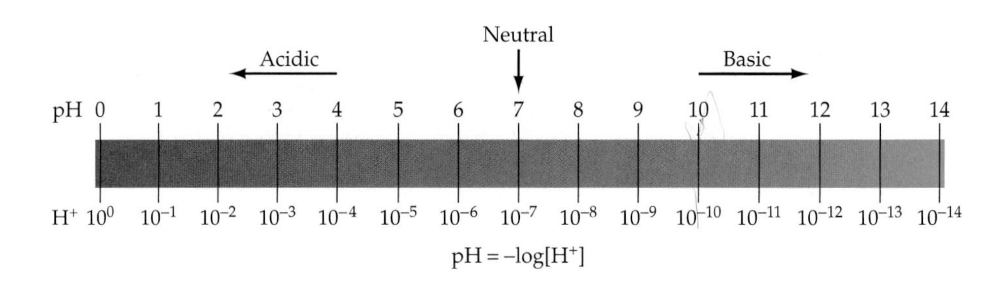

The figure above shows how the pH function scales small numbers on the H^+ axis into larger and more convenient numbers on the pH axis.

EXAMPLE 10 ■ Find the Hydronium-Ion Concentration of a Liquid

A sample of blood has a pH of 7.3. Find the hydronium-ion concentration of the blood.

Solution

$$\text{pH} = -\log[H^+]$$

$$7.3 = -\log[H^+] \qquad \text{• Substitute 7.3 for pH.}$$

$$-7.3 = \log[H^+] \qquad \text{• Multiply both sides by } -1.$$

$$10^{-7.3} = H^+ \qquad \text{• Change to exponential form.}$$

$$5.0 \times 10^{-8} \approx H^+ \qquad \text{• Evaluate } 10^{-7.3} \text{ and write the answer in scientific notation.}$$

The hydronium-ion concentration of the blood is about 5.0×10^{-8} mole per liter.

CHECK YOUR PROGRESS 10 The water in the Great Salt Lake in Utah has a pH of 10.0. Find the hydronium-ion concentration of the water.

Solution *See page S26.*

Excursion

Benford's Law

point of interest

Benford's Law has been used to identify fraudulent accountants. In most cases these accountants are unaware of Benford's Law and have replaced valid numbers with numbers selected at random. Their numbers do not conform to Benford's Law. Hence, an audit is warranted.

The authors of this text know some interesting details about your finances. For instance, of the last 100 checks you have written, about 30% are for amounts that start with a 1. Also, you have written about three times as many checks for amounts that start with a 2 as you have for amounts that start with a 7.

We are sure of these results because of a mathematical formula known as *Benford's Law.* This law was first discovered by the mathematician Simon Newcomb in 1881 and was then rediscovered by the physicist Frank Benford in 1938. Benford's Law states that the probability P that the first digit of a number selected from a wide range of numbers d is given by

$$P(d) = \log\left(1 + \frac{1}{d}\right)$$

1. Use Benford's Law to complete the table and bar graph shown below.

d	$P(d) = \log(1 + 1/d)$
1	0.301
2	0.176
3	0.125
4	______
5	______
6	______
7	______
8	______
9	______

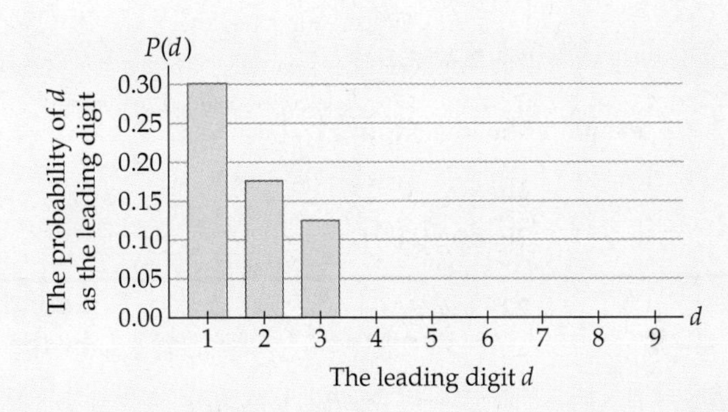

Benford's Law applies to most data with a wide range. For instance, it applies to

- the populations of the cities in the United States
- the numbers of dollars in the savings accounts at your local bank
- the numbers of miles driven during a month by each person in a state

2. Use Benford's Law to find the probability that in a U.S. city selected at random, the number of telephones in that city will be a number starting with a 6.

3. Use Benford's Law to find how many times as many purchases you have made for dollar amounts that start with a 1 than for dollar amounts that start with a 9.

4. Explain why Benford's Law would not apply to the set of telephone numbers in a small city such as Le Mars, Iowa.

5. Explain why Benford's Law would not apply to the set of all the ages, in years, of the students at a local high school.

Exercise Set 6.6

In Exercises 1–8, write the exponential equation in logarithmic form.

1. $7^2 = 49$
2. $10^3 = 1000$
3. $5^4 = 625$
4. $2^{-3} = \frac{1}{8}$
5. $10^{-4} = 0.0001$
6. $3^5 = 729$
7. $10^y = x$
8. $e^y = x$

In Exercises 9–16, write the logarithmic equation in exponential form.

9. $\log_3 81 = 4$
10. $\log_2 16 = 4$
11. $\log_5 125 = 3$
12. $\log_4 64 = 3$
13. $\log_4 \frac{1}{16} = -2$
14. $\log_2 \frac{1}{16} = -4$
15. $\ln x = y$
16. $\log x = y$

In Exercises 17–24, evaluate the logarithm.

17. $\log_3 81$
18. $\log_7 49$
19. $\log 100$
20. $\log 0.001$
21. $\log_3 \frac{1}{9}$
22. $\log_7 \frac{1}{7}$
23. $\log_2 64$
24. $\log 0.01$

In Exercises 25–32, solve the equation for x.

25. $\log_3 x = 2$
26. $\log_5 x = 1$
27. $\log_7 x = -1$
28. $\log_8 x = -2$
29. $\log_3 x = -2$
30. $\log_5 x = 3$
31. $\log_4 x = 0$
32. $\log_4 x = -1$

In Exercises 33–40, use a calculator to solve for x. Round to the nearest hundredth.

33. $\log x = 2.5$
34. $\log x = 3.2$
35. $\ln x = 2$
36. $\ln x = 4$
37. $\log x = 0.35$
38. $\log x = 0.127$
39. $\ln x = \frac{8}{3}$
40. $\ln x = \frac{1}{2}$

In Exercises 41–46, graph the function.

41. $g(x) = \log_2 x$
42. $g(x) = \log_4 x$
43. $f(x) = \log_3(2x - 1)$
44. $f(x) = -\log_2 x$
45. $f(x) = \log_2(x - 1)$
46. $f(x) = \log_3(x - 2)$

Light The percent of light that will pass through a material is given by the formula $\log P = -kd$, where P is the percent of light passing through the material, k is a constant that depends on the material, and d is the thickness of the material in centimeters. Use this formula for Exercises 47 and 48.

47. The constant k for a piece of opaque glass that is 0.5 centimeters thick is 0.2. Find the percent of light that will pass through the glass. Round to the nearest percent.
48. The constant k for a piece of tinted glass is 0.5. How thick is a piece of this glass that allows 60% of the light incident to the glass to pass through it? Round to the nearest hundredth of a centimeter.

Sound The number of decibels, D, of a sound can be given by the equation $D = 10(\log I + 16)$, where I is the power of the sound measured in watts. Use this formula for Exercises 49 and 50. Round to the nearest decibel.

49. Find the number of decibels of normal conversation. The power of the sound of normal conversation is approximately 3.2×10^{-10} watts.
50. The loudest sound made by an animal is made by the blue whale and can be heard over 500 miles away. The power of the sound is 630 watts. Find the number of decibels of the sound emitted by the blue whale.

pH of a Solution For Exercises 51 and 52, use the equation

$$\text{pH} = -\log(\text{H}^+),$$

where H^+ is the hydronium-ion concentration of a solution. Round to the nearest hundredth.

51. Find the pH of the digestive solution of the stomach, for which the hydronium-ion concentration is 0.045 mole per liter.
52. Find the pH of a morphine solution used to relieve pain, for which the hydronium-ion concentration is 3.2×10^{-10} mole per liter.

Earthquakes For Exercises 53–57, use the Richter scale equation $M = \log \frac{I}{I_0}$, where M is the magnitude of an earthquake, I is the intensity of the shock waves, and I_0 is the measure of the intensity of a zero-level earthquake.

53. On May 21, 2003, an earthquake struck Northern Algeria. The earthquake had an intensity of $I = 6{,}309{,}573I_0$. Find the Richter scale magnitude of the earthquake. Round to the nearest tenth.

54. The earthquake on November 17, 2003, in the Aleutian Islands of Moska had an intensity of $I = 63{,}095{,}734I_0$. Find the Richter scale magnitude of the earthquake. Round to the nearest tenth.

55. An earthquake in Japan on March 2, 1933 measured 8.9 on the Richter scale. Find the intensity of the earthquake in terms of I_0. Round to the nearest whole number.

56. An earthquake that occurred in China in 1978 measured 8.2 on the Richter scale. Find the intensity of the earthquake in terms of I_0. Round to the nearest whole number.

57. How many times as strong is an earthquake whose magnitude is 8 than one whose magnitude is 6?

Astronomy Astronomers use the distance modulus function $M(r) = 5 \log r - 5$, where M is the distance modulus and r is the distance of a star from Earth in parsecs. (One parsec is approximately 1.92×10^{13} miles, or approximately 20 trillion miles.) Use this function for Exercises 58 and 59. Round to the nearest tenth.

58. The distance modulus of the star Betelgeuse is 5.89. How many parsecs from Earth is this star?

59. The distance modulus of Alpha Centauri is -1.11. How many parsecs from Earth is this star?

World's Oil Supply One model for the time it will take for the world's oil supply to be depleted is given by the function $T(r) = 14.29 \ln(0.00411r + 1)$, where r is the estimated oil reserves in billions of barrels and T is the time in years before that amount of oil is depleted. Use this function for Exercises 60 and 61. Round to the nearest tenth of a year.

60. How many barrels of oil are necessary to last 20 years?

61. How many barrels of oil are necessary to last 50 years?

Extensions

CRITICAL THINKING

62. As mentioned in this section, the main motivation for the development of logarithms was to aid astronomers and other scientists with arithmetic calculations. The idea was to allow scientists to multiply large numbers by adding the logarithms of the numbers. Dividing large numbers was accomplished by subtracting the logarithms of the numbers. Follow through the exercises below to see how this was accomplished. For simplicity, we will use small numbers (2 and 3) to illustrate the procedure.

a. Write each of the equations $\log 2 = 0.30103$ and $\log 3 = 0.47712$ in exponential form.

b. Replace 2 and 3 in $x = 2 \cdot 3$ with the exponential expressions from part a. (In this exercise, we know that x is 6. However, if the two numbers were very large, the value of x would not be obvious.)

c. Simplify the exponential expression. Recall that to multiply two exponential expressions with the same base, *add* the exponents.

d. If you completed part c correctly, you should have

$$x = 10^{0.77815}$$

Write this expression in logarithmic form.

e. Using a calculator, verify that the solution of the equation you created in part d is 6. *Note:* When tables of logarithms were used, a scientist would have looked through the table to find 0.77815 and observed that it was the logarithm of 6.

63. Replace the denominator 2 and the numerator 3 in $x = \frac{3}{2}$ by the exponential expressions from part a of Exercise 62. Simplify the expression by *subtracting* the exponents. Write the answer in logarithmic form and verify that $x = 1.5$.

64. Write a few sentences explaining how adding the logarithms of two numbers can be used to find the product of the two numbers.

65. Write a few sentences explaining how subtracting the logarithms of two numbers can be used to find the quotient of the two numbers.

EXPLORATIONS

Earthquakes Seismologists generally determine the Richter scale magnitude of an earthquake by examining a *seismogram,* an example of which is shown below.

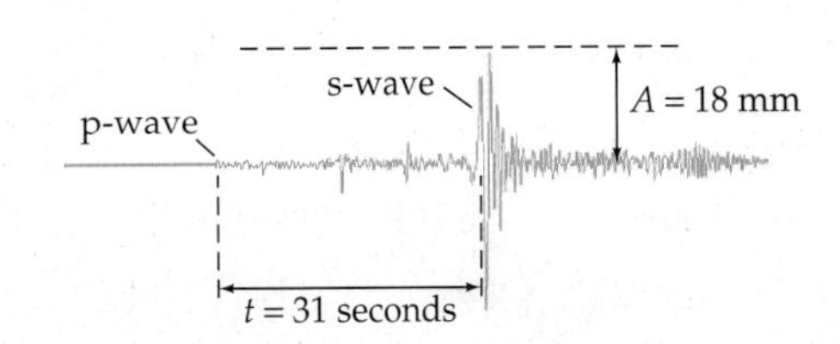

The magnitude of an earthquake cannot be determined just by examining the amplitude of a seismogram, because this amplitude decreases as the distance between the epicenter of the earthquake and the observation station increases. To account for the distance between the epicenter and the observation station, a seismologist examines a seismogram for both small waves, called *p-waves,* and larger waves, called *s-waves.* The Richter scale magnitude M of the earthquake is a function of both the amplitude A of the s-waves and the time t between the occurrence of the s-waves and the occurrence of the p-waves. In the 1950's, Charles Richter developed the formula on the right to determine the magnitude M of an earthquake from the data in a seismogram.

The Amplitude-Time-Difference Formula

The Richter scale magnitude of an earthquake is given by

$$M = \log A + 3 \log 8t - 2.92$$

where A is the amplitude, in millimeters, of the s-waves on a seismogram and t is the time, in seconds, between the s-waves and the p-waves.

66. Find the Richter scale magnitude of the earthquake that produced the seismogram shown on page 401. Round to the nearest tenth.

67. Find the Richter scale magnitude of the earthquake that produced the seismograph shown on the right. Round to the nearest tenth.

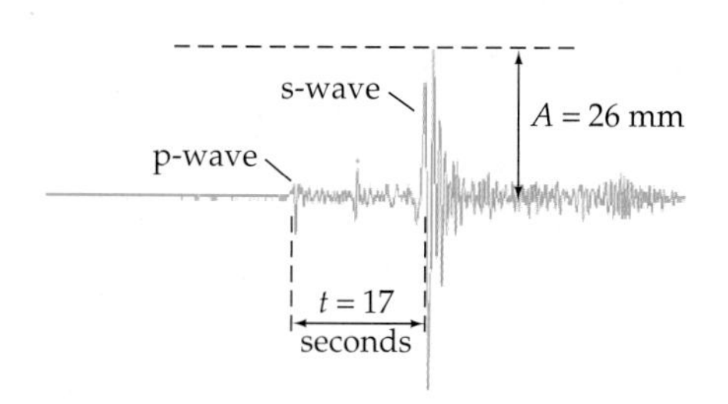

CHAPTER 6 Summary

Key Terms

abscissa [p. 329]
acid [p. 398]
acidity and alkalinity [p. 397]
axis of symmetry [p. 365]
base [p. 398]
coefficient of determination [p. 359]
common logarithm [p. 392]
common logarithmic function [p. 392]
coordinate axes [p. 328]
coordinates of a point [p. 329]
correlation coefficient [p. 359]
dependent variable [p. 333]
domain [p. 332]
e [p. 380]
equation in two variables [p. 329]
evaluating a function [p. 333]
exponential decay function [p. 377]
exponential function [p. 377]
exponential growth function [p. 377]
first coordinate [p. 329]
function [p. 332]
functional notation [p. 333]
graph a point [p. 329]
graph of an equation in two variables [p. 330]
graph of an ordered pair [p. 329]
independent variable [p. 332]
interval notation [p. 354]
linear function [p. 340]
logarithm [p. 390]
logarithmic function [p. 390]
maximum of a quadratic function [p. 368]
minimum of a quadratic function [p. 368]
natural exponential function [p. 381]
natural logarithm [p. 392]
natural logarithmic function [p. 392]
ordered pair [p. 329]
ordinate [p. 329]
origin [p. 328]
parabola [p. 364]
plot a point [p. 329]
quadrants [p. 328]
quadratic function [p. 364]
range [p. 332]
rectangular coordinate system [p. 328]
regression line [p. 358]
scatter diagram [p. 357]
second coordinate [p. 329]
slope [p. 344]
solution of an equation in two variables [p. 329]
value of a function [p. 333]
vertex [p. 365]
x-coordinate [p. 329]
x-intercept [pp. 341 and 366]
xy-plane [p. 328]
y-coordinate [p. 329]
y-intercept [p. 341]
zero-level earthquake [p. 395]

Essential Concepts

- **Functional Notation**
  The symbol $f(x)$ is the value of the function and represents the value of the dependent variable for a given value of the independent variable.
- **Linear Function**
  A linear function is one that can be written in the form $f(x) = mx + b$, where m is the coefficient of x and b is a constant. The slope of the graph of the function is m, the coefficient of x.
- **y-Intercept of a Linear Function**
  The y-intercept of the graph of $f(x) = mx + b$ is $(0, b)$.
- **Slope of a Line**
  Let (x_1, y_1) and (x_2, y_2) be two points on a line. Then the slope of the line through the two points is the ratio of the change in the y-coordinates to the change in the x-coordinates.
  $$m = \frac{\text{change in } y}{\text{change in } x} = \frac{y_2 - y_1}{x_2 - x_1}, x_1 \neq x_2$$
- **Slope–Intercept Form of the Equation of a Line**
  The graph of $f(x) = mx + b$ is a straight line with slope m and y-intercept $(0, b)$.
- **Point–Slope Formula of a Straight Line**
  Let (x_1, y_1) be a point on a line and let m be the slope of the line. Then the equation of the line can be found using the point–slope formula
  $y - y_1 = m(x - x_1)$.
- **Vertex of a Parabola**
  Let $f(x) = ax^2 + bx + c, a \neq 0$, be the equation of a parabola. Then the coordinates of the vertex are
  $\left(-\frac{b}{2a}, f\left(-\frac{b}{2a}\right)\right)$.
- **Definition of Logarithm**
  For $b > 0, b \neq 1, y = \log_b x$ is equivalent to $x = b^y$.
- **Equality of Exponents Property**
  If $b > 0$ and $b^x = b^y$, then $x = y$.
- **Richter Scale**
  An earthquake with an intensity of I has a Richter scale magnitude of $M = \log\left(\frac{I}{I_0}\right)$, where I_0 is the measure of the intensity of a zero-level earthquake.
- **The pH of a Solution**
  The pH of a solution with a hydronium-ion concentration of H^+ moles per liter is given by $\text{pH} = -\log[H^+]$.

CHAPTER 6 Review Exercises

1. Draw a line through all points with an x-coordinate of 4.
2. Draw a line through all points with a y-coordinate of 3.
3. Graph the ordered-pair solutions of $y = 2x^2$ when $x = -2, -1, 0, 1,$ and 2.
4. Graph the ordered-pair solutions of $y = 2x^2 - 5$ when $x = -2, -1, 0, 1,$ and 2.
5. Graph the ordered-pair solutions of $y = -2x + 1$ when $x = -2, -1, 0, 1,$ and 2.
6. Graph the ordered-pair solutions of $y = |x + 1|$ when $x = -5, -3, 0, 3,$ and 5.

In Exercises 7–16, graph the equation.

7. $y = -2x + 1$
8. $y = 3x + 2$
9. $f(x) = x^2 + 2$
10. $f(x) = x^2 - 3x + 1$
11. $y = |x + 4|$
12. $f(x) = 2|x| - 1$
13. $f(x) = 2^x - 3$
14. $f(x) = 3^{-x+2}$
15. $f(x) = e^{0.5x}$
16. $f(x) = \log_2(x - 2)$

In Exercises 17–23, evaluate the function for the given value.

17. $f(x) = 4x - 5; x = -2$
18. $g(x) = 2x^2 - x - 2; x = 3$
19. $s(t) = \frac{4}{3t - 5}; t = -1$
20. $R(s) = s^3 - 2s^2 + s - 3; s = -2$
21. $f(x) = 2^{x-3}; x = 5$
22. $g(x) = \left(\frac{2}{3}\right)^x; x = 2$
23. $T(r) = 2e^r + 1;\ r = 2$. Round to the nearest hundredth.

24. **Geometry** The volume of a sphere is a function of its radius and is given by $V(r) = \frac{4\pi r^3}{3}$, where r is the radius of the sphere.
   a. Find the volume of a sphere whose radius is 3 inches. Round to the nearest tenth of a cubic inch.
   b. Find the volume of a sphere whose radius is 12 centimeters. Round to the nearest tenth of a cubic centimeter.

25. **Sports** The height h of a ball that is released 5 feet above the ground with an initial velocity of 96 feet per second is a function of the time t, in seconds, the ball is in the air and is given by $h(t) = -16t^2 + 96t + 5$.
   a. Find the height of the ball above the ground 2 seconds after it is released.
   b. Find the height of the ball above the ground 4 seconds after it is released.

26. **Mixtures** The percent concentration P of sugar in a water solution depends on the amount of sugar that is added to the solution and is given by $P(x) = \frac{100x + 100}{x + 20}$, where x is the number of grams of sugar that are added.
   a. What is the original percent concentration of sugar?
   b. What is the percent concentration of sugar after 10 more grams of sugar are added? Round to the nearest tenth of a percent.

In Exercises 27–30, find the x- and y-intercepts of the graph of the function.

27. $f(x) = 2x + 10$
28. $f(x) = \frac{3}{4}x - 9$
29. $3x - 5y = 15$
30. $4x + 3y = 24$

31. **Depreciation** The accountant for a small business uses the model $V(t) = 25{,}000 - 5000t$ to approximate the value, $V(t)$, of a small truck t years after its purchase. Find and discuss the meaning of the intercepts, on the vertical and horizontal axes, in the context of this application.

In Exercises 32–35, find the slope of the line containing the given points.

32. $(3, 2), (2, -3)$
33. $(-1, 4), (-3, -1)$
34. $(2, -5), (-4, -5)$
35. $(5, 2), (5, 7)$

36. 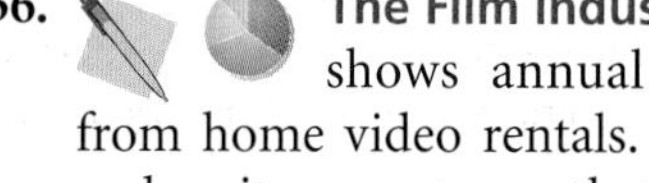 **The Film Industry** The following graph shows annual projections for revenue from home video rentals. Find the slope of the line and write a sentence that explains the meaning of the slope in the context of this application. (*Source:* Forrester Research)

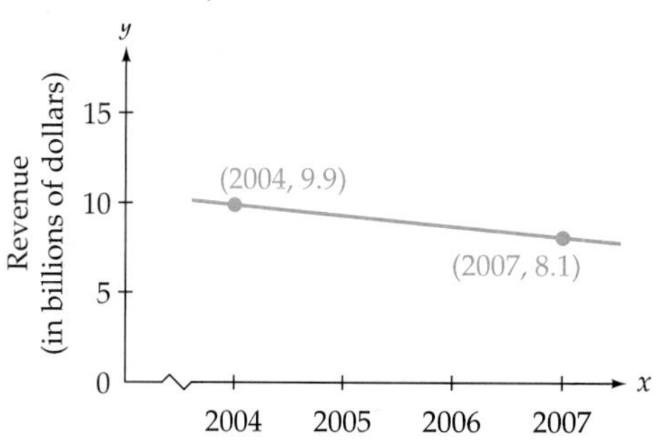

37. Graph the line that passes through the point $(3, -2)$ and has slope -2.

38. Graph the line that passes through the point $(-1, -3)$ and has slope $\frac{3}{4}$.

In Exercises 39–42, find the equation of the line that contains the given point and has the given slope.

39. $(-2, 3), m = 2$
40. $(1, -4), m = 1$
41. $(-3, 1), m = \frac{2}{3}$
42. $(4, 1), m = -\frac{1}{4}$

43. Graph $f(x) = \frac{3}{2}x - 1$ using the slope and y-intercept.

44. **Interior Decorating** A dentist's office is being recarpeted. The cost to install the new carpet is \$100 plus \$25 per square yard of carpeting.
   a. Determine a linear function for the cost to carpet the office.
   b. Use this function to determine the cost to carpet 32 square yards of floor space.

45. **Gasoline Sales** The manager of Valley Gas Mart has determined that 10,000 gallons of regular unleaded gasoline can be sold each week if the price is the same as that of Western QuickMart, a gas station across the street. If the manager increases the price \$.02 above Western QuickMart's price, the manager will sell 500 less gallons per week. If the manager decreases the price \$.02 below that of Western QuickMart, 500 more gallons of gasoline per week will be sold.
   a. Determine a linear function that will predict the number of gallons of gas per week that Valley Gas Mart can sell as a function of the price relative to that of Western QuickMart.
   b. Use the model to predict the number of gallons of gasoline Valley Gas Mart will sell if its price is \$.03 below that of Western QuickMart.

46. **BMI** The body mass index (BMI) of a person is a measure of the person's ideal body weight. The data in the table below show the BMI for different weights for a person 5′6″ tall.

BMI Data for Person 5'6" Tall			
Weight (in pounds)	BMI	Weight (in pounds)	BMI
110	18	160	26
120	19	170	27
125	20	180	29
135	22	190	31
140	23	200	32
145	23	205	33
150	24	215	35

Source: Centers for Disease Control and Prevention

a. Find the linear regression equation for this data. Round to the nearest hundred thousandth.

b. Use the regression equation to estimate the BMI for a person 5′6″ tall whose weight is 158 pounds. Round to the nearest whole number.

In Exercises 47–50, find the vertex of the graph of the function.

47. $y = x^2 + 2x + 4$

48. $y = -2x^2 - 6x + 1$

49. $f(x) = -3x^2 + 6x - 1$

50. $f(x) = x^2 + 5x - 1$

In Exercises 51–54, find the x-intercepts of the parabola given by the equation.

51. $y = x^2 + x - 20$

52. $y = x^2 + 2x - 1$

53. $f(x) = 2x^2 + 9x + 4$

54. $f(x) = x^2 + 4x + 6$

In Exercises 55–58, find the minimum or maximum value of the quadratic function. State whether the value is a maximum or a minimum.

55. $y = -x^2 + 4x + 1$

56. $y = 2x^2 + 6x - 3$

57. $f(x) = x^2 - 4x - 1$

58. $f(x) = -2x^2 + 3x - 1$

59. **Physics** The height s, in feet, of a rock thrown upward at an initial speed of 80 feet per second from a cliff 25 feet above an ocean beach is given by the function $s(t) = -16t^2 + 80t + 25$, where t is the time in seconds after the rock is released. Find the maximum height above the beach that the rock will attain.

60. **Manufacturing** A manufacturer of rewritable CDs (CD-RW) estimates that the average daily cost C of producing a single CD-RW is given by $C(x) = 0.01x^2 - 40x + 50{,}000$, where x is the number of CD-RWs produced each day. Find the number of CD-RWs the company should produce in order to minimize the average daily cost.

61. **Investments** A \$5000 certificate of deposit (CD) earns 6% annual interest compounded daily. What is the value of the investment after 15 years? Use the compound interest formula $A = P\left(1 + \frac{r}{n}\right)^{nt}$, where P is the amount deposited, A is the value of the money after t years, r is the annual interest rate as a decimal, and n is the number of compounding periods per year.

62. **Isotopes** An isotope of technetium has a half-life of approximately 6 hours. If a patient is injected with 10 milligrams of this isotope, what will be the technetium level in the patient after 2 hours? Use the function $A(t) = 10\left(\frac{1}{2}\right)^{t/6}$, where A is the technetium level, in milligrams, in the patient after t hours. Round to the nearest hundredth of a milligram.

63. **Isotopes** Iodine-131 has a half-life of approximately 8 days. If a patient is given an injection that contains 8 micrograms of iodine-131, what will be the iodine level in the patient after 10 days? Use the function $A(t) = 8\left(\frac{1}{2}\right)^{t/8}$, where A is the amount of iodine-131, in micrograms, in the patient after t days. Round to the nearest hundredth of a milligram.

64. **Golf** A golf ball is dropped from a height of 6 feet. On each successive bounce, the ball rebounds to a height that is $\frac{2}{3}$ of the previous height.

a. Find an exponential model for the height of the ball after the nth bounce.

b. What is the height of the ball after the fifth bounce? Round to the nearest hundredth of a foot.

65. **The Film Industry** When a new movie is released, there is initially a surge in the number of people who go to see the movie. After 2 weeks, attendance generally begins to drop off. The data in the following table show the number of people attending a certain movie; $t = 0$ represents the number of patrons 2 weeks after the movie's initial release.

Number of weeks two weeks after release, t	0	1	2	3	4
Number of people (in thousands), N	250	162	110	65	46

a. Find an exponential regression equation for these data. Round to the nearest ten-thousandth.

b. Use the equation to predict the number of people who will attend the movie 8 weeks after it has been released.

In Exercises 66–69, evaluate the logarithm.

66. $\log_3 243$

67. $\log_2 \frac{1}{16}$

68. $\log_4 \frac{1}{4}$

69. $\log_2 64$

In Exercises 70–73, solve for x. Round to the nearest ten-thousandth.

70. $\log_4 x = 3$

71. $\log_3 x = \frac{1}{3}$

72. $\ln x = 2.5$

73. $\log x = 2.4$

74. Astronomy Use the distance modulus function $M(r) = 5 \log r - 5$, where M is the distance modulus and r is the distance of a star from Earth in parsecs, to find the distance in parsecs to a star whose distance modulus is 3.2. Round to the nearest tenth of a parsec.

75. Sound The number of decibels, D, of a sound can be given by the function $D(I) = 10(\log I + 16)$, where I is the power of the sound measured in watts. The pain threshold for a sound for most humans is approximately 0.01 watt. Find the number of decibels of this sound.

CHAPTER 6 Test

In Exercises 1 and 2, evaluate the function for the given value of the independent variable.

1. $s(t) = -3t^2 + 4t - 1$; $t = -2$

2. $f(x) = 3^{x-4}$; $x = 2$

3. Evaluate: $\log_5 125$

4. Solve for x: $\log_6 x = 2$

In Exercises 5–8, graph the function.

5. $f(x) = 2x - 3$

6. $f(x) = x^2 + 2x - 3$

7. $f(x) = 2^x - 5$

8. $f(x) = \log_3(x - 1)$

9. Find the slope of the line that passes through $(3, -1)$ and $(-2, -4)$.

10. Find the equation of the line that passes through $(3, 5)$ and has slope $\frac{2}{3}$.

11. Find the vertex of the graph of $f(x) = x^2 + 6x - 1$.

12. Find the x-intercepts of the parabola given by the equation $y = x^2 + 2x - 8$.

13. Find the minimum or maximum value of the quadratic function $y = -x^2 - 3x + 10$ and state whether the value is a minimum or a maximum.

14. Travel The distance d, in miles, a small plane is from its final destination is given by $d(t) = 250 - 100t$, where t is the time, in hours, remaining for the flight. Find and discuss the meaning of the intercepts of the graph of the function.

15. Sports The height h, in feet, of a ball that is thrown straight up and released 4 feet above the ground is given by $h(t) = -16t^2 + 96t + 4$, where t is the time in seconds. Find the maximum height the ball attains.

16. Isotopes A radioactive isotope has a half-life of 5 hours. If a chemist has a 10-gram sample of this isotope, what amount will remain after 8 hours? Use the function $A(t) = 10\left(\frac{1}{2}\right)^{t/5}$, where A is the amount of the isotope, in grams, remaining after t hours. Round to the nearest hundredth of a gram.

17. Earthquakes Two earthquakes struck Papua, Indonesia, in February 2004. One had a magnitude of 7.0 on the Richter scale, and the second one had a magnitude of 7.3 on the Richter scale. How many times as great was the intensity of the second earthquake than that of the first? Round to the nearest tenth.

18. Farming The manager of an orange grove has determined that when there are 320 trees per acre, the average yield per tree is 260 pounds of oranges. If the number of trees is increased to 330 trees per acre, the average yield per tree decreases to 245 pounds. Find a linear model for the average yield per tree as a function of the number of trees per acre.

19. Farming Giant pumpkin contests are popular at many state fairs. Suppose the data below show the weights, in pounds, of the winning pumpkins at recent fairs, where $x = 0$ represents 2000.

Year, x	0	1	2	3	4
Weight (in pounds), y	650	715	735	780	820

a. Find a linear regression equation for these data.

b. Use the regression equation to predict the winning weight in 2008. Round to the nearest pound.

20. Light The equation $\log P = -kd$ gives the relationship between the percent P, as a decimal, of light passing through a substance of thickness d. The value of k for a swimming pool is approximately 0.05. At what depth, in meters, will the percent of light be 75% of the light at the surface of the pool? Round to the nearest tenth.

CHAPTER

8 Geometry

Look closely at each of the four rectangles shown below. Which rectangle seems to be the most pleasant to look at?

Rectangle A

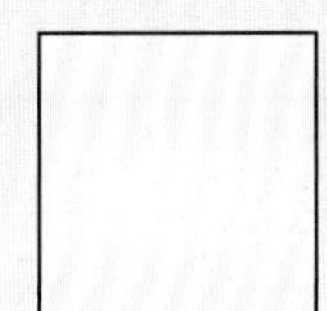

Rectangle B

Rectangle C

Rectangle D

The length of rectangle A is three times its width; the ratio of the length to the width is 3:1. Rectangle B is a square, so its length is equal to its width; the ratio of the length to the width is 1:1. The ratio of the length of rectangle C to its width is 1.2:1. The shape of rectangle D is referred to as a *golden rectangle*. In a **golden rectangle,** the ratio of the length to the width is approximately 1.6 to 1. The ratio is exactly

$$\frac{\text{length}}{\text{width}} = \frac{1 + \sqrt{5}}{2}$$

The early Greeks considered golden rectangles to be the most pleasing to the eye. Therefore, they used it extensively in their art and architecture. The Parthenon in Athens, Greece, exhibits many golden rectangles. It was built around 435 B.C. as a temple to the goddess Athena.

The World Wide Web is a wonderful source of information about golden rectangles. Simply enter "golden rectangle" in a search engine. It will respond with thousands of websites where you can learn more about the use of golden rectangles in art, architecture, music, nature, and mathematics. Be sure to research Leonardo da Vinci's use of the golden rectangle in his painting *Mona Lisa*.

Mona Lisa by Leonardo da Vinci

For online student resources, visit this textbook's website at **math.college.hmco.com/students.**

SECTION 8.1 Basic Concepts of Euclidean Geometry

Lines and Angles

historical note

Geometry is one of the oldest branches of mathematics. Around 350 B.C., Euclid (yo͞o′klĭd) of Alexandria wrote *Elements,* which contained all of the known concepts of geometry. Euclid's contribution to geometry was to unify various concepts into a single deductive system that was based on a set of postulates. ■

The word *geometry* comes from the Greek words for "earth" and "measure". In ancient Egypt, geometry was used by the Egyptians to measure land and to build structures such as the pyramids.

Today geometry is used in many fields, such as physics, medicine, and geology. Geometry is also used in applied fields such as mechanical drawing and astronomy. Geometric forms are used in art and design.

If you play a musical instrument, you know the meaning of the words *measure, rest, whole note,* and *time signature.* If you are a football fan, you have learned the terms *first down, sack, punt,* and *touchback.* Every field has its associated vocabulary. Geometry is no exception. We will begin by introducing two basic geometric concepts: point and line.

A **point** is symbolized by drawing a dot. A **line** is determined by two distinct points and extends indefinitely in both directions, as the arrows on the line at the right indicate. This line contains points A and B and is represented by $\overleftrightarrow{AB}$. A line can also be represented by a single letter, such as ℓ.

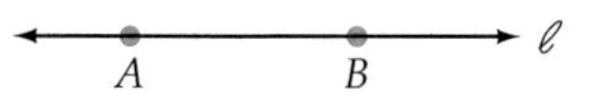

A **ray** starts at a point and extends indefinitely in *one* direction. The point at which a ray starts is called the **endpoint** of the ray. The ray shown at the right is denoted $\overrightarrow{AB}$. Point A is the endpoint of the ray.

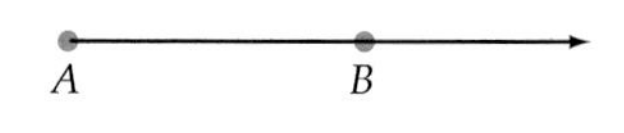

A **line segment** is part of a line and has two endpoints. The line segment shown at the right is denoted by $\overline{AB}$.

QUESTION *Classify each diagram as a line, a ray, or a line segment.*

a.

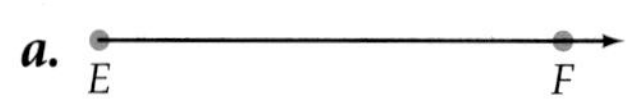

b. C ——— D

c. ←— J ——— K —→

The distance between the endpoints of $\overline{AC}$ is denoted by AC. If B is a point on $\overline{AC}$, then AC (the distance from A to C) is the sum of AB (the distance from A to B) and BC (the distance from B to C).

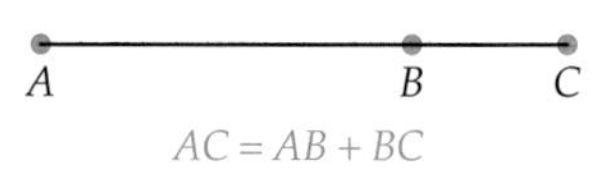

ANSWER **a.** *Ray* **b.** *Line segment* **c.** *Line*

Given $AB = 22$ cm and $BC = 13$ cm, find AC.

Make a drawing and write an equation to represent the distances between the points on the line segment.

22 cm 13 cm
A B C

$$AB + BC = AC$$

Substitute the given distances for AB and BC into the equation. Solve for AC.

$$22 + 13 = AC$$
$$35 = AC$$

$AC = 35$ cm

EXAMPLE 1 ■ Find a Distance on a Line Segment

Given $MN = 14$ mm, $NO = 17$ mm, and $OP = 15$ mm on $\overline{MP}$, find MP.

M N O P

Solution

14 m 17 m 15 cm
M N O P

• Make a drawing.

$$MN + NO + OP = MP$$

• Write an equation to represent the distances on the line segment.

$$14 + 17 + 15 = MP$$

• Replace *MN* by **14**, *NO* by **17**, and *OP* by **15**.

$$46 = MP$$

• Solve for *MP*.

$MP = 46$ mm

CHECK YOUR PROGRESS 1 Given $QR = 28$ cm, $RS = 16$ cm, and $ST = 10$ cm on $\overline{QT}$, find QT.

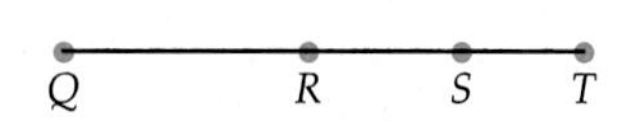

Solution *See page S29.*

EXAMPLE 2 ■ Use an Equation to Find a Distance on a Line Segment

X, Y, and Z are all on line ℓ. Given $XY = 9$ m and YZ is twice XY, find XZ.

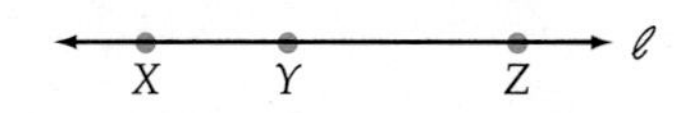

Solution

$$XZ = XY + YZ$$
$$XZ = XY + 2(XY)$$
• *YZ* is twice *XY*.
$$XZ = 9 + 2(9)$$
• Replace *XY* by 9.
$$XZ = 9 + 18$$
• Solve for *XZ*.
$$XZ = 27$$

$XZ = 27$ m

CHECK YOUR PROGRESS 2 *A*, *B*, and *C* are all on line ℓ. Given $BC = 16$ ft and $AB = \frac{1}{4}(BC)$, find *AC*.

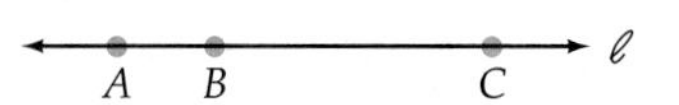

Solution *See page S29.*

In this section, we are discussing figures that lie in a plane. A **plane** is a flat surface with no thickness and no boundaries. It can be pictured as a desktop or blackboard that extends forever. Figures that lie in a plane are called **plane figures.**

Lines in a plane can be intersecting or parallel. **Intersecting lines** cross at a point in the plane.

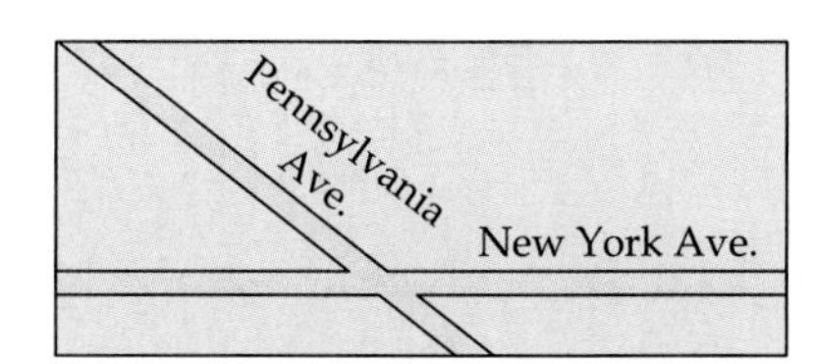

Parallel lines never intersect. The distance between them is always the same.

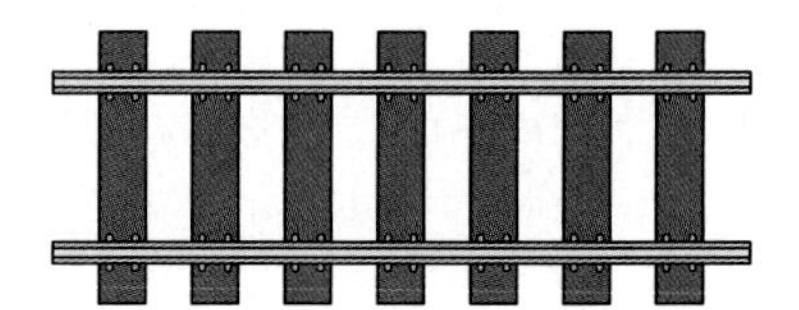

The symbol $\|$ means "is parallel to." In the figure at the right, $j \| k$ and $\overline{AB} \| \overline{CD}$. Note that *j* contains $\overline{AB}$ and *k* contains $\overline{CD}$. Parallel lines contain parallel line segments.

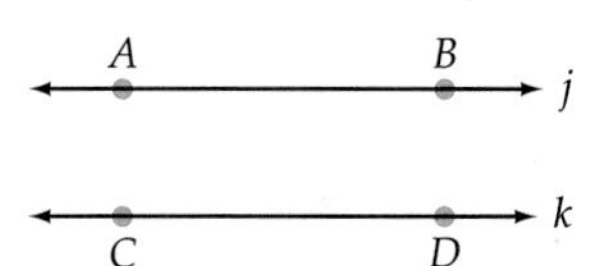

An **angle** is formed by two rays with the same endpoint. The **vertex** of the angle is the point at which the two rays meet. The rays are called the **sides** of the angle.

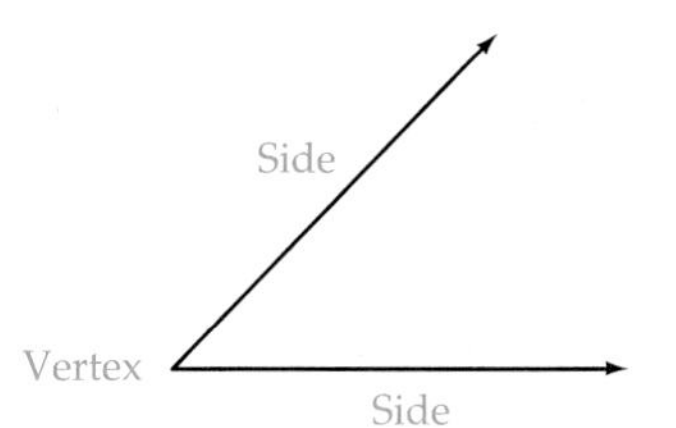

If *A* is a point on one ray of an angle, *C* is a point on the other ray, and *B* is the vertex, then the angle is called $\angle B$ or $\angle ABC$, where $\angle$ is the symbol for angle. Note that an angle can be named by the vertex, or by giving three points, where the second point listed is the vertex. $\angle ABC$ could also be called $\angle CBA$.

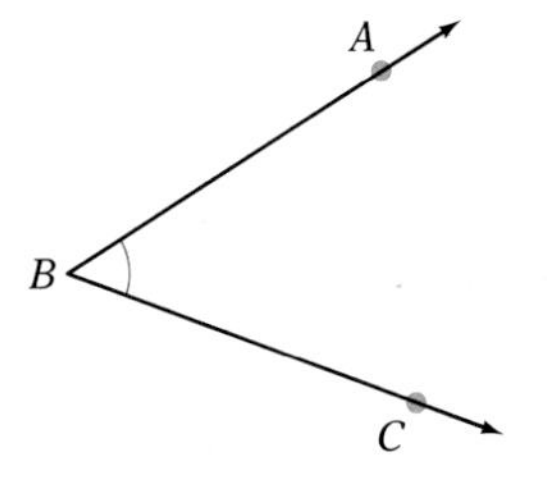

An angle can also be named by a variable written between the rays close to the vertex.

In the figure at the right, $\angle x$ and $\angle QRS$ are two different names for the same angle. $\angle y$ and $\angle SRT$ are two different names for the same angle. Note that in this figure, more than two rays meet at R. In this case, the vertex alone cannot be used to name $\angle QRT$.

historical note

The Babylonians knew that Earth was in approximately the same position in the sky every 365 days. Historians suggest that the reason one complete revolution of a circle is 360° is that 360 is the closest number to 365 that is divisible by many natural numbers. ■

An angle is often measured in **degrees.** The symbol for degrees is a small raised circle, °. The angle formed by rotating a ray through a complete circle has a measure of 360°.

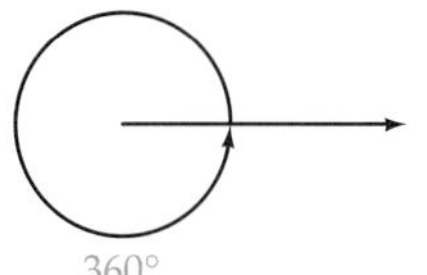

A **protractor** is often used to measure an angle. Place the line segment near the bottom edge of the protractor on BC as shown in the figure below. Make sure the center of the line segment on the protractor is directly over the vertex. $\angle ABC$ shown in the figure below measures 58°.

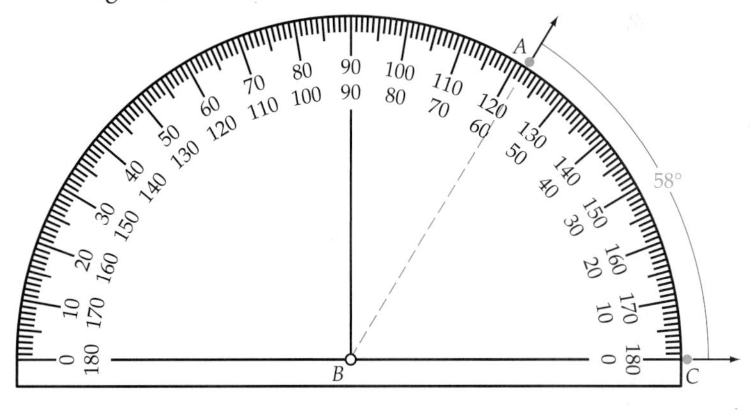

point of interest

The Leaning Tower of Pisa is the bell tower of the Cathedral in Pisa, Italy. Its construction began on August 9, 1173 and continued for about 200 years. The Tower was designed to be vertical, but it started to lean during its construction. By 1350 it was 2.5° off from the vertical; by 1817, it was 5.1° off; and by 1990, it was 5.5° off. In 2001, work on the structure that returned its list to 5° was completed. (*Source: Time* magazine, June 25, 2001, pp. 34–35.)

A 90° angle is called a **right angle.** The symbol ∟ represents a right angle.

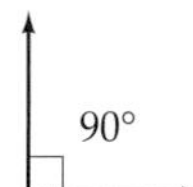

Perpendicular lines are intersecting lines that form right angles.

The symbol $\perp$ means "is perpendicular to." In the figure at the right, $p \perp q$ and $\overline{AB} \perp \overline{CD}$. Note that line p contains $\overline{AB}$ and line q contains $\overline{CD}$. Perpendicular lines contain perpendicular line segments.

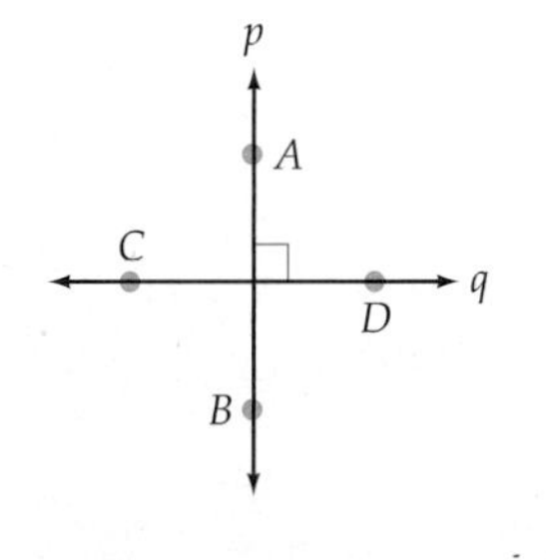

A **straight angle** is an angle whose measure is 180°. $\angle AOB$ is a straight angle.

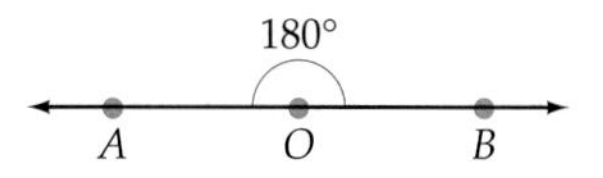

Straight angle

Complementary angles are two angles whose measures have the sum 90°.

$\angle A$ and $\angle B$ at the right are complementary angles.

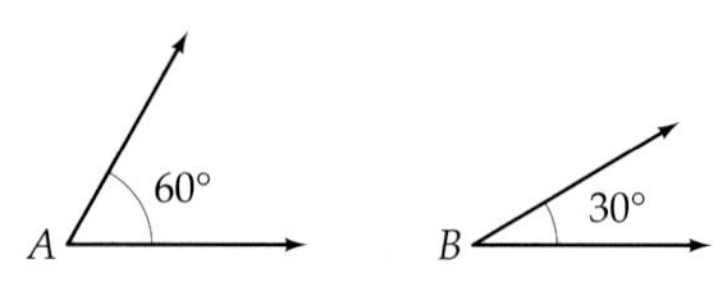

Supplementary angles are two angles whose measures have the sum 180°.

$\angle C$ and $\angle D$ at the right are supplementary angles.

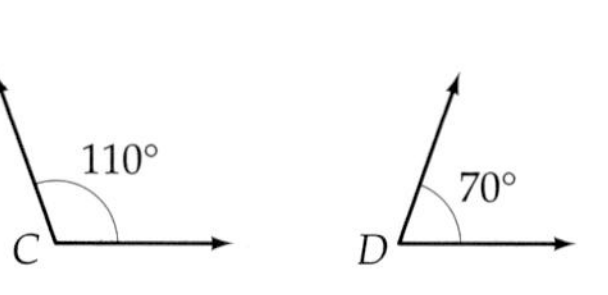

An **acute angle** is an angle whose measure is between 0° and 90°. $\angle D$ above is an acute angle. An **obtuse angle** is an angle whose measure is between 90° and 180°. $\angle C$ above is an obtuse angle.

The measure of $\angle C$ is 110°. This is often written as $m\angle C = 110°$, where m is an abbreviation for "the measure of." This notation is used in Example 3.

EXAMPLE 3 ■ Determine If Two Angles are Complementary

Are angles E and F complementary angles?

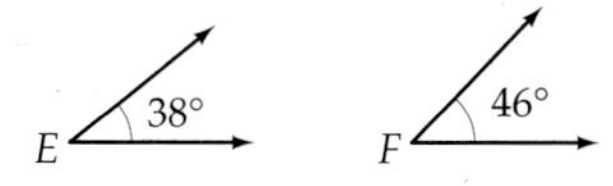

Solution

$$m\angle E + m\angle F = 38° + 46° = 84°$$

The sum of the measures of $\angle E$ and $\angle F$ is not 90°. Therefore, angles E and F are not complementary angles.

CHECK YOUR PROGRESS 3 Are angles G and H supplementary angles?

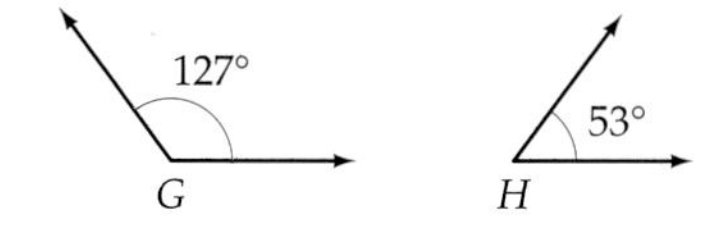

Solution *See page S29.*

EXAMPLE 4 ■ Find the Measure of the Complement of an Angle

Find the measure of the complement of a 38° angle.

Solution

Complementary angles are two angles the sum of whose measures is 90°. To find the measure of the complement, let x represent the complement of a 38° angle. Write an equation and solve for x.

$$x + 38° = 90°$$
$$x = 52°$$

CHECK YOUR PROGRESS 4 Find the measure of the supplement of a 129° angle.

Solution *See page S29.*

Adjacent angles are two angles that have a common vertex and a common side but have no interior points in common. In the figure at the right, $\angle DAC$ and $\angle CAB$ are adjacent angles.

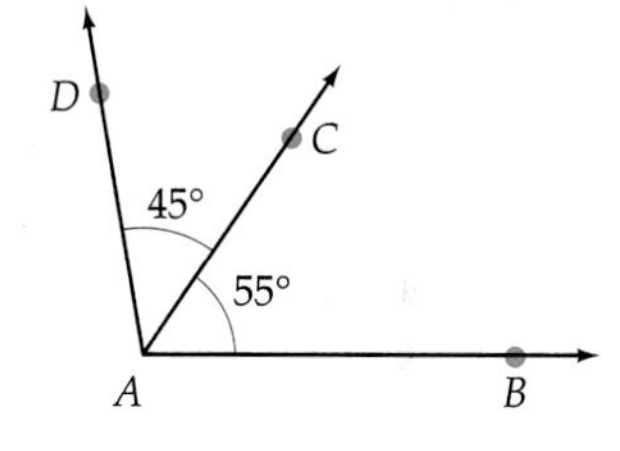

$m\angle DAC = 45°$ and $m\angle CAB = 55°$.

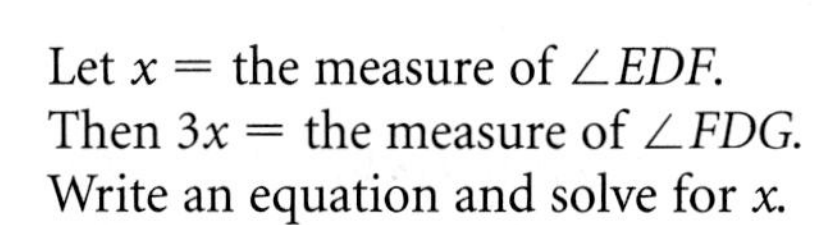

$$m\angle DAB = m\angle DAC + m\angle CAB$$
$$= 45° + 55° = 100°$$

In the figure at the right, $m\angle EDG = 80°$. The measure of $\angle FDG$ is three times the measure of $\angle EDF$. Find the measure of $\angle EDF$.

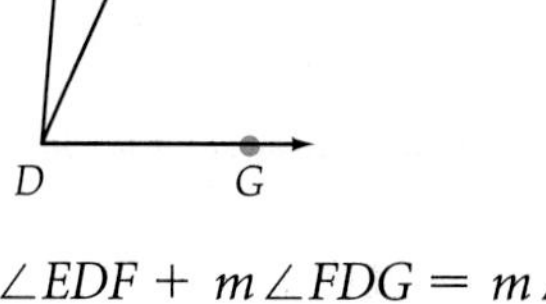

Let $x =$ the measure of $\angle EDF$.
Then $3x =$ the measure of $\angle FDG$.
Write an equation and solve for x.

$$m\angle EDF + m\angle FDG = m\angle EDG$$
$$x + 3x = 80°$$
$$4x = 80°$$
$$x = 20°$$

$$m\angle EDF = 20°$$

EXAMPLE 5 ■ Solve a Problem Involving Adjacent Angles

Find the measure of $\angle x$.

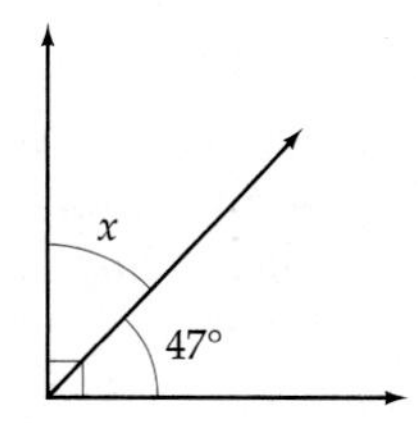

Solution

To find the measure of $\angle x$, write an equation using the fact that the sum of the measures of $\angle x$ and 47° is 90°. Solve for $m\angle x$.

$$m\angle x + 47° = 90°$$
$$m\angle x = 43°$$

CHECK YOUR PROGRESS 5 Find the measure of $\angle a$.

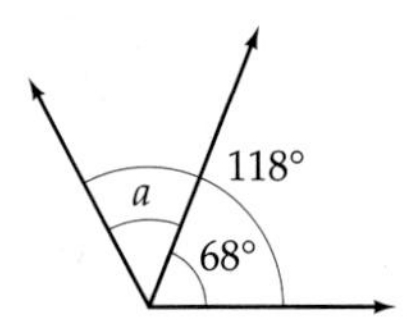

Solution *See page S29.*

Angles Formed by Intersecting Lines

Four angles are formed by the intersection of two lines. If the two lines are perpendicular, each of the four angles is a right angle.

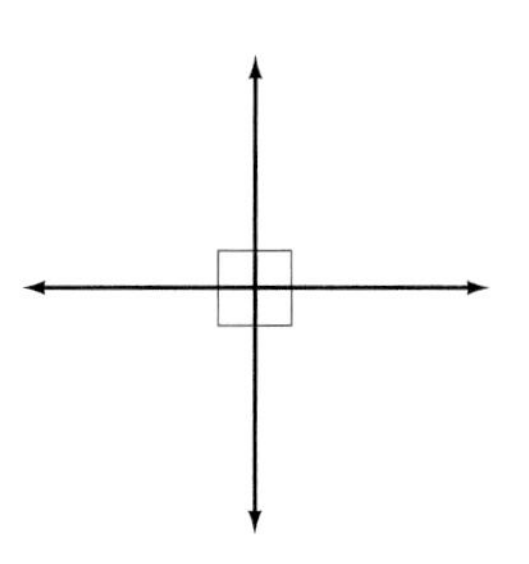

If the two lines are not perpendicular, then two of the angles formed are acute angles and two of the angles formed are obtuse angles. The two acute angles are always opposite each other, and the two obtuse angles are always opposite each other.

point of interest

Many cities in the New World, unlike those in Europe, were designed using rectangular street grids. Washington, D.C. was designed this way, except that diagonal avenues were added, primarily for the purpose of enabling troop movement in the event the city required defense. As an added precaution, monuments were constructed at major intersections so that attackers would not have a straight shot down a boulevard.

In the figure at the right, $\angle w$ and $\angle y$ are acute angles. $\angle x$ and $\angle z$ are obtuse angles.

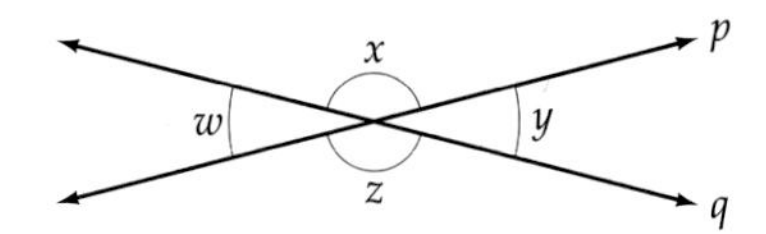

Two angles that are on opposite sides of the intersection of two lines are called **vertical angles.** Vertical angles have the same measure. $\angle w$ and $\angle y$ are vertical angles. $\angle x$ and $\angle z$ are vertical angles.

Vertical angles have the same measure.

$$m\angle w = m\angle y$$
$$m\angle x = m\angle z$$

Recall that two angles that have a common vertex and a common side, but have no interior points in common, are called adjacent angles. For the figure shown above, $\angle x$ and $\angle y$ are adjacent angles, as are $\angle y$ and $\angle z$, $\angle z$ and $\angle w$, and $\angle w$ and $\angle x$. Adjacent angles of intersecting lines are supplementary angles.

Adjacent angles formed by intersecting lines are supplementary angles.

$$m\angle x + m\angle y = 180°$$
$$m\angle y + m\angle z = 180°$$
$$m\angle z + m\angle w = 180°$$
$$m\angle w + m\angle x = 180°$$

EXAMPLE 6 ■ Solve a Problem Involving Intersecting Lines

In the diagram at the right, $m\angle b = 115°$. Find the measures of angles a, c, and d.

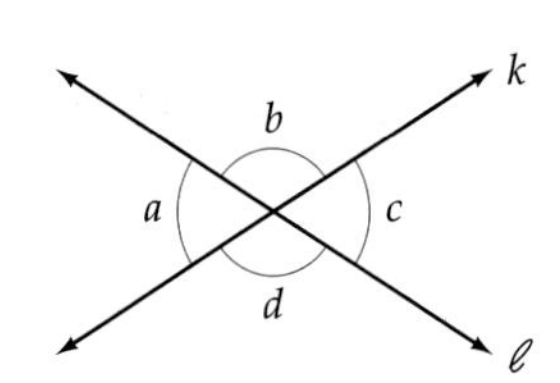

Solution

$m\angle a + m\angle b = 180°$ • $\angle a$ is supplementary to $\angle b$ because $\angle a$ and $\angle b$ are adjacent angles of intersecting lines.

$m\angle a + 115° = 180°$ • Replace $m\angle b$ with **115°**.

$m\angle a = 65°$ • Subtract 115° from each side of the equation.

$m\angle c = 65°$ • $m\angle c = m\angle a$ because $\angle c$ and $\angle a$ are vertical angles.

$m\angle d = 115°$ • $m\angle d = m\angle b$ because $\angle d$ and $\angle b$ are vertical angles.

$m\angle a = 65°$, $m\angle c = 65°$, and $m\angle d = 115°$.

CHECK YOUR PROGRESS 6

In the diagram at the right, $m\angle a = 35°$. Find the measures of angles b, c, and d.

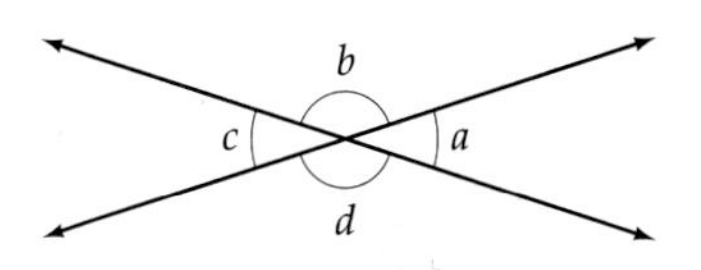

Solution *See page S29.*

A line that intersects two other lines at different points is called a **transversal.**

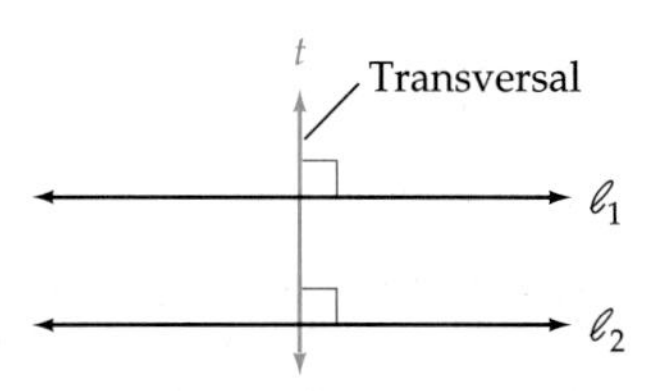

If the lines cut by a transversal t are parallel lines and the transversal is perpendicular to the parallel lines, all eight angles formed are right angles.

If the lines cut by a transversal t are parallel lines and the transversal is *not* perpendicular to the parallel lines, all four acute angles have the same measure and all four obtuse angles have the same measure. For the figure at the right:

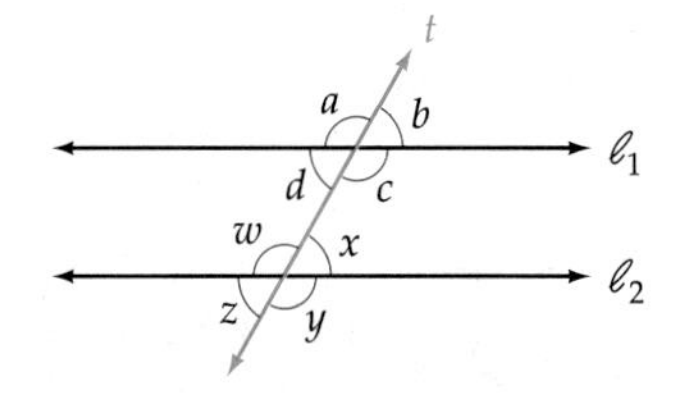

$$m\angle b = m\angle d = m\angle x = m\angle z$$

$$m\angle a = m\angle c = m\angle w = m\angle y$$

If two lines in a plane are cut by a transversal, then any two non-adjacent angles that are on opposite sides of the transversal and between the parallel lines are **alternate interior angles.** In the figure above, $\angle c$ and $\angle w$ are alternate interior angles; $\angle d$ and $\angle x$ are alternate interior angles. Alternate interior angles formed by two parallel lines cut by a transversal have the same measure.

Two alternate interior angles formed by two parallel lines cut by a transversal have the same measure.

$$m\angle c = m\angle w$$

$$m\angle d = m\angle x$$

If two lines in a plane are cut by a transversal, then any two non-adjacent angles that are on opposite sides of the transversal and outside the parallel lines are **alternate exterior angles.** In the figure at the top of the following page,

Two alternate exterior angles formed by two parallel lines cut by a transversal have the same measure.

$m\angle a = m\angle y$

$m\angle b = m\angle z$

$\angle a$ and $\angle y$ are alternate exterior angles; $\angle b$ and $\angle z$ are alternate exterior angles. Alternate exterior angles formed by two parallel lines cut by a transversal have the same measure.

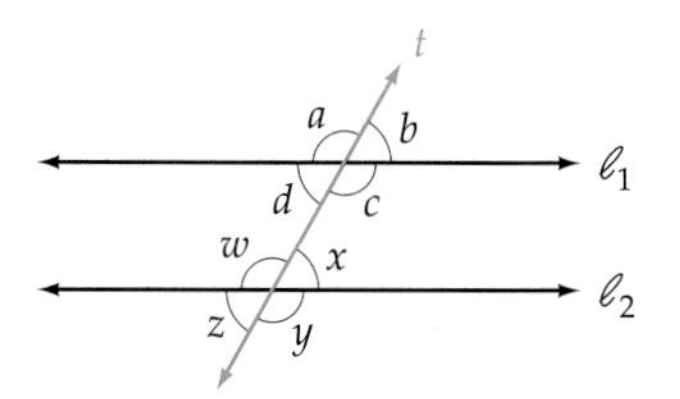

If two lines in a plane are cut by a transversal, then any two angles that are on the same side of the transversal and are both acute angles or both obtuse angles are **corresponding angles.** For the figure at the right above, the following pairs of angles are corresponding angles: $\angle a$ and $\angle w$, $\angle d$ and $\angle z$, $\angle b$ and $\angle x$, $\angle c$ and $\angle y$. Corresponding angles formed by two parallel lines cut by a transversal have the same measure.

Two corresponding angles formed by two parallel lines cut by a transversal have the same measure.

$m\angle a = m\angle w$

$m\angle d = m\angle z$

$m\angle b = m\angle x$

$m\angle c = m\angle y$

QUESTION *Which angles in the diagram at the right above have the same measure as angle a? Which angles have the same measure as angle b?*

EXAMPLE 7 ■ Solve a Problem Involving Parallel Lines Cut by a Transversal

In the diagram at the right, $\ell_1 \| \ell_2$ and $m\angle f = 58°$. Find the measures of $\angle a$, $\angle c$, and $\angle d$.

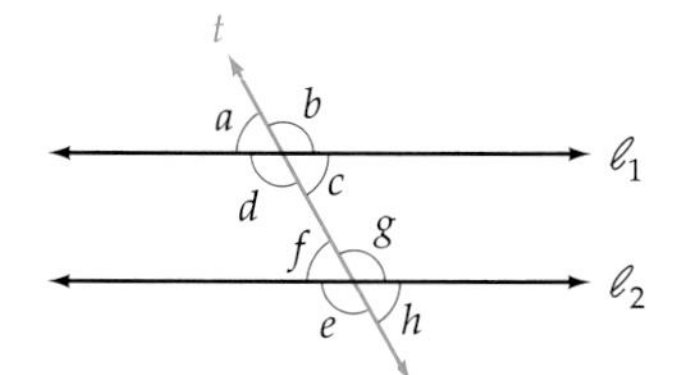

Solution

$m\angle a = m\angle f = 58°$ • $\angle a$ and $\angle f$ are corresponding angles.

$m\angle c = m\angle f = 58°$ • $\angle c$ and $\angle f$ are alternate interior angles.

$m\angle d + m\angle a = 180°$ • $\angle d$ is supplementary to $\angle a$.

$m\angle d + 58° = 180°$ • Replace $m\angle a$ with **58°**.

$m\angle d = 122°$ • Subtract 58° from each side of the equation.

$m\angle a = 58°$, $m\angle c = 58°$, and $m\angle d = 122°$.

ANSWER *Angles c, w, and y have the same measure as angle a. Angles d, x, and z have the same measure as angle b.*

CHECK YOUR PROGRESS 7

In the diagram at the right, $\ell_1 \| \ell_2$ and $m\angle g = 124°$. Find the measures of $\angle b$, $\angle c$, and $\angle d$.

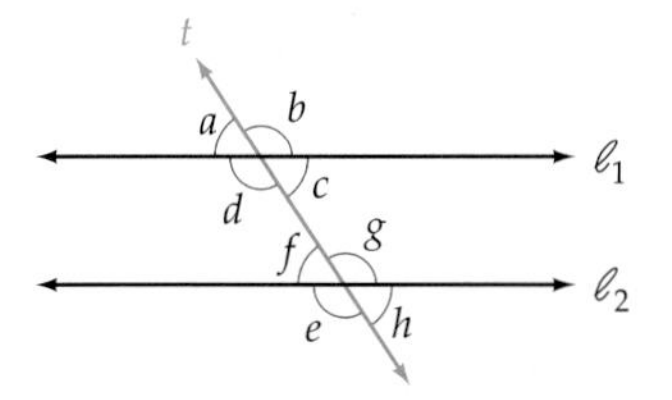

Solution *See page S29.*

Math Matters The Principle of Reflection

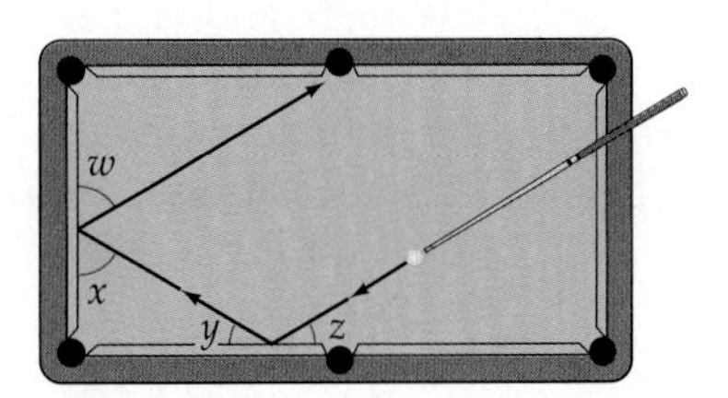

When a ray of light hits a flat surface, such as a mirror, the light is reflected at the same angle at which it hit the surface. For example, in the diagram at the left, $m\angle x = m\angle y$.

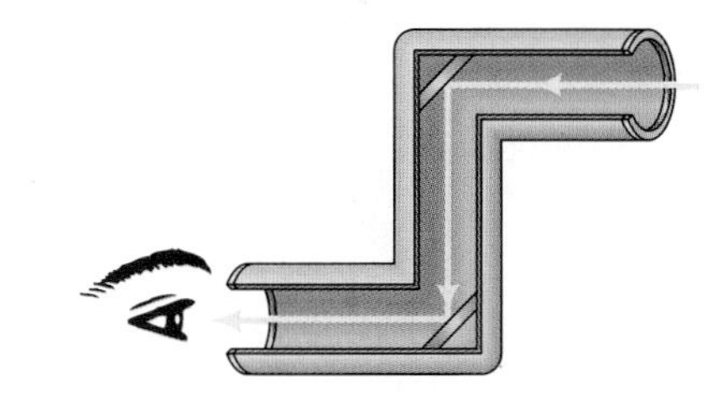

This principle of reflection is in operation in a simple periscope. In a periscope light is reflected twice, with the result that light rays entering the periscope are parallel to the light rays at eye level.

The same principle is in operation on a billiard table. Assuming it has no "side spin," a ball bouncing off the side of the table will bounce off at the same angle at which it hit the side. In the figure below, $m\angle w = m\angle x$ and $m\angle y = m\angle z$.

In the miniature golf shot illustrated below, $m\angle w = m\angle x$ and $m\angle y = m\angle z$.

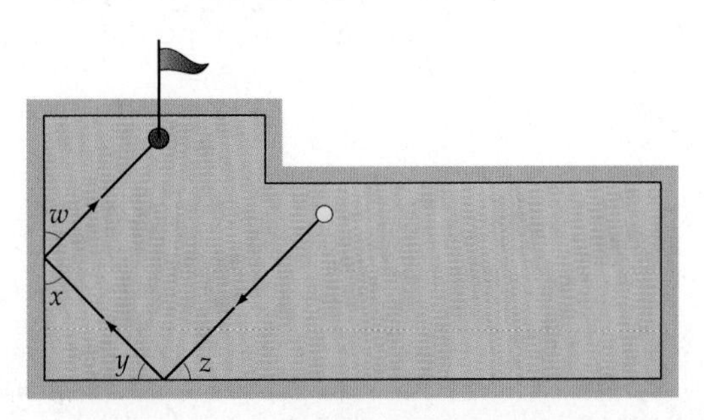

Angles of a Triangle

If the lines cut by a transversal are not parallel lines, the three lines will intersect at three points. In the figure at the right, the transversal t intersects lines p and q. The three lines intersect at points A, B, and C. These three points define three line segments, $\overline{AB}$, $\overline{BC}$, and $\overline{AC}$. The plane figure formed by these three line segments is called a **triangle.**

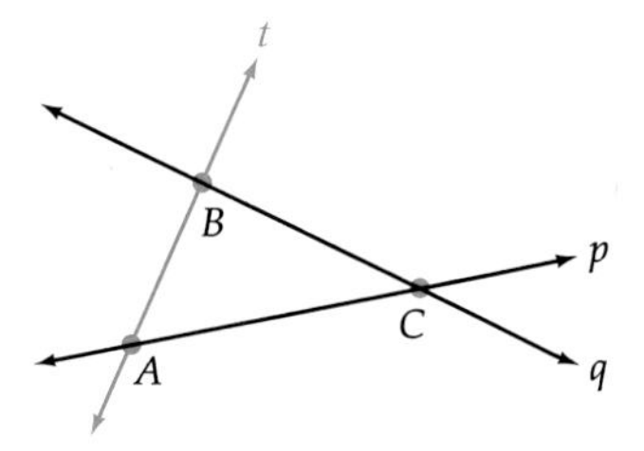

Each of the three points of intersection is the vertex of four angles. The angles within the region enclosed by the triangle are called **interior angles.** In the figure at the right, angles a, b, and c are interior angles. The sum of the measures of the interior angles of a triangle is 180°.

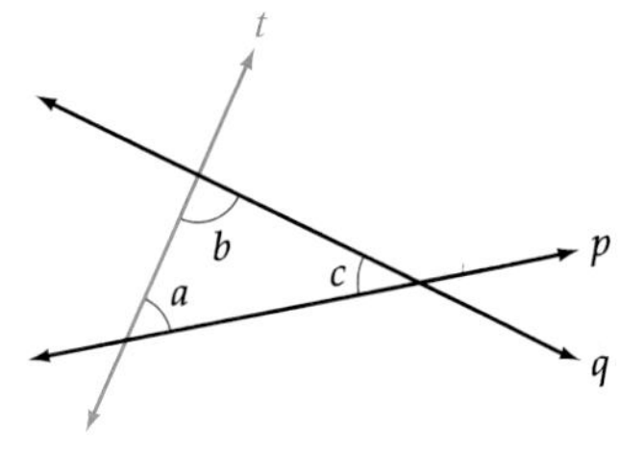

$$m\angle a + m\angle b + m\angle c = 180°$$

The Sum of the Measures of the Interior Angles of a Triangle

The sum of the measures of the interior angles of a triangle is 180°.

QUESTION *Can the measures of the three interior angles of a triangle be 87°, 51°, and 43°?*

An **exterior angle of a triangle** is an angle which is adjacent to an interior angle of the triangle and is a supplement of the interior angle. In the figure at the right, angles m and n are exterior angles for angle a. **The sum of the measures of an interior angle and one of its exterior angles is 180°.**

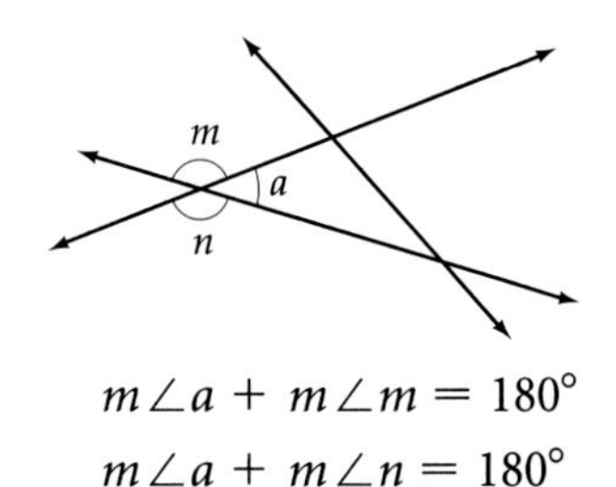

$$m\angle a + m\angle m = 180°$$
$$m\angle a + m\angle n = 180°$$

EXAMPLE 8 ■ Solve a Problem Involving the Angles of a Triangle

In the diagram at the right, $m\angle c = 40°$ and $m\angle e = 60°$. Find the measure of $\angle d$.

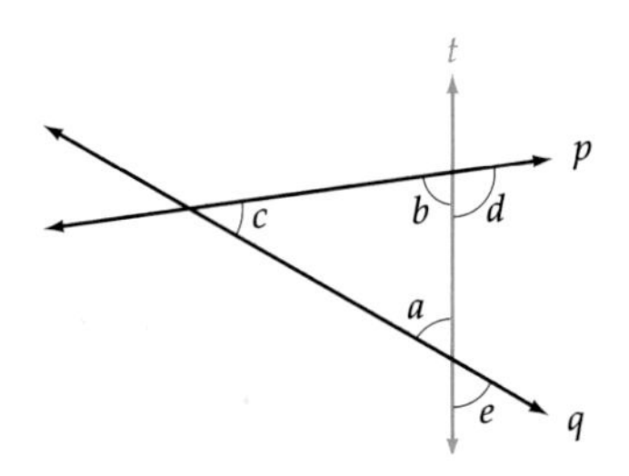

ANSWER *No, because 87° + 51° + 43° = 181°, and the sum of the measures of the three interior angles of a triangle must be 180°.*

Solution

$m\angle a = m\angle e = 60°$ • $\angle a$ and $\angle e$ are vertical angles.

$m\angle c + m\angle a + m\angle b = 180°$ • The sum of the interior angles is **180°**.

$40° + 60° + m\angle b = 180°$ • Replace $m\angle c$ with **40°** and $m\angle a$ with **60°**.

$100° + m\angle b = 180°$ • Add $40° + 60°$.

$m\angle b = 80°$ • Subtract 100° from each side of the equation.

$m\angle b + m\angle d = 180°$ • $\angle b$ and $\angle d$ are supplementary angles.

$80° + m\angle d = 180°$ • Replace $m\angle b$ with **80°**.

$m\angle d = 100°$ • Subtract 80° from each side of the equation.

$m\angle d = 100°$

CHECK YOUR PROGRESS 8

In the diagram at the right, $m\angle c = 35°$ and $m\angle d = 105°$. Find the measure of $\angle e$.

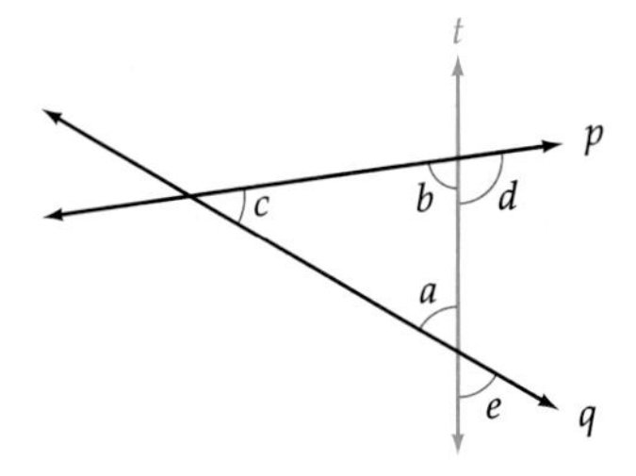

Solution *See page S29.*

EXAMPLE 9 ■ Find the Measure of the Third Angle of a Triangle

Two angles of a triangle measure 43° and 86°. Find the measure of the third angle.

TAKE NOTE

In this text, when we refer to the angles of a triangle, we mean the interior angles of the triangle unless specifically stated otherwise.

Solution

Use the fact that the sum of the measures of the interior angles of a triangle is 180°. Write an equation using x to represent the measure of the third angle. Solve the equation for x.

$x + 43° + 86° = 180°$

$x + 129° = 180°$ • Add $43° + 86°$.

$x = 51°$ • Subtract 129° from each side of the equation.

The measure of the third angle is 51°.

CHECK YOUR PROGRESS 9 One angle in a triangle is a right angle, and one angle measures 27°. Find the measure of the third angle.

Solution *See page S29.*

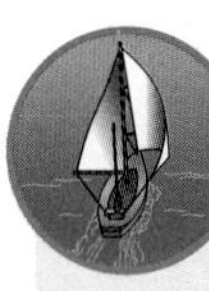

Excursion

Preparing a Circle Graph

On page 453, a protractor was used to measure angles. Preparing a circle graph requires the ability to use a protractor to draw angles.

To draw an angle of 142°, first draw a ray. Place a dot at the endpoint of the ray. This dot will be the vertex of the angle.

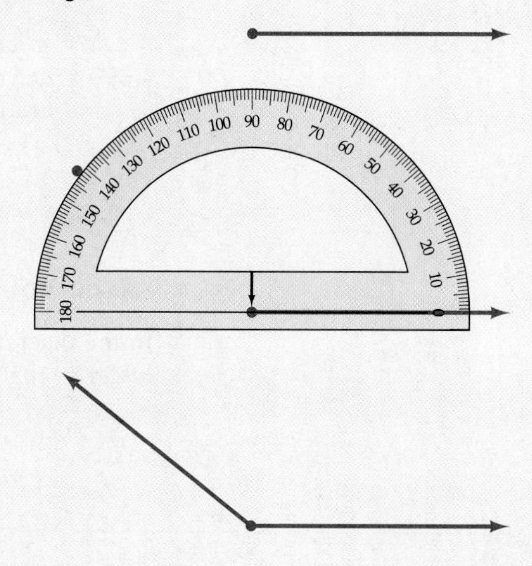

Place the line segment near the bottom edge of the protractor on the ray as shown in the figure at the right. Make sure the center of the bottom edge of the protractor is located directly over the vertex point. Locate the position of the 142° mark. Place a dot next to the mark.

Remove the protractor and draw a ray from the vertex to the dot at the 142° mark.

Here is an example of how to prepare a circle graph.

The revenues (in thousands of dollars) from four segments of a car dealership for the first quarter of a recent year were

New car sales:	\$2100	Used car/truck sales:	\$1500
New truck sales:	\$1200	Parts/service:	\$700

To draw a circle graph to represent the percent that each segment contributed to the total revenue from all four segments, proceed as follows:

Find the total revenue from all four segments.

$2100 + 1200 + 1500 + 700 = 5500$

Find what percent each segment is of the total revenue of \$5500 thousand.

New car sales: $\dfrac{2100}{5500} \approx 38.2\%$

New truck sales: $\dfrac{1200}{5500} \approx 21.8\%$

Used car/truck sales: $\dfrac{1500}{5500} \approx 27.3\%$

Parts/service: $\dfrac{700}{5500} \approx 12.7\%$

(continued)

Each percent represents the part of the circle for that sector. Because the circle contains 360°, multiply each percent by 360° to find the measure of the angle for each sector. Round to the nearest degree.

New car sales:	$0.382 \times 360° \approx 138°$
New truck sales:	$0.218 \times 360° \approx 78°$
Used car/truck sales:	$0.273 \times 360° \approx 98°$
Parts/service:	$0.127 \times 360° \approx 46°$

Draw a circle and use a protractor to draw the sectors representing the percents that each segment contributed to the total revenue from all four segments.

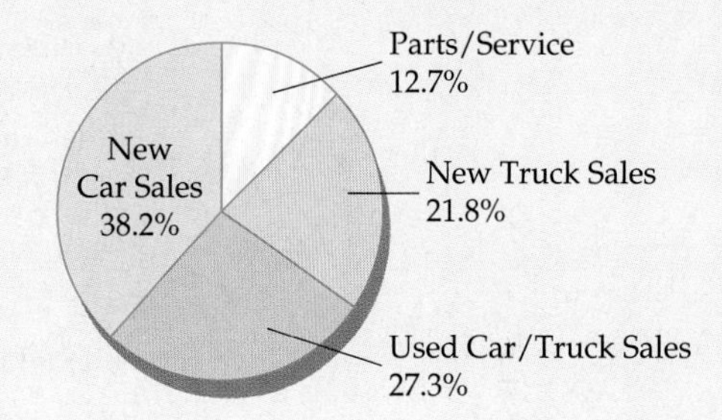

Excursion Exercises

Prepare a circle graph for the data provided in each exercise.

1. Shown below are American adults' favorite pizza toppings. (*Source:* Market Facts for Bolla wines)

Pepperoni:	43%
Sausage:	19%
Mushrooms:	14%
Vegetables:	13%
Other:	7%
Onions:	4%

2. According to a Pathfinder Research Group survey, more than 94% of American adults have heard of the Three Stooges. The choices of a favorite among those who have one are

Curly:	52%
Moe:	31%
Larry:	12%
Curly Joe:	3%
Shemp:	2%

(continued)

3. Only 18 million of the 67 million bicyclists in the United States own or have use of a helmet. The list below shows how often helmets are worn. (*Source:* U.S. Consumer Product Safety Commission)

Never or almost never:	13,680,000
Always or almost always:	2,412,000
Less than half the time:	1,080,000
More than half the time:	756,000
Unknown:	72,000

4. In a recent year, ten million tons of glass containers were produced. Given below is a list of the number of tons used by various industries. The category "Other" includes containers used for items such as drugs and toiletries. (*Source:* Salomon Bros.)

Beer:	4,600,000 tons
Food:	3,500,000 tons
Wine and Liquor:	900,000 tons
Soft Drinks:	500,000 tons
Other:	500,000 tons

Exercise Set 8.1

1. Provide three names for the angle below.

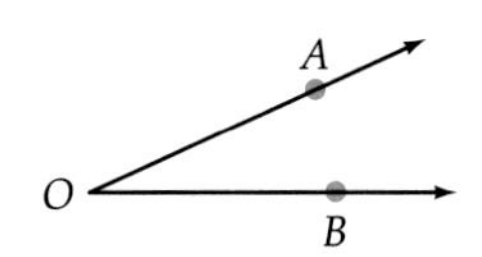

2. State the number of degrees in a full circle, a straight angle, and a right angle.

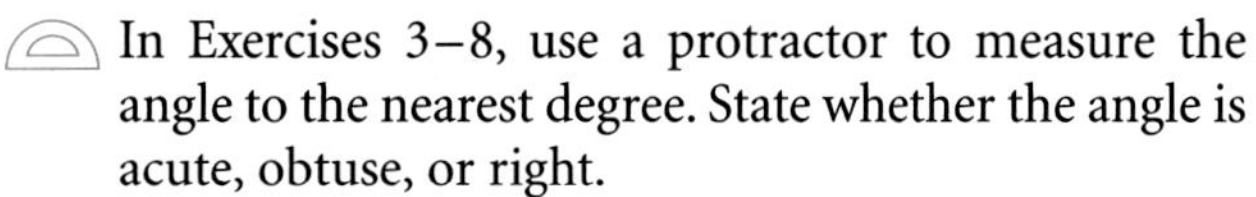

In Exercises 3–8, use a protractor to measure the angle to the nearest degree. State whether the angle is acute, obtuse, or right.

3. **4.** **5.** **6.**

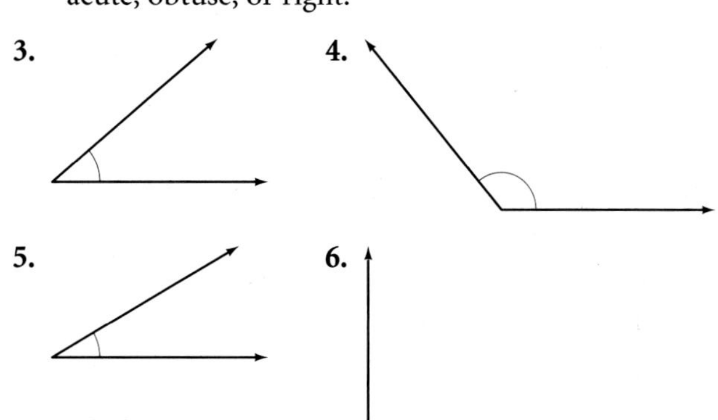

7. **8.**

9. Are angles A and B complementary angles?

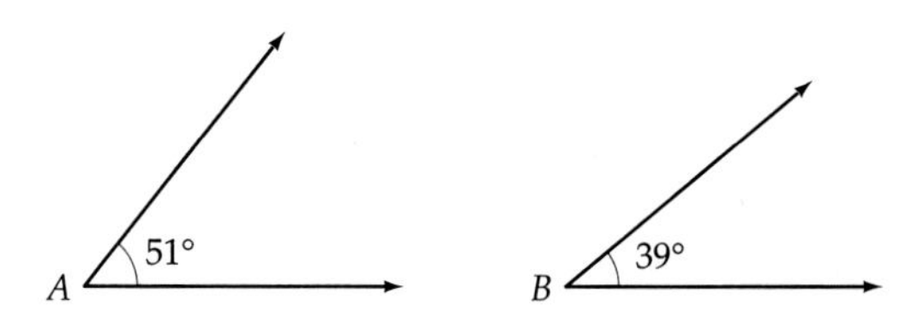

10. Are angles E and F complementary angles?

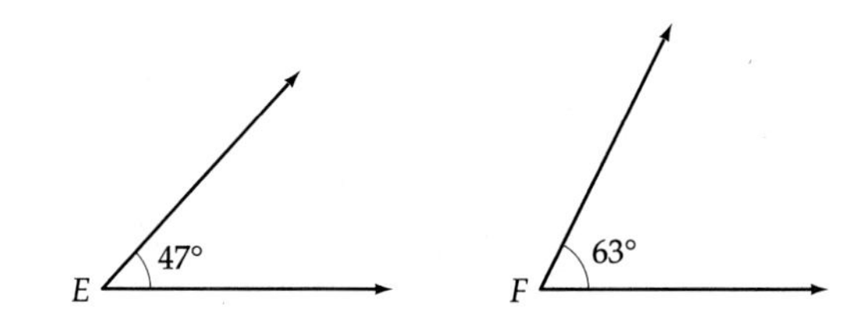

11. Are angles C and D supplementary angles?

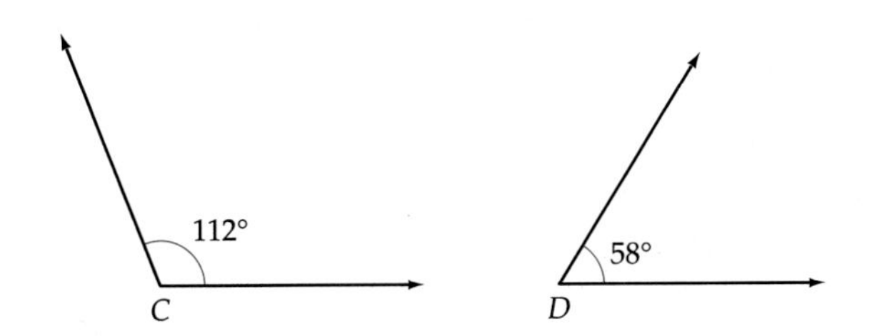

12. Are angles G and H supplementary angles?

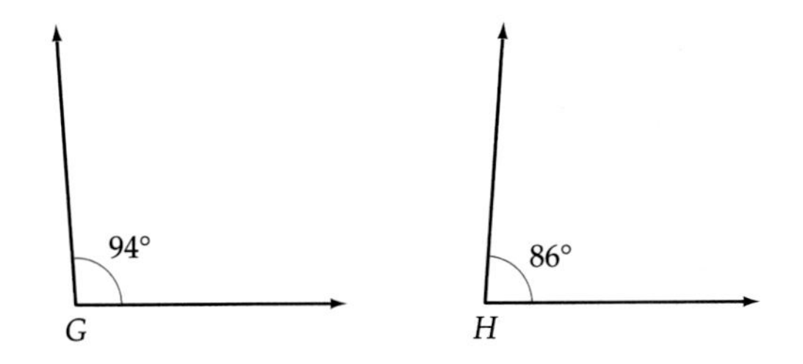

13. Find the complement of a 62° angle.
14. Find the complement of a 31° angle.
15. Find the supplement of a 162° angle.
16. Find the supplement of a 72° angle.
17. Given $AB = 12$ cm, $CD = 9$ cm, and $AD = 35$ cm, find the length of $\overline{BC}$.

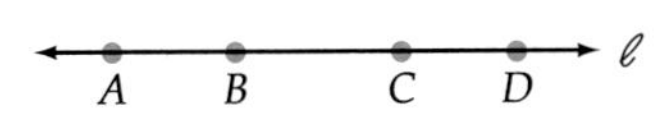

18. Given $AB = 21$ mm, $BC = 14$ mm, and $AD = 54$ mm, find the length of $\overline{CD}$.

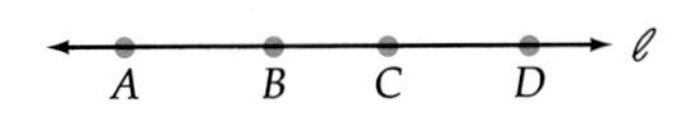

19. Given $QR = 7$ ft and RS is three times the length of $\overline{QR}$, find the length of $\overline{QS}$.

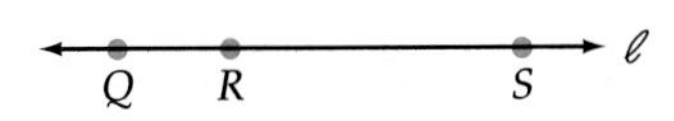

20. Given $QR = 15$ in. and RS is twice the length of $\overline{QR}$, find the length of $\overline{QS}$.

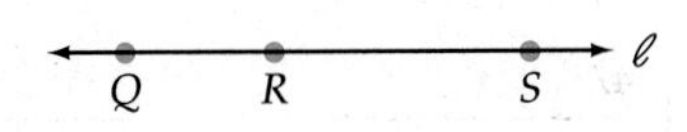

21. Given $EF = 20$ m and FG is $\frac{1}{2}$ the length of $\overline{EF}$, find the length of $\overline{EG}$.

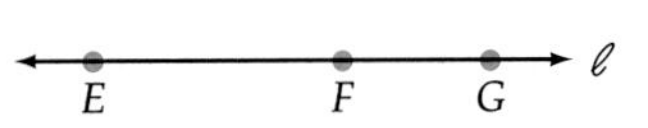

22. Given $EF = 18$ cm and FG is $\frac{1}{3}$ the length of $\overline{EF}$, find the length of $\overline{EG}$.

E F G ℓ

23. Given $m\angle LOM = 53°$ and $m\angle LON = 139°$, find the measure of $\angle MON$.

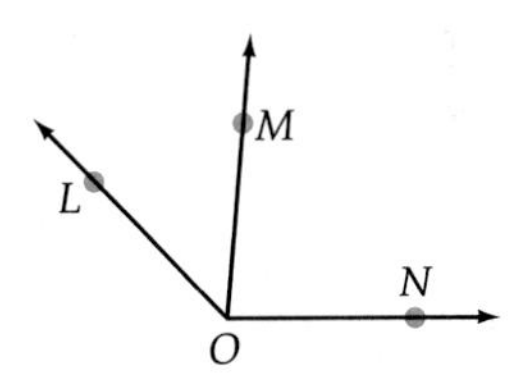

24. Given $m\angle MON = 38°$ and $m\angle LON = 85°$, find the measure of $\angle LOM$.

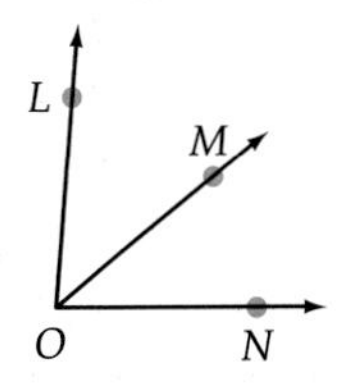

In Exercises 25 and 26, find the measure of $\angle x$.

25.

145° x 74°

26.

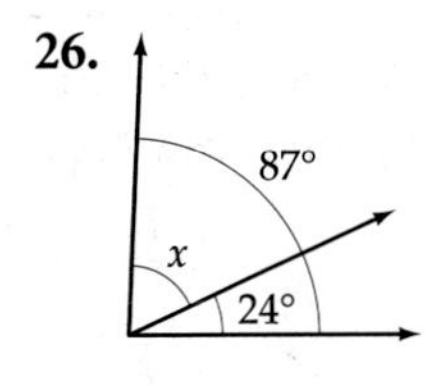

In Exercises 27–30, given that $\angle LON$ is a right angle, find the measure of $\angle x$.

27.

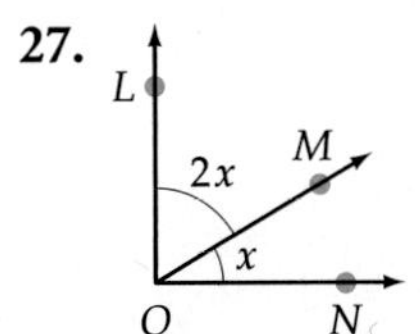

28.

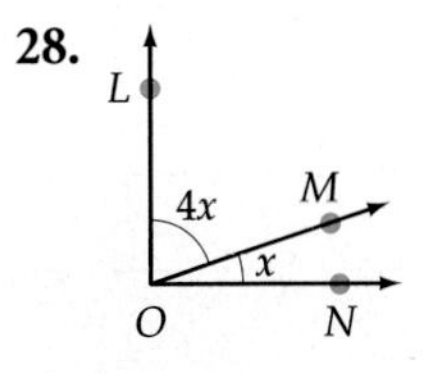

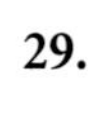

29.

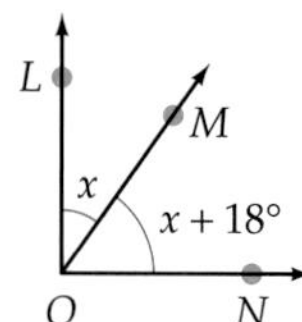

30.

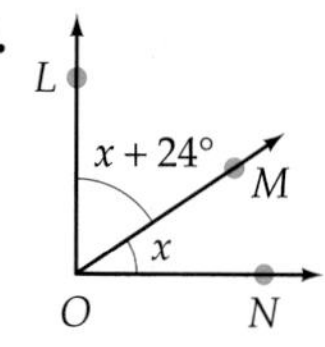

In Exercises 31–34, find the measure of $\angle a$.

31.

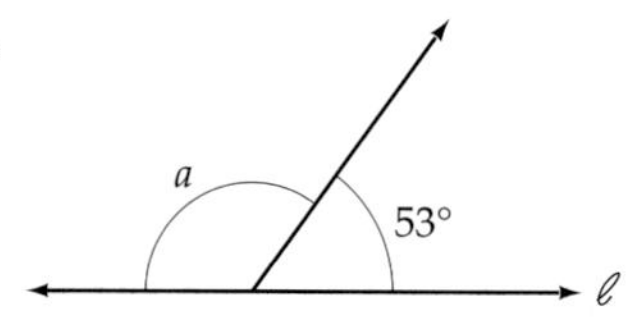

32.

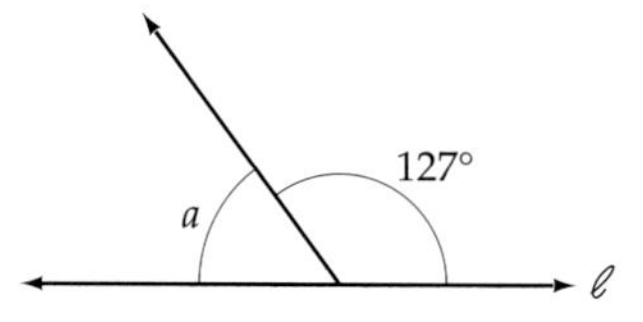

33.

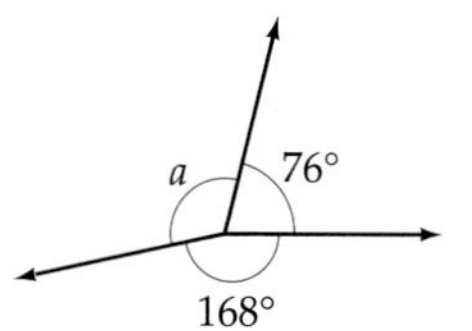

34.

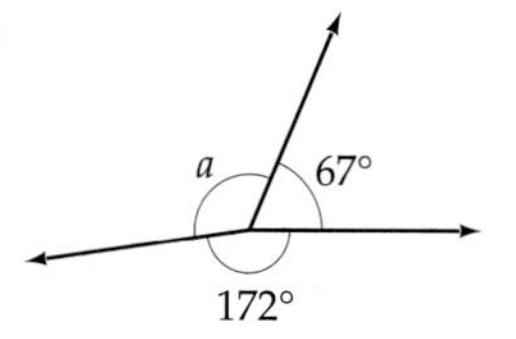

In Exercises 35–40, find the value of x.

35.

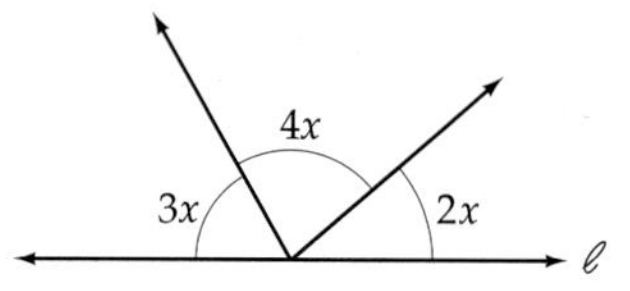

36.

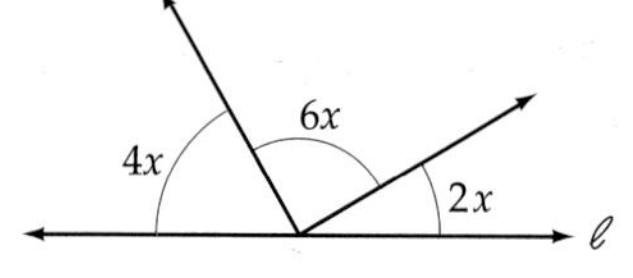

37.

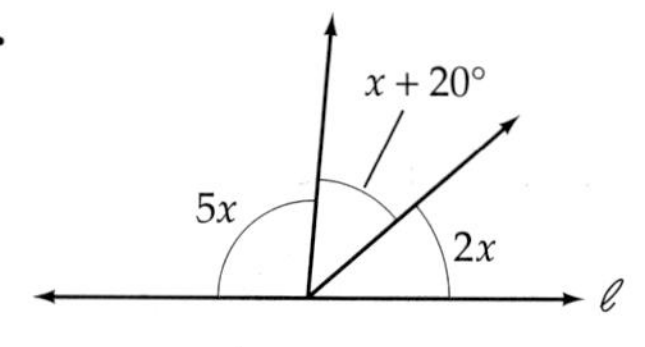

38.

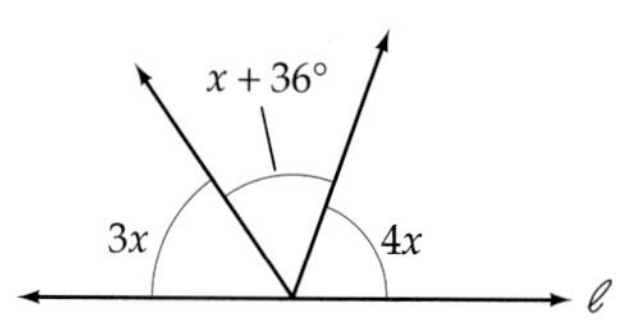

39.

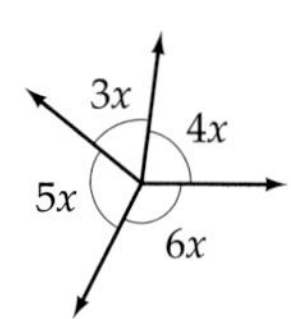

40.

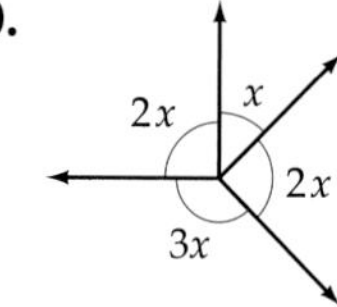

41. Given $m\angle a = 51°$, find the measure of $\angle b$.

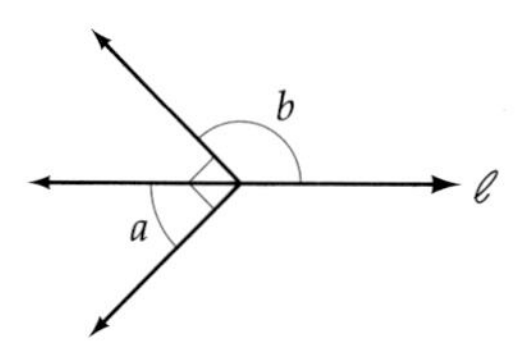

42. Given $m\angle a = 38°$, find the measure of $\angle b$.

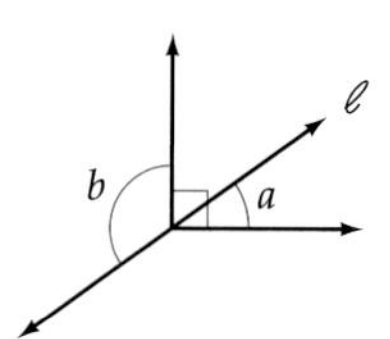

In Exercises 43 and 44, find the measure of $\angle x$.

43.

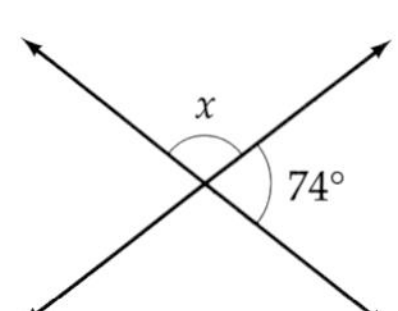

44.

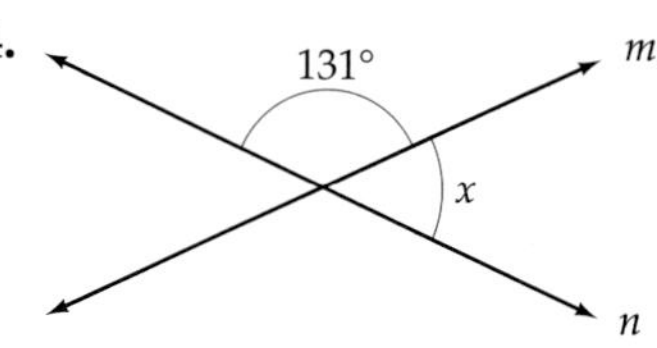

In Exercises 45 and 46, find the value of x.

45.

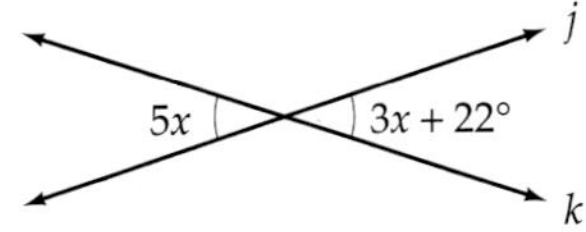

46.

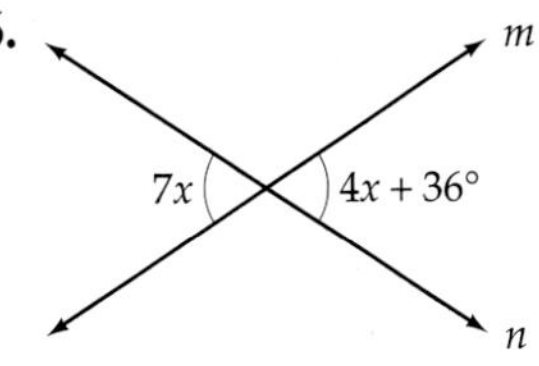

In Exercises 47–50, given that $\ell_1 \parallel \ell_2$, find the measures of angles a and b.

47.

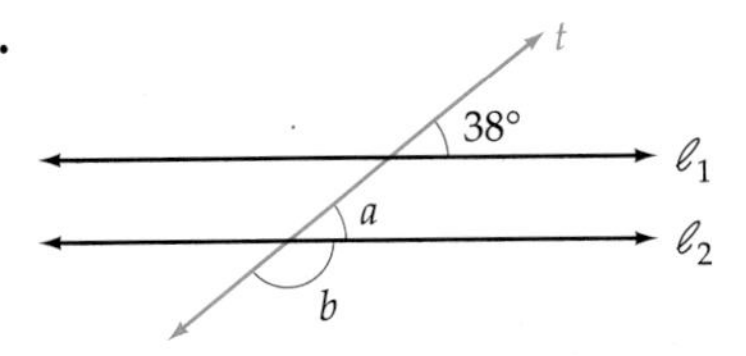

48.

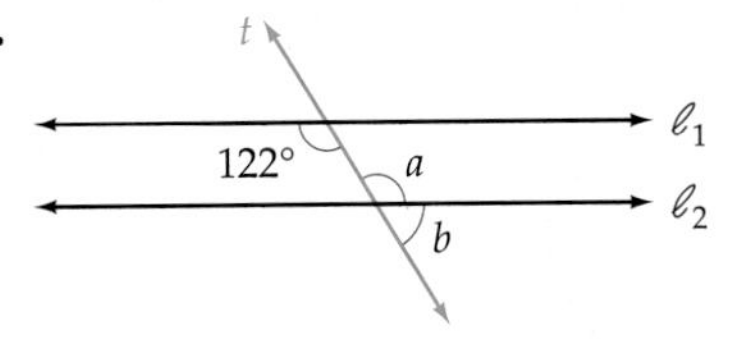

49.

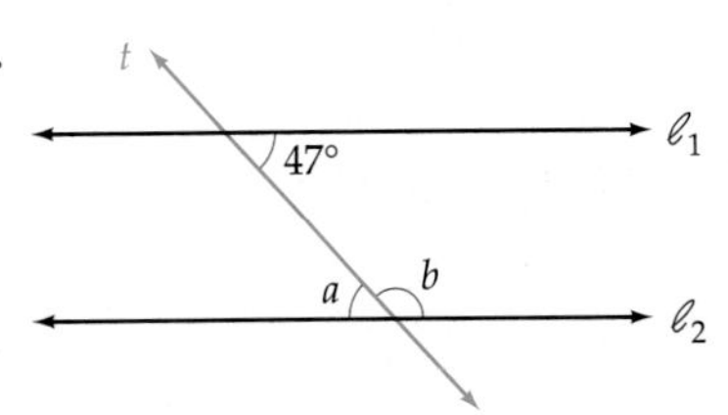

50.

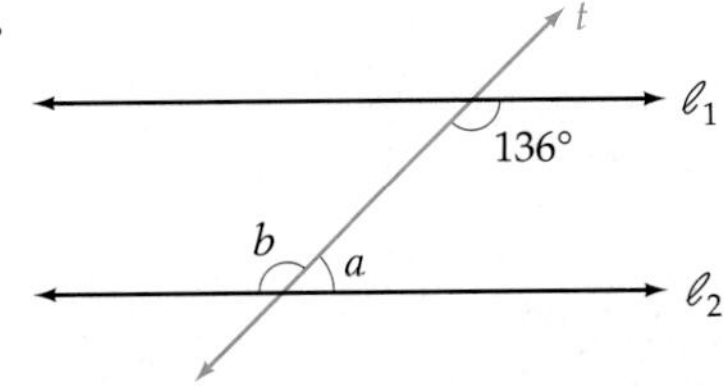

In Exercises 51–54, given that $\ell_1 \parallel \ell_2$, find x.

51.

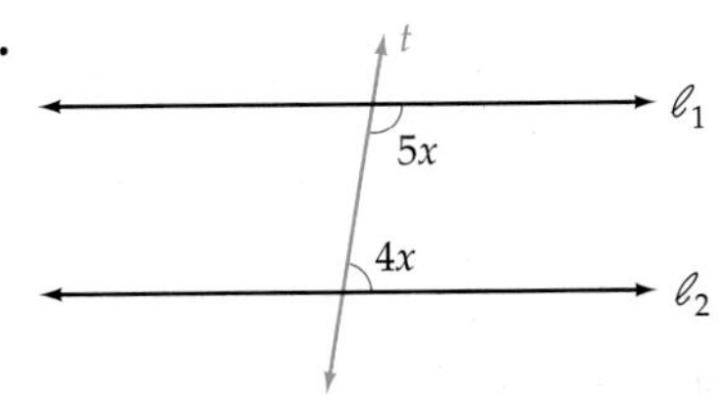

52.

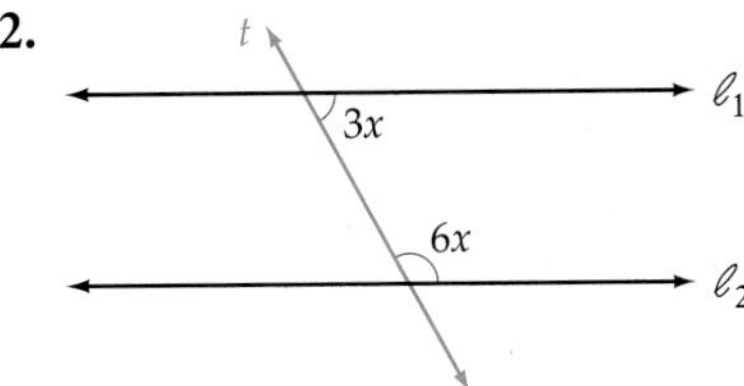

53.

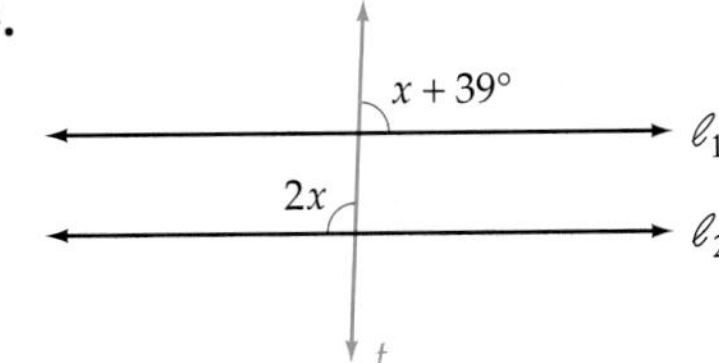

54.

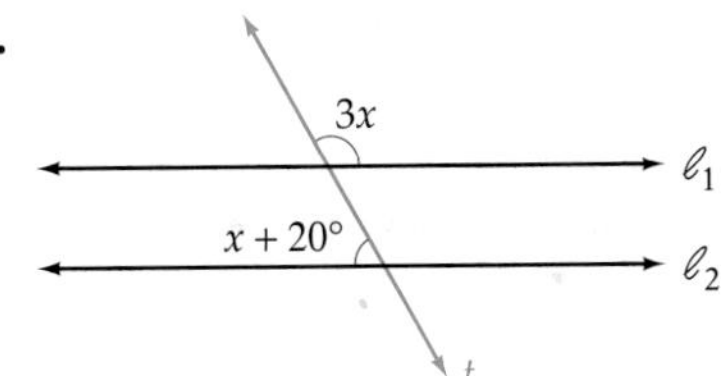

55. Given that $m\angle a = 95°$ and $m\angle b = 70°$, find the measures of angles x and y.

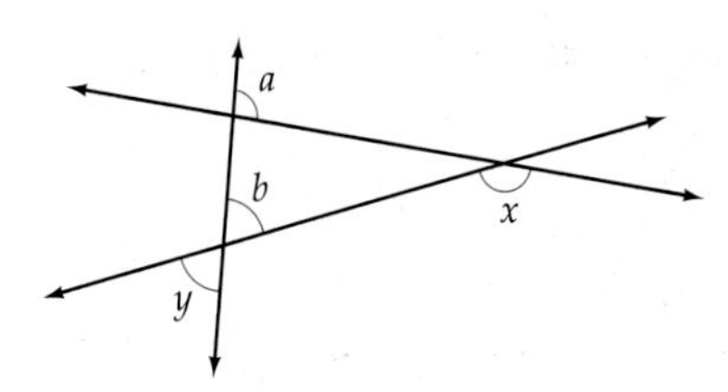

56. Given that $m\angle a = 35°$ and $m\angle b = 55°$, find the measures of angles x and y.

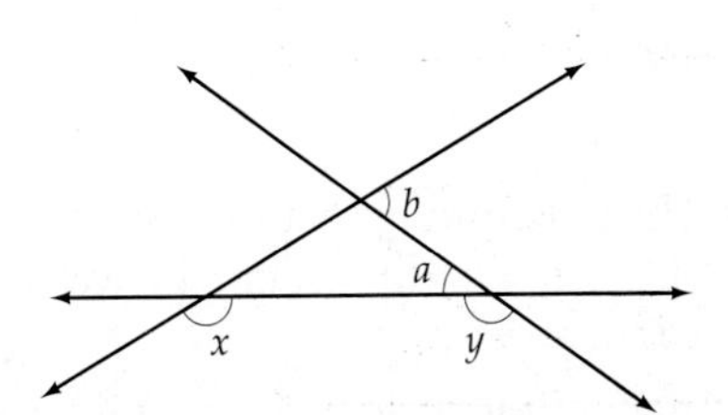

57. Given that $m\angle y = 45°$, find the measures of angles a and b.

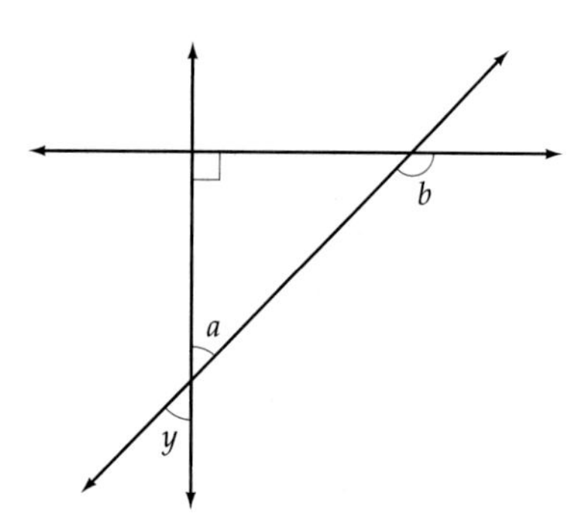

58. Given that $m\angle y = 130°$, find the measures of angles a and b.

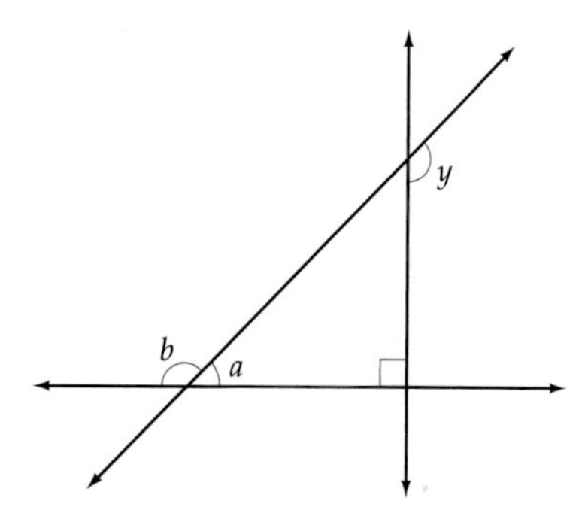

59. Given that $\overline{AO} \perp \overline{OB}$, express in terms of x the number of degrees in $\angle BOC$.

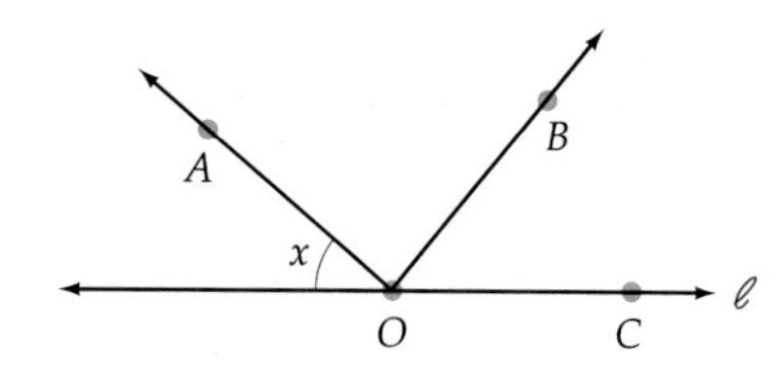

60. Given that $\overline{AO} \perp \overline{OB}$, express in terms of x the number of degrees in $\angle AOC$.

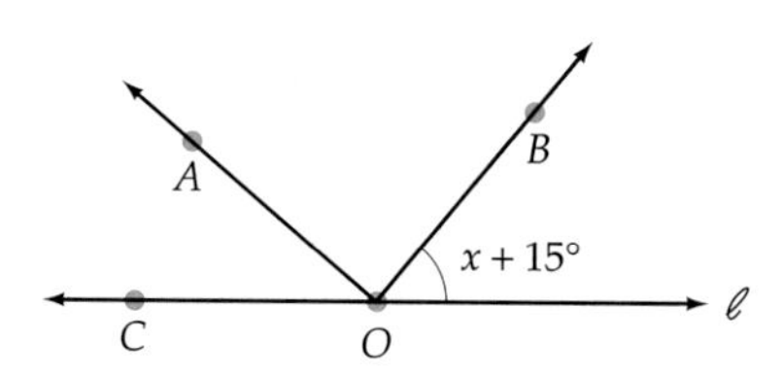

61. One angle in a triangle is a right angle, and one angle is equal to 30°. What is the measure of the third angle?

62. A triangle has a 45° angle and a right angle. Find the measure of the third angle.

63. Two angles of a triangle measure 42° and 103°. Find the measure of the third angle.

64. Two angles of a triangle measure 62° and 45°. Find the measure of the third angle.

65. A triangle has a 13° angle and a 65° angle. What is the measure of the third angle?

66. A triangle has a 105° angle and a 32° angle. What is the measure of the third angle?

67. Cut out a triangle and then tear off two of the angles, as shown below. Position the pieces you tore off so that angle a is adjacent to angle b and angle c is adjacent to angle b. Describe what you observe. What does this demonstrate?

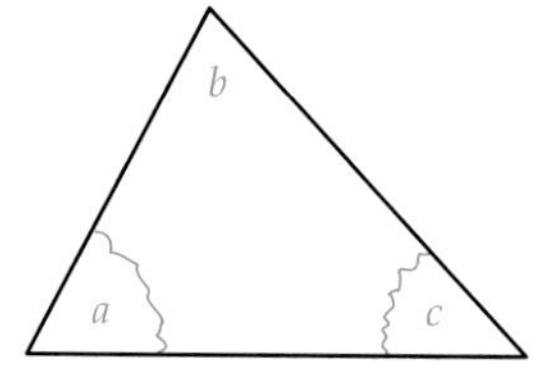

Extensions

CRITICAL THINKING

68. The road mileage between San Francisco, California and New York City is 3036 mi. The air distance between these two cities is 2571 mi. Why do the distances differ?

69. How many dimensions does a point have? a line? a line segment? a ray? an angle?

70. Which line segment is longer, $\overline{AB}$ or $\overline{CD}$?

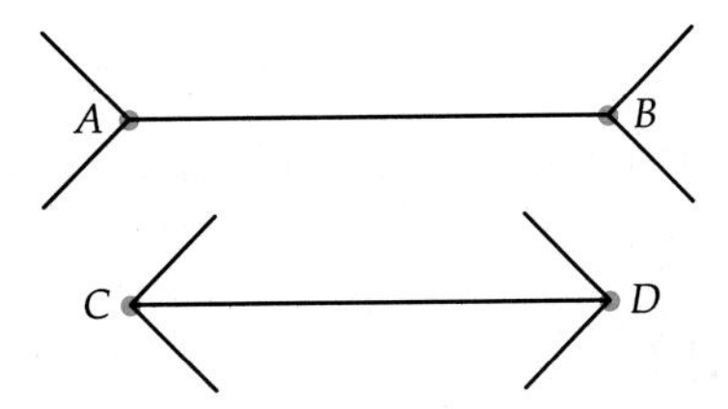

COOPERATIVE LEARNING

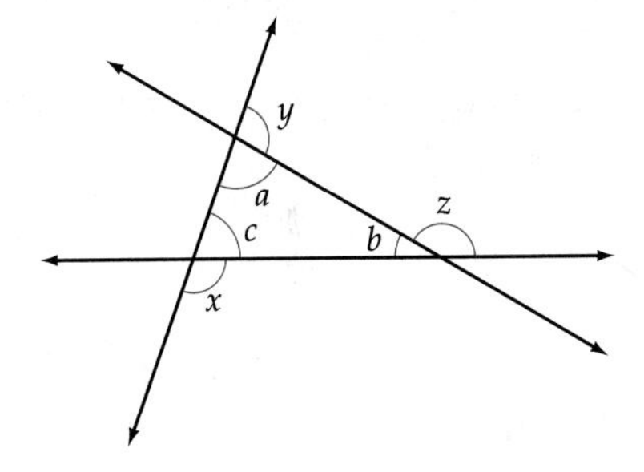

71. For the figure at the right, find the sum of the measures of angles x, y, and z.

72. For the figure at the right, explain why $m\angle a + m\angle b = m\angle x$. Write a rule that describes the relationship between the measure of an exterior angle of a triangle and the sum of the measures of its two opposite interior angles (the interior angles that are non-adjacent to the exterior angle). Use the rule to write an equation involving angles a, c, and z.

73. If $\overline{AB}$ and $\overline{CD}$ intersect at point O, and $m\angle AOC = m\angle BOC$, explain why $\overline{AB} \perp \overline{CD}$.

SECTION 8.2 Perimeter and Area of Plane Figures

Perimeter of Plane Geometric Figures

A **polygon** is a closed figure determined by three or more line segments that lie in a plane. The line segments that form the polygon are called its **sides.** The figures below are examples of polygons.

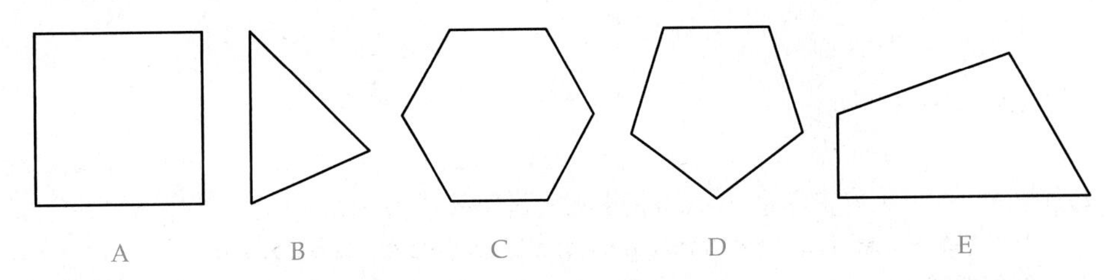

point of interest

Although a polygon is described in terms of the number of its sides, the word actually comes from the Latin word *polygonum,* meaning "many *angles.*"

A **regular polygon** is one in which each side has the same length and each angle has the same measure. The polygons in Figures A, C, and D above are regular polygons.

The name of a polygon is based on the number of its sides. The table below lists the names of polygons that have from 3 to 10 sides.

The Pentagon in Arlington, Virginia

Number of Sides	Name of Polygon
3	Triangle
4	Quadrilateral
5	Pentagon
6	Hexagon
7	Heptagon
8	Octagon
9	Nonagon
10	Decagon

Triangles and quadrilaterals are two of the most common types of polygons. Triangles are distinguished by the number of equal sides and also by the measures of their angles.

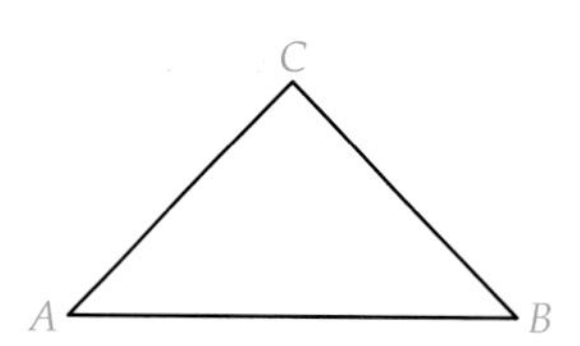

An **isosceles triangle** has exactly two sides of equal length. The angles opposite the equal sides are of equal measure.
$AC = BC$
$m\angle A = m\angle B$

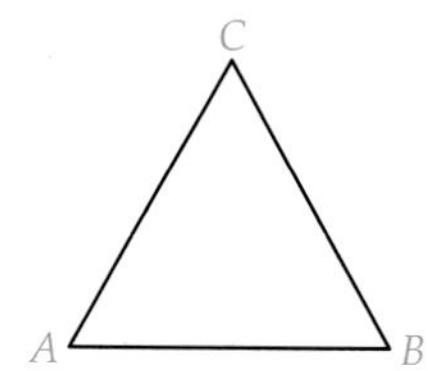

The three sides of an **equilateral triangle** are of equal length. The three angles are of equal measure.
$AB = BC = AC$
$m\angle A = m\angle B$
$= m\angle C$

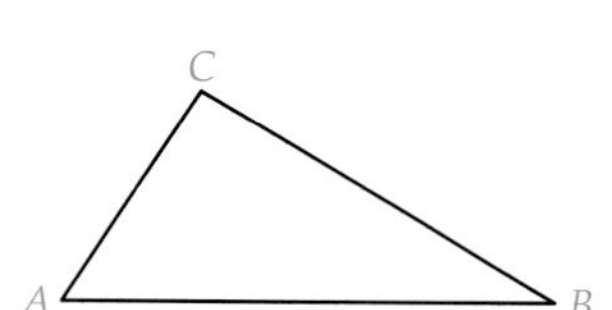

A **scalene triangle** has no two sides of equal length. No two angles are of equal measure.

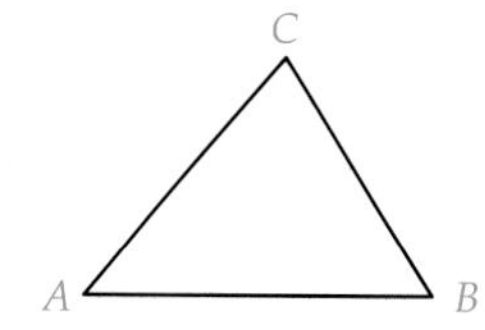

An **acute triangle** has three acute angles.

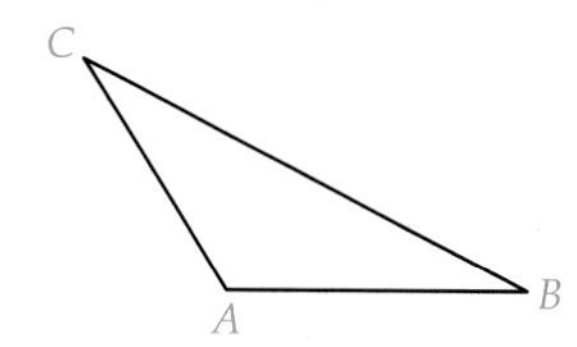

An **obtuse triangle** has one obtuse angle.

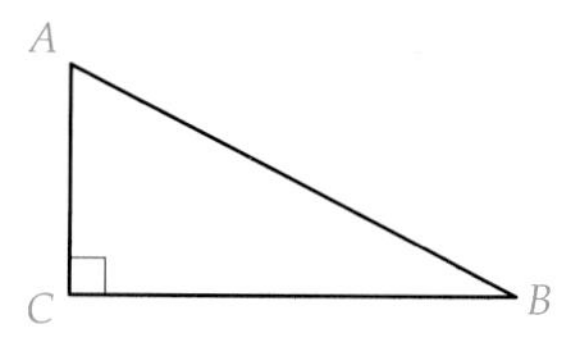

A **right triangle** has a right angle.

A **quadrilateral** is a four-sided polygon. Quadrilaterals are also distinguished by their sides and angles, as shown on the following page. Note that a rectangle, a square, and a rhombus are different forms of a parallelogram.

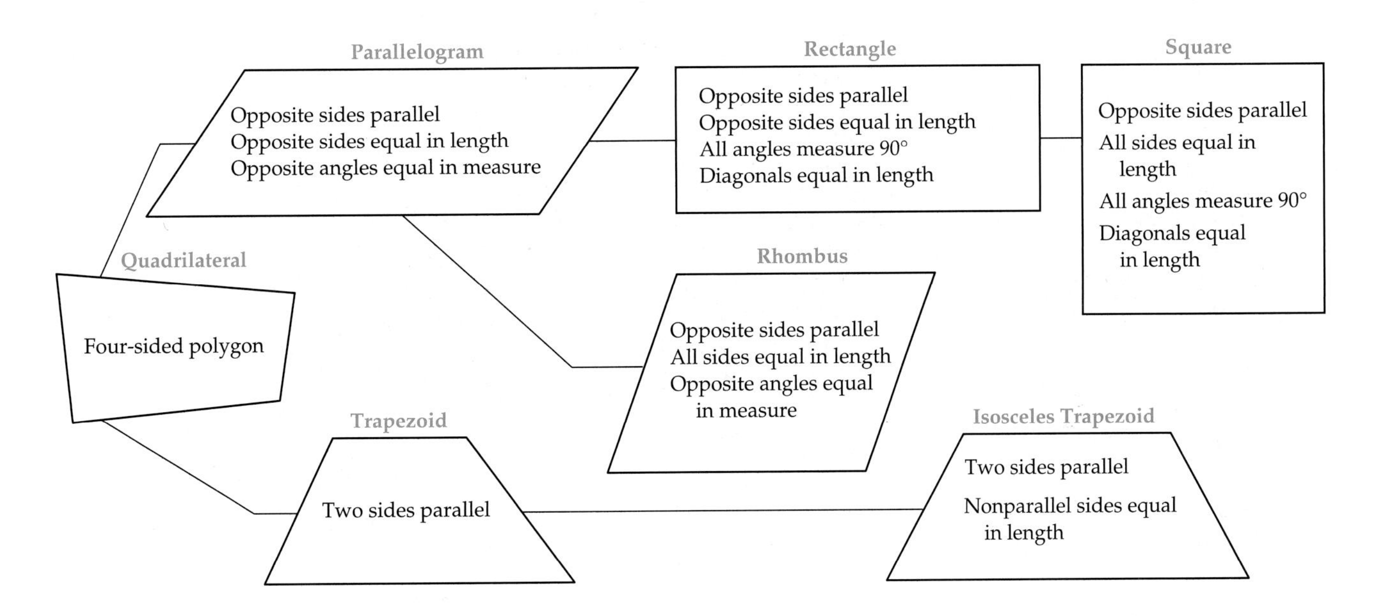

QUESTION *a. What distinguishes a rectangle from other parallelograms?*
b. What distinguishes a square from other rectangles?

The **perimeter** of a plane geometric figure is a measure of the distance around the figure. Perimeter is used when buying fencing for a garden or determining how much baseboard is needed for a room.

The perimeter of a triangle is the sum of the lengths of the three sides.

Perimeter of a Triangle

Let a, b, and c be the lengths of the sides of a triangle. The perimeter, P, of the triangle is given by $P = a + b + c$.

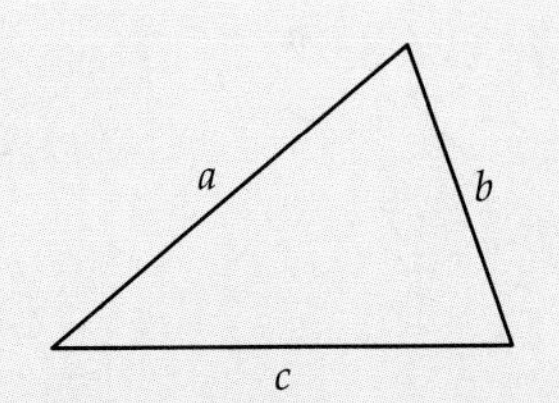

$P = a + b + c$

To find the perimeter of the triangle shown at the right, add the lengths of the three sides.

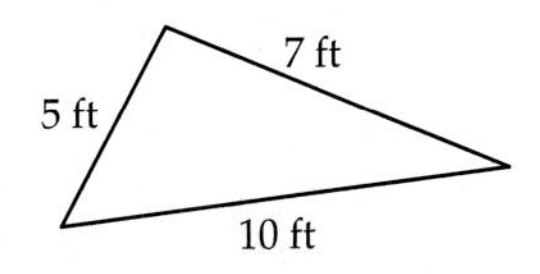

$P = 5 + 7 + 10 = 22$

The perimeter is 22 ft.

EXAMPLE 1 ■ Find the Perimeter of a Triangle

You want to sew bias binding along the edges of a cloth flag that has sides that measure $1\frac{1}{4}$ ft, $3\frac{1}{2}$ ft, and $3\frac{3}{4}$ ft. Find the length of bias binding needed.

ANSWER *a. In a rectangle, all angles measure 90°. b. In a square, all sides are equal in length.*

Solution

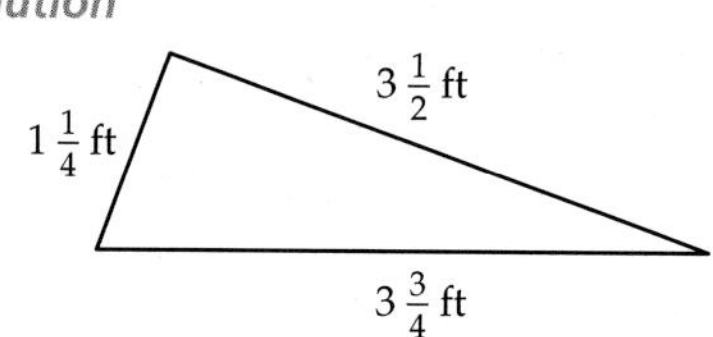

• Draw a diagram.

$P = a + b + c$

• Use the formula for the perimeter of a triangle.

$P = 1\frac{1}{4} + 3\frac{1}{2} + 3\frac{3}{4}$

$P = 1\frac{1}{4} + 3\frac{2}{4} + 3\frac{3}{4}$

$P = 7\frac{6}{4}$

• Replace a, b, and c with $1\frac{1}{4}$, $3\frac{1}{2}$, and $3\frac{3}{4}$. You can replace a with the length of any of the three sides, and then replace b and c with the lengths of the other two sides. The order does not matter. The result will be the same.

$P = 8\frac{1}{2}$

• $7\frac{6}{4} = 7\frac{3}{2} = 7 + \frac{3}{2} = 7 + 1\frac{1}{2} = 8\frac{1}{2}$

You need $8\frac{1}{2}$ ft of bias binding.

CHECK YOUR PROGRESS 1 A bicycle trail in the shape of a triangle has sides that measure $4\frac{3}{10}$ mi, $2\frac{1}{10}$ mi, and $6\frac{1}{2}$ mi. Find the total length of the bike trail.

Solution *See page S29.*

The perimeter of a quadrilateral is the sum of the lengths of its four sides.

A **rectangle** is a quadrilateral with all right angles and opposite sides of equal length. Usually the length, L, of a rectangle refers to the length of one of the longer sides of the rectangle and the width, W, refers to the length of one of the shorter sides. The perimeter can then be represented as $P = L + W + L + W$.

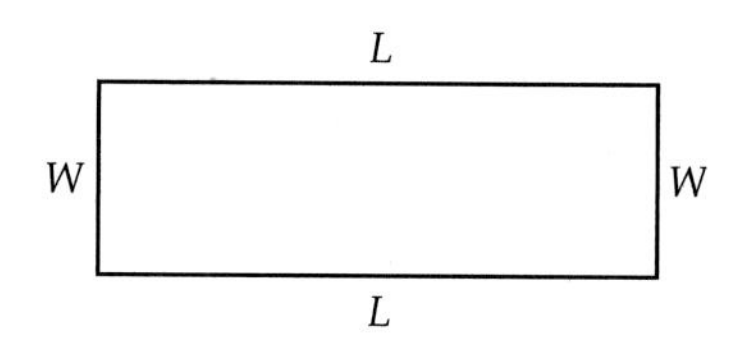

The formula for the perimeter of a rectangle is derived by combining like terms.

$P = L + W + L + W$

$P = 2L + 2W$

Perimeter of a Rectangle

Let L represent the length and W the width of a rectangle. The perimeter, P, of the rectangle is given by $P = 2L + 2W$.

EXAMPLE 2 ■ Find the Perimeter of a Rectangle

You want to trim a rectangular frame with a metal strip. The frame measures 30 in. by 20 in. Find the length of metal strip you will need to trim the frame.

Solution

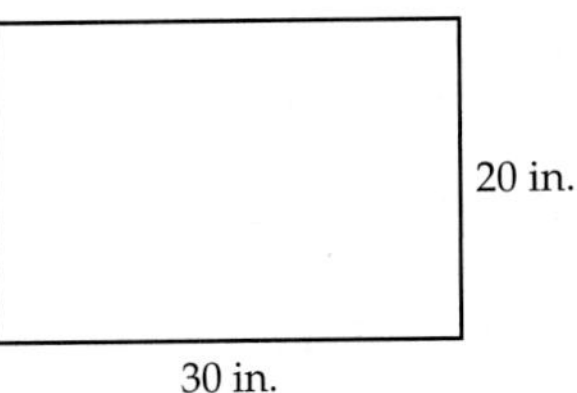

• Draw a diagram.

$P = 2L + 2W$ • Use the formula for the perimeter of a rectangle.

$P = 2(30) + 2(20)$ • The length is **30** in. Substitute **30** for *L*. The width is **20** in. Substitute **20** for *W*.

$P = 60 + 40$

$P = 100$

You will need 100 in. of the metal strip.

CHECK YOUR PROGRESS 2 Find the length of decorative molding needed to edge the top of the walls in a rectangular room that is 12 ft long and 8 ft wide.

Solution *See page S29.*

A **square** is a rectangle in which each side has the same length. Letting s represent the length of each side of a square, the perimeter of the square can be represented $P = s + s + s + s$.

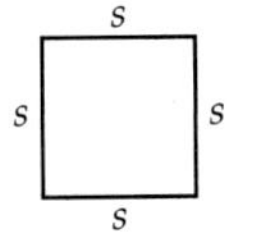

The formula for the perimeter of a square is derived by combining like terms.

$P = s + s + s + s$

$P = 4s$

Perimeter of a Square

Let s represent the length of a side of a square. The perimeter, P, of the square is given by $P = 4s$.

EXAMPLE 3 ■ **Find the Perimeter of a Square**

Find the length of fencing needed to surround a square corral that measures 60 ft on each side.

Solution

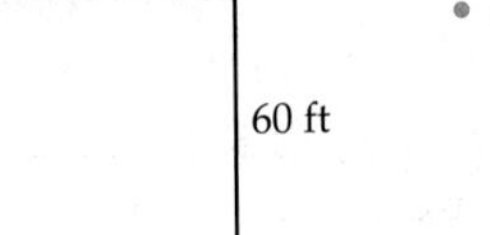

• Draw a diagram.

$P = 4s$ • Use the formula for the perimeter of a square.

$P = 4(60)$ • The length of a side is **60** ft. Substitute **60** for *s*.

$P = 240$

240 ft of fencing is needed.

CHECK YOUR PROGRESS 3 A homeowner plans to fence in the area around the swimming pool in the back yard. The area to be fenced in is a square measuring 24 ft on each side. How many feet of fencing should the homeowner purchase?

Solution *See page S30.*

historical note

Benjamin Banneker (băn′ĭ-kər) (1731–1806), a noted American scholar who was largely self-taught, was both a surveyor and an astronomer. As a surveyor, he was a member of the commission that defined the boundary lines and laid out the streets of the District of Columbia. (See the Point of Interest on page 456.) ■

Figure *ABCD* is a parallelogram. $\overline{BC}$ is the **base** of the parallelogram. Opposite sides of a parallelogram are equal in length, so $\overline{AD}$ is the same length as $\overline{BC}$, and $\overline{AB}$ is the same length as $\overline{CD}$.

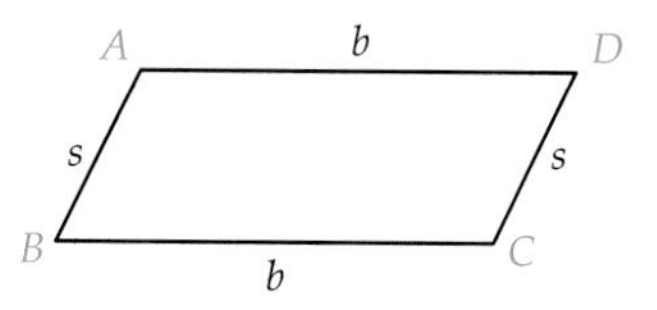

Let b represent the length of the base and s the length of an adjacent side. Then the perimeter of a parallelogram can be represented as $P = b + s + b + s$.

$$P = b + s + b + s$$

The formula for the perimeter of a parallelogram is derived by combining like terms.

$$P = 2b + 2s$$

Perimeter of a Parallelogram

Let b represent the length of the base of a parallelogram and s the length of a side adjacent to the base. The perimeter, P, of the parallelogram is given by $P = 2b + 2s$.

EXAMPLE 4 ■ Find the Perimeter of a Parallelogram

You plan to trim the edge of a kite with a strip of red fabric. The kite is in the shape of a parallelogram with a base measuring 40 in. and a side measuring 28 in. Find the length of the fabric needed to trim the kite.

Solution

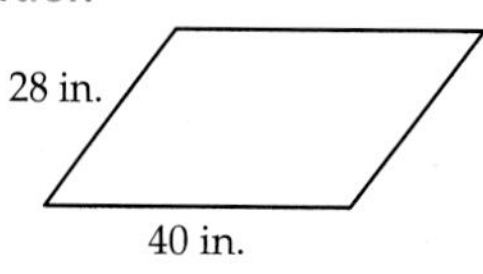

• Draw a diagram.

$P = 2b + 2s$ • Use the formula for the perimeter of a parallelogram.

$P = 2(40) + 2(28)$ • The base is **40** in. Substitute **40** for b. The length of a side is **28** in. Substitute **28** for s.

$P = 80 + 56$ • Simplify using the Order of Operations Agreement.

$P = 136$

To trim the kite, 136 in. of fabric is needed.

CHECK YOUR PROGRESS 4 A flower bed is in the shape of a parallelogram that has a base of length 5 m and a side of length 7 m. Wooden planks are used to edge the garden. Find the length of wooden planks needed to surround the garden.

Solution *See page S30.*

point of interest

A **glazier** is a person who cuts, fits, and installs glass, generally in doors and windows. Of particular challenge to a glazier are intricate stained glass window designs.

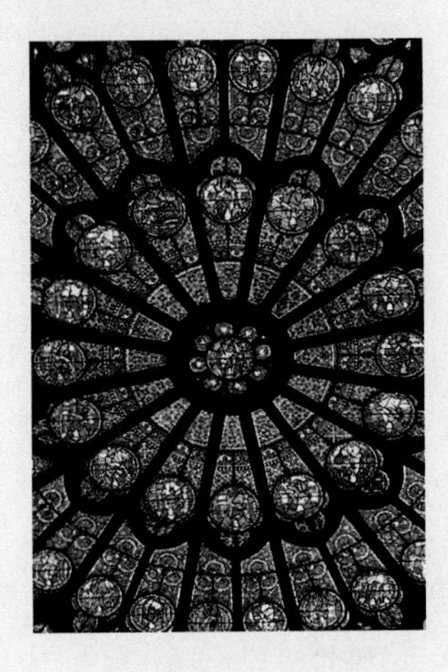

A **circle** is a plane figure in which all points are the same distance from point O, called the **center** of the circle.

A **diameter** of a circle is a line segment with endpoints on the circle and passing through the center. $\overline{AB}$ is a diameter of the circle at the right. The variable d is used to designate the length of a diameter of a circle.

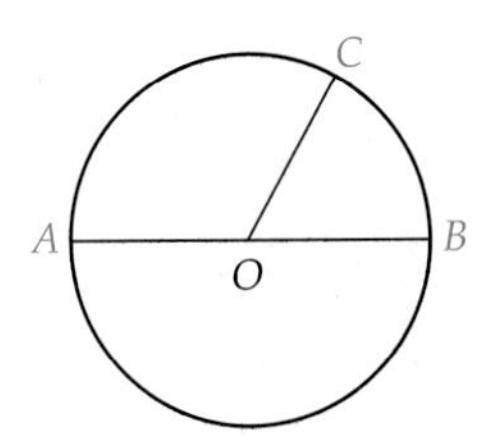

A **radius** of a circle is a line segment from the center of the circle to a point on the circle. $\overline{OC}$ is a radius of the circle at the right above. The variable r is used to designate the length of a radius of a circle.

The length of the diameter is twice the length of the radius. $\quad d = 2r \quad \text{or} \quad r = \frac{1}{2}d$

The distance around a circle is called the **circumference.** The formula for the circumference, C, of a circle is: $\quad C = \pi d$

Because $d = 2r$, the formula for the circumference can be written: $\quad C = 2\pi r$

Circumference of a Circle

The circumference, C, of a circle with diameter d and radius r is given by $C = \pi d$ or $C = 2\pi r$.

TAKE NOTE

Recall that an irrational number is a number whose decimal representation never terminates and does not have a pattern of numerals that keep repeating.

The formula for circumference uses the number π (pi), which is an irrational number. The value of π can be approximated by a fraction or by a decimal. $\quad \pi \approx 3\frac{1}{7} \quad \text{or} \quad \pi \approx 3.14$

The π key on a scientific calculator gives a closer approximation of π than 3.14. A scientific calculator is used in this section to find approximate values in calculations involving π.

CALCULATOR NOTE

The π key on your calculator can be used to find decimal approximations to expressions that contain π. To perform the calculation at the right, enter

6 [×] [π] [=].

Find the circumference of a circle with a diameter of 6 m.

The diameter of the circle is given. Use the circumference formula that involves the diameter. $d = 6$.

$$C = \pi d$$
$$C = \pi(6)$$

The exact circumference of the circle is 6π m.

$$C = 6\pi$$

An approximate measure can be found by using the π key on a calculator.

An approximate circumference is 18.85 m.

$$C \approx 18.85$$

point of interest

Archimedes (ar-kə-mē′dēz) (c. 287–212 B.C.) is the person who calculated that $\pi \approx 3\frac{1}{7}$. He actually showed that $3\frac{10}{71} < \pi < 3\frac{1}{7}$. The approximation $3\frac{10}{71}$ is a more accurate approximation of π than $3\frac{1}{7}$, but it is more difficult to use without a calculator.

EXAMPLE 5 ■ Find the Circumference of a Circle

Find the circumference of a circle with a radius of 15 cm. Round to the nearest hundredth of a centimeter.

Solution

$C = 2\pi r$ • The radius is given. Use the circumference formula that involves the radius.

$C = 2\pi(15)$ • Replace *r* with **15**.

$C = 30\pi$ • Multiply 2 times 15.

$C \approx 94.25$ • An approximation is asked for. Use the π key on a calculator.

The circumference of the circle is approximately 94.25 cm.

CHECK YOUR PROGRESS 5 Find the circumference of a circle with a diameter of 9 km. Give the exact measure.

Solution *See page S30.*

EXAMPLE 6 ■ Application of Finding the Circumference of a Circle

A bicycle tire has a diameter of 24 in. How many feet does the bicycle travel when the wheel makes 8 revolutions? Round to the nearest hundredth of a foot.

Solution

24 in. = 2 ft • The diameter is given in inches, but the answer must be expressed in feet. Convert the diameter (24 in.) to feet. There are 12 in. in 1 ft. Divide 24 by 12.

$C = \pi d$ • The diameter is given. Use the circumference formula that involves the diameter.

$C = \pi(2)$ • Replace *d* with **2**.

$C = 2\pi$ • This is the distance traveled in 1 revolution.

$8C = 8(2\pi) = 16\pi \approx 50.27$ • Find the distance traveled in **8** revolutions.

The bicycle will travel about 50.27 ft when the wheel makes 8 revolutions.

CALCULATOR NOTE

To approximate $8(2\pi)$ on your calculator, enter

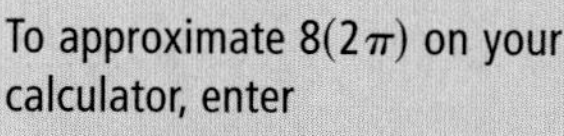

CHECK YOUR PROGRESS 6 A tricycle tire has a diameter of 12 in. How many feet does the tricycle travel when the wheel makes 12 revolutions? Round to the nearest hundredth of a foot.

Solution *See page S30.*

Area of Plane Geometric Figures

Area is the amount of surface in a region. Area can be used to describe, for example, the size of a rug, a parking lot, a farm, or a national park. Area is measured in square units.

A square that measures 1 inch on each side has an area of 1 square inch, written 1 in^2.

A square that measures 1 centimeter on each side has an area of 1 square centimeter, written 1 cm^2.

1 in^2

1 cm^2

Larger areas are often measured in square feet (ft^2), square meters (m^2), square miles (mi^2), acres (43,560 ft^2), or any other square unit.

QUESTION

a. *What is the area of a square that measures 1 yard on each side?*

b. *What is the area of a square that measures 1 kilometer on each side?*

The area of a geometric figure is the number of squares (each of area 1 square unit) that are necessary to cover the figure. In the figures below, two rectangles have been drawn and covered with squares. In the figure on the left, 12 squares, each of area 1 cm^2, were used to cover the rectangle. The area of the rectangle is 12 cm^2. In the figure on the right, six squares, each of area 1 in^2, were used to cover the rectangle. The area of the rectangle is 6 in^2.

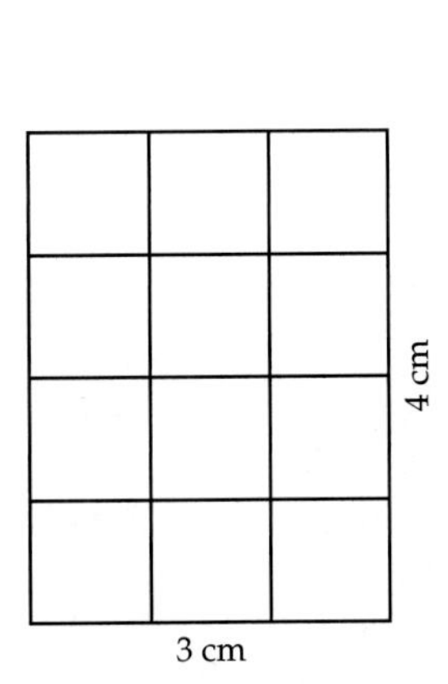

The area of the rectangle is 12 cm^2.

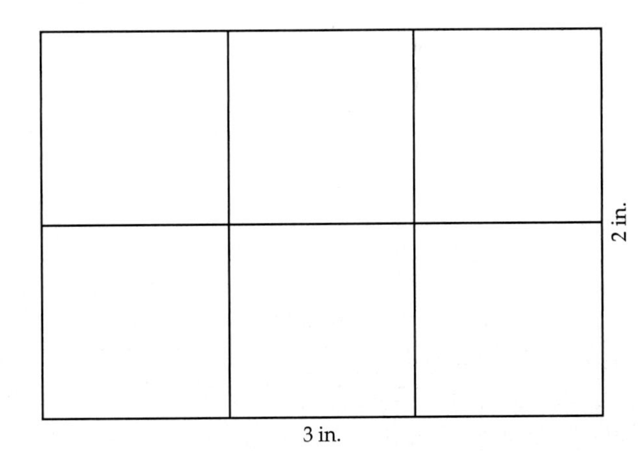

The area of the rectangle is 6 in^2.

Note from these figures that the area of a rectangle can be found by multiplying the length of the rectangle by its width.

ANSWER ***a.*** *The area is 1 square yard, written 1 yd^2.* ***b.*** *The area is 1 square kilometer, written 1 km^2.*

Area of a Rectangle

Let L represent the length and W the width of a rectangle. The area, A, of the rectangle is given by $A = LW$.

QUESTION *How many squares, each 1 inch on a side, are needed to cover a rectangle that has an area of 18 in^2?*

EXAMPLE 7 ■ Find the Area of a Rectangle

How many square feet of sod are needed to cover a football field? A football field measures 360 ft by 160 ft.

Solution

160 ft
360 ft

• Draw a diagram.

$A = LW$ • Use the formula for the area of a rectangle.

$A = 360(160)$ • The length is **360** ft. Substitute **360** for *L*. The width is **160** ft. Substitute **160** for *W*. Remember that *LW* means "*L* times *W*."

$A = 57{,}600$

57,600 ft^2 of sod is needed. • Area is measured in square units.

CHECK YOUR PROGRESS 7 Find the amount of fabric needed to make a rectangular flag that measures 308 cm by 192 cm.

Solution *See page S30.*

TAKE NOTE

Recall that the rules of exponents state that when multiplying variables with like bases, we add the exponents.

A square is a rectangle in which all sides are the same length. Therefore, both the length and the width of a square can be represented by s, and $A = LW = s \cdot s = s^2$.

s

$A = s \cdot s$

$A = s^2$

Area of a Square

Let s represent the length of a side of a square. The area, A, of the square is given by $A = s^2$.

ANSWER *18 squares, each 1 inch on a side, are needed to cover the rectangle.*

EXAMPLE 8 ■ Find the Area of a Square

A homeowner wants to carpet the family room. The floor is square and measures 6 m on each side. How much carpet should be purchased?

Solution

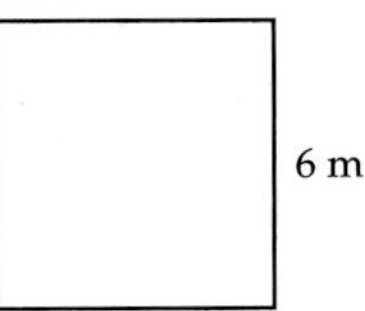

• Draw a diagram.

$A = s^2$ • Use the formula for the area of a square.

$A = 6^2$ • The length of a side is 6 m. Substitute 6 for *s*.

$A = 36$

36 m² of carpet should be purchased. • Area is measured in square units.

CHECK YOUR PROGRESS 8 Find the area of the floor of a two-car garage that is in the shape of a square that measures 24 ft on a side.

Solution *See page S30.*

Figure *ABCD* is a parallelogram. $\overline{BC}$ is the **base** of the parallelogram. $\overline{AE}$, perpendicular to the base, is the **height** of the parallelogram.

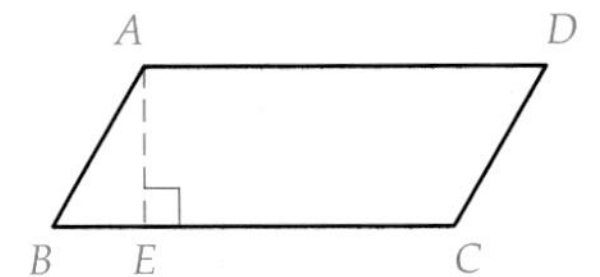

Any side of a parallelogram can be designated as the base. The corresponding height is found by drawing a line segment perpendicular to the base from the opposite side. In the figure at the right, $\overline{CD}$ is the base and $\overline{AE}$ is the height.

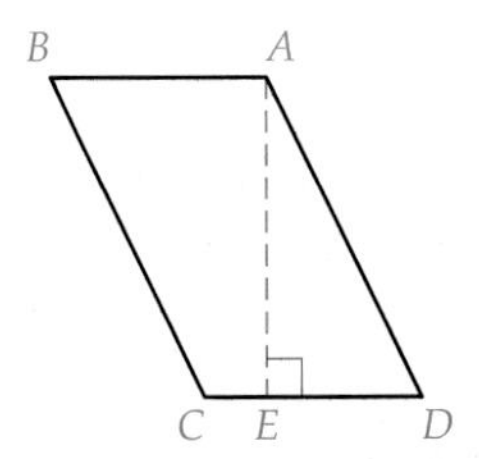

A rectangle can be formed from a parallelogram by cutting a right triangle from one end of the parallelogram and attaching it to the other end. The area of the resulting rectangle will equal the area of the original parallelogram.

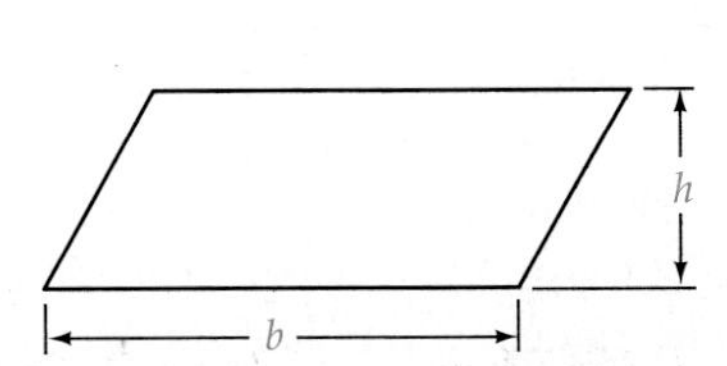

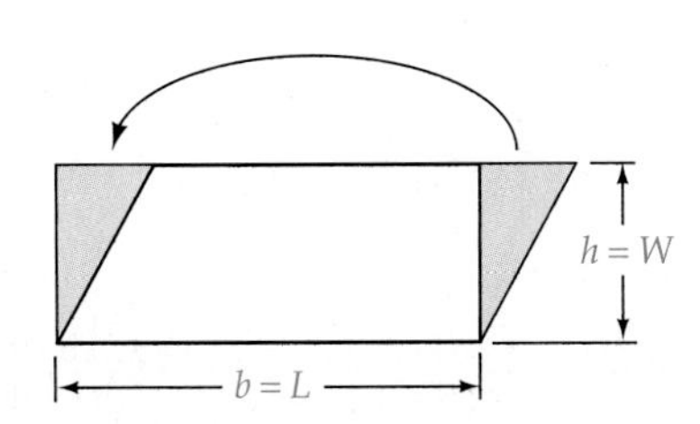

Area of a Parallelogram

Let b represent the length of the base and h the height of a parallelogram. The area, A, of the parallelogram is given by $A = bh$.

EXAMPLE 9 ■ Find the Area of a Parallelogram

A solar panel is in the shape of a parallelogram that has a base of 2 ft and a height of 3 ft. Find the area of the solar panel.

Solution

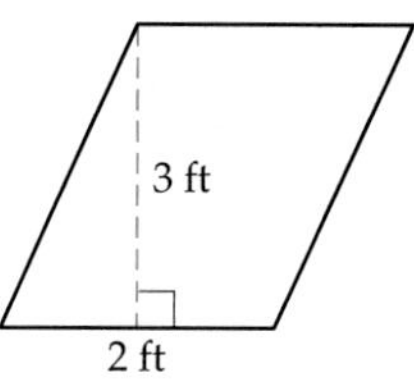

• Draw a diagram.

$A = bh$ • Use the formula for the area of a parallelogram.

$A = 2(3)$ • The base is **2** ft. Substitute **2** for b. The height is **3** ft. Substitute **3** for h. Remember that bh means "b times h."

$A = 6$

The area is 6 ft^2. • Area is measured in square units.

CHECK YOUR PROGRESS 9 A fieldstone patio is in the shape of a parallelogram that has a base measuring 14 m and a height measuring 8 m. What is the area of the patio?

Solution *See page S30.*

Figure ABC is a triangle. $\overline{AB}$ is the **base** of the triangle. $\overline{CD}$, perpendicular to the base, is the **height** of the triangle.

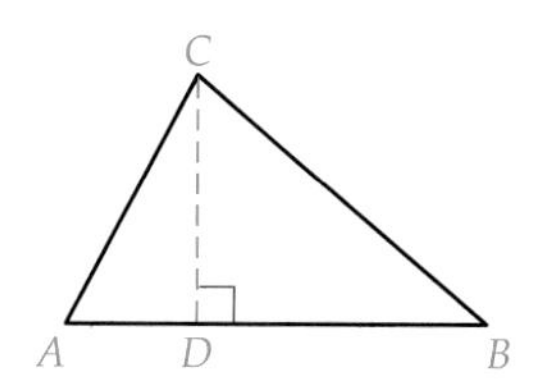

Any side of a triangle can be designated as the base. The corresponding height is found by drawing a line segment perpendicular to the base from the vertex opposite the base.

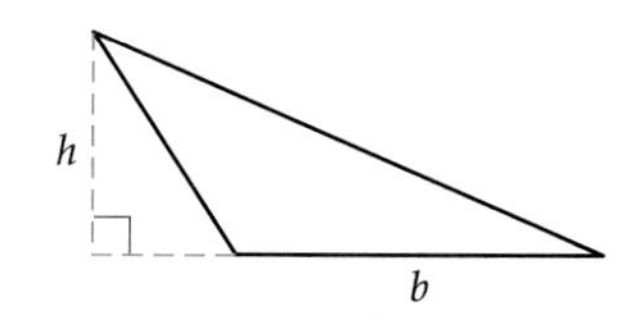

Consider triangle ABC with base b and height h shown at the right. By extending a line segment from C parallel to the base $\overline{AB}$ and equal in length to the base, a parallelogram is formed. The area of the parallelogram is bh and is twice the area of the original triangle. Therefore, the area of the triangle is one half the area of the parallelogram, or $\frac{1}{2}bh$.

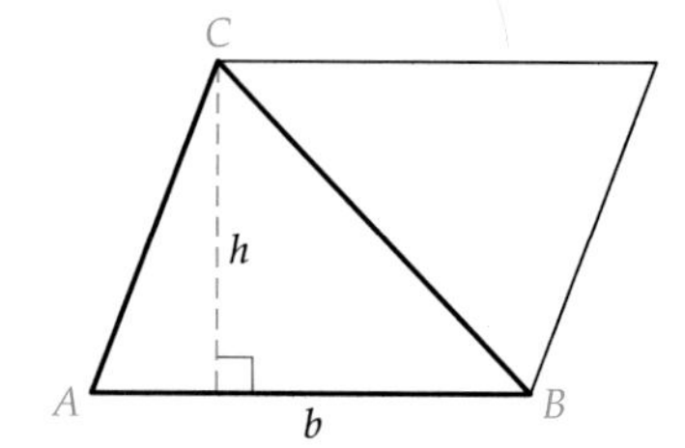

Area of a Triangle

Let b represent the length of the base and h the height of a triangle. The area, A, of the triangle is given by $A = \frac{1}{2}bh$.

EXAMPLE 10 ■ Find the Area of a Triangle

A riveter uses metal plates that are in the shape of a triangle with a base of 12 cm and a height of 6 cm. Find the area of one metal plate.

Solution

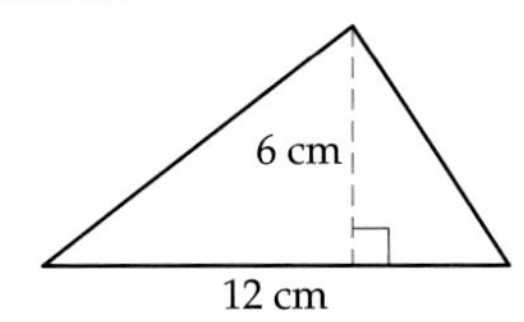

• Draw a diagram.

$$A = \frac{1}{2}bh$$

• Use the formula for the area of a triangle.

$$A = \frac{1}{2}(12)(6)$$

• The base is **12** cm. Substitute **12** for *b*. The height is **6** cm. Substitute **6** for *h*. Remember that *bh* means "*b* times *h*."

$$A = 6(6)$$
$$A = 36$$

The area is 36 cm^2.

• Area is measured in square units.

CHECK YOUR PROGRESS 10 Find the amount of felt needed to make a banner that is in the shape of a triangle with a base of 18 in. and a height of 9 in.

Solution *See page S30.*

TAKE NOTE

The bases of a trapezoid are the parallel sides of the figure.

Figure $ABCD$ is a *trapezoid*. $\overline{AB}$, with length b_1, is one **base** of the trapezoid and $\overline{CD}$, with length b_2, is the other base. $\overline{AE}$, perpendicular to the two bases, is the **height.**

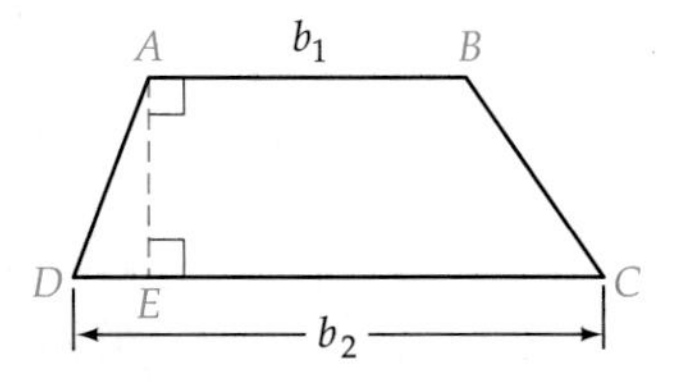

In the trapezoid at the right, the line segment $\overline{BD}$ divides the trapezoid into two triangles, ABD and BCD. In triangle ABD, b_1 is the base and h is the height. In triangle BCD, b_2 is the base and h is the height. The area of the trapezoid is the sum of the areas of the two triangles.

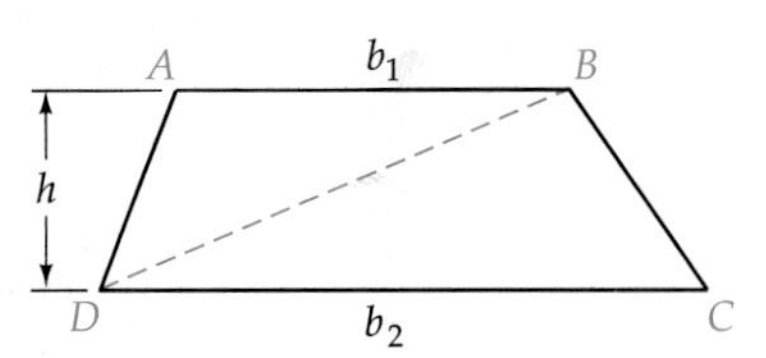

$$\text{Area of trapezoid } ABCD = \text{Area of triangle } ABD + \text{area of triangle } BCD$$
$$= \frac{1}{2}b_1h + \frac{1}{2}b_2h = \frac{1}{2}h(b_1 + b_2)$$

Area of a Trapezoid

Let b_1 and b_2 represent the lengths of the bases and h the height of a trapezoid. The area, A, of the trapezoid is given by $A = \frac{1}{2}h(b_1 + b_2)$.

EXAMPLE 11 ■ Find the Area of a Trapezoid

A boat dock is built in the shape of a trapezoid with bases measuring 14 ft and 6 ft and a height of 7 ft. Find the area of the dock.

Solution

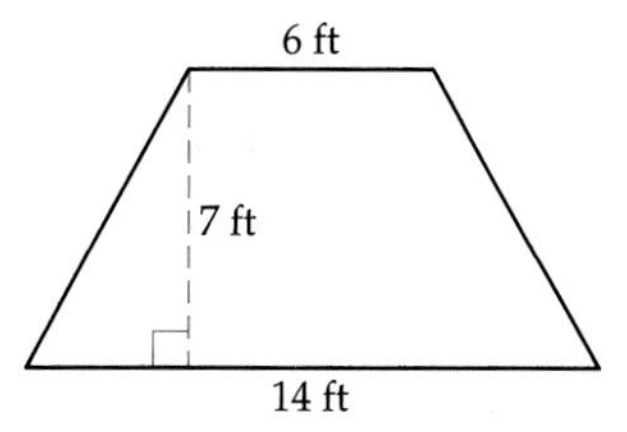

• Draw a diagram.

$$A = \frac{1}{2}h(b_1 + b_2)$$
• Use the formula for the area of a trapezoid.

$$A = \frac{1}{2} \cdot 7(14 + 6)$$
• The height is **7** ft. Substitute **7** for h. The bases measure **14** ft and **6** ft. Substitute **14** and **6** for b_1 and b_2.

$$A = \frac{1}{2} \cdot 7(20)$$
$$A = 70$$

The area is 70 ft^2.
• Area is measured in square units.

CHECK YOUR PROGRESS 11 Find the area of a patio that has the shape of a trapezoid with a height of 9 ft and bases measuring 12 ft and 20 ft.

Solution *See page S30.*

The area of a circle is the product of π and the square of the radius.

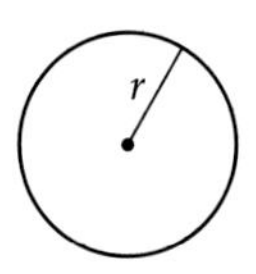

$A = \pi r^2$

The Area of a Circle

The area, A, of a circle with radius of length r is given by $A = \pi r^2$.

TAKE NOTE

For your reference, all of the formulas for the perimeters and areas of the geometric figures presented in this section are listed in the Chapter Summary at the end of this chapter.

Find the area of a circle that has a radius of 6 cm.

Use the formula for the area of a circle. $r = 6$.

$$A = \pi r^2$$
$$A = \pi(6)^2$$
$$A = \pi(36)$$

The exact area of the circle is 36π cm^2.

$$A = 36\pi$$

An approximate measure can be found by using the π key on a calculator.

$$A \approx 113.10$$

The approximate area of the circle is 113.10 cm^2.

EXAMPLE 12 ■ Find the Area of a Circle

Find the area of a circle with a diameter of 10 m. Round to the nearest hundredth of a square meter.

Solution

$r = \frac{1}{2}d = \frac{1}{2}(10) = 5$ • Find the radius of the circle.

$A = \pi r^2$ • Use the formula for the area of a circle.

$A = \pi(5)^2$ • Replace r with **5**.

$A = \pi(25)$ • Square 5.

$A \approx 78.54$ • An approximation is asked for. Use the π key on a calculator.

The area of the circle is approximately 78.54 m^2.

CALCULATOR NOTE

To evaluate the expression $\pi(5)^2$ on your calculator, enter

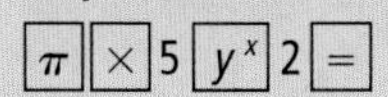

or

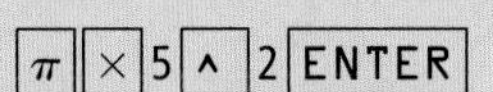

CHECK YOUR PROGRESS 12 Find the area of a circle with a diameter of 12 km. Give the exact measure.

Solution *See page S30.*

EXAMPLE 13 ■ Application of Finding the Area of a Circle

How large a cover is needed for a circular hot tub that is 8 ft in diameter? Round to the nearest tenth of a square foot.

Solution

$r = \frac{1}{2}d = \frac{1}{2}(8) = 4$ • Find the radius of a circle with a diameter of 8 ft.

$A = \pi r^2$ • Use the formula for the area of a circle.

$A = \pi(4)^2$ • Replace r with **4**.

$A = \pi(16)$ • Square 4.

$A \approx 50.3$ • Use the π key on a calculator.

The cover for the hot tub must be 50.3 ft^2.

CHECK YOUR PROGRESS 13 How much material is needed to make a circular tablecloth that is to have a diameter of 4 ft? Round to the nearest hundredth of a square foot.

Solution *See page S30.*

Math Matters Möbius Bands

Cut out a long, narrow rectangular strip of paper.

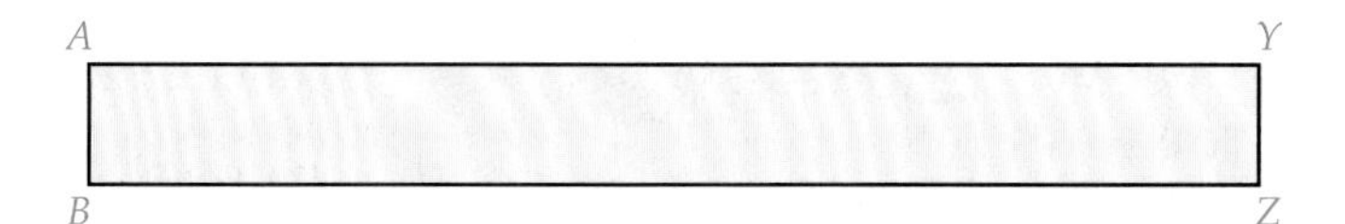

Give the strip of paper a half-twist.

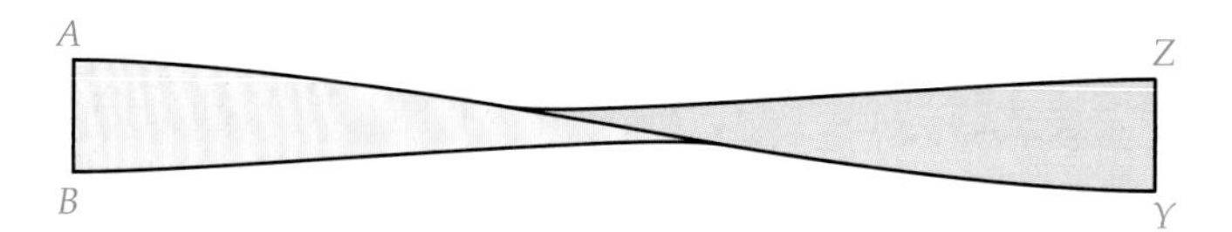

Put the ends together so that A meets Z and B meets Y. Tape the ends together. The result is a *Möbius band.*

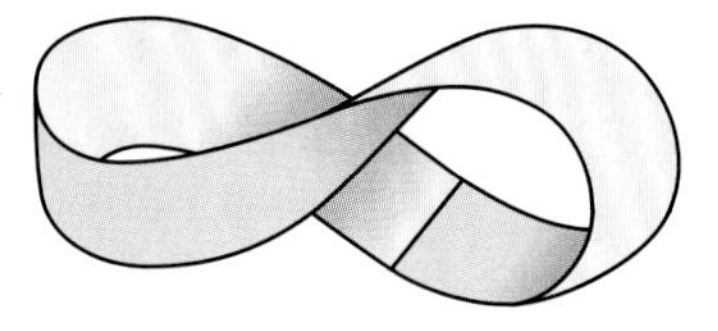

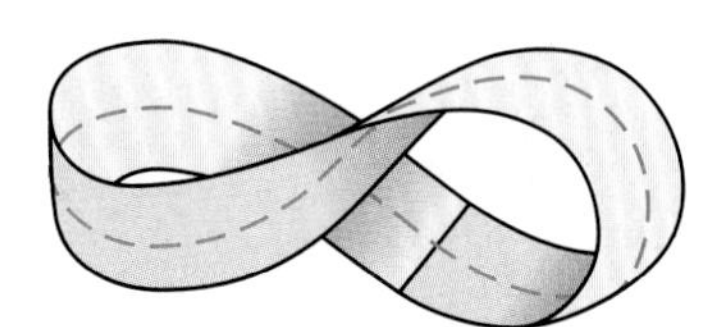

Make a Möbius band that is $1\frac{1}{2}$ in. wide. Use a pair of scissors to cut the Möbius band lengthwise down the middle, staying $\frac{3}{4}$ in. from each edge. Describe the result.

Make a Möbius band from plain white paper and then shade one side. Describe what remains unshaded on the Möbius band, and state the number of sides a Möbius band has.

Excursion

Slicing Polygons into Triangles[1]

Shown at the right is a triangle with three "slices" through it. Notice that the resulting pieces are six quadrilaterals and a triangle.

Excursion Exercises

1. Determine how you can slice a triangle so that all the resulting pieces are triangles. Use three slices. Each slice must cut all the way through the triangle.
2. Determine how you can slice each of the following polygons so that all the resulting pieces are triangles. Each slice must cut all the way through the figure. Record the number of slices required for each polygon. *Note:* There may be more than one solution for a figure.

   a. Quadrilateral **b.** Pentagon **c.** Hexagon **d.** Heptagon

Using Patterns in Experimentation

3. Try to cut a pie into the greatest number of pieces with only five straight cuts of a knife. An illustration showing how five cuts can produce 13 pieces is shown below. The correct answer, however, yields more than thirteen pieces.

   A reasonable question is "How do I know when I have the maximum number of pieces?" To determine the answer, we suggest that you start with one cut, then two cuts, then three cuts, and so on. Try to discover a pattern for the greatest number of pieces that each successive cut can produce.

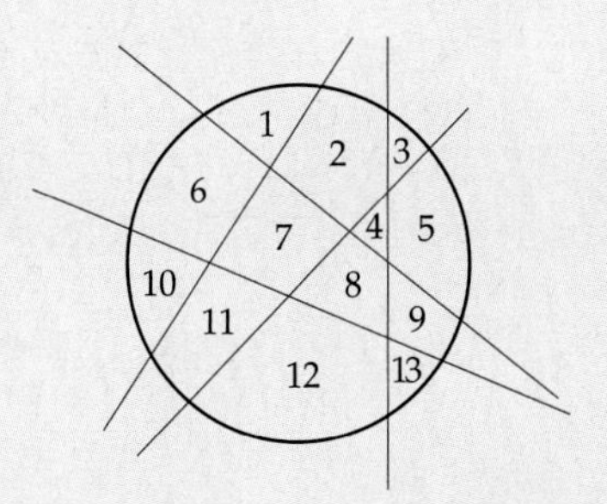

1. This activity is adapted from James Gary Propp's "The Slicing Game," *American Mathematical Monthly,* April 1996.

Exercise Set 8.2

1. Label the length of the rectangle L and the width of the rectangle W.

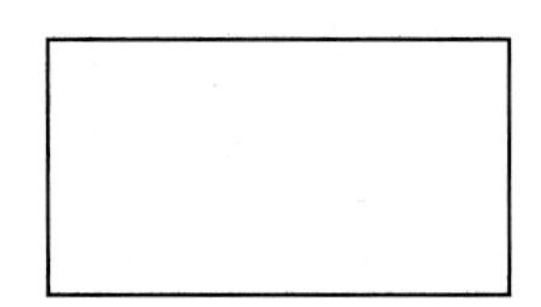

2. Label the base of the parallelogram b, the side s, and the height h.

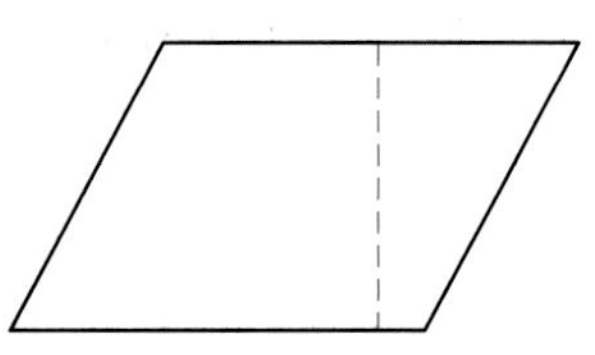

3. What is wrong with each statement?

a. The perimeter is 40 m^2.

b. The area is 120 ft.

In Exercises 4–7, name each polygon.

4.

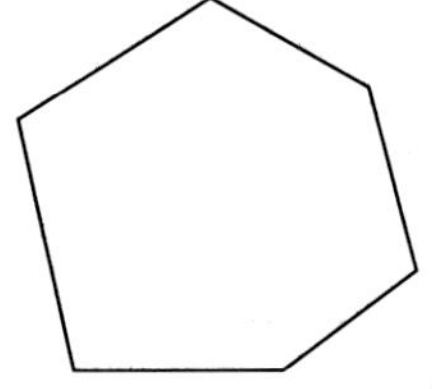

5.

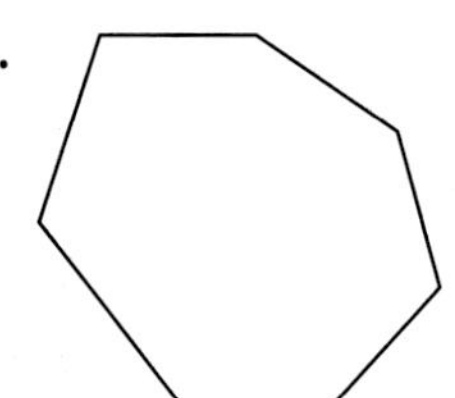

6.

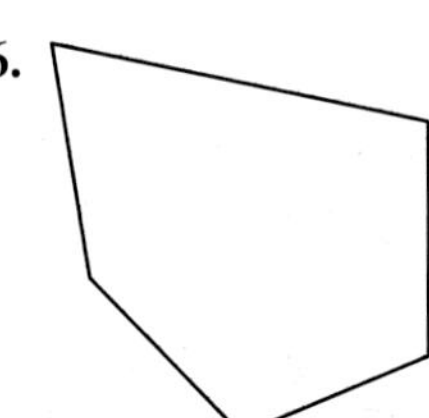

7.

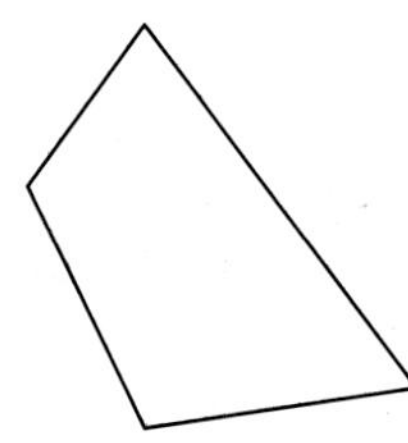

In Exercises 8–11, classify the triangle as isosceles, equilateral, or scalene.

8.

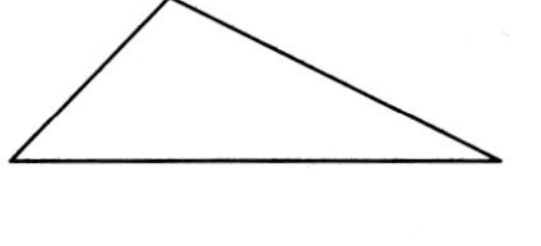

9.

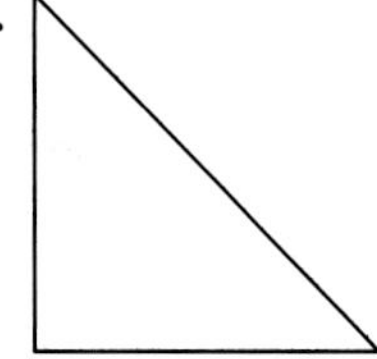

10.

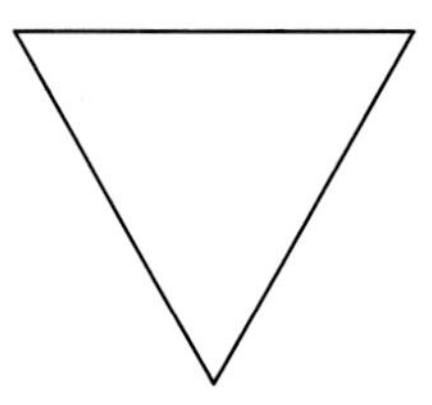

11.

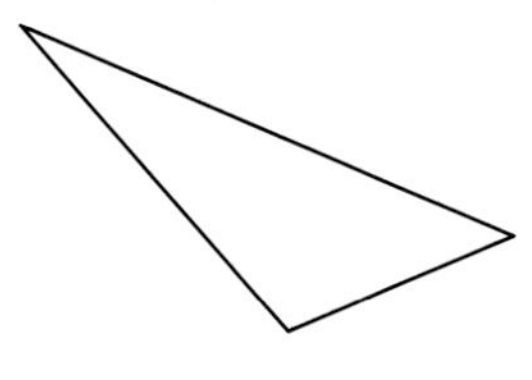

In Exercises 12–15, classify the triangle as acute, obtuse, or right.

12.

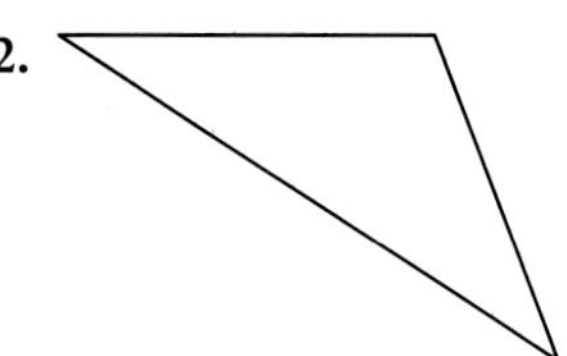

13.

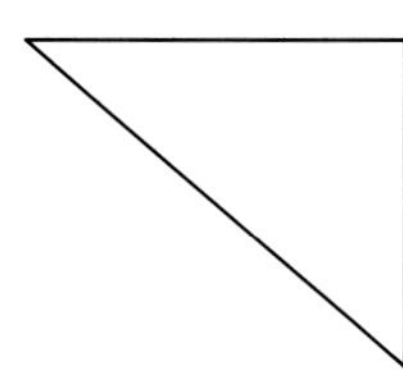

14.

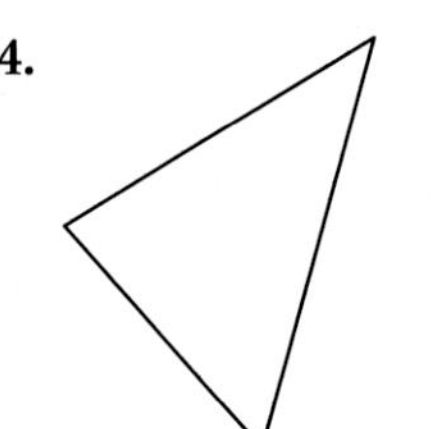

15.

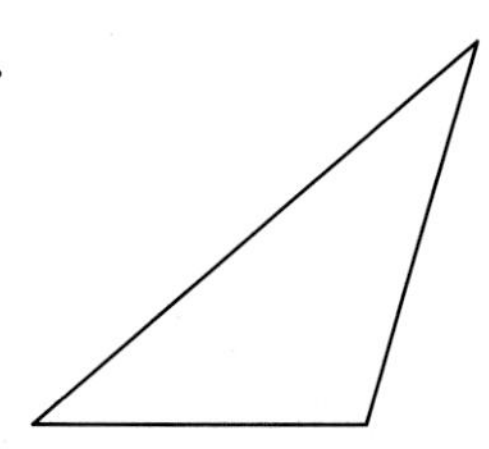

In Exercises 16–24, find (**a**) the perimeter and (**b**) the area of the figure.

16.

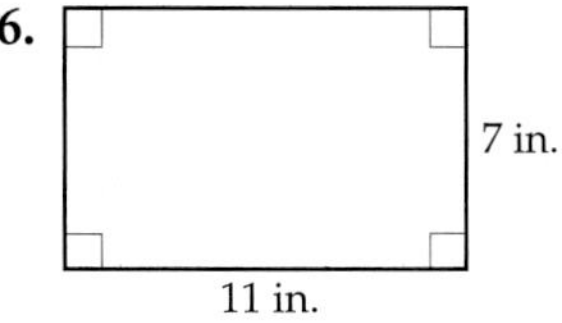

17.

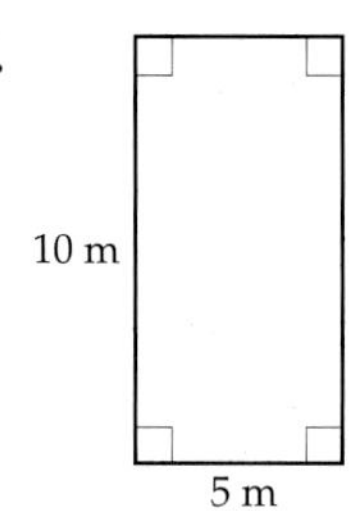

18.

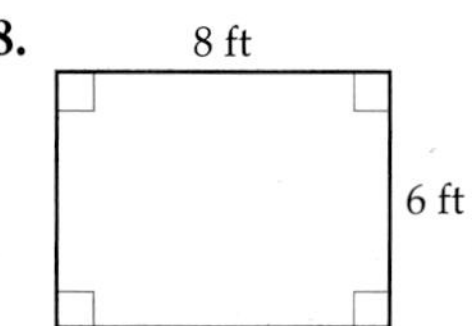

19.

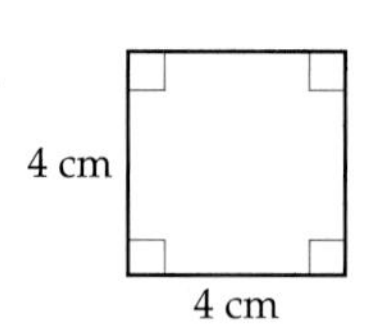

20.

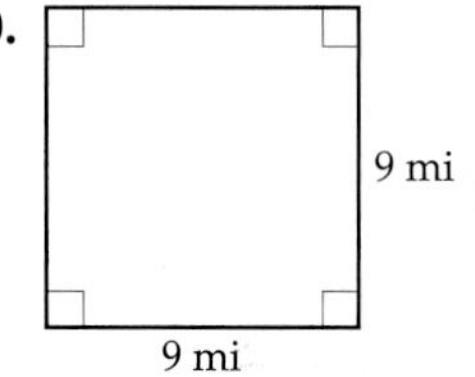

21.

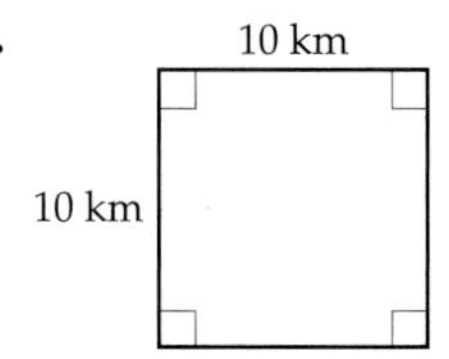

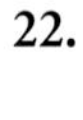

22.

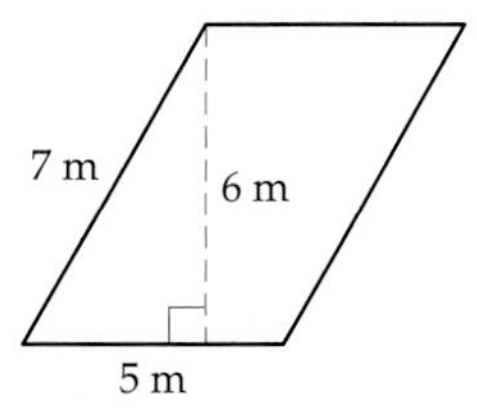

23.

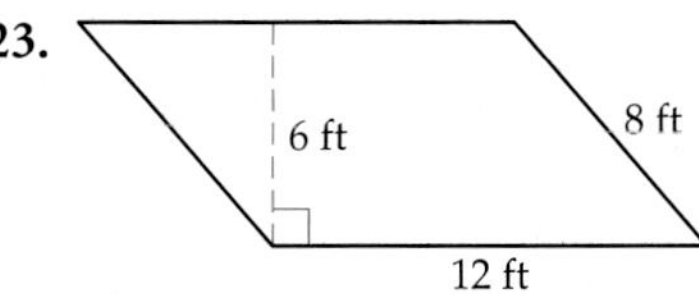

24.

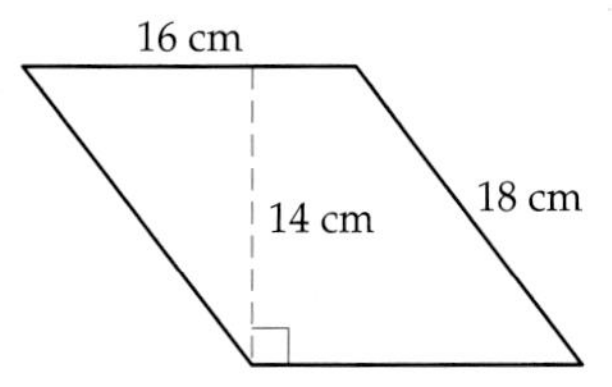

In Exercises 25–30, find (**a**) the circumference and (**b**) the area of the figure. Give both exact values and approximations to the nearest hundredth.

25.

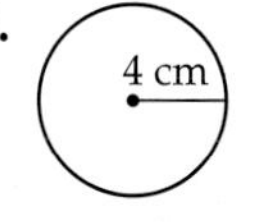

26.

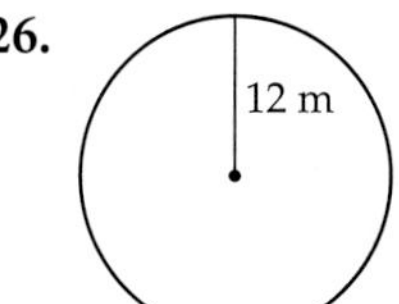

27.

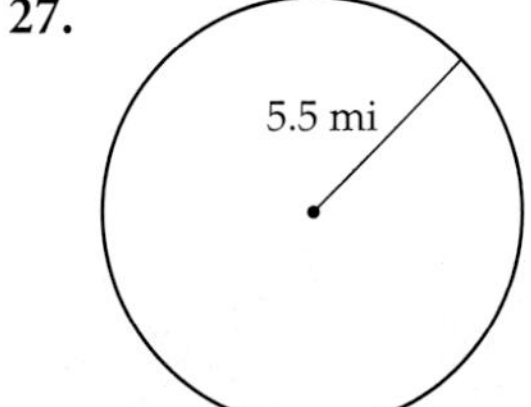

28.

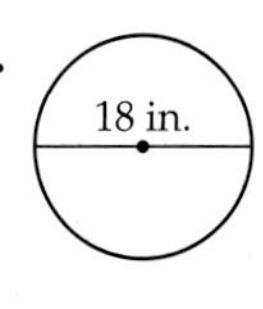

29.

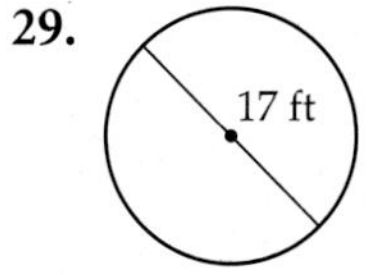

30.

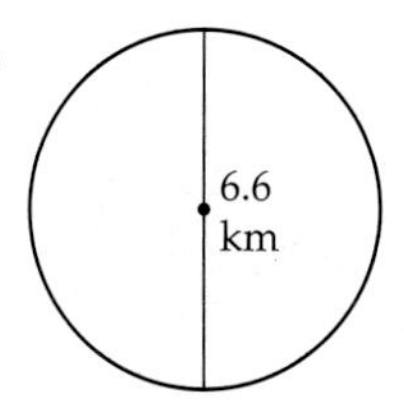

31. Fencing You need to fence in the triangular plot of land shown below. How many feet of fencing do you need?

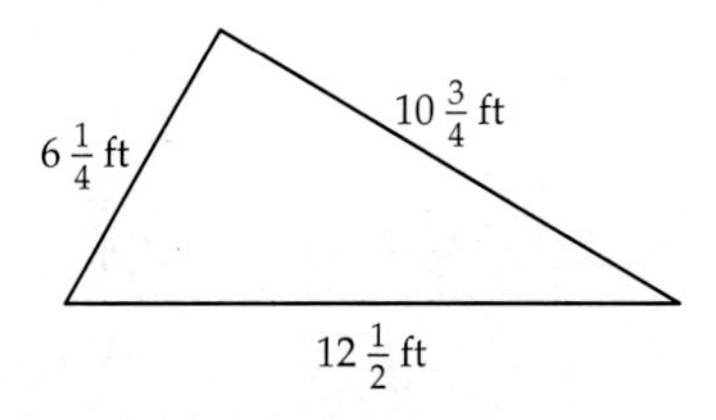

32. Gardens A flower garden in the yard of a historical home is in the shape of a triangle, as shown below. The wooden beams lining the edge of the garden need to be replaced. Find the total length of wood beams that must be purchased in order to replace the old beams.

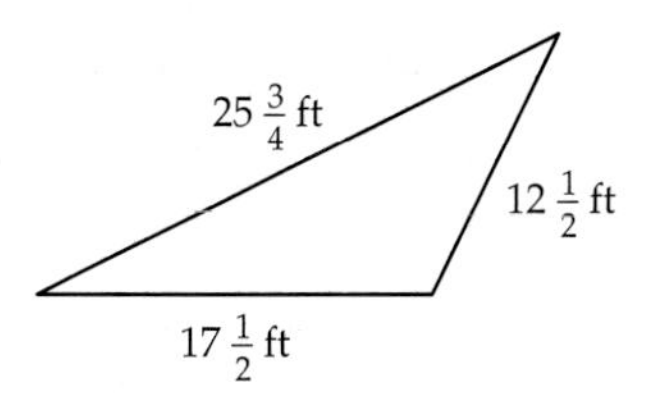

33. Yacht Racing The course of a yachting race is in the shape of a triangle with sides that measure $4\frac{3}{10}$ mi, $3\frac{7}{10}$ mi, and $2\frac{1}{2}$ mi. Find the total length of the course.

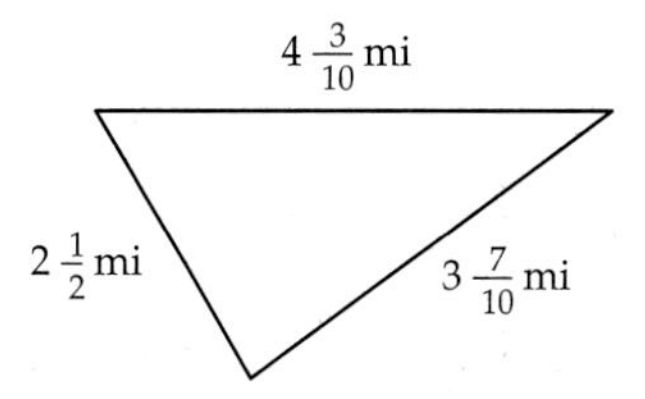

34. Physical Fitness An exercise course has stations set up along a path that is in the shape of a triangle with sides that measure $12\frac{1}{12}$ yd, $29\frac{1}{3}$ yd, and $26\frac{3}{4}$ yd. What is the entire length of the exercise course?

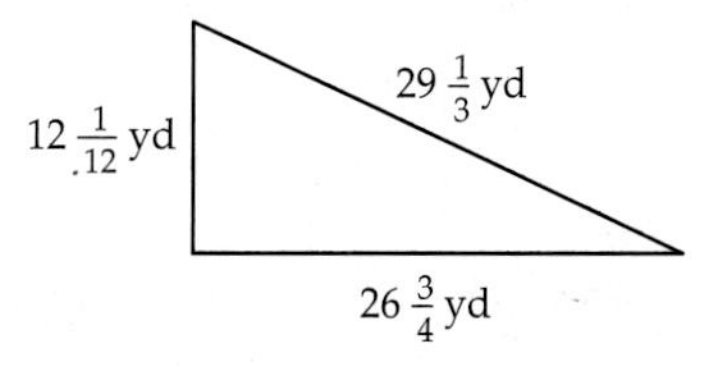

35. Fencing How many feet of fencing should be purchased to enclose a rectangular garden that is 20 ft long and 14 ft wide?

36. Sewing How many meters of binding are required to bind the edge of a rectangular quilt that measures 4 m by 6 m?

37. Perimeter Find the perimeter of a regular pentagon that measures 4 in. on each side.

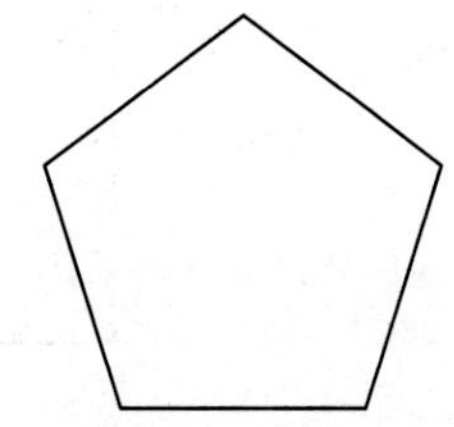

38. **Interior Decorating** Wall-to-wall carpeting is installed in a room that is 15 ft long and 10 ft wide. The edges of the carpet are held down by tack strips. How many feet of tack-strip material are needed?

39. **Parks and Recreation** The length of a rectangular park is 62 yd. The width is 45 yd. How many yards of fencing are needed to surround the park?

40. **Perimeter** What is the perimeter of a regular hexagon that measures 9 cm on each side?

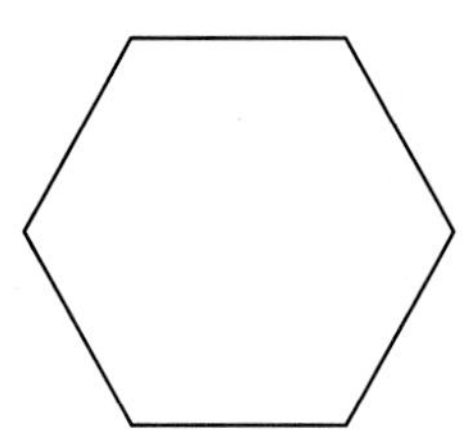

41. **Cross-Country** A cross-country course is in the shape of a parallelogram with a base of length 3 mi and a side of length 2 mi. What is the total length of the cross-country course?

42. **Parks and Recreation** A rectangular playground has a length of 160 ft and a width of 120 ft. Find the length of hedge that surrounds the playground.

43. **Sewing** Bias binding is to be sewn around the edge of a rectangular tablecloth measuring 68 in. by 42 in. If the bias binding comes in packages containing 15 ft of binding, how many packages of bias binding are needed for the tablecloth?

44. **Gardens** Find the area of a rectangular flower garden that measures 24 ft by 18 ft.

45. **Construction** What is the area of a square patio that measures 12 m on each side?

46. **Athletic Fields** Artificial turf is being used to cover a playing field. If the field is rectangular with a length of 110 yd and a width of 80 yd, how much artificial turf must be purchased to cover the field?

47. **Framing** The perimeter of a square picture frame is 36 in. Find the length of each side of the frame.

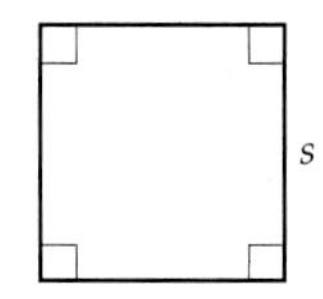

48. **Carpeting** A square rug has a perimeter of 24 ft. Find the length of each edge of the rug.

49. **Area** The area of a rectangle is 400 in^2. If the length of the rectangle is 40 in., what is the width?

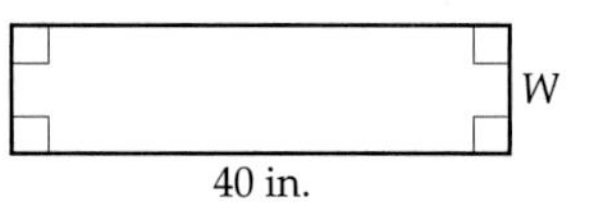

50. **Area** The width of a rectangle is 8 ft. If the area is 312 ft^2, what is the length of the rectangle?

51. **Area** The area of a parallelogram is 56 m^2. If the height of the parallelogram is 7 m, what is the length of the base?

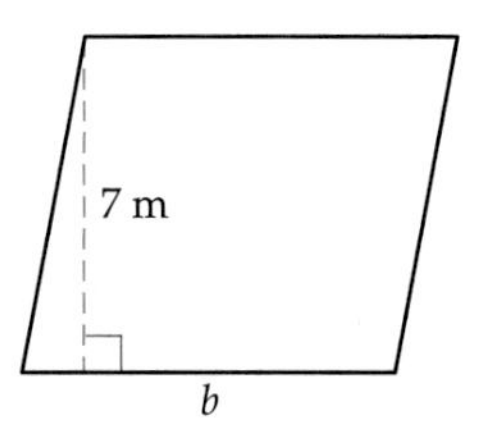

52. **Storage Units** You want to rent a storage unit. You estimate that you will need 175 ft^2 of floor space. In the Yellow Pages, you see the ad shown below. You want to rent the smallest possible unit that will hold everything you want to store. Which of the six units pictured in the ad should you select?

53. **Sailing** A sail is in the shape of a triangle with a base of 12 m and a height of 16 m. How much canvas was needed to make the body of the sail?

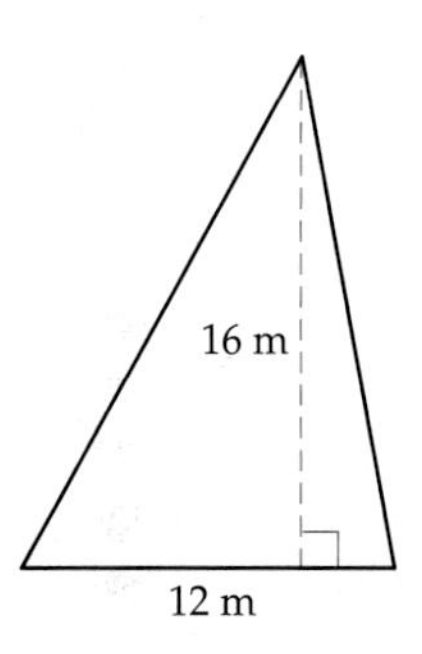

54. Gardens A vegetable garden is in the shape of a triangle with a base of 21 ft and a height of 13 ft. Find the area of the vegetable garden.

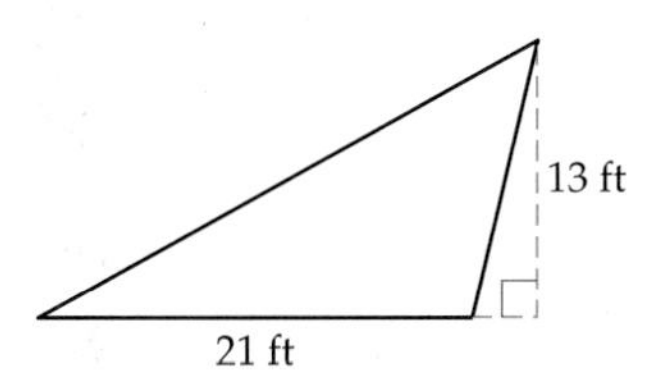

55. Athletic Fields How much artificial turf should be purchased to cover an athletic field that is in the shape of a trapezoid with a height of 15 m and bases that measure 45 m and 36 m?

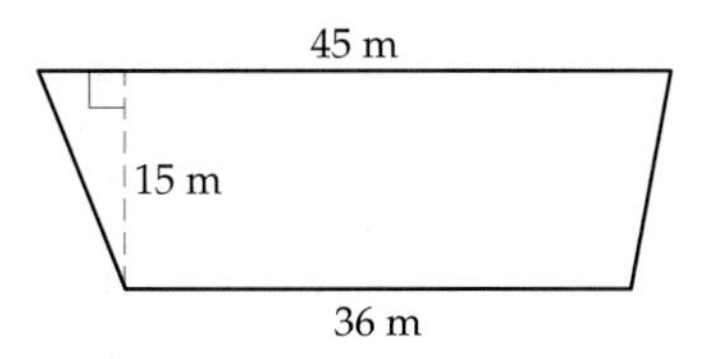

56. Land Area A township is in the shape of a trapezoid with a height of 10 km and bases measuring 9 km and 23 km. What is the land area of the township?

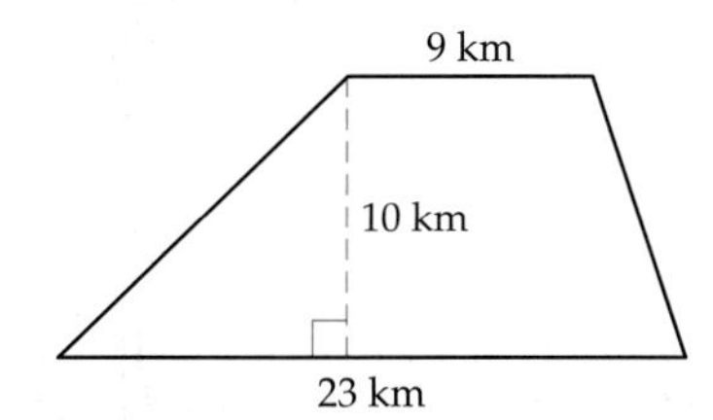

57. Parks and Recreation A city plans to plant grass seed in a public playground that has the shape of a triangle with a height of 24 m and a base of 20 m. Each bag of grass seed will seed 120 m². How many bags of seed should be purchased?

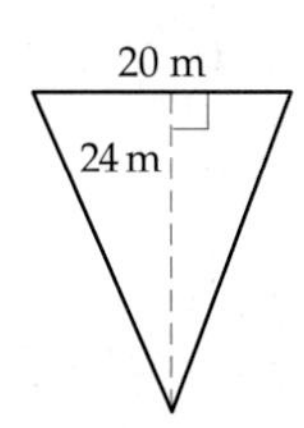

58. Interior Decorating The family room of a home is on the third floor of the house. Because of the pitch of the roof, two walls of the room are in the shape of a trapezoid with bases that measure 16 ft and 24 ft and a height of 8 ft. You plan to wallpaper the two walls. One roll of wallpaper will cover 40 ft². How many rolls of wallpaper should you purchase?

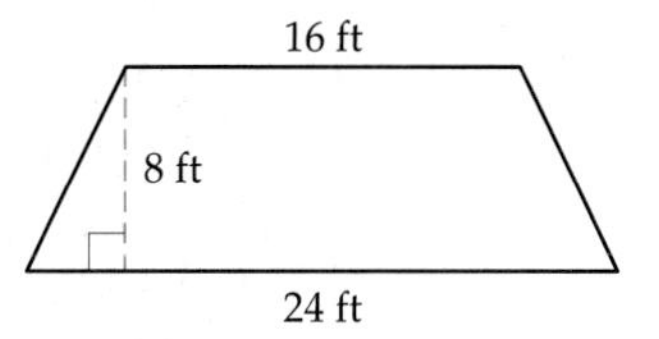

59. Home Maintenance You plan to stain the wooden deck at the back of your house. The deck is in the shape of a trapezoid with bases that measure 10 ft and 12 ft and a height of 10 ft. A quart of stain will cover 55 ft². How many quarts of stain should you purchase?

60. Interior Decorating A fabric wall hanging is in the shape of a triangle that has a base of 4 ft and a height of 3 ft. An additional 1 ft² of fabric is needed for hemming the material. How much fabric should be purchased to make the wall hanging?

61. Interior Decorating You want to tile your kitchen floor. The floor measures 10 ft by 8 ft. How many square tiles that measure 2 ft along each side should you purchase for the job?

62. Interior Decorating You are wallpapering two walls of a den, one measuring 10 ft by 8 ft and the other measuring 12 ft by 8 ft. The wallpaper costs $24 per roll, and each roll will cover 40 ft². What is the cost to wallpaper the two walls?

63. Gardens An urban renewal project involves reseeding a garden that is in the shape of a square, 80 ft on each side. Each bag of grass seed costs $8 and will seed 1500 ft². How much money should be budgeted for buying grass seed for the garden?

64. Carpeting You want to install wall-to-wall carpeting in the family room. The floor plan is drawn below. If the cost of the carpet you would like to purchase is $19 per square yard, what is the cost of carpeting your family room? Assume there is no waste. *Hint:* $9 \text{ ft}^2 = 1 \text{ yd}^2$

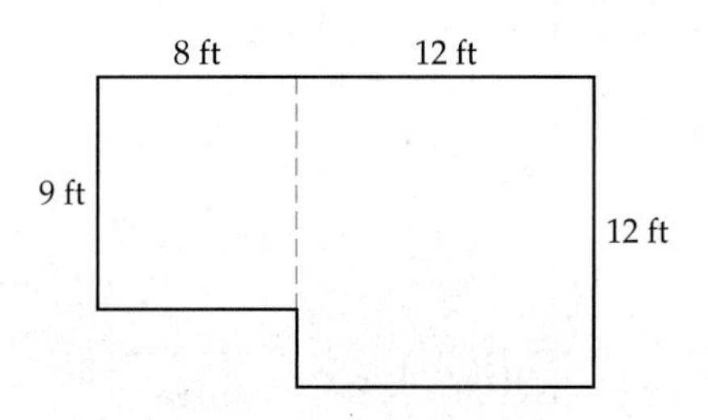

65. **Interior Decorating** You want to paint the walls of your bedroom. Two walls measure 16 ft by 8 ft, and the other two walls measure 12 ft by 8 ft. The paint you wish to purchase costs \$17 per gallon, and each gallon will cover 400 ft^2 of wall. Find the total amount you will spend on paint.

66. **Landscaping** A walkway 2 m wide surrounds a rectangular plot of grass. The plot is 25 m long and 15 m wide. What is the area of the walkway?

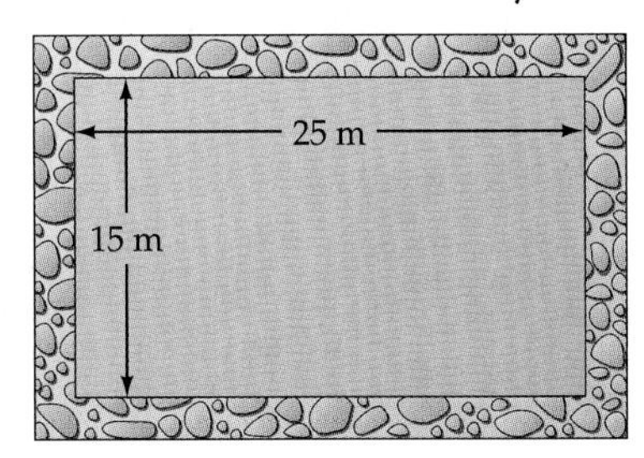

67. **Draperies** The material used to make pleated draperies for a window must be twice as wide as the width of the window. Draperies are being made for four windows, each 3 ft wide and 4 ft high. Because the drapes will fall slightly below the window sill and extra fabric is needed for hemming the drapes, 1 ft must be added to the height of the window. How much material must be purchased to make the drapes?

68. **Fencing** How many feet of fencing should be purchased to enclose a circular flower garden that has a diameter of 18 ft? Round to the nearest tenth of a foot.

69. **Carpentry** Find the length of molding needed to put around a circular table that is 4.2 ft in diameter. Round to the nearest hundredth of a foot.

70. **Sewing** How much binding is needed to bind the edge of a circular rug that is 3 m in diameter? Round to the nearest hundredth of a meter.

71. **Gardens** Find the area of a circular flower garden that has a radius of 20 ft. Round to the nearest tenth of a square foot.

72. **Pulleys** A pulley system is diagrammed below. If pulley B has a diameter of 16 in. and is rotating at 240 revolutions per minute, how far does the belt travel each minute that the pulley system is in operation? Assume the belt does not slip as the pulley rotates. Round to the nearest inch.

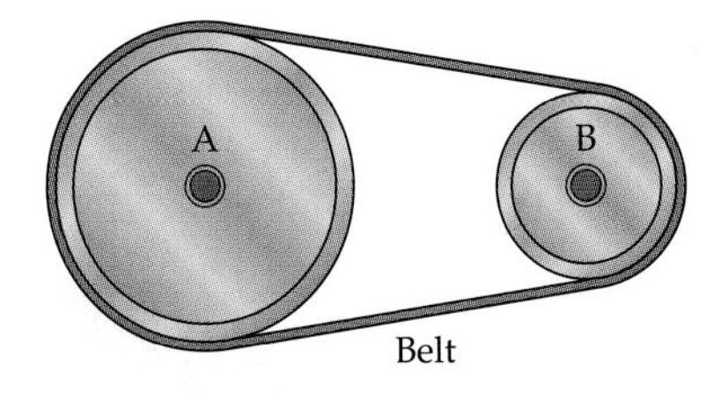

73. **Bicycles** A bicycle tire has a diameter of 18 in. How many feet does the bicycle travel when the wheel makes 20 revolutions? Round to the nearest hundredth of a foot.

74. **Tricycles** The front wheel of a tricycle has a diameter of 16 in. How many feet does the tricycle travel when the wheel makes 15 revolutions? Round to the nearest hundredth of a foot.

75. **Telescopes** The lens located on an astronomical telescope has a diameter of 24 in. Find the exact area of the lens.

76. **Irrigation** An irrigation system waters a circular field that has a 50-foot radius. Find the exact area watered by the irrigation system.

77. **Pizza** How much greater is the area of a pizza that has a radius of 10 in. than the area of a pizza that has a radius of 8 in.? Round to the nearest hundredth of a square inch.

78. **Pizza** A restaurant serves a small pizza that has a radius of 6 in. The restaurant's large pizza has a radius that is twice the radius of the small pizza. How much larger is the area of the large pizza? Round to the nearest hundredth of a square inch. Is the area of the large pizza more or less than twice the area of the small pizza?

79. **Satellites** There are two general types of satellite systems that orbit our Earth: geostationary Earth orbit (GEO) and non-geostationary, primarily low Earth orbit (LEO). Geostationary satellite systems orbit at a distance of 36,000 km above Earth. An orbit at this altitude allows the satellite to maintain a fixed position in relation to Earth. What is the distance traveled by a GEO satellite in one orbit around Earth? The radius of Earth at the equator is 6380 km. Round to the nearest kilometer.

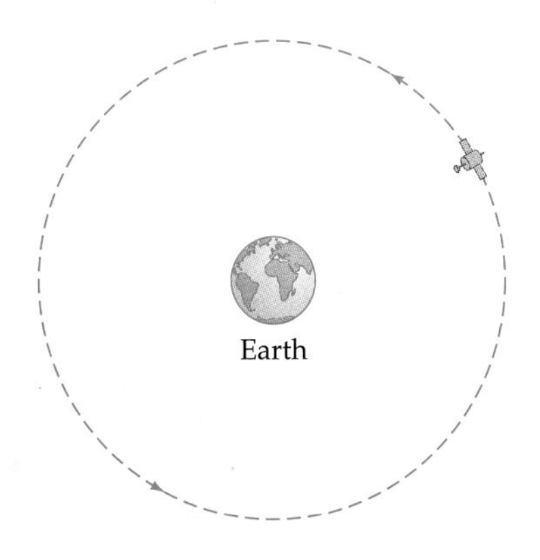

80. Ball Fields How much farther is it around the bases of a baseball diamond than around the bases of a softball diamond? *Hint:* Baseball and softball diamonds are squares.

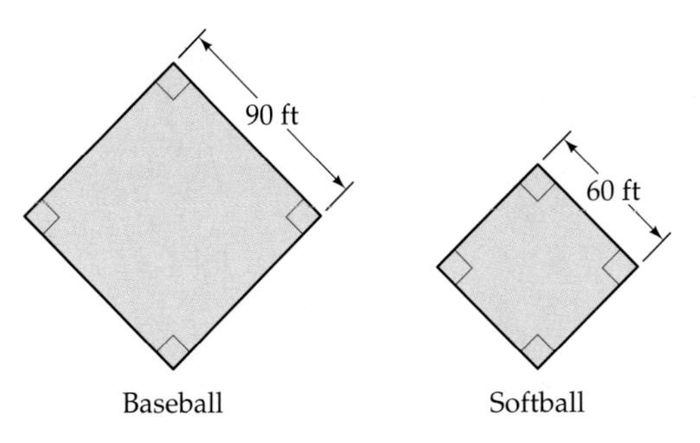

81. Area Write an expression for the area of the shaded portion of the diagram. Leave the answer in terms of π and r.

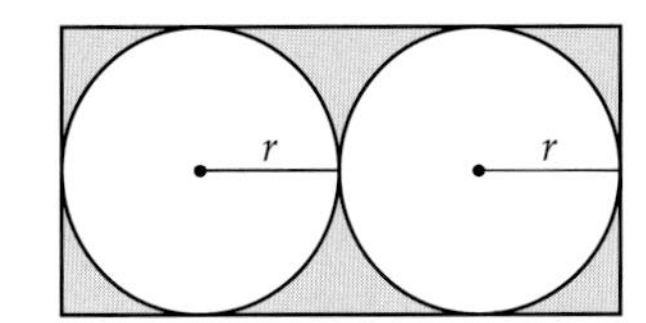

82. Area Write an expression for the area of the shaded portion of the diagram. Leave the answer in terms of π and r.

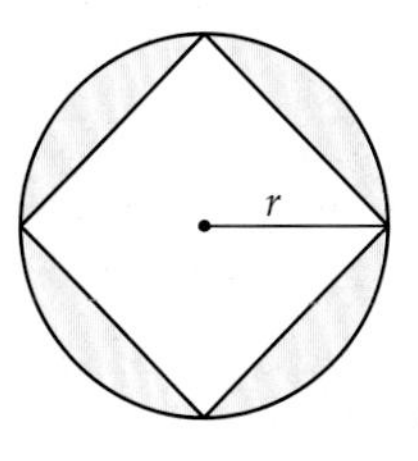

83. Area If both the length and width of a rectangle are doubled, how many times as large is the area of the resulting rectangle?

Extensions

CRITICAL THINKING

84. Determine whether the statement is always true, sometimes true, or never true.

a. Two triangles that have the same perimeter have the same area.

b. Two rectangles that have the same area have the same perimeter.

c. If two squares have the same area, then the sides of the squares have the same length.

85. Find the dimensions of a rectangle that has the same area as the shaded region in the diagram below. Write the dimensions in terms of the variable a.

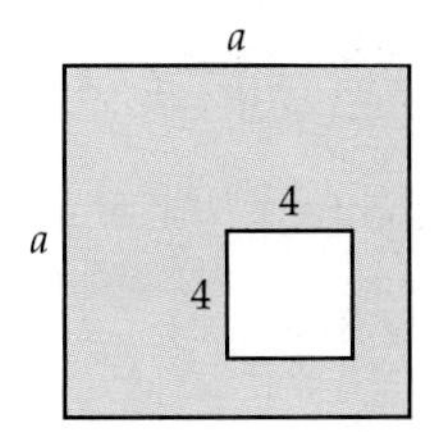

86. Find the dimensions of a rectangle that has the same area as the shaded region in the diagram below. Write the dimensions in terms of the variable x.

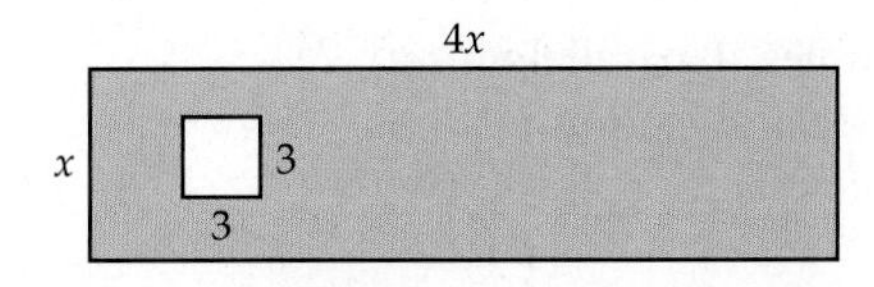

87. In the diagrams below, the length of one side of a square is 1 cm. Find the perimeter and the area of the eighth figure in the pattern.

a. ...

b. ...

c. ...

EXPLORATIONS

88. The perimeter of the square at the right is 4 units.

If two such squares are joined along one of the sides, the perimeter is 6 units. Note that it does not matter which sides are joined; the perimeter is still 6 units.

If three squares are joined, the perimeter of the resulting figure is 8 units for each possible placement of the squares.

Four squares can be joined in five different ways as shown. There are two possible perimeters: 10 units for A, B, C, and D, and 8 units for E.

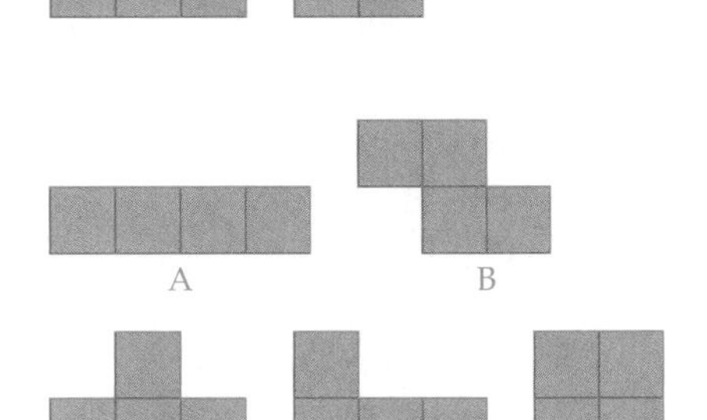

a. If five squares are joined, what is the maximum perimeter possible?

b. If five squares are joined, what is the minimum perimeter possible?

c. If six squares are joined, what is the maximum perimeter possible?

d. If six squares are joined, what is the minimum perimeter possible?

89. Suppose a circle is cut into 16 equal pieces, which are then arranged as shown at the right. The figure formed resembles a parallelogram. What variable expression could describe the base of the parallelogram? What variable could describe its height? Explain how the formula for the area of a circle is derived from this approach.

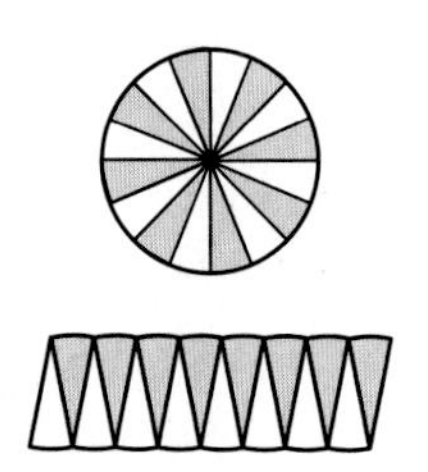

SECTION 8.3 Properties of Triangles

point of interest

Model trains are similar to actual trains. They come in a variety of sizes that manufacturers have agreed upon, so that, for example, an engine made by manufacturer A is able to run on a track made by manufacturer B. Listed below are three model railroad sizes by name, along with the ratio of model size to actual size.

Name	*Ratio*
Z	1:220
N	1:160
HO	1:87

Z Scale

N Scale

HO Scale

HO's ratio of 1:87 means that in every dimension, an HO scale model railroad car is $\frac{1}{87}$ the size of the real railroad car.

Similar Triangles

Similar objects have the same shape but not necessarily the same size. A tennis ball is similar to a basketball. A model ship is similar to an actual ship.

Similar objects have corresponding parts; for example, the rudder on the model ship corresponds to the rudder on the actual ship. The relationship between the sizes of each of the corresponding parts can be written as a ratio, and each ratio will be the same. If the rudder on the model ship is $\frac{1}{100}$ the size of the rudder on the actual ship, then the model wheelhouse is $\frac{1}{100}$ of the size of the actual wheelhouse, the width of the model is $\frac{1}{100}$ the width of the actual ship, and so on.

The two triangles ABC and DEF shown at the right are similar. Side $\overline{AB}$ corresponds to side $\overline{DE}$, side $\overline{BC}$ corresponds to side $\overline{EF}$, and side $\overline{AC}$ corresponds to side $\overline{DF}$. The ratios of the lengths of corresponding sides are equal.

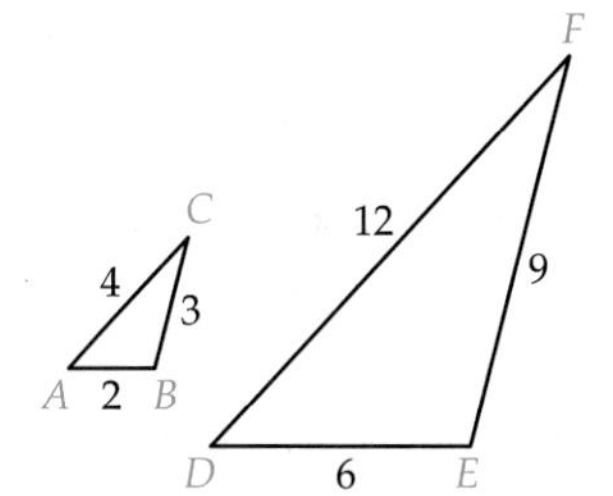

$$\frac{AB}{DE} = \frac{2}{6} = \frac{1}{3}, \quad \frac{BC}{EF} = \frac{3}{9} = \frac{1}{3}, \quad \text{and}$$
$$\frac{AC}{DF} = \frac{4}{12} = \frac{1}{3}$$

Because the ratios of corresponding sides are equal, several proportions can be formed.

$$\frac{AB}{DE} = \frac{BC}{EF}, \quad \frac{AB}{DE} = \frac{AC}{DF}, \quad \text{and} \quad \frac{BC}{EF} = \frac{AC}{DF}$$

The measures of corresponding angles in similar triangles are equal. Therefore,

$$m\angle A = m\angle D, \quad m\angle B = m\angle E, \quad \text{and}$$
$$m\angle C = m\angle F$$

Triangles ABC and DEF at the right are similar triangles. AH and DK are the heights of the triangles. The ratio of the heights of similar triangles equals the ratio of the lengths of corresponding sides.

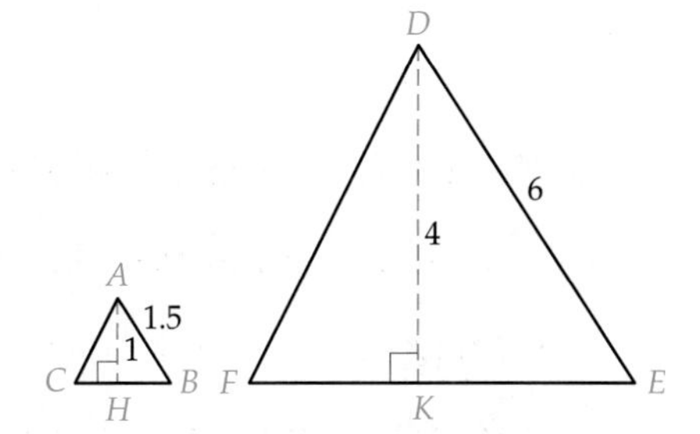

Ratio of corresponding sides $= \frac{1.5}{6} = \frac{1}{4}$

Ratio of heights $= \frac{1}{4}$

Properties of Similar Triangles

For similar triangles, the ratios of corresponding sides are equal. The ratio of corresponding heights is equal to the ratio of corresponding sides.

> **TAKE NOTE**
>
> The notation $\triangle ABC \sim \triangle DEF$ is used to indicate that triangle ABC is similar to triangle DEF.

The two triangles at the right are similar triangles. Find the length of side $\overline{EF}$. Round to the nearest tenth of a meter.

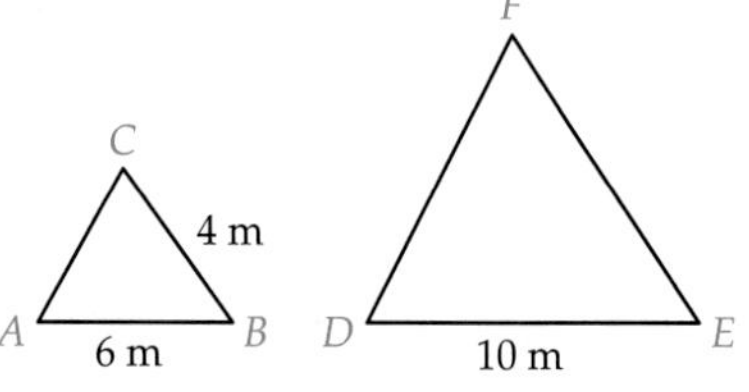

The triangles are similar, so the ratios of the lengths of corresponding sides are equal.

$$\frac{EF}{BC} = \frac{DE}{AB}$$
$$\frac{EF}{4} = \frac{10}{6}$$
$$6(EF) = 4(10)$$
$$6(EF) = 40$$
$$EF \approx 6.7$$

The length of side EF is approximately 6.7 m.

QUESTION *What are two other proportions that can be written for the similar triangles shown above?*

EXAMPLE 1 ■ Use Similar Triangles to Find the Unknown Height of a Triangle

Triangles ABC and DEF are similar. Find FG, the height of triangle DEF.

Solution

$$\frac{AB}{DE} = \frac{CH}{FG}$$
• For similar triangles, the ratio of corresponding sides equals the ratio of corresponding heights.

$$\frac{8}{12} = \frac{4}{FG}$$
• Replace AB, DE, and CH with their values.

$$8(FG) = 12(4)$$
• The cross products are equal.

$$8(FG) = 48$$
$$FG = 6$$
• Divide both sides of the equation by 8.

The height FG of triangle DEF is 6 cm.

CHECK YOUR PROGRESS 1

Triangles ABC and DEF are similar. Find FG, the height of triangle DEF.

Solution *See page S30.*

ANSWER *In addition to* $\frac{EF}{BC} = \frac{DE}{AB}$, *we can write the proportions* $\frac{DE}{AB} = \frac{DF}{AC}$ *and* $\frac{EF}{BC} = \frac{DF}{AC}$. *These three proportions can also be written using the reciprocal of each fraction:* $\frac{BC}{EF} = \frac{AB}{DE}$, $\frac{AB}{DE} = \frac{AC}{DF}$, *and* $\frac{BC}{EF} = \frac{AC}{DF}$. *Also, the right and left sides of each proportion can be interchanged.*

Triangles ABC and DEF are similar triangles. Find the area of triangle ABC.

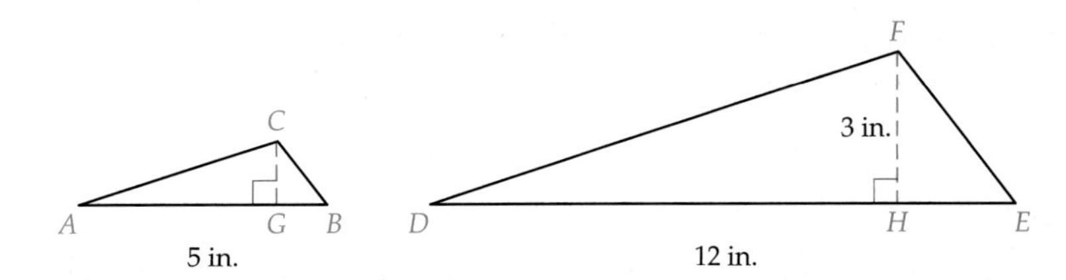

Solve a proportion to find the height of triangle ABC.

$$\frac{AB}{DE} = \frac{CG}{FH}$$

$$\frac{5}{12} = \frac{CG}{3}$$

$$12(CG) = 5(3)$$

$$12(CG) = 15$$

$$CG = 1.25$$

Use the formula for the area of a triangle.

$$A = \frac{1}{2}bh$$

The base is 5 in. The height is 1.25 in.

$$A = \frac{1}{2}(5)(1.25)$$

$$A = 3.125$$

The area of triangle ABC is 3.125 in^2.

If the three angles of one triangle are equal in measure to the three angles of another triangle, then the triangles are similar.

In triangle ABC at the right, line segment $\overline{DE}$ is drawn parallel to the base $\overline{AB}$. Because the measures of corresponding angles are equal, $m\angle x = m\angle r$ and $m\angle y = m\angle n$. We know that $m\angle C = m\angle C$. Thus the measures of the three angles of triangle ABC are equal, respectively, to the measures of the three angles of triangle DEC. Therefore, triangles ABC and DEC are similar triangles.

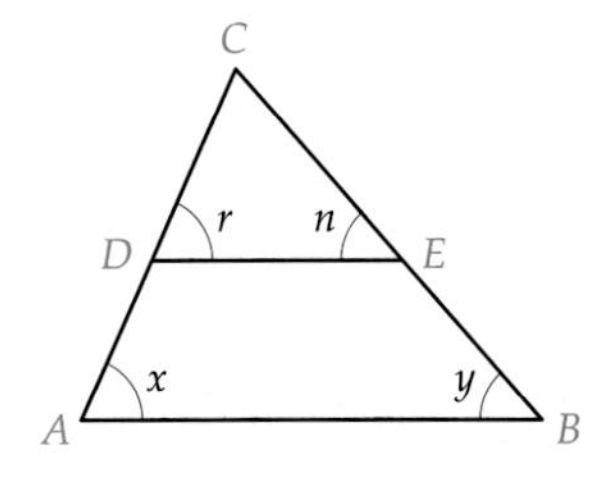

TAKE NOTE

You can always create similar triangles by drawing a line segment inside the original triangle parallel to one side of the triangle. In the triangle below, $\overline{ST} \parallel \overline{QR}$ and triangle PST is similar to triangle PQR.

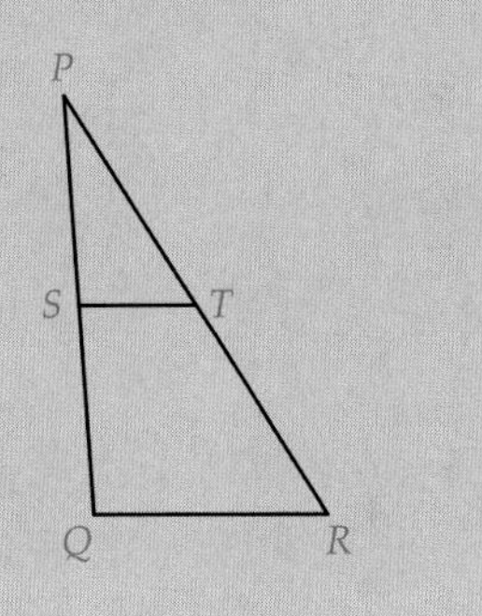

The sum of the measures of the three angles of a triangle is 180°. If two angles of one triangle are equal in measure to two angles of another triangle, then the third angles must be equal. Thus, we can say that if two angles of one triangle are equal in measure to two angles of another triangle, then the two triangles are similar.

In the figure at the right, $\overline{AB}$ intersects $\overline{CD}$ at point O. Angles C and D are right angles. Find the length of $\overline{DO}$.

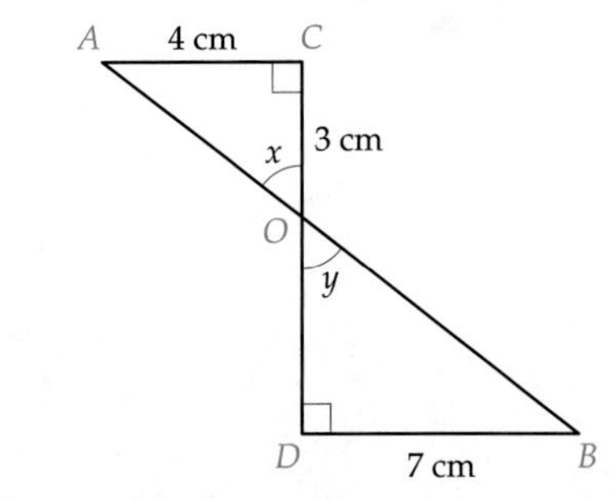

First determine whether triangles AOC and BOD are similar.

$m\angle C = m\angle D$ because they are both right angles.

$m\angle x = m\angle y$ because vertical angles have the same measure.

historical note

Many mathematicians have studied similar objects. Thales of Miletus (c. 624 B.C.–547 B.C.) discovered that he could determine the heights of pyramids and other objects by measuring a small object and the length of its shadow and then making use of similar triangles.

Fibonacci (1170–1250) also studied similar objects. The second section of Fibonacci's text *Liber abaci* contains practical information for merchants and surveyors. There is a chapter devoted to using similar triangles to determine the heights of tall objects. ■

Because two angles of triangle AOC are equal in measure to two angles of triangle BOD, triangles AOC and BOD are similar.

Use a proportion to find the length of the unknown side.

$$\frac{AC}{BD} = \frac{CO}{DO}$$

$$\frac{4}{7} = \frac{3}{DO}$$

$$4(DO) = 7(3)$$

$$4(DO) = 21$$

$$DO = 5.25$$

The length of $\overline{DO}$ is 5.25 cm.

EXAMPLE 2 ■ Solve a Problem Involving Similar Triangles

In the figure at the right, $\overline{AB}$ is parallel to $\overline{DC}$, $\angle B$ and $\angle D$ are right angles, $AB = 12$ m, $DC = 4$ m, and $AC = 18$ m. Find the length of $\overline{CO}$.

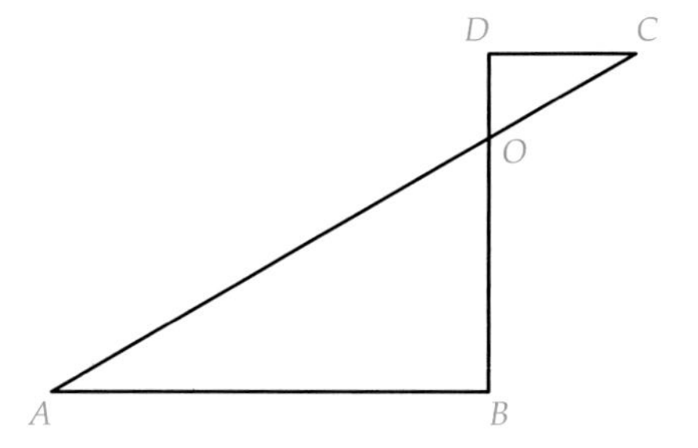

Solution

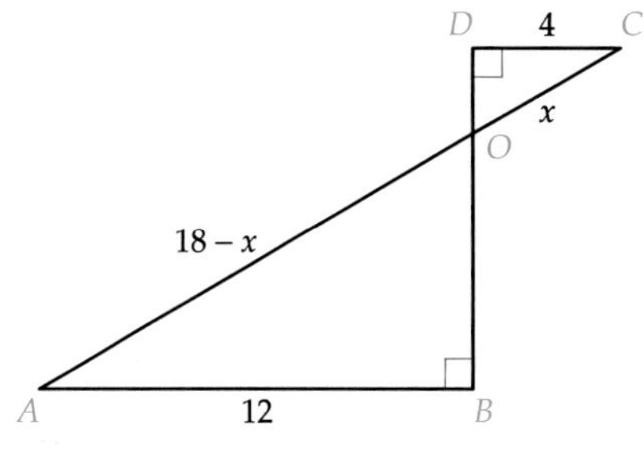

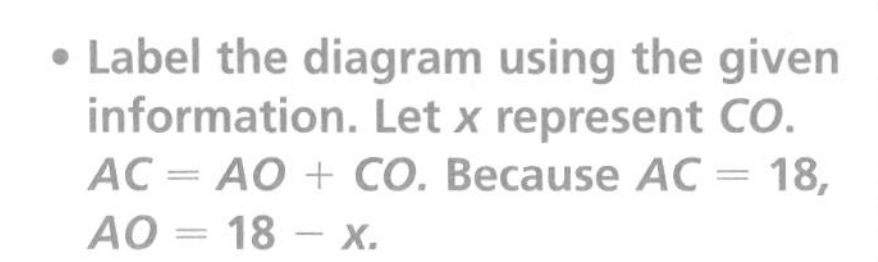

• Label the diagram using the given information. Let x represent CO. $AC = AO + CO$. Because $AC = 18$, $AO = 18 - x$.

$$\frac{DC}{BA} = \frac{CO}{AO}$$

• Triangles AOB and COD are similar triangles. The ratios of corresponding sides are equal.

$$\frac{4}{12} = \frac{x}{18 - x}$$

$$12x = 4(18 - x)$$

$$12x = 72 - 4x$$

• Use the Distributive Property.

$$16x = 72$$

$$x = 4.5$$

The length of $\overline{CO}$ is 4.5 m.

CHECK YOUR PROGRESS 2

In the figure at the right, $\overline{AB}$ is parallel to $\overline{DC}$, $\angle A$ and $\angle D$ are right angles, $AB = 10$ cm, $CD = 4$ cm, and $DO = 3$ cm. Find the area of triangle AOB.

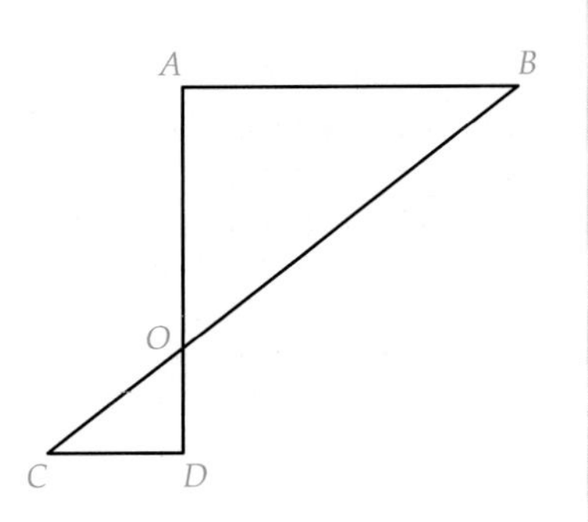

Solution *See page S31.*

Math Matters **Similar Polygons**

For similar triangles, the measures of corresponding angles are equal and the ratios of the lengths of corresponding sides are equal. The same is true for similar polygons: the measures of corresponding angles are equal and the lengths of corresponding sides are in proportion.

Quadrilaterals $ABCD$ and $LMNO$ are similar.

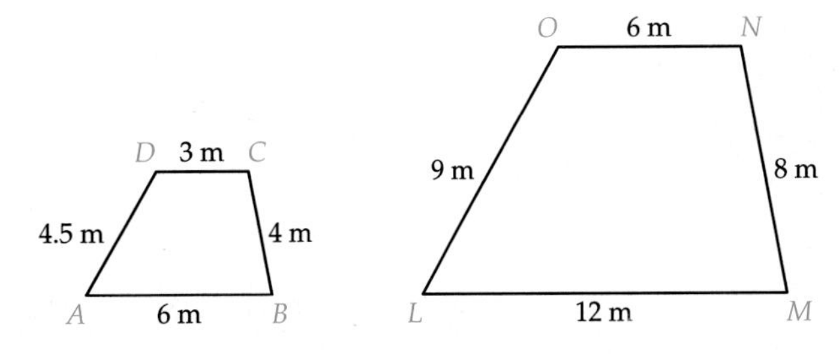

The ratio of the lengths of corresponding sides is: $\dfrac{AB}{LM} = \dfrac{6}{12} = \dfrac{1}{2}$

The ratio of the perimeter of $ABCD$ to the perimeter of $LMNO$ is:

$$\frac{\text{perimeter of } ABCD}{\text{perimeter of } LMNO} = \frac{17.5}{35} = \frac{1}{2}$$

Note that this ratio is the same as the ratio of corresponding sides. This is true for all similar polygons: If two polygons are similar, the ratio of their perimeters is equal to the ratio of the lengths of any pair of corresponding sides.

Congruent Triangles

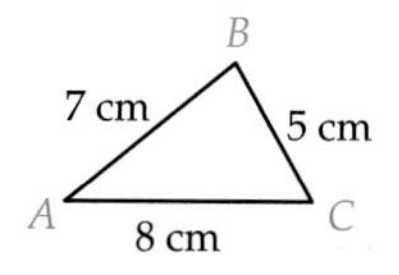

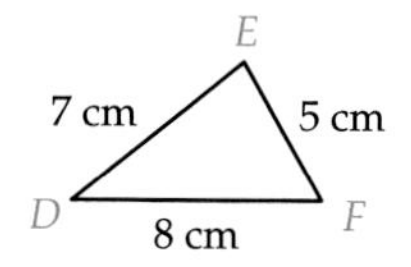

The two triangles at the right are **congruent.** They have the same shape and the same size.

TAKE NOTE

The notation $\triangle ABC \cong \triangle DEF$ is used to indicate that triangle *ABC* is congruent to triangle *DEF*.

Congruent and similar triangles differ in that the corresponding angles of congruent triangles have the same measure and the corresponding sides are equal in length, whereas for similar triangles, corresponding angles have the same measure but corresponding sides are not necessarily the same length.

The three major theorems used to determine whether two triangles are congruent are given below.

Side-Side-Side Theorem (SSS)

If the three sides of one triangle are equal in measure to the corresponding three sides of a second triangle, the two triangles are congruent.

In the triangles at the right, $AC = DE$, $AB = EF$, and $BC = DF$. The corresponding sides of triangles *ABC* and *DEF* are equal in measure. The triangles are congruent by the SSS Theorem.

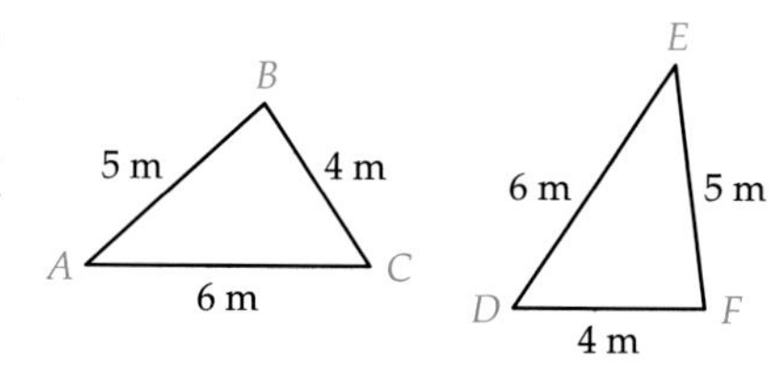

Side-Angle-Side Theorem (SAS)

If two sides and the included angle of one triangle are equal in measure to two sides and the included angle of a second triangle, the two triangles are congruent.

TAKE NOTE

$\angle A$ is *included* between sides $\overline{AB}$ and $\overline{AC}$. $\angle E$ is *included* between sides $\overline{DE}$ and $\overline{EF}$.

In the two triangles at the right, $AB = EF$, $AC = DE$, and $m\angle BAC = m\angle DEF$. The triangles are congruent by the SAS Theorem.

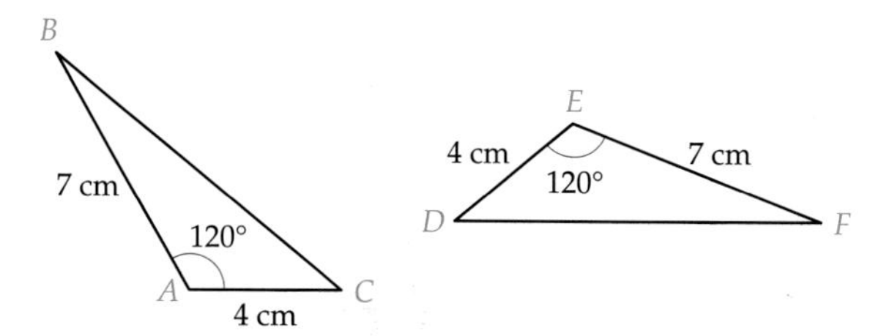

Angle-Side-Angle Theorem (ASA)

If two angles and the included side of one triangle are equal in measure to two angles and the included side of a second triangle, the two triangles are congruent.

TAKE NOTE

Side $\overline{AC}$ is *included* between $\angle A$ and $\angle C$. Side $\overline{EF}$ is *included* between $\angle E$ and $\angle F$.

For triangles *ABC* and *DEF* at the right, $m\angle A = m\angle F$, $m\angle C = m\angle E$, and $AC = EF$. The triangles are congruent by the ASA Theorem.

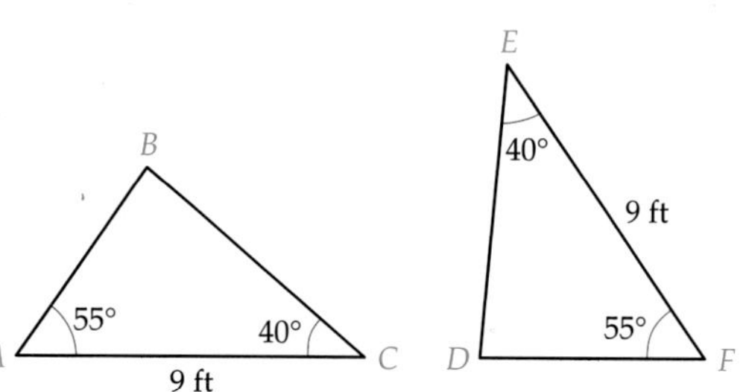

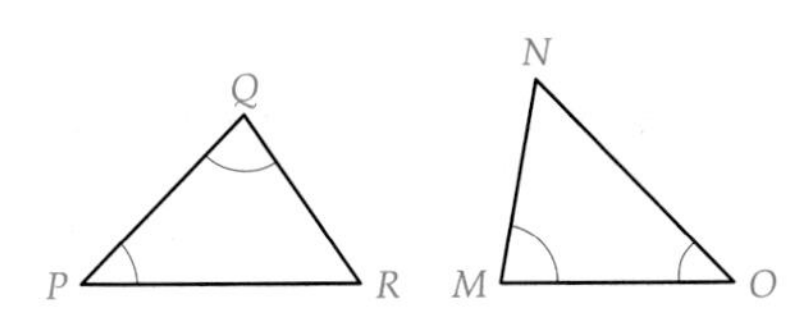

Given triangles PQR and MNO, do the conditions $m\angle P = m\angle O$, $m\angle Q = m\angle M$, and $PQ = MO$ guarantee that triangle PQR is congruent to triangle MNO?

Draw a sketch of the two triangles and determine whether one of the theorems for congruence is satisfied. See the figures at the left.

Because two angles and the included side of one triangle are equal in measure to two angles and the included side of the second triangle, the triangles are congruent by the ASA Theorem.

EXAMPLE 3 ■ Determine Whether Two Triangles Are Congruent

In the figure at the right, is triangle ABC congruent to triangle DEF?

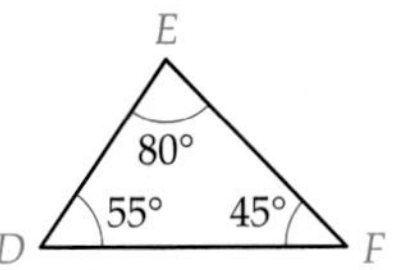

Solution

To determine whether the triangles are congruent, determine whether one of the theorems for congruence is satisfied.

The triangles do not satisfy the SSS Theorem, the SAS Theorem, or the ASA Theorem.

The triangles are not necessarily congruent.

CHECK YOUR PROGRESS 3 In the figure at the right, is triangle PQR congruent to triangle MNO?

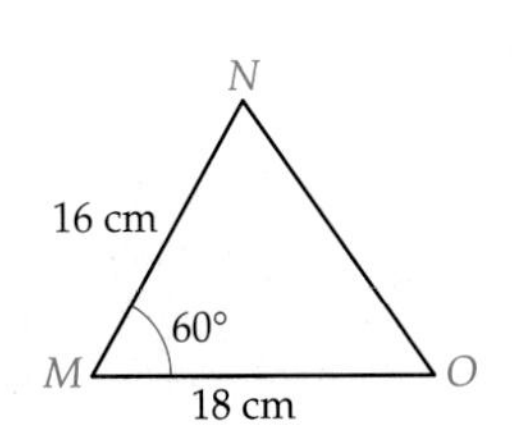

Solution *See page S31.*

The Pythagorean Theorem

Recall that a right triangle contains one right angle. The side opposite the right angle is called the **hypotenuse.** The other two sides are called **legs.**

The angles in a right triangle are usually labeled with the capital letters A, B, and C, with C reserved for the right angle. The side opposite angle A is side a, the side opposite angle B is side b, and c is the hypotenuse.

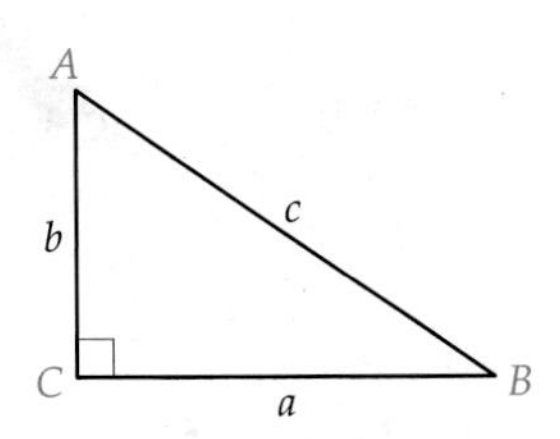

The figure at the right is a right triangle with legs measuring 3 units and 4 units and a hypotenuse measuring 5 units. Each side of the triangle is also the side of a square. The number of square units in the area of the largest square is equal to the sum of the numbers of square units in the areas of the smaller squares.

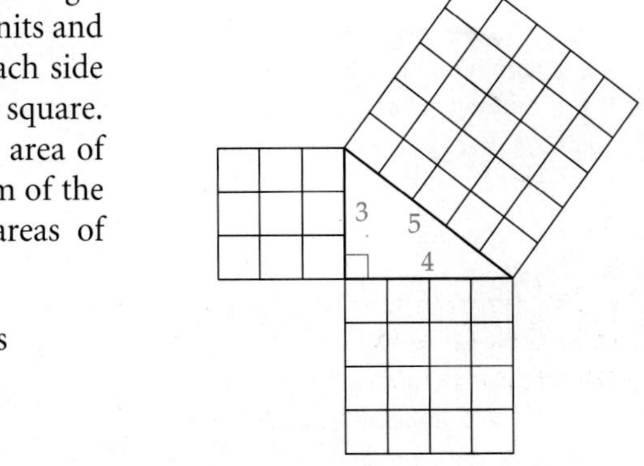

Square of the hypotenuse = sum of the squares of the two legs

$$5^2 = 3^2 + 4^2$$
$$25 = 9 + 16$$
$$25 = 25$$

point of interest

The first known proof of the Pythagorean Theorem is in a Chinese textbook that dates from 150 B.C. The book is called *Nine Chapters on the Mathematical Art*. The diagram below is from that book and was used in the proof of the theorem.

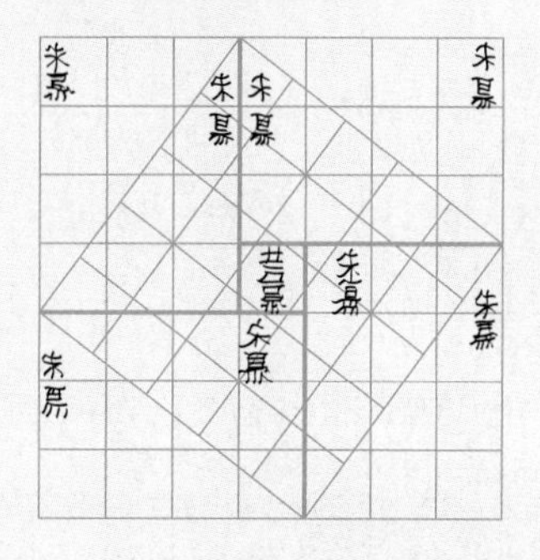

TAKE NOTE

The length of the side of a triangle cannot be negative. Therefore, we take only the principal, or positive, square root of 169.

CALCULATOR NOTE

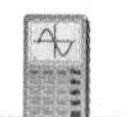

The way in which you evaluate the square root of a number depends on the type of calculator you have. Here are two possible keystrokes to find $\sqrt{80}$:

80 [√] [=]

or

[√] 80 [ENTER]

The first method is used on many scientific calculators. The second method is used on many graphing calculators.

The Greek mathematician Pythagoras is generally credited with the discovery that the square of the hypotenuse of a right triangle is equal to the sum of the squares of the two legs. This is called the **Pythagorean Theorem.**

Pythagorean Theorem

If a and b are the lengths of the legs of a right triangle and c is the length of the hypotenuse, then $c^2 = a^2 + b^2$.

If the lengths of two sides of a right triangle are known, the Pythagorean Theorem can be used to find the length of the third side.

Consider a right triangle with legs that measure 5 cm and 12 cm. Use the Pythagorean Theorem, with $a = 5$ and $b = 12$, to find the length of the hypotenuse. (If you let $a = 12$ and $b = 5$, the result will be the same.) Take the square root of each side of the equation.

The length of the hypotenuse is 13 cm.

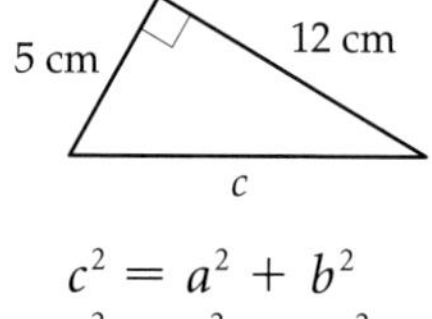

$$c^2 = a^2 + b^2$$
$$c^2 = 5^2 + 12^2$$
$$c^2 = 25 + 144$$
$$c^2 = 169$$
$$\sqrt{c^2} = \sqrt{169}$$
$$c = 13$$

EXAMPLE 4 ■ Determine the Length of the Unknown Side of a Right Triangle

The length of one leg of a right triangle is 8 in. The length of the hypotenuse is 12 in. Find the length of the other leg. Round to the nearest hundredth of an inch.

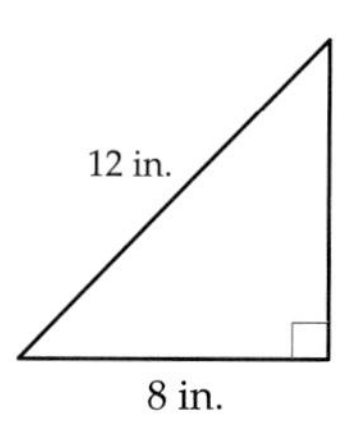

Solution

$a^2 + b^2 = c^2$ • Use the Pythagoran Theorem.

$8^2 + b^2 = 12^2$ • $a = 8$, $c = 12$

$64 + b^2 = 144$

$b^2 = 80$ • Solve for b^2. Subtract 64 from each side.

$\sqrt{b^2} = \sqrt{80}$ • Take the square root of each side of the equation.

$b \approx 8.94$ • Use a calculator to approximate $\sqrt{80}$.

The length of the other leg is approximately 8.94 in.

CHECK YOUR PROGRESS 4 The hypotenuse of a right triangle measures 6 m, and one leg measures 2 m. Find the measure of the other leg. Round to the nearest hundredth of a meter.

Solution *See page S31.*

Excursion

Topology

In this section, we discussed similar figures—that is, figures with the same shape. The branch of geometry called *topology* is the study of even more basic properties of figures than their sizes and shapes. For example, look at the figures below. We could take a rubber band and stretch it into any one of these shapes.

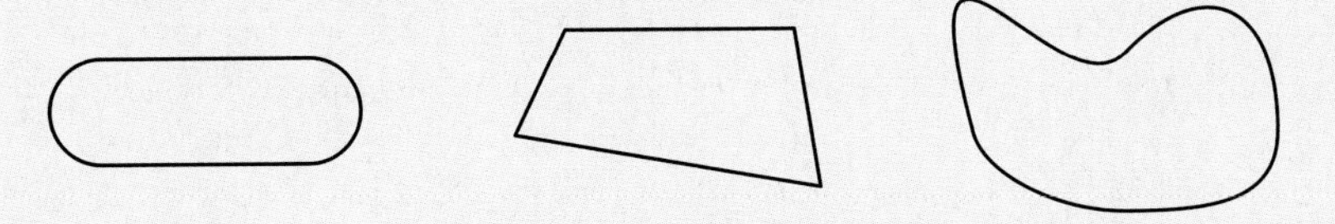

All three of these figures are different shapes, but each can be turned into one of the others by stretching the rubber band.

In topology, figures that can be stretched, molded, or bent into the same shape *without puncturing or cutting* belong to the same family. They are called **topologically equivalent.**

Rectangles, triangles, and circles are topologically equivalent.

Line segments and wavy curves are topologically equivalent.

Note that the figures formed from a rubber band and those formed from a line segment are not topologically equivalent; to form a line segment from a rubber band, we would have to cut the rubber band.

In the following plane figures, the lines are joined where they cross. The figures are topologically equivalent. They are not topologically equivalent to any of the figures shown above.

A **topologist** (a person who studies topology) is interested in identifying and describing different families of equivalent figures. Topology applies to solids as well as plane figures.

(continued)

For example, a topologist considers a brick, a potato, and a cue ball to be topologically equivalent to each other. Think of using modeling clay to form each of these shapes.

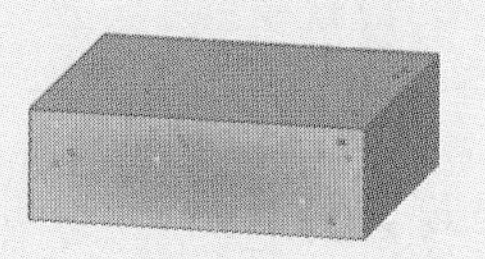

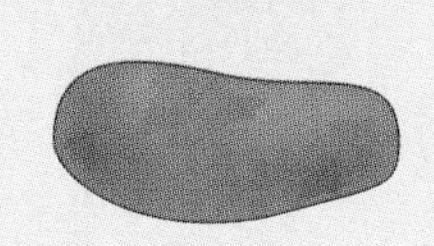

Excursion Exercises

Which of the figures listed is not topologically equivalent to the others?

1. **a.** Parallelogram **b.** Square **c.** Ray **d.** Trapezoid
2. **a.** Wedding ring **b.** Doughnut **c.** Fork **d.** Sewing needle
3. **a.** A **b.** D **c.** O **d.** P **e.** T

Exercise Set 8.3

In Exercises 1–4, find the ratio of the lengths of corresponding sides for the similar triangles.

1.

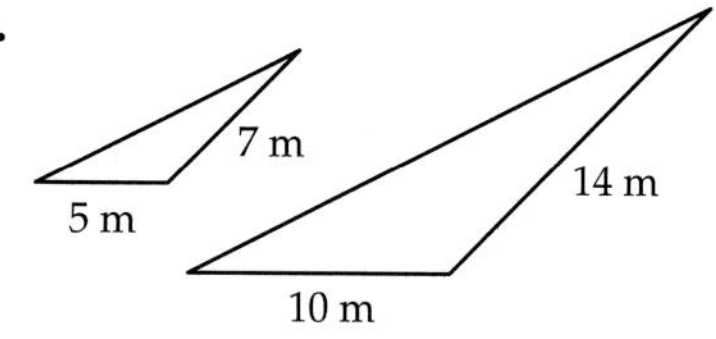

2.

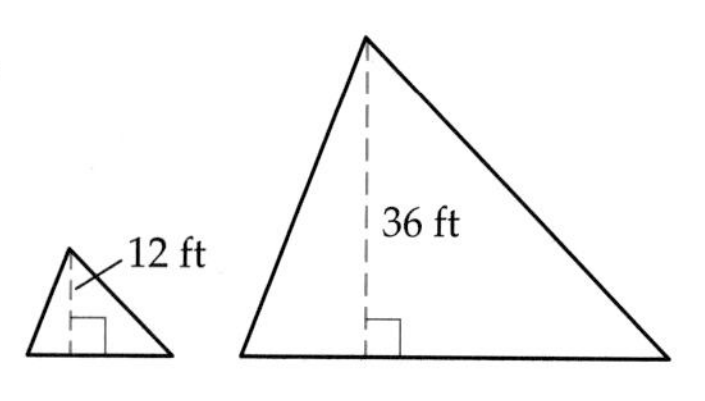

3.

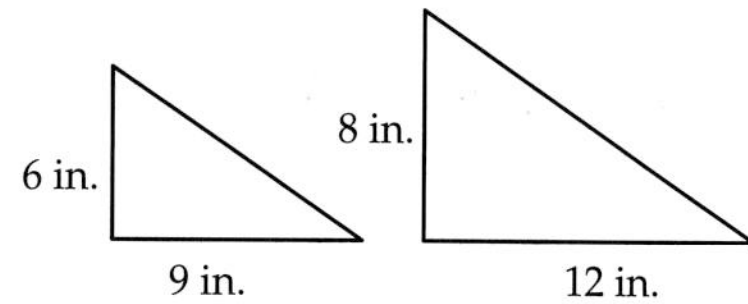

4.

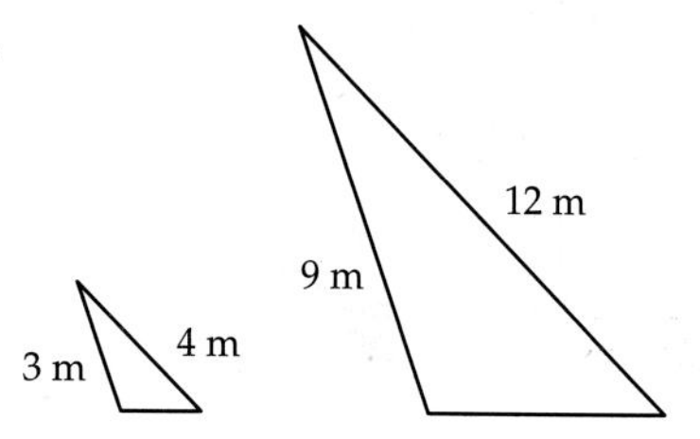

In Exercises 5–12, triangles *ABC* and *DEF* are similar triangles. Solve. Round to the nearest tenth.

5. Find side *DE*.

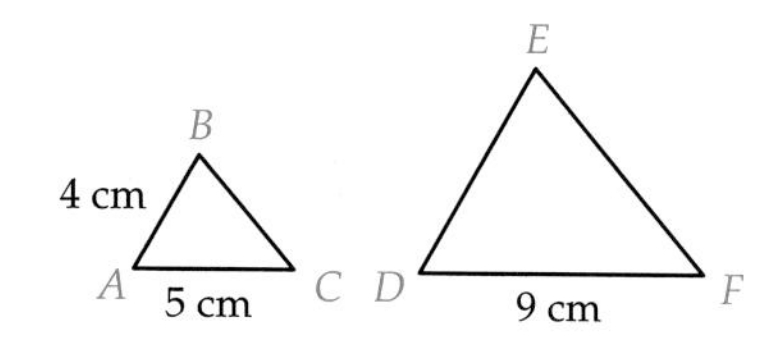

6. Find side *DE*.

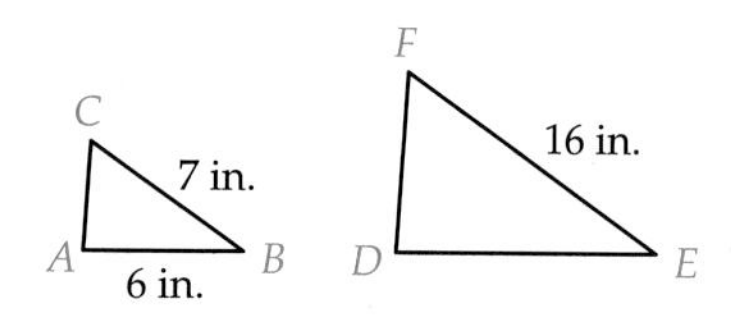

7. Find the height of triangle *DEF*.

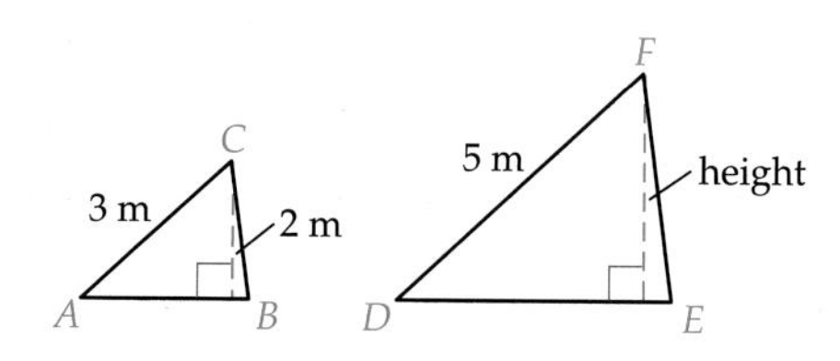

8. Find the height of triangle *ABC*.

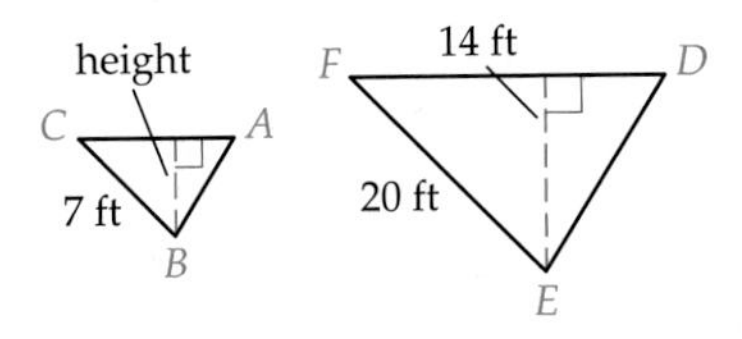

9. Find the perimeter of triangle *ABC*.

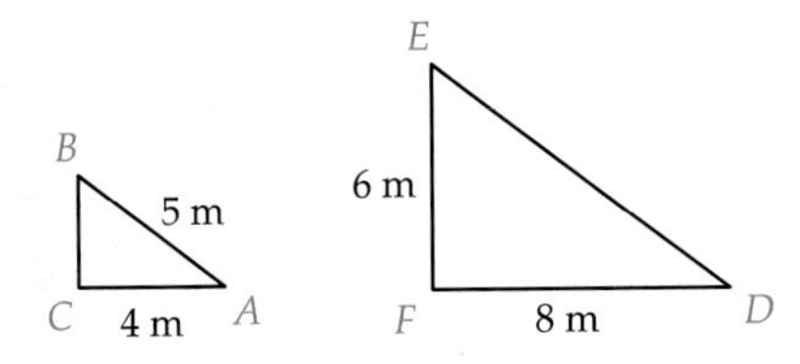

10. Find the perimeter of triangle *DEF*.

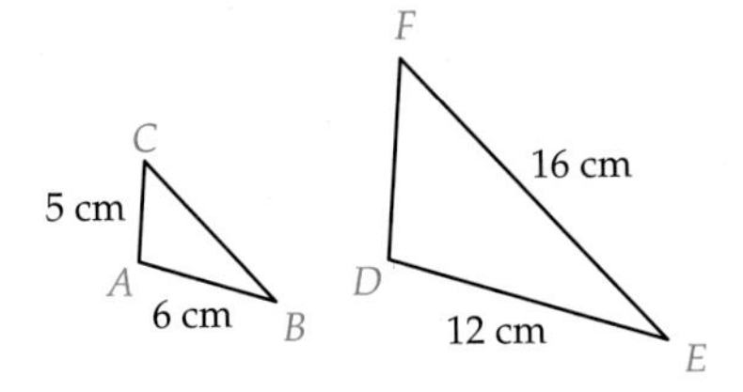

11. Find the perimeter of triangle *ABC*.

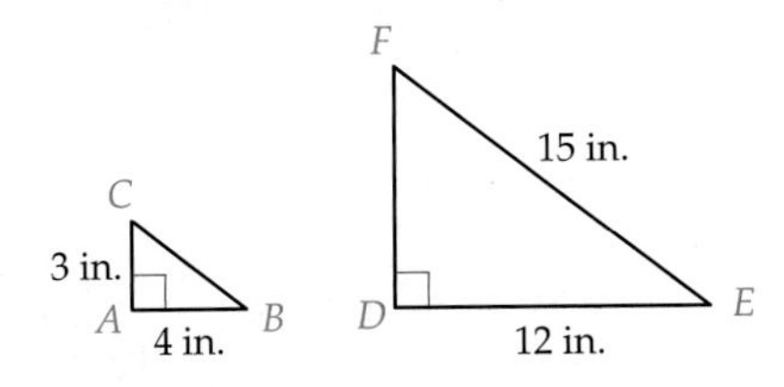

12. Find the area of triangle *DEF*.

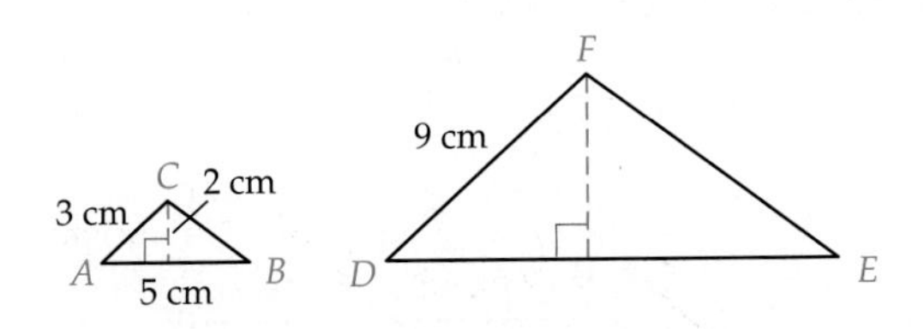

In Exercises 13 and 14, triangles *ABC* and *DEF* are similar triangles. Solve. Round to the nearest tenth.

13. Find the area of triangle *ABC*.

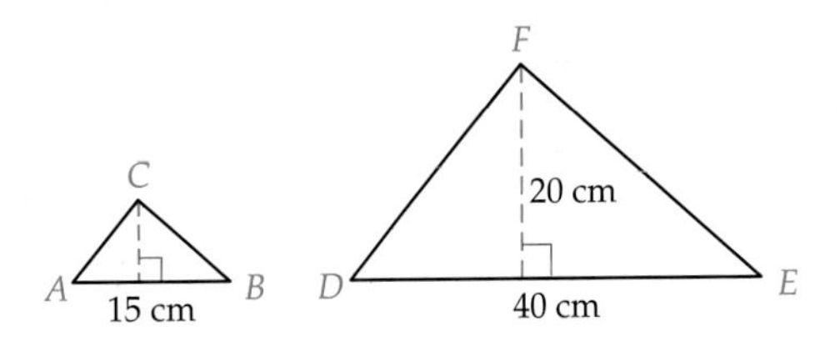

14. Find the area of triangle *DEF*.

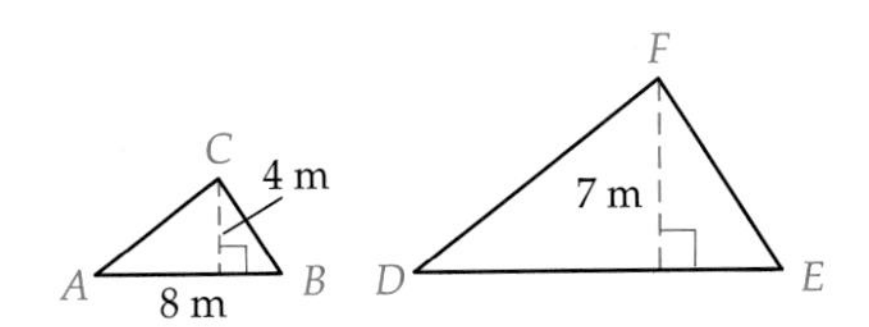

In Exercises 15–19, the given triangles are similar triangles. Use this fact to solve each exercise.

15. Find the height of the flagpole.

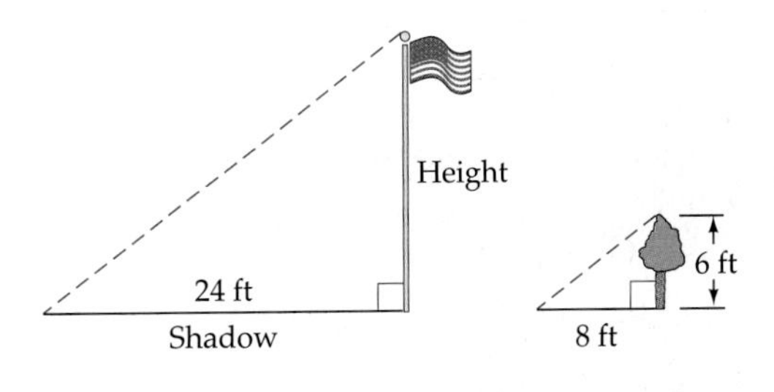

16. Find the height of the flagpole.

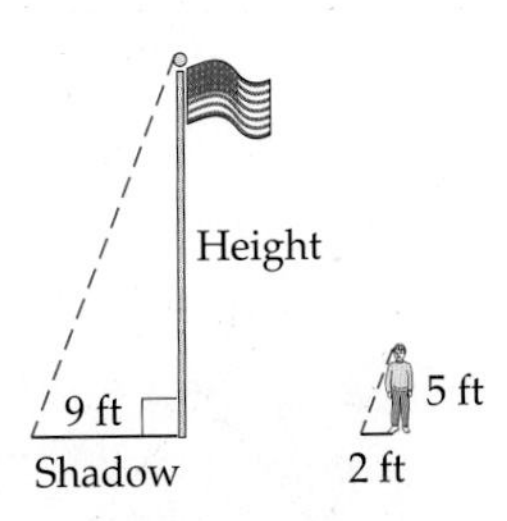

17. Find the height of the building.

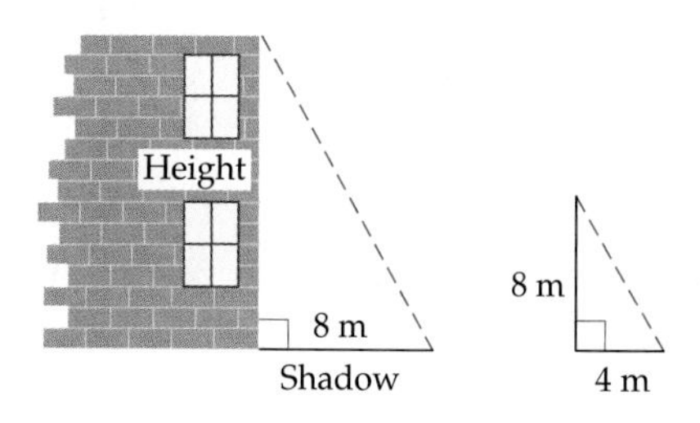

18. Find the height of the building.

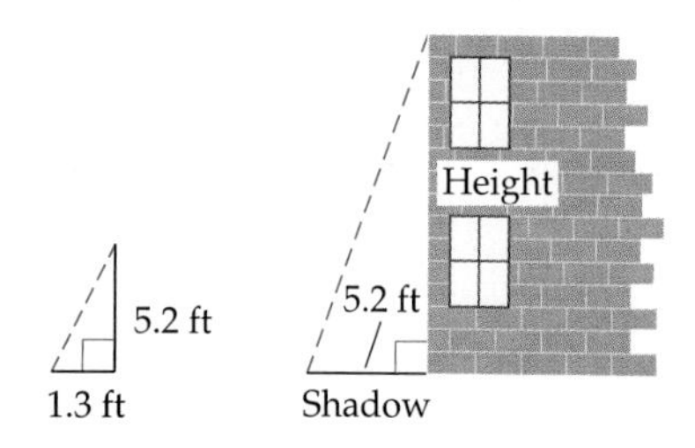

19. Find the height of the flagpole.

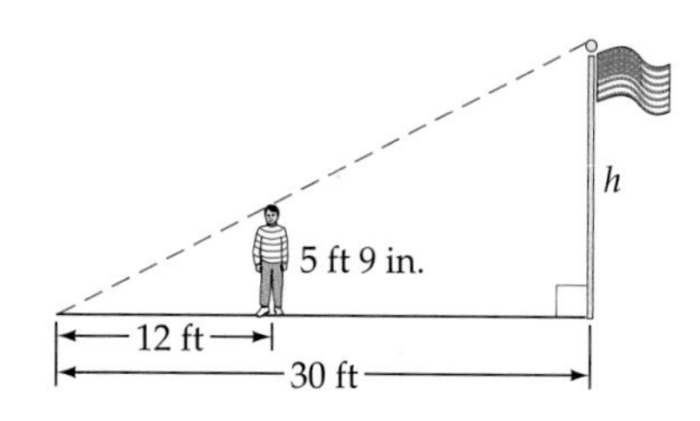

20. In the figure below, $\overline{BD} \parallel \overline{AE}$, BD measures 5 cm, AE measures 8 cm, and AC measures 10 cm. Find the length of $\overline{BC}$.

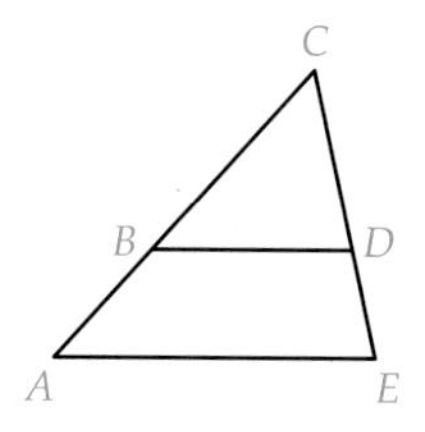

21. In the figure below, $\overline{AC} \parallel \overline{DE}$, BD measures 8 m, AD measures 12 m, and BE measures 6 m. Find the length of $\overline{BC}$.

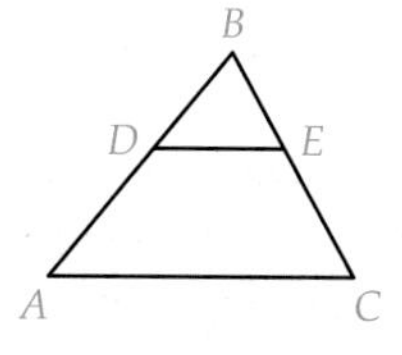

22. In the figure below, $\overline{DE} \parallel \overline{AC}$, DE measures 6 in., AC measures 10 in., and AB measures 15 in. Find the length of $\overline{DA}$.

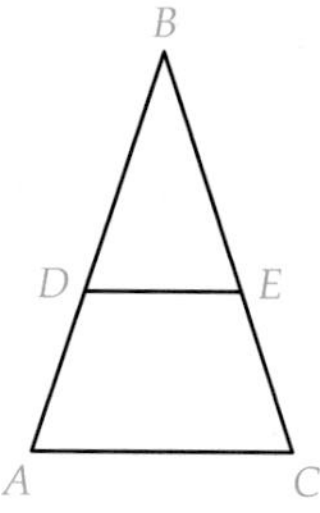

23. In the figure below, $\overline{AE} \parallel \overline{BD}$, $AB = 3$ ft, $ED = 4$ ft, and $BC = 3$ ft. Find the length of $\overline{CE}$.

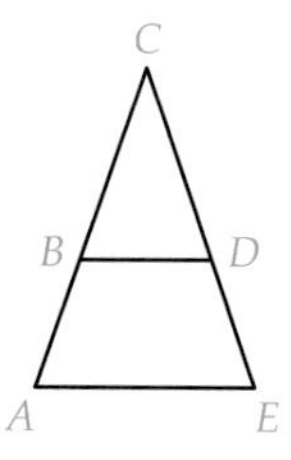

24. In the figure below, $\overline{MP}$ and $\overline{NQ}$ intersect at O, $NO = 25$ ft, $MO = 20$ ft, and $PO = 8$ ft. Find the length of $\overline{QO}$.

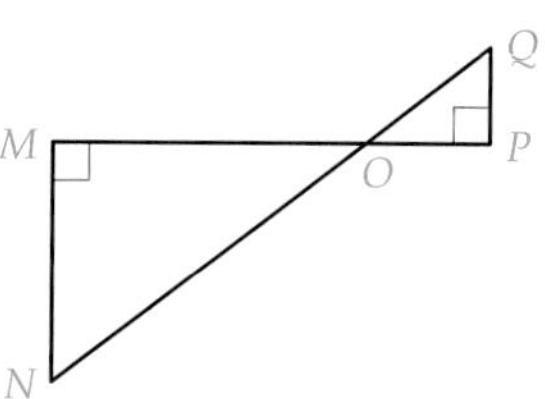

25. In the figure below, $\overline{MP}$ and $\overline{NQ}$ intersect at O, $NO = 24$ cm, $MN = 10$ cm, $MP = 39$ cm, and $QO = 12$ cm. Find the length of $\overline{OP}$.

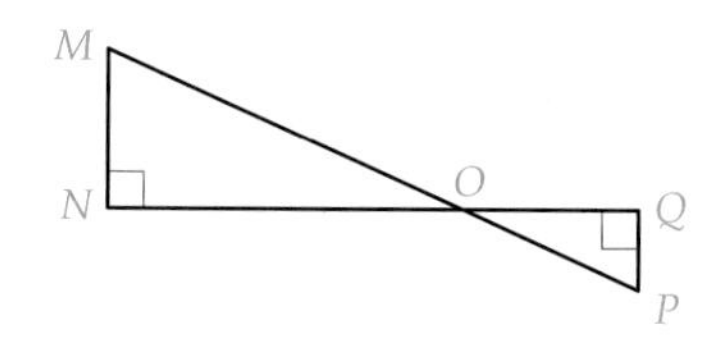

26. In the figure below, $\overline{MQ}$ and $\overline{NP}$ intersect at O, $NO = 12$ m, $MN = 9$ m, $PQ = 3$ m, and $MQ = 20$ m. Find the perimeter of triangle OPQ.

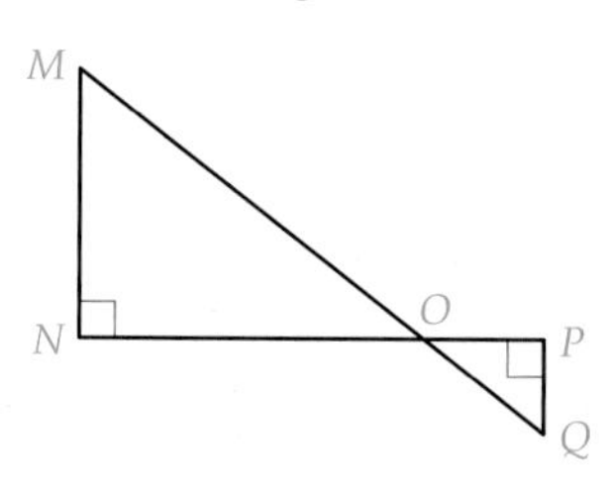

Surveying Surveyors use similar triangles to measure distances that cannot be measured directly. This is illustrated in Exercises 27 and 28.

27. The diagram below represents a river of width CD. Triangles AOB and DOC are similar. The distances AB, BO, and OC were measured and found to have the lengths given in the diagram. Find CD, the width of the river.

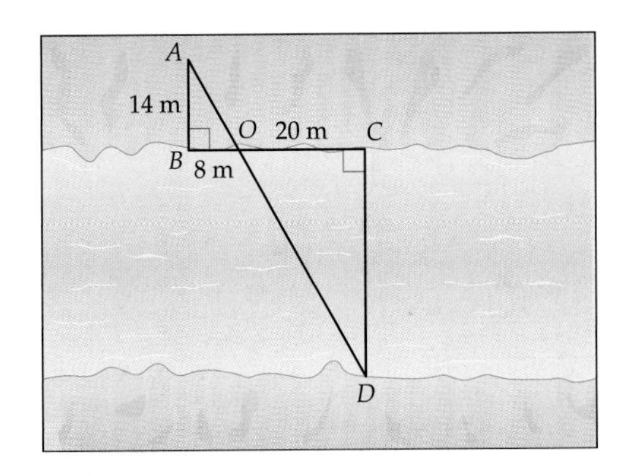

28. The diagram below shows how surveyors laid out similar triangles along the Winnepaugo River. Find the width, d, of the river.

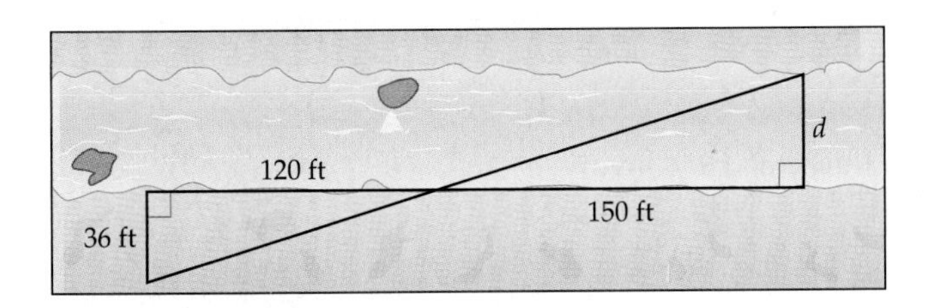

In Exercises 29–36, determine whether the two triangles are congruent. If they are congruent, state by what theorem (SSS, SAS, or ASA) they are congruent.

29.

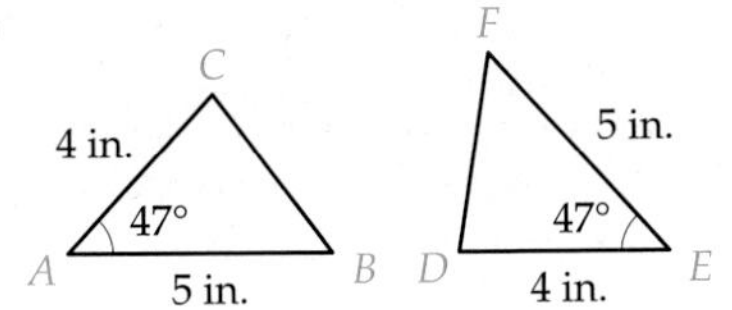

30.

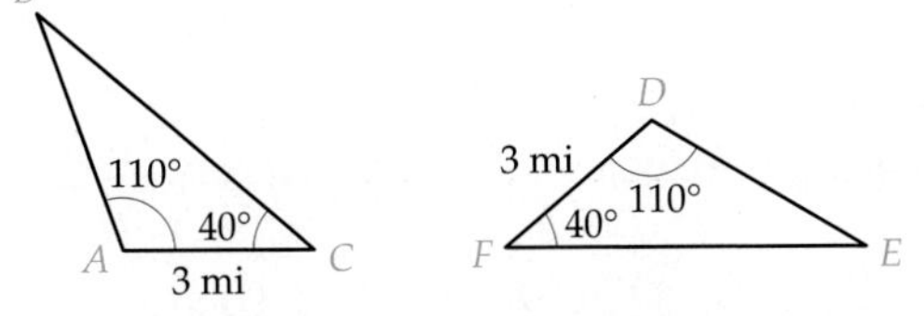

31.

C
9 ft
6 ft
A
10 ft
B
F
6 ft
9 ft
D
10 ft
E

32.

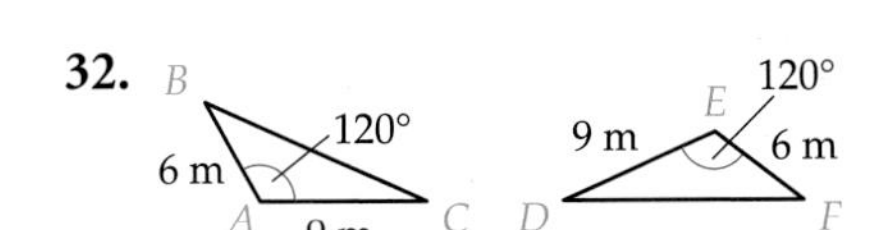

33.

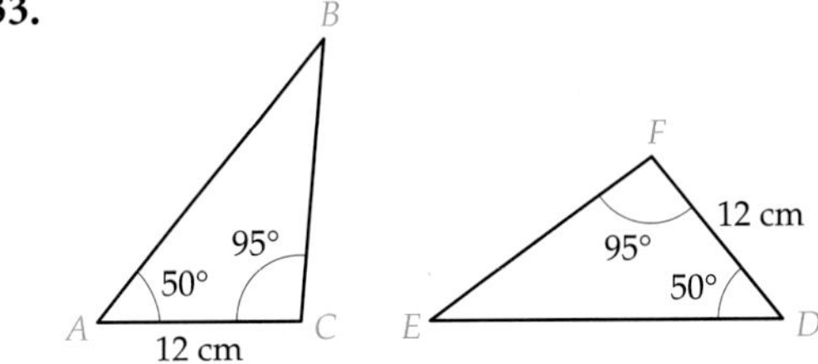

34.

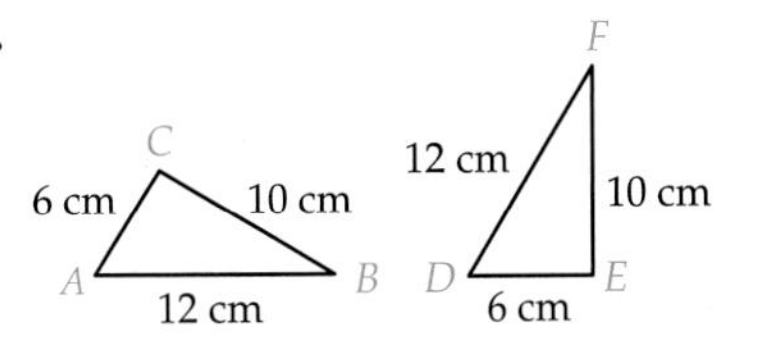

35.

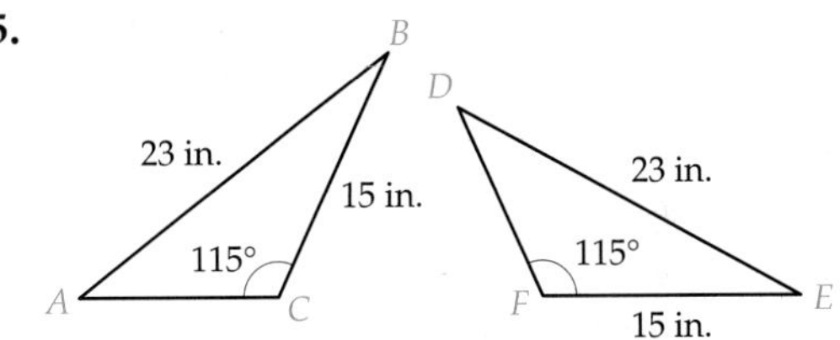

36.

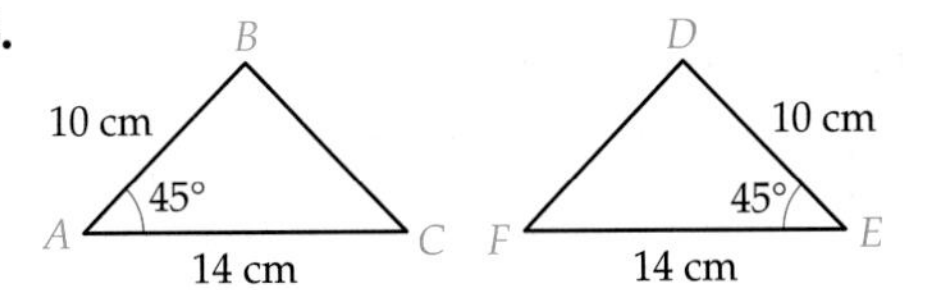

37. Given triangle ABC and triangle DEF, do the conditions $m\angle C = m\angle E$, $AC = EF$, and $BC = DE$ guarantee that triangle ABC is congruent to triangle DEF? If they are congruent, by what theorem are they congruent?

38. Given triangle PQR and triangle MNO, do the conditions $PR = NO$, $PQ = MO$, and $QR = MN$ guarantee that triangle PQR is congruent to triangle MNO? If they are congruent, by what theorem are they congruent?

39. Given triangle LMN and triangle QRS, do the conditions $m\angle M = m\angle S$, $m\angle N = m\angle Q$, and $m\angle L = m\angle R$ guarantee that triangle LMN is congruent to triangle QRS? If they are congruent, by what theorem are they congruent?

40. Given triangle DEF and triangle JKL, do the conditions $m\angle D = m\angle K$, $m\angle E = m\angle L$, and $DE = KL$ guarantee that triangle DEF is congruent to triangle JKL? If they are congruent, by what theorem are they congruent?

41. Given triangle ABC and triangle PQR, do the conditions $m\angle B = m\angle P$, $BC = PQ$, and $AC = QR$ guarantee that triangle ABC is congruent to triangle PQR? If they are congruent, by what theorem are they congruent?

In Exercises 42–50, find the unknown side of the triangle. Round to the nearest tenth.

42.

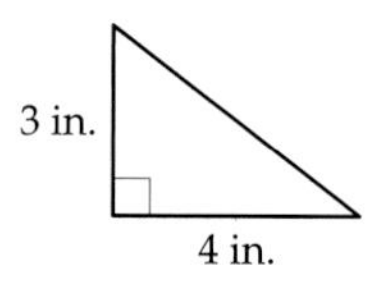

43.

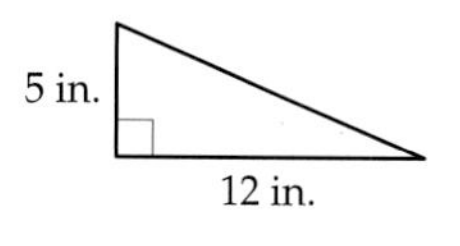

44.

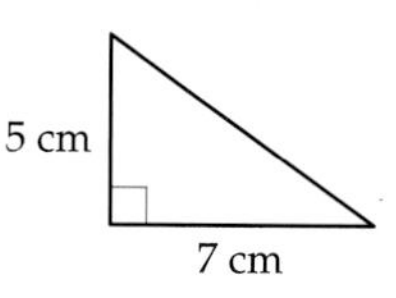

45.

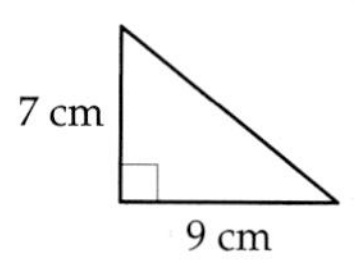

46.

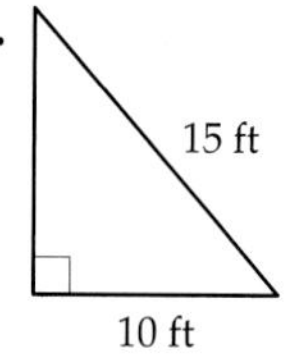

47.

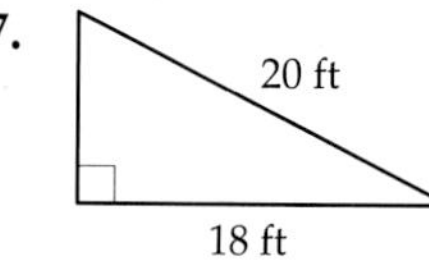

48.

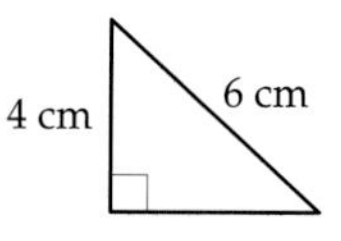

49.

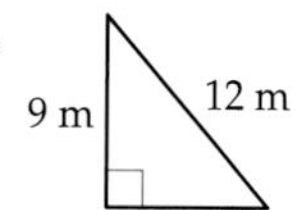

50.

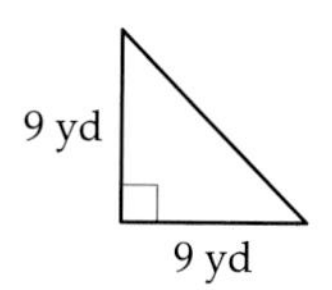

In Exercises 51–55, solve. Round to the nearest tenth.

51. Home Maintenance A ladder 8 m long is leaning against a building. How high on the building will the ladder reach when the bottom of the ladder is 3 m from the building?

52. Mechanics Find the distance between the centers of the holes in the metal plate.

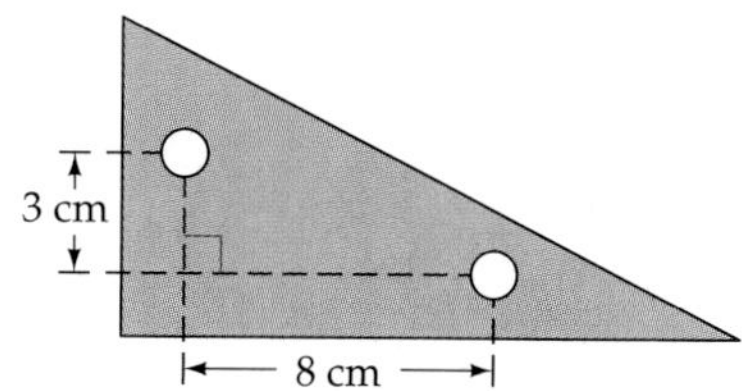

53. Travel If you travel 18 mi east and then 12 mi north, how far are you from your starting point?

54. Perimeter Find the perimeter of a right triangle with legs that measure 5 cm and 9 cm.

55. Perimeter Find the perimeter of a right triangle with legs that measure 6 in. and 8 in.

Extensions

CRITICAL THINKING

56. Determine whether the statement is always true, sometimes true, or never true.

a. If two angles of one triangle are equal to two angles of a second triangle, then the triangles are similar triangles.

b. Two isosceles triangles are similar triangles.

c. Two equilateral triangles are similar triangles.

d. If an acute angle of a right triangle is equal to an acute angle of another right triangle, then the triangles are similar triangles.

57. **Home Maintenance** You need to clean the gutters of your home. The gutters are 24 ft above the ground. For safety, the distance a ladder reaches up a wall should be four times the distance from the bottom of the ladder to the base of the side of the house. Therefore, the ladder must be 6 ft from the base of the house. Will a 25-foot ladder be long enough to reach the gutters? Explain how you found your answer.

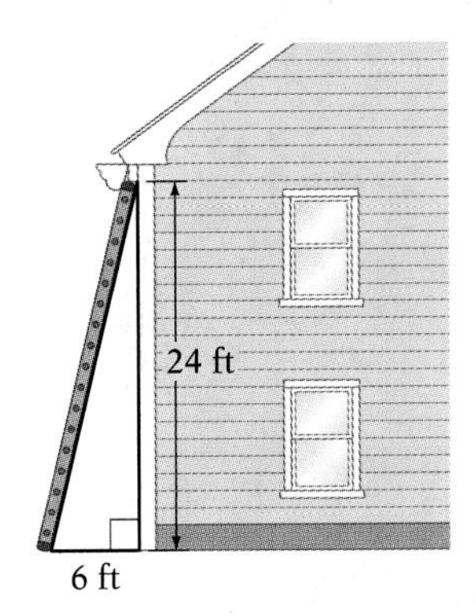

COOPERATIVE LEARNING

58. In the figure below, the height of a right triangle is drawn from the right angle perpendicular to the hypotenuse. (Recall that the hypotenuse of a right triangle is the side opposite the right angle.) Verify that the two smaller triangles formed are similar to the original triangle and similar to each other.

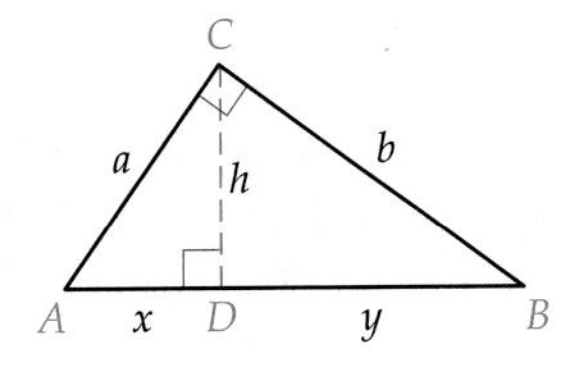

SECTION 8.4 Volume and Surface Area

Volume

In Section 2 of this chapter, we developed the geometric concepts of perimeter and area. Perimeter and area refer to plane figures (figures that lie in a plane). We are now ready to introduce *volume* of geometric solids.

Geometric solids are figures in space. Figures in space include baseballs, ice cubes, milk cartons, and trucks.

Volume is a measure of the amount of space occupied by a geometric solid. Volume can be used to describe, for example, the amount of trash in a land fill, the amount of concrete poured for the foundation of a house, or the amount of water in a town's reservoir.

Five common geometric solids are rectangular solids, spheres, cylinders, cones, and pyramids.

A **rectangular solid** is one in which all six sides, called **faces,** are rectangles. The variable L is used to represent the length of a rectangular solid, W is used to represent its width, and H is used to represent its height. A shoe box is an example of a rectangular solid.

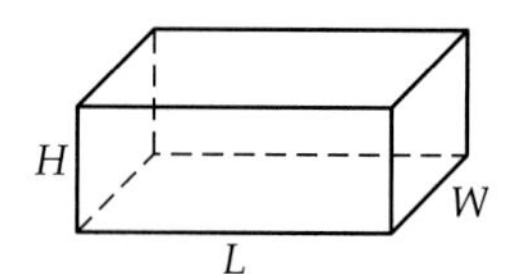

A **cube** is a special type of rectangular solid. Each of the six faces of a cube is a square. The variable s is used to represent the length of one side of a cube. A baby's block is an example of a cube.

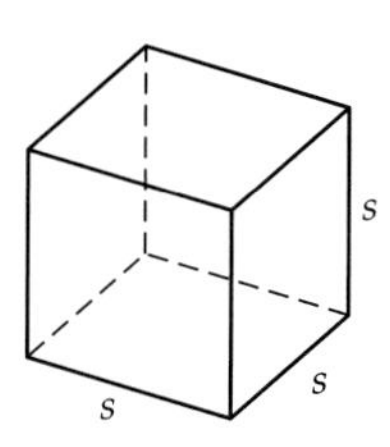

point of interest

Originally, the human body was used as the standard of measure. A mouthful was used as a unit of measure in ancient Egypt; it was later referred to as a *half jigger.* In French, the word for inch is *pouce,* which means thumb. A *span* was the distance from the tip of the outstretched thumb to the tip of the little finger. A *cubit* referred to the distance from the elbow to the end of the fingers. A *fathom* was the distance from the tip of the fingers on one hand to the tip of the fingers on the other hand when standing with arms fully extended out from the sides. The *hand* is still used today to measure the height of horses.

A cube that is 1 ft on each side has a volume of 1 cubic foot, which is written 1 ft^3. A cube that measures 1 cm on each side has a volume of 1 cubic centimeter, written 1 cm^3.

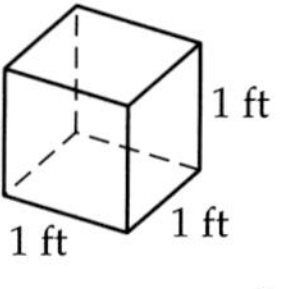

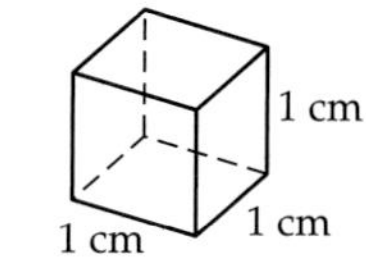

The volume of a solid is the number of cubes, each of volume 1 cubic unit, that are necessary to exactly fill the solid. The volume of the rectangular solid at the right is 24 cm^3 because it will hold exactly 24 cubes, each 1 cm on a side. Note that the volume can be found by multiplying the length times the width times the height.

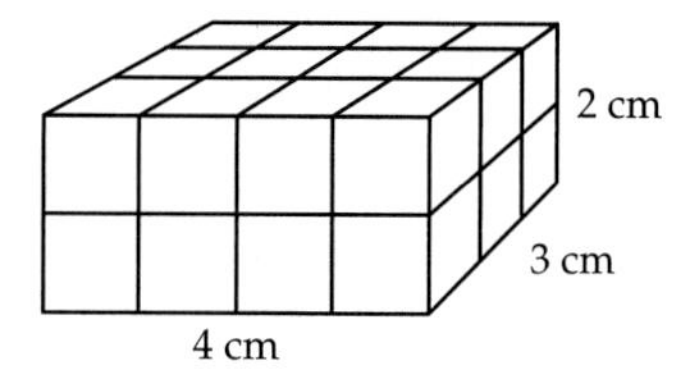

$$4 \cdot 3 \cdot 2 = 24$$

The volume of the solid is 24 cm^3.

Volume of a Rectangular Solid

The volume, V, of a rectangular solid with length L, width W, and height H is given by $V = LWH$.

Volume of a Cube

The volume, V, of a cube with side of length s is given by $V = s^3$.

QUESTION *Which of the following are rectangular solids: a juice box, a milk carton, a can of soup, a compact disc, a jewel box (plastic container) a compact disc is packaged in?*

A **sphere** is a solid in which all points are the same distance from a point O, called the **center** of the sphere. A **diameter** of a sphere is a line segment with endpoints on the sphere and passing through the center. A **radius** is a line segment from the center to a point on the sphere. $\overline{AB}$ is a diameter and $\overline{OC}$ is a radius of the sphere shown at the right. A basketball is an example of a sphere.

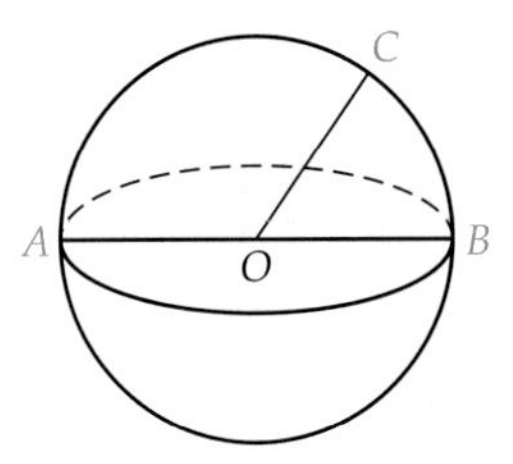

If we let d represent the length of a diameter and r represent the length of a radius, then $d = 2r$ or $r = \frac{1}{2}d$.

$$d = 2r \quad \text{or} \quad r = \frac{1}{2}d$$

historical note

Al-Sijzi was born in Persia around A.D. 945. He was both an astrologer and a geometer. It has been said that his papers related to mathematics were less numerous but more significant than those related to astrology.

One of Al-Sijzi's mathematical works was *Book of the Measurement of Spheres by Spheres.* It contains 12 theorems regarding a large sphere containing smaller spheres. The small spheres are mutually tangent and tangent to the big sphere. In his book Al-Sijzi calculates the volume of the large sphere using the radii of the smaller spheres. ■

Volume of a Sphere

The volume, V, of a sphere with radius of length r is given by $V = \frac{4}{3}\pi r^3$.

Find the volume of a rubber ball that has a diameter of 6 in.

First find the length of a radius of the sphere. $\quad r = \frac{1}{2}d = \frac{1}{2}(6) = 3$

Use the formula for the volume of a sphere. $\quad V = \frac{4}{3}\pi r^3$

Replace r with 3. $\quad V = \frac{4}{3}\pi(3)^3$

$$V = \frac{4}{3}\pi(27)$$

The exact volume of the sphere is 36π in³. $\quad V = 36\pi$

An approximate measure can be found by using the π key on a calculator. $\quad V \approx 113.10$

The volume of the rubber ball is approximately 113.10 in³.

CALCULATOR NOTE

To approximate 36π on your calculator, enter

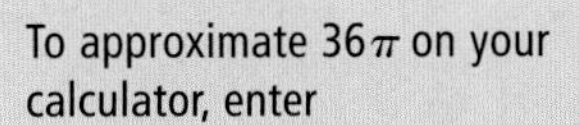

ANSWER *A juice box and a jewel box are rectangular solids.*

The most common cylinder, called a **right circular cylinder,** is one in which the bases are circles and are perpendicular to the height of the cylinder. The variable r is used to represent the length of the radius of a base of a cylinder, and h represents the height of the cylinder. In this text, only right circular cylinders are discussed.

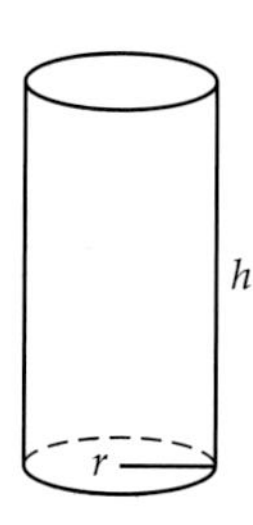

Volume of a Right Circular Cylinder

The volume, V, of a right circular cylinder is given by $V = \pi r^2 h$, where r is the radius of the base and h is the height of the cylinder.

A **right circular cone** is obtained when one base of a right circular cylinder is shrunk to a point, called the **vertex,** V. The variable r is used to represent the radius of the base of the cone, and h represents the height of the cone. The variable l is used to represent the **slant height,** which is the distance from a point on the circumference of the base to the vertex. In this text, only right circular cones are discussed. An ice cream cone is an example of a right circular cone.

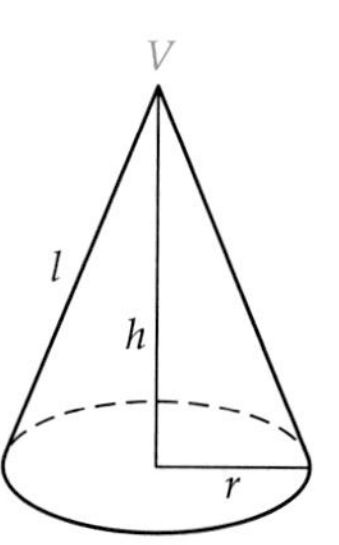

TAKE NOTE

Note that πr^2 appears in the formula for the volume of a right circular cylinder and in the formula for a right circular cone. This is because, in each case, the base of the figure is a circle.

Volume of a Right Circular Cone

The volume, V, of a right circular cone is given by $V = \frac{1}{3}\pi r^2 h$, where r is the length of a radius of the circular base and h is the height of the cone.

The base of a **regular pyramid** is a regular polygon and the sides are isosceles triangles (two sides of the triangle are the same length). The height, h, is the distance from the vertex, V, to the base and is perpendicular to the base. The variable l is used to represent the **slant height,** which is the height of one of the isosceles triangles on the face of the pyramid. The regular square pyramid at the right has a square base. This is the only type of pyramid discussed in this text. Many Egyptian pyramids are regular square pyramids.

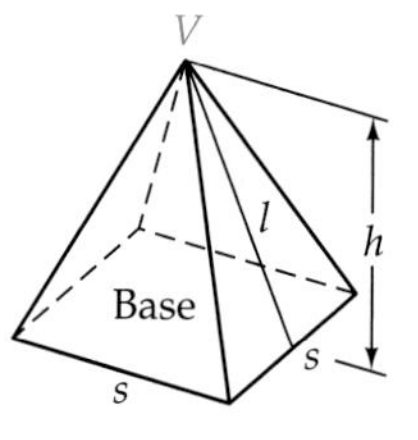

TAKE NOTE

Recall that an isosceles triangle has two sides of equal length.

Pyramid at Giza

Volume of a Regular Square Pyramid

The volume, V, of a regular square pyramid is given by $V = \frac{1}{3}s^2 h$, where s is the length of a side of the base and h is the height of the pyramid.

QUESTION *Which of the following units could not be used to measure the volume of a regular square pyramid?*

a. ft^3 **b.** m^3 **c.** yd^2 **d.** cm^3 **e.** mi

EXAMPLE 1 ■ Find the Volume of a Geometric Solid

Find the volume of a cube that measures 1.5 m on a side.

Solution

$V = s^3$ • Use the formula for the volume of a cube.

$V = 1.5^3$ • Replace s with 1.5.

$V = 3.375$

The volume of the cube is 3.375 m^3.

CHECK YOUR PROGRESS 1 The length of a rectangular solid is 5 m, the width is 3.2 m, and the height is 4 m. Find the volume of the solid.

Solution *See page S31.*

EXAMPLE 2 ■ Find the Volume of a Geometric Solid

The radius of the base of a cone is 8 cm. The height of the cone is 12 cm. Find the volume of the cone. Round to the nearest hundredth of a cubic centimeter.

Solution

$V = \frac{1}{3}\pi r^2 h$ • Use the formula for the volume of a cone.

$V = \frac{1}{3}\pi(8)^2(12)$ • Replace r with 8 and h with 12.

$V = \frac{1}{3}\pi(64)(12)$

$V = 256\pi$

$V \approx 804.25$ • Use the π key on a calculator.

The volume of the cone is approximately 804.25 cm^3.

CHECK YOUR PROGRESS 2 The length of a side of the base of a regular square pyramid is 15 m and the height of the pyramid is 25 m. Find the volume of the pyramid.

Solution *See page S31.*

ANSWER *Volume is measured in cubic units. Therefore, the volume of a regular square pyramid could be measured in ft^3, m^3, or cm^3, but not in yd^2 or mi.*

EXAMPLE 3 ■ Find the Volume of a Geometric Solid

An oil storage tank in the shape of a cylinder is 4 m high and has a diameter of 6 m. The oil tank is two-thirds full. Find the number of cubic meters of oil in the tank. Round to the nearest hundredth of a cubic meter.

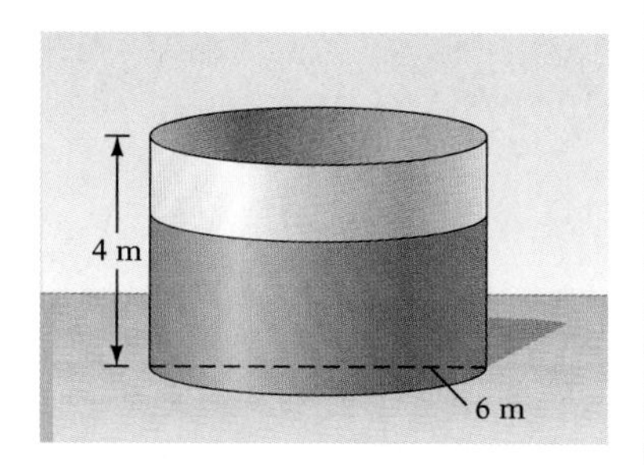

Solution

$$r = \frac{1}{2}d = \frac{1}{2}(6) = 3$$ • Find the radius of the base.

$$V = \pi r^2 h$$ • Use the formula for the volume of a cylinder.

$$V = \pi(3)^2(4)$$ • Replace r with **3** and h with **4**.

$$V = \pi(9)(4)$$

$$V = 36\pi$$

$$\frac{2}{3}(36\pi) = 24\pi$$ • Multiply the volume by $\frac{2}{3}$.

$$\approx 75.40$$ • Use the π key on a calculator.

There are approximately 75.40 m^3 of oil in the storage tank.

CHECK YOUR PROGRESS 3 A silo in the shape of a cylinder is 16 ft in diameter and has a height of 30 ft. The silo is three-fourths full. Find the volume of the portion of the silo that is not being used for storage. Round to the nearest hundredth of a cubic foot.

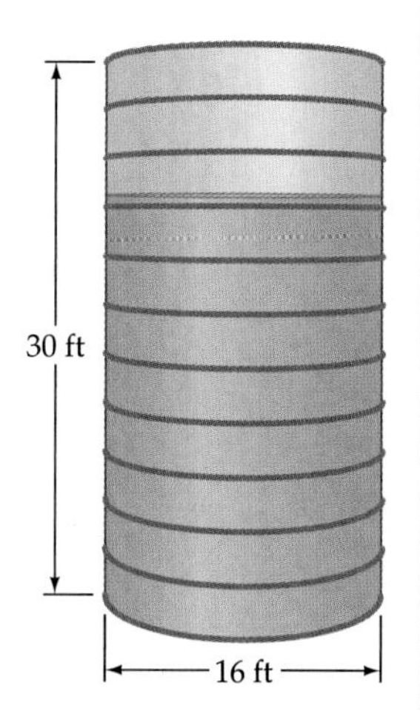

Solution *See page S31.*

Surface Area

The **surface area** of a solid is the total area on the surface of the solid. Suppose you want to cover a geometric solid with wallpaper. The amount of wallpaper needed is equal to the surface area of the figure.

When a rectangular solid is cut open and flattened out, each face is a rectangle. The surface area, S, of the rectangular solid is the sum of the areas of the six rectangles:

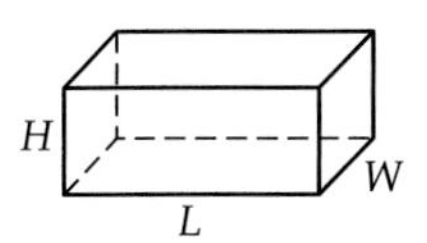

$$S = LW + LH + WH + LW + WH + LH$$

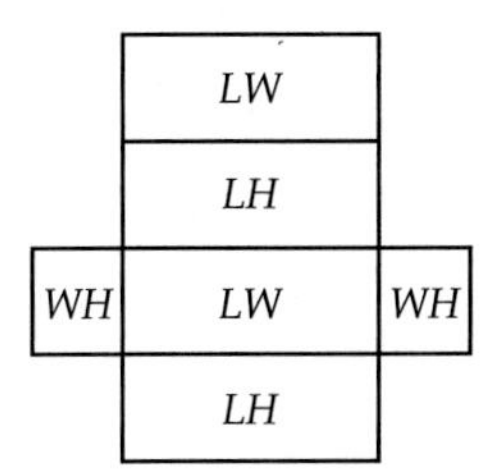

which simplifies to

$$S = 2LW + 2LH + 2WH$$

The surface area of a cube is the sum of the areas of the six faces of the cube. The area of each face is s^2. Therefore, the surface area S, of a cube is given by the formula $S = 6s^2$.

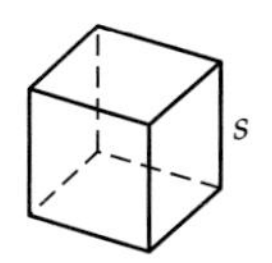

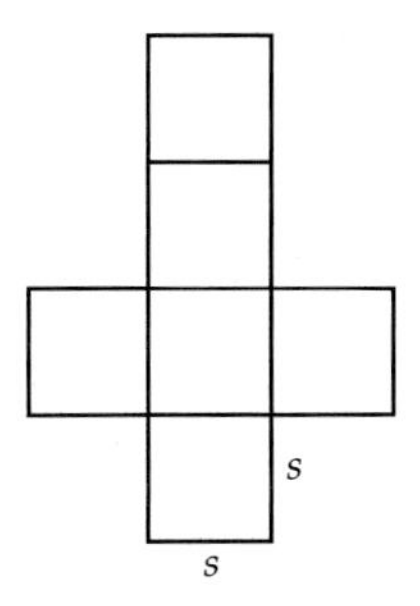

historical note

Pappus (păp′ŭs) of Alexandria (c. 290–350) was born in Egypt. We know the time period in which he lived because he wrote about his observation of the eclipse of the sun that took place in Alexandria on October 18, 320.

Pappas has been called the last of the great Greek geometers. His major work in geometry is *Synagoge,* or the *Mathematical Collection.* It consists of eight books. In Book V, Pappas proves that the sphere has a greater volume than any regular geometric solid with equal surface area. He also proves that if two regular solids have equal surface areas, the solid with the greater number of faces has the greater volume. ■

When a cylinder is cut open and flattened out, the top and bottom of the cylinder are circles. The side of the cylinder flattens out to a rectangle. The length of the rectangle is the circumference of the base, which is $2\pi r$; the width is h, the height of the cylinder. Therefore, the area of the rectangle is $2\pi rh$. The surface area, S, of the cylinder is

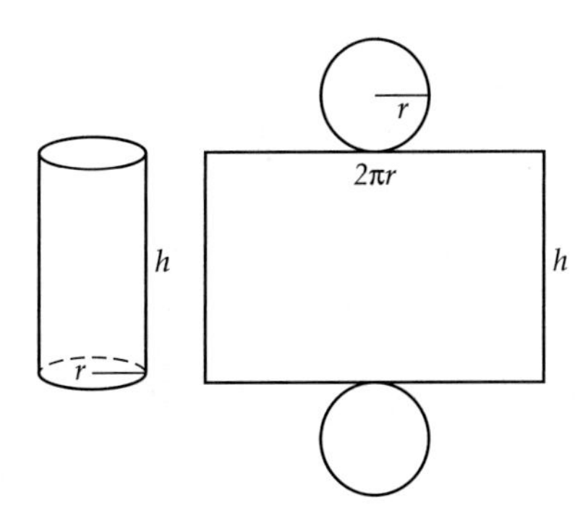

$$S = \pi r^2 + 2\pi rh + \pi r^2$$

which simplifies to

$$S = 2\pi r^2 + 2\pi rh$$

The surface area of a regular square pyramid is the area of the base plus the area of the four isosceles triangles. The length of a side of the square base is s; therefore, the area of the base is s^2. The slant height, l, is the height of each triangle, and s is the length of the base of each triangle. The surface area, S, of a regular square pyramid is

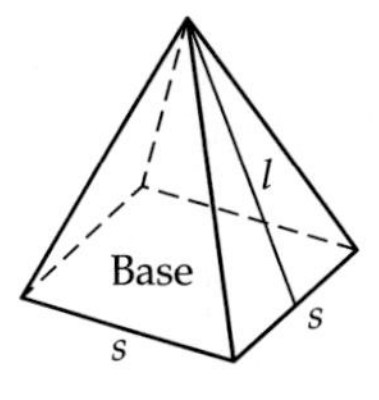

$$S = s^2 + 4\left(\frac{1}{2}sl\right)$$

which simplifies to

$$S = s^2 + 2sl$$

Formulas for the surface areas of geometric solids are given below.

Surface Areas of Geometric Solids

The surface area, S, of a **rectangular solid** with length L, width W, and height H is given by $S = 2LW + 2LH + 2WH$.

The surface area, S, of a **cube** with sides of length s is given by $S = 6s^2$.

The surface area, S, of a **sphere** with radius r is given by $S = 4\pi r^2$.

The surface area, S, of a **right circular cylinder** is given by $S = 2\pi r^2 + 2\pi rh$, where r is the radius of the base and h is the height.

The surface area, S, of a **right circular cone** is given by $S = \pi r^2 + \pi rl$, where r is the radius of the circular base and l is the slant height.

The surface area, S, of a **regular square pyramid** is given by $S = s^2 + 2sl$, where s is the length of a side of the base and l is the slant height.

QUESTION *Which of the following units could not be used to measure the surface area of a rectangular solid?*

a. in^2 **b.** m^3 **c.** cm^2 **d.** ft^3 **e.** yd

Find the surface area of a sphere with a diameter of 18 cm.

First find the radius of the sphere. $r = \frac{1}{2}d = \frac{1}{2}(18) = 9$

Use the formula for the surface area of a sphere.

$$S = 4\pi r^2$$
$$S = 4\pi(9)^2$$
$$S = 4\pi(81)$$
$$S = 324\pi$$

The exact surface area of the sphere is 324π cm^2.

An approximate measure can be found by using the π key on a calculator. $S \approx 1017.88$

The approximate surface area is 1017.88 cm^2.

ANSWER *Surface area is measured in square units. Therefore, the surface area of a rectangular solid could be measured in in^2 or cm^2, but not in m^3, ft^3 or yd.*

EXAMPLE 4 ■ Find the Surface Area of a Geometric Solid

The diameter of the base of a cone is 5 m and the slant height is 4 m. Find the surface area of the cone. Round to the nearest hundredth of a square meter.

Solution

$r = \frac{1}{2}d = \frac{1}{2}(5) = 2.5$ • Find the radius of the cone.

$S = \pi r^2 + \pi rl$ • Use the formula for the surface area of a cone.

$S = \pi(2.5)^2 + \pi(2.5)(4)$ • Replace r with **2.5** and l with **4**.

$S = \pi(6.25) + \pi(2.5)(4)$

$S = 6.25\pi + 10\pi$

$S = 16.25\pi$

$S \approx 51.05$

The surface area of the cone is approximately 51.05 m^2.

CHECK YOUR PROGRESS 4 The diameter of the base of a cylinder is 6 ft and the height is 8 ft. Find the surface area of the cylinder. Round to the nearest hundredth of a square foot.

Solution *See page S31.*

EXAMPLE 5 ■ Find the Surface Area of a Geometric Solid

Find the area of a label used to cover a soup can that has a radius of 4 cm and a height of 12 cm. Round to the nearest hundredth of a square centimeter.

Solution
The surface area of the side of a cylinder is given by $2\pi rh$.

$$\begin{aligned}\text{Area of the label} &= 2\pi rh\\ &= 2\pi(4)(12)\\ &= 96\pi\\ &\approx 301.59\end{aligned}$$

The area of the label is approximately 301.59 cm^2.

CHECK YOUR PROGRESS 5 Which has a larger surface area, a cube with a side measuring 8 cm or a sphere with a diameter measuring 10 cm?

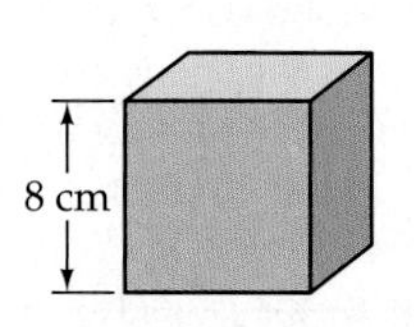

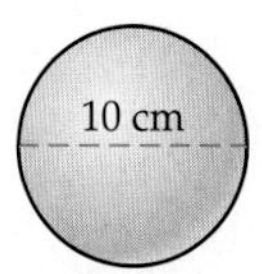

Solution *See page S31.*

Math Matters Survival of the Fittest

The ratio of an animal's surface area to the volume of its body is a crucial factor in its survival. The more square units of skin for every cubic unit of volume, the more rapidly the animal loses body heat. Therefore, animals living in a warm climate benefit from a higher ratio of surface area to volume, whereas those living in a cool climate benefit from a lower ratio.

Excursion

Water Displacement

A recipe for peanut butter cookies calls for 1 cup of peanut butter. Peanut butter is difficult to measure. If you have ever used a measuring cup to measure peanut butter, you know that there tend to be pockets of air at the bottom of the cup. And trying to scrape all of the peanut butter out of the cup and into the mixing bowl is a challenge.

A more convenient method of measuring 1 cup of peanut butter is to fill a 2-cup measuring cup with 1 cup of water. Then add peanut butter to the water until the water reaches the 2-cup mark. (Make sure all the peanut butter is below the top of the water.) Drain off the water, and the one cup of peanut butter drops easily into the mixing bowl.

This method of measuring peanut butter works because when an object sinks below the surface of the water, the object displaces an amount of water that is equal to the volume of the object.

(continued)

point of interest

King Hiero of Syracuse commissioned a new crown from his goldsmith. When the crown was completed, Hiero suspected that the goldsmith stole some of the gold and replaced it with lead. Hiero asked Archimedes to determine, without defacing the surface of the crown, whether the crown was pure gold.

Archimedes knew that the first problem he had to solve was how to determine the volume of the crown. Legend has it that one day he was getting into a bathtub that was completely full of water, and the water splashed over the side. He surmised that the volume of water that poured over the side was equal to the volume of his submerged body. Supposedly, he was so excited that he jumped from the tub and went running through the streets yelling "Eureka! Eureka!", meaning "I found it!, I found it!"—referring to the solution of the problem of determining the volume of the crown.

A sphere with a diameter of 4 in. is placed in a rectangular tank of water that is 6 in. long and 5 in. wide. How much does the water level rise? Round to the nearest hundredth of an inch.

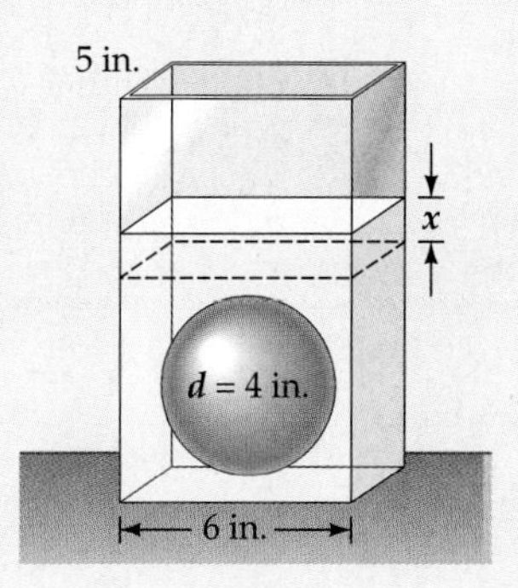

$V = \frac{4}{3}\pi r^3$ • **Use the formula for the volume of a sphere.**

$V = \frac{4}{3}\pi(2^3) = \frac{32}{3}\pi$ • $r = \frac{1}{2}d = \frac{1}{2}(4) = 2$

Let x represent the amount of the rise in water level. The volume of the sphere will equal the volume of the water displaced. As shown above, this volume is the rectangular solid with width 5 in., length 6 in., and height x in.

$V = LWH$ • **Use the formula for the volume of a rectangular solid.**

$\frac{32}{3}\pi = (6)(5)x$ • **Substitute $\frac{32}{3}\pi$ for V, 6 for L, 5 for W, and x for H.**

$\frac{32}{90}\pi = x$ • **The exact height that the water will rise is $\frac{32}{90}\pi$.**

$1.12 \approx x$ • **Use a calculator to find an approximation.**

The water will rise approximately 1.12 in.

Excursion Exercises

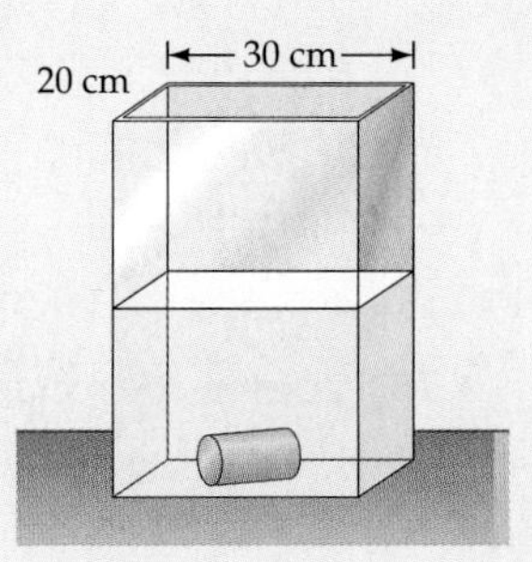

Figure 1

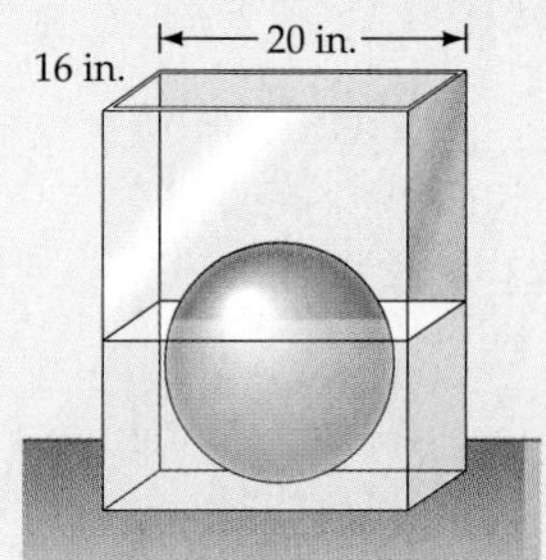

Figure 2

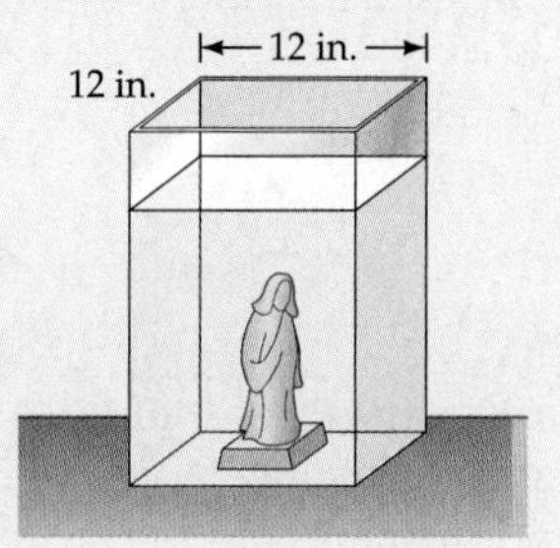

Figure 3

1. A cylinder with a 2-centimeter radius and a height of 10 cm is submerged in a tank of water that is 20 cm wide and 30 cm long (see Figure 1). How much does the water level rise? Round to the nearest hundredth of a centimeter.

(continued)

2. A sphere with a radius of 6 in. is placed in a rectangular tank of water that is 16 in. wide and 20 in. long (see Figure 2). The sphere displaces water until two-thirds of the sphere, with respect to its volume, is submerged. How much does the water level rise? Round to the nearest hundredth of an inch.

3. A chemist wants to know the density of a statue that weighs 15 lb. The statue is placed in a rectangular tank of water that is 12 in. long and 12 in. wide (see Figure 3). The water level rises 0.42 in. Find the density of the statue. Round to the nearest hundredth of a pound per cubic inch. *Hint:* Density = weight ÷ volume.

Exercise Set 8.4

In Exercises 1–6, find the volume of the figure. For calculations involving π, give both the exact value and an approximation to the nearest hundredth of a unit.

1.

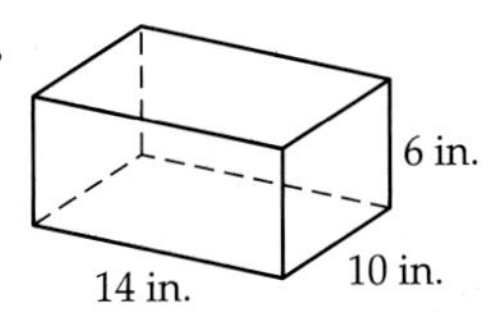

2.

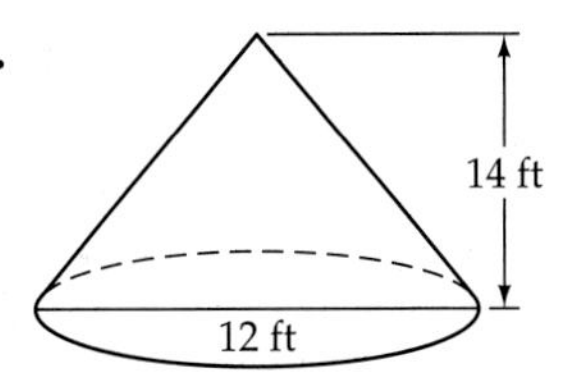

3.

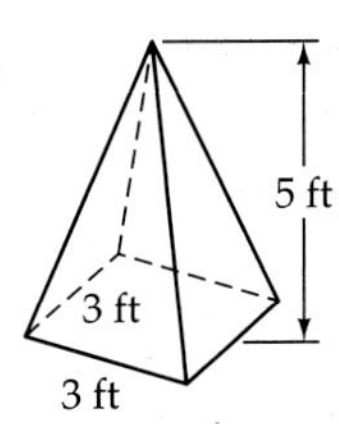

4.

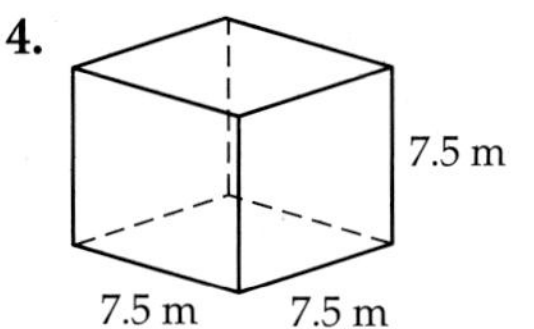

5.

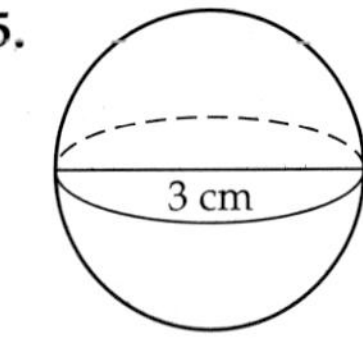

6.

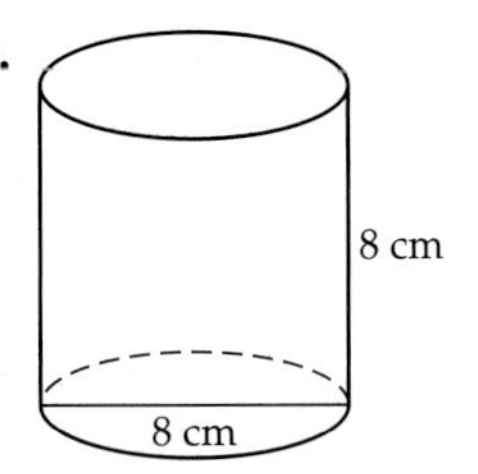

In Exercises 7–12, find the surface area of the figure. For calculations involving π, give both the exact value and an approximation to the nearest hundredth of a unit.

7.

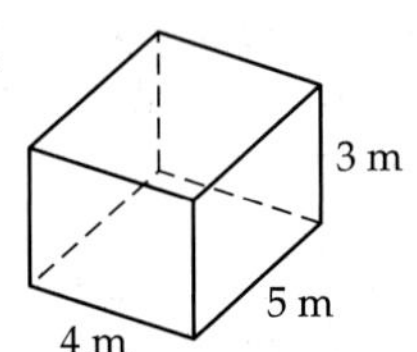

8.

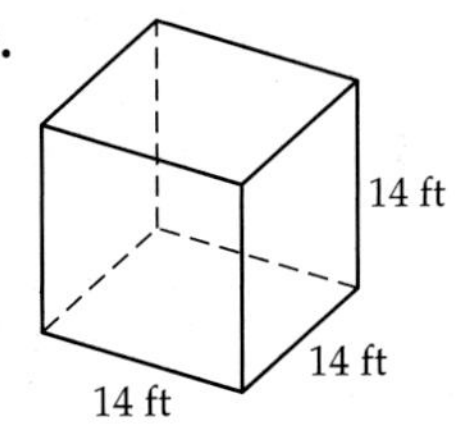

9.

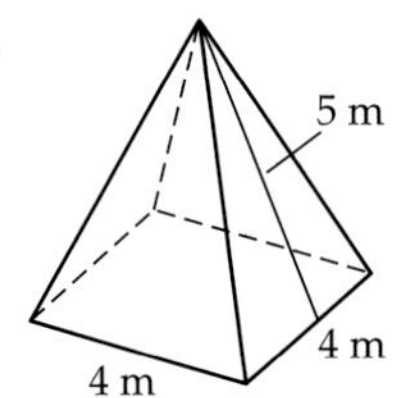

10.

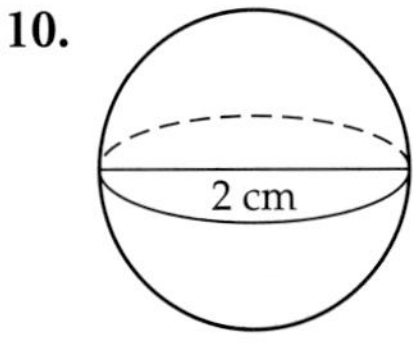

11.

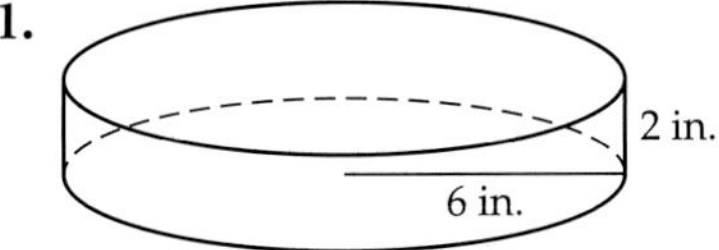

12.

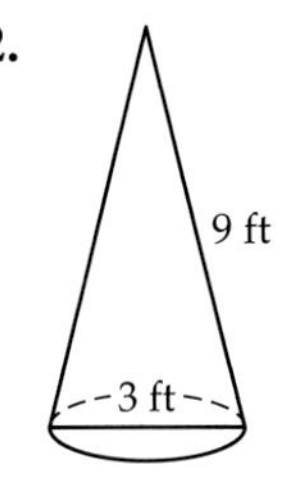

In Exercises 13–46, solve.

13. **Volume** A rectangular solid has a length of 6.8 m, a width of 2.5 m, and a height of 2 m. Find the volume of a solid.

14. **Volume** Find the volume of a rectangular solid that has a length of 4.5 ft, a width of 3 ft, and a height of 1.5 ft.

15. **Volume** Find the volume of a cube whose side measures 2.5 in.

16. **Volume** The length of a side of a cube is 7 cm. Find the volume of the cube.

17. **Volume** The diameter of a sphere is 6 ft. Find the exact volume of the sphere.

18. **Volume** Find the volume of a sphere that has a radius of 1.2 m. Round to the nearest hundredth of a cubic meter.

19. **Volume** The diameter of the base of a cylinder is 24 cm. The height of the cylinder is 18 cm. Find the volume of the cylinder. Round to the nearest hundredth of a cubic centimeter.

20. **Volume** The height of a cylinder is 7.2 m. The radius of the base is 4 m. Find the exact volume of the cylinder.

21. **Volume** The radius of the base of a cone is 5 in. The height of the cone is 9 in. Find the exact volume of the cone.

22. **Volume** The height of a cone is 15 cm. The diameter of the cone is 10 cm. Find the volume of the cone. Round to the nearest hundredth of a cubic centimeter.

23. **Volume** The length of a side of the base of a regular square pyramid is 6 in. and the height of the pyramid is 10 in. Find the volume of the pyramid.

24. **Volume** The height of a regular square pyramid is 8 m and the length of a side of the base is 9 m. What is the volume of the pyramid?

25. **The Statue of Liberty** The index finger of the Statue of Liberty is 8 ft long. The circumference at the second joint is 3.5 ft. Use the formula for the volume of a cylinder to approximate the volume of the index finger on the Statue of Liberty. Round to the nearest hundredth of a cubic foot.

26. **Surface Area** The height of a rectangular solid is 5 ft, the length is 8 ft, and the width is 4 ft. Find the surface area of the solid.

27. **Surface Area** The width of a rectangular solid is 32 cm, the length is 60 cm, and the height is 14 cm. What is the surface area of the solid?

28. **Surface Area** The side of a cube measures 3.4 m. Find the surface area of the cube.

29. **Surface Area** Find the surface area of a cube with a side measuring 1.5 in.

30. **Surface Area** Find the exact surface area of a sphere with a diameter of 15 cm.

31. **Surface Area** The radius of a sphere is 2 in. Find the surface area of the sphere. Round to the nearest hundredth of a square inch.

32. **Surface Area** The radius of the base of a cylinder is 4 in. The height of the cylinder is 12 in. Find the surface area of the cylinder. Round to the nearest hundredth of a square inch.

33. **Surface Area** The diameter of the base of a cylinder is 1.8 m. The height of the cylinder is 0.7 m. Find the exact surface area of the cylinder.

34. **Surface Area** The slant height of a cone is 2.5 ft. The radius of the base is 1.5 ft. Find the exact surface area of the cone. The formula for the surface area of a cone is given on page 514.

35. **Surface Area** The diameter of the base of a cone is 21 in. The slant height is 16 in. What is the surface area of the cone? The formula for the surface area of a cone is given on page 514. Round to the nearest hundredth of a square inch.

36. **Surface Area** The length of a side of the base of a pyramid is 9 in., and the pyramid's slant height is 12 in. Find the surface area of the pyramid.

37. **Surface Area** The slant height of a regular square pyramid is 18 m, and the length of a side of the base is 16 m. What is the surface area of the pyramid?

38. **Appliances** The volume of a freezer that is a rectangular solid with a length of 7 ft and a height of 3 ft is 52.5 ft^3. Find the width of the freezer.

39. **Aquariums** The length of an aquarium is 18 in. and the width is 12 in. If the volume of the aquarium is 1836 in^3, what is the height of the aquarium?

40. Surface Area The surface area of a rectangular solid is 108 cm^2. The height of the solid is 4 cm, and the length is 6 cm. Find the width of the rectangular solid.

41. Surface Area The length of a rectangular solid is 12 ft and the width is 3 ft. If the surface area is 162 ft^2, find the height of the rectangular solid.

42. Paint A can of paint will cover 300 ft^2 of surface. How many cans of paint should be purchased to paint a cylinder that has a height of 30 ft and a radius of 12 ft?

43. Ballooning A hot air balloon is in the shape of a sphere. Approximately how much fabric was used to construct the balloon if its diameter is 32 ft? Round to the nearest square foot.

44. Aquariums How much glass is needed to make a fish tank that is 12 in. long, 8 in. wide, and 9 in. high? The fish tank is open at the top.

45. Food Labels Find the area of a label used to completely cover the side of a cylindrical can of juice that has a diameter of 16.5 cm and a height of 17 cm. Round to the nearest hundredth of a square centimeter.

46. Surface Area The length of a side of the base of a regular square pyramid is 5 cm and the slant height of the pyramid is 8 cm. How much larger is the surface area of this pyramid than the surface area of a cone with a diameter of 5 cm and a slant height of 8 cm? Round to the nearest hundredth of a square centimeter.

In Exercises 47–52, find the volume of the figure. Round to the nearest hundredth of a unit.

47.

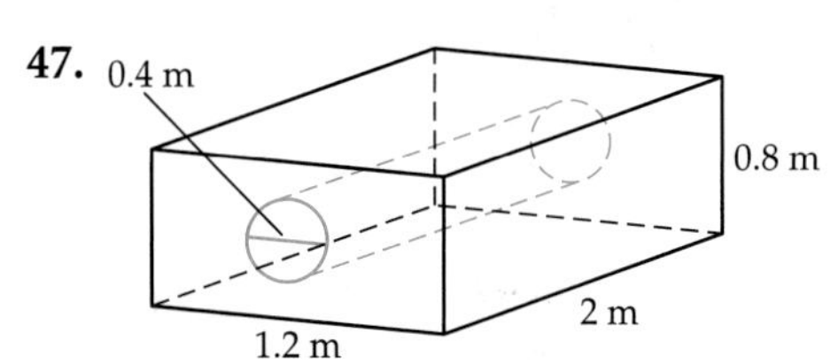

48.

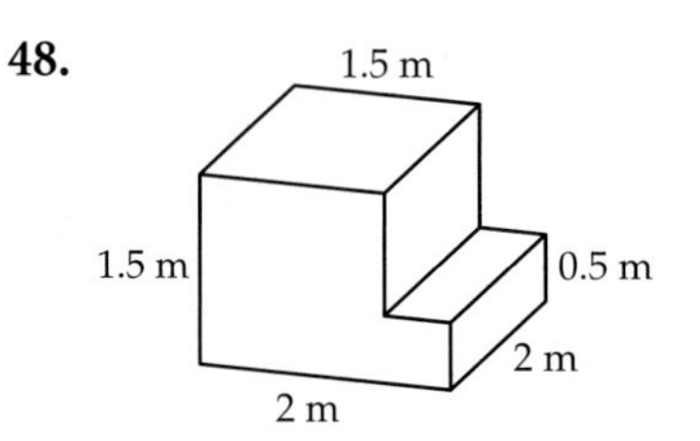

49.

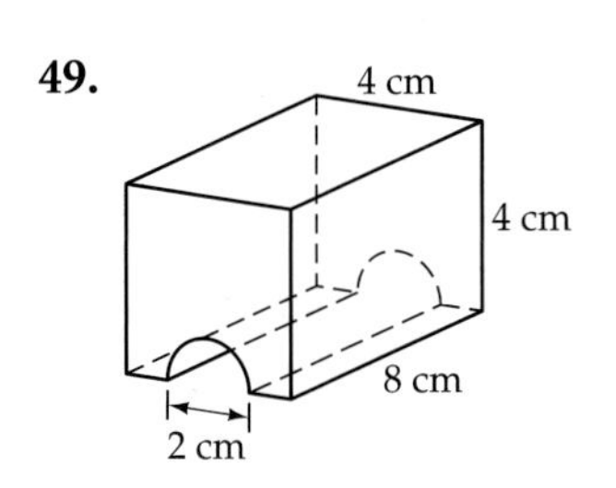

50.

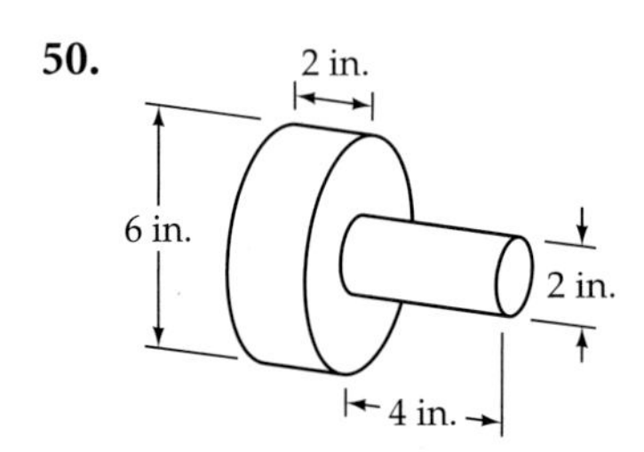

51.

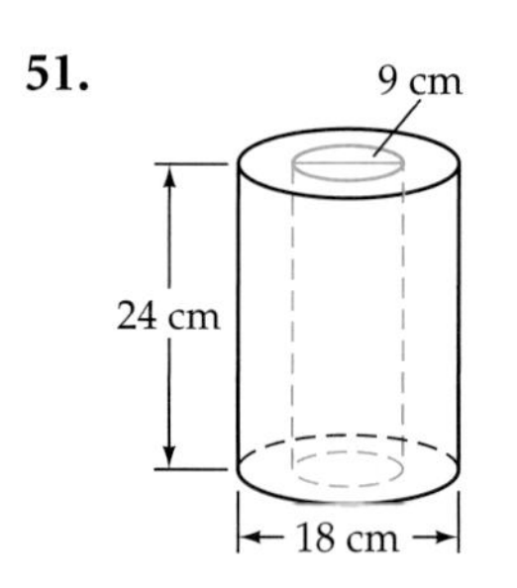

52.

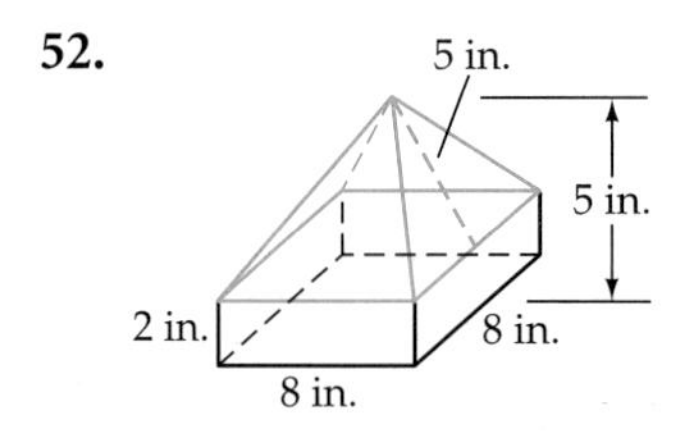

In Exercises 53–58, find the surface area of the figure. Round to the nearest hundredth of a unit.

53.

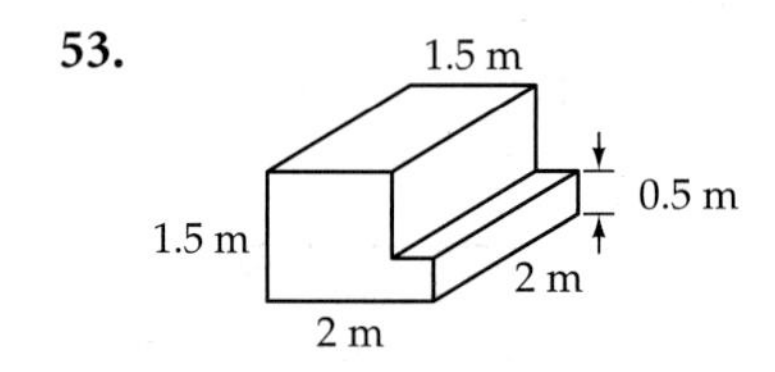

54.

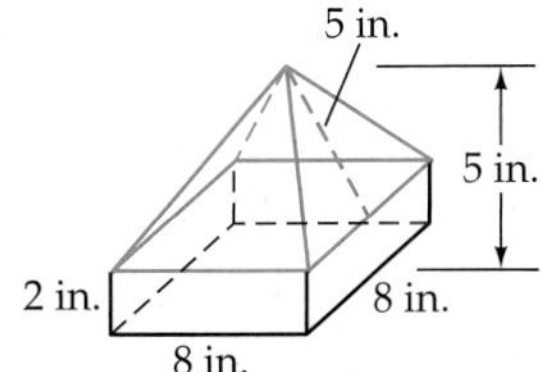

55.

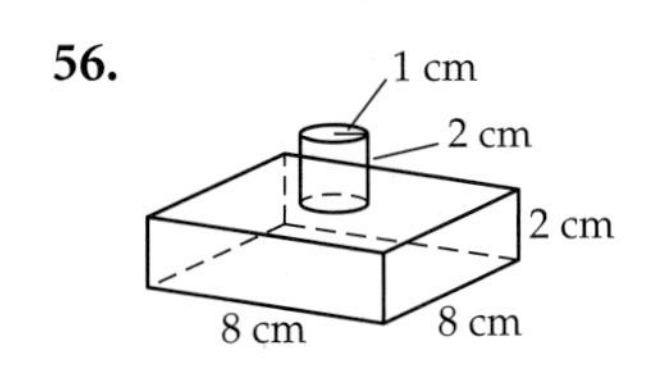

56.

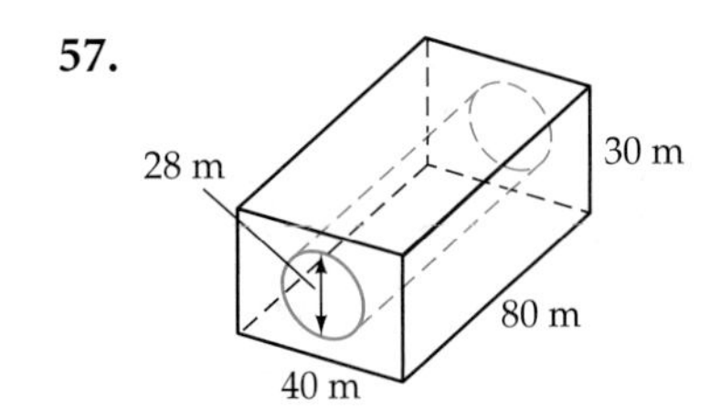

57.

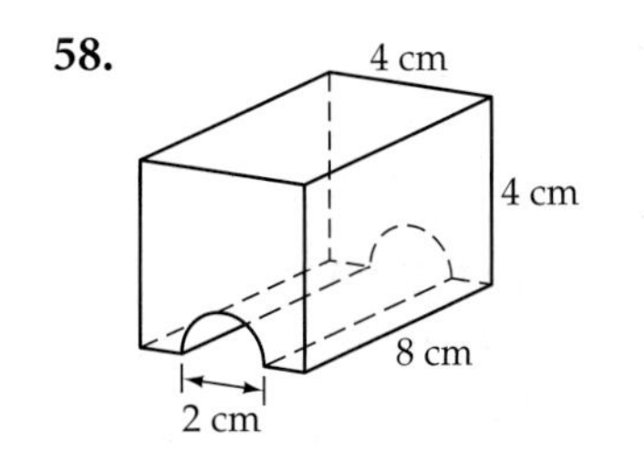

58.

59. Oil Tanks A truck is carrying an oil tank, as shown in the figure below. If the tank is half full, how many cubic feet of oil is the truck carrying? Round to the nearest hundredth of a cubic foot.

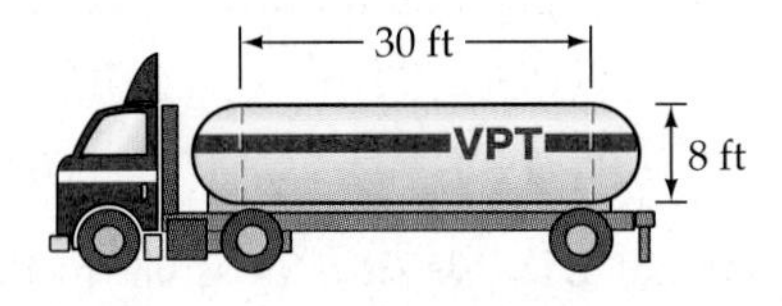

60. Concrete The concrete floor of a building is shown in the following figure. At a cost of \$3.15 per cubic foot, find the cost of having the floor poured. Round to the nearest cent.

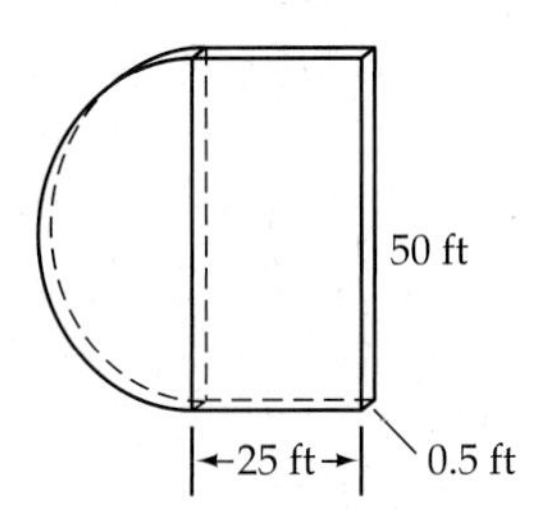

61. Swimming Pools How many liters of water are needed to fill the swimming pool shown below? (1 m^3 contains 1000 L.)

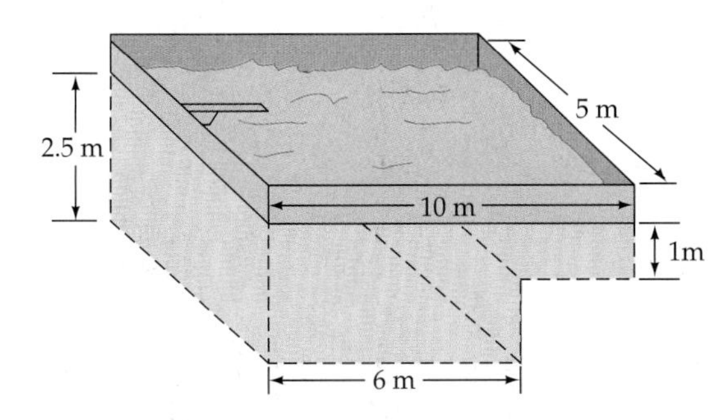

62. Paint A can of paint will cover 250 ft^2 of surface. Find the number of cans of paint that should be purchased to paint the exterior of the auditorium shown in the figure below.

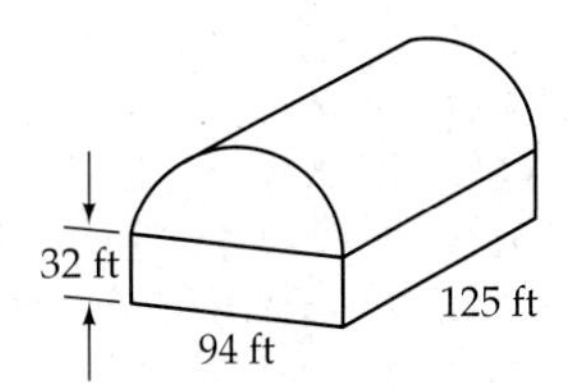

63. Metallurgy A piece of sheet metal is cut and formed into the shape shown below. Given that there are 0.24 g in 1 cm^2 of the metal, find the total number of grams of metal used. Round to the nearest hundredth of a gram.

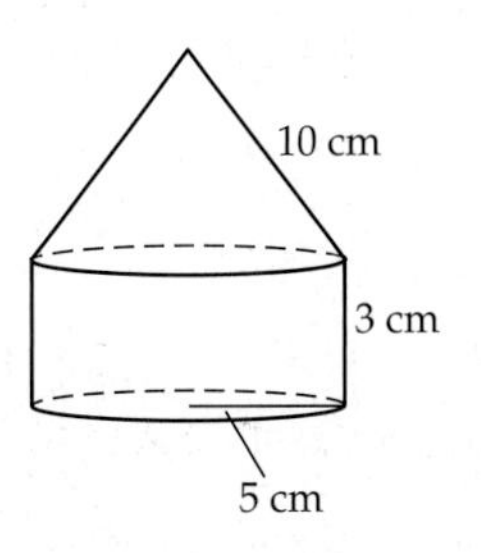

64. **Plastering** The walls of a room that is 25.5 ft long, 22 ft wide, and 8 ft high are being plastered. There are two doors in the room, each measuring 2.5 ft by 7 ft. Each of the six windows in the room measures 2.5 ft by 4 ft. At a cost of $.75 per square foot, find the cost of plastering the walls of the room.

65. **Gold** A solid sphere of gold with a radius of 0.5 cm has a value of $180. Find the value of a sphere of the same type of gold with a radius of 1.5 cm.

66. **Swimming Pools** A swimming pool is built in the shape of a rectangular solid. It holds 32,000 gal of water. If the length, width, and height of the pool are each doubled, how many gallons of water will be needed to fill the pool?

Extensions

CRITICAL THINKING

67. Half of a sphere is called a **hemisphere.** Derive formulas for the volume and surface area of a hemisphere.

68. **a.** Draw a two-dimensional figure that can be cut out and made into a right circular cone.

    b. Draw a two-dimensional figure that can be cut out and made into a regular square pyramid.

69. A sphere fits inside a cylinder as shown in the following figure. The height of the cylinder equals the diameter of the sphere. Show that the surface area of the sphere equals the surface area of the side of the cylinder.

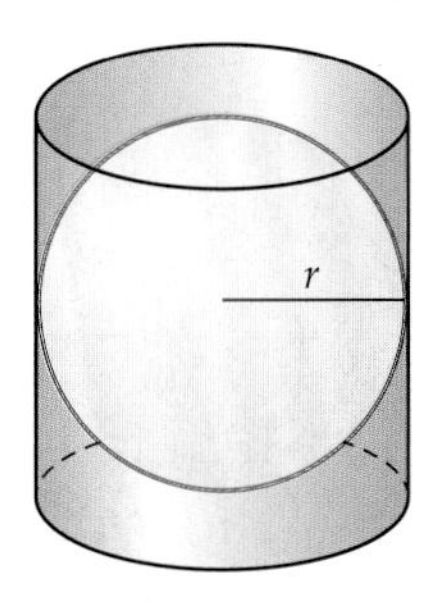

70. Determine whether the statement is always true, sometimes true, or never true.

    a. The slant height of a regular square pyramid is longer than the height.

    b. The slant height of a cone is shorter than the height.

    c. The four triangular faces of a regular square pyramid are equilateral triangles.

71. **a.** What is the effect on the surface area of a rectangular solid of doubling the width and height?

    b. What is the effect on the volume of a rectangular solid of doubling the length and width?

    c. What is the effect on the volume of a cube of doubling the length of each side of the cube?

    d. What is the effect on the surface area of a cylinder of doubling the radius and height?

EXPLORATIONS

72. Explain how you could cut through a cube so that the face of the resulting solid is

    a. a square. **b.** an equilateral triangle.

    c. a trapezoid. **d.** a hexagon.

SECTION 8.5 Introduction to Trigonometry

Trigonometric Functions of an Acute Angle

Given the lengths of two sides of a right triangle, it is possible to determine the length of the third side by using the Pythagorean Theorem. In some situations, however, it may not be practical or possible to know the lengths of two of the sides of a right triangle.

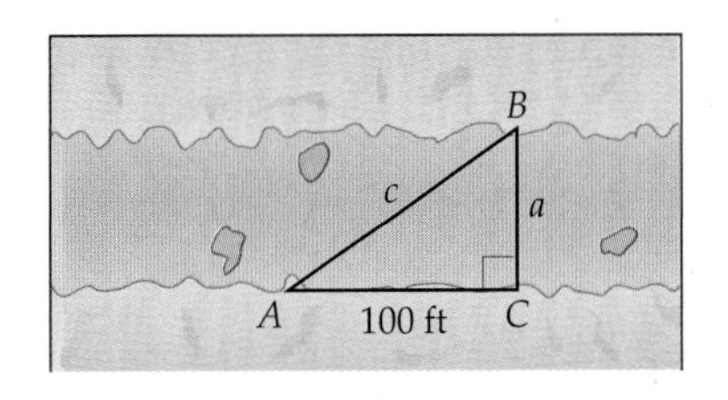

Consider, for example, the problem of engineers trying to determine the distance across a ravine so that they can design a bridge that can be built connecting the two sides. Look at the triangle to the left.

It is fairly easy to measure the length of the side of the triangle that is on the land (100 feet), but the lengths of sides a and c cannot be measured easily because of the ravine.

The study of *trigonometry*, a term that comes from two Greek words meaning "triangle measurement," began about 2000 years ago, partially as a means of solving surveying problems such as the one above. In this section, we will examine *right triangle* trigonometry—that is, trigonometry that applies only to right triangles.

When working with right triangles, it is convenient to refer to the side *opposite* an angle and to the side *adjacent* to (next to) an angle. The hypotenuse of a right triangle is not adjacent to or opposite either of the acute angles in a right triangle.

TAKE NOTE

In trigonometry, it is common practice to use Greek letters for angles of a triangle. Here are some frequently used letters: α (alpha), β (beta), and θ (theta). The word *alphabet* is derived from the first two letters of the Greek alphabet, α and β.

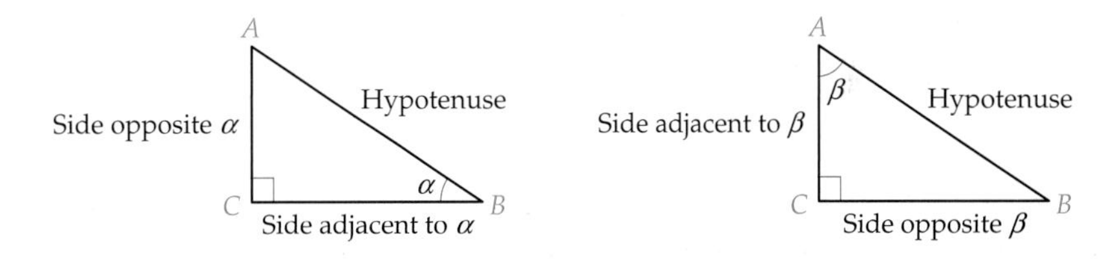

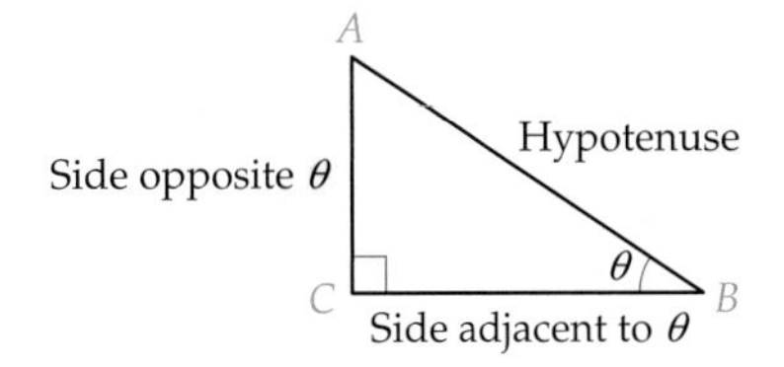

Figure 8.1

Consider the right triangle in Figure 8.1 shown at the left. Six possible ratios can be formed using the lengths of the sides of the triangle.

$$\frac{\text{length of opposite side}}{\text{length of hypotenuse}} \qquad \frac{\text{length of hypotenuse}}{\text{length of opposite side}}$$

$$\frac{\text{length of adjacent side}}{\text{length of hypotenuse}} \qquad \frac{\text{length of hypotenuse}}{\text{length of adjacent side}}$$

$$\frac{\text{length of opposite side}}{\text{length of adjacent side}} \qquad \frac{\text{length of adjacent side}}{\text{length of opposite side}}$$

Each of these ratios defines a value of a trigonometric function of the acute angle θ. The functions are **sine** (sin), **cosine** (cos), **tangent** (tan), **cosecant** (csc), **secant** (sec), and **cotangent** (cot).

The Trigonometric Functions of an Acute Angle of a Right Triangle

If θ is an acute angle of a right triangle ABC, then

$$\sin \theta = \frac{\text{length of opposite side}}{\text{length of hypotenuse}} \qquad \csc \theta = \frac{\text{length of hypotenuse}}{\text{length of opposite side}}$$

$$\cos \theta = \frac{\text{length of adjacent side}}{\text{length of hypotenuse}} \qquad \sec \theta = \frac{\text{length of hypotenuse}}{\text{length of adjacent side}}$$

$$\tan \theta = \frac{\text{length of opposite side}}{\text{length of adjacent side}} \qquad \cot \theta = \frac{\text{length of adjacent side}}{\text{length of opposite side}}$$

As a convenience, we will write opp, adj, and hyp as abbreviations for *the length of the* opposite side, adjacent side, and hypotenuse, respectively. Using this convention, the definitions of the trigonometric functions are written

$$\sin\theta = \frac{\text{opp}}{\text{hyp}} \qquad \csc\theta = \frac{\text{hyp}}{\text{opp}}$$

$$\cos\theta = \frac{\text{adj}}{\text{hyp}} \qquad \sec\theta = \frac{\text{hyp}}{\text{adj}}$$

$$\tan\theta = \frac{\text{opp}}{\text{adj}} \qquad \cot\theta = \frac{\text{adj}}{\text{opp}}$$

All of the trigonometric functions have applications, but the sine, cosine, and tangent functions are used most frequently. For the remainder of this section, we will focus on those functions.

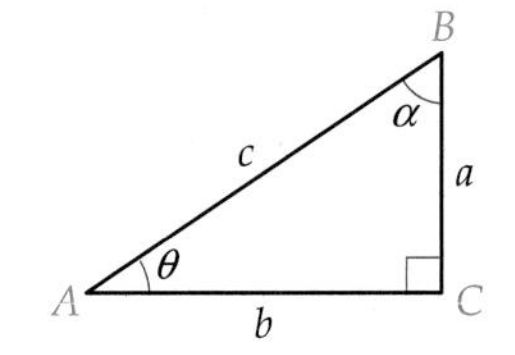

When working with trigonometric functions, be sure to draw a diagram and label the adjacent and opposite sides of an angle. For instance, in the definition above, if we had placed θ at angle A, then the triangle would have been labeled as shown at the left. The definitions of the functions remain the same.

$$\sin\theta = \frac{\text{opp}}{\text{hyp}} \qquad \cos\theta = \frac{\text{adj}}{\text{hyp}} \qquad \tan\theta = \frac{\text{opp}}{\text{adj}}$$

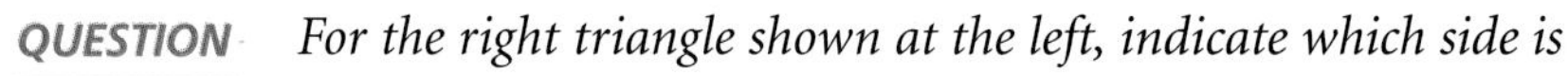

QUESTION *For the right triangle shown at the left, indicate which side is*

a. *adjacent to $\angle A$* ***b.*** *opposite θ*
c. *adjacent to α* ***d.*** *opposite $\angle B$*

EXAMPLE 1 ■ Find the Value of Trigonometric Functions

For the right triangle at the right, find the values of $\sin\theta$, $\cos\theta$, and $\tan\theta$.

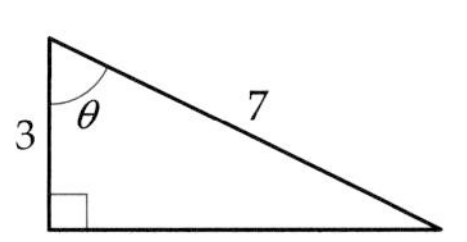

Solution

Use the Pythagorean Theorem to find the length of the side opposite θ.

$$a^2 + b^2 = c^2$$
$$3^2 + b^2 = 7^2 \qquad \text{• } a = 3, c = 7$$
$$9 + b^2 = 49$$
$$b^2 = 40$$
$$b = \sqrt{40} = 2\sqrt{10}$$

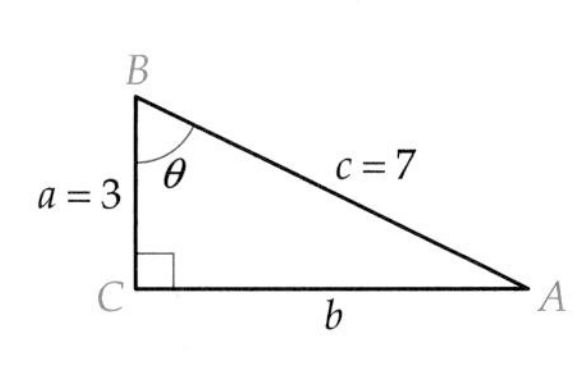

Using the definition of the trigonometric functions, we have

$$\sin\theta = \frac{\text{opp}}{\text{hyp}} = \frac{2\sqrt{10}}{7} \qquad \cos\theta = \frac{\text{adj}}{\text{hyp}} = \frac{3}{7} \qquad \tan\theta = \frac{\text{opp}}{\text{adj}} = \frac{2\sqrt{10}}{3}$$

CHECK YOUR PROGRESS 1 For the right triangle at the right, find the values of $\sin\theta$, $\cos\theta$, and $\tan\theta$.

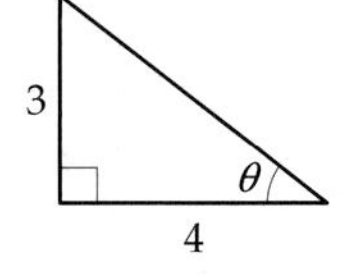

Solution *See page S31.*

ANSWER ***a.*** *b* ***b.*** *a* ***c.*** *a* ***d.*** *b*

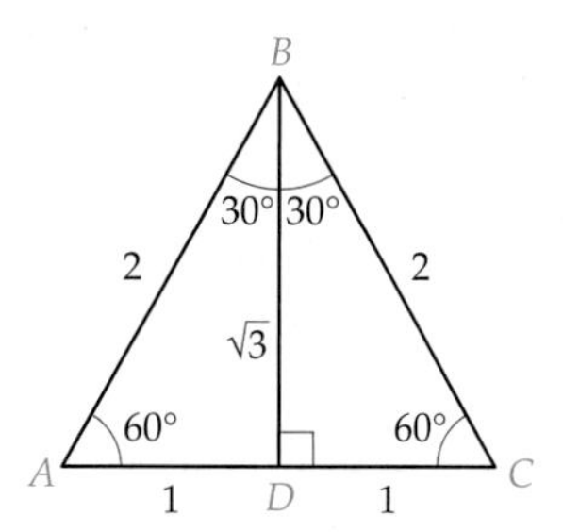

In Example 1, we gave the exact answers. In many cases, approximate values of trigonometric functions are given. The answers to Example 1, rounded to the nearest ten-thousandth, are

$$\sin\theta = \frac{2\sqrt{10}}{7} \approx 0.9035 \quad \cos\theta = \frac{3}{7} \approx 0.4286 \quad \tan\theta = \frac{2\sqrt{10}}{3} \approx 2.1082$$

There are many occasions when we will want to know the value of a trigonometric function for a given angle. Triangle ABC at the left is an equilateral triangle with sides of length 2 units and angle bisector $\overline{BD}$. Because $\overline{BD}$ bisects $\angle ABC$, the measures of $\angle ABD$ and $\angle DBC$ are both 30°. The angle bisector $\overline{BD}$ also bisects $\overline{AC}$. Therefore, $AD = 1$ and $DC = 1$. Using the Pythagorean Theorem, we can find the measure of BD.

$$\begin{aligned}(DC)^2 + (BD)^2 &= (BC)^2\\ 1^2 + (BD)^2 &= 2^2\\ 1 + (BD)^2 &= 4\\ (BD)^2 &= 3\\ BD &= \sqrt{3}\end{aligned}$$

Using the definitions of the trigonometric functions and triangle BCD, we can find the values of the sine, cosine, and tangent of 30° and 60°.

$$\sin 30^\circ = \frac{\text{opp}}{\text{hyp}} = \frac{1}{2} = 0.5 \qquad \sin 60^\circ = \frac{\text{opp}}{\text{hyp}} = \frac{\sqrt{3}}{2} \approx 0.8660$$

$$\cos 30^\circ = \frac{\text{adj}}{\text{hyp}} = \frac{\sqrt{3}}{2} \approx 0.8660 \qquad \cos 60^\circ = \frac{\text{adj}}{\text{hyp}} = \frac{1}{2} = 0.5$$

$$\tan 30^\circ = \frac{\text{opp}}{\text{adj}} = \frac{1}{\sqrt{3}} \approx 0.5774 \qquad \tan 60^\circ = \frac{\text{opp}}{\text{adj}} = \sqrt{3} \approx 1.732$$

The properties of an equilateral triangle enabled us to calculate the values of the trigonometric functions for 30° and 60°. Calculating values of the trigonometric functions for most other angles, however, would be quite difficult. Fortunately, many calculators have been programmed to allow us to estimate these values.

To use a TI-83/84 calculator to find tan 30°, confirm that your calculator is in "degree mode." Press the tan button and key in 30. Then press ENTER.

$\tan 30^\circ \approx 0.5774$

Despite the fact that the values of many trigonometric functions are approximate, it is customary to use the equals sign rather than the approximately equals sign when writing these function values. Thus we write $\tan 30^\circ = 0.5774$.

CALCULATOR NOTE

Just as distances can be measured in feet, miles, meters, and other units, angles can be measured in various units: degrees, radians, and grads. In this section, we use only degree measurements for angles, so be sure your calculator is in degree mode.

On a TI-83/84, press the MODE key to determine whether the calculator is in degree mode.

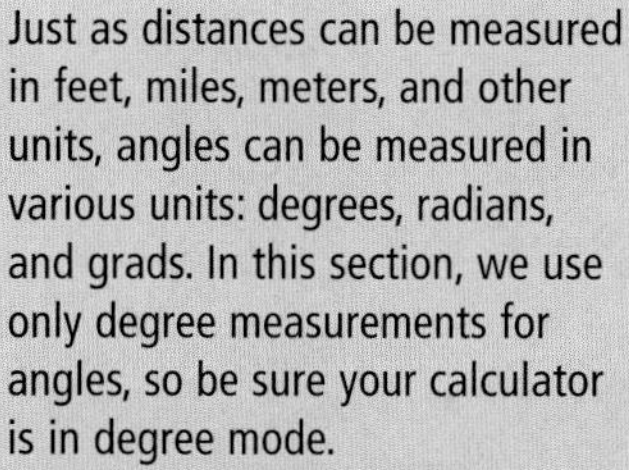

EXAMPLE 2 ■ Use a Calculator to Find the Value of a Trigonometric Function

Use a calculator to find sin 43.8° to the nearest ten-thousandth.

Solution $\sin 43.8^\circ = 0.6921$

CHECK YOUR PROGRESS 2 Use a calculator to find tan 37.1° to the nearest ten-thousandth.

Solution *See page S32.*

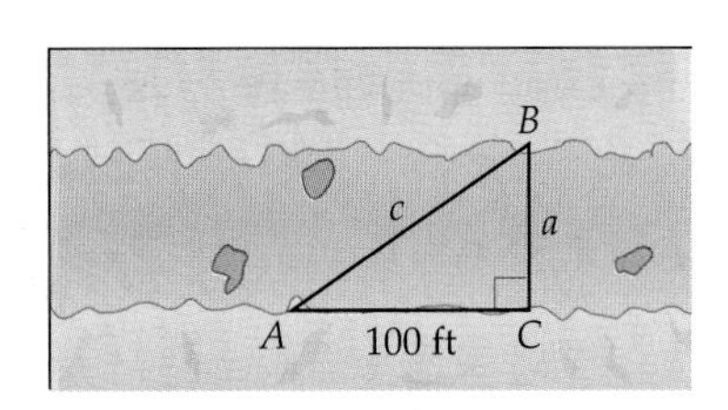

Using trigonometry, the engineers mentioned at the beginning of this section could determine the distance across the ravine after determining the measure of $\angle A$. Suppose the engineers measure the angle as 33.8°. Now the engineers would ask, "Which trigonometric function, sine, cosine, or tangent, involves the side opposite an angle and the side adjacent to that angle?" Knowing that the tangent function is the required function, the engineers could write and solve the equation $\tan 33.8° = \frac{a}{100}$.

$$\tan 33.8° = \frac{a}{100}$$

$$100(\tan 33.8°) = a$$ • Multiply each side of the equation by **100**.

$$66.9 \approx a$$ • Use a calculator to find tan 33.8°. Multiply the result in the display by 100.

The distance across the ravine is approximately 66.9 feet.

EXAMPLE 3 ■ Find the Length of a Side of a Triangle

For the right triangle shown at the left, find the length of side a. Round to the nearest hundredth of a meter.

Solution

We are given the measure of $\angle A$ and the hypotenuse. We want to find the length of side a. Side a is opposite $\angle A$. The sine function involves the side opposite an angle and the hypotenuse.

$$\sin A = \frac{\text{opp}}{\text{hyp}}$$

$$\sin 26° = \frac{a}{24}$$ • A = **26°**, hypotenuse = **24** meters.

$$24(\sin 26°) = a$$ • Multiply each side by **24**.

$$10.52 \approx a$$ • Use a calculator to find sin 26°. Multiply the result in the display by 24.

The length of side a is approximately 10.52 meters.

CHECK YOUR PROGRESS 3 For the right triangle shown at the right, find the length of side a. Round to the nearest hundredth of a foot.

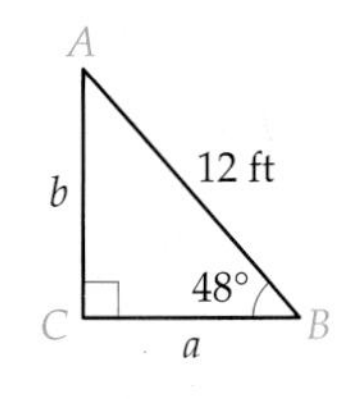

Solution *See page S32.*

Inverse Trigonometric Functions

Sometimes it is necessary to find one of the acute angles in a right triangle. For instance, suppose it is necessary to find the measure of $\angle A$ in the figure at the left. Because the side adjacent to $\angle A$ is known and the hypotenuse is known, we can write

$$\cos A = \frac{\text{adj}}{\text{hyp}}$$

$$\cos A = \frac{25}{27}$$

The solution of this equation is the angle whose cosine is $\frac{25}{27}$. This angle can be found by using the $\cos^{-1}$ key on a calculator.

$$\cos^{-1}\left(\frac{25}{27}\right) \approx 22.19160657$$

To the nearest tenth, the measure of $\angle A$ is 22.2°.
The function $\cos^{-1}$ is called the *inverse cosine function.*

Definitions of the Inverse Sine, Inverse Cosine, and Inverse Tangent Functions

For $0° < x < 90°$:
$y = \sin^{-1}(x)$ can be read "y is the angle whose sine is x."
$y = \cos^{-1}(x)$ can be read "y is the angle whose cosine is x."
$y = \tan^{-1}(x)$ can be read "y is the angle whose tangent is x."

TAKE NOTE

The expression $y = \sin^{-1}(x)$ is sometimes written $y = \arcsin(x)$. The two expressions are equivalent. The expressions $y = \cos^{-1}(x)$ and $y = \arccos(x)$ are equivalent, as are $y = \tan^{-1}(x)$ and $y = \arctan(x)$.

Note that $\sin^{-1}(x)$ is used to denote the inverse of the sine function. It is not the reciprocal of $\sin x$ but the notation used for its inverse. The same is true for $\cos^{-1}$ and $\tan^{-1}$.

CALCULATOR NOTE 

To find an inverse function on a calculator, usually the INV or 2nd key is pressed prior to pushing the function key. Some calculators have $\sin^{-1}$, $\cos^{-1}$, and $\tan^{-1}$ keys. Consult the instruction manual for your calculator.

EXAMPLE 4 ■ Evaluate an Inverse Trigonometric Function

Use a calculator to find $\sin^{-1}(0.9171)$. Round to the nearest tenth of a degree.

Solution

$\sin^{-1}(0.9171) \approx 66.5°$ • The calculator must be in degree mode. Press the keys for the inverse sine function followed by .9171. Press ENTER.

CHECK YOUR PROGRESS 4 Use a calculator to find $\tan^{-1}(0.3165)$. Round to the nearest tenth of a degree.

Solution *See page S32.*

EXAMPLE 5 ■ Find the Measure of an Angle Using the Inverse of a Trigonometric Function

Given $\sin\theta = 0.7239$, find θ. Use a calculator. Round to the nearest tenth of a degree.

TAKE NOTE

If
$\sin\theta = 0.7239$,
then
$\theta = \sin^{-1}(0.7239)$.

Solution
This is equivalent to finding $\sin^{-1}(0.7239)$. The calculator must be in the degree mode.

$$\sin^{-1}(0.7239) \approx 46.4°$$
$$\theta \approx 46.4°$$

CHECK YOUR PROGRESS 5 Given $\tan\theta = 0.5681$, find θ. Use a calculator. Round to the nearest tenth of a degree.

Solution *See page S32.*

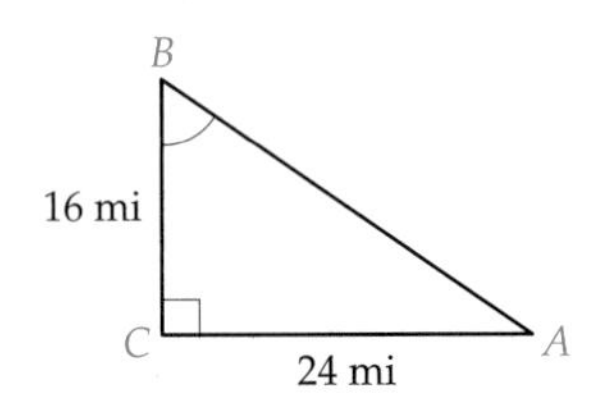

EXAMPLE 6 ■ Find the Measure of an Angle in a Right Triangle

For the right triangle shown at the left, find the measure of $\angle B$. Round to the nearest tenth of a degree.

Solution
We want to find the measure of $\angle B$, and we are given the lengths of the sides opposite $\angle B$ and adjacent to $\angle B$. The tangent function involves the side opposite an angle and the side adjacent to that angle.

$$\tan B = \frac{\text{opposite}}{\text{adjacent}}$$

$$\tan B = \frac{24}{16}$$

$$B = \tan^{-1}\left(\frac{24}{16}\right)$$

$$B \approx 56.3^\circ$$ • Use the $\tan^{-1}$ key on a calculator.

The measure of $\angle B$ is approximately 56.3°.

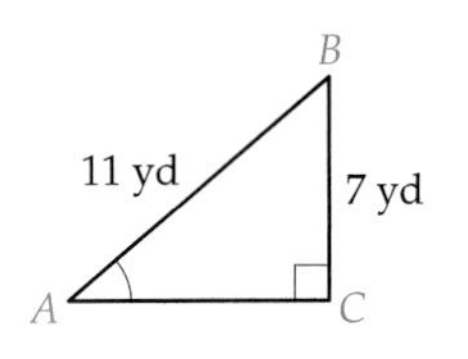

CHECK YOUR PROGRESS 6 For the right triangle shown at the left, find the measure of $\angle A$. Round to the nearest tenth of a degree.

Solution *See page S32.*

Angles of Elevation and Depression

The use of trigonometry is necessary in a variety of situations. One application, called **line-of-sight problems,** concerns an observer looking at an object.

Angles of elevation and depression are measured with respect to a horizontal line. If the object being sighted is above the observer, the acute angle formed by the line of sight and the horizontal line is an **angle of elevation.** If the object being sighted is below the observer, the acute angle formed by the line of sight and the horizontal line is an **angle of depression.**

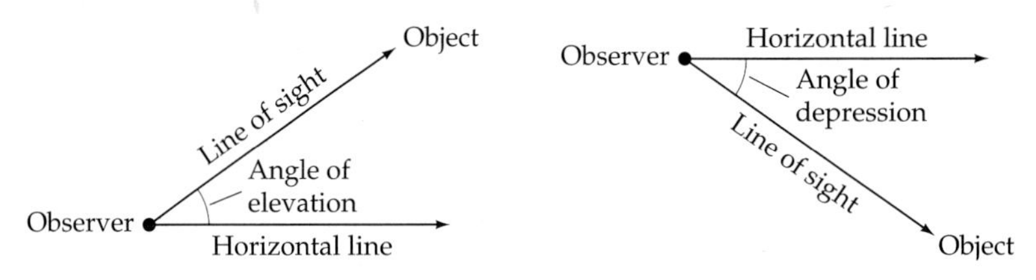

EXAMPLE 7 ■ Solve an Angle of Elevation Problem

The angle of elevation of the top of a flagpole 62 feet away is 34°. Find the height of the flagpole. Round to the nearest tenth of a foot.

Solution
Draw a diagram. To find the height, h, write a trigonometric function that relates the given information and the unknown side of the triangle.

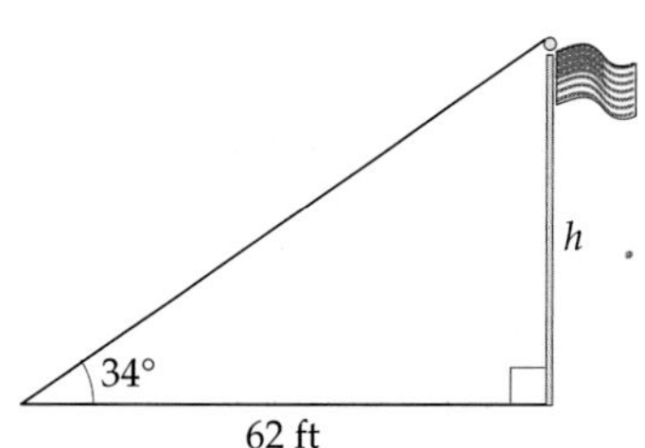

$$\tan 34^\circ = \frac{h}{62}$$

$$62(\tan 34^\circ) = h$$ • **Multiply each side by 62.**

$$41.8 \approx h$$ • **Use a calculator to find $\tan 34^\circ$. Multiply the result in the display by 62.**

The height of the flagpole is approximately 41.8 feet.

CHECK YOUR PROGRESS 7 The angle of depression from the top of a lighthouse that is 20 meters high to a boat on the water is 25°. How far is the boat from the base of the lighthouse? Round to the nearest tenth of a meter.

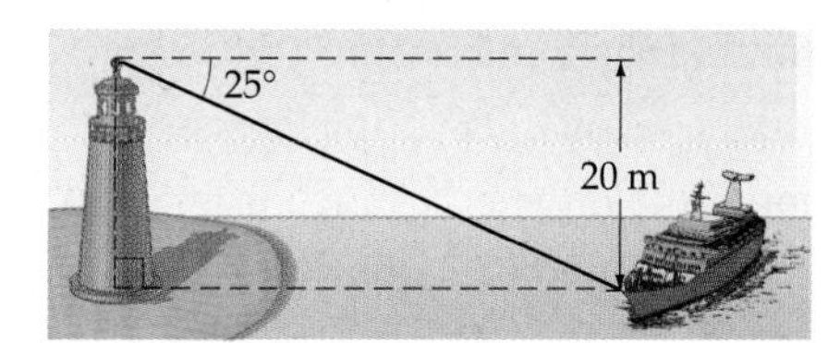

Solution *See page S32.*

Excursion

Approximating the Value of Trigonometric Functions

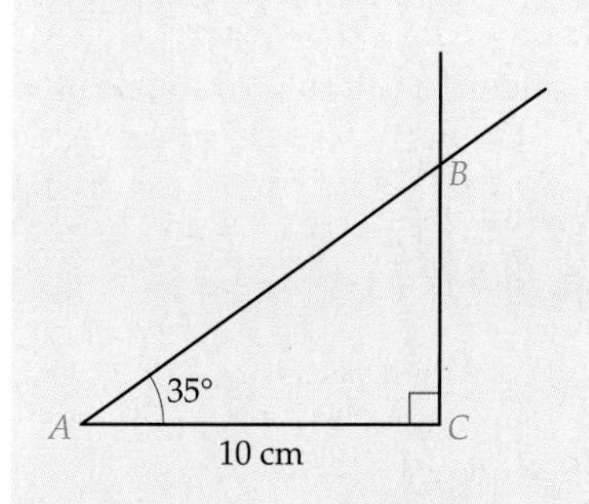

The value of a trigonometric function can be approximated by drawing a triangle with a given angle. To illustrate, we will choose an angle of 35°.

To find the tangent of 35° using the definitions given in this section, we can carefully construct a right triangle containing an angle of 35°. Because any two right triangles containing an angle of 35° are similar, *the value for* tan 35° *is the same no matter what triangle we draw.*

Excursion Exercises

1. Draw a horizontal line segment 10 cm long with left endpoint A and right endpoint C. See the diagram at the left.
2. Using a protractor, construct at A a 35° angle.
3. Draw at C a vertical line that intersects the terminal side of angle A at B. Your drawing should be similar to the one at the left.
4. Measure line segment BC.
5. What is the approximate value of tan 35°?
6. Using your value for BC and the Pythagorean Theorem, estimate AB.
7. Estimate sin 35° and cos 35°.
8. What are the values of sin 35°, cos 35°, and tan 35° as produced by a calculator? Round to the nearest ten-thousandth.

Exercise Set 8.5

1. Use the right triangle at the right and sides a, b, and c to do the following:

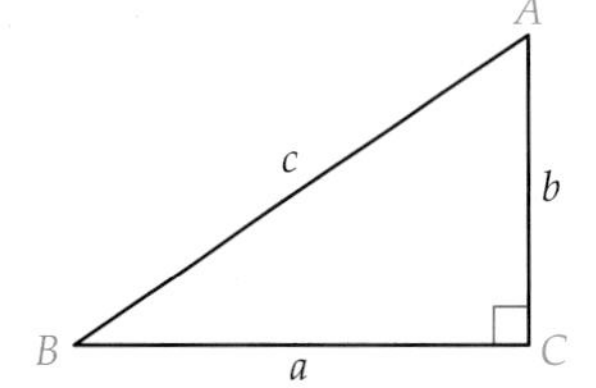

   a. Name the ratio for the trigonometric function sin A.
   b. Name the ratio for the trigonometric function sin B.
   c. Name the ratio for the trigonometric function cos A.
   d. Name the ratio for the trigonometric function cos B.
   e. Name the ratio for the trigonometric function tan A.
   f. Name the ratio for the trigonometric function tan B.

2. 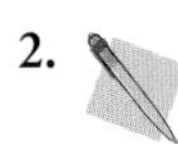 Explain the meaning of the notation $y = \sin^{-1}(x)$, $y = \cos^{-1}(x)$, and $y = \tan^{-1}(x)$.

In Exercises 3–10, find the values of sin θ, cos θ, and tan θ for the given right triangle. Give the exact values.

3. 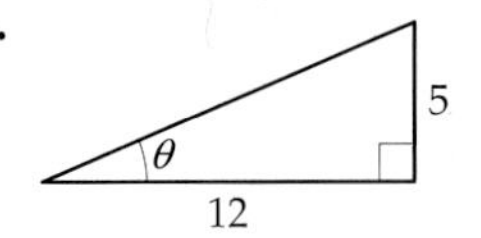

4. 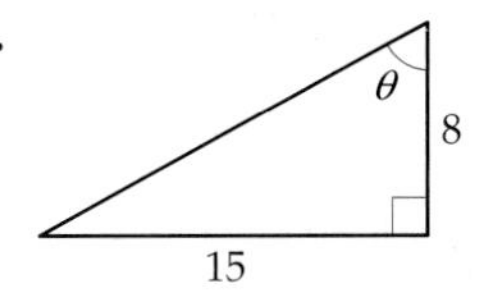

5. 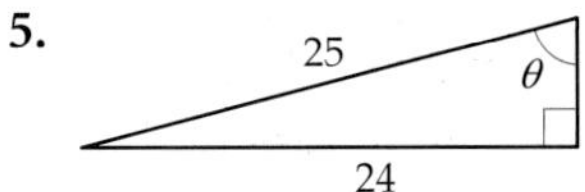

6. 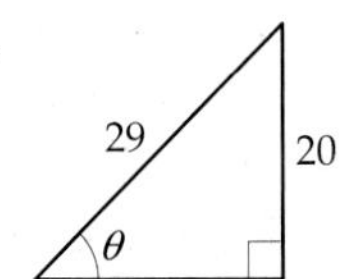

7. 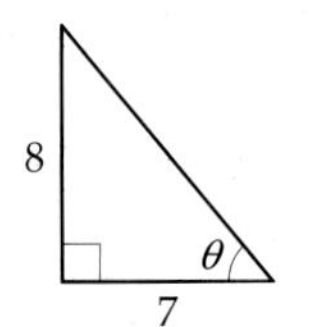

8. 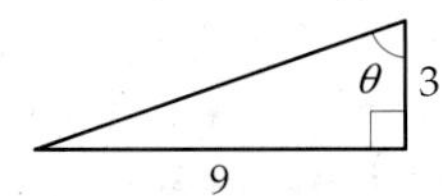

9. θ, 80, 40

10. 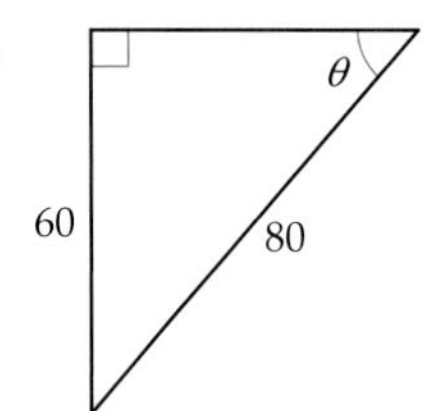

In Exercises 11–26, use a calculator to estimate the value of each of the following. Round to the nearest ten-thousandth.

11. cos 47°
12. sin 62°
13. tan 55°
14. cos 11°
15. sin 85.6°
16. cos 21.9°
17. tan 63.4°
18. sin 7.8°
19. tan 41.6°
20. cos 73°
21. sin 57.7°
22. tan 39.2°
23. sin 58.3°
24. tan 35.1°
25. cos 46.9°
26. sin 50°

In Exercises 27–42, use a calculator. Round to the nearest tenth.

27. Given $\sin\theta = 0.6239$, find θ.
28. Given $\cos\beta = 0.9516$, find β.
29. Find $\cos^{-1}(0.7536)$.
30. Find $\sin^{-1}(0.4478)$.
31. Given $\tan\alpha = 0.3899$, find α.
32. Given $\sin\beta = 0.7349$, find β.
33. Find $\tan^{-1}(0.7815)$.
34. Find $\cos^{-1}(0.6032)$.
35. Given $\cos\theta = 0.3007$, find θ.
36. Given $\tan\alpha = 1.588$, find α.
37. Find $\sin^{-1}(0.0105)$.
38. Find $\tan^{-1}(0.2438)$.
39. Given $\sin\beta = 0.9143$, find β.
40. Given $\cos\theta = 0.4756$, find θ.
41. Find $\cos^{-1}(0.8704)$.
42. Find $\sin^{-1}(0.2198)$.

For Exercises 43–56, draw a picture and label it. Set up an equation and solve it. Show all your work. Round an angle to the nearest tenth of a degree. Round the length of a side to the nearest hundredth of a unit.

43. Ballooning A balloon, tethered by a cable 997 feet long, was blown by a wind so that the cable made an angle of 57.6° with the ground. Find the height of the balloon off the ground.

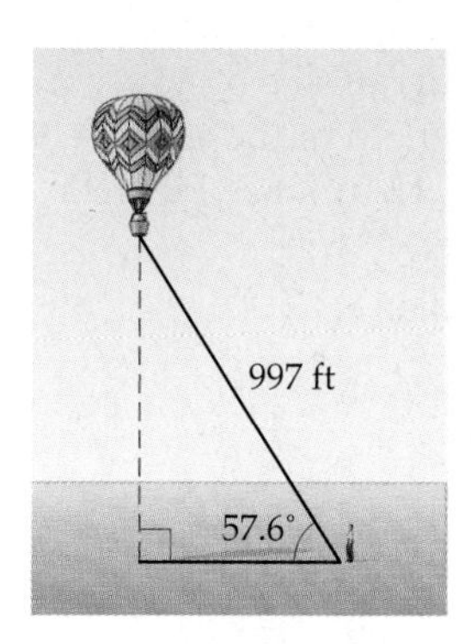

44. Roadways A road is inclined at an angle of 9.8° with the horizontal. Find the distance that one must drive on this road in order to be elevated 14.8 feet above the horizontal.

45. Home Maintenance A ladder 30.8 feet long leans against a building. If the foot of the ladder is 7.25 feet from the base of the building, find the angle the top of the ladder makes with the building.

46. Aviation A plane takes off from a field and rises at an angle of 11.4° with the horizontal. Find the height of the plane after it has traveled a distance of 1250 feet.

47. Guy Wires A guy wire whose grounded end is 16 feet from the telephone pole it supports makes an angle of 56.7° with the ground. How long is the wire?

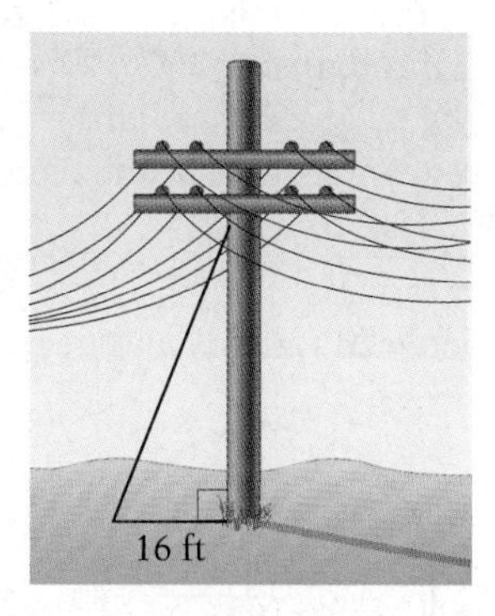

48. Angle of Depression A lighthouse built at sea level is 169 feet tall. From its top, the angle of depression to a boat below measures 25.1°. Find the distance from the boat to the foot of the lighthouse.

49. Angle of Elevation At a point 39.3 feet from the base of a tree, the angle of elevation of its top measures 53.4°. Find the height of the tree.

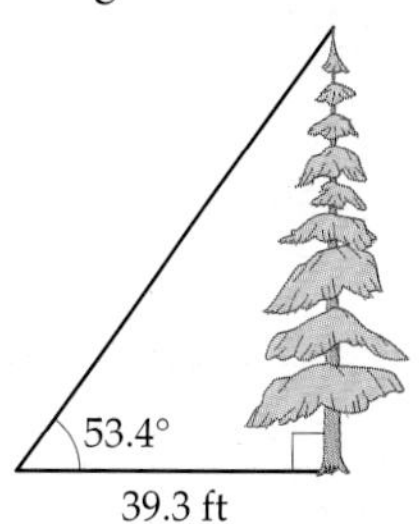

50. Angle of Depression An artillery spotter in a plane that is at an altitude of 978 feet measures the angle of depression of an enemy tank as 28.5°. How far is the enemy tank from the point on the ground directly below the spotter?

51. Home Maintenance A 15-foot ladder leans against a house. The ladder makes an angle of 65° with the ground. How far up the side of the house does the ladder reach?

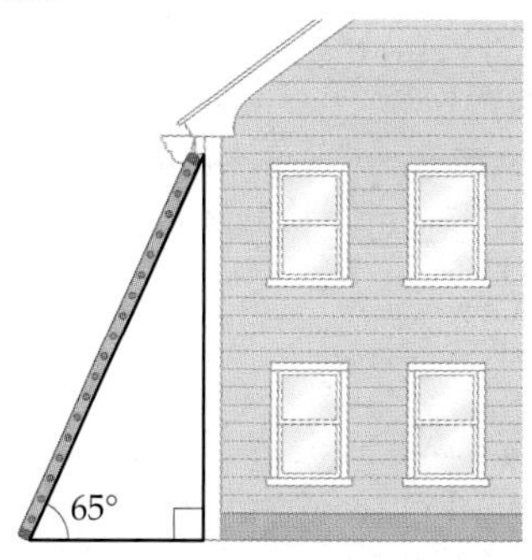

52. Angle of Elevation Find the angle of elevation of the sun when a tree 40.5 feet high casts a shadow 28.3 feet long.

53. Guy Wires A television transmitter tower is 600 feet high. If the angle between the guy wire (attached at the top) and the tower is 55.4°, how long is the guy wire?

54. Ramps A ramp used to load a racing car onto a flatbed carrier is 5.25 meters long, and its upper end is 1.74 meters above the lower end. Find the angle between the ramp and the road.

55. Angle of Elevation The angle of elevation of the sun is 51.3° at a time when a tree casts a shadow 23.7 yards long. Find the height of the tree.

56. Angle of Depression From the top of a building 312 feet tall, the angle of depression to a flower bed on the ground below is 12.0°. What is the distance between the base of the building and the flower bed?

Extensions

CRITICAL THINKING

57. Can the value of sin θ or cos θ ever be greater than 1? Explain your answer.

58. Can the value of tan θ ever be greater than 1? Explain your answer.

59. Let $\sin\theta = \frac{2}{3}$. Find the exact value of $\cos\theta$.

60. Let $\tan\theta = \frac{5}{4}$. Find the exact value of $\sin\theta$.

61. Let $\cos\theta = \frac{3}{4}$. Find the exact value of $\tan\theta$.

62. Let $\sin\theta = \frac{\sqrt{5}}{4}$. Find the exact value of $\tan\theta$.

63. Let $\sin\theta = a$, $a > 0$. Find $\cos\theta$.

64. Let $\tan\theta = a$, $a > 0$. Find $\sin\theta$.

EXPLORATIONS

As we noted in this section, angles can also be measured in *radians.* To define a radian, first consider a circle of radius r and two radii $\overline{OA}$ and $\overline{OB}$. The angle θ formed by the two radii is a **central angle**. The portion of the circle between A and B is an **arc** of the circle and is written $\overset{\frown}{AB}$. We say that $\overset{\frown}{AB}$ *subtends* the angle θ. The length of the arc is s. (See Figure 1 below.)

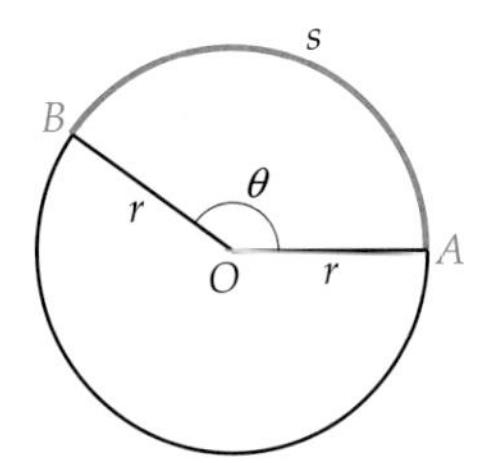

Figure 1

Definition of Radian

One **radian** is the measure of the central angle subtended by an arc of length r. The measure of θ in Figure **2** is 1 radian.

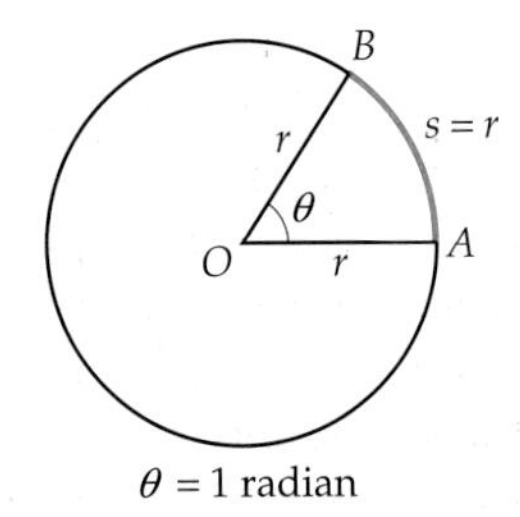

Figure 2

To find the radian measure of an angle subtended by an arc of length s, use the following formula.

Radian Measure

Given an arc of length s on a circle of radius r, the measure of the central angle subtended by the arc is $\theta = \frac{s}{r}$ radians.

For example, to find the measure in radians of the central angle subtended by an arc of 9 in. in a circle of radius 12 in., divide the length of the arc ($s = 9$ in.) by the length of the radius ($r = 12$ in.). See Figure 3.

$$\theta = \frac{9 \text{ in.}}{12 \text{ in.}} \text{ radian}$$
$$= \frac{3}{4} \text{ radian}$$

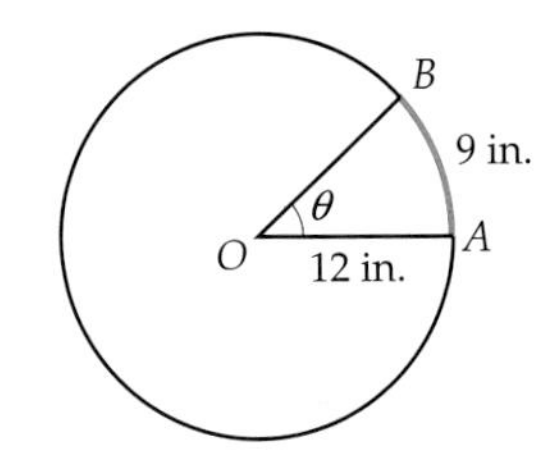

Figure 3

65. Find the measure in radians of the central angle subtended by an arc of 12 cm in a circle of radius 3 cm.

66. Find the measure in radians of the central angle subtended by an arc of 4 cm in a circle of radius 8 cm.

67. Find the measure in radians of the central angle subtended by an arc of 6 in. in a circle of radius 9 in.

68. Find the measure in radians of the central angle subtended by an arc of 12 ft in a circle of radius 10 ft.

Recall that the circumference of a circle is given by $C = 2\pi r$. Therefore, the radian measure of the central angle subtended by the circumference is $\theta = \frac{2\pi r}{r} = 2\pi$.

In degree measure, the central angle has a measure of 360°. Thus we have 2π radians $= 360°$. Dividing each side of the equation by 2 gives π radians $= 180°$. From the last equation, we can establish the conversion factors $\frac{\pi \text{ radians}}{180°}$ and $\frac{180°}{\pi \text{ radians}}$. These conversion factors are used to convert between radians and degrees.

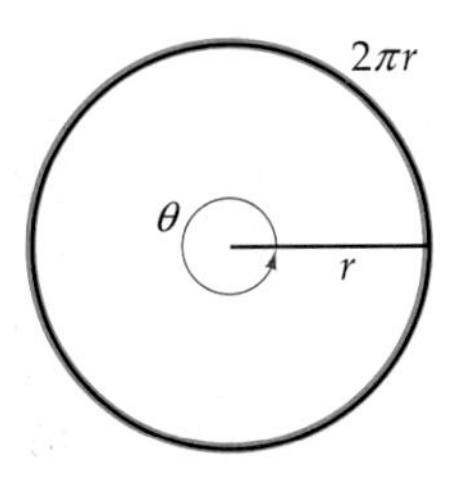

Conversion between Radians and Degrees

- To convert from degrees to radians, multiply by $\frac{\pi \text{ radians}}{180°}$.
- To convert from radians to degrees, multiply by $\frac{180°}{\pi \text{ radians}}$.

For instance, to convert 30° to radians, multiply 30° by $\frac{\pi \text{ radians}}{180°}$.

$$30° = 30°\left(\frac{\pi \text{ radians}}{180°}\right)$$

$$= \frac{\pi}{6} \text{ radian} \quad \text{• Exact answer}$$

$$\approx 0.5236 \text{ radian} \quad \text{• Approximate answer}$$

To convert 2 radians to degrees, multiply 2 by $\frac{180°}{\pi \text{ radians}}$.

$$2 \text{ radians} = 2\left(\frac{180°}{\pi \text{ radians}}\right)$$

$$= \left(\frac{360}{\pi}\right)^{\circ} \quad \text{• Exact answer}$$

$$\approx 114.5916° \quad \text{• Approximate answer}$$

69. What is the measure in degrees of 1 radian?

70. Is the measure of 1 radian larger or smaller than the measure of 1°?

In Exercises 71–76, convert degree measure to radian measure. Find an exact answer and an answer rounded to the nearest ten-thousandth.

71. 45° **72.** 180° **73.** 315°

74. 90° **75.** 210° **76.** 18°

In Exercises 77–82, convert radian measure to degree measure. For Exercises 80–82, find an exact answer and an answer rounded to the nearest ten-thousandth.

77. $\frac{\pi}{3}$ radians **78.** $\frac{11\pi}{6}$ radians

79. $\frac{4\pi}{3}$ radians **80.** 1.2 radians

81. 3 radians **82.** 2.4 radians

SECTION 8.6 Non-Euclidean Geometry

Euclidean Geometry versus Non-Euclidean Geometry

Learning to play a game by reading the rules can be frustrating. If the game has too many rules, you may decide that learning to play is not worth the effort. Most popular games are based on a few rules that are easy to learn but that still allow the game to develop into complex situations. The ancient Greek mathematician Euclid was immersed in the mathematical "game" of geometry. It was his desire to construct a substantial body of geometrical knowledge based on the *fewest* possible number of rules, which he called **postulates.** Euclid based his study of geometry on the following five postulates.

TAKE NOTE

In addition to postulates, Euclid's geometry, referred to as *Euclidean geometry,* involves definitions and some *undefined terms.* For instance, the term *point* is an undefined term in Euclidean geometry. To define a point would require a description involving additional undefined terms. Because every term cannot be defined by simpler defined terms, Euclid chose to use *point* as an undefined term. In Euclidean geometry, the term *line* is also an undefined term.

Euclid's Postulates

P1: A line segment can be drawn from any point to any other point.

P2: A line segment can be extended continuously in a straight line.

P3: A circle can be drawn with any center and any radius.

P4: All right angles are equal to one another.

P5: ***The Parallel Postulate*** Through a given point not on a given line, exactly one line can be drawn parallel to the given line.

For many centuries, the truth of these postulates was felt to be self-evident. However, a few mathematicians suspected that the fifth postulate, known as the Parallel Postulate, could be deduced from the other postulates. Many individuals took on the task of proving the Parallel Postulate, but after more than 2000 years, the proof of the Parallel Postulate still remained unsolved.

Carl Friedrich Gauss (gaus′) (1777–1855) became interested in the problem of proving the Parallel Postulate as a teenager. After many failed attempts to establish the Parallel Postulate as a theorem, Gauss decided to take a different route. He came to the conclusion that the Parallel Postulate was an independent postulate and that it could be changed to produce a new type of geometry. This is analogous to changing one of the rules of a game to create a new game.

The diaries and letters written by Gauss show that he proposed the following alternative postulate.

Gauss's Alternative to the Parallel Postulate

Through a given point not on a given line, there are *at least two* lines parallel to the given line.

Thus Gauss was the first person to realize that non-Euclidean geometries could be created by merely changing or excluding the Parallel Postulate. It has been speculated that Gauss did not publish this result because he felt it would not be readily accepted by other mathematicians.

historical note

Nikolai Lobachevsky (lō′bə-chĕf′skē) (1793–1856), was a noted Russian mathematician. Lobachevski's concept of a non-Euclidean geometry was so revolutionary that he is called the Copernicus of Geometry. ■

The idea of using an alternative postulate in place of the Parallel Postulate was first proposed *openly* by the Russian mathematician Nikolai Lobachevsky. In a lecture given in 1826, and in a series of monthly articles that appeared in the academic journal of the University of Kazan in 1829, Lobachevsky provided a detailed investigation into the problem of the Parallel Postulate. He proposed that a consistent new geometry could be developed by replacing the Parallel Postulate with the alternative postulate, which assumes that *more than one* parallel line can be drawn through a point not on a given line.

Lobachevsky described his geometry as *imaginary geometry,* because he could not comprehend a model of such a geometry. Independently of each other, Gauss and Lobachevsky had each conceived of the same idea; but whereas Gauss chose to keep the idea to himself, Lobachevsky professed the idea to the world. It is for this reason that we now refer to the geometry they each conceived as *Lobachevskian geometry.* Today Lobachevskian geometry is often called *hyperbolic geometry.*

The year 1826, in which Lobachevsky first lectured about a new non-Euclidean geometry, also marks the birth of the mathematician Bernhard Riemann. Although Riemann died of tuberculosis at age 39, he made major contributions in several areas of mathematics and physics, including the theory of functions of complex

historical note

Bernhard Riemann (rē′mən) (1826–1866). "Riemann's achievement has taught mathematicians to disbelieve in *any* geometry, or any space, as a necessary mode of human perception."[2] ■

variables, electrodynamics, and non-Euclidean geometry. Riemann was the first person to consider a geometry in which the Parallel Postulate was replaced with the following postulate.

Riemann's Alternative to the Parallel Postulate

Through a given point not on a given line, there exist *no* lines parallel to the given line.

Unlike the geometry developed by Lobachevsky, which was not based on a physical model, the non-Euclidean geometry of Riemann was closely associated with a sphere and the remarkable idea that because a line is an undefined term, a line on the surface of a sphere can be different from a line on a plane. Even though a line on a sphere can be different from a line on a plane, it seems reasonable that "spherical lines" should retain some of the properties of lines on a plane. For example, on a plane, the shortest distance between two points is measured along the line that connects the points. The line that connects the points is an example of what is called a *geodesic.*

Definition of a Geodesic

A **geodesic** is a curve C on a surface S such that for any two points on C, the portion of C between these points is the shortest path on S that joins these points.

On a sphere, the geodesic between two points is a *great circle* that connects the points.

Definition of a Great Circle

A **great circle** of a sphere is a circle on the surface of the sphere whose center is at the center of the sphere. Any two points on a great circle divide the circle into two arcs. The shorter arc is the **minor arc** and the longer arc is the **major arc.**

In *Riemannian geometry,* which is also called *spherical geometry* or *elliptical geometry,* great circles, which are the geodesics of a sphere, are thought of as lines. Figure 8.2 shows a sphere and two of its great circles. Because all great circles of a sphere intersect, a sphere provides us with a model of a geometry in which there are no parallel lines.

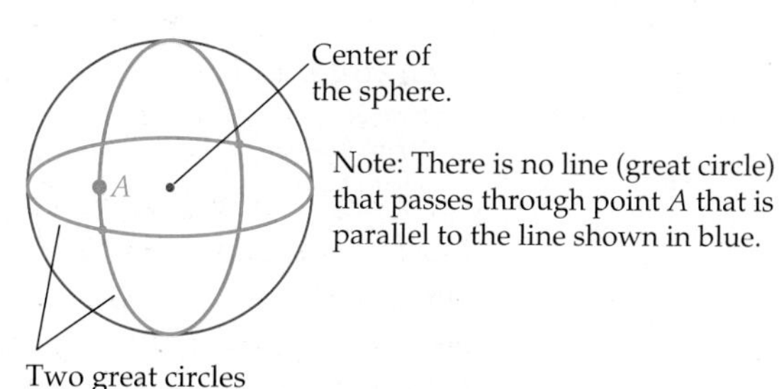

Figure 8.2 *A sphere and its great circles serve as a physical model for Riemannian geometry.*

2. Bell, E. T. *Men of Mathematics.* New York: Touchstone Books, Simon and Schuster, 1986.

TAKE NOTE

Riemannian geometry has many applications. For instance, because Earth is nearly spherical, navigation over long distances is facilitated by the use of Riemannian geometry as opposed to Euclidean geometry.

In Riemannian geometry, a triangle may have as many as three right angles. Figure 8.3 illustrates a spherical triangle with one right angle, a spherical triangle with two right angles, and a spherical triangle with three right angles.

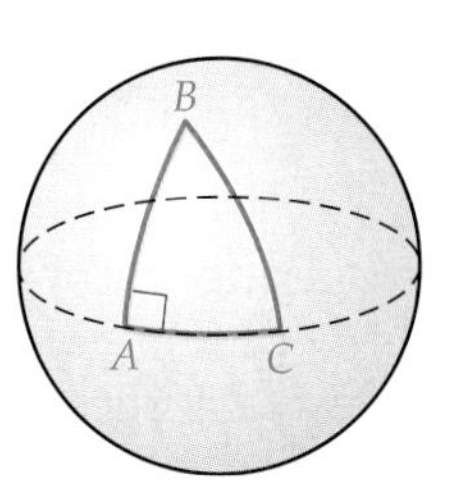

a. A spherical triangle with one right angle

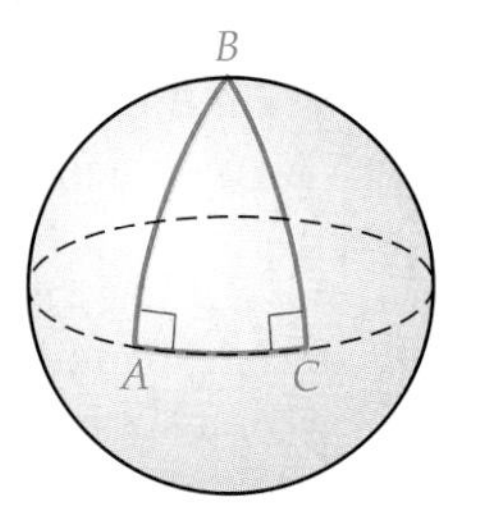

b. A spherical triangle with two right angles

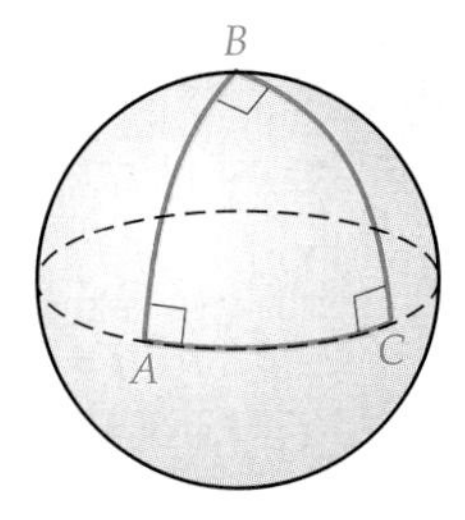

c. A spherical triangle with three right angles

Figure 8.3

In Euclidean geometry it is not possible to determine the area of a triangle from the measures of its angles alone.

In spherical geometry, however, any two triangles on the same sphere that are similar are also congruent. Therefore, in spherical geometry, the area of a triangle depends on the size of its angles and the radius of the sphere.

QUESTION *In Euclidean geometry, are two similar triangles necessarily congruent triangles?*

The Spherical Triangle Area Formula

The area S of the spherical triangle ABC on a sphere with radius r is given by

$$S = (m\angle A + m\angle B + m\angle C - 180°) \cdot \left(\frac{\pi}{180°}\right) r^2$$

where each angle is measured in degrees.

EXAMPLE 1 ■ Find the Area of a Spherical Triangle

Find the area of a spherical triangle with three right angles on a sphere with a radius of 1 ft. Find both the exact area and the approximate area rounded to the nearest hundredth of a square foot.

ANSWER *No.*

TAKE NOTE

To check the result in Example 1, use the fact that the given triangle covers $\frac{1}{8}$ of the surface of the sphere. See Figure 8.3c. The total surface area of a sphere is $4\pi r^2$. In Example 1, $r = 1$ ft, so the sphere has a surface area of 4π ft². Thus the area of the spherical triangle in Example 1 should be

$$\left(\frac{1}{8}\right) \cdot (4\pi) = \frac{\pi}{2}\text{ ft}^2$$

Solution

Apply the spherical triangle area formula.

$$S = (m\angle A + m\angle B + m\angle C - 180°) \cdot \left(\frac{\pi}{180°}\right)r^2$$

$$= (90° + 90° + 90° - 180°) \cdot \left(\frac{\pi}{180°}\right)(1)^2$$

$$= (90°) \cdot \left(\frac{\pi}{180°}\right)$$

$$= \frac{\pi}{2}\text{ ft}^2 \quad \text{• Exact area}$$

$$\approx 1.57\text{ ft}^2 \quad \text{• Approximate area}$$

CHECK YOUR PROGRESS 1 Find the area of the spherical triangle whose angles measure 200°, 90°, and 90° on a sphere with a radius of 6 in. Find both the exact area and the approximate area rounded to the nearest hundredth of a square inch.

Solution *See page S32.*

Mathematicians have not been able to create a three-dimensional model that *perfectly* illustrates all aspects of hyperbolic geometry. However, an infinite saddle surface can be used to visualize some of the basic aspects of hyperbolic geometry. Figure 8.4 shows a portion of an infinite saddle surface.

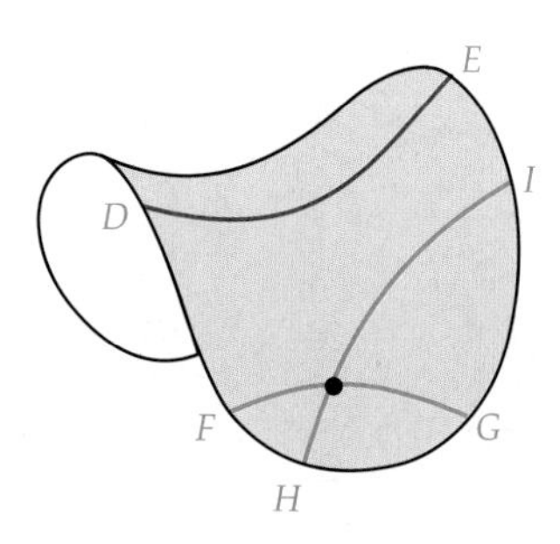

Figure 8.4 *A portion of an infinite saddle surface*

A line (geodesic) can be drawn through any two points on the saddle surface. Most lines on the saddle surface have a concave curvature as shown by $\overleftrightarrow{DE}$, $\overleftrightarrow{FG}$, and $\overleftrightarrow{HI}$. Keep in mind that the saddle surface is an infinite surface. Figure 8.4 shows only a portion of the surface.

Parallel lines on an infinite saddle surface are defined as two lines that do not intersect. In Figure 8.4, $\overleftrightarrow{FG}$ and $\overleftrightarrow{HI}$ are *not* parallel because they intersect at a point. The lines $\overleftrightarrow{DE}$ and $\overleftrightarrow{FG}$ are parallel because they do not intersect. The lines $\overleftrightarrow{DE}$ and $\overleftrightarrow{HI}$ are also parallel lines. Figure 8.4 provides a geometric model of a hyperbolic geometry because for a given line, *more than one* parallel line exists through a point not on the given line.

TAKE NOTE

Several different physical models have been invented to illustrate hyperbolic geometry, but most of these models require advanced mathematics beyond the scope of this text.

Figure 8.5 shows a triangle drawn on a saddle surface. The triangle is referred to as a *hyperbolic triangle.* Due to the curvature of the sides of the hyperbolic triangle, the sum of the measures of the angles of the triangle is less than 180 degrees. This is true for all hyperbolic triangles.

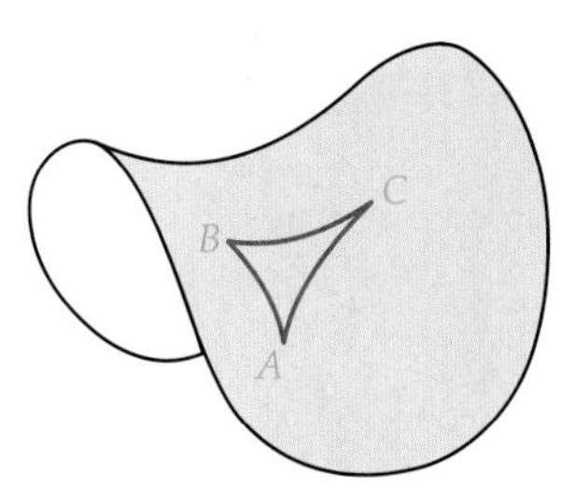

Figure 8.5 *A hyperbolic triangle*

The following chart compares and contrasts some of the properties of plane, hyperbolic, and spherical geometries.

Euclidean Geometry	Non-Euclidean Geometries	
Euclidean or Plane Geometry (circa 300 B.C.):	**Lobachevskian or Hyperbolic Geometry (1826):**	**Riemannian or Spherical Geometry (1855):**
Through a given point not on a given line, exactly one line can be drawn parallel to the given line.	***Through a given point not on a given line, there are at least two lines parallel to the given line.***	***Through a given point not on a given line, there exist no lines parallel to the given line.***
Geometry on a plane	Geometry on an infinite saddle surface	Geometry on a sphere
Parallel lines; Plane triangle	Parallel lines; Hyperbolic triangle	Spherical triangle
For any triangle ABC, $m\angle A + m\angle B + m\angle C = 180°$	For any triangle ABC, $m\angle A + m\angle B + m\angle C < 180°$	For any triangle ABC, $180° < m\angle A + m\angle B + m\angle C < 540°$
A triangle can have at most one right angle.	A triangle can have at most one right angle.	A triangle can have one, two, or three right angles.
The shortest path between two points is the line segment that connects the points.	The curves shown in the above figure illustrate some of the geodesics of an infinite saddle surface.	The shortest path between two points is the minor arc of a great circle that passes through the points.

Math Matters Curved Space

In 1915, Albert Einstein proposed a revolutionary theory. This theory is now called the *general theory of relativity.* One of the major ideas of this theory is that space is curved, or "warped," by the mass of stars and planets. The greatest curvature occurs around those stars with the largest mass. Light rays in space do not travel in a straight path, but rather follow the geodesics of this curved space. Recall that the shortest path that joins two points on a given surface is on a geodesic of the surface.

It is interesting to consider the paths of light rays in space from a different perspective, in which the light rays travel along straight paths, in a space that is non-Euclidean. To better understand this concept, consider an airplane that flies from Los Angeles to London. We say that the airplane flies on a straight path between the two cities. The path is not really a straight path if we use Euclidean geometry as our frame of reference, but if we think in terms of Reimannian geometry, then the path *is* straight. We can use Euclidean geometry or a non-Euclidean geometry as our frame of reference. It does not matter which we choose, but it does change our concept about what is a straight path and what is a curved path.

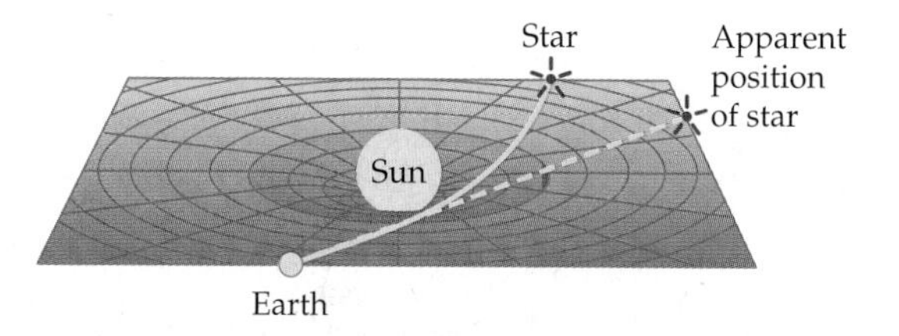

Light rays from a star follow the geodesics of space. Any light rays that pass near the sun are slightly bent. This causes some stars to appear to an observer on Earth to be in different positions than their actual positions.

City Geometry: A Contemporary Non-Euclidean Geometry

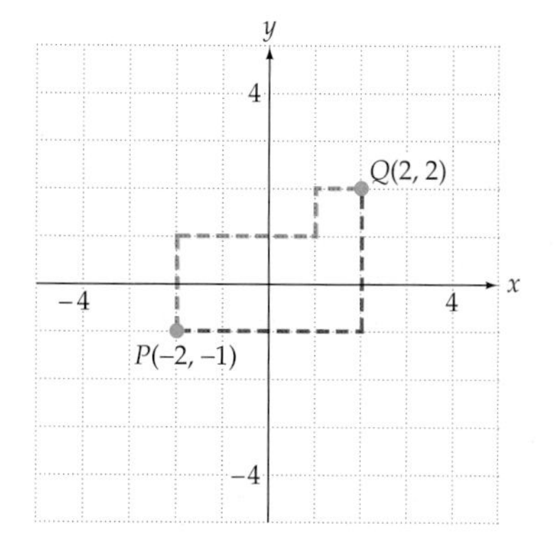

Figure 8.6 *Two city paths from P to Q*

Consider the geometric model of a city shown in Figure 8.6. In this city, all of the streets run either straight north and south or straight east and west. The distance between adjacent north–south streets is 1 block, and the distance between adjacent east–west streets is 1 block. In a city it is generally not possible to travel from P to Q along a straight path. Instead, one must travel between P and Q by traveling along the streets. **As you travel from P to Q, we assume that you always travel in a direction that gets you closer to point Q.** Two such paths are shown by the red and the green dashed line segments in Figure 8.6.

We will use the notation $d_C(P, Q)$ to represent the *city distance* between the points P and Q. For P and Q as shown in Figure 8.6, $d_C(P, Q) = 7$ blocks. This distance can be determined by counting the number of blocks needed to travel along the streets from P to Q or by using the following formula.

point of interest

The city distance formula is also called the *Manhattan metric*. It first appeared in an article published by the mathematician Hermann Minkowski (1864–1909).

The City Distance Formula

If $P(x_1, y_1)$ and $Q(x_2, y_2)$ are two points in a city, then the **city distance** between P and Q is given by

$$d_C(P, Q) = |x_2 - x_1| + |y_2 - y_1|$$

In Euclidean geometry, the distance between the points P and Q is defined as the length of $\overline{PQ}$. To determine a *Euclidean distance formula* for the distance between $P(x_1, y_1)$ and $Q(x_2, y_2)$, we first locate the point $R(x_2, y_1)$. See Figure 8.7.

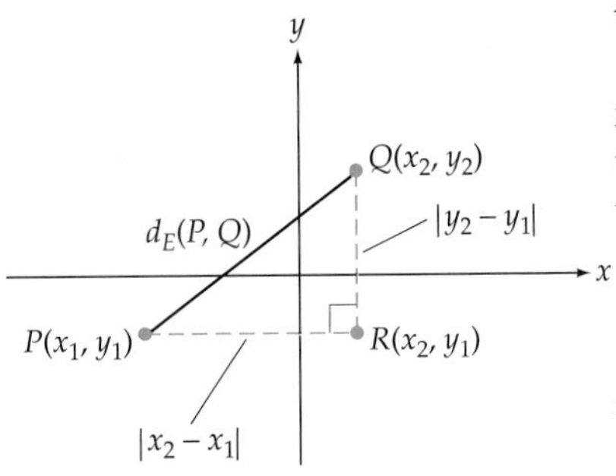

Figure 8.7

Note that R has the same x-coordinate as Q and that R has the same y-coordinate as P. The horizontal distance between P and R is $|x_2 - x_1|$, and the vertical distance between R and Q is $|y_2 - y_1|$. Apply the Pythagorean Theorem to the right triangle PRQ to produce

$$[d_E(P, Q)]^2 = |x_2 - x_1|^2 + |y_2 - y_1|^2$$

Because the square of a number cannot be negative, the absolute value signs are not necessary.

$$[d_E(P, Q)]^2 = (x_2 - x_1)^2 + (y_2 - y_1)^2$$

Take the square root of each side of the equation to produce

$$d_E(P, Q) = \sqrt{(x_2 - x_1)^2 + (y_2 - y_1)^2}$$

The Euclidean Distance Formula

If $P(x_1, y_1)$ and $Q(x_2, y_2)$ are two points in a plane, then the **Euclidean distance** between P and Q is given by

$$d_E(P, Q) = \sqrt{(x_2 - x_1)^2 + (y_2 - y_1)^2}$$

TAKE NOTE

In Example 2 we have calculated the city distances by using the city distance formula. These distances can also be determined by counting the number of blocks needed to travel along the streets from P to Q on a rectangular coordinate grid.

EXAMPLE 2 ■ Find the Euclidean Distance and the City Distance Between Two Points

For each of the following, find $d_E(P, Q)$ and $d_C(P, Q)$. Assume that both $d_E(P, Q)$ and $d_C(P, Q)$ are measured in blocks. Round approximate results to the nearest tenth of a block.

a. $P(-4, -3), Q(2, -1)$ **b.** $P(2, -3), Q(-5, 4)$

Solution

a.
$$\begin{aligned} d_E(P, Q) &= \sqrt{(x_2 - x_1)^2 + (y_2 - y_1)^2} \\ &= \sqrt{[2 - (-4)]^2 + [(-1) - (-3)]^2} \\ &= \sqrt{6^2 + 2^2} \\ &= \sqrt{40} \approx 6.3 \text{ blocks} \end{aligned}$$

$$\begin{aligned} d_C(P, Q) &= |x_2 - x_1| + |y_2 - y_1| \\ &= |2 - (-4)| + |(-1) - (-3)| \\ &= |6| + |2| \\ &= 8 \text{ blocks} \end{aligned}$$

b. $d_E(P, Q) = \sqrt{(x_2 - x_1)^2 + (y_2 - y_1)^2}$
$= \sqrt{[(-5) - 2]^2 + [4 - (-3)]^2}$
$= \sqrt{(-7)^2 + 7^2}$
$= \sqrt{98} \approx 9.9$ blocks

$d_C(P, Q) = |x_2 - x_1| + |y_2 - y_1|$
$= |(-5) - 2| + |4 - (-3)|$
$= |-7| + |7|$
$= 7 + 7$
$= 14$ blocks

CHECK YOUR PROGRESS 2 For each of the following, find $d_E(P, Q)$ and $d_C(P, Q)$. Assume that both $d_E(P, Q)$ and $d_C(P, Q)$ are measured in blocks. Round approximate results to the nearest tenth of a block.
a. $P(-1, 4)$, $Q(3, 2)$ **b.** $P(3, -4)$, $Q(-1, 5)$

Solution *See page S32.*

Recall that a circle is a plane figure in which all points are the same distance from a given center point and the length of the radius r of the circle is the distance from the center point to a point on the circle. Figure 8.8 shows a *Euclidean circle* centered at (0, 0) with a radius of 3 blocks. Figure 8.9 shows all the points in a city that are 3 blocks from the center point (0, 0). These points form a *city circle* with a radius of 3 blocks.

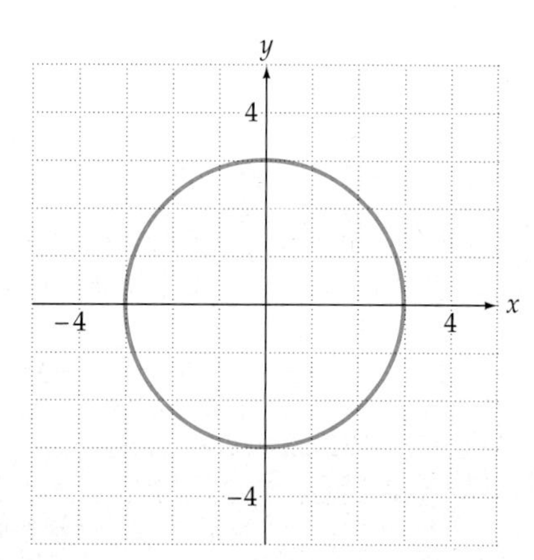

Figure 8.8 *A Euclidean circle with center (0, 0) and a radius of 3 blocks*

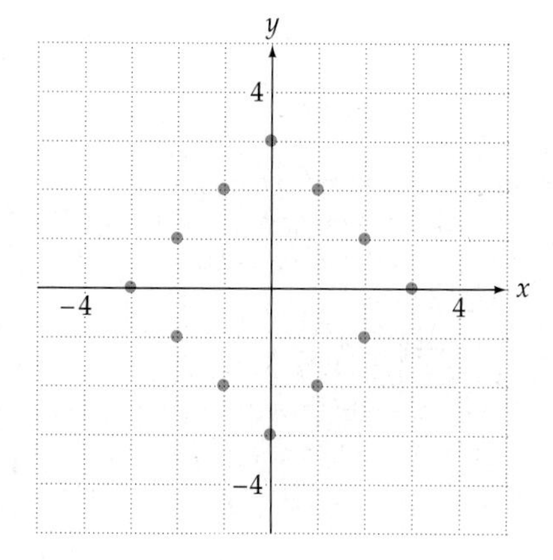

Figure 8.9 *A city circle with center (0, 0) and a radius of 3 blocks*

It is interesting to observe that the *city circle* shown in Figure 8.9 consists of just 12 points and that these points all lie on a square with vertices (3, 0), (0, 3), (−3, 0), and (0, −3).

Excursion

Finding Geodesics

Form groups of three or four students. Each group needs a roll of narrow tape or a ribbon and the two geometrical models shown at the right. The purpose of this Excursion is to use the tape (ribbon) to determine the geodesics of a surface.

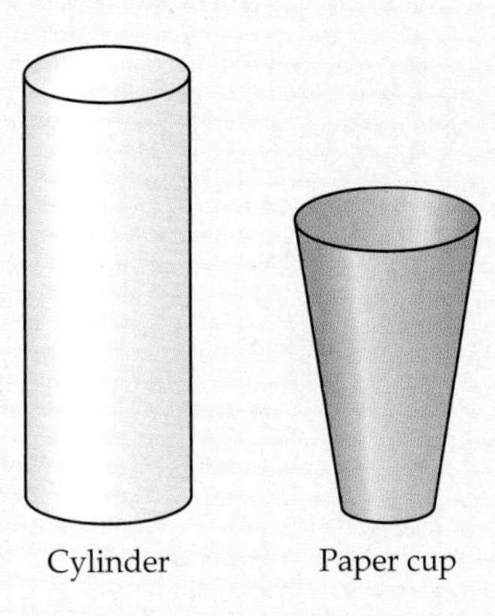

The following three theorems can be used to determine the geodesics of a surface.

Geodesic Theorems

Theorem 1 If a surface is smooth with no edges or holes, then the shortest path between any two points is on a geodesic of the surface.

Theorem 2 ***The Tape Test*** If a piece of tape is placed so that it lies flat on a smooth surface, then the center line of the tape is on a geodesic of the surface.

Theorem 3 ***Inverse of the Tape Test*** If a piece of tape does not lie flat on a smooth surface, then the center line of the tape is not on a geodesic of the surface.

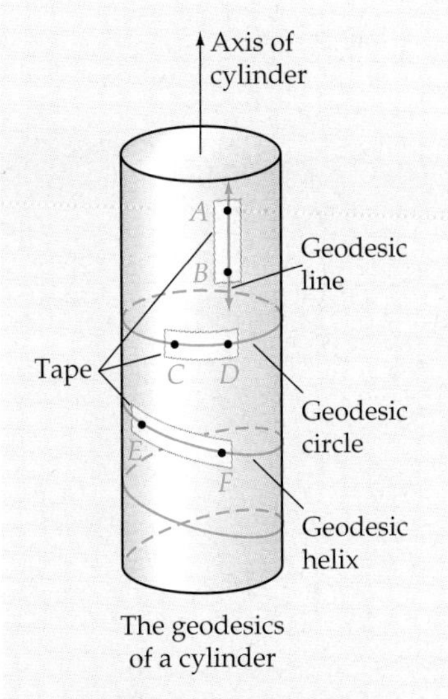

Figure 8.10 *The geodesics of a cylinder*

What are the geodesics of a cylinder? If two points A and B are as shown in Figure 8.10, then the vertical line segment between A and B is the shortest path between the points. A piece of tape can be placed so that it covers point A and point B and lies flat on the cylinder. Thus, by Theorem 2, line segment $\overline{AB}$ is on a geodesic of the cylinder.

If two points C and D are as shown in Figure 8.10, then the minor arc of a circle is the shortest path between the points. Once again we see that a piece of tape can be placed so that it covers points C and D and lies flat on the cylinder. Theorem 2 indicates that the arc $\overset{\frown}{CD}$ is on a geodesic of the cylinder.

To find a geodesic that passes through the two points E and F, start your tape at E and proceed slightly downward and to the right, toward point F. If your tape lies flat against the cylinder, you have found the geodesic for the two points. If your tape does not lie flat against the surface, then your path is not a geodesic and you need to experiment further. Eventually you will find the *circular helix* curve, shown in Figure 8.10, that allows the tape to lie flat on the cylinder.[3]

(continued)

3. The tread of a bolt is an example of a circular helix.

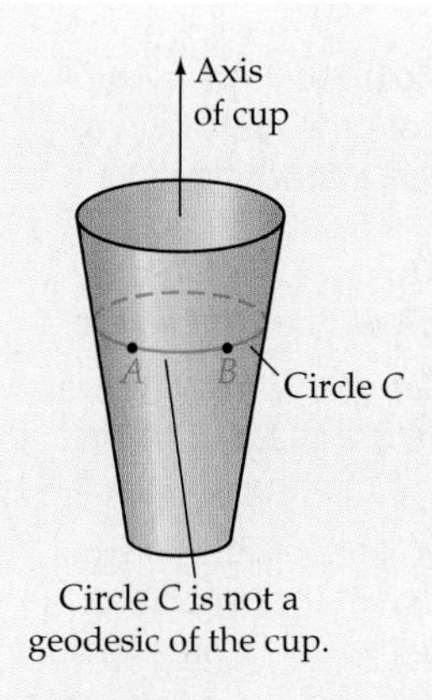

Circle C is not a geodesic of the cup.

Figure 8.11

Additional experiments with the tape and the cylinder should convince you that a geodesic of a cylinder is either (a) a line parallel to the axis of the cylinder, (b) a circle with center on the axis of the cylinder and diameter perpendicular to the axis, or (c) a circular helix curve that has a constant slope and a center on the axis of the cylinder.

Excursion Exercises

1. a. Place two points A and B on a paper cup so that A and B are both at the same height, as in Figure 8.11. We know circle C that passes through A and B is *not* a geodesic of the cup because a piece of tape will not lie flat when placed directly on top of circle C. Experiment with a piece of tape to determine the *actual* geodesic that passes through A and B. Make a drawing that shows this geodesic and illustrate how it differs from circle C.

b. Use a cup similar to the one in Figure 8.11 and a piece of tape to determine two other types of geodesics of the cup. Make a drawing that shows each of these two additional types of geodesics.

2.  The only geodesics of a sphere are great circles. Write a sentence that explains how you can use a piece of tape to show that circle D in Figure 8.12 is not a geodesic of the sphere.

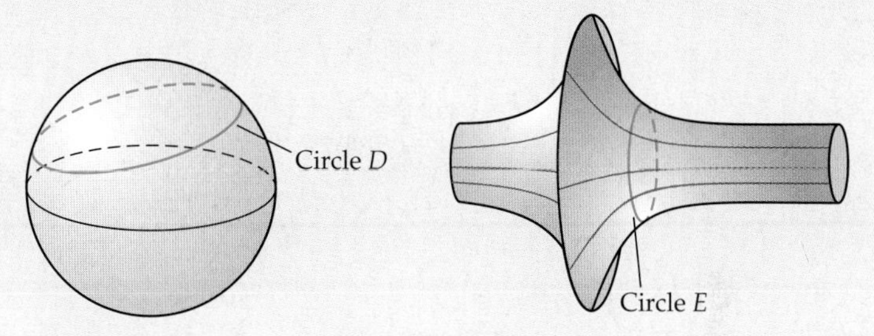

Figure 8.12 **Figure 8.13**

3.  Write a sentence that explains how you know that circle E in Figure 8.13 is not a geodesic of the figure.

Excursion Exercises 4 to 6 require a world globe that shows the locations of major cities. We suggest that you use a thin ribbon, instead of tape, to determine the great circle routes in the following exercises, because tape may damage the globe.

4. A pilot flies a great circle route from Miami, Florida to Hong Kong. Which one of the following states will the plane fly over?

a. California **b.** Oregon
c. Washington **d.** Alaska

5. A pilot flies a great circle route from Los Angeles to London. Which one of the following cities will the plane fly over?

a. New York **b.** Chicago
c. Godthaab, Greenland **d.** Vancouver, Canada

(continued)

6. Washington, D.C. and Seoul, Korea both have a latitude of about 38°. How many miles (to the nearest 100 miles) will a pilot save by flying a great circle route between the cities as opposed to the route that follows the 38th parallel? *Hint:* Use a cloth measuring tape to measure the minor arc of the great circle that passes through the two cities. Use the scale on the globe to convert this distance to miles. Then use the measuring tape to determine the distance of the route that follows the 38th parallel.

7. **a.** The surface at the left below is called a *hyperboloid of one sheet.* Explain how you can determine that the blue circle is a geodesic of the hyperboloid, but the red circles are not geodesics of the hyperboloid.

   b. The saddle surface at the right below is called a *hyperbolic paraboloid.* Explain how you can determine that the blue parabola is a geodesic of the hyperbolic paraboloid, but the red parabolas are not geodesics of the hyperbolic paraboloid.

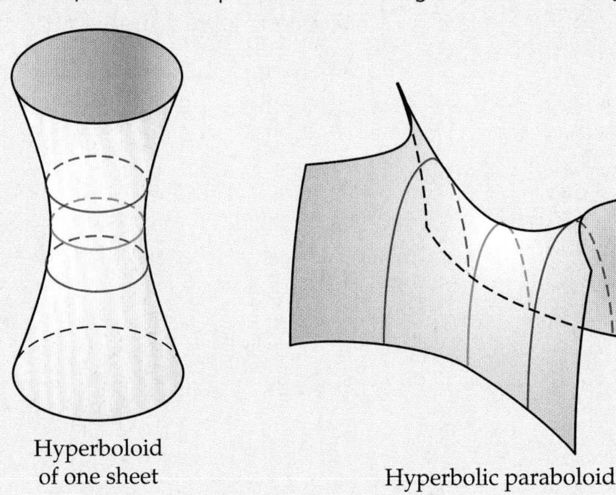

Exercise Set 8.6

1. State the parallel postulate for each of the following.
   a. Euclidean geometry
   b. Lobachevskian geometry
   c. Riemannian geometry
2. Name the mathematician who is called the Copernicus of Geometry.
3. Name the mathematician who was the first to consider a geometry in which Euclid's Parallel Postulate was replaced with "Through a given point not on a given line, there are *at least two* lines parallel to the given line."
4. Name the mathematician who was the first to create a non-Euclidean geometry using the postulate "Through a given point not on a given line, there exist no lines parallel to the given line."
5. What can be stated about the sum of the measures of the angles of a triangle in
   a. Euclidean geometry?
   b. Lobachevskian geometry?
   c. Riemannian geometry?
6. What is the maximum number of right angles a triangle can have in
   a. Euclidean geometry?
   b. Lobachevskian geometry?
   c. Riemannian geometry?

7. What name did Lobachevsky give to the geometry that he created?

8. Explain why great circles in Riemannian geometry are thought of as lines.

9. What is a geodesic?

10. In which geometry can two distinct lines be parallel to a third line but not parallel to each other?

11. What model was used in this text to illustrate hyperbolic geometry?

12. In which geometry do all perpendiculars to a given line intersect each other?

13. Find the exact area of a spherical triangle with angles of 150°, 120°, and 90° on a sphere with a radius of 1.

14. Find the area of a spherical triangle with three right angles on a sphere with a radius of 1980 mi. Round to the nearest ten thousand square miles.

A **spherical polygon** is a polygon on a sphere. Each side of a spherical polygon is an arc of a great circle. The area A of a spherical polygon of n sides on a sphere of radius r is given by

$$A = [\Sigma\theta - (n - 2) \cdot 180°] \cdot \left(\frac{\pi}{180°}\right)r^2$$

where $\Sigma\theta$ is the sum of the measures of the angles of the polygon in degrees.

15. Use the above spherical polygon area formula to determine the area of a spherical quadrilateral with angles of 90°, 90°, 100°, and 100° on the planet Earth, which has a radius of 1980 mi. Round to the nearest ten thousand square miles.

16. Use the above spherical polygon area formula to determine the area of a spherical quadrilateral with angles of 90°, 90°, 120°, and 120° on the planet Earth, which has a radius of 1980 mi. Round to the nearest ten thousand square miles.

City Geometry In Exercises 17–24, find the Euclidean distance between the points and the city distance between the points. Assume that both $d_E(P, Q)$ and $d_C(P, Q)$ are measured in blocks. Round approximate results to the nearest tenth of a block.

17. $P(-3, 1), Q(4, 1)$
18. $P(-2, 4), Q(3, -1)$
19. $P(2, -3), Q(-3, 5)$
20. $P(-2, 0), Q(3, 7)$
21. $P(-1, 4), Q(5, -2)$
22. $P(-5, 2), Q(3, -4)$
23. $P(2, 0), Q(3, -6)$
24. $P(2, -2), Q(5, -2)$

A Distance Conversion Formula The following formula can be used to convert the Euclidean distance between the points P and Q to the city distance between P and Q. In this formula, the variable m represents the slope of the line segment $\overline{PQ}$.

$$d_C(P, Q) = \frac{1 + |m|}{\sqrt{1 + m^2}}\, d_E(P, Q)$$

In Exercises 25–30, use the above formula to find the city distance between P and Q.

25. $d_E(P, Q) = 5$ blocks, slope of $\overline{PQ} = \frac{3}{4}$

26. $d_E(P, Q) = \sqrt{29}$ blocks, slope of $\overline{PQ} = \frac{2}{5}$

27. $d_E(P, Q) = \sqrt{13}$ blocks, slope of $\overline{PQ} = -\frac{2}{3}$

28. $d_E(P, Q) = 2\sqrt{10}$ blocks, slope of $\overline{PQ} = -3$

29. $d_E(P, Q) = \sqrt{17}$ blocks, slope of $\overline{PQ} = \frac{1}{4}$

30. $d_E(P, Q) = 4\sqrt{2}$ blocks, slope of $\overline{PQ} = -1$

31. Explain why there is no formula that can be used to convert $d_C(P, Q)$ to $d_E(P, Q)$. Assume that no additional information is given other than the value of $d_C(P, Q)$.

32. a. If $d_E(P, Q) = d_E(R, S)$, must $d_C(P, Q) = d_C(R, S)$? Explain.
 b. If $d_C(P, Q) = d_C(R, S)$, must $d_E(P, Q) = d_E(R, S)$? Explain.

33. Plot the points in the city circle with center $(-2, -1)$ and radius $r = 2$ blocks.

34. Plot the points in the city circle with center $(1, -1)$ and radius $r = 3$ blocks.

35. Plot the points in the city circle with center $(0, 0)$ and radius $r = 2.5$ blocks.

36. Plot the points in the city circle with center $(0, 0)$ and radius $r = 3.5$ blocks.

37. How many points are on the city circle with center $(0, 0)$ and radius $r = n$ blocks, where n is a natural number?

38. Which of the following city circles has the most points, a city circle with center (0, 0) and radius 4.5 blocks or a city circle with center (0, 0) and radius 5 blocks?

Extensions

CRITICAL THINKING

Apartment Hunting Use the following information to answer each of the questions in Exercises 39 and 40.

Amy and her husband Ryan are looking for an apartment located adjacent to a city street. Amy works at $P(-3, -1)$ and Ryan works at $Q(2, 3)$. Both Amy and Ryan plan to walk from their apartment to work along routes that follow the north–south and the east–west streets.

39. **a.** Plot the points where Amy and Ryan should look for an apartment if they wish the sum of the city distances they need to walk to work to be a minimum.

b. Plot the points where Amy and Ryan should look for an apartment if they wish the sum of the city distances they need to walk to work to be a minimum and they both will walk the same distance. *Hint:* Find the intersection of the city circle with center P and radius of 4.5 blocks, and the city circle with center Q and radius of 4.5 blocks.

40. **a.** Plot the points where Amy and Ryan should look for an apartment if they wish the sum of the city distances they need to walk to work to be less than or equal to 10 blocks.

b. Plot the points where Amy and Ryan should look for an apartment if they wish the sum of the city distances they need to walk to work to be less than or equal to 10 blocks and they both will walk the same distance. *Hint:* Find the intersection of the city circle with center P and radius of 5.5 blocks, and the city circle with center Q and radius of 5.5 blocks.

41. **A Finite Geometry** Consider a finite geometry with the five points A, B, C, D, and E. In this geometry a line is any two of the five points. For example, the two points A and B together form the line denoted by AB. Parallel lines are defined as two lines that do not share a common point.

a. How may lines are in this geometry?

b. How many of the lines are parallel to line AB?

42. **A Finite Geometry** Consider a finite geometry with the six points A, B, C, D, E, and F. In this geometry a line is any two of the six points. Parallel lines are defined as two lines that do not share a common point.

a. How many lines are in this geometry?

b. How many of the lines are parallel to line AB?

EXPLORATIONS

43. **Janos Bolyai** The mathematician Janos Bolyai (bōl′yoi) (1802–1860) proposed a geometry that excluded Euclid's Parallel Postulate. Concerning the geometry he created, he remarked, "I have made such wonderful discoveries that I am myself lost in astonishment: Out of nothing I have created a strange new world." Use the Internet or a library to find information about Janos Bolyai and the mathematics he created. Write a report that summarizes the pertinent details concerning his work in the area of non-Euclidean geometry.

44. **Girolamo Saccheri** The mathematician Girolamo Saccheri (säk′kā-rē) (1667–1733) did important early work in non-Euclidean geometry. Use the Internet or a library to find information about Saccheri's work. Write a report that summarizes the pertinent details.

SECTION 8.7 Fractals

Fractals—Endlessly Repeated Geometric Figures

Have you ever used a computer program to enlarge a portion of a photograph? Sometimes the result is a satisfactory enlargement; however, if the photograph is enlarged too much, the image may become blurred. For example, the photograph in Figure 8.14 on the following page is shown at its original size. The image in

Figure 8.15 is an enlarged portion of Figure 8.14. A computer monitor displays an image using small dots called *pixels.* If a computer image is enlarged using a program such as Adobe Photoshop™, the program must determine the color of each pixel in the enlargement. If the image file for the photograph cannot supply the needed color information for each pixel, then Photoshop determines the color of some pixels by *averaging* the numerical color values of neighboring pixels for which the image file has the color information. In the enlarged photograph in Figure 8.15, the image is somewhat blurred because some of the pixels do not provide accurate information about the original photograph. Figure 8.16 is an enlargement of a portion of Figure 8.15. The blurred image in Figure 8.16 provides even less detail than does Figure 8.15, because the colors of many of the pixels in this image were determined by averaging. If we continue to make enlargements of enlargements we will produce extremely blurred images that provide little information about the original photograph.

Figure 8.14

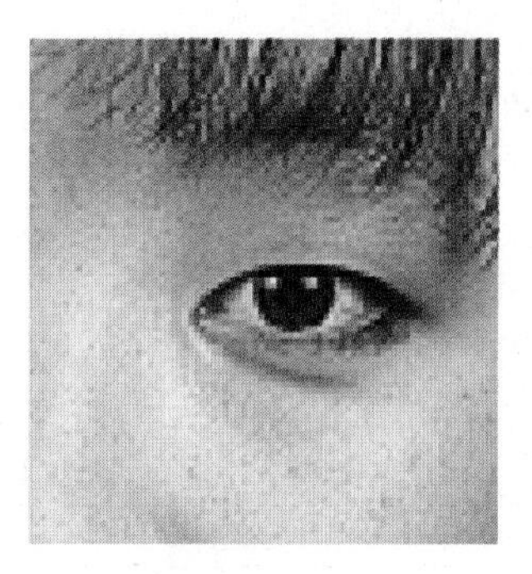

Figure 8.15

Figure 8.16

In the 1970s, the mathematician Benoit Mandelbrot discovered some remarkable methods that enable us to create geometric figures with a special property: if any portion of the figure is enlarged repeatedly, then additional details (not fewer details, as with the enlargement of a photograph) of the figure are displayed. Mandelbrot called these endlessly repeated geometric figures *fractals.* At the present time there is no universal agreement on the precise definition of a fractal, but the fractals that we will study in this lesson can be defined as follows. A **fractal** is a geometric figure in which a self-similar motif repeats itself on an ever-diminishing scale.

Fractals are generally constructed by using **iterative processes** in which the fractal is more closely approximated as a repeated cycle of procedures is performed. For example, a fractal known as the *Koch curve* is constructed as follows.

Construction of the Koch Curve

Step 0: Start with a line segment. This initial segment is shown as stage 0 in Figure 8.17. Stage 0 of a fractal is called the **initiator** of the fractal.

Step 1: On the middle third of the line segment, draw an equilateral triangle and remove its base. The resulting curve is stage 1 in Figure 8.17. Stage 1 of a fractal is called the **generator** of the fractal.

Step 2: Replace each initiator shape (line segment, in this example) with a *scaled version* of the generator to produce the next stage of the Koch curve. The width of the scaled version of the generator is the same as the width of the line segment it replaces. Continue to repeat this step ad infinitum to create additional stages of the Koch curve.

Three applications of step 2 produce stage 2, stage 3, and stage 4 of the Koch curve, as shown in Figure 8.17.

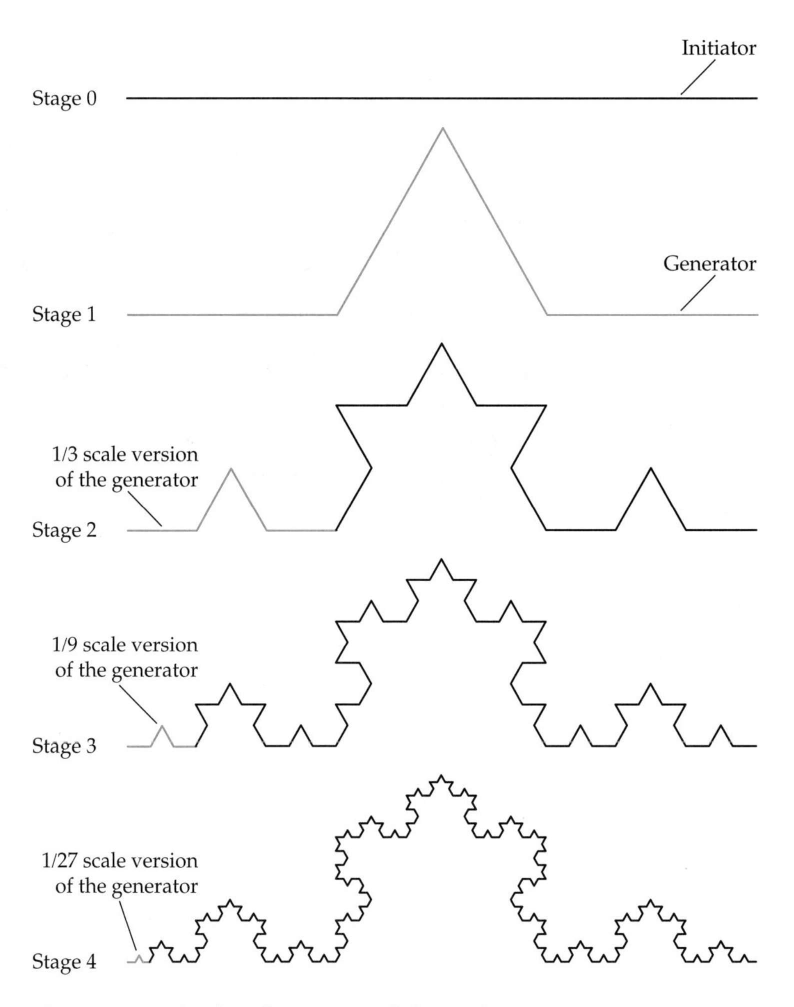

Figure 8.17 *The first five stages of the Koch curve*

TAKE NOTE

The line segments at any stage of the Koch curve are $\frac{1}{3}$ the length of the line segments at the previous stage. To create the next stage of the Koch curve (after stage 1), replace each line segment with a scaled version of the generator.

At any stage after stage 2, the scaled version of the generator is $\frac{1}{3}$ the size of the preceding scaled generator.

historical note

The Koch curve was created by the mathematician Niels Fabian Helge von Koch (kôkh) (1870–1924). Von Koch attended Stockholm University and in 1911 became a professor of mathematics at Stockholm University. In addition to his early pioneering work with fractals, he wrote several papers on number theory. ■

None of the curves shown in Figure 8.17 is the Koch curve. The Koch curve is the curve that would be produced if step 2 in the above construction process were repeated ad infinitum. No one has ever seen the Koch curve, but we know that it is a very jagged curve in which the self-similar motif shown in Figure 8.17 repeats itself on an ever-diminishing scale.

The curves shown in Figure 8.18 are the first five stages of the *Koch snowflake.* Each stage of the Koch snowflake is composed of three congruent sides, each of which is a stage of the Koch curve.

point of interest

It can be shown that the Koch snowflake has a finite area but an infinite perimeter. Thus it has the remarkable property that you could paint the surface of the snowflake, but you could never get enough paint to paint its boundary.

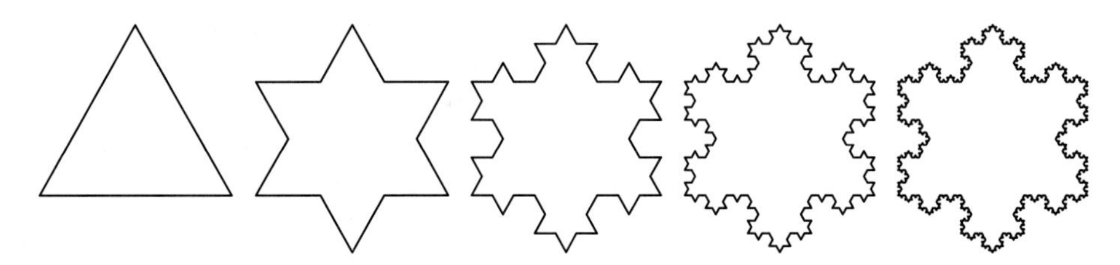

Figure 8.18 *The first five stages of the Koch snowflake*

EXAMPLE 1 ■ Draw Stages of a Fractal

Draw stage 2 and stage 3 of the *box curve,* which is defined by the following iterative process.

Step 0: Start with a line segment as the initiator. See stage 0 in Figure 8.19.

Step 1: On the middle third of the line segment, draw a square and remove its base. This produces the generator of the box curve. See stage 1 in Figure 8.19.

Step 2: Replace each initiator shape with a scaled version of the generator to produce the next stage.

Solution

Two applications of step 2 yield stage 2 and stage 3 of the box curve, as shown in Figure 8.19.

TAKE NOTE

If your drawing of stage 0 of a fractal is small, it will be difficult to draw additional stages of the fractal, because each additional stage must display more details of the fractal. However, if you start by making a *large* drawing of stage 0 on a sheet of graph paper, you will be able to use the grid lines on the graph paper to make accurate drawings of a few additional stages. For instance, if you draw stage 0 of the box curve as a 9-inch line segment, then stage 1 will consist of five line segments, each 3 inches in length. Stage 2 will consist of 25 line segments, each 1 inch in length. The 125 line segments of stage 3 will each be $\frac{1}{3}$ of an inch in length.

Instead of erasing and drawing each new stage of the fractal on top of the previous stage, it is advantageous to draw each new stage directly below the previous stage, as shown in Figure 8.19.

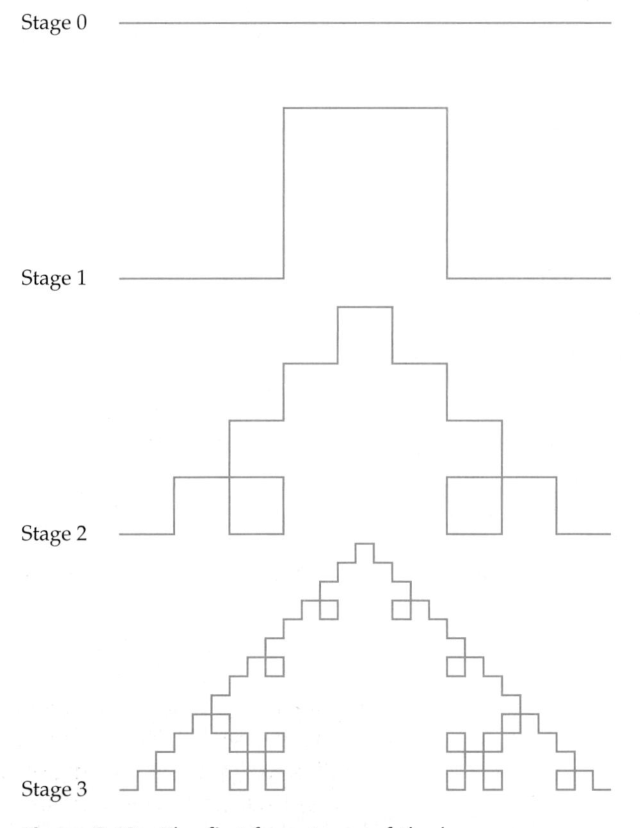

Figure 8.19 *The first four stages of the box curve*

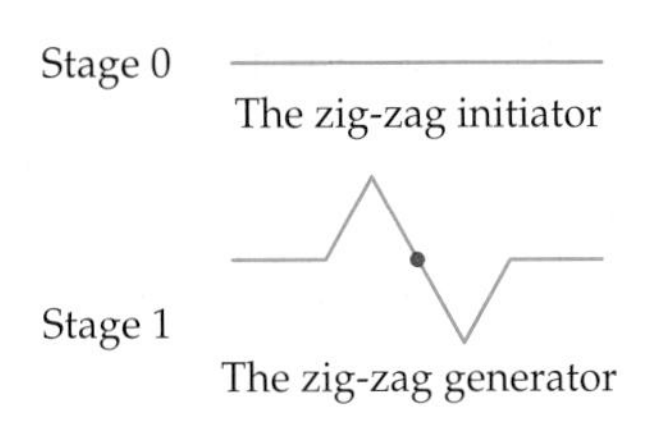

Figure 8.20

CHECK YOUR PROGRESS 1 Draw stage 2 of the *zig-zag curve*, which is defined by the following iterative process.

Step 0: Start with a line segment. See stage 0 of Figure 8.20.

Step 1: Remove the middle half of the line segment and draw a zig-zag, as shown in stage 1 of Figure 8.20. Each of the six line segments in the generator is a $\frac{1}{4}$-scale replica of the initiator.

Step 2: Replace each initiator shape with the scaled version of the generator to produce the next stage. Repeat this step to produce additional stages.

Solution *See page S32.*

In each of the previous examples, the initiator was a line segment. In Example 2, we use a triangle and its interior as the initiator.

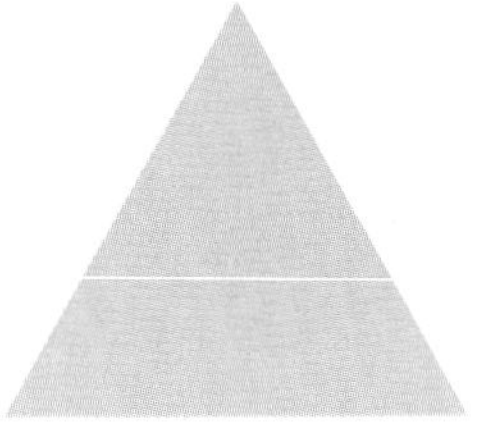
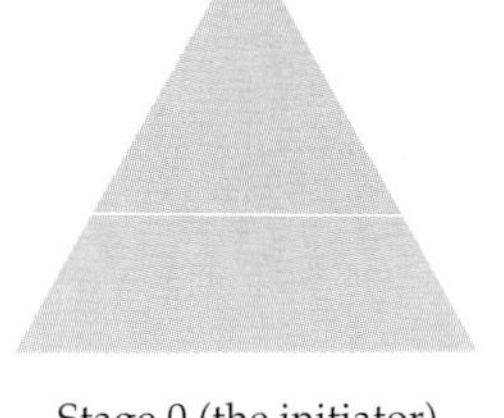

Stage 0 (the initiator)

Stage 1 (the generator)

Figure 8.21

EXAMPLE 2 ■ Draw Stages of a Fractal

Draw stage 2 and stage 3 of the *Sierpinski gasket* (also known as the *Sierpinski triangle*), which is defined by the following iterative process.

Step 0: Start with an equilateral triangle and its interior. This is stage 0 of the Sierpinski gasket. See Figure 8.21.

Step 1: Form a new triangle by connecting the midpoints of the sides of the triangle. Remove this center triangle. The result is the three blue triangles shown in stage 1 in Figure 8.21.

Step 2: Replace each initiator (green triangle) with a scaled version of the generator.

Solution

Two applications of step 2 of the above process produce stage 2 and stage 3 of the Sierpinski gasket, as shown in Figure 8.22.

Stage 2

Stage 3

Figure 8.22

CHECK YOUR PROGRESS 2 Draw stage 2 of the *Sierpinski carpet*, which is defined by the following process.

Step 0: Start with a square and its interior. See stage 0 in Figure 8.23.

Step 1: Subdivide the square into nine smaller congruent squares and remove the center square. This yields stage 1 (the generator) shown in Figure 8.23.

Step 2: Replace each initiator (tan square) with a scaled version of the generator. Repeat this step to create additional stages of the Sierpinski carpet.

Stage 0

Stage 1

Figure 8.23

Solution *See page S33.*

point of interest

Fractal geometry is not just a chapter of mathematics, but one that helps Everyman to see the same old world differently.— Benoit Mandelbrot, IBM Research (*Source:* **http://php.iupui.edu/~wijackso/fractal3.htm**)

Math Matters Benoit Mandelbrot (1924–)

Benoit Mandelbrot is often called the Father of Fractal Geometry. He was not the first person to create a fractal, but he was the first person to discover how some of the ideas of earlier mathematicians such as Georg Cantor, Giuseppe Peano, Helge von Koch, Waclaw Sierpinski, and Gaston Julia could be united to form a new type of geometry. Mandelbrot also recognized that many fractals share characteristics with shapes and curves found in nature. For instance, the leaves of a fern, when compared with the whole fern, are almost identical in shape, only smaller. This self-similarity characteristic is evident (to some degree) in all fractals. The following quote by Mandelbrot is from his 1983 book, *The Fractal Geometry of Nature.*[4]

> Clouds are not spheres, mountains are not cones, coastlines are not circles, and bark is not smooth, nor does lightning travel in a straight line. More generally, I claim that many patterns of Nature are so irregular and fragmented, that, compared with Euclid—a term used in this work to denote all of standard geometry—Nature exhibits not simply a higher degree but an altogether different level of complexity.

4. Mandelbrot, Benoit B. *The Fractal Geometry of Nature.* New York: W. H. Freeman and Company, 1983, p. 1.

Strictly Self-Similar Fractals

All fractals show a self-similar motif on an ever-diminishing scale; however, some fractals are *strictly self-similar* fractals, according to the following definition.

Definition of a Strictly Self-Similar Fractal

A fractal is said to be **strictly self-similar** if any arbitrary portion of the fractal contains a replica of the entire fractal.

EXAMPLE 3 ■ Determine Whether a Fractal is Strictly Self-Similar

Determine whether the following fractals are strictly self-similar.

a. The Koch snowflake **b.** The Koch curve

Solution

a. The Koch snowflake is a closed figure. Any portion of the Koch snowflake (like the portion circled in Figure 8.24) is not a closed figure. Thus the Koch snowflake is *not* a strictly self-similar fractal.

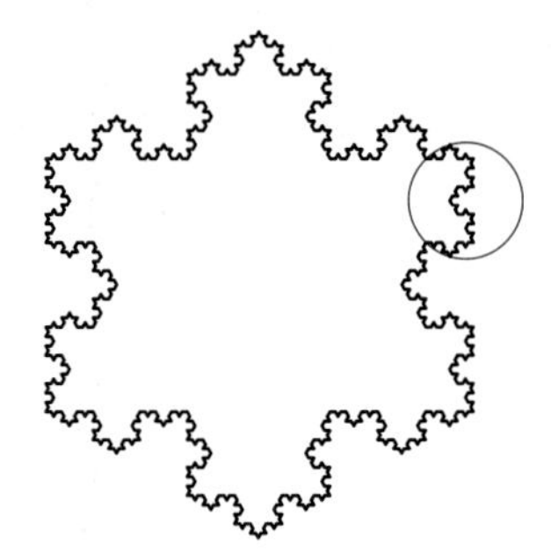

Figure 8.24 *The portion of the Koch snowflake shown in the circle is not a replica of the entire snowflake.*

b. Because any portion of the Koch curve replicates the entire fractal, the Koch curve is a strictly self-similar fractal. See Figure 8.25.

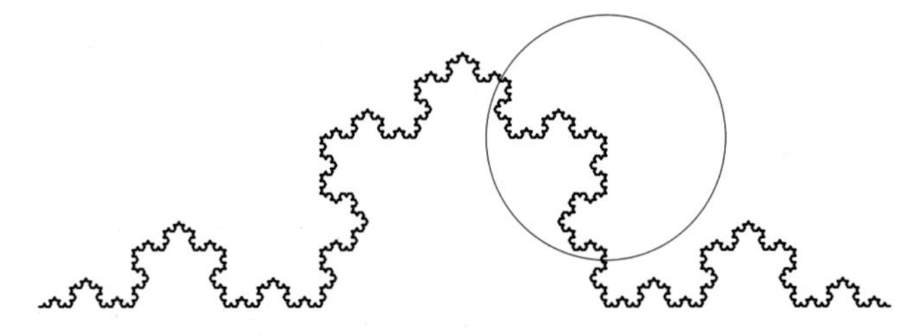

Figure 8.25 *Any portion of the Koch curve is a replica of the entire Koch curve.*

CHECK YOUR PROGRESS 3 Determine whether the following fractals are strictly self-similar.

a. The box curve **b.** The Sierpinski gasket

Solution *See page S33.*

Replacement Ratio and Scaling Ratio

Mathematicians like to assign numbers to fractals so that they can objectively compare fractals. Two numbers that are associated with many fractals are the *replacement ratio* and the *scaling ratio.*

Replacement Ratio and Scaling Ratio of a Fractal

- If the generator of a fractal consists of N replicas of the initiator, then the **replacement ratio** of the fractal is N.
- If the initiator of a fractal has linear dimensions that are r times the corresponding linear dimensions of its replicas in the generator, then the **scaling ratio** of the fractal is r.

EXAMPLE 4 ■ Find the Replacement Ratio and the Scaling Ratio of a Fractal

Find the replacement ratio and the scaling ratio of the

a. box curve.

b. Sierpinski gasket.

Solution

a. Figure 8.19 on page 549 shows that the generator of the box curve consists of five line segments and that the initiator consists of only one line segment. Thus the replacement ratio of the box curve is 5 : 1, or 5.

The initiator of the box curve is a line segment that is 3 times as long as the replica line segments in the generator. Thus the scaling ratio of the box curve is 3 : 1, or 3.

b. Figure 8.21 on page 550 shows that the generator of the Sierpinski gasket consists of three triangles and that the initiator consists of only one triangle. Thus the replacement ratio of the Sierpinski gasket is 3 : 1, or 3.

The initiator triangle of the Sierpinski gasket has a width (height) that is 2 times the width (height) of the replica triangles in the generator. Thus the scaling ratio of the Sierpinski gasket is 2 : 1, or 2.

CHECK YOUR PROGRESS 4 Find the replacement ratio and the scaling ratio of the

a. Koch curve. **b.** zig-zag curve.

Solution *See page S33.*

Similarity Dimension

A number called the *similarity dimension* is used to quantify how densely a strictly self-similar fractal fills a region.

The **similarity dimension** D of a strictly self-similar fractal is given by

$$D = \frac{\log N}{\log r}$$

where N is the replacement ratio of the fractal and r is the scaling ratio.

EXAMPLE 5 ■ Find the Similarity Dimension of a Fractal

Find the similarity dimension, to the nearest thousandth, of the

a. Koch curve. **b.** Sierpinski gasket.

Solution

a. Because the Koch curve is a strictly self-similar fractal, we can find its similarity dimension. Figure 8.17 on page 548 shows that stage 1 of the Koch curve consists of four line segments and stage 0 consists of only one line segment. Hence the replacement ratio is 4 : 1, or 4. The line segment in stage 0 is 3 times as long as each of the replica line segments in stage 1, so the scaling ratio is 3. Thus the Koch curve has a similarity dimension of

$$D = \frac{\log 4}{\log 3} \approx 1.262$$

b. In Example 4, we found that the Sierpinski gasket has a replacement ratio of 3 and a scaling ratio of 2. Thus the Sierpinski gasket has a similarity dimension of

$$D = \frac{\log 3}{\log 2} \approx 1.585$$

TAKE NOTE

Because the Koch snowflake is not a strictly self-similar fractal, we cannot compute its similarity dimension.

CHECK YOUR PROGRESS 5 Compute the similarity dimension, to the nearest thousandth, of the

a. box curve. **b.** Sierpinski carpet.

Solution *See page S33.*

The results of Example 5 show that the Sierpinski gasket has a larger similarity dimension than the Koch curve. This means that the Sierpinski gasket fills a flat two-dimensional surface more densely than does the Koch curve.

Computers are used to generate fractals such as those shown in Figure 8.26. These fractals were *not* rendered by using an initiator and a generator, but they were rendered using iterative procedures.

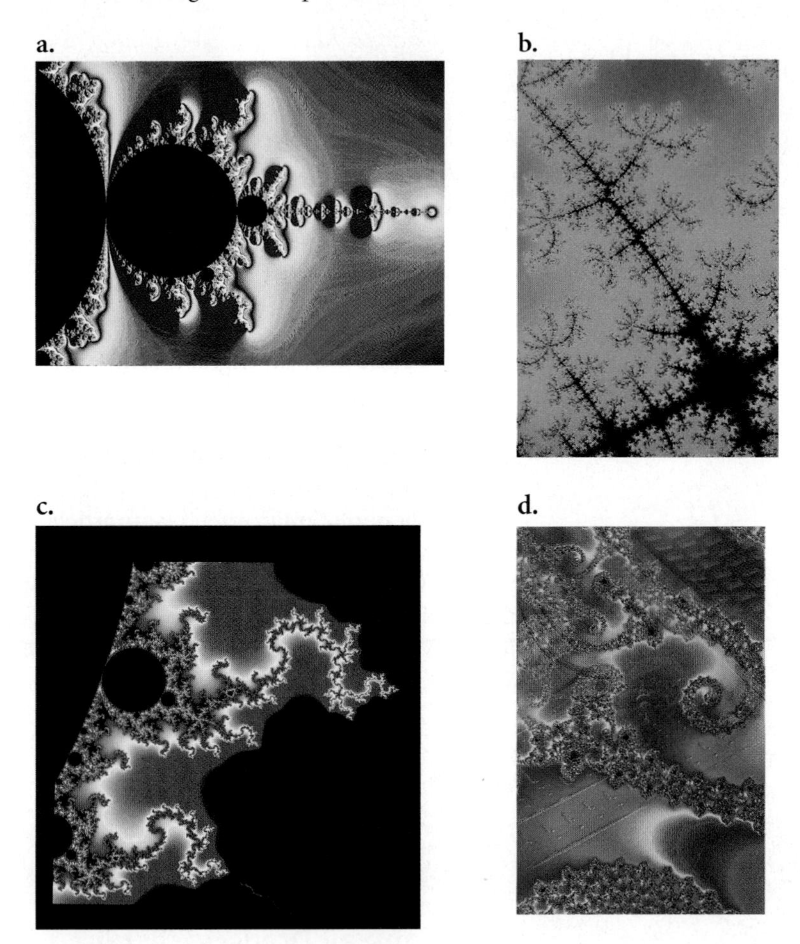

Figure 8.26 *Computer-generated fractals*

Fractals have other applications in addition to being used to produce intriguing images. For example, computer scientists have recently developed fractal image compression programs based on self-transformations of an image. An image compression program is a computer program that converts an image file to a smaller file that requires less computer memory. In some situations these fractal compression programs outperform standardized image compression programs such as JPEG (*jay-peg*), which was developed by the Joint Photographic Experts Group.

During the past few years, some cellular telephones have been manufactured with internal antennas that are fractal in design. Figure 8.27 shows a cellular telephone with an internal antenna in the shape of a stage on the Sierpinski carpet fractal. The antenna in Figure 8.28 is in the shape of a stage of the Koch curve.

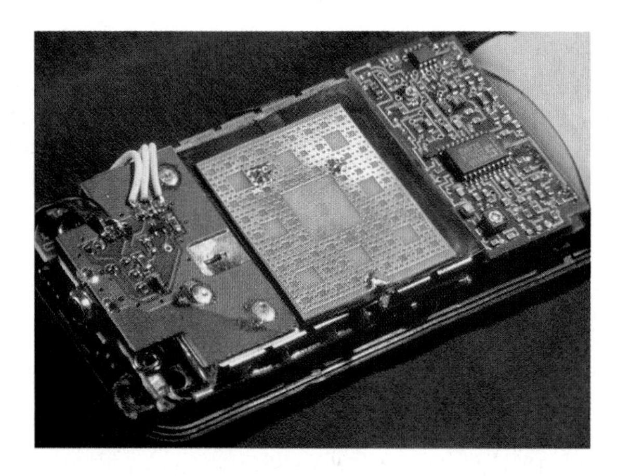

Figure 8.27 *A Sierpinski carpet fractal antenna hidden inside a cellular telephone*

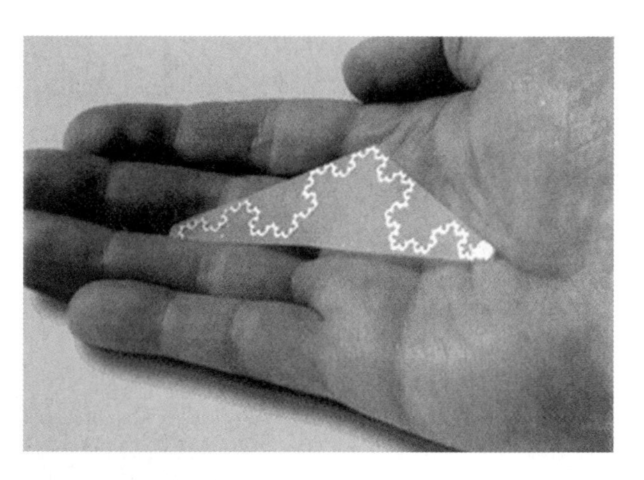

Figure 8.28 *A small Koch curve antenna designed for use in wireless communication devices*

John Chenoweth, an engineer with T&W Antennas claimed, in the journal *Scientific American*, that "fractal antennas are 25 percent more efficient than the rubbery 'stubby' found on most phones. In addition, they are cheaper to manufacture, operate on multiple bands—allowing, for example, a Global Positioning System receiver to be built into the phone—and can be tucked inside the phone body."[5]

Another quotation from the same article states, "Just why these fractal antennas work so well was answered in part in the March issue of the journal *Fractals.* [Nathan] Cohen and his colleague Robert Hohlfeld proved mathematically that for an antenna to work equally well at all frequencies, it must satisfy two criteria. It must be symmetrical about a point. And it must be self-similar, having the same basic appearance at every scale—that is, it has to be fractal."[6]

Excursion

The Heighway Dragon Fractal

In this Excursion we illustrate two methods of constructing the stages of a fractal known as the *Heighway dragon.*

The Heighway Dragon via Paper Folding

The first few stages of the Heighway dragon fractal can be constructed by the repeated folding of a strip of paper. In the following discussion we use a 1-inch-wide strip of paper that is 14 inches in length as stage 0. To create stage 1 of the dragon fractal, just fold

(continued)

5. Musser, George. Technology and Business: Wireless Communications. *Scientific American,* July 1999. **http://www.sciam.com/1999/0799issue/0799techbus3.html**

6. Ibid.

TAKE NOTE

The easiest way to construct stage 4 is to make all four folds before you open the paper. All folds must be made in the same direction.

point of interest

The Heighway dragon is sometimes called the "Jurassic Park fractal" because it was used in the book *Jurassic Park* to illustrate some of the surprising results that can be obtained by iterative processes.

the strip in half and open it so that the fold forms a right angle (see Figure 8.29). To create stage 2, fold the strip twice. The second fold should be in the same direction as the first fold. Open the paper so that each of the folds forms a right angle. Continue the iterative process of making an additional fold in the same direction as the first fold and then forming a right angle at each fold to produce additional stages. See Figure 8.29.

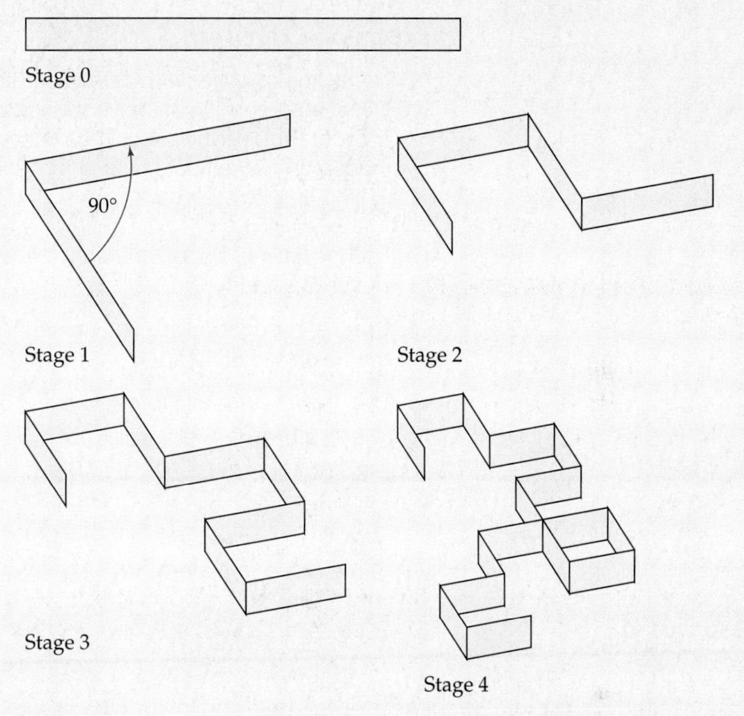

Figure 8.29 *The first five stages of the Heighway dragon via paper folding*

The Heighway Dragon via the Left-Right Rule

The nth stage of the Heighway dragon can also be created by the following drawing procedure.

Step 0: Draw a small vertical line segment. Label the bottom point of this segment as vertex $v = 0$ and label the top as vertex $v = 1$.

Step 1: Use the following left-right rule to determine whether to make a left turn or a right turn.

The Left-Right Rule

At vertex v, where v is an *odd* number, go

- **right** if the remainder of v divided by 4 is 1.
- **left** if the remainder of v divided by 4 is 3.

At vertex 2, go to the right. At vertex v, where v is an *even* number greater than 2, go in the same direction in which you went at vertex $\frac{v}{2}$.

(continued)

Draw another line segment of the same length as the original segment. Label the endpoint of this segment with a number that is 1 larger than the number used for the preceding vertex.

Step 2: Continue to repeat step 1 until you reach the last vertex. The last vertex of an n-stage Heighway dragon is the vertex numbered 2^n.

Excursion Exercises

1. Use a strip of paper and the folding procedure explained on pages 556 and 557 to create models of the first five stages (stage 0 through stage 4) of the Heighway dragon. Explain why it would be difficult to create the 10th stage of the Heighway dragon using the paper-folding procedure.
2. Use the left-right rule to draw stage 2 of the Heighway dragon.
3. Use the left-right rule to determine the direction in which to turn at vertex 7 of the Heighway dragon.
4. Use the left-right rule to determine the direction in which to turn at vertex 50 of the Heighway dragon.
5. Use the left-right rule to determine the direction in which to turn at vertex 64 of the Heighway dragon.

Exercise Set 8.7

In Exercises 1 and 2, use an iterative process to draw stage 2 and stage 3 of the fractal with the given initiator (stage 0) and the given generator (stage 1).

1. The Cantor point set

Stage 0

Stage 1

2. Lévy's curve

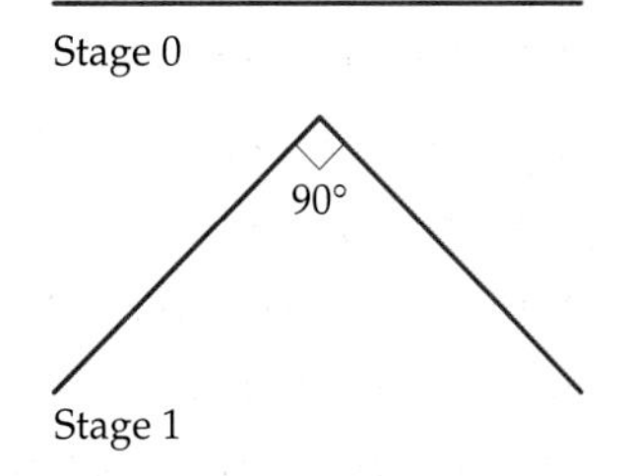

In Exercises 3–8, use an iterative process to draw stage 2 of the fractal with the given initiator (stage 0) and the given generator (stage 1).

3. The Sierpinski carpet, variation 1

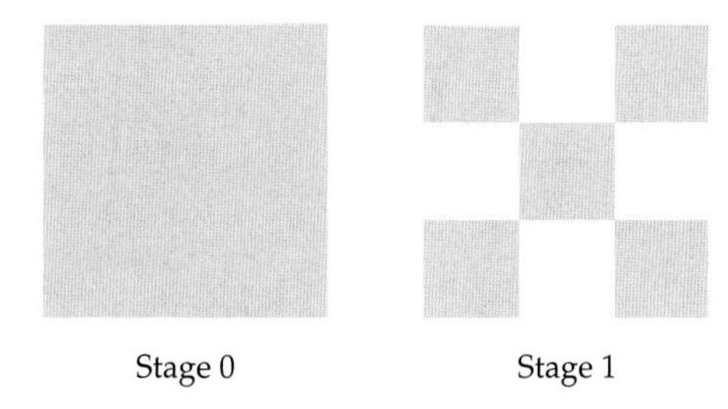

4. The Sierpinski carpet, variation 2

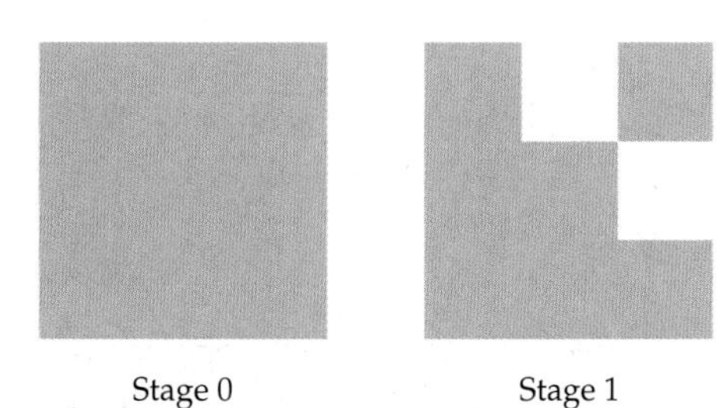

5. The river tree of Peano Cearo

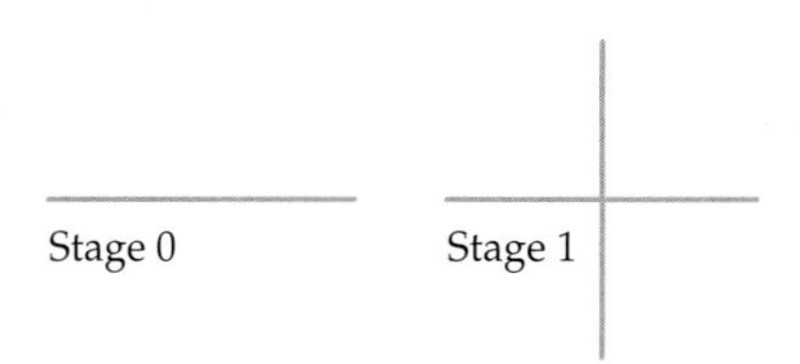

6. Minkowski's fractal

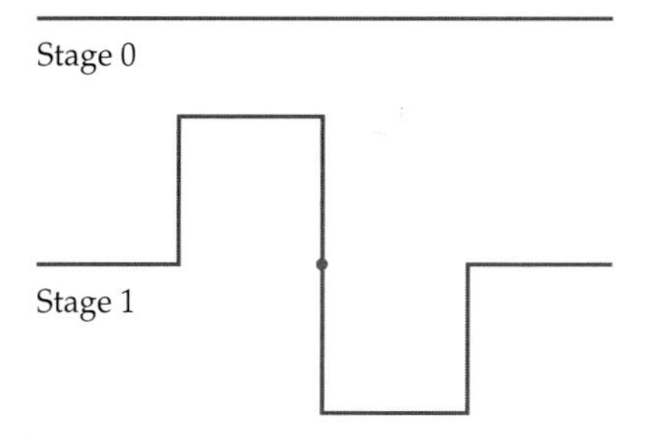

7. The square fractal

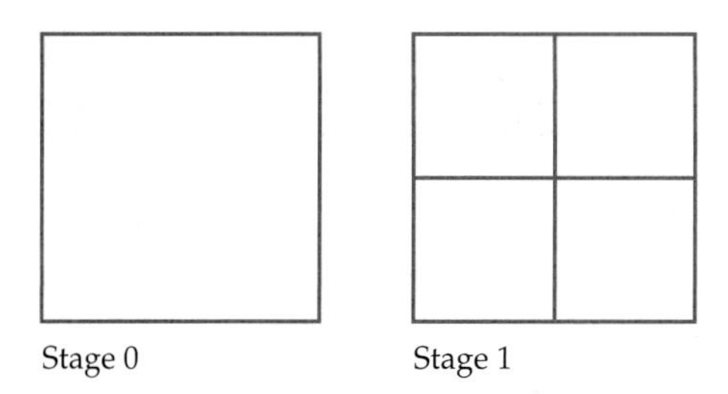

8. The cube fractal

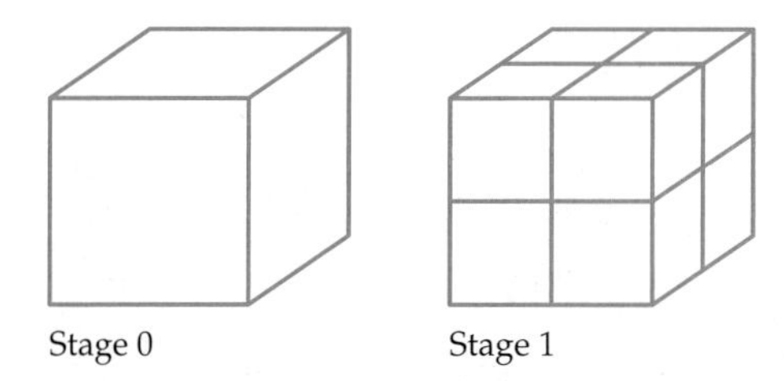

In Exercises 9 and 10, draw stage 3 and stage 4 of the fractal defined by the given iterative process.

9. The binary tree

Step 0: Start with a "T". This is stage 0 of the binary tree. The vertical line segment is the trunk of the tree and the horizontal line segment is the branch of the tree. The branch is half the length of the trunk.

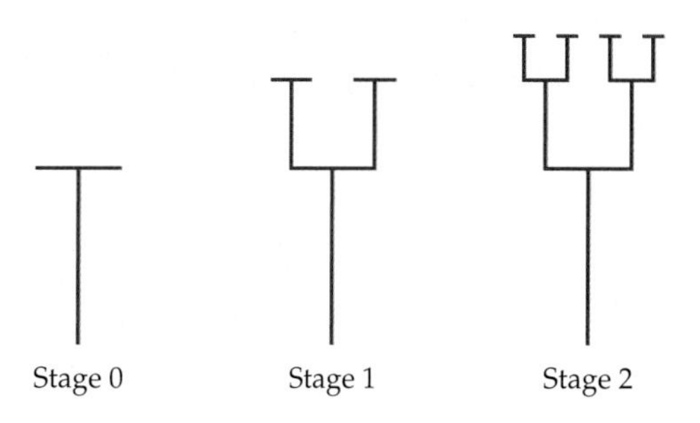

Step 1: At the ends of each branch, draw an upright T that is half the size of the T in the preceding stage.

Step 2: Continue to repeat step 1 to generate additional stages of the binary tree.

10. The I-fractal

Step 0: Start with the line segment shown as stage 0 below.

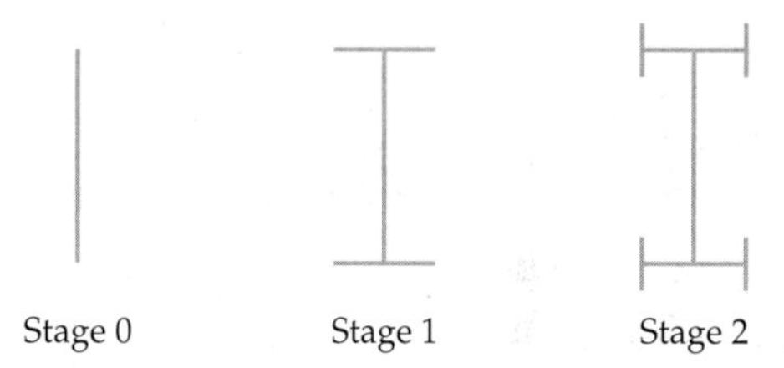

Step 1: At each end of the line segment, draw a crossbar that is half the length of the line segment it contacts. (This produces stage 1 of the I-fractal.)

Step 2: Use each crossbar from the preceding step as the connecting segment of a new "I." Attach new crossbars that are half the length of the connecting segment. Continue to repeat this step to generate additional stages of the I-fractal.

In Exercises 11–20, compute, if possible, the similarity dimension of the fractal. Round to the nearest thousandth.

11. The Cantor point set (See Exercise 1.)
12. Lévy's curve (See Exercise 2.)
13. The Sierpinski carpet, variation 1 (See Exercise 3.)
14. The Sierpinski carpet, variation 2 (See Exercise 4.)
15. The river tree of Peano Cearo (See Exercise 5.)

16. Minkowski's fractal (See Exercise 6.)
17. The square fractal (See Exercise 7.)
18. The cube fractal (See Exercise 8.)
19. The quadric Koch curve, defined by the following stages.

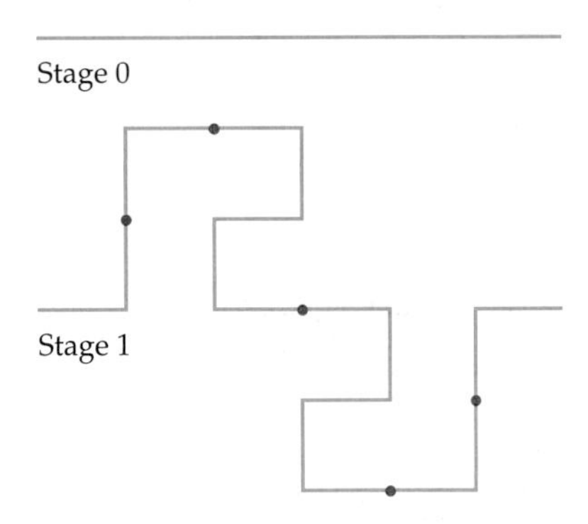

20. The Menger sponge, defined by the following stages.

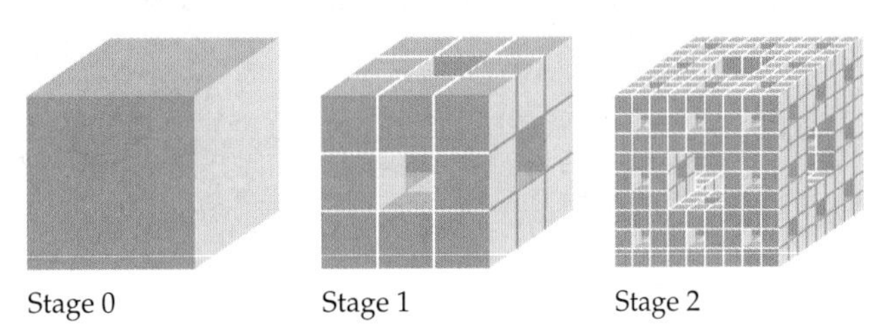

Extensions

CRITICAL THINKING

21. **a.** Rank, from largest to smallest, the similarity dimensions of the Sierpinski carpet; the Sierpinski carpet, variation 1 (see Exercise 3); and the Sierpinski carpet, variation 2 (see Exercise 4).
    b. Which of the three fractals is the most dense?
22. The *Peano curve* is defined by the following stages.

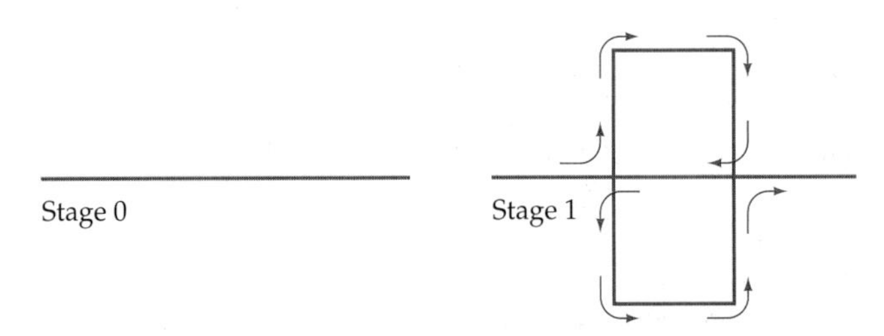

The arrows show the route used to trace the generator.

a. What is the similarity dimension of the Peano curve?

b. Explain why the Peano curve is referred to as a plane-filling curve.

23. Explain why the similarity dimension formula cannot be used to find the similarity dimension of the binary tree defined in Exercise 9.
24. Stage 0 and stage 1 of Lévy's curve (see Exercise 2) and the Heighway dragon (see page 556) are identical, but the fractals start to differ at stage 2. Make two drawings that illustrate how they differ at stage 2.

EXPLORATIONS

25. Create a strictly self-similar fractal that is different from any of the fractals in this lesson.
    a. Draw the first four stages of your fractal.
    b. Compute the replacement ratio, the scaling ratio, and the similarity dimension of your fractal.
26. **The Sierpinski Pyramid** The following figure shows three stages of the *Sierpinski pyramid.* This fractal has been called a three-dimensional model of the Sierpinski gasket; however, its similarity dimension is 2, not 3. Build scale models of stage 0 and stage 1 of the Sierpinski pyramid. Use the models to demonstrate to your classmates why the Sierpinski pyramid fractal has a similarity dimension of 2.

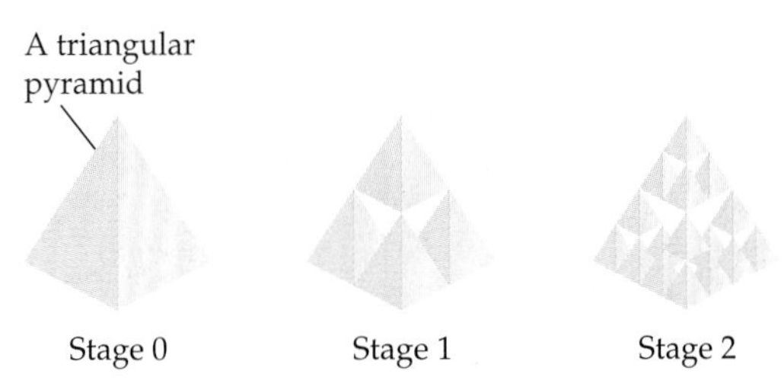

The first three stages of the Sierpinski pyramid

CHAPTER 8 Summary

Key Terms

acute angle [p. 454]
acute triangle [p. 470]
adjacent angles [p. 455]
adjacent side [p. 523]
alternate exterior angles [p. 457]
alternate interior angles [p. 457]
angle [p. 452]
angle of depression [p. 528]
angle of elevation [p. 528]
area [p. 477]
base of a parallelogram [p. 474]
base of a trapezoid [p. 481]
box curve [p. 549]
center of a circle [p. 475]
center of a sphere [p. 509]
circle [p. 475]
circumference [p. 475]
city circle [p. 541]
complementary angles [p. 454]
corresponding angles [p. 458]
cosine [p. 523]
cosecant [p. 523]
cotangent [p. 523]
cube [p. 508]
degree [p. 453]
diameter [p. 475]
diameter of a sphere [p. 509]
endpoint [p. 450]
equilateral triangle [p. 470]
Euclidean circle [p. 541]
Euclidean geometry [p. 534]
exterior angle of a triangle [p. 460]
face of a rectangular solid [p. 508]
fractal [p. 547]
generator [p. 547]
geodesic [p. 535]
geometric solids [p. 507]
great circle [p. 535]
height of a parallelogram [p. 479]
height of a trapezoid [p. 481]
hyperbolic geometry [p. 534]
hyperbolic triangle [p. 538]
hypotenuse [p. 499]
imaginary geometry [p. 534]
initiator [p. 547]
interior angle of a triangle [p. 460]
intersecting lines [p. 452]
isosceles triangle [p. 470]
iterative process [p. 547]
Koch curve [p. 547]
Koch snowflake [p. 548]
legs of a right triangle [p. 499]
line [p. 450]
line segment [p. 450]
line of sight [p. 528]
Lobachevskian geometry [p. 534]
major/minor arc [p. 535]
non-Euclidean geometry [p. 534]
obtuse angle [p. 454]
obtuse triangle [p. 470]
opposite side [p. 523]
parallel lines [p. 452]
parallel lines on an infinite saddle surface [p. 537]
parallelogram [p. 471]
perimeter [p. 471]
perpendicular lines [p. 453]
plane [p. 452]
plane figure [p. 452]
point [p. 450]
polygon [p. 469]
postulate [p. 533]
protractor [p. 453]
quadrilateral [p. 470]
radius [p. 475]
radius of a sphere [p. 509]
ray [p. 450]
rectangle [p. 472]
rectangular solid [p. 508]
regular polygon [p. 469]
regular pyramid [p. 510]
replacement ratio [p. 553]
Riemmanian geometry [p. 535]
right angle [p. 453]
right circular cone [p. 510]
right circular cylinder [p. 510]
right triangle [p. 470]
scalene triangle [p. 470]
scaling ratio [p. 553]
secant [p. 523]
sides of an angle [p. 452]
sides of a polygon [p. 469]
Sierpinski carpet [p. 551]

Essential Concepts

- **Triangles**
  Sum of the measures of the interior angles = 180°
  Sum of the measures of an interior and a corresponding exterior angle = 180°

- **Perimeter Formulas**

Triangle:	$P = a + b + c$
Rectangle:	$P = 2L + 2W$
Square:	$P = 4s$
Parallelogram:	$P = 2b + 2s$
Circle:	$C = \pi d$ or $C = 2\pi r$

- **Area Formulas**

Triangle:	$A = \frac{1}{2}bh$
Rectangle:	$A = LW$
Square:	$A = s^2$
Circle:	$A = \pi r^2$
Parallelogram:	$A = bh$
Trapezoid:	$A = \frac{1}{2}h(b_1 + b_2)$

- **Volume Formulas**

Rectangular solid:	$V = LWH$
Cube:	$V = s^3$
Sphere:	$V = \frac{4}{3}\pi r^3$
Right circular cylinder:	$V = \pi r^2 h$
Right circular cone:	$V = \frac{1}{3}\pi r^2 h$
Regular pyramid:	$V = \frac{1}{3}s^2 h$

- **Surface Area Formulas**

Rectangular solid:	$S = 2LW + 2LH + 2WH$
Cube:	$S = 6s^2$
Sphere:	$S = 4\pi r^2$
Right circular cylinder:	$S = 2\pi r^2 + 2\pi rh$
Right circular cone:	$S = \pi r^2 + \pi rl$
Regular pyramid:	$S = s^2 + 2sl$

- **Similar Triangles**
  The ratios of corresponding sides are equal. The ratio of corresponding heights is equal to the ratio of corresponding sides.

- **Congruent Triangles**

  Side-Side-Side Theorem (SSS)

  If the three sides of one triangle are equal in measure to the corresponding three sides of a second triangle, the two triangles are congruent.

  Side-Angle-Side Theorem (SAS)

  If two sides and the included angle of one triangle are equal in measure to two sides and the included angle of a second triangle, the two triangles are congruent.

  Angle-Side-Angle Theorem (ASA)

  If two angles and the included side of one triangle are equal in measure to two angles and the included side of a second triangle, the two triangles are congruent.

- **Pythagorean Theorem**

  If a and b are the lengths of the legs of a right triangle and c is the length of the hypotenuse, then

  $$c^2 = a^2 + b^2.$$

- **The Trigonometric Functions of an Acute Angle of a Right Triangle**
  If θ is an acute angle of a right triangle ABC, then

$$\sin \theta = \frac{\text{length of opposite side}}{\text{length of hypotenuse}}$$

$$\cos \theta = \frac{\text{length of adjacent side}}{\text{length of hypotenuse}}$$

$$\tan \theta = \frac{\text{length of opposite side}}{\text{length of adjacent side}}$$

$$\csc \theta = \frac{\text{length of hypotenuse}}{\text{length of opposite side}}$$

$$\sec \theta = \frac{\text{length of hypotenuse}}{\text{length of adjacent side}}$$

$$\cot \theta = \frac{\text{length of adjacent side}}{\text{length of opposite side}}$$

- **Definitions of the Inverse Sine, Inverse Cosine, and Inverse Tangent Functions**
  For $0° < x < 90°$:

  $y = \sin^{-1}(x)$ can be read "y is the angle whose sine is x."

  $y = \cos^{-1}(x)$ can be read "y is the angle whose cosine is x."

  $y = \tan^{-1}(x)$ can be read "y is the angle whose tangent is x."

- **Parallel Postulates**

  The Euclidean Parallel Postulate (*Plane Geometry*) Through a given point not on a given line, exactly one line can be drawn parallel to the given line.

  Gauss's Alternative to the Parallel Postulate (*Hyperbolic Geometry*) Through a given point not on a given line, there are *at least two* lines parallel to the given line.

  Riemann's Alternative to the Parallel Postulate (*Spherical Geometry*) Through a given point not on a given line, there exist *no* lines parallel to the given line.

- **The Spherical Triangle Area Formula**
  The area S of the spherical triangle ABC on a sphere with radius r is

$$S = (m\angle A + m\angle B + m\angle C - 180°) \cdot \left(\frac{\pi}{180°}\right) r^2$$

- **City Distance Formula** The city distance between $P(x_1, y_1)$ and $Q(x_2, y_2)$ is given by

$$d_C(P, Q) = |x_2 - x_1| + |y_2 - y_1|$$

- **Euclidean Distance Formula** The Euclidean distance between $P(x_1, y_1)$ and $Q(x_2, y_2)$ is given by

$$d_E(P, Q) = \sqrt{(x_2 - x_1)^2 + (y_2 - y_1)^2}$$

- **Similarity Dimension of a Fractal**
  The similarity dimension D of a strictly self-similar fractal is $D = \frac{\log N}{\log r}$, where N is the replacement ratio of the fractal and r is the scaling ratio.

CHAPTER 8 Review Exercises

1. Given that $m\angle a = 74°$ and $m\angle b = 52°$, find the measures of angles x and y.

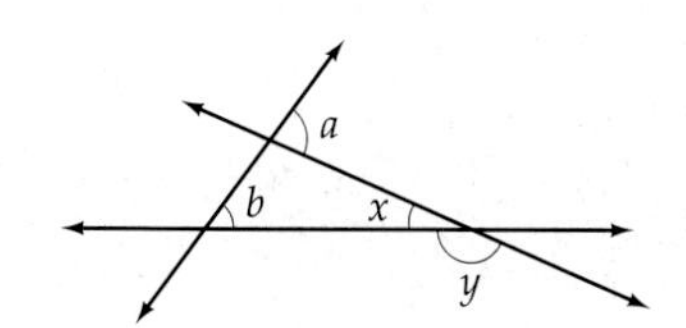

2. Triangles ABC and DEF are similar. Find the perimeter of triangle ABC.

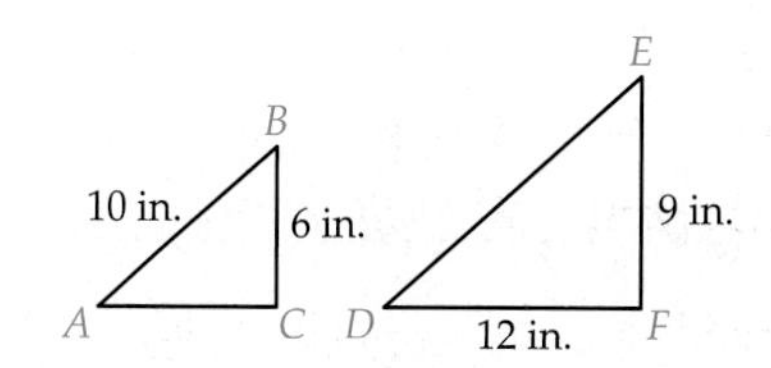

3. Find the volume of the geometric solid.

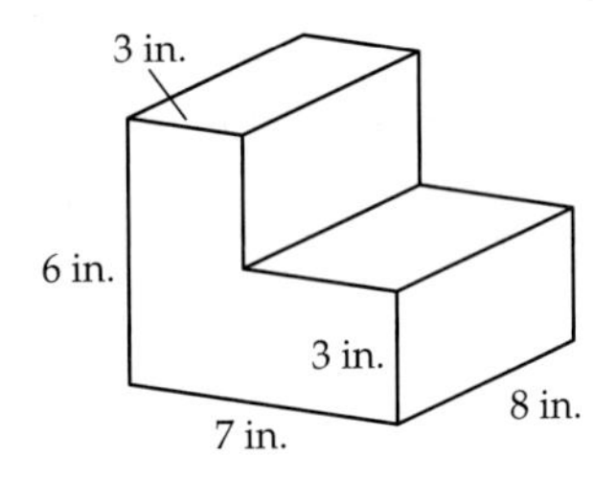

4. Find the measure of $\angle x$.

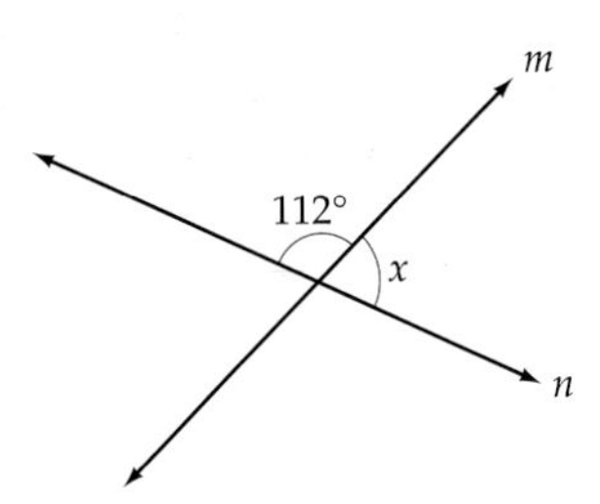

5. Find the surface area of the rectangular solid.

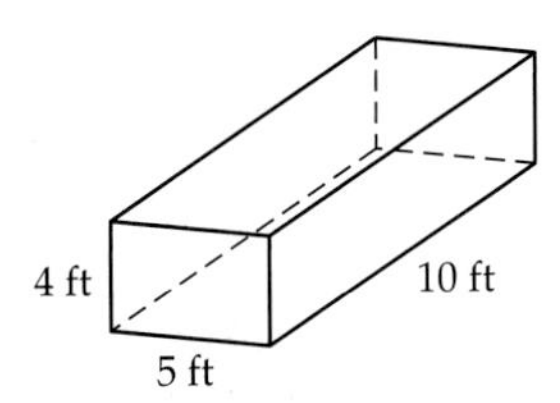

6. The length of a diameter of the base of a cylinder is 4 m, and the height of the cylinder is 8 m. Find the surface area of the cylinder. Give the exact value.

7. Given that $BC = 11$ cm and that AB is three times the length of BC, find the length of AC.

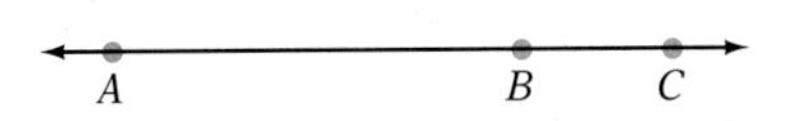

8. Find x.

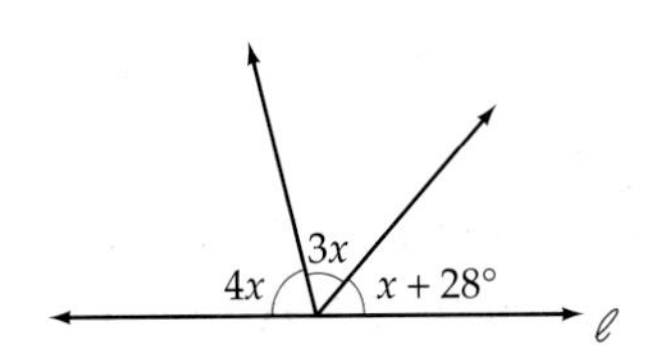

9. Find the area of a parallelogram that has a base of 6 in. and a height of 4.5 in.

10. Find the volume of the square pyramid.

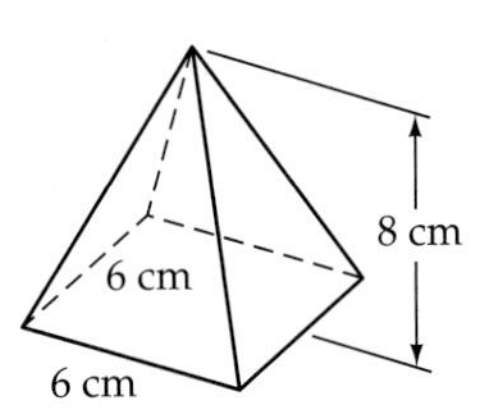

11. Find the circumference of a circle that has a diameter of 4.5 m. Round to the nearest hundredth of a meter.

12. Given that $\ell_1 \parallel \ell_2$, find the measures of angles a and b.

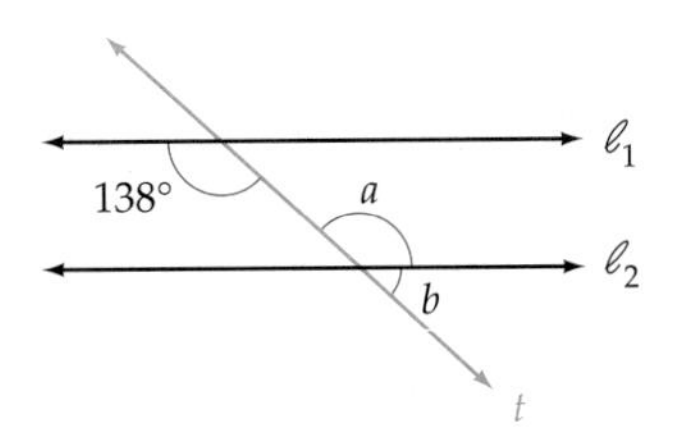

13. Find the supplement of a 32° angle.

14. Find the volume of a rectangular solid with a length of 6.5 ft, a width of 2 ft, and a height of 3 ft.

15. Two angles of a triangle measure 37° and 48°. Find the measure of the third angle.

16. The height of a triangle is 7 cm. The area of the triangle is 28 cm². Find the length of the base of the triangle.

17. Find the volume of a sphere that has a diameter of 12 mm. Give the exact value.

18. **Framing** The perimeter of a square picture frame is 86 cm. Find the length of each side of the frame.

19. **Paint** A can of paint will cover 200 ft² of surface. How many cans of paint should be purchased to paint a cylinder that has a height of 15 ft and a radius of 6 ft?

20. **Parks and Recreation** The length of a rectangular park is 56 yd. The width is 48 yd. How many yards of fencing are needed to surround the park?

21. **Patios** What is the area of a square patio that measures 9.5 m on each side?

22. **Landscaping** A walkway 2 m wide surrounds a rectangular plot of grass. The plot is 40 m long and 25 m wide. What is the area of the walkway?

23. Name the mathematician who was the first person to develop a non-Euclidean geometry, but chose not to publish his work.

24. Determine whether the two triangles are congruent. If they are congruent, state by what theorem they are congruent.

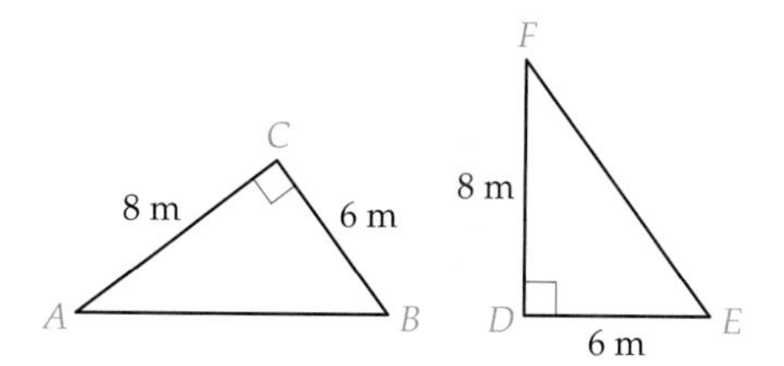

25. Find the unknown side of the triangle. Round to the nearest hundredth of a foot.

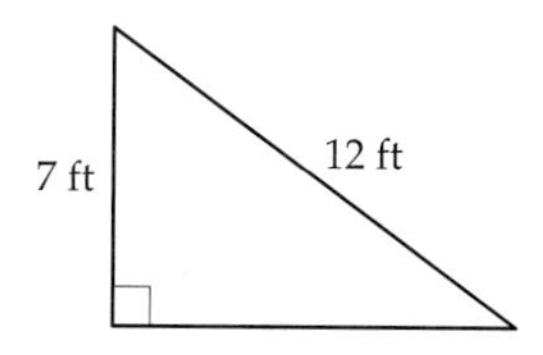

In Exercises 26 and 27, find the values of $\sin \theta$, $\cos \theta$, and $\tan \theta$ for the given right triangle.

26.

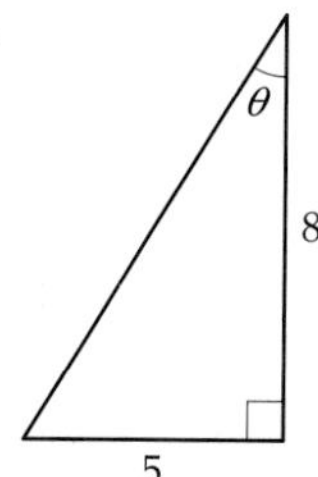

27.

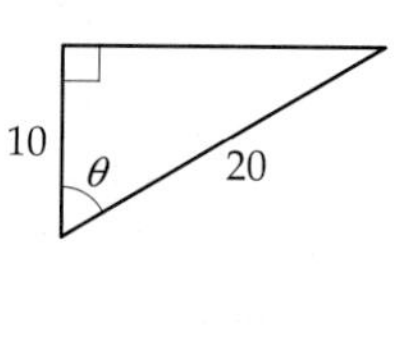

In Exercises 28–31, use a calculator. Round to the nearest tenth of a degree.

28. Find $\cos^{-1}(0.9013)$.

29. Find $\sin^{-1}(0.4871)$.

30. Given $\tan \beta = 1.364$, find β.

31. Given $\sin \theta = 0.0325$, find θ.

32. **Surveying** Find the distance across the marsh in the following figure. Round to the nearest tenth of a foot.

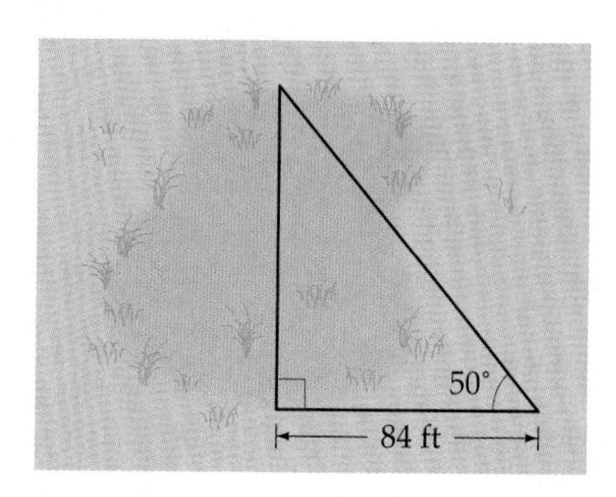

33. **Angle of Depression** The distance from a plane to a radar station is 200 miles, and the angle of depression is 40°. Find the number of ground miles from a point directly under the plane to the radar station. Round to the nearest tenth of a mile.

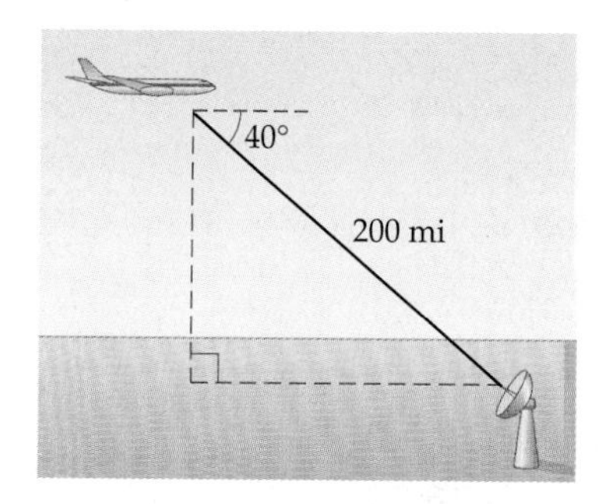

34. **Angle of Elevation** The angle of elevation from a point A on the ground to the top of a space shuttle is 27°. If point A is 110 feet from the base of the space shuttle, how tall is the space shuttle? Round to the nearest tenth of a foot.

35. Name the mathematician who called the geometry he developed "imaginary geometry."

36. What is another name for Riemannian geometry?

37. What is another name for Lobachevskian geometry?

38. Name a geometry in which the sum of the measures of the interior angles of a triangle is less than 180°.

39. Name a geometry in which there are no parallel lines.

In Exercises 40 and 41, determine the exact area of the spherical triangle.

40. Radius: 12 in.; angles: 90°, 150°, 90°

41. Radius: 5 ft; angles: 90°, 60°, 90°

City Geometry In Exercises 42–45, find the Euclidean distance and the city distance between the points. Assume that the distances are measured in blocks. Round approximate results to the nearest tenth of a block.

42. $P(-1, 1)$, $Q(3, 4)$

43. $P(-5, -2)$, $Q(2, 6)$

44. $P(2, 8)$, $Q(3, 2)$

45. $P(-3, 3)$, $Q(5, -2)$

46. Consider the points $P(1, 1)$, $Q(4, 5)$, and $R(-4, 2)$.

a. Which two points are closest together if you use the Euclidean distance formula to measure distance?

b. Which two points are closest together if you use the city distance formula to measure distance?

47. Draw stage 0, stage 1, and stage 2 of the Koch curve.

48. Draw stage 2 of the fractal with the following initiator and generator.

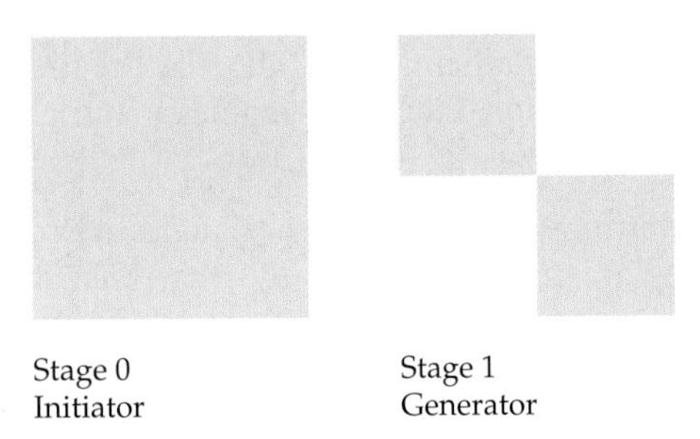

49. Compute the similarity dimension of a strictly self-similar fractal with a replacement ratio of 5 and a scaling ratio of 4. Round to the nearest thousandth.

50. Determine the similarity dimension of the fractal defined in Exercise 48.

CHAPTER 8 Test

1. Find the volume of a cylinder with a height of 6 m and a radius of 3 m. Round to the nearest hundredth of a cubic meter.

2. Find the perimeter of a rectangle that has a length of 2 m and a width of 1.4 m.

3. Find the complement of a 32° angle.

4. Find the area of a circle that has a diameter of 2 m. Round to the nearest hundredth of a square meter.

5. In the figure below, lines ℓ_1 and ℓ_2 are parallel. Angle x measures 30°. Find the measure of angle y.

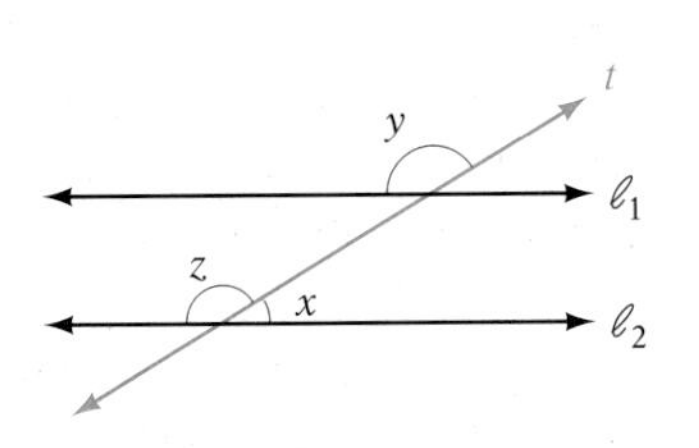

6. In the figure below, lines ℓ_1 and ℓ_2 are parallel. Angle x measures 45°. Find the measures of angles a and b.

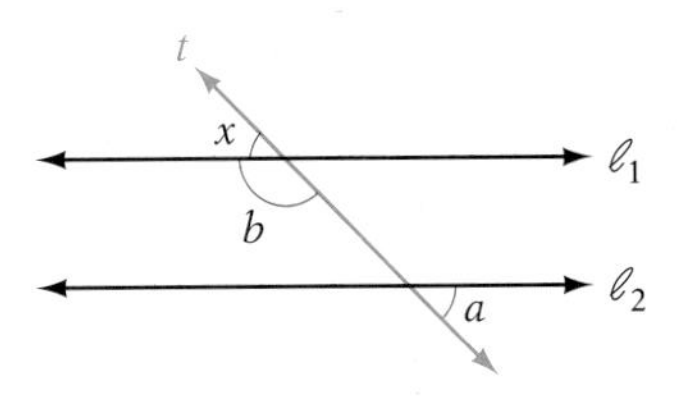

7. Find the area of a square that measures 2.25 ft on each side.

8. Find the volume of the figure. Give the exact value.

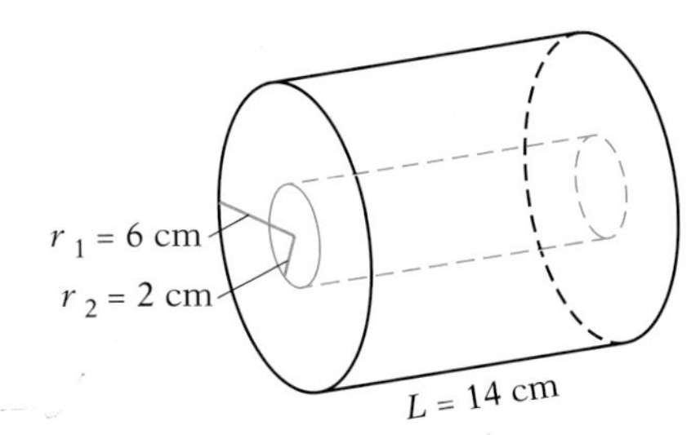

9. Triangles ABC and DEF are similar. Find side BC.

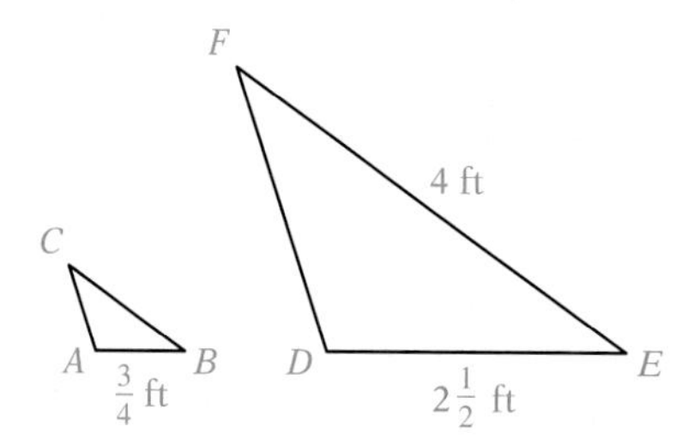

10. A right triangle has a 40° angle. Find the measures of the other two angles.

11. Find the measure of $\angle x$.

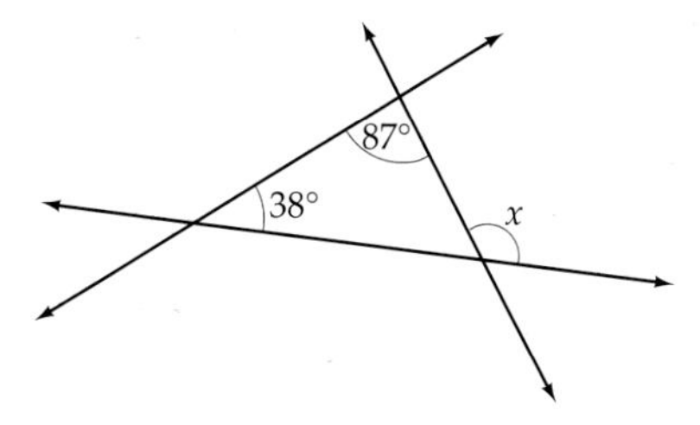

12. Find the area of the parallelogram shown below.

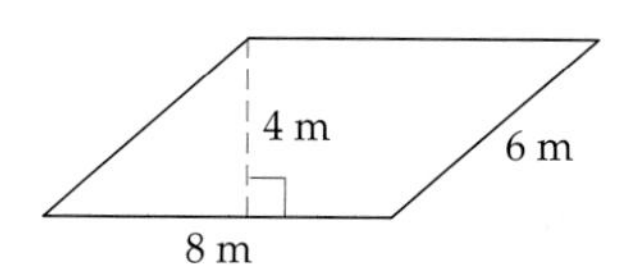

13. **Surveying** Find the width of the canal shown in the figure below.

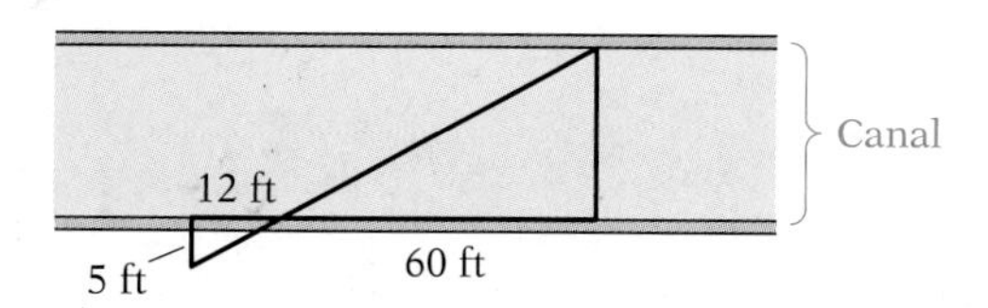

14. **Pizza** How much more area is in a pizza with radius 10 in. than in a pizza with radius 8 in.? Round to the nearest hundredth of a square inch.

15. Determine whether the two triangles are congruent. If they are congruent, state by what theorem they are congruent.

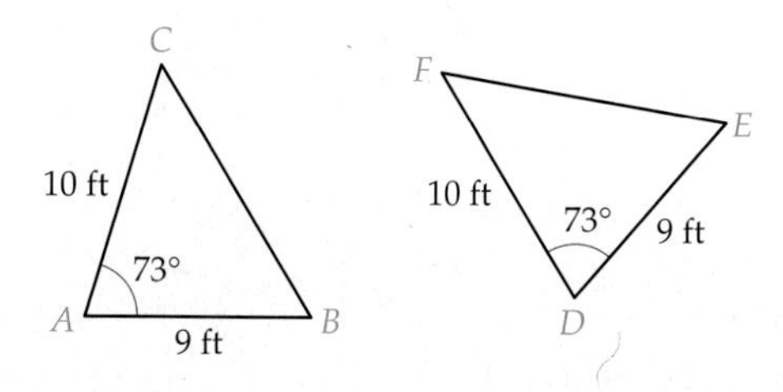

16. For the right triangle shown below, determine the length of side BC. Round to the nearest hundredth of a centimeter.

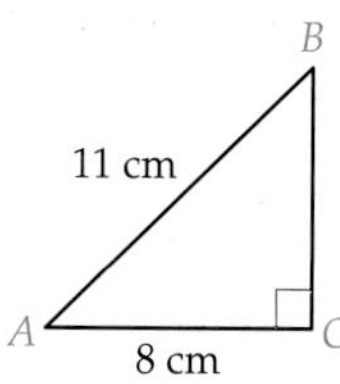

17. Find the values of $\sin\theta$, $\cos\theta$, and $\tan\theta$ for the given right triangle.

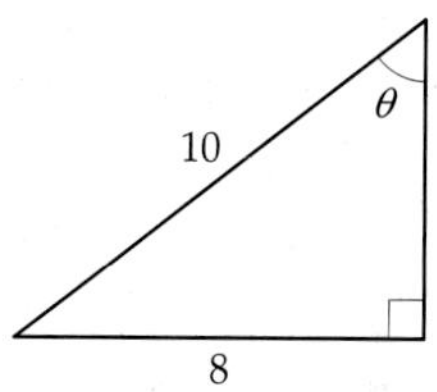

18. **Angle of Elevation** From a point 27 feet from the base of a Roman aqueduct, the angle of elevation to the top of the aqueduct is 78°. Find the height of the aqueduct. Round to the nearest foot.

19. **Trees** Find the cross-sectional area of a redwood tree that is 11 ft 6 in. in diameter. Round to the nearest hundredth of a square foot.

20. **Toolbox** A toolbox is 14 in. long, 9 in. wide, and 8 in. high. The sides and bottom of the toolbox are $\frac{1}{2}$ in. thick. The toolbox is open at the top. Find the volume of the interior of the toolbox in cubic inches.

21. **a.** State the Euclidean parallel postulate.

    b. State the parallel postulate used in Riemannian geometry.

22. What is the maximum number of right angles a triangle can have in

    a. Lobachevskian geometry?

    b. Riemannian geometry?

23. What is a great circle?

24. Find the area of a spherical triangle with a radius of 12 ft and angles of 90°, 100°, and 90°. Give the exact area and the area rounded to the nearest tenth of a square foot.

25. **City Geometry** Find the Euclidean distance and the city distance between the points $P(-4, 2)$ and $Q(5, 1)$. Assume that the distances are measured in blocks. Round approximate results to the nearest tenth of a block.

26. **City Geometry** How many points are on the city circle with center $(0, 0)$ and radius $r = 4$ blocks?

In Exercises 27 and 28, draw stage 2 of the fractal with the given initiator and generator.

27.

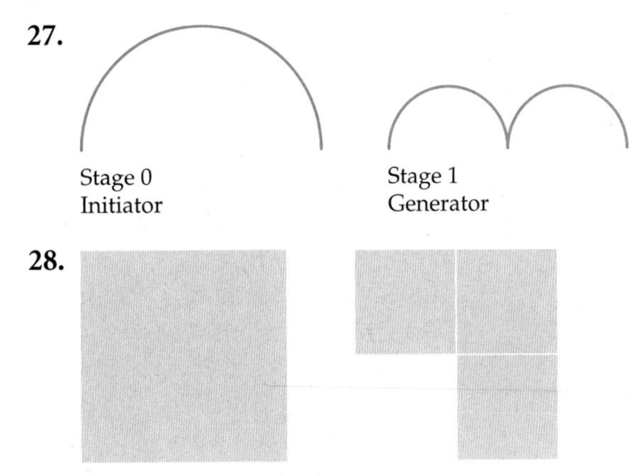

28.

Stage 0
Initiator

Stage 1
Generator

29. Compute the replacement ratio, scale ratio, and similarity dimension of the fractal defined by the initiator and generator in Exercise 27.

30. Compute the replacement ratio, scale ratio, and similarity dimension of the fractal defined by the initiator and generator in Exercise 28.

APPENDIX | The Metric System of Measurement

International trade, or trade between nations, is a vital and growing segment of business in the world today. The opening of McDonald's restaurants around the globe is testimony to the expansion of international business.

The United States, as a nation, is dependent on world trade. And world trade is dependent on internationally standardized units of measurement. Almost all countries use the metric system as their sole system of measurement. The United States is one of only a few countries that has not converted to the metric system as its official system of measurement. The International Mathematics and Science Study compared the performances of half a million students from 41 countries at five different grade levels on tests of their mathematics and science knowledge. One area of mathematics in which the U.S. average was below the international average was measurement, largely because the units cited in the questions were metric units. Because the United States has not yet converted to the metric system, its citizens are less familiar with it.

In this Appendix we will present the metric system of measurement and explain how to convert between different units.

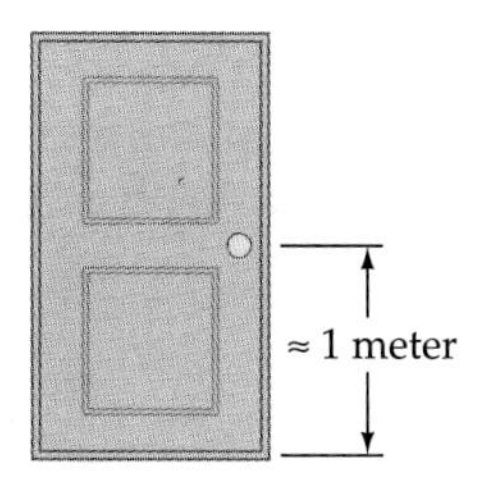

The basic unit of *length,* or distance, in the metric system is the **meter** (m). One meter is approximately the distance from a doorknob to the floor. All units of length in the metric system are derived from the meter. Prefixes added to the basic unit denote the length of the unit. For example, the prefix *centi-* means "one hundredth;" therefore, 1 centimeter is 1 one hundredth of a meter (0.01 m).

kilo-	= 1 000	1 kilometer (km)	= 1 000 meters (m)
hecto-	= 100	1 hectometer (hm)	= 100 m
deca-	= 10	1 decameter (dam)	= 10 m
		1 meter (m)	= 1 m
deci-	= 0.1	1 decimeter (dm)	= 0.1 m
centi-	= 0.01	1 centimeter (cm)	= 0.01 m
milli-	= 0.001	1 millimeter (mm)	= 0.001 m

Notice that in this list 1000 is written as 1 000, with a space between the 1 and the zeros. When writing numbers using metric units, each group of three numbers is separated by a space instead of a comma. A space is also used after each group of three numbers to the right of a decimal. For example, 31,245.2976 is written 31 245.297 6 in metric notation.

point of interest

Originally the meter (spelled *metre* in some countries) was defined as $\frac{1}{10,000,000}$ of the distance from the equator to the North Pole. Modern scientists have redefined the meter as 1,650,753.73 wavelengths of the orange-red light given off by the element krypton.

QUESTION *Which unit in the metric system is one thousandth of a meter?*

Mass and weight are closely related. *Weight* is a measure of how strongly gravity is pulling on an object. Therefore, an object's weight is less in space than on Earth's surface. However, the amount of material in the object, its *mass,* remains the same. On the surface of Earth, the terms *mass* and *weight* can be used interchangeably.

ANSWER *The millimeter is one thousandth of a meter.*

The basic unit of mass in the metric system is the **gram** (g). If a box 1 centimeter long on each side is filled with pure water, the mass of that water is 1 gram.

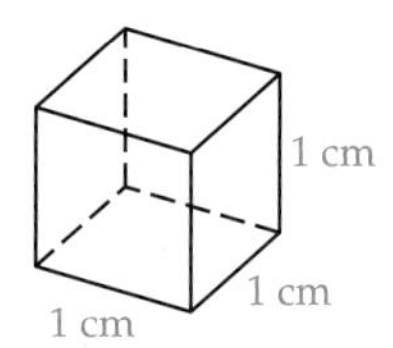

1 gram = the mass of water in a box that is 1 centimeter long on each side

The units of mass in the metric system have the same prefixes as the units of length.

1 kilogram (kg) = 1 000 grams (g)
1 hectogram (hg) = 100 g
1 decagram (dag) = 10 g
1 gram (g) = 1 g
1 decigram (dg) = 0.1 g
1 centigram (cg) = 0.01 g
1 milligram (mg) = 0.001 g

The gram is a small unit of mass. A paperclip weighs about 1 gram. In many applications, the kilogram (1 000 grams) is a more useful unit of mass. This textbook weighs about 1 kilogram.

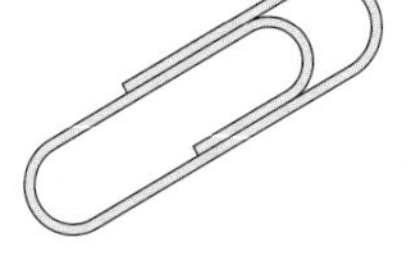

Weight ≈ 1 gram

QUESTION *Which unit in the metric system is equal to 1 000 grams?*

Liquid substances are measured in units of *capacity*.

The basic unit of capacity in the metric system is the **liter** (L). One liter is defined as the capacity of a box that is 10 centimeters long on each side.

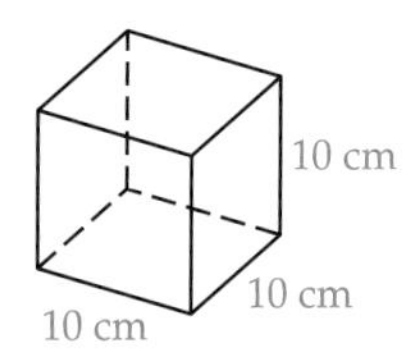

1 liter = the capacity of a box that is 10 centimeters long on each side

The units of capacity in the metric system have the same prefixes as the units of length.

1 kiloliter (kl) = 1 000 liters (L)
1 hectoliter (hl) = 100 L
1 decaliter (dal) = 10 L
1 liter (L) = 1 L
1 deciliter (dl) = 0.1 L
1 centiliter (cl) = 0.01 L
1 milliliter (ml) = 0.001 L

ANSWER *The kilogram is equal to 1 000 grams.*

Converting between units in the metric system involves moving the decimal point to the right or to the left. Listing the units in order from largest to smallest will indicate how many places to move the decimal point and in which direction.

To convert 3 800 centimeters to meters, write the units of length in order from largest to smallest.

km hm dam m dm cm mm

2 positions

- Converting from cm to m requires moving 2 positions to the left.

3 800 cm = 38.00 m

2 places

- Move the decimal point the same number of places and in the same direction.

Convert 27 kilograms to grams.

kg hg dag g dg cg mg

3 positions

- Write the units of mass in order from largest to smallest.
- Converting kg to g requires moving 3 positions to the right.

27 kg = 27 000 g

3 places

- Move the decimal point the same number of places and in the same direction.

point of interest

The definition of 1 inch has been changed as a consequence of the wide acceptance of the metric system. One inch is now exactly 25.4 millimeters.

EXAMPLE 1 ■ Convert Units in the Metric System of Measurement

Convert.

a. 4.08 meters to centimeters **b.** 5.93 grams to milligrams

c. 82 milliliters to liters **d.** 9 kiloliters to liters

Solution

a. Write the units of length from largest to smallest.

km hm dam (m) dm (cm) mm

Converting m to cm requires moving 2 positions to the right.

4.08 m = 408 cm

b. Write the units of mass from largest to smallest.

kg hg dag (g) dg cg (mg)

Converting g to mg requires moving 3 positions to the right.

5.93 g = 5 930 mg

c. Write the units of capacity from largest to smallest.

kl hl dal (L) dl cl (ml)

Converting ml to L requires moving 3 positions to the left.

82 ml = 0.082 L

d. Write the units of capacity from largest to smallest.

(kl) hl dal (L) dl cl ml

Converting kl to L requires moving 3 positions to the right.

9 kl = 9 000 L

CHECK YOUR PROGRESS 1 Convert.

a. 1 295 meters to kilometers **b.** 7 543 grams to kilograms

c. 6.3 liters to milliliters **d.** 2 kiloliters to liters

Solution *See page S54.*

Other prefixes in the metric system are becoming more common as a result of technological advances in the computer industry. For example:

tera- = 1 000 000 000 000
giga- = 1 000 000 000
mega- = 1 000 000

micro- = 0.000 001
nano- = 0.000 000 001
pico- = 0.000 000 000 001

A **bit** is the smallest unit of code that computers can read; it is a binary digit, either a 0 or a 1. Usually bits are grouped into bytes of 8 bits. Each byte stands for a letter, number, or any other symbol we might use in communicating information. For example, the letter W can be represented by 01010111. The amount of memory in a computer hard drive is measured in terabytes, gigabytes, and megabytes. The speed of a computer was measured in microseconds, then in nanoseconds, and then in picoseconds.

Here are a few more examples of how these prefixes are used.

The mass of Earth gains 40 Gg (gigagrams) each year from captured meteorites and cosmic dust.

The average distance from Earth to the moon is 384.4 Mm (megameters) and the average distance from Earth to the sun is 149.5 Gm (gigameters).

The wavelength of yellow light is 590 nm (nanometers).

The diameter of a hydrogen atom is about 70 pm (picometers).

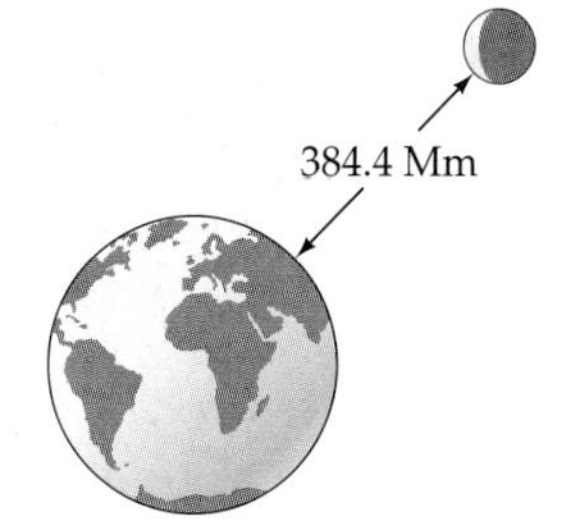

There are additional prefixes in the metric system, representing both larger and smaller units. We may hear them more and more often as computer chips hold more and more information, as computers get faster and faster, and as we learn more and more about objects in our universe and beyond that are great distances away.

Exercises

1. In the metric system, what is the basic unit of length? of liquid measure? of weight?

2. a. Explain how to convert meters to centimeters.

   b. Explain how to convert milliliters to liters.

In Exercises 3–26, name the unit in the metric system that can most conveniently be used to measure each of the following.

3. The distance from New York to London
4. The weight of a truck
5. A person's waist
6. The amount of coffee in a mug
7. The weight of a thumbtack
8. The amount of water in a swimming pool
9. The distance a baseball player hits a baseball
10. A person's hat size
11. The amount of fat in a slice of cheddar cheese
12. A person's weight
13. The maple syrup served with pancakes
14. The amount of water in a water cooler
15. The amount of vitamin C in a vitamin tablet
16. A serving of cereal
17. The width of a hair
18. A person's height
19. The amount of medication in an aspirin
20. The weight of a lawnmower
21. The weight of a slice of bread
22. The contents of a bottle of salad dressing
23. The amount of water a family uses monthly
24. The newspapers collected at a recycling center
25. The amount of liquid in a bowl of soup
26. The distance to the bank

27. a. Complete the table.

Metric system prefix	Symbol	Magnitude	Means multiply the basic unit by:
tera-	T	10^{12}	1 000 000 000 000
giga-	G	?	1 000 000 000
mega-	M	10^6	?
kilo-	?	?	1 000
hecto-	h	?	100
deca-	da	10^1	?
deci-	d	$\frac{1}{10}$	?
centi-	?	$\frac{1}{10^2}$	?
milli-	?	?	0.001
micro-	μ (mu)	$\frac{1}{10^6}$	?
nano-	n	$\frac{1}{10^9}$	?
pico-	p	?	0.000 000 000 001

b. How can the magnitude column in the table above be used to determine how many places to move the decimal point when converting to the basic unit in the metric system?

In Exercises 28–57, convert the given measure.

28. 42 cm = ______ mm
29. 91 cm = ______ mm
30. 360 g = ______ kg
31. 1 856 g = ______ kg
32. 5 194 ml = ______ L

33. 7 285 ml = ______ L

34. 2 m = ______ mm

35. 8 m = ______ mm

36. 217 mg = ______ g

37. 34 mg = ______ g

38. 4.52 L = ______ ml

39. 0.029 7 L = ______ ml

40. 8 406 m = ______ km

41. 7 530 m = ______ km

42. 2.4 kg = ______ g

43. 9.2 kg = ______ g

44. 6.18 kl = ______ L

45. 0.036 kl = ______ L

46. 9.612 km = ______ m

47. 2.35 km = ______ m

48. 0.24 g = ______ mg

49. 0.083 g = ______ mg

50. 298 cm = ______ m

51. 71.6 cm = ______ m

52. 2 431 L = ______ kl

53. 6 302 L = ______ kl

54. 0.66 m = ______ cm

55. 4.58 m = ______ cm

56. 243 mm = ______ cm

57. 92 mm = ______ cm

58. The Olympics **a.** One of the events in the summer Olympics is the 50 000-meter walk. How many kilometers do the entrants in this event walk?

b. One of the events in the winter Olympic games is the 10 000-meter speed skating event. How many kilometers do the entrants in this event skate?

59. Gemology A carat is a unit of weight equal to 200 milligrams. Find the weight in grams of a 10-carat precious stone.

60. Sewing How many pieces of material, each 75 centimeters long, can be cut from a bolt of fabric that is 6 meters long?

61. Water Treatment An athletic club uses 800 milliliters of chlorine each day for its swimming pool. How many liters of chlorine are used in a month of 30 days?

62. Carpentry Each of the four shelves in a bookcase measures 175 centimeters. Find the cost of the shelves when the price of lumber is $15.75 per meter.

63. Consumerism The printed label from a container of milk is shown below. To the nearest whole number, how many 230-milliliter servings are in the container?

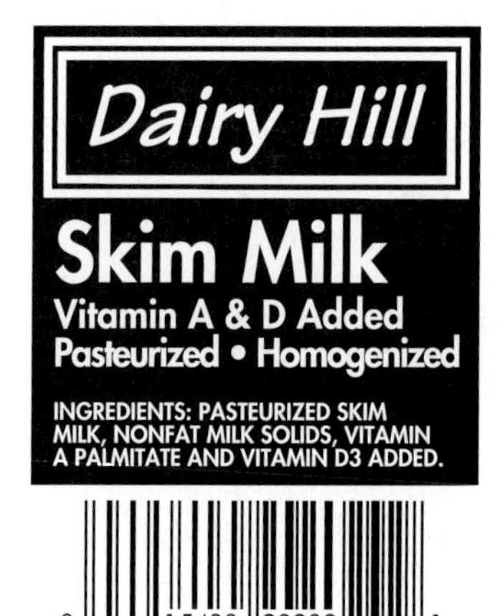

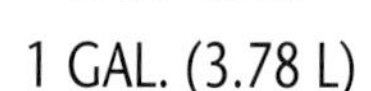

64. Consumerism A 1.19-kilogram container of Quaker Oats contains 30 servings. Find the number of grams in one serving of the oatmeal. Round to the nearest gram.

65. Nutrition A patient is advised to supplement her diet with 2 grams of calcium per day. The calcium tablets she purchases contain 500 milligrams of calcium per tablet. How many tablets per day should the patient take?

66. Education A laboratory assistant is in charge of ordering acid for three chemistry classes of 30 students each. Each student requires 80 milliliters of acid. How many liters of acid should be ordered? The assistant must order by the whole liter.

67. Consumerism A case of 12 one-liter bottles of apple juice costs $19.80. A case of 24 cans, each can containing 340 milliliters of apple juice, costs $14.50. Which case of apple juice costs less per milliliter?

68. **Construction** A column assembly is being constructed in a building. The components are shown in the diagram below. What height column must be cut?

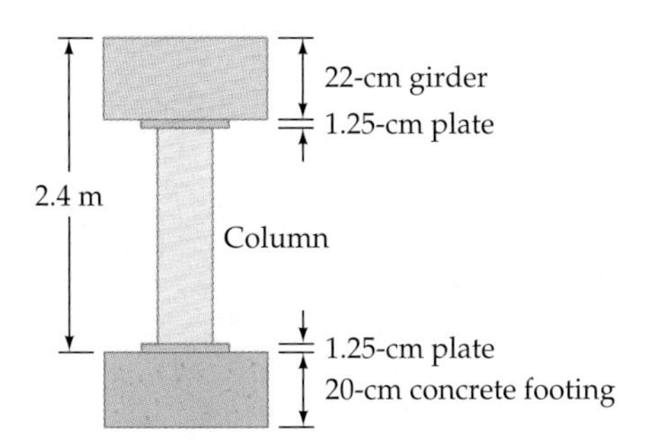

69. **Light** The distance between Earth and the sun is 150 000 000 kilometers. Light travels 300 000 000 meters in 1 second. How many seconds does it take for light to reach Earth from the sun?

70. Explain why is it advantageous to have internationally standardized units of measure.

Extensions

CRITICAL THINKING

71. **Business** A service station operator bought 85 kl of gasoline for $38,500. The gasoline was sold for $.658 per liter. Find the profit on the 85 kl of gasoline.

72. **Business** For $149.50, a cosmetician buys 5 L of moisturizer and repackages it in 125-milliliter jars. Each jar costs the cosmetician $.55. Each jar of moisturizer is sold for $8.95. Find the profit on the 5 L of moisturizer.

73. **Business** A health food store buys nuts in 10-kilogram containers and repackages the nuts for resale. The store packages the nuts in 200-gram bags, costing $.06 each, and sells them for $2.89 per bag. Find the profit on a 10-kilogram container of nuts costing $75.

COOPERATIVE LEARNING

74. Form two debating teams. One team should argue in favor of changing to the metric system in the United States, and the other should argue against it.

SOLUTIONS TO *CHECK YOUR PROGRESS* PROBLEMS

CHAPTER 1

SECTION 1.1

CHECK YOUR PROGRESS 1, *page 2*

a. Each successive number is 5 larger than the preceding number. Thus we predict that the next number in the list is 5 larger than 25, which is 30.

b. The first two numbers differ by 3. The second and third numbers differ by 5. It appears that the difference between any two numbers is always 2 more than the preceding difference. Thus we predict that the next number will be 11 more than 26, which is 37.

CHECK YOUR PROGRESS 2, *page 3*

If the original number is 2, then $\frac{2 \times 9 + 15}{3} - 5 = 6$, which is three times the original number.

If the original number is 7, then $\frac{7 \times 9 + 15}{3} - 5 = 21$, which is three times the original number.

If the original number is -12, then $\frac{-12 \times 9 + 15}{3} - 5 = -36$, which is three times the original number.

It appears, by inductive reasoning, that the procedure produces a number that is three times the original number.

CHECK YOUR PROGRESS 3, *page 4*

a. It appears that when the velocity of a tsunami is doubled, its height is quadrupled.

b. A tsunami with a velocity of 30 feet per second will have a height that is four times that of a tsunami with a speed of 15 feet per second. Thus, we predict a height of $4 \times 25 = 100$ feet for a tsunami with a velocity of 30 feet per second.

CHECK YOUR PROGRESS 4, *page 5*

a. Let $x = 0$. Then $\frac{x}{x} \neq 1$, because division by 0 is undefined.

b. Let $x = 1$. Then $\frac{x + 3}{3} = \frac{1 + 3}{3} = \frac{4}{3}$, whereas $x + 1 = 1 + 1 = 2$.

c. Let $x = 3$. Then $\sqrt{x^2 + 16} = \sqrt{3^2 + 16} = \sqrt{25} = 5$, whereas $x + 4 = 3 + 4 = 7$.

CHECK YOUR PROGRESS 5, *page 6*

Let n represent the original number.

Multiply the number by 6: $6n$

Add 10 to the product: $6n + 10$

Divide the sum by 2: $\frac{6n + 10}{2} = 3n + 5$

Subtract 5: $3n + 5 - 5 = 3n$

The procedure always produces a number that is three times the original number.

CHECK YOUR PROGRESS 6, *page 7*

a. The conclusion is a specific case of a general assumption, so the argument is an example of deductive reasoning.

b. The argument reaches a conclusion based on specific examples, so the argument is an example of inductive reasoning.

CHECK YOUR PROGRESS 7, *page 9*

From clue 1, we know that Ashley is not the president or the treasurer. In the following chart, write X1 (which stands for "ruled out by clue 1") in the President and Treasurer columns of Ashley's row.

	Pres.	V. P.	Sec.	Treas.
Brianna				
Ryan				
Tyler				
Ashley	X1			X1

From clue 2, Brianna is not the secretary. We know from clue 1 that the president is not the youngest, and we know from clue 2 that Brianna and the secretary are the youngest members of the group. Thus Brianna is not the president. In the chart, write X2 for these two conditions. Also we know from clues 1 and 2 that Ashley is not the secretary, because she is older than the treasurer. Write an X2 in the Secretary column of Ashley's row.

	Pres.	V. P.	Sec.	Treas.
Brianna	X2		X2	
Ryan				
Tyler				
Ashley	X1		X2	X1

At this point we see that Ashley must be the vice president and that none of the other members is the vice president. Thus we can update the chart as shown below.

	Pres.	V. P.	Sec.	Treas.
Brianna	X2	X2	X2	
Ryan		X2		
Tyler		X2		
Ashley	X1	√	X2	X1

Now we can see that Brianna must be the treasurer and that neither Ryan nor Tyler is the treasurer. Update the chart as shown below.

	Pres.	V. P.	Sec.	Treas.
Brianna	X2	X2	X2	√
Ryan		X2		X2
Tyler		X2		X2
Ashley	X1	√	X2	X1

From clue 3, we know that Tyler is not the secretary. Thus we can conclude that Tyler is the president and Ryan must be the secretary. See the chart below.

	Pres.	V. P.	Sec.	Treas.
Brianna	X2	X2	X2	√
Ryan	X3	X2	√	X2
Tyler	√	X2	X3	X2
Ashley	X1	√	X2	X1

Tyler is the president, Ashley is the vice president, Ryan is the secretary, and Brianna is the treasurer.

SECTION 1.2

CHECK YOUR PROGRESS 1, *page 17*

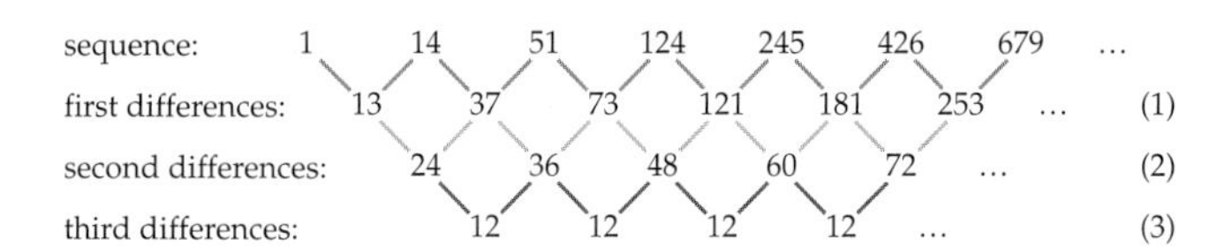

Using the method of extending the difference table, we predict that 679 is the next term in the sequence.

CHECK YOUR PROGRESS 2, *page 19*

a. Each figure after the first figure consists of a square region and a "tail." The number of tiles in the square region is n^2 and the number of tiles in the tail is $n - 1$. Thus the nth term formula for the number of tiles in the nth figure is $a_n = n^2 + n - 1$.

b. Let $n = 10$. Then $n^2 + n - 1 = (10)^2 + (10) - 1 = 109$.

c.
$$n^2 + n - 1 = 419$$
$$n^2 + n - 420 = 0$$
$$(n + 21)(n - 20) = 0$$

$$n + 21 = 0 \quad \text{or} \quad n - 20 = 0$$
$$n = -21 \qquad\qquad n = 20$$

We consider only the positive result. The 20th figure will consist of 419 tiles.

CHECK YOUR PROGRESS 3, *page 21*

$$F_9 = F_8 + F_7$$
$$= 21 + 13$$
$$= 34$$

CHECK YOUR PROGRESS 4, *page 22*

a. The inequality $2F_n > F_{n+1}$ is true for $n = 3, 4, 5, 6, \ldots, 10$. Thus, by inductive reasoning, we conjecture that the statement is a true statement.

b. The equality $2F_n + 3 = F_{n+2}$ is not true for $n = 4$, because $2F_4 + 3 = 2(3) + 3 = 9$ and $F_{4+2} = F_6 = 8$. Thus the statement is a false statement.

SECTION 1.3

CHECK YOUR PROGRESS 1, *page 32*

Understand the Problem In order to go past Starbucks, Allison must walk along Third Avenue from Boardwalk to Park Avenue.

Devise a Plan Label each intersection that Allison can pass through with the number of routes to that intersection. If she can reach an intersection from two different routes, then the number of routes to that intersection is the sum of the numbers of routes to the two adjacent intersections.

Carry Out the Plan The following figure shows the number of routes to each of the intersections that Allison could pass through. Thus there are a total of nine routes that Allison can take if she wishes to walk directly from point A to point B and pass by Starbucks.

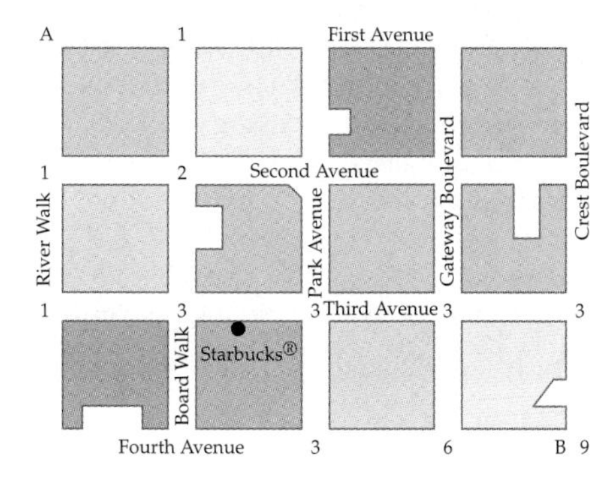

Review the Solution The total of nine routes seems reasonable. We know from Example 1 that if Allison can take any route, the total number of routes is 35. Requiring Allison to go past Starbucks eliminates several routes.

CHECK YOUR PROGRESS 2, *page 33*

Understand the Problem There are several ways to answer the questions so that two answers are "false" and three answers are "true." One way is TTTFF and another is FFTTT.

Devise a Plan Make an organized list. Try the strategy of listing a T unless doing so will produce too many T's or a duplicate of one of the previous orders in your list.

Carry Out the Plan (Start with 3 T's in a row.)

TTTFF (1)
TTFTF (2)
TTFFT (3)
TFTTF (4)
TFTFT (5)
TFFTT (6)
FTTTF (7)
FTTFT (8)
FTFTT (9)
FFTTT (10)

Review the Solution Each entry in the list has two F's and three T's. Since the list is complete and has no duplications, we know that there are 10 ways for a student to mark two questions with "false" and the other three with "true."

CHECK YOUR PROGRESS 3, *page 34*

Understand the Problem There are six people, and each person shakes hands with each of the other people.

Devise a Plan Each person will shake hands with five other people (a person won't shake his or her own hand; that would be silly). Since there are six people, we could multiply 6 times 5 to get the total number of handshakes. However, this procedure would count each handshake exactly twice, so we must divide this product by 2 for the actual answer.

Carry Out the Plan 6 times 5 is 30. 30 divided by 2 is 15.

Review the Solution Denote the people by the letters A, B, C, D, E, and F. Make an organized list. Remember that AB and BA represent the same people shaking hands, so do not list both AB and BA.

AB AC AD AE AF
BC BD BE BF
CD CE CF
DE DF
EF

The method of making an organized list verifies that if six people shake hands with each other there will be a total of 15 handshakes.

CHECK YOUR PROGRESS 4, *page 35*

Understand the Problem We need to find the ones digit of 4^{200}.

Devise a Plan Compute a few powers of 4 to see if there are any patterns. $4^1 = 4$, $4^2 = 16$, $4^3 = 64$, and $4^4 = 256$. It appears that the last digit (ones digit) of 4^{200} must be either a 4 or a 6.

Carry Out the Plan If the exponent n is an even number, then 4^n has a ones digit of 6. If the exponent n is an odd number, then 4^n has a ones digit of 4. Because 200 is an even number, we conjecture that 4^{200} has a ones digit of 6.

Review the Solution You could try to check the answer by using a calculator, but you would find that 4^{200} is too large to be displayed. Thus we need to rely on the patterns we have observed to conclude that 6 is indeed the ones digit of 4^{200}.

CHECK YOUR PROGRESS 5, *page 35*

Understand the Problem We are asked to find the possible numbers that Melody could have started with.

Devise a Plan Work backward from 18 and do the inverse of each operation that Melody performed.

Carry Out the Plan To get 18, Melody subtracted 30 from a number, so that number was $18 + 30 = 48$. To get 48, she divided a number by 3, so that number was $48 \times 3 = 144$. To get 144, she squared a number. She could have squared either 12 or -12 to produce 144. If the number she squared was 12, then she must have doubled 6 to get 12. If the number she squared was -12, then the number she doubled was -6.

Review the Solution We can check by starting with 6 or −6. If we do exactly as Melody did, we end up with 18. The operation that prevents us from knowing with 100% certainty which number she started with is the squaring operation. We have no way of knowing whether the number she squared was a positive number or a negative number.

CHECK YOUR PROGRESS 6, *page 36*

Understand the Problem We need to find Diophantus's age when he died.

Devise a Plan Read the hint and then look for clues that will help you make an educated guess. You know from the given information that Diophantus's age must be divisible by 6, 12, 7, and 2. Find a number divisible by all of these numbers and check to see if it is a possible solution to the problem.

Carry Out the Plan All multiples of 12 are divisible by 6 and 2, but the smallest multiple of 12 that is divisible by 7 is $12 \times 7 = 84$. Thus we conjecture that Diophantus's age when he died was $x = 84$ years. If $x = 84$, then $\frac{1}{6}x = 14$, $\frac{1}{12}x = 7$, $\frac{1}{7}x = 12$, and $\frac{1}{2}x = 42$. Then $\frac{1}{6}x + \frac{1}{12}x + \frac{1}{7}x + 5 + \frac{1}{2}x + 4 = 14 + 7 + 12 + 5 + 42 + 4 = 84$. It seems that 84 years is a correct solution to the problem.

Review the Solution After 84, the next multiple of 12 that is divisible by 7 is 168. The number 168 also satisfies all the conditions of the problem, but it is unlikely that Diophantus died at the age of 168 years or at any age older than 168 years. Hence the only reasonable solution is 84 years.

CHECK YOUR PROGRESS 7, *page 38*

Understand the Problem We need to determine two U.S. coins that have a total value of 35¢, given that one of the coins is not a quarter.

Devise a Plan Experiment with different coins to try to produce 35¢. After a few attempts, you should conclude that one of the coins must be a quarter. Consider that the problem may be a *deceptive problem.*

Carry Out the Plan A total of 35¢ can be produced by using a dime and a quarter. One of the coins is a quarter, but it is also true that *one of the coins, the dime, is not a quarter.*

Review the Solution A dime and a quarter satisfy all the conditions of the problem. No other combination of coins satisfies the conditions of the problem. Thus the only solution is a dime and a quarter.

CHECK YOUR PROGRESS 8, *page 40*

a. The maximum of the average yearly ticket prices is displayed by the tallest vertical bar in Figure 1.3. Thus the maximum of the average yearly U.S. movie theatre ticket price for the years from 1996 to 2003 was $6.03, in the year 2003.

b. Figure 1.4 indicates that in 2005, about 9% of the automobile accidents in Twin Falls were accidents involving lane changes. Thus $0.09 \cdot 4300 = 387$ of the accidents were accidents involving lane changes.

c. To estimate the average age at which women married for the first time in the year 1975, locate 1975 on the horizontal axis of Figure 1.5 and then move directly upward to the point on the green broken-line graph. The height of this point represents the average age at first marriage for women in the year 1975, and it can be estimated by moving horizontally to the vertical axis on the left. Thus the average age at first marriage for women in the year 1975 was 21 years, rounded to the nearest quarter of a year. This same procedure shows that in the year 1975, the average age at which men first married was 23.5 years, rounded to the nearest quarter of a year.

CHAPTER 2

SECTION 2.1

CHECK YOUR PROGRESS 1, *page 53*

The only months that start with the letter A are April and August. When we use the roster method, the set is given by {April, August}.

CHECK YOUR PROGRESS 2, *page 54*

The set {March, May} is the set of all months that start with the letter M.

CHECK YOUR PROGRESS 3, *page 55*

a. $\{0, 1, 2, 3\}$ **b.** $\{12, 13, 14, 15, 16, 17, 18, 19\}$

c. $\{-4, -3, -2, -1\}$

CHECK YOUR PROGRESS 4, *page 55*

a. False **b.** True **c.** True **d.** True

CHECK YOUR PROGRESS 5, *page 56*

a. $\{x \mid x \in I \text{ and } x < 9\}$ **b.** $\{x \mid x \in N \text{ and } x > 4\}$

CHECK YOUR PROGRESS 6, *page 57*

a. $n(C) = 5$ **b.** $n(D) = 1$ **c.** $n(E) = 0$

CHECK YOUR PROGRESS 7, *page 57*

a. The sets are not equal but they both contain six elements. Thus the sets are equivalent.

b. The sets are not equal but they both contain 16 elements. Thus the sets are equivalent.

SECTION 2.2

CHECK YOUR PROGRESS 1, *page 65*

a. $M = \{0, 4, 6, 17\}$. The set of elements in $U = \{0, 2, 3, 4, 6, 7, 17\}$ but not in M is $M' = \{2, 3, 7\}$.

b. $P = \{2, 4, 6\}$. The set of elements in $U = \{0, 2, 3, 4, 6, 7, 17\}$ but not in P is $P' = \{0, 3, 7, 17\}$.

CHECK YOUR PROGRESS 2, *page 66*

a. False. The number 3 is an element of the first set but not an element of the second set. Therefore, the first set is not a subset of the second set.

b. True. The set of counting numbers is the same set as the set of natural numbers, and every set is a subset of itself.

c. True. The empty set is a subset of every set.

d. True. Each element of the first set is an integer.

CHECK YOUR PROGRESS 3, *page 67*

a. Yes, because every natural number is a whole number, and the whole numbers include 0, which is not a natural number.

b. The first set is not a proper subset of the second set because the sets are equal.

CHECK YOUR PROGRESS 4, *page 68*

Subsets with zero elements: { }

Subsets with one element: $\{a\}, \{b\}, \{c\}, \{d\}, \{e\}$

Subsets with two elements: $\{a, b\}, \{a, c\}, \{a, d\}, \{a, e\}, \{b, c\}, \{b, d\}, \{b, e\}, \{c, d\}, \{c, e\}, \{d, e\}$

Subsets with three elements: $\{a, b, c\}, \{a, b, d\}, \{a, b, e\}, \{a, c, d\}, \{a, c, e\}, \{a, d, e\}, \{b, c, d\}, \{b, c, e\}, \{b, d, e\}, \{c, d, e\}$

Subsets with four elements: $\{a, b, c, d\}, \{a, b, c, e\}, \{a, b, d, e\}, \{a, c, d, e\}, \{b, c, d, e\}$

Subsets with five elements: $\{a, b, c, d, e\}$

CHECK YOUR PROGRESS 5, *page 68*

a. $2^3 = 8$ **b.** $2^{15} = 32{,}768$ **c.** $2^{10} = 1024$

SECTION 2.3

CHECK YOUR PROGRESS 1, *page 74*

a. $D \cap E = \{0, 3, 8, 9\} \cap \{3, 4, 8, 9, 11\}$
$= \{3, 8, 9\}$

b. $D \cap F = \{0, 3, 8, 9\} \cap \{0, 2, 6, 8\}$
$= \{0, 8\}$

CHECK YOUR PROGRESS 2, *page 75*

a. $D \cup E = \{0, 4, 8, 9\} \cup \{1, 4, 5, 7\}$
$= \{0, 1, 4, 5, 7, 8, 9\}$

b. $D \cup F = \{0, 4, 8, 9\} \cup \{2, 6, 8\}$
$= \{0, 2, 4, 6, 8, 9\}$

CHECK YOUR PROGRESS 3, *page 76*

a. The set $D \cap (E' \cup F)$ can be described as "the set of all elements that are in D, and in F or not E."

b. The set $L' \cup M$ can be described as "the set of all elements that are in M or are not in L."

CHECK YOUR PROGRESS 4, *page 76*

The following Venn diagrams show that $(A \cap B)'$ is equal to $A' \cup B'$.

The white region represents $(A \cap B)$.
The shaded region represents $(A \cap B)'$.

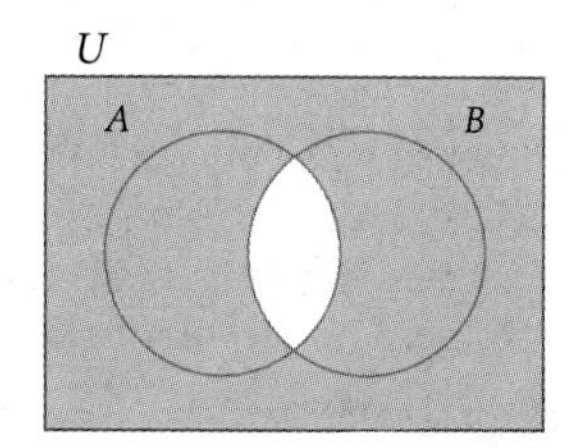

The grey shaded region below represents A'.
The diagonal patterned region below represents B'.
$A' \cup B'$ is the union of the grey shaded region and the diagonal patterned region.

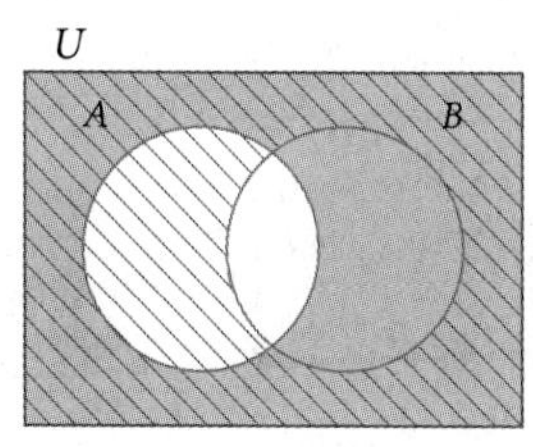

CHECK YOUR PROGRESS 5, *page 78*

The following Venn diagrams show that $A \cup (B \cap C) = (A \cup B) \cap (A \cup C)$.

The grey shaded region represents $A \cup (B \cap C)$.

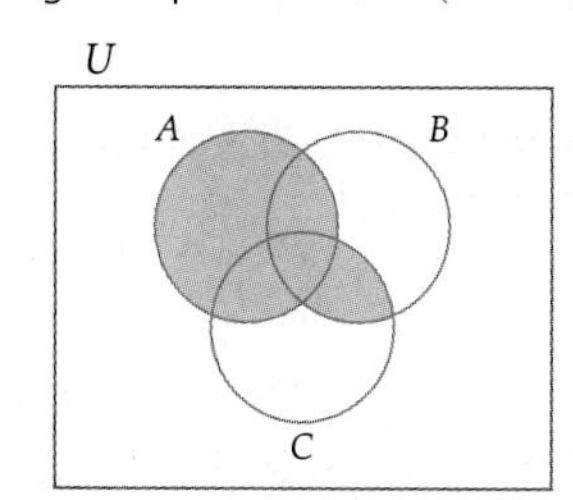

The grey shaded region below represents $A \cup B$.
The diagonal patterned region below represents $A \cup C$.
The intersection of the grey shaded region and the diagonal patterned region represents $(A \cup B) \cap (A \cup C)$.

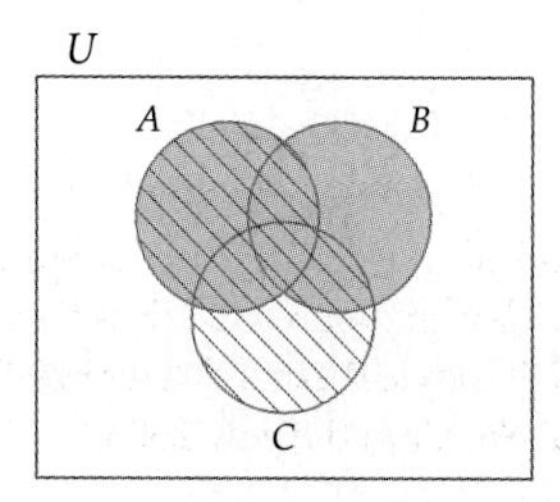

CHECK YOUR PROGRESS 6, *page 79*

a. Because Alex is in blood group A, not in blood group B, and is Rh+, his blood type is A+.

b. Roberto is in both blood group A and blood group B. Roberto is not Rh+. Thus Roberto's blood type is AB−.

CHECK YOUR PROGRESS 7, *page 80*

a. Alex's blood type is A+. The blood transfusion table shows that a person with blood type A+ can safely receive type A− blood.

b. The blood transfusion table shows that a person with type AB+ blood can safely receive each of the eight different types of blood. Thus a person with AB+ blood is classified as a universal recipient.

SECTION 2.4

CHECK YOUR PROGRESS 1, *page 87*
The intersection of the two sets includes the 85 students who like both volleyball and basketball.

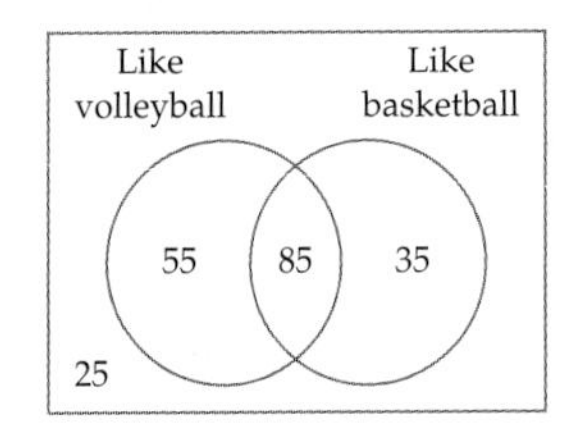

a. Because 140 students like volleyball and 85 like both sports, there must be $140 - 85 = 55$ students who like only volleyball.

b. Because 120 students like basketball and 85 like both sports, there must be $120 - 85 = 35$ students who like only basketball.

c. The Venn diagram shows that the number of students who like only volleyball plus the number who like only basketball plus the number who like both sports is $55 + 35 + 85 = 175$. Thus of the 200 students surveyed, only $200 - 175 = 25$ do not like either of the sports.

CHECK YOUR PROGRESS 2, *page 88*
The intersection of the three sets includes the 15 people who like all three activities.

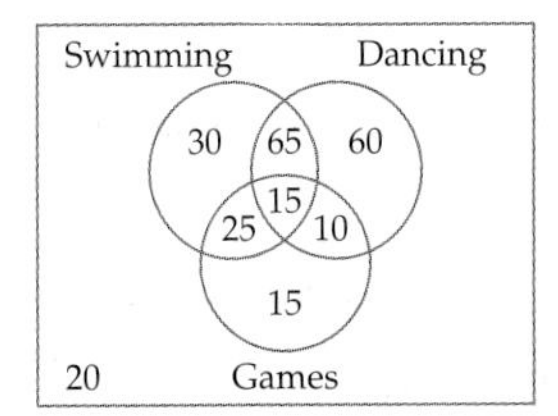

a. There are 25 people who like dancing and games. This includes the 15 people who like all three activities. Thus there must be another $25 - 15 = 10$ people who like only dancing and games. There are 40 people who like swimming and games. Thus there must be another $40 - 15 = 25$ people who like only swimming and games. There are 80 people who like swimming and dancing. Thus there must be another $80 - 15 = 65$ people who like only swimming and dancing. Hence $10 + 25 + 65 = 100$ people who like exactly two of the three activities.

b. There are 135 people who like swimming. We have determined that 15 people like all three activities, 25 like only swimming and games, and 65 like only swimming and dancing. This means that $135 - (15 + 25 + 65) = 30$ people like only swimming.

c. There are a total of 240 passengers surveyed. The Venn diagram shows that $15 + 25 + 10 + 15 + 30 + 65 + 60 = 220$ passengers like at least one of the activities. Thus $240 - 220 = 20$ passengers like none of the activities.

CHECK YOUR PROGRESS 3, *page 90*
Let B = {the set of students who play basketball}.
Let S = {the set of students who play soccer}.

$$\begin{aligned} n(B \cup S) &= n(B) + n(S) - n(B \cap S) \\ &= 80 + 60 - 24 \\ &= 116 \end{aligned}$$

Using the inclusion-exclusion principle, we see that 116 students play either basketball or soccer.

CHECK YOUR PROGRESS 4, *page 90*

$$\begin{aligned} n(A \cup B) &= n(A) + n(B) - n(A \cap B) \\ 852 &= 785 + 162 - n(A \cap B) \\ 852 &= 947 - n(A \cap B) \\ n(A \cap B) &= 947 - 852 \\ n(A \cap B) &= 95 \end{aligned}$$

CHECK YOUR PROGRESS 5, *page 91*

$$\begin{aligned} p(A \cup \text{Rh+}) &= p(A) + p(\text{Rh+}) - p(A \cap \text{Rh+}) \\ 91\% &= 44\% + 84\% - p(A \cap \text{Rh+}) \\ 91\% &= 128\% - p(A \cap \text{Rh+}) \\ p(A \cap \text{Rh+}) &= 128\% - 91\% \\ p(A \cap \text{Rh+}) &= 37\% \end{aligned}$$

Thus about 37% of the U.S. population has the A antigen and is Rh+.

CHECK YOUR PROGRESS 6, *page 92*

a. The number 410 appears in both the column labeled "Yahoo!" and the row labeled "children." Thus the table shows that 410 children surveyed use Yahoo! as a search engine. Thus $n(Y \cap C) = 410$.

b. The set $L \cap M'$ is the set of surveyed Lycos users who are women or children. The number in this set is $325 + 40 = 365$. Thus $n(L \cap M') = 365$.

c. The set $G \cap M$ represents the set of surveyed Google users who are men. The table shows that this set includes 440 people. The set $G \cap W$ represents the set of surveyed Google users who are women. The table shows that this set includes 390 people. Thus $n((G \cap M) \cup (G \cap W)) = 440 + 390 = 830$.

SECTION 2.5

CHECK YOUR PROGRESS 1, *page 99*

Write the sets so that one is aligned below the other. Draw arrows to show how you wish to pair the elements of each set. One possible method is shown in the following figure.

$$\begin{array}{l} N = \{1, 2, 3, 4, \ldots, \quad n, \quad \ldots\} \\ \quad\;\; \updownarrow\;\updownarrow\;\updownarrow\;\updownarrow \qquad\quad \updownarrow \\ D = \{1, 3, 5, 7, \ldots, 2n - 1, \ldots\} \end{array}$$

In the preceding correspondence, each natural number $n \in N$ is paired with the odd number $(2n - 1) \in D$. The *general correspondence* $n \leftrightarrow (2n - 1)$ enables us to determine exactly which element of D will be paired with any given element of N, and vice versa. For instance, under this correspondence, $8 \in N$ is paired with the odd number $2 \cdot 8 - 1 = 15$ and $21 \in D$ is paired with the natural number $\frac{21 + 1}{2} = 11$. The general correspondence $n \leftrightarrow (2n - 1)$ establishes a one-to-one correspondence between the sets.

CHECK YOUR PROGRESS 2, *page 100*

One proper subset of V is $P = \{41, 42, 43, 44, \ldots, 40 + n \ldots\}$, which was produced by deleting 40 from set V. To establish a one-to-one correspondence between V and P, consider the following diagram.

$$\begin{array}{l} V = \{40, 41, 42, 43, \ldots, 39 + n, \ldots\} \\ \qquad\;\; \updownarrow\;\;\;\updownarrow\;\;\;\updownarrow\;\;\;\updownarrow \qquad\quad \updownarrow \\ P = \{41, 42, 43, 44, \ldots, 40 + n, \ldots\} \end{array}$$

In the above correspondence, each element of the form $39 + n$ from set V is paired with an element of the form $40 + n$ from set P. The general correspondence $(39 + n) \leftrightarrow (40 + n)$ establishes a one-to-one correspondence between V and P. Because V can be placed in a one-to-one correspondence with a proper subset of itself, V is an infinite set.

CHECK YOUR PROGRESS 3, *page 101*

The following figure shows that we can establish a one-to-one correspondence between M and the set of natural numbers N by pairing $\frac{1}{n + 1}$ of set M with n of set N.

$$\begin{array}{l} M = \left\{\frac{1}{2}, \frac{1}{3}, \frac{1}{4}, \frac{1}{5}, \ldots, \frac{1}{n + 1}, \ldots\right\} \\ \qquad\;\; \updownarrow\;\;\updownarrow\;\;\updownarrow\;\;\updownarrow \qquad\quad \updownarrow \\ N = \{\,1,\; 2,\; 3,\; 4,\; \ldots, \quad n, \quad \ldots\} \end{array}$$

Thus the cardinality of M must be the same as the cardinality of N, which is $\aleph_0$.

CHAPTER 3

SECTION 3.1

CHECK YOUR PROGRESS 1, *page 115*

a. The sentence "Open the door." is a command. It is not a statement.

b. The word *large* is not a precise term. It is not possible to determine whether the sentence "7055 is a large number" is true or false and thus the sentence is not a statement.

c. The sentence $4 + 5 = 8$ is a false statement.

d. At this time we do not know whether the given sentence is true or false, but we know that the sentence is either true or false and that it is not both true and false. Thus the sentence is a statement.

e. The sentence $x > 3$ is a statement because for any given value of x, the inequality $x > 3$ is true or false, but not both.

CHECK YOUR PROGRESS 2, *page 117*

a. 1001 is not divisible by 7.

b. 5 is not an even number.

c. That fire engine is red.

CHECK YOUR PROGRESS 3, *page 118*

a. $\sim p \wedge r$

b. $\sim s \wedge \sim r$

c. $r \leftrightarrow q$

d. $p \rightarrow \sim r$

CHECK YOUR PROGRESS 4, *page 118*

$e \wedge \sim t$: All men are created equal and I am not trading places.

$a \vee \sim t$: I get Abe's place or I am not trading places.

$e \rightarrow t$: If all men are created equal, then I am trading places.

$t \leftrightarrow g$: I am trading places if and only if I get George's place.

CHECK YOUR PROGRESS 5, *page 119*

a. True. A conjunction is true provided both components are true.

b. True. A disjunction is true provided at least one component is true.

c. False. If both components of a disjunction are false, then the disjunction is false.

CHECK YOUR PROGRESS 6, *page 120*

a. Some bears are not brown.

b. Some math classes are fun.

c. All vegetables are green.

SECTION 3.2

CHECK YOUR PROGRESS 1, *page 127*

a.

p	q	$\sim p$	$\sim q$	$p \wedge \sim q$	$\sim p \vee q$	$(p \wedge \sim q) \vee (\sim p \vee q)$	
T	T	F	F	F	T	T	Row 1
T	F	F	T	T	F	T	Row 2
F	T	T	F	F	T	T	Row 3
F	F	T	T	F	T	T	Row 4
		1	2	3	4	5	

b. p is true and q is false in row 2 of the above truth table. The truth value of $(p \wedge \sim q) \vee (\sim p \vee q)$ in row 2 is T (true).

CHECK YOUR PROGRESS 2, *page 127*

a.

p	q	r	$\sim p$	$\sim r$	$\sim p \wedge r$	$q \wedge \sim r$	$(\sim p \wedge r) \vee (q \wedge \sim r)$	
T	T	T	F	F	F	F	F	Row 1
T	T	F	F	T	F	T	T	Row 2
T	F	T	F	F	F	F	F	Row 3
T	F	F	F	T	F	F	F	Row 4
F	T	T	T	F	T	F	T	Row 5
F	T	F	T	T	F	T	T	Row 6
F	F	T	T	F	T	F	T	Row 7
F	F	F	T	T	F	F	F	Row 8
			1	2	3	4	5	

b. p is false, q is true, and r is false in row 6 of the above truth table. The truth value of $(\sim p \wedge r) \vee (q \wedge \sim r)$ in row 6 is T (true).

CHECK YOUR PROGRESS 3, *page 129*

The given statement has two simple statements. Thus you should use a standard form that has $2^2 = 4$ rows.

Step 1 Enter the truth values for each simple statement and their negations. See columns 1, 2, and 3 in the table on the right.

Step 2 Use the truth values in columns 2 and 3 to determine the truth values to enter under the "and" connective. See column 4 in the table on the right.

Step 3 Use the truth values in columns 1 and 4 to determine the truth values to enter under the "or" connective. See column 5 in the table on the right.

p	q	$\sim p$	$\vee$	$(p$	$\wedge$	$q)$
T	T	F	T	T	T	T
T	F	F	F	T	F	F
F	T	T	T	F	F	T
F	F	T	T	F	F	F
		1	5	2	4	3

The truth table for $\sim p \vee (p \wedge q)$ is displayed in column 5.

CHECK YOUR PROGRESS 4, *page 130*

p	q	p	$\vee$	$(p$	$\wedge$	$\sim q)$
T	T	T	T	T	F	F
T	F	T	T	T	T	T
F	T	F	F	F	F	F
F	F	F	F	F	F	T
		1	5	2	4	3

The above truth table shows that $p \equiv p \vee (p \wedge \sim q)$.

CHECK YOUR PROGRESS 5, *page 131*

Let d represent "I am going to the dance." Let g represent "I am going to the game." The original sentence in symbolic form is $\sim(d \wedge g)$. Applying one of De Morgan's laws, we find that $\sim(d \wedge g) \equiv \sim d \vee \sim g$. Thus an equivalent form of "It is not true that I am going to the dance and I am going to the game" is "I am not going to the dance or I am not going to the game."

CHECK YOUR PROGRESS 6, *page 131*

The following truth table shows that $p \wedge (\sim p \wedge q)$ is always false. Thus $p \wedge (\sim p \wedge q)$ is a self-contradiction.

p	q	p	$\wedge$	$(\sim p$	$\wedge$	$q)$
T	T	T	F	F	F	T
T	F	T	F	F	F	F
F	T	F	F	T	T	T
F	F	F	F	T	F	F
		1	5	2	4	3

SECTION 3.3

CHECK YOUR PROGRESS 1, *page 137*

a. *Antecedent:* I study for at least 6 hours
Consequent: I will get an A on the test

b. *Antecedent:* I get the job
Consequent: I will buy a new car

c. *Antecedent:* you can dream it
Consequent: you can do it

CHECK YOUR PROGRESS 2, *page 139*

a. Because the antecedent is true and the consequent is false, the statement is a false statement.

b. Because the antecedent is false, the statement is a true statement.

c. Because the consequent is true, the statement is a true statement.

CHECK YOUR PROGRESS 3, *page 139*

p	q	$[p$	$\wedge$	$(p$	$\rightarrow$	$q)]$	$\rightarrow$	q
T	T	T	T	T	T	T	T	T
T	F	T	F	T	F	F	T	F
F	T	F	F	F	T	T	T	T
F	F	F	F	F	T	F	T	F
		1	6	2	5	3	7	4

CHECK YOUR PROGRESS 4, *page 140*

a. I will move to Georgia or I will live in Houston.

b. The number is not divisible by 2 or the number is even.

CHECK YOUR PROGRESS 5, *page 141*

a. I finished the report and I did not go to the concert.

b. The square of n is 25 and n is not 5 or -5.

CHECK YOUR PROGRESS 6, *page 141*

a. Let $x = 6.5$. Then the first component of the biconditional is false and the second component of the biconditional is true. Thus the given biconditional statement is false.

b. Both components of the biconditional are true for $x > 2$, and both components are false for $x \leq 2$. Because both components have the same truth value for any real number x, the given biconditional is true.

SECTION 3.4

CHECK YOUR PROGRESS 1, *page 146*

a. If a geometric figure is a square, then it is a rectangle.

b. If I am older than 30, then I am at least 21.

CHECK YOUR PROGRESS 2, *page 147*

Converse: If we are not going to have a quiz tomorrow, then we will have a quiz today.

Inverse: If we don't have a quiz today, then we will have a quiz tomorrow.

Contrapositive: If we have a quiz tomorrow, then we will not have a quiz today.

CHECK YOUR PROGRESS 3, *page 148*

a. The second statement is the inverse of the first statement. Thus the statements are not equivalent. This can also be demonstrated by the fact that the first statement is true for $c = 0$ and the second statement is false for $c = 0$.

b. The second statement is the contrapositive of the first statement. Thus the statements are equivalent.

CHECK YOUR PROGRESS 4, *page 148*

a. *Contrapositive:* If x is an odd integer, then $3 + x$ is an even integer. The contrapositive is true and so the original statement is also true.

b. *Contrapositive:* If two triangles are congruent triangles, then the two triangles are similar triangles. The contrapositive is true and so the original statement is also true.

c. *Contrapositive:* If tomorrow is Thursday, then today is Wednesday. The contrapositive is true and so the original statement is also true.

SECTION 3.5

CHECK YOUR PROGRESS 1, *page 153*
Let p represent the statement "She got on the plane." Let r represent the statement "She will regret it." Then the symbolic form of the argument is

$$\begin{array}{l} \sim p \rightarrow r \\ \underline{\sim r} \\ \therefore p \end{array}$$

CHECK YOUR PROGRESS 2, *page 155*
Let r represent the statement "The stock market rises." Let f represent the statement "The bond market will fall." Then the symbolic form of the argument is

$$\begin{array}{l} r \rightarrow f \\ \underline{\sim f} \\ \therefore \sim r \end{array}$$

The truth table for this argument is as follows:

		First premise	Second premise	Conclusion	
r	f	$r \rightarrow f$	$\sim f$	$\sim r$	
T	T	T	F	F	Row 1
T	F	F	T	F	Row 2
F	T	T	F	T	Row 3
F	F	T	T	T	Row 4

Row 4 is the only row in which all the premises are true, so it is the only row that we examine. Because the conclusion is true in row 4, the argument is valid.

CHECK YOUR PROGRESS 3, *page 156*
Let a represent the statement "I arrive before 8 A.M." Let f represent the statement "I will make the flight." Let p represent the statement "I will give the presentation." Then the symbolic form of the argument is

$$\begin{array}{l} a \rightarrow f \\ \underline{f \rightarrow p} \\ \therefore a \rightarrow p \end{array}$$

The truth table for this argument is as follows:

			First premise	Second premise	Conclusion	
a	f	p	$a \rightarrow f$	$f \rightarrow p$	$a \rightarrow p$	
T	T	T	T	T	T	Row 1
T	T	F	T	F	F	Row 2
T	F	T	F	T	T	Row 3
T	F	F	F	T	F	Row 4
F	T	T	T	T	T	Row 5
F	T	F	T	F	T	Row 6
F	F	T	T	T	T	Row 7
F	F	F	T	T	T	Row 8

The only rows in which all the premises are true are rows 1, 5, 7, and 8. In each of these rows the conclusion is also true. Thus the argument is a valid argument.

CHECK YOUR PROGRESS 4, *page 157*
Let f represent "I go to Florida for spring break." Let $\sim s$ represent "I will not study." Then the symbolic form of the argument is

$$\begin{array}{l} f \rightarrow \sim s \\ \underline{\sim f} \\ \therefore s \end{array}$$

This argument has the form of the fallacy of the inverse. Thus the argument is invalid.

CHECK YOUR PROGRESS 5, *page 158*
Let r represent "I read a math book." Let f represent "I start to fall asleep." Let d represent "I drink a soda." Let e represent "I eat a candy bar." Then the symbolic form of the argument is

$$\begin{array}{l} r \rightarrow f \\ f \rightarrow d \\ \underline{d \rightarrow e} \\ \therefore r \rightarrow e \end{array}$$

The argument has the form of the extended law of syllogism. Thus the argument is valid.

CHECK YOUR PROGRESS 6, *page 159*
We are given the following premises:

$$\begin{array}{l} \sim m \vee t \\ t \rightarrow \sim d \\ e \vee g \\ \underline{e \rightarrow d} \\ \therefore ? \end{array}$$

The first premise can be written as $m \rightarrow t$, the third premise can be written as $\sim e \rightarrow g$, and the fourth premise can be written as $\sim d \rightarrow \sim e$. Thus the argument can be expressed in the following equivalent form.

$$\begin{array}{l} m \rightarrow t \\ t \rightarrow \sim d \\ \sim e \rightarrow g \\ \underline{\sim d \rightarrow \sim e} \\ \therefore ? \end{array}$$

If we switch the order of the third and fourth premises, then we have the following equivalent form.

$$\begin{array}{l} m \rightarrow t \\ t \rightarrow \sim d \\ \sim d \rightarrow \sim e \\ \underline{\sim e \rightarrow g} \\ \therefore ? \end{array}$$

An application of the extended law of syllogism produces $m \rightarrow g$ as a valid conclusion for the argument. *Note:* Although $m \rightarrow \sim e$ is also a valid conclusion for the argument, we do not list it as our answer because it can be obtained without using all of the given premises.

SECTION 3.6

CHECK YOUR PROGRESS 1, *page 165*
The following Euler diagram shows that the argument is valid.

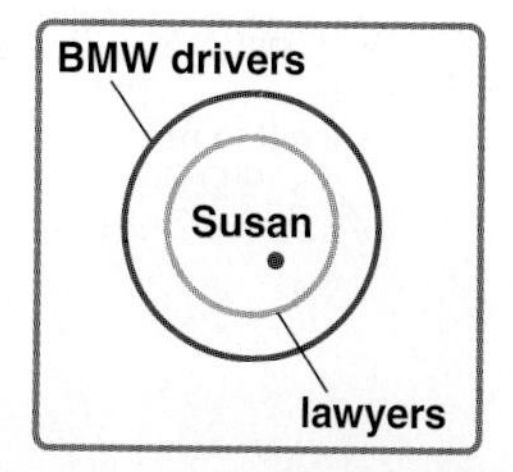

CHECK YOUR PROGRESS 2, *page 166*
From the given premises we can conclude that 7 may or may not be a prime number. Thus the argument is invalid.

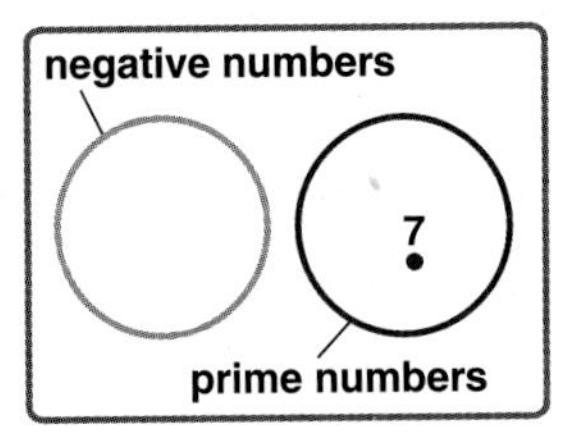

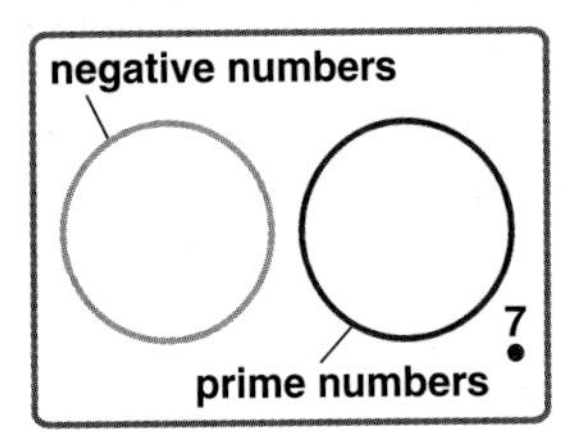

CHECK YOUR PROGRESS 3, *page 167*
From the given premises we can construct two possible Euler diagrams.

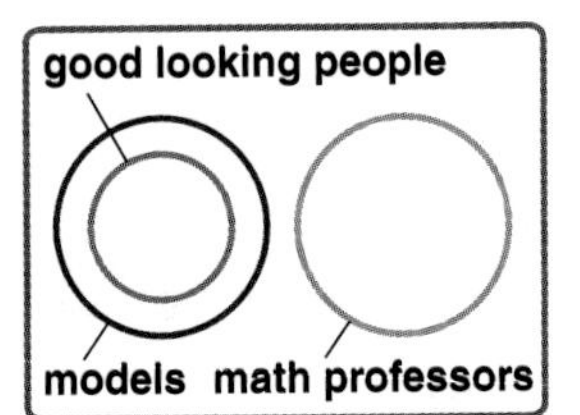

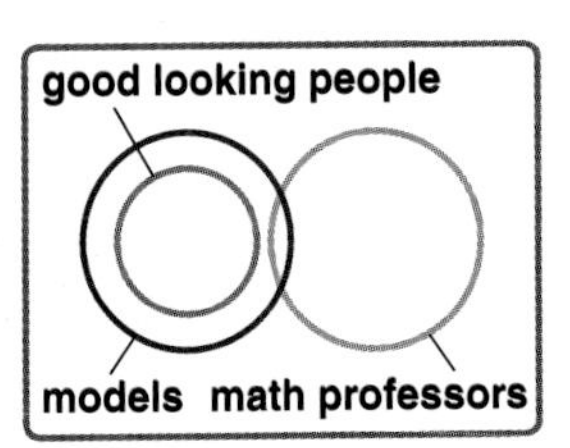

From the rightmost Euler diagram we can determine that the argument is invalid.

CHECK YOUR PROGRESS 4, *page 168*
The following Euler diagram illustrates that all squares are quadrilaterals, so the argument is a valid argument.

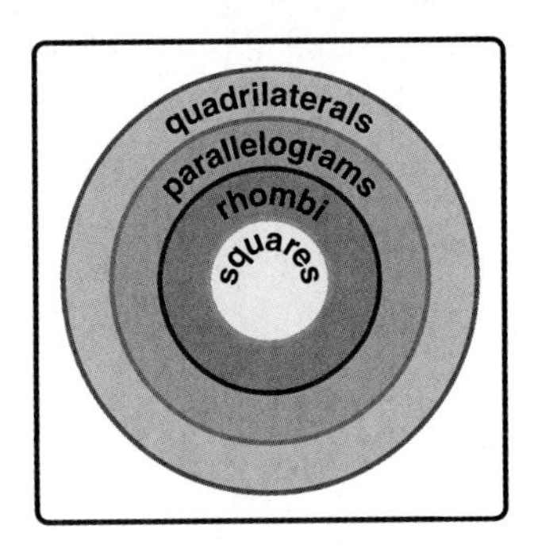

CHECK YOUR PROGRESS 5, *page 169*
The following Euler diagrams illustrate two possible cases. In both cases we see that all white rabbits like tomatoes.

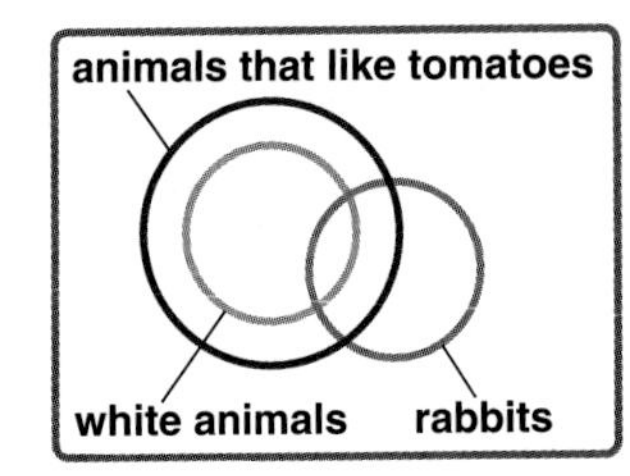

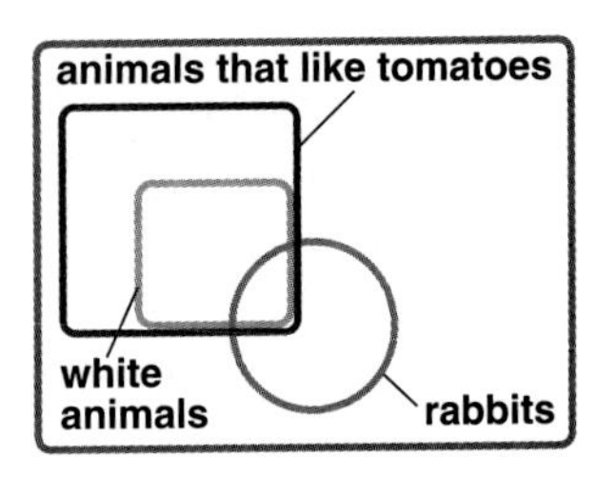

CHAPTER 4

SECTION 4.1

CHECK YOUR PROGRESS 1, *page 179*

𓍢𓍢𓎆𓎆𓎆
𓆐𓆼𓍢𓍢𓎆𓎆𓎆𓎆𓏺𓏺𓏺

CHECK YOUR PROGRESS 2, *page 179*

$(1 \times 1{,}000{,}000) + (3 \times 100{,}000) + (1 \times 10{,}000) + (4 \times 1000)$
$+ (3 \times 100) + (2 \times 10) + (1 \times 6) = 1{,}314{,}326$

CHECK YOUR PROGRESS 3, *page 180*

23,341	𓂭𓂭𓆼𓆼𓆼𓍢𓍢𓍢𓎆𓎆𓎆𓎆𓏺
+ 10,562	+ 𓂭𓍢𓍢𓍢𓍢𓍢𓎆𓎆𓎆𓎆𓎆𓎆𓏺𓏺

𓍢𓍢𓍢 𓎆𓎆𓎆𓎆
𓂭𓂭𓂭𓆼𓆼𓆼𓍢𓍢𓍢𓍢𓍢𓎆𓎆𓎆𓎆𓎆𓎆𓏺𓏺𓏺

Replace 10 heel bones with one scroll to produce:

𓍢𓍢𓍢𓍢
𓂭𓂭𓂭𓆼𓆼𓆼𓍢𓍢𓍢𓍢𓍢𓏺𓏺𓏺

which is 33,903.

CHECK YOUR PROGRESS 4, *page 180*

61,432	𓂭𓂭𓂭𓂭𓂭𓂭𓆼𓍢𓍢𓍢𓍢𓎆𓎆𓎆𓏺𓏺
− 45,121	− 𓂭𓂭𓂭𓂭𓆼𓆼𓆼𓆼𓆼𓍢𓎆𓎆𓏺

Replace one pointing finger with 10 lotus flowers to produce:

𓂭𓂭𓂭𓂭𓂭𓆼𓆼𓆼𓆼𓆼𓆼𓆼𓆼𓆼𓆼𓆼𓍢𓍢𓍢𓍢𓎆𓎆𓎆𓏺𓏺
− 𓂭𓂭𓂭𓂭𓆼𓆼𓆼𓆼𓆼𓍢𓎆𓎆𓏺
𓂭𓆼𓆼𓆼𓆼𓆼𓆼𓍢𓍢𓍢𓎆𓏺

which is 16,311.

CHECK YOUR PROGRESS 5, *page 182*

MCDXLV = M + (CD) + (XL) + V
= 1000 + 400 + 40 + 5 = 1445

CHECK YOUR PROGRESS 6, *page 182*

473 = 400 + 70 + 3 = CD + LXX + III = CDLXXIII

CHECK YOUR PROGRESS 7, *page 183*

a. $\overline{\text{VII}}$CCLIV = $\overline{\text{VII}}$ + CCLIV = 7000 + 254 = 7254

b. 8070 = 8000 + 70 = $\overline{\text{VIII}}$ + LXX = $\overline{\text{VIII}}$LXX

SECTION 4.2

CHECK YOUR PROGRESS 1, *page 187*

$17{,}325 = 10{,}000 + 7000 + 300 + 20 + 5$
$= (1 \times 10{,}000) + (7 \times 1000) + (3 \times 100) + (2 \times 10) + 5$
$= (1 \times 10^4) + (7 \times 10^3) + (3 \times 10^2) + (2 \times 10^1) + (5 \times 10^0)$

CHECK YOUR PROGRESS 2, *page 188*

$(5 \times 10^4) + (9 \times 10^3) + (2 \times 10^2) + (7 \times 10^1) + (4 \times 10^0)$
$= (5 \times 10{,}000) + (9 \times 1000) + (2 \times 100) + (7 \times 10) + (4 \times 1)$
$= 50{,}000 + 9000 + 200 + 70 + 4$
$= 59{,}274$

CHECK YOUR PROGRESS 3, *page 188*

$$\begin{array}{rl} 152 = & (1 \times 100) + (5 \times 10) + 2 \\ +\ 234 = & (2 \times 100) + (3 \times 10) + 4 \\ \hline & (3 \times 100) + (8 \times 10) + 6 = 386 \end{array}$$

CHECK YOUR PROGRESS 4, *page 189*

$$\begin{array}{rl} 147 = & (1 \times 100) + (4 \times 10) + 7 \\ +\ 329 = & (3 \times 100) + (2 \times 10) + 9 \\ \hline & (4 \times 100) + (6 \times 10) + 16 \end{array}$$

Replace 16 with $(1 \times 10) + 6$

$= (4 \times 100) + (6 \times 10) + (1 \times 10) + 6$
$= (4 \times 100) + (7 \times 10) + 6$
$= 476$

CHECK YOUR PROGRESS 5, *page 189*

$$\begin{array}{rl} 382 = & (3 \times 100) + (8 \times 10) + 2 \\ -\ 157 = & (1 \times 100) + (5 \times 10) + 7 \end{array}$$

Because $7 > 2$, it is necessary to borrow by rewriting (8×10) as $(7 \times 10) + 10$.

$$\begin{array}{rl} 382 = & (3 \times 100) + (7 \times 10) + 12 \\ -\ 157 = & (1 \times 100) + (5 \times 10) + 7 \\ \hline & (2 \times 100) + (2 \times 10) + 5 = 225 \end{array}$$

CHECK YOUR PROGRESS 6, *page 190*

<<Y YYYYY <<<YYYY

$= (21 \times 60^2) + (5 \times 60) + (34 \times 1)$
$= 75{,}600 + 300 + 34 = 75{,}934$

CHECK YOUR PROGRESS 7, *page 191*

$$\begin{array}{r} 3 \\ 3600\overline{)12578} \\ \underline{10800} \\ 1778 \end{array} \qquad \begin{array}{r} 29 \\ 60\overline{)1778} \\ \underline{120} \\ 578 \\ \underline{540} \\ 38 \end{array}$$

Thus $12{,}578 = (3 \times 60^2) + (29 \times 60) + (38 \times 1) =$

YYY <<YYYYYYYYY <<<YYYYYYYY

CHECK YOUR PROGRESS 8, *page 191*

Combine the symbols for each place value.

Replace ten Ys in the 1s' place with a <.

<<<<<<<YYYYYY

Take away 60 from the ones' place and add 1 to the 60s' place.

<<<<<<<YYYYYYY

Take away 60 from the 60s' place and add 1 to the 60^2 place.

Y <YYYYYYY <<<<<YYY

Thus

	<<<YY	<<<<<YYYYYY
+	<<<<YYYY	<<<<<YYYYYYYY
Y	<YYYYYYY	<<<<<YYY

CHECK YOUR PROGRESS 9, *page 193*

a. $(16 \times 360) + (0 \times 20) + (1 \times 1) = 5761$

b. $(9 \times 7200) + (1 \times 360) + (10 \times 20) + (4 \times 1) = 65{,}364$

CHECK YOUR PROGRESS 10, *page 194*

$$\begin{array}{r} 1 \\ 7200\overline{)11480} \\ \underline{7200} \\ 4280 \end{array} \qquad \begin{array}{r} 11 \\ 360\overline{)4280} \\ \underline{360} \\ 680 \\ \underline{360} \\ 320 \end{array} \qquad \begin{array}{r} 16 \\ 20\overline{)320} \\ \underline{20} \\ 120 \\ \underline{120} \\ 0 \end{array}$$

Thus $11{,}480 = (1 \times 7200) + (11 \times 360) + (16 \times 20) + (0 \times 1)$. In Mayan numerals this is

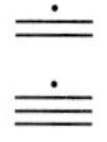

SECTION 4.3

CHECK YOUR PROGRESS 1, *page 198*

$3156_{\text{seven}} = (3 \times 7^3) + (1 \times 7^2) + (5 \times 7^1) + (6 \times 7^0)$
$= (3 \times 343) + (1 \times 49) + (5 \times 7) + (6 \times 1)$
$= 1029 + 49 + 35 + 6$
$= 1119$

CHECK YOUR PROGRESS 2, *page 199*

$111000101_{\text{two}} = (1 \times 2^8) + (1 \times 2^7) + (1 \times 2^6) + (0 \times 2^5) + (0 \times 2^4) + (0 \times 2^3) + (1 \times 2^2) + (0 \times 2^1) + (1 \times 2^0)$
$= (1 \times 256) + (1 \times 128) + (1 \times 64) + (0 \times 32) + (0 \times 16) + (0 \times 8) + (1 \times 4) + (0 \times 2) + (1 \times 1)$
$= 256 + 128 + 64 + 0 + 0 + 0 + 4 + 0 + 1$
$= 453$

CHECK YOUR PROGRESS 3, *page 199*

$$\begin{aligned} A5B_{twelve} &= (10 \times 12^2) + (5 \times 12^1) + (11 \times 12^0) \\ &= 1440 + 60 + 11 \\ &= 1511 \end{aligned}$$

CHECK YOUR PROGRESS 4, *page 200*

$$\begin{aligned} C24F_{sixteen} &= (12 \times 16^3) + (2 \times 16^2) + (4 \times 16^1) + (15 \times 16^0) \\ &= 49{,}152 + 512 + 64 + 15 \\ &= 49{,}743 \end{aligned}$$

CHECK YOUR PROGRESS 5, *page 201*

a.

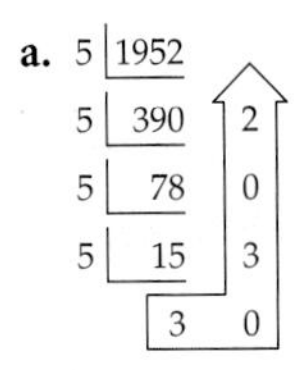

$1952 = 30302_{five}$

b.

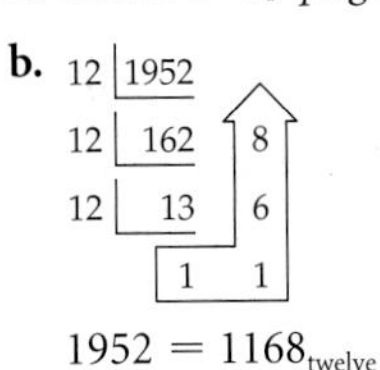

$1952 = 1168_{twelve}$

CHECK YOUR PROGRESS 6, *page 202*

$$\begin{array}{ccccc} 6 & 3 & 2 & 1 & 0_{eight} \\ \| & \| & \| & \| & \| \\ 110 & 011 & 010 & 001 & 000_{two} \end{array}$$

$63210_{eight} = 110011010001000_{two}$

CHECK YOUR PROGRESS 7, *page 202*

$$\begin{array}{cccc} 111 & 010 & 011 & 100_{two} \\ \| & \| & \| & \| \\ 7 & 2 & 3 & 4_{eight} \end{array}$$

$111010011100_{two} = 7234_{eight}$

CHECK YOUR PROGRESS 8, *page 203*

$$\begin{array}{ccc} C & 5 & A_{sixteen} \\ \| & \| & \| \\ 1100 & 0101 & 1010_{two} \end{array}$$

$C5A_{sixteen} = 110001011010_{two}$

CHECK YOUR PROGRESS 9, *page 203*

Insert a zero to make a group of four.

$$\begin{array}{cccc} 0101 & 0001 & 1101 & 0010_{two} \\ \| & \| & \| & \| \\ 5 & 1 & D & 2_{sixteen} \end{array}$$

$101000111010010_{two} = 51D2_{sixteen}$

CHECK YOUR PROGRESS 10, *page 204*

	dabble 3	dabble 7	double 14	double 28	dabble 57	double 114
1	1	1	0	0	1	0_{two}

$1110010_{two} = 114$

SECTION 4.4

CHECK YOUR PROGRESS 1, *page 209*

$$\begin{array}{rr} & 1\,1\quad 1 \\ & 1\,1\,0\,0\,1_{two} \\ + & 1\,1\,0\,1_{two} \\ \hline & 1\,0\,0\,1\,1\,0_{two} \end{array}$$

CHECK YOUR PROGRESS 2, *page 210*

$$\begin{array}{rr} & 1\,1 \\ & 3\,2_{four} \\ + & 1\,2_{four} \\ \hline & 1\,1\,0_{four} \end{array}$$

CHECK YOUR PROGRESS 3, *page 210*

$$\begin{array}{rr} & 1\,2 \\ & 3\,5_{seven} \\ & 4\,6_{seven} \\ + & 2\,4_{seven} \\ \hline & 1\,4\,1_{seven} \end{array}$$

CHECK YOUR PROGRESS 4, *page 211*

$$\begin{array}{rr} & 1\,1 \\ & A\,C\,4_{sixteen} \\ + & 6\,E\,8_{sixteen} \\ \hline & 1\,1\,A\,C_{sixteen} \end{array}$$

CHECK YOUR PROGRESS 5, *page 212*

$$\begin{array}{rr} & 2 + 1 \\ & \not{3}\,6\,5_{nine} \\ - & 1\,8\,3_{nine} \\ \hline \end{array}$$

Because $8_{nine} > 6_{nine}$, it is necessary to borrow from the 3 in the first column at the left.

$$\begin{array}{rr} & 10 \\ & 2\,6\,5_{nine} \\ - & 1\,8\,3_{nine} \\ \hline \end{array}$$

Borrow 1 nine from the first column and add $9 = 10_{nine}$ to the 6 in the middle column.

$$\begin{array}{rr} & 16 \\ & 2\,\not{6}\,5_{nine} \\ - & 1\,8\,3_{nine} \\ \hline & 1\,7\,2_{nine} \end{array}$$

$$\begin{aligned} 16_{nine} - 8_{nine} &= 15 - 8 \\ &= 7 \\ &= 7_{nine} \end{aligned}$$

CHECK YOUR PROGRESS 6, *page 213*

$$\begin{array}{rccc} & 7 & 10 & \\ & \not{8} & 3 & A_{twelve} \\ - & 4 & 6 & 7_{twelve} \\ \hline & 3 & 9 & 3_{twelve} \end{array}$$

- $A_{twelve} - 7_{twelve} = 10 - 7 = 3 = 3_{twelve}$
- $10_{twelve} + 3_{twelve} = 13_{twelve} = 15$ $15 - 6 = 9 = 9_{twelve}$
- $7_{twelve} - 4_{twelve} = 3_{twelve}$

CHECK YOUR PROGRESS 7, *page 213*

$$\begin{array}{rcccc} & & & 1 & \\ & & 2 & 1 & 3_{four} \\ \times & & & & 2_{four} \\ \hline & 1 & 0 & 3 & 2_{four} \end{array}$$

- $2_{four} \times 3_{four} = 12_{four}$
- $2_{four} \times 1_{four} + 1_{four} = 3_{four}$
- $2_{four} \times 2_{four} = 10_{four}$

CHECK YOUR PROGRESS 8, *page 215*

$$\begin{array}{rrrl} & & 2 & \\ & & 3 & 4_{\text{eight}} \\ \times & & 2 & 5_{\text{eight}} \\ \hline & 2 & 1 & 4_{\text{eight}} \\ \hline \end{array}$$

- $5_{\text{eight}} \times 4_{\text{eight}} = 20 = 24_{\text{eight}}$
- $5_{\text{eight}} \times 3_{\text{eight}} + 2_{\text{eight}} = 15 + 2 = 17 = 21_{\text{eight}}$

$$\begin{array}{rrrl} & & 1 & \\ & & 3 & 4_{\text{eight}} \\ \times & & 2 & 5_{\text{eight}} \\ \hline & 2 & 1 & 4_{\text{eight}} \\ & 7 & 0_{\text{eight}} & \\ \hline 1 & 1 & 1 & 4_{\text{eight}} \end{array}$$

- $2_{\text{eight}} \times 4_{\text{eight}} = 8 = 10_{\text{eight}}$
- $2_{\text{eight}} \times 3_{\text{eight}} + 1_{\text{eight}} = 6 + 1 = 7 = 7_{\text{eight}}$

CHECK YOUR PROGRESS 9, *page 216*

First list a few multiples of 3_{five}.

$3_{\text{five}} \times 0_{\text{five}} = 0_{\text{five}}$

$3_{\text{five}} \times 1_{\text{five}} = 3_{\text{five}}$

$3_{\text{five}} \times 2_{\text{five}} = 11_{\text{five}}$

$3_{\text{five}} \times 3_{\text{five}} = 14_{\text{five}}$

$3_{\text{five}} \times 4_{\text{five}} = 22_{\text{five}}$

$$\begin{array}{r} 1 \\ 3_{\text{five}}\overline{)3\,2\,4_{\text{five}}} \\ \underline{3} \\ 2 \end{array} \qquad \begin{array}{r} 1\,0 \\ 3_{\text{five}}\overline{)3\,2\,4_{\text{five}}} \\ \underline{3} \\ 2 \\ \underline{0} \\ 2\,4 \end{array} \qquad \begin{array}{r} 1\,0\,4 \\ 3_{\text{five}}\overline{)3\,2\,4_{\text{five}}} \\ \underline{3} \\ 2 \\ \underline{0} \\ 2\,4 \\ \underline{2\,2} \\ 2 \end{array}$$

Thus $324_{\text{five}} \div 3_{\text{five}} = 104_{\text{five}}$ with a remainder of 2_{five}.

CHECK YOUR PROGRESS 10, *page 216*

The divisor is 10_{two}. The multiples of the divisor are $10_{\text{two}} \times 0_{\text{two}} = 0_{\text{two}}$ and $10_{\text{two}} \times 1_{\text{two}} = 10_{\text{two}}$.

$$\begin{array}{r} 1\,1\,1\,0\,0\,1_{\text{two}} \\ 10_{\text{two}}\overline{)1\,1\,1\,0\,0\,1\,1_{\text{two}}} \\ \underline{1\,0} \\ 1\,1 \\ \underline{1\,0} \\ 1\,0 \\ \underline{1\,0} \\ 0\,0 \\ \underline{0} \\ 0\,1 \\ \underline{0} \\ 1\,1 \\ \underline{1\,0} \\ 1 \end{array}$$

Thus $1110011_{\text{two}} \div 10_{\text{two}} = 111001_{\text{two}}$ with a remainder of 1_{two}.

SECTION 4.5

CHECK YOUR PROGRESS 1, *page 220*

a. Divide 9 by 1, 2, 3, . . . , 9 to determine that the only natural number divisors of 9 are 1, 3, and 9.

b. Divide 11 by 1, 2, 3, . . . , 11 to determine that the only natural number divisors of 11 are 1 and 11.

c. Divide 24 by 1, 2, 3, . . . , 24 to determine that the only natural number divisors of 24 are 1, 2, 3, 4, 6, 8, 12, and 24.

CHECK YOUR PROGRESS 2, *page 221*

a. The only divisors of 47 are 1 and 47. Thus 47 is a prime number.

b. 171 is divisible by 3, 9, 19, and 57. Thus 171 is a composite number.

c. The divisors of 91 are 1, 7, 13, and 91. Thus 91 is a composite number.

CHECK YOUR PROGRESS 3, *page 223*

a. The sum of the digits of 341,565 is 24; therefore, 341,565 is divisible by 3.

b. The number 341,565 is not divisible by 4 because the number formed by last two digits, 65, is not divisible by 4.

c. The number 341,565 is not divisible by 10 because it does not end in 0.

d. The sum of the digits with even place-value powers is 14. The sum of the digits with odd place-value powers is 10. The difference of these sums is 4. Thus 341,565 is not divisible by 11.

CHECK YOUR PROGRESS 4, *page 224*

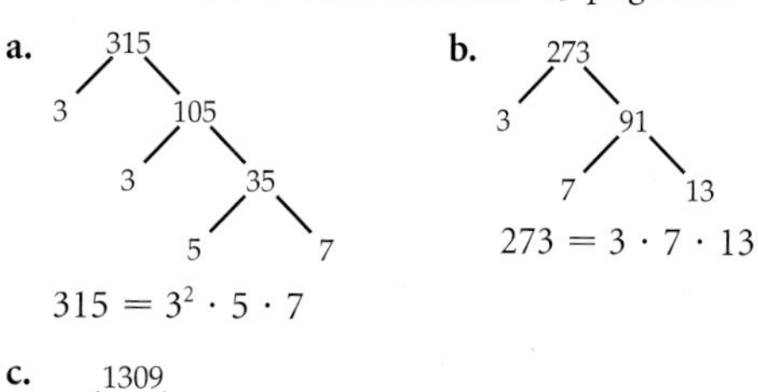

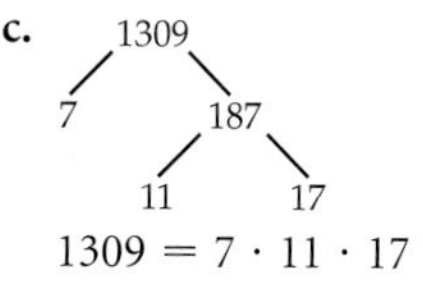

SECTION 4.6

CHECK YOUR PROGRESS 1, *page 232*

a. The proper factors of 24 are 1, 2, 3, 4, 6, 8, and 12. The sum of these proper factors is 36. Because 24 is less than the sum of its proper factors, 24 is an abundant number.

b. The proper factors of 28 are 1, 2, 4, 7, and 14. The sum of these proper factors is 28. Because 28 equals the sum of its proper factors, 28 is a perfect number.

c. The proper factors of 35 are 1, 5, and 7. The sum of these proper factors is 13. Because 35 is larger than the sum of its proper factors, 35 is a deficient number.

CHECK YOUR PROGRESS 2, *page 232*

$2^7 - 1 = 127$, which is a prime number.

CHECK YOUR PROGRESS 3, *page 233*

The exponent $n = 61$ is a prime number and we are given that $2^{61} - 1$ is a prime number, so the perfect number we seek is $2^{60}(2^{61} - 1)$.

CHECK YOUR PROGRESS 4, *page 236*

First consider $2^{2976221}$. The base b is 2. The exponent x is 2,976,221.

$$\begin{aligned}(x \log b) + 1 &= (2{,}976{,}221 \log 2) + 1 \\ &\approx 895{,}931.8 + 1 \\ &= 895{,}932.8\end{aligned}$$

The greatest integer of 895,932.8 is 895,932. Thus $2^{2976221}$ has 895,932 digits. The number $2^{2976221}$ is not a power of 10, so the Mersenne number $2^{2976221} - 1$ also has 895,932 digits.

CHECK YOUR PROGRESS 5, *page 237*

Substituting 9 for x, 11 for y, and 4 for n in $x^n + y^n = z^n$ yields

$$\begin{aligned}9^4 + 11^4 &= z^4 \\ 6561 + 14{,}641 &= z^4 \\ 21{,}202 &= z^4\end{aligned}$$

The real solution of $z^4 = 21{,}202$ is $\sqrt[4]{21{,}202} \approx 12.066858$, which is not a natural number. Thus $x = 9$, $y = 11$, and $n = 4$ do not satisfy the equation $x^n + y^n = z^n$, where z is a natural number.

CHAPTER 5

SECTION 5.1

CHECK YOUR PROGRESS 1, *page 249*

a.
$$\begin{aligned}c - 6 &= -13 \\ c - 6 + 6 &= -13 + 6 \\ c &= -7\end{aligned}$$

The solution is -7.

b.
$$\begin{aligned}4 &= -8z \\ \frac{4}{-8} &= \frac{-8z}{-8} \\ -\frac{1}{2} &= z\end{aligned}$$

The solution is $-\frac{1}{2}$.

c.
$$\begin{aligned}22 + m &= -9 \\ 22 - 22 + m &= -9 - 22 \\ m &= -31\end{aligned}$$

The solution is -31.

d.
$$\begin{aligned}5x &= 0 \\ \frac{5x}{5} &= \frac{0}{5} \\ x &= 0\end{aligned}$$

The solution is 0.

CHECK YOUR PROGRESS 2, *page 251*

a.
$$\begin{aligned}4x + 3 &= 7x + 9 \\ 4x - 7x + 3 &= 7x - 7x + 9 \\ -3x + 3 &= 9 \\ -3x + 3 - 3 &= 9 - 3 \\ -3x &= 6 \\ \frac{-3x}{-3} &= \frac{6}{-3} \\ x &= -2\end{aligned}$$

The solution is -2.

b.
$$\begin{aligned}7 - (5x - 8) &= 4x + 3 \\ 7 - 5x + 8 &= 4x + 3 \\ 15 - 5x &= 4x + 3 \\ 15 - 5x - 4x &= 4x - 4x + 3 \\ 15 - 9x &= 3 \\ 15 - 15 - 9x &= 3 - 15 \\ -9x &= -12 \\ \frac{-9x}{-9} &= \frac{-12}{-9} \\ x &= \frac{4}{3}\end{aligned}$$

The solution is $\frac{4}{3}$.

c.
$$\frac{3x-1}{4}+\frac{1}{3}=\frac{7}{3}$$
$$12\left(\frac{3x-1}{4}+\frac{1}{3}\right)=12\left(\frac{7}{3}\right)$$
$$12\cdot\frac{3x-1}{4}+12\cdot\frac{1}{3}=12\cdot\frac{7}{3}$$
$$9x-3+4=28$$
$$9x+1=28$$
$$9x+1-1=28-1$$
$$9x=27$$
$$\frac{9x}{9}=\frac{27}{9}$$
$$x=3$$

The solution is 3.

CHECK YOUR PROGRESS 3, *page 252*

a. $P=0.05Y-95$
$P=0.05(1990)-95$
$P=99.5-95$
$P=4.5$

The amount of garbage was about 4.5 pounds per day.

b.
$$P=0.05Y-95$$
$$5.6=0.05Y-95$$
$$5.6+95=0.05Y-95+95$$
$$100.6=0.05Y$$
$$\frac{100.6}{0.05}=\frac{0.05Y}{0.05}$$
$$2012=Y$$

The year will be 2012.

CHECK YOUR PROGRESS 4, *page 253*

\$17.50 for the first three lines + \$2.50 for each additional line	=	**\$30**

Let $L=$ the number of lines in the ad.

$$17.50+2.50(L-3)=30$$
$$17.50+2.50L-7.50=30$$
$$10.00+2.50L=30$$
$$10.00-10.00+2.50L=30-10.00$$
$$2.50L=20$$
$$\frac{2.50L}{2.50}=\frac{20}{2.50}$$
$$L=8$$

You can place an eight-line ad.

CHECK YOUR PROGRESS 5, *page 254*

Let $n=$ the number of years.

The 1990 population of Vermont plus an annual increase times *n*	=	**The 1990 population of North Dakota minus an annual decrease times *n***

$$562{,}576+5116n=638{,}800-1370n$$
$$562{,}576+5116n+1370n=638{,}800-1370n+1370n$$
$$562{,}576+6486n=638{,}800$$
$$562{,}576-562{,}576+6486n=638{,}800-562{,}576$$
$$6486n=76{,}224$$
$$\frac{6486n}{6486}=\frac{76{,}224}{6486}$$
$$n\approx 12$$

$1990+12=2002$

The populations would be the same in 2002.

CHECK YOUR PROGRESS 6, *page 256*

a.
$$s=\frac{A+L}{2}$$
$$2\cdot s=2\cdot\frac{A+L}{2}$$
$$2s=A+L$$
$$2s-A=A-A+L$$
$$2s-A=L$$

b.
$$L=a(1+ct)$$
$$\frac{L}{a}=\frac{a(1+ct)}{a}$$
$$\frac{L}{a}=1+ct$$
$$\frac{L}{a}-1=1-1+ct$$
$$\frac{L}{a}-1=ct$$
$$\frac{\frac{L}{a}-1}{t}=\frac{ct}{t}$$
$$\frac{\frac{L}{a}-1}{t}=c$$
$$\left(\frac{L}{a}-1\right)\left(\frac{1}{t}\right)=c$$
$$\frac{L}{at}-\frac{1}{t}=c$$

SECTION 5.2

CHECK YOUR PROGRESS 1, *page 264*

$4.92 \div 1.5 = 3.28$

$$\frac{\$4.92}{1.5 \text{ pounds}} = \frac{\$3.28}{1 \text{ pound}} = \$3.28/\text{pound}$$

The hamburger costs \$3.28 per pound.

CHECK YOUR PROGRESS 2, *page 265*

Find the difference in the hourly wage.

$\$6.75 - \$5.15 = \$1.60$

Multiply the difference in the hourly wage by 35.

$\$1.60(35) = \$56$

An employee's pay for working 35 hours and earning the California minimum wage is \$56 greater.

CHECK YOUR PROGRESS 3, *page 265*

$$\frac{\$2.99}{32 \text{ ounces}} \approx \frac{\$.093}{1 \text{ ounce}} \qquad \frac{\$3.99}{48 \text{ ounces}} \approx \frac{\$.083}{1 \text{ ounce}}$$

$\$.093 > \$.083$

The more economical purchase is 48 ounces of detergent for \$3.99.

CHECK YOUR PROGRESS 4, *page 267*

a. $20{,}000(1.3240) = 26{,}480$
26,480 Canadian dollars would be needed to pay for an order costing \$20,000.

b. $25{,}000(0.8110) = 20{,}275$
20,275 euros would be exchanged for \$25,000.

CHECK YOUR PROGRESS 5, *page 268*

a. $\frac{24 \text{ hours}}{1 \text{ day}} \cdot 7 \text{ days} = (24 \text{ hours})(7) = 168 \text{ hours}$

$$\frac{120 \text{ hours}}{1 \text{ week}} = \frac{120 \text{ hours}}{168 \text{ hours}} = \frac{120}{168} = \frac{5}{7}$$

The ratio is $\frac{5}{7}$.

b. $\frac{60 \text{ hours}}{(168 - 60) \text{ hours}} = \frac{60 \text{ hours}}{108 \text{ hours}} = \frac{60}{108} = \frac{5}{9}$

The ratio is 5 to 9.

CHECK YOUR PROGRESS 6, *page 269*

$6742 + 7710 = 14{,}452$

$$\frac{14{,}452}{798} \approx \frac{18.11}{1} \approx \frac{18}{1}$$

The ratio is 18 to 1.

CHECK YOUR PROGRESS 7, *page 271*

$$\frac{42}{x} = \frac{5}{8}$$
$$42 \cdot 8 = x \cdot 5$$
$$336 = 5x$$
$$\frac{336}{5} = \frac{5x}{5}$$
$$67.2 = x$$

The solution is 67.2.

CHECK YOUR PROGRESS 8, *page 272*

$$\frac{15 \text{ kilometers}}{2 \text{ centimeters}} = \frac{x \text{ kilometers}}{7 \text{ centimeters}}$$
$$\frac{15}{2} = \frac{x}{7}$$
$$15 \cdot 7 = 2 \cdot x$$
$$105 = 2x$$
$$\frac{105}{2} = \frac{2x}{2}$$
$$52.5 = x$$

The distance between the two cities is 52.5 kilometers.

CHECK YOUR PROGRESS 9, *page 273*

$$\frac{7}{5} = \frac{\$28{,}000}{x \text{ dollars}}$$
$$\frac{7}{5} = \frac{28{,}000}{x}$$
$$7 \cdot x = 5 \cdot 28{,}000$$
$$7x = 140{,}000$$
$$\frac{7x}{7} = \frac{140{,}000}{7}$$
$$x = 20{,}000$$

The other partner receives \$20,000.

CHECK YOUR PROGRESS 10, *page 274*

$$\frac{10.1 \text{ deaths}}{1{,}000{,}000 \text{ people}} = \frac{d \text{ deaths}}{4{,}000{,}000 \text{ people}}$$
$$10.1(4{,}000{,}000) = 1{,}000{,}000 \cdot d$$
$$40{,}400{,}000 = 1{,}000{,}000d$$
$$\frac{40{,}400{,}000}{1{,}000{,}000} = \frac{1{,}000{,}000d}{1{,}000{,}000}$$
$$40.4 = d$$

Approximately 40 people aged 5 to 34 die from asthma each year in New York City.

SECTION 5.3

CHECK YOUR PROGRESS 1, *page 284*

a. $74\% = 0.74$

b. $152\% = 1.52$

c. $8.3\% = 0.083$

d. $0.6\% = 0.006$

CHECK YOUR PROGRESS 2, *page 284*

a. $0.3 = 30\%$

b. $1.65 = 165\%$

c. $0.072 = 7.2\%$

d. $0.004 = 0.4\%$

CHECK YOUR PROGRESS 3, *page 285*

a. $8\% = 8\left(\frac{1}{100}\right) = \frac{8}{100} = \frac{2}{25}$

b. $180\% = 180\left(\frac{1}{100}\right) = \frac{180}{100} = 1\frac{80}{100} = 1\frac{4}{5}$

c. $2.5\% = 2.5\left(\frac{1}{100}\right) = \frac{2.5}{100} = \frac{25}{1000} = \frac{1}{40}$

d. $66\frac{2}{3}\% = \frac{200}{3}\% = \frac{200}{3}\left(\frac{1}{100}\right) = \frac{2}{3}$

CHECK YOUR PROGRESS 4, *page 286*

a. $\frac{1}{4} = 0.25 = 25\%$

b. $\frac{3}{8} = 0.375 = 37.5\%$

c. $\frac{5}{6} = 0.83\overline{3} = 83.\overline{3}\%$

d. $1\frac{2}{3} = 1.66\overline{6} = 166.\overline{6}\%$

CHECK YOUR PROGRESS 5, *page 287*

$$\frac{\text{Percent}}{100} = \frac{\text{amount}}{\text{base}}$$

$$\frac{70}{100} = \frac{22{,}400}{B}$$

$$70 \cdot B = 100(22{,}400)$$

$$70B = 2{,}240{,}000$$

$$\frac{70B}{70} = \frac{2{,}240{,}000}{70}$$

$$B = 32{,}000$$

The Blazer cost $32,000 when it was new.

CHECK YOUR PROGRESS 6, *page 288*

$$\frac{\text{Percent}}{100} = \frac{\text{amount}}{\text{base}}$$

$$\frac{p}{100} = \frac{416{,}000}{1{,}300{,}000}$$

$$p \cdot 1{,}300{,}000 = 100(416{,}000)$$

$$1{,}300{,}000p = 41{,}600{,}000$$

$$\frac{1{,}300{,}000p}{1{,}300{,}000} = \frac{41{,}600{,}000}{1{,}300{,}000}$$

$$p = 32$$

32% of the enlisted people are over the age of 30.

CHECK YOUR PROGRESS 7, *page 289*

$$\frac{\text{Percent}}{100} = \frac{\text{amount}}{\text{base}}$$

$$\frac{3.5}{100} = \frac{A}{32{,}500}$$

$$3.5(32{,}500) = 100(A)$$

$$113{,}750 = 100A$$

$$\frac{113{,}750}{100} = \frac{100A}{100}$$

$$1137.5 = A$$

The customer would receive a rebate of $1137.50.

CHECK YOUR PROGRESS 8, *page 289*

$$PB = A$$

$$0.05(32{,}685) = A$$

$$1634.25 = A$$

The teacher contributes $1634.25.

CHECK YOUR PROGRESS 9, *page 290*

$$PB = A$$

$$0.03B = 14{,}370$$

$$\frac{0.03B}{0.03} = \frac{14{,}370}{0.03}$$

$$B = 479{,}000$$

The selling price of the home was $479,000.

CHECK YOUR PROGRESS 10, *page 290*

$$PB = A$$

$$P \cdot 90 = 63$$

$$\frac{P \cdot 90}{90} = \frac{63}{90}$$

$$P = 0.7$$

$$P = 70\%$$

You answered 70% of the questions correctly.

CHECK YOUR PROGRESS 11, *page 291*

$$PB = A$$
$$0.90(21{,}262) = A$$
$$19{,}135.80 = A$$

$21{,}262 - 19{,}135.80 = 2126.20$

The difference between the cost of the remodeling and the increase in value of your home is \$2126.20.

CHECK YOUR PROGRESS 12, *page 293*

$5.67 - 1.82 = 3.85$

$$PB = A$$
$$P \cdot 1.82 = 3.85$$
$$\frac{P \cdot 1.82}{1.82} = \frac{3.85}{1.82}$$
$$P \approx 2.115$$

The percent increase in the federal debt from 1985 to 2000 was 211.5%.

CHECK YOUR PROGRESS 13, *page 295*

$$\frac{\text{Percent}}{100} = \frac{\text{amount}}{\text{base}}$$
$$\frac{3.81}{100} = \frac{A}{20{,}416}$$
$$3.81(20{,}416) = 100(A)$$
$$77{,}784.96 = 100A$$
$$\frac{77{,}784.96}{100} = \frac{100A}{100}$$
$$778 \approx A$$

$20{,}416 - 778 = 19{,}638$

There were 19,638 passenger car fatalities in the United States in 2003.

SECTION 5.4

CHECK YOUR PROGRESS 1, *page 307*

$$2s^2 = 6 - 4s$$
$$2s^2 + 4s = 6 - 4s + 4s$$
$$2s^2 + 4s = 6$$
$$2s^2 + 4s - 6 = 6 - 6$$
$$2s^2 + 4s - 6 = 0$$

CHECK YOUR PROGRESS 2, *page 308*

$$(n + 5)(2n - 3) = 0$$

$$n + 5 = 0 \qquad\qquad 2n - 3 = 0$$
$$n = -5 \qquad\qquad 2n = 3$$
$$n = \frac{3}{2}$$

Check:

$$\begin{array}{r|l} \multicolumn{2}{c}{(n + 5)(2n - 3) = 0} \\ \hline (-5 + 5)[2(-5) - 3] & 0 \\ 0(-13) & 0 \\ \multicolumn{2}{c}{0 = 0} \end{array} \qquad \begin{array}{r|l} \multicolumn{2}{c}{(n + 5)(2n - 3) = 0} \\ \hline \left(\frac{3}{2} + 5\right)\left(2 \cdot \frac{3}{2} - 3\right) & 0 \\ \frac{13}{2}(3 - 3) & 0 \\ \multicolumn{2}{c}{0 = 0} \end{array}$$

The solutions are -5 and $\frac{3}{2}$.

CHECK YOUR PROGRESS 3, *page 309*

$$2x^2 = x + 1$$
$$2x^2 - x = x - x + 1$$
$$2x^2 - x = 1$$
$$2x^2 - x - 1 = 1 - 1$$
$$2x^2 - x - 1 = 0$$

$$(2x + 1)(x - 1) = 0$$

$$2x + 1 = 0 \qquad\qquad x - 1 = 0$$
$$2x = -1 \qquad\qquad x = 1$$
$$x = -\frac{1}{2}$$

Check:

$$\begin{array}{r|l} \multicolumn{2}{c}{2x^2 = x + 1} \\ \hline 2\left(-\frac{1}{2}\right)^2 & -\frac{1}{2} + 1 \\ 2\left(\frac{1}{4}\right) & \frac{1}{2} \\ \multicolumn{2}{c}{\frac{1}{2} = \frac{1}{2}} \end{array} \qquad \begin{array}{r|l} \multicolumn{2}{c}{2x^2 = x + 1} \\ \hline 2(1)^2 & 1 + 1 \\ 2(1) & 2 \\ \multicolumn{2}{c}{2 = 2} \end{array}$$

The solutions are $-\frac{1}{2}$ and 1.

CHECK YOUR PROGRESS 4, *page 310*

$$2x^2 = 8x - 5$$
$$2x^2 - 8x + 5 = 0$$
$$a = 2,\ b = -8,\ c = 5$$
$$x = \frac{-b \pm \sqrt{b^2 - 4ac}}{2a}$$

$$x = \frac{-(-8) \pm \sqrt{(-8)^2 - 4(2)(5)}}{2(2)} = \frac{8 \pm \sqrt{64 - 40}}{4}$$
$$= \frac{8 \pm \sqrt{24}}{4} = \frac{8 \pm 2\sqrt{6}}{4} = \frac{2(4 \pm \sqrt{6})}{2(2)} = \frac{4 \pm \sqrt{6}}{2}$$

The exact solutions are $\frac{4 + \sqrt{6}}{2}$ and $\frac{4 - \sqrt{6}}{2}$.

$\frac{4 + \sqrt{6}}{2} \approx 3.225 \qquad \frac{4 - \sqrt{6}}{2} \approx 0.775$

To the nearest thousandth, the solutions are 3.225 and 0.775.

CHECK YOUR PROGRESS 5, *page 311*

$z^2 = -6 - 2z$

$z^2 + 2z + 6 = 0$

$a = 1, b = 2, c = 6$

$$z = \frac{-b \pm \sqrt{b^2 - 4ac}}{2a}$$
$$z = \frac{-(2) \pm \sqrt{(2)^2 - 4(1)(6)}}{2(1)}$$
$$= \frac{-2 \pm \sqrt{4 - 24}}{2} = \frac{-2 \pm \sqrt{-20}}{2}$$

$\sqrt{-20}$ is not a real number.

The equation has no real number solutions.

CHECK YOUR PROGRESS 6, *page 312*

$h = 64t - 16t^2$

$0 = 64t - 16t^2$

$16t^2 - 64t = 0$

$16t(t - 4) = 0$

$16t = 0 \qquad t - 4 = 0$

$t = 0 \qquad t = 4$

The object will be on the ground at 0 seconds and after 4 seconds.

CHECK YOUR PROGRESS 7, *page 313*

$h = -16t^2 + 32t + 6.5$

$10 = -16t^2 + 32t + 6.5$

$16t^2 - 32t + 3.5 = 0$

$a = 16, b = -32, c = 3.5$

$$t = \frac{-b \pm \sqrt{b^2 - 4ac}}{2a}$$
$$t = \frac{-(-32) \pm \sqrt{(-32)^2 - 4(16)(3.5)}}{2(16)} = \frac{32 \pm \sqrt{800}}{32}$$

$t = \frac{32 + \sqrt{800}}{32} \approx 1.88 \qquad t = \frac{32 - \sqrt{800}}{32} \approx 0.12$

The solution $t \approx 0.12$ second is not reasonable. The ball hits the basket 1.88 seconds after the ball is released.

CHAPTER 6

SECTION 6.1

CHECK YOUR PROGRESS 1, *page 330*

x	$-2x + 3 = y$	(x, y)
−2	−2(−2) + 3 = 7	(−2, 7)
−1	−2(−1) + 3 = 5	(−1, 5)
0	−2(0) + 3 = 3	(0, 3)
1	−2(1) + 3 = 1	(1, 1)
2	−2(2) + 3 = −1	(2, −1)
3	−2(3) + 3 = −3	(3, −3)

CHECK YOUR PROGRESS 2, *page 331*

x	$-x^2 + 1 = y$	(x, y)
−3	$-(-3)^2 + 1 = -8$	(−3, −8)
−2	$-(-2)^2 + 1 = -3$	(−2, −3)
−1	$-(-1)^2 + 1 = 0$	(−1, 0)
0	$-(0)^2 + 1 = 1$	(0, 1)
1	$-(1)^2 + 1 = 0$	(1, 0)
2	$-(2)^2 + 1 = -3$	(2, −3)
3	$-(3)^2 + 1 = -8$	(3, −8)

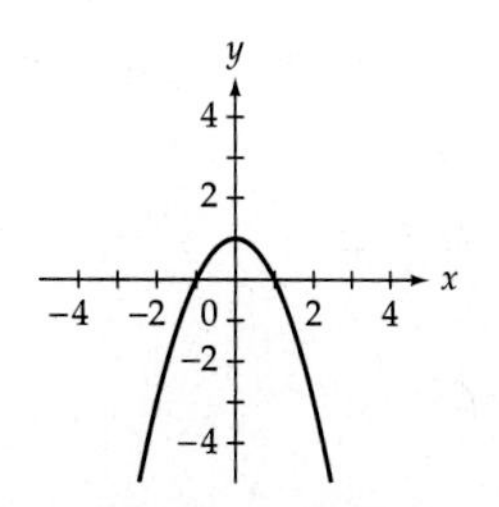

CHECK YOUR PROGRESS 3, *page 334*

$$f(z) = z^2 - z$$
$$f(-3) = (-3)^2 - (-3)$$
$$= 12$$

The value of the function is 12 when $z = -3$.

CHECK YOUR PROGRESS 4, *page 334*

$$N(s) = \frac{s^2 - 3s}{2}$$
$$N(12) = \frac{(12)^2 - 3(12)}{2}$$
$$= \frac{144 - 36}{2}$$
$$= 54$$

A polygon with 12 sides has 54 diagonals.

CHECK YOUR PROGRESS 5, *page 336*

x	$f(x) = 2 - \frac{3}{4}x$	(x, y)
−3	$f(-3) = 2 - \frac{3}{4}(-3) = 4\frac{1}{4}$	$\left(-3, 4\frac{1}{4}\right)$
−2	$f(-2) = 2 - \frac{3}{4}(-2) = 3\frac{1}{2}$	$\left(-2, 3\frac{1}{2}\right)$
−1	$f(-1) = 2 - \frac{3}{4}(-1) = 2\frac{3}{4}$	$\left(-1, 2\frac{3}{4}\right)$
0	$f(0) = 2 - \frac{3}{4}(0) = 2$	$(0, 2)$
1	$f(1) = 2 - \frac{3}{4}(1) = 1\frac{1}{4}$	$\left(1, 1\frac{1}{4}\right)$
2	$f(2) = 2 - \frac{3}{4}(2) = \frac{1}{2}$	$\left(2, \frac{1}{2}\right)$
3	$f(3) = 2 - \frac{3}{4}(3) = -\frac{1}{4}$	$\left(3, -\frac{1}{4}\right)$

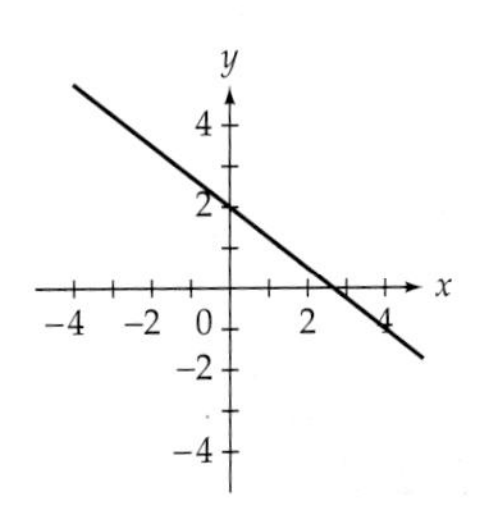

SECTION 6.2

CHECK YOUR PROGRESS 1, *page 342*

$$f(x) = \frac{1}{2}x + 3$$
$$0 = \frac{1}{2}x + 3$$
$$-3 = \frac{1}{2}x$$
$$-6 = x$$

The x-intercept is $(-6, 0)$.

$$f(x) = \frac{1}{2}x + 3$$
$$f(0) = \frac{1}{2}(0) + 3$$
$$= 3$$

The y-intercept is $(0, 3)$.

CHECK YOUR PROGRESS 2, *page 343*

$$g(t) = -20t + 8000$$
$$g(0) = -20(0) + 8000 = 8000$$

The intercept on the vertical axis is (0, 8000). This means that the plane is at an altitude of 8000 feet when it begins its descent.

$$g(t) = -20t + 8000$$
$$0 = -20t + 8000$$
$$-8000 = -20t$$
$$400 = t$$

The intercept on the horizontal axis is (400, 0). This means that the plane reaches the ground 400 seconds after beginning its descent.

CHECK YOUR PROGRESS 3, *page 346*

a. $(x_1, y_1) = (-6, 5), (x_2, y_2) = (4, -5)$

$$m = \frac{y_2 - y_1}{x_2 - x_1} = \frac{-5 - 5}{4 - (-6)} = \frac{-10}{10} = -1$$

The slope is -1.

b. $(x_1, y_1) = (-5, 0), (x_2, y_2) = (-5, 7)$

$$m = \frac{y_2 - y_1}{x_2 - x_1} = \frac{7 - 0}{-5 - (-5)} = \frac{7}{0}$$

The slope is undefined.

c. $(x_1, y_1) = (-7, -2), (x_2, y_2) = (8, 8)$

$$m = \frac{y_2 - y_1}{x_2 - x_1} = \frac{8 - (-2)}{8 - (-7)} = \frac{10}{15} = \frac{2}{3}$$

The slope is $\frac{2}{3}$.

d. $(x_1, y_1) = (-6, 7), (x_2, y_2) = (1, 7)$

$$m = \frac{y_2 - y_1}{x_2 - x_1} = \frac{7 - 7}{1 - (-6)} = \frac{0}{7} = 0$$

The slope is 0.

CHECK YOUR PROGRESS 4, *page 347*

For the linear function $d(t) = 50t$, the slope is the coefficient of t. Therefore, the slope is 50. This means that a homing pigeon can fly 50 miles for each 1 hour of flight time.

CHECK YOUR PROGRESS 5, *page 348*

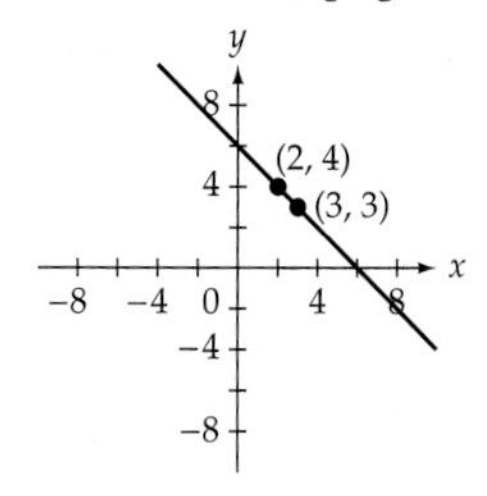

CHECK YOUR PROGRESS 6, *page 348*

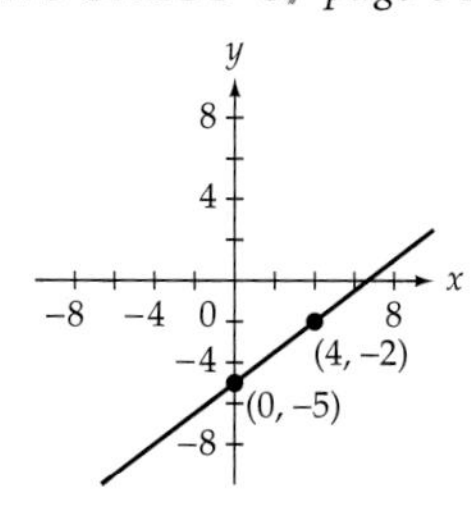

SECTION 6.3

CHECK YOUR PROGRESS 1, *page 355*

$$f(a) = ma + b$$
$$f(a) = -3.5a + 100$$

The linear function is $f(a) = -3.5a + 100$, where $f(a)$ is the boiling point of water at an altitude of a kilometers above sea level.

CHECK YOUR PROGRESS 2, *page 355*

$$y - y_1 = m(x - x_1)$$
$$y - 2 = -\frac{1}{2}[x - (-2)]$$
$$y - 2 = -\frac{1}{2}x - 1$$
$$y = -\frac{1}{2}x + 1$$

CHECK YOUR PROGRESS 3, *page 356*

$$C - C_1 = m(t - t_1)$$
$$C - 191 = 3.8(t - 50)$$
$$C - 191 = 3.8t - 190$$
$$C = 3.8t + 1$$

A linear function that models the number of calories burned is $C(t) = 3.8t + 1$.

CHECK YOUR PROGRESS 4, *page 356*

$$m = \frac{y_2 - y_1}{x_2 - x_1} = \frac{1 - 3}{4 - (-2)} = \frac{-2}{6} = -\frac{1}{3}$$

$$y - y_1 = m(x - x_1)$$
$$y - 3 = -\frac{1}{3}[x - (-2)]$$
$$y - 3 = -\frac{1}{3}x - \frac{2}{3}$$
$$y = -\frac{1}{3}x + \frac{7}{3}$$

CHECK YOUR PROGRESS 5, *page 359*

The regression equation is $y = 5.63x - 252.86$.

The estimated weight of a woman swimmer who is 63 inches tall is approximately 102 pounds.

SECTION 6.4

CHECK YOUR PROGRESS 1, *page 365*

$$a = 1, b = 0; -\frac{b}{2a} = -\frac{0}{2(1)} = 0$$

$$y = x^2 - 2$$
$$y = (0)^2 - 2$$
$$y = -2$$

The vertex is $(0, -2)$.

CHECK YOUR PROGRESS 2, *page 367*

a. $y = 2x^2 - 5x + 2$
$0 = 2x^2 - 5x + 2$
$0 = (2x - 1)(x - 2)$

$2x - 1 = 0 \qquad x - 2 = 0$
$x = \frac{1}{2} \qquad x = 2$

The x-intercepts are $\left(\frac{1}{2}, 0\right)$ and $(2, 0)$.

b. $y = x^2 + 4x + 4$
$0 = x^2 + 4x + 4$
$0 = (x + 2)(x + 2)$

$x + 2 = 0 \qquad x + 2 = 0$
$x = -2 \qquad x = -2$

The x-intercept is $(-2, 0)$.

CHECK YOUR PROGRESS 3, *page 368*

$a = 2, b = -3; -\frac{b}{2a} = -\frac{-3}{2(2)} = \frac{3}{4}$

$f(x) = 2x^2 - 3x + 1$

$f\left(\frac{3}{4}\right) = 2\left(\frac{3}{4}\right)^2 - 3\left(\frac{3}{4}\right) + 1$

$f\left(\frac{3}{4}\right) = -\frac{1}{8}$

The vertex is $\left(\frac{3}{4}, -\frac{1}{8}\right)$. The minimum value of the function is $-\frac{1}{8}$, the y-coordinate of the vertex.

CHECK YOUR PROGRESS 4, *page 369*

$a = -16, b = 64; -\frac{b}{2a} = -\frac{64}{2(-16)} = 2$

The ball reaches its maximum height in 2 seconds.

$s(t) = -16t^2 + 64t + 4$

$s(2) = -16(2)^2 + 64(2) + 4$

$s(2) = 68$

The maximum height of the ball is 68 feet.

CHECK YOUR PROGRESS 5, *page 370*

Perimeter: $w + l + w + l = 44$

$2w + 2l = 44$

$w + l = 22$

$l = -w + 22$

Area: $A = lw$

$= (-w + 22)w$

$A = -w^2 + 22w$

$w = -\frac{b}{2a} = -\frac{22}{2(-1)} = 11$

The width is 11 feet.

$l = -w + 22$

$l = -(11) + 22 = 11$

The length is 11 feet.

The dimensions of the rectangle with maximum area are 11 feet by 11 feet.

SECTION 6.5

CHECK YOUR PROGRESS 1, *page 378*

$g(x) = \left(\frac{1}{2}\right)^x$

$g(3) = \left(\frac{1}{2}\right)^3 = \frac{1}{8}$

$g(-1) = \left(\frac{1}{2}\right)^{-1} = \frac{1}{\frac{1}{2}} = 2$

$g(\sqrt{3}) = \left(\frac{1}{2}\right)^{\sqrt{3}} \approx \left(\frac{1}{2}\right)^{1.732} \approx 0.301$

CHECK YOUR PROGRESS 2, *page 380*

Because the base $\frac{3}{2}$ is greater than 1, f is an exponential growth function.

x	$f(x) = \left(\frac{3}{2}\right)^x$	(x, y)
-3	$f(-3) = \left(\frac{3}{2}\right)^{-3} = \frac{8}{27}$	$\left(-3, \frac{8}{27}\right)$
-2	$f(-2) = \left(\frac{3}{2}\right)^{-2} = \frac{4}{9}$	$\left(-2, \frac{4}{9}\right)$
-1	$f(-1) = \left(\frac{3}{2}\right)^{-1} = \frac{2}{3}$	$\left(-1, \frac{2}{3}\right)$
0	$f(0) = \left(\frac{3}{2}\right)^{0} = 1$	$(0, 1)$
1	$f(1) = \left(\frac{3}{2}\right)^{1} = \frac{3}{2}$	$\left(1, \frac{3}{2}\right)$
2	$f(2) = \left(\frac{3}{2}\right)^{2} = \frac{9}{4}$	$\left(2, \frac{9}{4}\right)$
3	$f(3) = \left(\frac{3}{2}\right)^{3} = \frac{27}{8}$	$\left(3, \frac{27}{8}\right)$

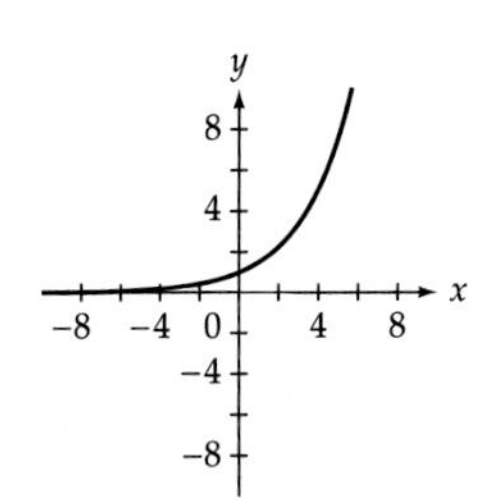

CHECK YOUR PROGRESS 3, *page 381*

x	-2	-1	0	1	2
$f(x) = e^{-x} + 2$	9.4	4.7	3	2.4	2.1

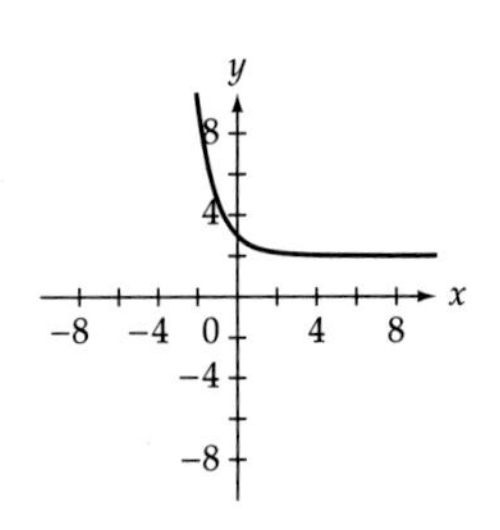

CHECK YOUR PROGRESS 4, *page 382*

$$N(t) = 1.5\left(\frac{1}{2}\right)^{t/193.7}$$
$$N(24) = 1.5\left(\frac{1}{2}\right)^{24/193.7}$$
$$\approx 1.5(0.9177) \approx 1.3766$$

After 24 hours, there are approximately 1.3766 grams of the isotope in the body.

CHECK YOUR PROGRESS 5, *page 383*

$$A(t) = 200e^{-0.014t}$$
$$A(45) = 200e^{-0.014(45)}$$
$$\approx 107$$

After 45 minutes, there are approximately 107 milligrams of aspirin in the patient's bloodstream.

CHECK YOUR PROGRESS 6, *pages 384–385*

The regression equation is $P(a) \approx 10.1468(0.8910)^a$.

The atmospheric pressure at an altitude of 24 kilometers is approximately 0.6 newton per square centimeter.

SECTION 6.6

CHECK YOUR PROGRESS 1, *page 391*

a. $2^{10} = 4x$

b. $\log_{10} 2x = 3$

CHECK YOUR PROGRESS 2, *page 392*

a.
$$\log_{10} 0.001 = x$$
$$10^x = 0.001$$
$$10^x = 10^{-3}$$
$$x = -3$$
$$\log_{10} 0.001 = -3$$

b.
$$\log_5 125 = x$$
$$5^x = 125$$
$$5^x = 5^3$$
$$x = 3$$
$$\log_5 125 = 3$$

CHECK YOUR PROGRESS 3, *page 392*

$$\log_2 x = 6$$
$$2^6 = x$$
$$64 = x$$

CHECK YOUR PROGRESS 4, *page 393*

a.
$$\log x = -2.1$$
$$10^{-2.1} = x$$
$$0.008 \approx x$$

b.
$$\ln x = 2$$
$$e^2 = x$$
$$7.389 \approx x$$

CHECK YOUR PROGRESS 5, *page 394*

$$y = \log_5 x$$
$$5^y = x$$

$x = 5^y$	$\frac{1}{25}$	$\frac{1}{5}$	1	5	25
y	−2	−1	0	1	2

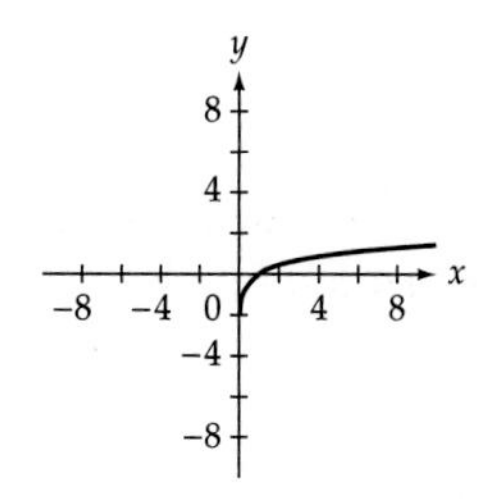

CHECK YOUR PROGRESS 6, *page 395*

a. $S(0) = 5 + 29\ln(0 + 1) = 5$
The average typing speed when the student first started to type was 5 words per minute.
$S(3) = 5 + 29\ln(3 + 1) \approx 45$
The average typing speed after 3 months was about 45 words per minute.

b. $S(3) - S(0) = 45 - 5 = 40$
The typing speed increased by 40 words per minute during the 3 months.

CHECK YOUR PROGRESS 7, *page 396*

$I = 2 \cdot (12{,}589{,}254I_0) = 25{,}178{,}508I_0$

$$M = \log\left(\frac{I}{I_0}\right) = \log\left(\frac{25{,}178{,}508I_0}{I_0}\right) = \log(25{,}178{,}508) \approx 7.4$$

The Richter scale magnitude of an earthquake whose intensity is twice that of the Amazonas, Brazil, earthquake is 7.4.

CHECK YOUR PROGRESS 8, *page 396*

$$\log\left(\frac{I}{I_0}\right) = 4.6$$
$$\frac{I}{I_0} = 10^{4.6}$$
$$I = 10^{4.6}I_0$$
$$I \approx 39{,}811I_0$$

The April 29, 2003 earthquake had an intensity that was approximately 40,000 times the intensity of a zero-level earthquake.

CHECK YOUR PROGRESS 9, *page 397*

a. $\text{pH} = -\log[\text{H}^+] = -\log(2.41 \times 10^{-13}) \approx 12.6$
The cleaning solution has a pH of 12.6.

b. $\text{pH} = -\log[\text{H}^+] = -\log(5.07 \times 10^{-4}) \approx 3.3$
The cola soft drink has a pH of 3.3.

c. $\text{pH} = -\log[\text{H}^+] = -\log(6.31 \times 10^{-5}) \approx 4.2$
The rainwater has a pH of 4.2.

CHECK YOUR PROGRESS 10, *page 398*

$$\begin{aligned} \text{pH} &= -\log[\text{H}^+] \\ 10.0 &= -\log[\text{H}^+] \\ -10.0 &= \log[\text{H}^+] \\ 10^{-10.0} &= \text{H}^+ \\ 1.0 \times 10^{-10} &= \text{H}^+ \end{aligned}$$

The hydronium-ion concentration of the water in the Great Salt Lake in Utah is 1.0×10^{-10} moles per liter.

CHAPTER 7

SECTION 7.1

CHECK YOUR PROGRESS 1, *page 408*

a. $6 \oplus 10 = 4$

b. $5 \oplus 9 = 2$

c. $7 \ominus 11 = 8$

d. $5 \ominus 10 = 7$

CHECK YOUR PROGRESS 2, *page 410*
The years 2008, 2012, and 2016 are leap years, so there are 3 years between the two dates with 366 days and 6 years with 365 days. The total number of days between the dates is $3 \cdot 366 + 6 \cdot 365 = 3288$. $3288 \div 7 = 469$ remainder 5, so $3288 \equiv 5 \bmod 7$. The day of the week 3288 days after Tuesday, February 12, 2008 will be the same as the day 5 days later, a Sunday.

CHECK YOUR PROGRESS 3, *page 411*
$51 + 72 = 123$, and $123 \div 3 = 41$ remainder 0, so $(51 + 72) \bmod 3 \equiv 0$.

CHECK YOUR PROGRESS 4, *page 411*
$21 - 43 = -22$, a negative number. Repeatedly add the modulus 7 to the difference until a whole number is reached.

$$\begin{aligned} -22 + 7 &= -15 \\ -15 + 7 &= -8 \\ -8 + 7 &= -1 \\ -1 + 7 &= 6 \end{aligned}$$

$(21 - 43) \bmod 7 \equiv 6$.

CHECK YOUR PROGRESS 5, *page 412*
Tuesday corresponds to 2 (see the chart on page 408), so the day of the week 93 days from now is represented by $(2 + 93) \bmod 7$. Because $95 \div 7 = 13$ remainder 4, $(2 + 93) \bmod 7 \equiv 4$, which corresponds to Thursday.

CHECK YOUR PROGRESS 6, *page 412*
$33 \cdot 41 = 1353$ and $1353 \div 17 = 79$ remainder 10, so $(33 \cdot 41) \bmod 17 \equiv 10$.

CHECK YOUR PROGRESS 7, *page 414*
Substitute each whole number from 0 to 11 into the congruence.

$4(0) + 1 \not\equiv 5 \bmod 12$	Not a solution
$4(1) + 1 \equiv 5 \bmod 12$	1 is a solution.
$4(2) + 1 \not\equiv 5 \bmod 12$	Not a solution
$4(3) + 1 \not\equiv 5 \bmod 12$	Not a solution
$4(4) + 1 \equiv 5 \bmod 12$	4 is a solution.
$4(5) + 1 \not\equiv 5 \bmod 12$	Not a solution
$4(6) + 1 \not\equiv 5 \bmod 12$	Not a solution
$4(7) + 1 \equiv 5 \bmod 12$	7 is a solution.
$4(8) + 1 \not\equiv 5 \bmod 12$	Not a solution
$4(9) + 1 \not\equiv 5 \bmod 12$	Not a solution
$4(10) + 1 \equiv 5 \bmod 12$	10 is a solution.
$4(11) + 1 \not\equiv 5 \bmod 12$	Not a solution

The solutions from 0 to 11 are 1, 4, 7, and 10. The remaining solutions are obtained by repeatedly adding the modulus 12 to these solutions. So the solutions are 1, 4, 7, 10, 13, 16, 19, 22, ...

CHECK YOUR PROGRESS 8, *page 414*
In mod 12 arithmetic, $6 + 6 = 12$, so the additive inverse of 6 is 6.

CHECK YOUR PROGRESS 9, *page 415*
Solve the congruence equation $5x \equiv 1 \bmod 11$ by substituting whole number values of x less than the modulus.

$$\begin{aligned} 5(1) &\not\equiv 1 \bmod 11 \\ 5(2) &\not\equiv 1 \bmod 11 \\ 5(3) &\not\equiv 1 \bmod 11 \\ 5(4) &\not\equiv 1 \bmod 11 \\ 5(5) &\not\equiv 1 \bmod 11 \\ 5(6) &\not\equiv 1 \bmod 11 \\ 5(7) &\not\equiv 1 \bmod 11 \\ 5(8) &\not\equiv 1 \bmod 11 \\ 5(9) &\equiv 1 \bmod 11 \end{aligned}$$

In mod 11 arithmetic, the multiplicative inverse of 5 is 9.

SECTION 7.2

CHECK YOUR PROGRESS 1, *page 420*

Check the ISBN congruence equation.

$$0(10) + 2(9) + 0(8) + 1(7) + 1(6) + 5(5) + 5(4) + 0(3) + 2(2) + 4 \equiv ? \bmod 11$$
$$84 \equiv 7 \bmod 11$$

Because $84 \not\equiv 0 \bmod 11$, the ISBN is invalid.

CHECK YOUR PROGRESS 2, *page 421*

Check the UPC congruence equation.

$$1(3) + 3(1) + 2(3) + 3(1) + 4(3) + 2(1) + 6(3) + 5(1) + 9(3) + 3(1) + 3(3) + 9 \equiv ? \bmod 10$$
$$100 \equiv 0 \bmod 10$$

Because $100 \equiv 0 \bmod 10$, the UPC is valid.

CHECK YOUR PROGRESS 3, *page 422*

Highlight every other digit, reading from right to left:

6 0 1 1 0 1 2 3 9 1 4 5 2 3 1 7

Double the highlighted digits:

12 0 2 1 0 1 4 3 18 1 8 5 4 3 2 7

Add all the digits, treating two-digit numbers as two single digits:

$$(1 + 2) + 0 + 2 + 1 + 0 + 1 + 4 + 3 + (1 + 8) + 1 + 8 + 5 + 4 + 3 + 2 + 7 = 53$$

Because $53 \not\equiv 0 \bmod 10$, this is not a valid credit card number.

CHECK YOUR PROGRESS 4, *page 425*

a. The encrypting congruence is $c \equiv (p + 17) \bmod 26$.

A	$c \equiv (1 + 17) \bmod 26 \equiv 18 \bmod 26 \equiv 18$	Code A as R.
L	$c \equiv (12 + 17) \bmod 26 \equiv 29 \bmod 26 \equiv 3$	Code L as C.
P	$c \equiv (16 + 17) \bmod 26 \equiv 33 \bmod 26 \equiv 7$	Code P as G.
I	$c \equiv (9 + 17) \bmod 26 \equiv 26 \bmod 26 \equiv 0$	Code I as Z.
N	$c \equiv (14 + 17) \bmod 26 \equiv 31 \bmod 26 \equiv 5$	Code N as E.
E	$c \equiv (5 + 17) \bmod 26 \equiv 22 \bmod 26 \equiv 22$	Code E as V.
S	$c \equiv (19 + 17) \bmod 26 \equiv 36 \bmod 26 \equiv 10$	Code S as J.
K	$c \equiv (11 + 17) \bmod 26 \equiv 28 \bmod 26 \equiv 2$	Code K as B.
G	$c \equiv (7 + 17) \bmod 26 \equiv 24 \bmod 26 \equiv 24$	Code G as X.

Thus the plaintext ALPINE SKIING is coded as RCGZEV JBZZEX.

b. To decode, because $m = 17$, $n = 26 - 17 = 9$, and the decoding congruence is $p \equiv (c + 9) \bmod 26$.

T	$c \equiv (20 + 9) \bmod 26 \equiv 29 \bmod 26 \equiv 3$	Decode T as C.
I	$c \equiv (9 + 9) \bmod 26 \equiv 18 \bmod 26 \equiv 18$	Decode I as R.
F	$c \equiv (6 + 9) \bmod 26 \equiv 15 \bmod 26 \equiv 15$	Decode F as O.
J	$c \equiv (10 + 9) \bmod 26 \equiv 19 \bmod 26 \equiv 19$	Decode J as S.

Continuing, the ciphertext TIFJJ TFLEKIP JBZZEX decodes as CROSS COUNTRY SKIING.

CHECK YOUR PROGRESS 5, *page 426*

The encrypting congruence is $c \equiv (3p + 1) \bmod 26$.

C	$c \equiv (3 \cdot 3 + 1) \bmod 26 \equiv 10 \bmod 26 = 10$	Code C as J.
O	$c \equiv (3 \cdot 15 + 1) \bmod 26 \equiv 46 \bmod 26 = 20$	Code O as T.
L	$c \equiv (3 \cdot 12 + 1) \bmod 26 \equiv 37 \bmod 26 \equiv 11$	Code L as K.
R	$c \equiv (3 \cdot 18 + 1) \bmod 26 \equiv 55 \bmod 26 \equiv 3$	Code R as C.

Continuing, the plaintext COLOR MONITOR is coded as JTKTC NTQBITC.

CHECK YOUR PROGRESS 6, *page 427*

Solve the congruence equation $c \equiv (7p + 1) \bmod 26$ for p.

$$c = 7p + 1$$

$$c - 1 = 7p$$

• Subtract 1 from each side of the equation.

$$15(c - 1) = 15(7p)$$

• Multiply each side of the equation by the multiplicative inverse of 7. Because $7 \cdot 15 \equiv 1 \bmod 26$, multiply each side by 15.

$$[15(c - 1)] \bmod 26 \equiv p$$

The decoding congruence is $p \equiv [15(c - 1)] \bmod 26$.

I	$p \equiv [15(9 - 1)] \bmod 26 \equiv 120 \bmod 26 \equiv 16$	Decode I as P.
G	$p \equiv [15(7 - 1)] \bmod 26 \equiv 90 \bmod 26 \equiv 12$	Decode G as L.
H	$p \equiv [15(8 - 1)] \bmod 26 \equiv 105 \bmod 26 \equiv 1$	Decode H as A.
T	$p \equiv [15(20 - 1)] \bmod 26 \equiv 285 \bmod 26 \equiv 25$	Decode T as Y.

Continuing, the ciphertext IGHT OHGG decodes as PLAY BALL.

SECTION 7.3

CHECK YOUR PROGRESS 1, *page 433*

Check to see whether the four properties of a group are satisfied.

1. The product of two integers is always an integer, so the integers are closed with respect to multiplication.
2. The associative property of multiplication is true for integers.
3. The integers have an identity element for multiplication, namely 1.
4. Not every integer has a multiplicative inverse that is also an integer. For instance, $\frac{1}{2}$ is the multiplicative inverse of 2, but $\frac{1}{2}$ is not an integer. There is no integer that can be multiplied by 2 that gives the identity element 1.

Because property 4 is not satisfied, the integers with multiplication do not form a group.

CHECK YOUR PROGRESS 2, *page 436*

Rotate the original triangle, I, about the line of symmetry through the bottom right vertex, followed by a clockwise rotation of 240°.

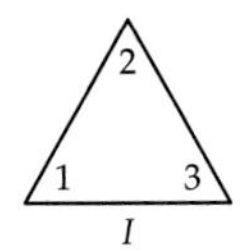

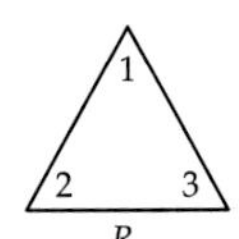

followed by R_{240}

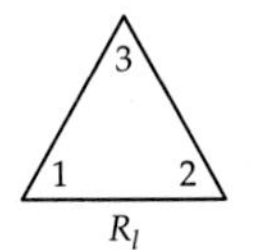

Therefore, $R_r \Delta R_{240} = R_l$.

CHECK YOUR PROGRESS 3, *page 437*

$$R_r \Delta R_{240} = \begin{pmatrix} 1 & 2 & 3 \\ 2 & 1 & 3 \end{pmatrix} \Delta \begin{pmatrix} 1 & 2 & 3 \\ 3 & 1 & 2 \end{pmatrix} = \begin{pmatrix} 1 & 2 & 3 \\ 1 & 3 & 2 \end{pmatrix} = R_l$$

• $1 \to 2 \to 1$. Thus $1 \to 1$.
• $2 \to 1 \to 3$. Thus $2 \to 3$.
• $3 \to 3 \to 2$. Thus $3 \to 2$.

CHECK YOUR PROGRESS 4, *page 438*

$E \Delta B = \begin{pmatrix} 1 & 2 & 3 \\ 2 & 1 & 3 \end{pmatrix} \Delta \begin{pmatrix} 1 & 2 & 3 \\ 3 & 1 & 2 \end{pmatrix}$. 1 is replaced by 2, which is then replaced by 1 in the second permutation. Thus 1 remains as 1. 2 is replaced by 1, which is then replaced by 3, so ultimately, 2 is replaced by 3. Finally, 3 remains as 3 in the first permutation but is replaced by 2 in the second, so ultimately 3 is replaced by 2. The result is $\begin{pmatrix} 1 & 2 & 3 \\ 1 & 3 & 2 \end{pmatrix}$, which is C. Thus $E \Delta B = C$.

CHECK YOUR PROGRESS 5, *page 439*

$D = \begin{pmatrix} 1 & 2 & 3 \\ 3 & 2 & 1 \end{pmatrix}$ replaces 1 with 3, 2 with 2, and 3 with 1. Reversing these, we need to replace 3 with 1, leave 2 alone, and replace 1 with 3. This is the element $\begin{pmatrix} 1 & 2 & 3 \\ 3 & 2 & 1 \end{pmatrix}$, which is D again.

Thus D is its own inverse.

CHAPTER 8

SECTION 8.1

CHECK YOUR PROGRESS 1, *page 451*

$QR + RS + ST = QT$

$28 + 16 + 10 = QT$

$54 = QT$

$QT = 54 \text{ cm}$

CHECK YOUR PROGRESS 2, *page 452*

$AB + BC = AC$

$\frac{1}{4}(BC) + BC = AC$

$\frac{1}{4}(16) + 16 = AC$

$4 + 16 = AC$

$20 = AC$

$AC = 20 \text{ ft}$

CHECK YOUR PROGRESS 3, *page 454*

$m\angle G + m\angle H = 127° + 53° = 180°$

The sum of the measures of $\angle G$ and $\angle H$ is 180°. Angles G and H are supplementary angles.

CHECK YOUR PROGRESS 4, *page 455*

Supplementary angles are two angles the sum of whose measures is 180°. To find the supplement, let x represent the supplement of a 129° angle.

$x + 129 = 180$

$x = 51$

The supplement of a 129° angle is a 51° angle.

CHECK YOUR PROGRESS 5, *page 456*

$m\angle a + 68° = 118°$

$m\angle a = 50°$

The measure of $\angle a$ is 50°.

CHECK YOUR PROGRESS 6, *page 457*

$m\angle b + m\angle a = 180°$
$m\angle b + 35° = 180°$
$m\angle b = 145°$

$m\angle c = m\angle a = 35°$

$m\angle d = m\angle b = 145°$

$m\angle b = 145°$, $m\angle c = 35°$, and $m\angle d = 145°$.

CHECK YOUR PROGRESS 7, *page 459*

$m\angle b = m\angle g = 124°$

$m\angle d = m\angle g = 124°$

$m\angle c + m\angle b = 180°$
$m\angle c + 124° = 180°$
$m\angle c = 56°$

$m\angle b = 124°$, $m\angle c = 56°$, and $m\angle d = 124°$.

CHECK YOUR PROGRESS 8, *page 461*

$m\angle b + m\angle d = 180°$
$m\angle b + 105° = 180°$
$m\angle b = 75°$

$m\angle a + m\angle b + m\angle c = 180°$
$m\angle a + 75° + 35° = 180°$
$m\angle a + 110° = 180°$
$m\angle a = 70°$

$m\angle e = m\angle a = 70°$

CHECK YOUR PROGRESS 9, *page 461*

Let x represent the measure of the third angle.

$x + 90° + 27° = 180°$

$x + 117° = 180°$

$x = 63°$

The measure of the third angle is 63°.

SECTION 8.2

CHECK YOUR PROGRESS 1, *page 472*

$P = a + b + c$

$P = 4\frac{3}{10} + 2\frac{1}{10} + 6\frac{1}{2}$

$P = 4\frac{3}{10} + 2\frac{1}{10} + 6\frac{5}{10}$

$P = 12\frac{9}{10}$

The total length of the bike trail is $12\frac{9}{10}$ mi.

CHECK YOUR PROGRESS 2, *page 473*

$P = 2L + 2W$
$P = 2(12) + 2(8)$
$P = 24 + 16$
$P = 40$

You will need 40 ft of molding to edge the top of the walls.

CHECK YOUR PROGRESS 3, *page 474*

$P = 4s$
$P = 4(24)$
$P = 96$

The homeowner should purchase 96 ft of fencing.

CHECK YOUR PROGRESS 4, *page 474*

$P = 2b + 2s$
$P = 2(5) + 2(7)$
$P = 10 + 14$
$P = 24$

24 m of plank is needed to surround the garden.

CHECK YOUR PROGRESS 5, *page 476*

$C = \pi d$
$C = 9\pi$

The circumference of the circle is 9π km.

CHECK YOUR PROGRESS 6, *page 476*

12 in. = 1 ft

$C = \pi d$
$C = \pi(1)$
$C = \pi$

$12C = 12\pi \approx 37.70$

The tricycle travels approximately 37.70 ft when the wheel makes 12 revolutions.

CHECK YOUR PROGRESS 7, *page 478*

$A = LW$
$A = 308(192)$
$A = 59{,}136$

59,136 cm^2 of fabric is needed.

CHECK YOUR PROGRESS 8, *page 479*

$A = s^2$
$A = 24^2$
$A = 576$

The area of the floor is 576 ft^2.

CHECK YOUR PROGRESS 9, *page 480*

$A = bh$
$A = 14(8)$
$A = 112$

The area of the patio is 112 m^2.

CHECK YOUR PROGRESS 10, *page 481*

$A = \frac{1}{2}bh$

$A = \frac{1}{2}(18)(9)$

$A = 9(9)$
$A = 81$

81 in^2 of felt is needed.

CHECK YOUR PROGRESS 11, *page 482*

$A = \frac{1}{2}h(b_1 + b_2)$

$A = \frac{1}{2} \cdot 9(12 + 20)$

$A = \frac{1}{2} \cdot 9(32)$

$A = \frac{9}{2} \cdot (32)$

$A = 144$

The area of the patio is 144 ft^2.

CHECK YOUR PROGRESS 12, *page 483*

$r = \frac{1}{2}d = \frac{1}{2}(12) = 6$

$A = \pi r^2$
$A = \pi(6)^2$
$A = 36\pi$

The area of the circle is 36π km^2.

CHECK YOUR PROGRESS 13, *page 484*

$r = \frac{1}{2}d = \frac{1}{2}(4) = 2$

$A = \pi r^2$
$A = \pi(2)^2$
$A = \pi(4)$
$A \approx 12.57$

Approximately 12.57 ft^2 of material is needed.

SECTION 8.3

CHECK YOUR PROGRESS 1, *page 494*

$\frac{AC}{DF} = \frac{CH}{FG}$

$\frac{10}{15} = \frac{7}{FG}$

$10(FG) = (15)7$
$10(FG) = 105$
$FG = 10.5$

The height *FG* of triangle *DEF* is 10.5 m.

CHECK YOUR PROGRESS 2, *page 497*

$$\frac{AO}{DO} = \frac{AB}{DC}$$
$$\frac{AO}{3} = \frac{10}{4}$$
$$4(AO) = 3(10)$$
$$4(AO) = 30$$
$$AO = 7.5$$

$$A = \frac{1}{2}bh$$
$$A = \frac{1}{2}(10)(7.5)$$
$$A = 5(7.5)$$
$$A = 37.5$$

The area of triangle AOB is 37.5 cm^2.

CHECK YOUR PROGRESS 3, *page 499*

Because two sides and the included angle of one triangle are equal in measure to two sides and the included angle of the second triangle, the triangles are congruent by the SAS Theorem.

CHECK YOUR PROGRESS 4, *page 500*

$a^2 + b^2 = c^2$ • Use the Pythagorean Theorem.

$2^2 + b^2 = 6^2$ • $a = 2$, $c = 6$

$4 + b^2 = 36$

$b^2 = 32$ • Solve for b^2. Subtract 4 from each side.

$\sqrt{b^2} = \sqrt{32}$ • Take the square root of each side of the equation.

$b \approx 5.66$ • Use a calculator to approximate $\sqrt{32}$.

The length of the other leg is approximately 5.66 m.

SECTION 8.4

CHECK YOUR PROGRESS 1, *page 511*

$$V = LWH$$
$$V = 5(3.2)(4)$$
$$V = 64$$

The volume of the solid is 64 m^3.

CHECK YOUR PROGRESS 2, *page 511*

$$V = \frac{1}{3}s^2h$$
$$V = \frac{1}{3}(15)^2(25)$$
$$V = \frac{1}{3}(225)(25)$$
$$V = 1875$$

The volume of the pyramid is 1875 m^3.

CHECK YOUR PROGRESS 3, *page 512*

$$r = \frac{1}{2}d = \frac{1}{2}(16) = 8$$

$$V = \pi r^2h$$
$$V = \pi(8)^2(30)$$
$$V = \pi(64)(30)$$
$$V = 1920\pi$$

$$\frac{1}{4}(1920\pi) = 480\pi$$
$$\approx 1507.96$$

Approximately 1507.96 ft^3 is not being used for storage.

CHECK YOUR PROGRESS 4, *page 515*

$$r = \frac{1}{2}d = \frac{1}{2}(6) = 3$$

$$S = 2\pi r^2 + 2\pi rh$$
$$S = 2\pi(3)^2 + 2\pi(3)(8)$$
$$S = 2\pi(9) + 2\pi(3)(8)$$
$$S = 18\pi + 48\pi$$
$$S = 66\pi$$
$$S \approx 207.35$$

The surface area of the cylinder is approximately 207.35 ft^2.

CHECK YOUR PROGRESS 5, *page 515*

$$\text{Surface area of the cube} = 6s^2 = 6(8)^2 = 6(64) = 384 \text{ cm}^2$$

$$\text{Surface area of the sphere} = 4\pi r^2 = 4\pi(5)^2 = 4\pi(25) \approx 314.16 \text{ cm}^2$$

The cube has a larger surface area.

SECTION 8.5

CHECK YOUR PROGRESS 1, *page 524*

Use the Pythagorean Theorem to find the length of the hypotenuse.

$$a^2 + b^2 = c^2$$
$$3^2 + 4^2 = c^2$$
$$9 + 16 = c^2$$
$$25 = c^2$$
$$\sqrt{25} = \sqrt{c^2}$$
$$5 = c$$

$$\sin\theta = \frac{\text{opp}}{\text{hyp}} = \frac{3}{5},$$
$$\cos\theta = \frac{\text{adj}}{\text{hyp}} = \frac{4}{5},$$
$$\tan\theta = \frac{\text{opp}}{\text{adj}} = \frac{3}{4}$$

CHECK YOUR PROGRESS 2, *page 525*

$\tan 37.1° = 0.7563$

CHECK YOUR PROGRESS 3, *page 526*

We are given the measure of $\angle B$ and the hypotenuse. We want to find the length of side a. The cosine function involves the side adjacent and the hypotenuse.

$$\cos B = \frac{\text{adj}}{\text{hyp}}$$

$$\cos 48° = \frac{a}{12}$$

$$12(\cos 48°) = a$$

$$8.03 \approx a$$

The length of side a is approximately 8.03 ft.

CHECK YOUR PROGRESS 4, *page 527*

$\tan^{-1}(0.3165) \approx 17.6°$

CHECK YOUR PROGRESS 5, *page 527*

$\theta \approx \tan^{-1}(0.5681)$

$\theta \approx 29.6°$

CHECK YOUR PROGRESS 6, *page 528*

We want to find the measure of $\angle A$, and we are given the length of the side opposite $\angle A$ and the hypotenuse. The sine function involves the side opposite an angle and the hypotenuse.

$$\sin A = \frac{\text{opp}}{\text{hyp}}$$

$$\sin A = \frac{7}{11}$$

$$A = \sin^{-1}\frac{7}{11}$$

$$A \approx 39.5°$$

The measure of $\angle A$ is approximately 39.5°.

CHECK YOUR PROGRESS 7, *page 529*

Let d be the distance from the base of the lighthouse to the boat.

$$\tan 25° = \frac{20}{d}$$

$$d(\tan 25°) = 20$$

$$d = \frac{20}{\tan 25°}$$

$$d \approx 42.9$$

The boat is approximately 42.9 m from the base of the lighthouse.

SECTION 8.6

CHECK YOUR PROGRESS 1, *page 537*

$$S = (m\angle A + m\angle B + m\angle C - 180°) \cdot \left(\frac{\pi}{180°}\right)r^2$$

$$= (200° + 90° + 90° - 180°) \cdot \left(\frac{\pi}{180°}\right)(6)^2$$

$$= (200°) \cdot \left(\frac{\pi}{180°}\right) \cdot (36)$$

$$= 40\pi \text{ in}^2 \quad \text{• Exact area}$$

$$\approx 125.66 \text{ in}^2 \quad \text{• Approximate area}$$

CHECK YOUR PROGRESS 2, *page 541*

a. $d_E(P, Q) = \sqrt{(x_2 - x_1)^2 + (y_2 - y_1)^2}$

$= \sqrt{[3 - (-1)]^2 + [2 - 4]^2}$

$= \sqrt{4^2 + (-2)^2}$

$= \sqrt{20} \approx 4.5$ blocks

$d_C(P, Q) = |x_2 - x_1| + |y_2 - y_1|$

$= |3 - (-1)| + |2 - 4|$

$= |4| + |-2|$

$= 6$ blocks

b. $d_E(P, Q) = \sqrt{(x_2 - x_1)^2 + (y_2 - y_1)^2}$

$= \sqrt{[(-1) - 3]^2 + [5 - (-4)]^2}$

$= \sqrt{(-4)^2 + 9^2}$

$= \sqrt{97} \approx 9.8$ blocks

$d_C(P, Q) = |x_2 - x_1| + |y_2 - y_1|$

$= |(-1) - 3| + |5 - (-4)|$

$= |-4| + 9$

$= 4 + 9$

$= 13$ blocks

SECTION 8.7

CHECK YOUR PROGRESS 1, *page 550*

Replace each line segment with a scaled version of the generator. As you move from left to right, your first zig should be to the left.

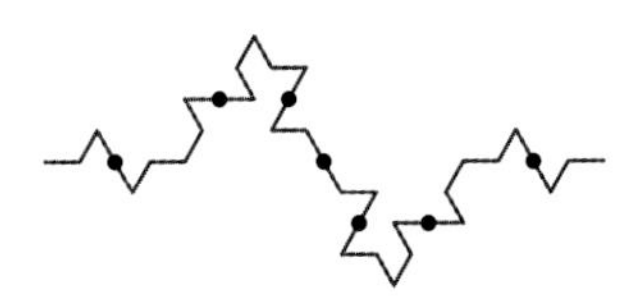

Stage 2 of the zig-zag curve

CHECK YOUR PROGRESS 2, *page 551*

Replace each square with a scaled version of the generator.

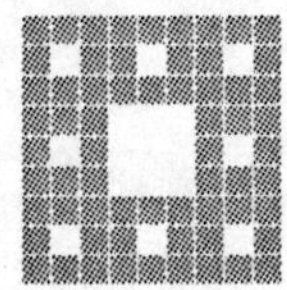

Stage 2 of the Sierpinski carpet

CHECK YOUR PROGRESS 3, *page 553*

a. Any portion of the box curve replicates the entire fractal, so t box curve is a strictly self-similar fractal.

b. Any portion of the Sierpinski gasket replicates the entire frac so the Sierpinski gasket is a strictly self-similar fractal.

CHECK YOUR PROGRESS 4, *page 554*

a. The generator of the Koch curve consists of four line segmer and the initiator consists of only one line segment. Thus the replacement ratio of the Koch curve is 4 : 1, or 4. The initiator of the Koch curve is a line segment that is 3 times as long as the replica line segments in the generator. Thus the scaling ratio of the Koch curve is 3 : 1, or 3.

b. The generator of the zig-zag curve consists of six line segments, and the initiator consists of only one line segment. Thus the replacement ratio of the zig-zag curve is 6 : 1, or 6. The initiator of the zig-zag curve is a line segment that is 4 times as long as the replica line segments in the generator. Thus the scaling ratio of the zig-zag curve is 4 : 1, or 4.

CHECK YOUR PROGRESS 5, *page 554*

a. In Example 4 we determined that the replacement ratio of the box curve is 5 and the scaling ratio of the box curve is 3. Thus the similarity dimension of the box curve is $D = \frac{\log 5}{\log 3} \approx 1.465$.

b. The replacement ratio of the Sierpinski carpet is 8 and the scaling ratio of the Sierpinski carpet is 3. Thus the similarity dimension of the Sierpinski carpet is $D = \frac{\log 8}{\log 3} \approx 1.893$.

APPENDIX

APPENDIX

CHECK YOUR PROGRESS 1, *page 936*

a. 1 295 m = 1.295 km

b. 7 543 g = 7.543 kg

c. 6.3 L = 6 300 ml

d. 2 kl = 2 000 L

ANSWERS TO SELECTED EXERCISES

CHAPTER 1

EXERCISE SET 1.1 *page 12*

1. 28 **3.** 45 **5.** 64 **7.** $\frac{15}{17}$ **9.** -13 **11.** correct **13.** correct **15.** incorrect
17. no effect **19.** 150 inches **21.** The distance is quadrupled. **23.** 0.5 second **25.** inductive
27. deductive **29.** deductive **31.** inductive
In Exercises 33–39, only one possible answer is given. Your answers may vary from the given answers.
33. $x = \frac{1}{2}$ **35.** $x = \frac{1}{2}$ **37.** $x = -3$ **39.** Consider 1 and 3. $1 + 3$ is even, but $1 \cdot 3$ is odd.
41. It does not work for 121. **43.** n

$6n + 8$

$\frac{6n + 8}{2} = 3n + 4$

$3n + 4 - 2n = n + 4$

$n + 4 - 4 = n$

45. Maria: the utility stock; Jose: the automotive stock; Anita: the technology stock; Tony: the oil stock
47. Atlanta: stamps; Chicago: baseball cards; Philadelphia: coins; Seattle: comic books **49.** Home, bookstore, supermarket, credit union, home; or home, credit union, supermarket, bookstore, home **51.** N, because the first letter of Nine is N. **53.** d

EXERCISE SET 1.2 *page 24*

1. 97 **3.** 329 **5.** 159 **7.** $\frac{3}{2}, 5, \frac{21}{2}, 18, \frac{55}{2}$ **9.** 2, 14, 36, 68, 110 **11.** $a_n = n^2 + n - 1$
13. $a_n = 2n$ **15. a.** There are 56 cannonballs in the sixth pyramid and 84 cannonballs in the seventh pyramid.
b. The eighth pyramid has eight levels of cannonballs. The number of cannonballs in the nth level is given by the nth number in the sequence 1, 3, 6, 10, 15, 21, 28, 36. Thus the total number of cannonballs in the eighth pyramid is $1 + 3 + 6 + 10 + 15 + 21 + 28 + 36 = 120$. **17. a.** Five cuts produce six pieces and six cuts produce 7 pieces. **b.** $a_n = n + 1$ **19. a.** 26
b. 7 **21.** $a_3 = 7, a_4 = 9, a_5 = 11$ **23.** $F_{20} = 6765, F_{30} = 832{,}040, F_{40} = 102{,}334{,}155$ **25.** n^2
27. a. 38.8 AU **b.** $38.8 - 30.6 = 8.2$ AU. The prediction is not close compared to the results obtained for the inner planets.
29. a. 154 AU **b.** Yes **31. a.** $F_n + 2F_{n+1} + F_{n+2} = F_{n+4}$ **b.** $F_n + F_{n+1} + F_{n+3} = F_{n+4}$
33. 1.615385; 1.617647; 1.617978; 1.618026. If n is an even number, then the ratio $\frac{F_n}{F_{n-1}}$ is slightly less than Φ.

EXERCISE SET 1.3 *page 41*

1. 195 **3.** 91 **5.** \$40 **7.** 18 **9.** $2^{12} = 4096$ **11. a.** B: 1, C: 9, D: 36, E: 84, F: 126, G: 126, H: 84
b. Region F is the fifth region from the left and region G is the fifth region from the right. For any path from A to F, the ball makes a number of left or right turns. If the turns are reversed, right turns instead of left turns and left turns instead of right turns, then the ball ends up in region G. Thus for each path from A to F, there is exactly one symmetrical path, about a central vertical line, from A to G. Also, from each path from A to G, there is exactly one symmetrical path from A to F. Thus there are the same numbers of paths to both regions.
13. 28 **15.** 21 ducks, 14 pigs **17.** 12 **19.** 6 **21.** 7 **23. a.** 80,200 **b.** 151,525 **c.** 1892

25. a. 121, 484, and 676 **b.** 1331 **27.** $1\frac{1}{2}$ inches **29. a.** 1.3 billion; 1.5 billion; 1.6 billion **b.** 1995 **c.** 2002 **d.** 2001 to 2002 **31. a.** 1994 **b.** 2002 **c.** The number of theatre admissions in 2003 was less than the number of admissions in 2002. **33.** 2601 tiles **35.** Four more sisters than brothers **37.** the 11th day **39.** 91 **41. a.** Place four coins on the left balance pan and the other four on the right balance pan. The pan that is the higher contains the fake coin. Take the four coins from the higher pan and use the balance scale to compare the weight of two of these coins to the weight of the other two. The pan that is the higher contains the fake coin. Take the two coins from the higher pan and use the balance scale to compare the weight of one of these coins to the weight of the other. The pan that is the higher contains the fake coin. This procedure enables you to determine the fake coin in three weighings. **b.** Place three of the coins on one of the balance pans and another three coins on the other. If the pans balance, then the fake coin is one of the two remaining coins. You can use the balance scale to determine which of the remaining coins is the fake coin because it will be lighter than the other coin. If the three coins on the left pan do not balance with the three coins on the right pan, then the fake coin must be one of the three coins on the higher pan. Pick any two coins from these three and place one on each balance pan. If these two coins do not balance, then the one that is the higher is the fake. If the coins balance, then the third coin (the one that you did not place on the balance pan) is the fake. In any case, this procedure enables you to determine the fake coin in two weighings. **43.** a. 1600. Sally likes perfect squares. **45.** d. 64. Each number is the cube of a term in the sequence 1, 2, 3, 4, 5, 6. **47. a.** People born in 1980 will be 45 in 2025. ($2025 = 45^2$) **b.** 2070, because people born in 2070 will be 46 in 2116. ($2116 = 46^2$) **49.** 612 **51.** Answers will vary.

CHAPTER 1 REVIEW EXERCISES *page 47*

1. deductive [Sec 1.1] **2.** inductive [Sec. 1.1] **3.** inductive [Sec. 1.1] **4.** deductive [Sec. 1.1] **5.** $x = 0$ provides a counterexample because $0^4 = 0$ and 0 is not greater than 0. [Sec. 1.1] **6.** $x = 4$ provides a counterexample because $\frac{(4)^3 + 5(4) + 6}{6} = 15$, which is not an even number. [Sec. 1.1] **7.** $x = 1$ provides a counterexample because $[(1) + 4]^2 = 25$, but $(1)^2 + 16 = 17$. [Sec. 1.1] **8.** Let $a = 1$ and $b = 1$. Then $(a + b)^3 = (1 + 1)^3 = 2^3 = 8$. However, $a^3 + b^3 = 1^3 + 1^3 = 2$. [Sec. 1.1] **9. a.** 112 **b.** 479 [Sec. 1.2] **10. a.** -72 **b.** -768 [Sec. 1.2] **11.** $a_1 = 1, a_2 = 12, a_3 = 31, a_4 = 58, a_5 = 93, a_{20} = 1578$ [Sec. 1.2] **12.** $a_1 = 3, a_2 = -6, a_3 = -39, a_4 = -108, a_5 = -225$ $a_{25} = -31{,}125$ [Sec. 1.2] **13.** $a_n = 3n$ [Sec. 1.2] **14.** $a_n = n^2 + 3n + 4$ [Sec. 1.2] **15.** $a_n = n^2 + 3n + 2$ [Sec. 1.2] **16.** $a_n = 5n - 1$ [Sec. 1.2] **17.** 320 feet by 1600 feet [Sec. 1.3] **18.** $3^{15} = 14{,}348{,}907$ [Sec. 1.3] **19.** 48 skyboxes [Sec. 1.3] **20.** On the first trip the rancher takes the rabbit across the river. The rancher returns alone. The rancher takes the dog across the river and returns with the rabbit. The rancher next takes the carrots across the river and returns alone. On the final trip the rancher takes the rabbit across the river. [Sec. 1.3] **21.** $300 [Sec. 1.3] **22.** 105 [Sec. 1.3] **23.** Answers will vary. [Sec. 1.3] **24.** Answers will vary. [Sec. 1.3] **25.** Michael: biology major; Clarissa: business major; Reggie: computer science major; Ellen: chemistry major [Sec. 1.1] **26.** Dodgers: drug store; Pirates: supermarket; Tigers: bank; Giants: service station [Sec. 1.1] **27. a.** Yes. Answers will vary. **b.** No. The countries of India, Bangladesh, and Myanmar all share borders with each of the other two countries. Thus at least three colors are needed to color the map. [Sec. 1.1] **28. a.** The following figure shows a route that passes over each bridge once and only once. **b.** No. [Sec. 1.3]

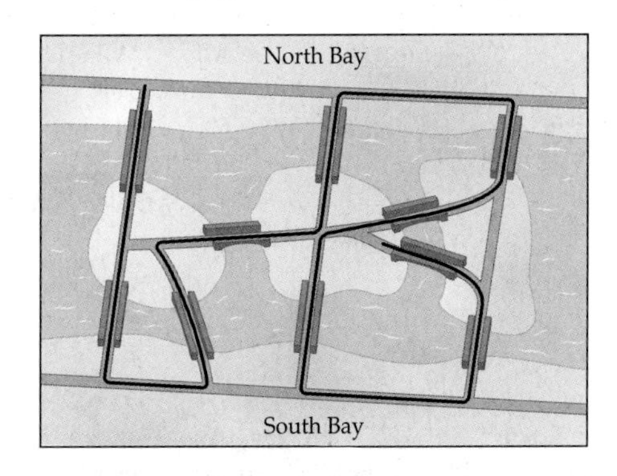

29. 1 square inch; 4 square inches; 25 square inches [Sec. 1.3] **30. a.** $2^{10} = 1024$ **b.** $2^{30} = 1{,}073{,}741{,}824$ [Sec. 1.2] **31.** A represents 1, B represents 9, and D represents 0. [Sec. 1.3] **32.** 5 [Sec. 1.3] **33.** 10 [Sec. 1.3] **34.** 1 [Sec. 1.3] **35.** 3 [Sec. 1.3]

36. n

$4n$

$4n + 12$

$\frac{4n + 12}{2} = 2n + 6$

$2n + 6 - 6 = 2n$ [Sec. 1.1]

37. Each nickel is worth 5 cents. Thus 2004 nickels are worth $2004 \times 5 = 10{,}020$ cents, or $100.20. [Sec. 1.3]

38. a. 1970 to 1980 **b.** 5.5% [Sec. 1.3] **39. a.** 16 times as many **b.** 61 times as many [Sec. 1.3] **40. a.** 2002 **b.** 2004 [Sec. 1.3] **41.** 5005 [Sec. 1.3] **42.** There are no narcissistic numbers. [Sec. 1.3] **43. a.** 10 **b.** yes [Sec. 1.2] **44. a.** 22 **b.** No. $9^9 = 387{,}420{,}489$. Thus $9^{(9^9)}$ is the product of 387,420,489 nines. At one multiplication per second, this computation would take about 12.3 years. [Sec. 1.1/1.3]

CHAPTER 1 TEST *page 50*

1. deductive [Sec. 1.1] **2.** inductive [Sec. 1.1] **3.** inductive [Sec. 1.1] **4.** deductive [Sec. 1.1] **5.** 384 [Sec. 1.2] **6.** 1, 1, 2, 3, 5, 8, 13, 21, 34, 55 [Sec. 1.2] **7. a.** $a_n = 4n$ **b.** $a_n = 3n + 1$ [Sec. 1.2] **8.** 0, 1, −3, 6, −10, and −5460 [Sec. 1.2] **9.** 131, 212, 343 [Sec. 1.2] **10.** Understand the problem. Devise a plan. Carry out the plan. Review the solution. [Sec. 1.3] **11.** 6 [Sec. 1.3] **12.** 15 [Sec. 1.3] **13.** 3 [Sec. 1.3] **14.** $672 [Sec. 1.3] **15.** 126 [Sec. 1.3] **16.** 36 [Sec. 1.3] **17.** Reynaldo is 13, Ramiro is 5, Shakira is 15, and Sasha is 7. [Sec. 1.1] **18.** 606 [Sec. 1.3] **19.** $x = 4$ provides a counterexample because division by zero is undefined. [Sec. 1.1] **20. a.** 2002 to 2003 **b.** 250,000 [Sec. 1.3]

CHAPTER 2

EXERCISE SET 2.1 *page 61*

1. {penny, nickel, dime, quarter} **3.** {Mercury, Mars} **5.** {Reagan, G. H. W. Bush, Clinton, G. W. Bush} **7.** $\{-5, -4, -3, -2, -1\}$ **9.** {7} **11.** { }

In Exercises 13–19, only one possible answer is given.

13. The set of days of the week that begin with the letter T. **15.** The set consisting of the two planets in our solar system that are closest to the sun. **17.** The set of single-digit natural numbers. **19.** The set of natural numbers less than or equal to 7. **21.** True **23.** False; $b \in \{a, b, c\}$, but $\{b\}$ is not an element of $\{a, b, c\}$. **25.** False; {0} has one element, whereas ∅ has no elements. **27.** False; the word "good" is subjective. **29.** False; 0 is an element of the first set, but 0 is not an element of the second set.

In Exercises 31–39, only one possible answer is given.

31. $\{x \mid x \in N \text{ and } x < 13\}$ **33.** $\{x \mid x \text{ is a multiple of 5 and } 4 < x < 16\}$ **35.** $\{x \mid x$ is the name of a month that has 31 days$\}$ **37.** $\{x \mid x$ is the name of a U.S. state that begins with the letter A$\}$ **39.** $\{x \mid x$ is a season that starts with the letter s$\}$ **41.** {California, Arizona} **43.** {California, Arizona, Florida, Texas} **45.** {2000, 2002, 2004} **47.** {1997, 1998} **49.** {June, October, November} **51.** {1985, 1986, 1987, 1989} **53.** {1988, 1990, 1991, 1992, 1993, 1995, 1996} **55.** 11 **57.** 0 **59.** 4 **61.** 16 **63.** 121 **65.** Neither **67.** Both **69.** Equivalent **71.** Equivalent **73.** Not well defined **75.** Not well defined **77.** Well defined **79.** Well defined **81.** Not well defined **83.** Not well defined **85.** $A = B$ **87.** $A \neq B$ **89.** Answers will vary; however, the set of all real numbers between 0 and 1 is one example of a set that cannot be written using the roster method.

EXERCISE SET 2.2 *page 71*

1. {0, 1, 3, 5, 8} **3.** $U = \{0, 1, 2, 3, 4, 5, 6, 7, 8\}$ **5.** {0, 7, 8} **7.** {0, 2, 4, 6, 8} **9.** $\subseteq$ **11.** $\not\subseteq$ **13.** $\subseteq$ **15.** $\subseteq$ **17.** $\subseteq$ **19.** True **21.** True **23.** True **25.** False **27.** True **29.** True **31.** False **33.** False **35.** False **37.** 18 hours **39.** $\emptyset, \{\alpha\}, \{\beta\}, \{\alpha, \beta\}$ **41.** ∅, {I}, {II}, {III}, {I, II}, {I, III}, {II, III}, {I, II, III} **43.** 4 **45.** 128 **47.** 2048 **49.** 1 **51. a.** 15 **b.** 10 **c.** Two different sets of coins can have the same value. **53. a.** 6 **b.** 2 **c.** 4 **d.** 1 **55. a.** 1024 **b.** 12

57. a. {1, 2, 3} has only three elements, namely 1, 2, and 3. Because {2} is not equal to 1, 2, or 3, {2} $\notin$ {1, 2, 3}. **b.** 1 is not a set, so it cannot be a subset. **c.** The given set has the elements 1 and {1}. Because 1 $\neq$ {1}, there are exactly two elements in {1, {1}}.
59. a. {A, B, C}, {A, B, D}, {A, B, E}, {A, C, D}, {A, C, E}, {A, D, E}, {B, C, D}, {B, C, E}, {B, D, E}, {C, D, E}, {A, B, C, D}, {A, B, C, E}, {A, B, D, E}, {A, C, D, E}, {B, C, D, E}, {A, B, C, D, E} **b.** {A}, {B}, {C}, {D}, {E}, {A, B}, {A, C}, {A, D}, {A, E}, {B, C}, {B, D}, {B, E}, {C, D}, {C, E}, {D, E}

EXERCISE SET 2.3 *page 83*

1. {1, 2, 4, 5, 6, 8} **3.** {4, 6} **5.** {3, 7} **7.** $U = \{1, 2, 3, 4, 5, 6, 7, 8\}$ **9.** $\emptyset$ **11.** $B = \{1, 2, 5, 8\}$
13. $U = \{1, 2, 3, 4, 5, 6, 7, 8\}$ **15.** {2, 5, 8} **17.** {1, 3, 4, 6, 7} **19.** {2, 5, 8}
In Exercises 21–27, one possible answer is given. Your answers may vary from the given answers.
21. The set of all elements that are not in *L* or are in *T*. **23.** The set of all elements that are in *A*, or are in *C* but not in *B*.
25. The set of all elements that are in *T*, and are also in *J* or not in *K*. **27.** The set of all elements that are in both *W* and *V*, or are in both *W* and *Z*. **29.** *U*

31. *U*

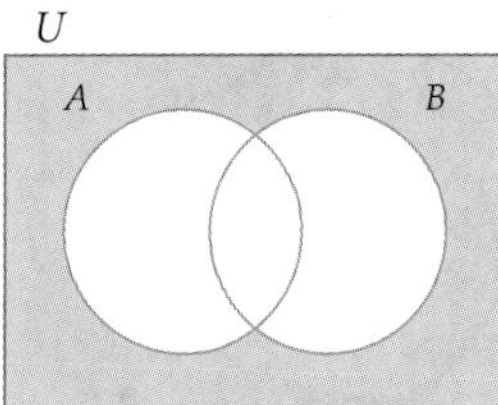

33. *U*

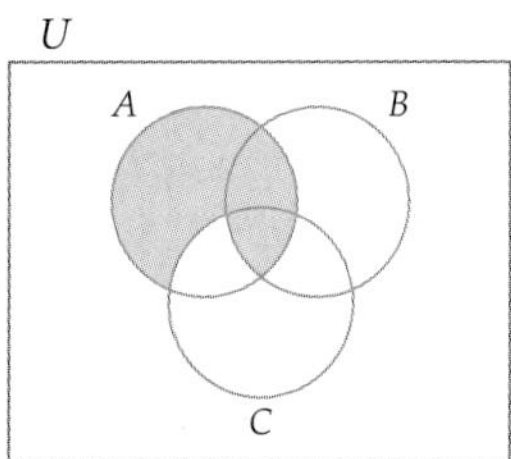

35. *U*

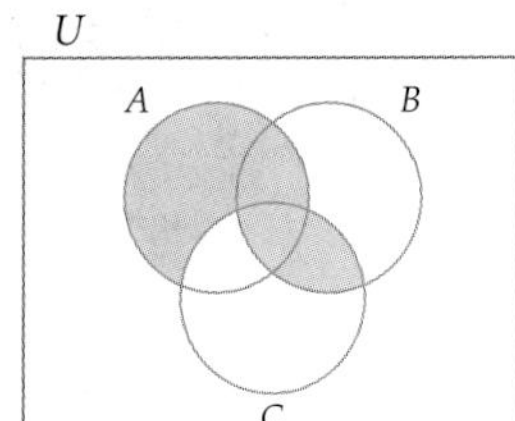

37. Not equal **39.** Equal **41.** Not equal **43.** Not equal

45. Equal **47.** Yellow **49.** Cyan **51.** Red
In Exercises 53–61, one possible answer is given. Your answers may vary from the given answers.
53. $A \cap B'$ **55.** $(A \cup B)'$ **57.** $B \cup C$ **59.** $C \cap (A \cup B)'$ **61.** $(A \cup B)' \cup (A \cap B \cap C)$
63. a. *U*

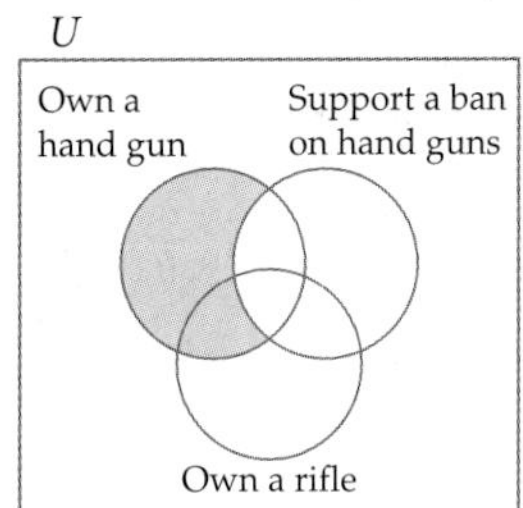

b. *U*

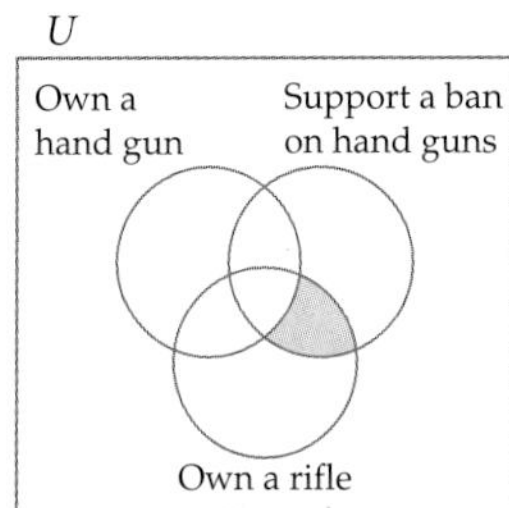

c. *U*

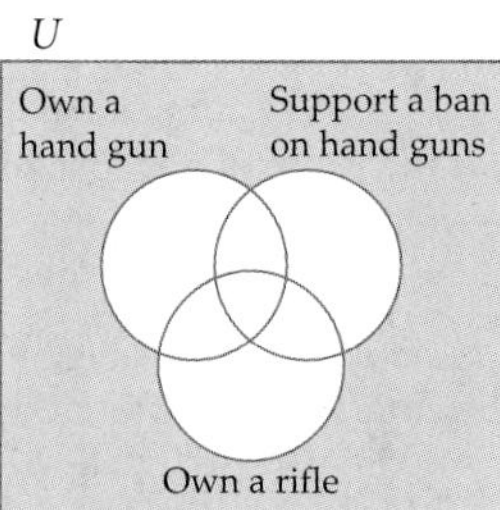

65. *U*

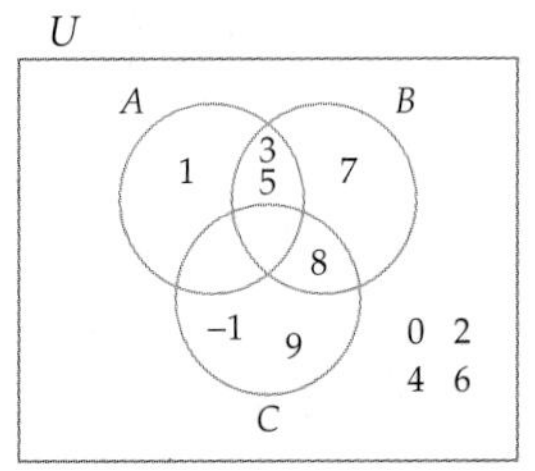

67. *U*

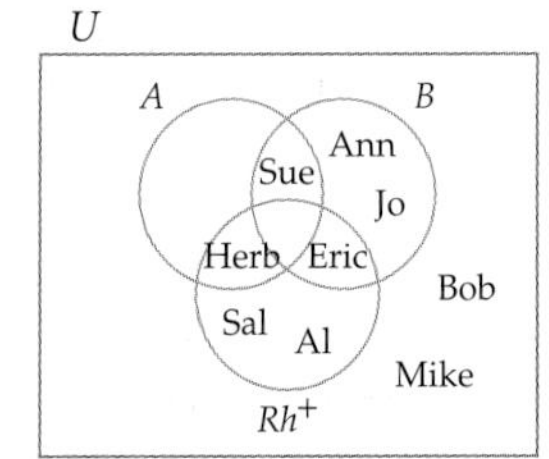

See the *Student Solutions Manual* for the verification for Exercise 69. **71.** {3, 9} **73.** {2, 8} **75.** {3, 9}
77. Responses will vary. **79.**

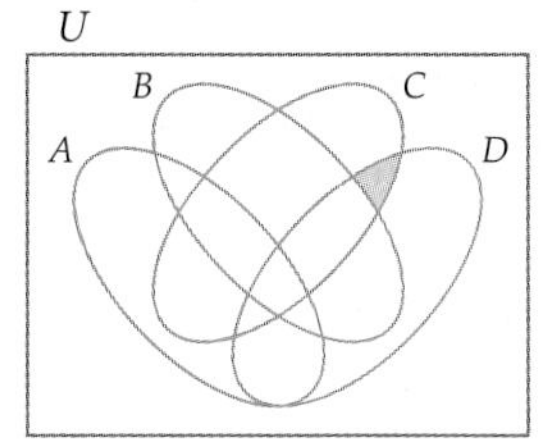

EXERCISE SET 2.4 *page 94*

1. 7 **3.** 8 **5.** 8 **7.** 12 **9.** $n(A \cup B) = 7$; $n(A) + n(B) - n(A \cap B) = 4 + 5 - 2 = 7$ **11.** 113
13. 1060 **15.** U

A B C
6 12 11
3
7 5
25
6

17. a. 180 **b.** 200 **19. a.** 15% **b.** 13%

21. a. 450 **b.** 140 **c.** 130 **23. a.** 109 **b.** 328 **c.** 104 **25. a.** 101 **b.** 370
c. 380 **d.** 373 **e.** 225 **f.** 530 **27. a.** 72 **b.** 47 **c.** 25 **d.** 0
29. a. 200 **b.** 271 **c.** 16

EXERCISE SET 2.5 *page 107*

1. a. One possible one-to-one correspondence between V and M is given by

$$V = \{a, e, i\}$$
$$\updownarrow \; \updownarrow \; \updownarrow$$
$$M = \{3, 6, 9\}$$

b. 6 **3.** Pair $(2n - 1)$ of D with $(3n)$ of M. **5.** $\aleph_0$ **7.** c

9. c **11.** Equivalent **13.** Equivalent **15.** Let $S = \{10, 20, 30, \ldots, 10n, \ldots\}$. Then S is a proper subset of A. A rule for a one-to-one correspondence between A and S is $(5n) \leftrightarrow (10n)$. Because A can be placed in a one-to-one correspondence with a proper subset of itself, A is an infinite set.

17. Let $R = \left\{\frac{3}{4}, \frac{5}{6}, \frac{7}{8}, \ldots, \frac{2n+1}{2n+2}, \ldots\right\}$. Then R is a proper subset of C. A rule for a one-to-one correspondence between C and R is $\left(\frac{2n-1}{2n}\right) \leftrightarrow \left(\frac{2n+1}{2n+2}\right)$. Because C can be placed in a one-to-one correspondence with a proper subset of itself, C is an infinite set.

In Exercises 19–25, let $N = \{1, 2, 3, 4, \ldots, n, \ldots\}$. Then a one-to-one correspondence between the given sets and the set of natural numbers N is given by the following general correspondences.

19. $(n + 49) \leftrightarrow n$ **21.** $\left(\frac{1}{3^{n-1}}\right) \leftrightarrow n$ **23.** $(10^n) \leftrightarrow n$ **25.** $(n^3) \leftrightarrow n$

27. a. For any natural number n, the two natural numbers preceding $3n$ are not multiples of 3. Pair these two numbers, $3n - 2$ and $3n - 1$, with the multiples of 3 given by $6n - 3$ and $6n$, respectively. Using the two general correspondences $(6n - 3) \leftrightarrow (3n - 2)$ and $(6n) \leftrightarrow (3n - 1)$ (as shown below), we can establish a one-to-one correspondence between the multiples of 3 (set M) and the set K of all natural numbers that are not multiples of 3.

$$M = \{3, 6, 9, 12, 15, 18, \ldots, 6n - 3, \; 6n, \; \ldots\}$$
$$\updownarrow \qquad \updownarrow$$
$$K = \{1, 2, 4, 5, 7, 8, \ldots, 3n - 2, 3n - 1, \ldots\}$$

The following answers in parts b and c were produced by using the correspondences established in part a. **b.** 302 **c.** 1800

29. The set of real numbers x such that $0 < x < 1$ is equivalent to the set of all real numbers.

CHAPTER 2 REVIEW EXERCISES *page 110*

1. {0, 1, 2, 3, 4, 5, 6, 7} [Sec. 2.1] **2.** {−8, 8} [Sec. 2.1] **3.** {1, 2, 3, 4} [Sec. 2.1] **4.** {1, 2, 3, 4, 5, 6} [Sec. 2.1]
5. $\{x \mid x \in I \text{ and } x > -6\}$ [Sec. 2.1] **6.** $\{x \mid x$ is the name of a month with exactly 30 days$\}$ [Sec. 2.1]
7. $\{x \mid x$ is the name of a U.S. state that begins with the letter K$\}$ [Sec. 2.1] **8.** $\{x^3 \mid x = 1, 2, 3, 4, \text{ or } 5\}$ [Sec. 2.1]
9. False [Sec. 2.1] **10.** True [Sec. 2.1] **11.** True [Sec. 2.1] **12.** False [Sec. 2.1]
13. {6, 10} [Sec. 2.3] **14.** {2, 6, 10, 16, 18} [Sec. 2.3] **15.** $C = \{14, 16\}$ [Sec. 2.3]
16. {2, 6, 8, 10, 12, 16, 18} [Sec. 2.3] **17.** {2, 6, 10, 16} [Sec. 2.3] **18.** {8, 12} [Sec. 2.3]
19. {6, 8, 10, 12, 14, 16, 18} [Sec. 2.3] **20.** {8, 12} [Sec. 2.3] **21.** Proper subset [Sec. 2.2]
22. Proper subset [Sec. 2.2] **23.** Not a proper subset [Sec. 2.2] **24.** Not a proper subset [Sec. 2.2]
25. ∅, {I}, {II}, {I, II} [Sec. 2.2] **26.** ∅, {s}, {u}, {n}, {s, u}, {s, n}, {u, n}, {s, u, n} [Sec. 2.2]
27. ∅, {penny}, {nickel}, {dime}, {quarter}, {penny, nickel}, {penny, dime}, {penny, quarter}, {nickel, dime}, {nickel, quarter}, {dime, quarter}, {penny, nickel, dime}, {penny, nickel, quarter}, {penny, dime, quarter}, {nickel, dime, quarter}, {penny, nickel, dime, quarter} [Sec. 2.2] **28.** ∅, {A}, {B}, {C}, {D}, {E}, {A, B}, {A, C}, {A, D}, {A, E}, {B, C}, {B, D}, {B, E}, {C, D}, {C, E}, {D, E}, {A, B, C}, {A, B, D}, {A, B, E}, {A, C, D}, {A, C, E}, {A, D, E}, {B, C, D}, {B, C, E}, {B, D, E}, {C, D, E}, {A, B, C, D}, {A, B, C, E}, {A, B, D, E}, {A, C, D, E}, {B, C, D, E}, {A, B, C, D, E} [Sec. 2.2] **29.** $2^4 = 16$ [Sec. 2.2]
30. $2^{26} = 67{,}108{,}864$ [Sec. 2.2] **31.** $2^{15} = 32{,}768$ [Sec. 2.2] **32.** $2^7 = 128$ [Sec. 2.2]

33.

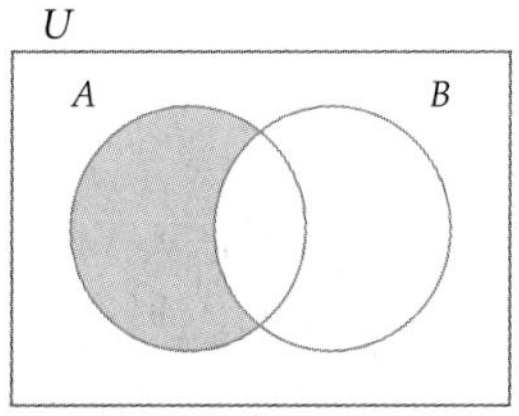

[Sec. 2.3]

34.

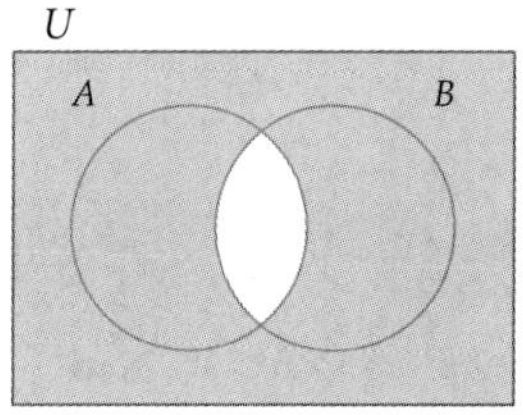

[Sec. 2.3]

35.

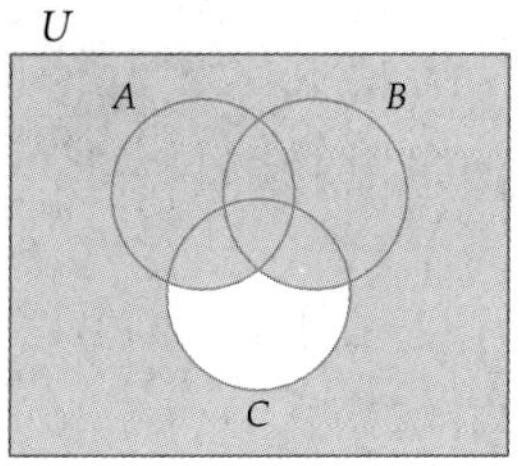

[Sec. 2.3]

36.

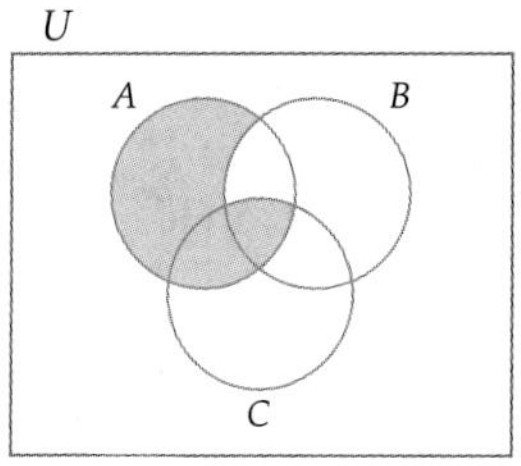

[Sec. 2.3]

37. Equal [Sec. 2.3]

38. Not equal [Sec. 2.3] **39.** Not equal [Sec. 2.3] **40.** Not equal [Sec. 2.3]
41. $(A \cup B)' \cap C$ or $C \cap (A' \cap B')$ [Sec. 2.3] **42.** $(A \cap B) \cup (B \cap C')$ [Sec. 2.3]

43.

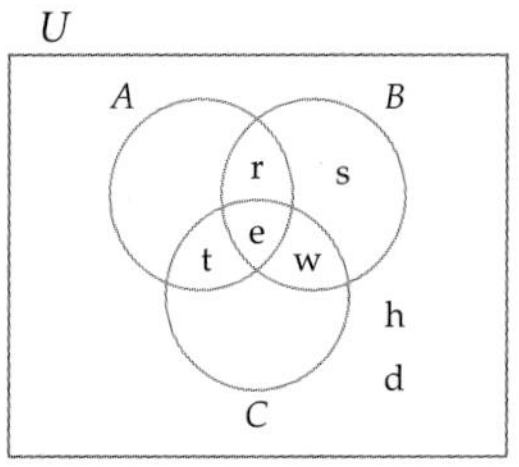

[Sec. 2.3]

44.

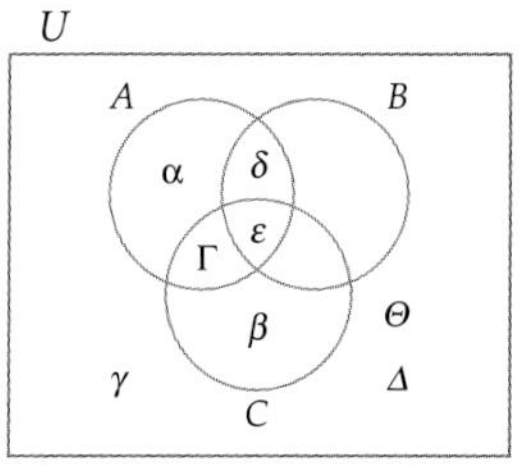

[Sec. 2.3]

45. 391 [Sec. 2.4]

46. a. 42 **b.** 31 **c.** 20 **d.** 142 [Sec. 2.4]

47. One possible one-to-one correspondence between {1, 3, 6, 10} and {1, 2, 3, 4} is given by
{1, 3, 6, 10}
↕ ↕ ↕ ↕
{1, 2, 3, 4} [Sec. 2.5]

48. $\{x \mid x > 10 \text{ and } x \in N\} = \{11, 12, 13, 14, \ldots, n + 10, \ldots\}$
Thus a one-to-one correspondence between the sets is given by
$\{11, 12, 13, 14, \ldots, n + 10, \ldots\}$
↕ ↕ ↕ ↕ ↕
$\{2, 4, 6, 8, \ldots, 2n, \ldots\}$ [Sec. 2.5]

49. One possible one-to-one correspondence between the sets is given by
$\{3, 6, 9, \ldots, 3n, \ldots\}$
↕ ↕ ↕ ↕
$\{10, 100, 1000, \ldots, 10^n, \ldots\}$ [Sec. 2.5]

50. In the following figure, the line from E that passes through $\overline{AB}$ and $\overline{CD}$ illustrates a method of establishing a one-to-one correspondence between $\{x \mid 0 \le x \le 1\}$ and $\{x \mid 0 \le x \le 4\}$.

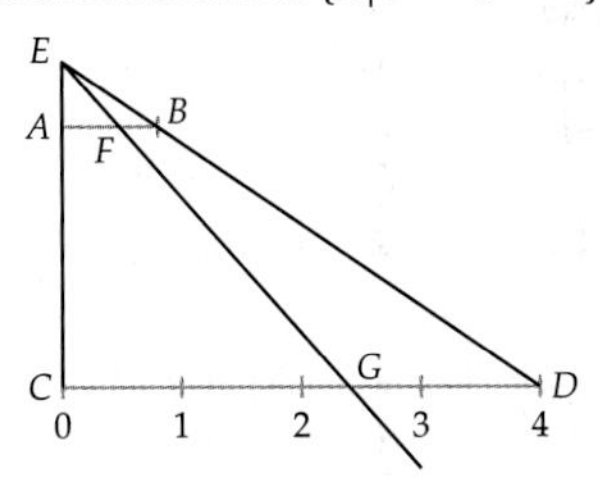

[Sec. 2.5]

51. A proper subset of A is $S = \{10, 14, 18, \ldots, 4n + 6, \ldots\}$. A one-to-one correspondence between A and S is given by

$$A = \{6, 10, 14, 18, \ldots, 4n + 2, \ldots\}$$
$$\updownarrow \quad \updownarrow \quad \updownarrow \quad \updownarrow \qquad \updownarrow$$
$$S = \{10, 14, 18, 22, \ldots, 4n + 6, \ldots\}$$

Because A can be placed in a one-to-one correspondence with a proper subset of itself, A is an infinite set. [Sec. 2.5]

52. A proper subset of B is $T = \left\{\frac{1}{2}, \frac{1}{4}, \frac{1}{8}, \frac{1}{16}, \ldots, \frac{1}{2^n}, \ldots\right\}$. A one-to-one correspondence between B and T is given by

$$B = \left\{1, \frac{1}{2}, \frac{1}{4}, \frac{1}{8}, \ldots, \frac{1}{2^{n-1}}, \ldots\right\}$$
$$\updownarrow \quad \updownarrow \quad \updownarrow \quad \updownarrow \qquad \updownarrow$$
$$T = \left\{\frac{1}{2}, \frac{1}{4}, \frac{1}{8}, \frac{1}{16}, \ldots, \frac{1}{2^n}, \ldots\right\}$$

Because B can be placed in a one-to-one correspondence with a proper subset of itself, B is an infinite set. [Sec. 2.5]

53. 5 [Sec. 2.1] **54.** 10 [Sec. 2.1] **55.** 2 [Sec. 2.1] **56.** 5 [Sec. 2.1] **57.** $\aleph_0$ [Sec. 2.5]
58. $\aleph_0$ [Sec. 2.5] **59.** c [Sec. 2.5] **60.** c [Sec. 2.5] **61.** $\aleph_0$ [Sec. 2.5] **62.** $\aleph_0$ [Sec. 2.5]
63. $\aleph_0$ [Sec. 2.5] **64.** c [Sec. 2.5] **65.** c [Sec. 2.5] **66.** c [Sec. 2.5] **67.** $\aleph_0$ [Sec. 2.5]
68. c [Sec. 2.5]

CHAPTER 2 TEST *page 112*

1. $\{2, 3, 5, 7, 8, 9, 10\}$ [Sec. 2.1/2.3] **2.** $\{2, 9, 10\}$ [Sec. 2.1/2.3] **3.** $\{1, 2, 4, 5, 6, 7, 9, 10\}$ [Sec. 2.1/2.3]
4. $\{2, 9, 10\}$ [Sec. 2.1/2.3] **5.** $\{1, 2, 3, 4, 6, 9, 10\}$ [Sec. 2.1/2.3] **6.** $\{5, 7, 8\}$ [Sec. 2.1/2.3]
7. $\{x \mid x \in W \text{ and } x < 7\}$ [Sec. 2.1] **8.** $\{x \mid x \in I \text{ and } -3 \le x \le 2\}$ [Sec. 2.1] **9. a.** 4
b. $\aleph_0$ **c.** $\aleph_0$ **d.** c [Sec. 2.1/2.5] **10. a.** Equivalent **b.** Equivalent [Sec. 2.5]
11. a. Neither **b.** Equivalent [Sec. 2.5]

12. $\emptyset, \{a\}, \{b\}, \{c\}, \{d\}$
$\{a, b\}, \{a, c\}, \{a, d\}, \{b, c\}, \{b, d\}, \{c, d\}$
$\{a, b, c\}, \{a, b, d\}, \{a, c, d\}, \{b, c, d\}$
$\{a, b, c, d\}$ [Sec. 2.2]

13. $2^{21} = 2{,}097{,}152$ [Sec. 2.2]

14. a. False **b.** True
c. False **d.** True [Sec. 2.1/2.2/2.5]

15. a.

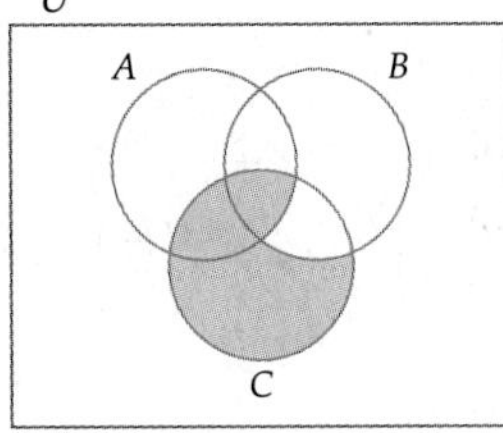
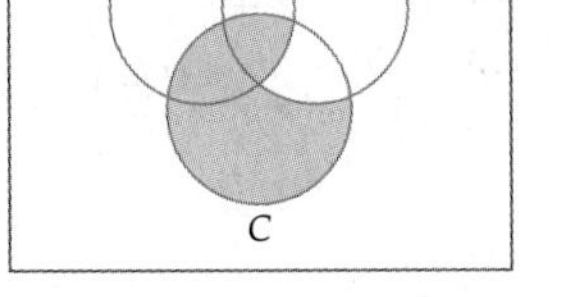

b.

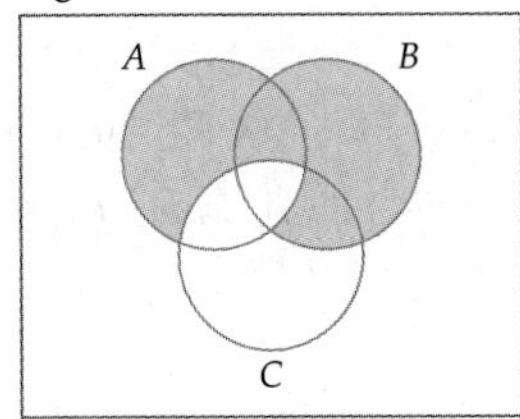

[Sec. 2.3]

16. $B \cup (A \cap C)$ [Sec. 2.3]

17. 541 [Sec. 2.4] **18. a.** 232 **b.** 102 **c.** 857 **d.** 79 [Sec. 2.4]

19.
$$\{5, 10, 15, 20, 25, \ldots, 5n, \ldots\}$$
$$\updownarrow \quad \updownarrow \quad \updownarrow \quad \updownarrow \quad \updownarrow \qquad \updownarrow$$
$$\{0, 1, 2, 3, 4, \ldots, n - 1, \ldots\}$$
$$(5n) \leftrightarrow (n - 1)$$
[Sec. 2.5]

20.

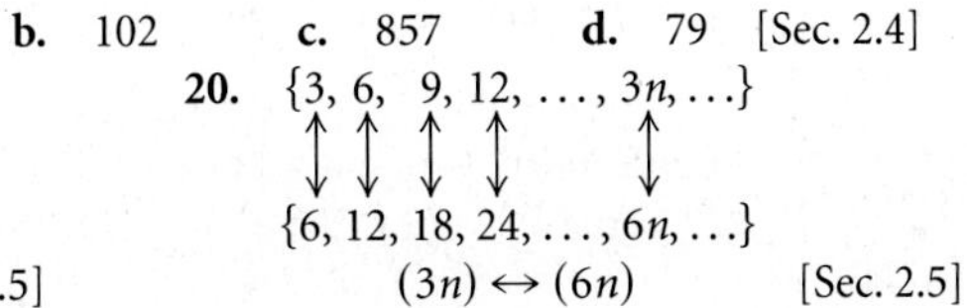

$$\{3, 6, 9, 12, \ldots, 3n, \ldots\}$$
$$\updownarrow \quad \updownarrow \quad \updownarrow \quad \updownarrow \qquad \updownarrow$$
$$\{6, 12, 18, 24, \ldots, 6n, \ldots\}$$
$$(3n) \leftrightarrow (6n)$$
[Sec. 2.5]

CHAPTER 3

EXERCISE SET 3.1 *page 123*

1. Statement **3.** Statement **5.** Not a statement **7.** Statement **9.** Not a statement
11. One component is "The principal will attend the class on Tuesday." The other component is "The principal will attend the class on Wednesday." **13.** One component is "A triangle is an acute triangle." The other component is "It has three acute angles."
15. One component is "I ordered a salad." The other component is "I ordered a cola." **17.** One component is $5 + 2 > 6$. The other component is $5 + 2 = 6$. **19.** The Giants did not lose the game. **21.** The game went into overtime.
23. $w \rightarrow t$; conditional **25.** $s \rightarrow r$; conditional **27.** $l \leftrightarrow a$; biconditional **29.** $d \rightarrow f$; conditional
31. $m \vee c$; disjunction **33.** The tour goes to Italy and the tour does not go to Spain. **35.** If we go to Venice, then we will not go to Florence. **37.** We will go to Florence if and only if we do not go to Venice. **39.** All cats have claws.
41. Some classic movies were not first produced in black and white. **43.** Some of the numbers were even numbers.
45. Some irrational numbers can be written as terminating decimals. **47.** Some cars do not run on gasoline.
49. Some items are not on sale. **51.** True **53.** True **55.** True **57.** True **59.** True **61.** True
63. True **65.** $p \rightarrow q$, where p represents "you can count your money" and q represents "you don't have a billion dollars."
67. $p \rightarrow q$, where p represents "you do not learn from history" and q represents "you are condemned to repeat it."
69. $p \rightarrow q$, where p represents "people concentrated on the really important things in life" and q represents "there'd be a shortage of fishing poles." **71.** $p \leftrightarrow q$, where p represents "an angle is a right angle" and q represents "its measure is 90°."
73. $p \rightarrow q$, where p represents "two sides of a triangle are equal in length" and q represents "the angles opposite those sides are congruent." **75.** $p \rightarrow q$, where p represents "it is a square" and q represents "it is a rectangle."

EXERCISE SET 3.2 *page 135*

1. True **3.** False **5.** False **7.** False **9.** False **11. a.** If p is false, then $p \wedge (q \vee r)$ must be a false statement. **b.** For a conjunctive statement to be true, it is necessary that all components of the statement be true. Because it is given that one of the components (p) is false, $p \wedge (q \vee r)$ must be a false statement.

p	q	13.	15.	17.
T	T	T	F	F
T	F	F	T	T
F	T	T	F	T
F	F	T	F	T

p	q	r	19.	21.	23.	25.	27.
T	T	T	F	T	T	T	F
T	T	F	F	T	F	T	T
T	F	T	F	T	F	T	T
T	F	F	F	T	F	T	F
F	T	T	F	F	T	T	F
F	T	F	F	F	F	T	F
F	F	T	F	T	T	F	T
F	F	F	T	T	F	T	F

See the *Student Solutions Manual* for the solutions to Exercises 29–35. **37.** It did not rain and it did not snow. **39.** She did not visit either France or Italy. **41.** She did not get a promotion and she did not receive a raise. **43.** Tautology
45. Tautology **47.** Tautology **49.** Self-contradiction **51.** Self-contradiction **53.** Not a self-contradiction
55. The symbol ≤ means "less than or equal to." **57.** $2^5 = 32$ **59.** F F F T T T F T F F F T F F F F

EXERCISE SET 3.3 *page 144*

1. *Antecedent:* I had the money
Consequent: I would buy the painting
3. *Antecedent:* they had a guard dog
Consequent: no one would trespass on their property
5. *Antecedent:* I change my major
Consequent: I must reapply for admission
7. True **9.** True **11.** True **13.** False

p	q	15.	17.
T	T	T	T
T	F	T	T
F	T	T	T
F	F	T	T

p	q	r	19.	21.	23.
T	T	T	T	T	T
T	T	F	T	F	T
T	F	T	T	T	T
T	F	F	T	T	T
F	T	T	T	T	T
F	T	F	T	F	T
F	F	T	T	T	T
F	F	F	T	T	T

25. She cannot sing or she would be perfect for the part. **27.** Either x is not an irrational number or x is not a terminating decimal. **29.** The fog must lift or our flight will be cancelled. **31.** They offered me the contract and I didn't accept. **33.** Pigs have wings and they still can't fly. **35.** She traveled to Italy and she didn't visit her relatives. **37.** False **39.** True **41.** False **43.** True **45.** True **47.** $p \rightarrow v$ **49.** $t \rightarrow \sim v$ **51.** $(\sim t \wedge p) \rightarrow v$ **53.** Not equivalent **55.** Not equivalent **57.** Equivalent **59.** If a number is a rational number, then it is a real number. **61.** If a number is a repeating decimal, then it is a rational number. **63.** If an animal is a Sauropod, then it is herbivorous.

EXERCISE SET 3.4 *page 150*

1. If we take the aerobics class, then we will be in good shape for the ski trip. **3.** If the number is an odd prime number, then it is greater than 2. **5.** If he has the talent to play a keyboard, then he can join the band. **7.** If I was able to prepare for the test, then I had the textbook. **9.** If you ran the Boston marathon, then you are in excellent shape. **11. a.** If I quit this job, then I am rich. **b.** If I were not rich, then I would not quit this job. **c.** If I would not quit this job, then I would not be rich. **13. a.** If we are not able to attend the party, then she did not return soon. **b.** If she returns soon, then we will be able to attend the party. **c.** If we are able to attend the party, then she returned soon. **15. a.** If a figure is a quadrilateral, then it is a parallelogram. **b.** If a figure is not a parallelogram, then it is not a quadrilateral. **c.** If a figure is not a quadrilateral, then it is not a parallelogram. **17. a.** If I am able to get current information about astronomy, then I have access to the Internet. **b.** If I do not have access to the Internet, then I will not be able to get current information about astronomy. **c.** If I am not able to get current information about astronomy, then I don't have access to the Internet. **19. a.** If we don't have enough money for dinner, then we took a taxi. **b.** If we did not take a taxi, then we will have enough money for dinner. **c.** If we have enough money for dinner, then we did not take a taxi. **21. a.** If she can extend her vacation for at least two days, then she will visit Kauai. **b.** If she does not visit Kauai, then she could not extend her vacation for at least two days. **c.** If she cannot extend her vacation for at least two days, then she will not visit Kauai. **23. a.** If two lines are parallel, then the two lines are perpendicular to a given line. **b.** If two lines are not perpendicular to a given line, then the two lines are not parallel. **c.** If two lines are not parallel, then the two lines are not both perpendicular to a given line. **25.** Not equivalent **27.** Equivalent **29.** Not equivalent **31.** If $x = 7$, then $3x - 7 \neq 11$. The original statement is true. **33.** If $|a| = 3$, then $a = 3$. The original statement is false. **35.** If $a + b = 25$, then $\sqrt{a + b} = 5$. The original statement is true. **37.** $p \rightarrow q$ **39. a. and b.** Answers will vary. **41.** If you can dream it, then you can do it. **43.** If I were a dancer, then I would not be a singer. **45.** A conditional statement and its contrapositive are equivalent. They always have the same truth values. **47.** The Hatter is telling the truth.

EXERCISE SET 3.5 *page 162*

1. $r \rightarrow c$, r, $\therefore c$ **3.** $g \rightarrow s$, $\sim g$, $\therefore \sim s$ **5.** $s \rightarrow i$, s, $\therefore i$ **7.** $\sim p \rightarrow \sim a$, a, $\therefore p$ **9.** Invalid **11.** Invalid **13.** Valid **15.** Invalid **17.** Invalid **19.** Invalid **21.** Valid **23.** Valid

25. $h \to r$
$\sim h$
$\therefore \sim r$
Invalid

27. $\sim b \to d$
$b \vee d$
$\therefore b$
Invalid

29. $c \to t$
t
$\therefore c$
Invalid

31. Valid argument; modus tollens
33. Invalid argument; fallacy of the inverse
35. Valid argument; law of syllogism
37. Valid argument; modus ponens
39. Valid argument; modus tollens

See the *Student Solutions Manual* for the solutions to Exercises 41–45.
47. q
49. it is not a theropod
51. Valid

EXERCISE SET 3.6 *page 171*

1. Valid 3. Valid 5. Valid 7. Valid 9. Invalid 11. Invalid 13. Invalid
15. Valid 17. Valid 19. Invalid 21. All Reuben sandwiches need mustard. 23. 1001 ends with a 5.
25. Some horses are grey. 27. a. Invalid b. Invalid c. Invalid d. Invalid e. Valid f. Valid

29.

1: 2	2: 1
3: 8	4: 4

CHAPTER 3 REVIEW EXERCISES *page 174*

1. Not a statement [Sec. 3.1] 2. Statement [Sec. 3.1] 3. Statement [Sec. 3.1] 4. Statement [Sec. 3.1]
5. Not a statement [Sec. 3.1] 6. Statement [Sec. 3.1] 7. $m \wedge b$; conjunction [Sec. 3.1]
8. $d \to e$; conditional [Sec. 3.1] 9. $g \leftrightarrow d$; biconditional [Sec. 3.1] 10. $t \to s$; conditional [Sec. 3.1]
11. No dogs bite. [Sec. 3.1] 12. Some desserts at the Cove restaurant are not good. [Sec. 3.1] 13. Some winners do not receive a prize. [Sec. 3.1] 14. All cameras use film. [Sec. 3.1] 15. Some of the students received an A. [Sec. 3.1]
16. Nobody enjoyed the story. [Sec. 3.1] 17. True [Sec. 3.1] 18. True [Sec. 3.1] 19. True [Sec. 3.1]
20. True [Sec. 3.1] 21. True [Sec. 3.1] 22. True [Sec. 3.1] 23. False [Sec. 3.2]
24. False [Sec. 3.2/3.3] 25. True [Sec. 3.2] 26. True [Sec. 3.2/3.3] 27. True [Sec. 3.2/3.3]
28. False [Sec. 3.2/3.3]

p	q	29. [Sec. 3.2/3.3]	30. [Sec. 3.2/3.3]	31. [Sec. 3.2/3.3]	32. [Sec. 3.2/3.3]
T	T	T	F	F	T
T	F	T	F	F	T
F	T	T	T	F	F
F	F	F	F	F	T

p	q	r	33. [Sec. 3.2/3.3]	34. [Sec. 3.2/3.3]	35. [Sec. 3.2/3.3]	36. [Sec. 3.2/3.3]
T	T	T	T	F	F	T
T	T	F	T	T	F	T
T	F	T	T	T	T	T
T	F	F	F	T	T	T
F	T	T	T	F	F	F
F	T	F	T	F	F	T
F	F	T	T	T	F	T
F	F	F	T	T	F	T

37. Bob passed the English proficiency test or he did not register for a speech course. [Sec. 3.2] **38.** It is not true that Ellen went to work this morning or she took her medication. [Sec. 3.2] **39.** It is not the case that Wendy will not go to the store this afternoon and she will be able to prepare her fettuccine al pesto recipe. [Sec. 3.2] **40.** It is not the case that Gina did not enjoy the movie or she enjoyed the party. [Sec. 3.2/3.3] See the *Student Solutions Manual* for solutions to Exercises 41–44. **45.** Self-contradiction [Sec. 3.2] **46.** Tautology [Sec. 3.2/3.3] **47.** Tautology [Sec. 3.2/3.3] **48.** Tautology [Sec. 3.2/3.3] **49.** *Antecedent:* he has talent *Consequent:* he will succeed [Sec. 3.3] **50.** *Antecedent:* I had a credential *Consequent:* I could get the job [Sec. 3.3] **51.** *Antecedent:* I join the fitness club *Consequent:* I will follow the exercise program [Sec. 3.3] **52.** *Antecedent:* I will attend *Consequent:* it is free [Sec. 3.3] **53.** She is not tall or she would be on the volleyball team. [Sec. 3.3] **54.** He cannot stay awake or he would finish the report. [Sec. 3.3] **55.** Rob is ill or he would start. [Sec. 3.3] **56.** Sharon will not be promoted or she closes the deal. [Sec. 3.3] **57.** I get my paycheck and I do not purchase a ticket. [Sec. 3.3] **58.** The tomatoes will get big and you did not provide plenty of water. [Sec. 3.3] **59.** You entered Cleggmore University and you did not have a high score on the SAT exam. [Sec. 3.3] **60.** Ryan enrolled at a university and he did not enroll at Yale. [Sec. 3.3] **61.** False [Sec. 3.3] **62.** True [Sec. 3.3] **63.** True [Sec. 3.3] **64.** False [Sec. 3.3] **65.** False [Sec. 3.3] **66.** False [Sec. 3.3] **67.** If a real number has a nonrepeating, nonterminating decimal form, then the real number is irrational. [Sec. 3.4] **68.** If you are a politician, then you are well known. [Sec. 3.4] **69.** If I can sell my condominium, then I can buy the house. [Sec. 3.4] **70.** If a number is divisible by 9, then the number is divisible by 3. [Sec. 3.4] **71. a.** *Converse:* If $x > 3$, then $x + 4 > 7$. **b.** *Inverse:* If $x + 4 \leq 7$, then $x \leq 3$. **c.** *Contrapositive:* If $x \leq 3$, then $x + 4 \leq 7$. [Sec. 3.4] **72. a.** *Converse:* If the recipe can be prepared in less than 20 minutes, then the recipe is in this book. **b.** *Inverse:* If the recipe is not in this book, then the recipe cannot be prepared in less than 20 minutes. **c.** *Contrapositive:* If the recipe cannot be prepared in less than 20 minutes, then the recipe is not in this book. [Sec. 3.4] **73. a.** *Converse:* If $(a + b)$ is divisible by 3, then a and b are both divisible by 3. **b.** *Inverse:* If a and b are not both divisible by 3, then $(a + b)$ is not divisible by 3. **c.** *Contrapositive:* If $(a + b)$ is not divisible by 3, then a and b are not both divisible by 3. [Sec. 3.4] **74. a.** *Converse:* If they come, then you built it. **b.** *Inverse:* If you do not build it, then they will not come. **c.** *Contrapositive:* If they do not come, then you did not build it. [Sec. 3.4] **75. a.** *Converse:* If it has exactly two parallel sides, then it is a trapezoid. **b.** *Inverse:* If it is not a trapezoid, then it does not have exactly two parallel sides. **c.** *Contrapositive:* If it does not have exactly two parallel sides, then it is not a trapezoid. [Sec. 3.4] **76. a.** *Converse:* If they returned, then they liked it. **b.** *Inverse:* If they do not like it, then they will not return. **c.** *Contrapositive:* If they do not return, then they did not like it. [Sec. 3.4] **77.** $q \rightarrow p$, the converse of the original statement [Sec. 3.4] **78.** $p \rightarrow q$, the original statement [Sec. 3.4] **79.** If x is an odd prime number, then $x > 2$. [Sec. 3.4] **80.** If the senator attends the meeting, then she will vote on the motion. [Sec. 3.4] **81.** If their manager contacts me, then I will purchase some of their products. [Sec. 3.4] **82.** If I can rollerblade, then Ginny can rollerblade. [Sec. 3.4] **83.** Valid [Sec. 3.5] **84.** Valid [Sec. 3.5] **85.** Invalid [Sec. 3.5] **86.** Valid [Sec. 3.5] **87.** Valid argument; disjunctive syllogism [Sec. 3.5] **88.** Valid argument; law of syllogism [Sec. 3.5] **89.** Invalid argument; fallacy of the inverse [Sec. 3.5] **90.** Valid argument; disjunctive syllogism [Sec. 3.5] **91.** Valid argument; modus tollens [Sec. 3.5] **92.** Invalid argument; fallacy of the inverse [Sec. 3.5] **93.** Valid [Sec. 3.6] **94.** Invalid [Sec. 3.6] **95.** Invalid [Sec. 3.6] **96.** Valid [Sec. 3.6]

CHAPTER 3 TEST *page 176*

1. a. Not a statement **b.** Statement [Sec. 3.1] **2. a.** All trees are green. **b.** Some of the kids had seen the movie. [Sec. 3.1] **3. a.** False **b.** True [Sec. 3.1] **4. a.** False **b.** True [Sec. 3.2/3.3]

p	q	5. [Sec. 3.2/3.3]
T	T	T
T	F	T
F	T	T
F	F	T

p	q	r	6. [Sec. 3.2/3.3]
T	T	T	F
T	T	F	T
T	F	T	F
T	F	F	F
F	T	T	F
F	T	F	T
F	F	T	T
F	F	F	F

7. It is not the case that Elle ate breakfast or took a lunch break. [Sec. 3.2]

8. A tautology is a statement that is always true. [Sec. 3.2] **9.** $\sim p \vee q$ [Sec. 3.3] **10. a.** False **b.** False [Sec. 3.2/3.3] **11. a.** *Converse:* If $x > 4$, then $x + 7 > 11$. **b.** *Inverse:* If $x + 7 \leq 11$, then $x \leq 4$. **c.** *Contrapositive:* If $x \leq 4$, then $x + 7 \leq 11$. [Sec. 3.4]

12. $p \rightarrow q$
$\underline{p\quad\quad}$
$\therefore q$ [Sec. 3.5]

13. $p \rightarrow q$
$\underline{q \rightarrow r}$
$\therefore p \rightarrow r$ [Sec. 3.5]

14. Valid [Sec. 3.5] **15.** Invalid [Sec. 3.5]

16. Invalid argument; the argument is a fallacy of the inverse. [Sec. 3.5] **17.** Valid argument; the argument is a disjunctive syllogism. [Sec. 3.5] **18.** Invalid argument, as shown by an Euler diagram. [Sec. 3.6] **19.** Invalid argument, as shown by an Euler diagram. [Sec. 3.6] **20.** Invalid argument; the argument is a fallacy of the converse. [Sec. 3.5]

CHAPTER 4

EXERCISE SET 4.1 *page 185*

1. 𓎆𓎆𓎆𓎆𓏺𓏺𓏺𓏺𓏺𓏺 **3.** 𓍢𓏺𓏺𓏺 **5.** 𓆼𓆼𓍢𓍢𓍢𓍢𓍢𓎆𓎆𓎆𓎆𓎆𓏺𓏺𓏺𓏺𓏺𓏺𓏺𓏺 **7.** 𓂭𓂭𓆼𓆼𓆼𓍢𓍢𓍢𓍢𓏺𓏺
9. 𓂭𓂭𓂭𓂭𓂭𓂭𓆼𓆼𓆼𓆼𓆼𓍢𓍢𓍢𓍢𓍢𓍢𓍢𓍢 **11.** 𓁨𓆐𓆐𓆐𓆐𓆼𓆼𓆼𓆼𓆼𓍢𓍢𓏺𓏺𓏺 **13.** 2134 **15.** 845 **17.** 1232
19. 221,011 **21.** 65,769 **23.** 5,122,406 **25.** 94 **27.** 666 **29.** 32 **31.** 56 **33.** 650
35. 1409 **37.** 1240 **39.** 840 **41.** 9044 **43.** 11,461 **45.** CLVII **47.** DXLII
49. MCXCVII **51.** DCCLXXXVII **53.** DCLXXXIII **55.** $\overline{\text{VI}}$DCCCXCVIII **57.** 504 **59.** 203
61. 595 **63.** 2484 **65. a. and b.** Answers will vary.

EXERCISE SET 4.2 *page 196*

1. $(4 \times 10^1) + (8 \times 10^0)$ **3.** $(4 \times 10^2) + (2 \times 10^1) + (0 \times 10^0)$ **5.** $(6 \times 10^3) + (8 \times 10^2) + (0 \times 10^1) + (3 \times 10^0)$
7. $(1 \times 10^4) + (0 \times 10^3) + (2 \times 10^2) + (0 \times 10^1) + (8 \times 10^0)$ **9.** 456 **11.** 5076 **13.** 35,407
15. 683,040 **17.** 76 **19.** 395 **21.** 2481 **23.** 27 **25.** 3363 **27.** 10,311 **29.** 23
31. 97 **33.** 72,133 **35.** 2,171,466 **37.** 𒌋𒌋𒌋𒌋𒁹𒁹 **39.** 𒁹𒁹 𒁹𒁹𒁹𒁹𒁹𒁹𒁹𒁹
41. 𒁹 𒌋𒌋𒌋𒁹𒁹𒁹𒁹 𒌋𒌋𒌋𒁹𒁹𒁹𒁹𒁹𒁹𒁹𒁹𒁹 **43.** 𒁹𒁹 𒌋𒌋𒌋𒌋𒌋𒁹𒁹𒁹𒁹𒁹𒁹 𒌋𒌋𒁹𒁹𒁹𒁹 **45.** 𒁹𒁹𒁹𒁹𒁹 𒌋𒌋𒌋𒌋𒌋𒁹𒁹𒁹𒁹𒁹 𒌋𒌋𒌋𒌋𒁹𒁹𒁹𒁹𒁹
47. 𒁹 𒁹𒁹𒁹𒁹𒁹𒁹𒁹𒁹 **49.** 𒁹 𒁹𒁹𒁹𒁹𒁹𒁹 𒁹𒁹𒁹 **51.** 𒌋𒁹 𒌋𒌋𒌋𒌋𒌋𒁹𒁹𒁹 𒌋𒁹𒁹𒁹𒁹𒁹
53. 194 **55.** 1803 **57.** 14,492 **59.** 36,103 **61.** 𝋦
𝋱 **63.** 𝋢
𝋫
𝋨 **65.** 𝋤
𝋬
𝋭
67. 𝋡
𝋠
𝋫
𝋬 **69. a. and b.** Answers will vary.

EXERCISE SET 4.3 *page 205*

1. 73 **3.** 61 **5.** 718 **7.** 485 **9.** 181 **11.** 2032_{five} **13.** 12540_{six} **15.** 22886_{nine} **17.** $111111011100_{\text{two}}$ **19.** $1B7_{\text{twelve}}$ **21.** 13 **23.** 27 **25.** 100 **27.** 139 **29.** 41 **31.** 90 **33.** 1338 **35.** 26_{eight} **37.** 23033_{four} **39.** 24_{five} **41.** 2446_{nine} **43.** 126_{eight} **45.** $7C_{\text{sixteen}}$ **47.** 11101010_{two} **49.** 312_{eight} **51.** 151_{sixteen} **53.** $1011111011110011_{\text{two}}$ **55.** $10111010010111001111_{\text{two}}$ **57.** Answers will vary. **59.** 54 **61.**  **63.** 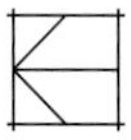 **65.** 256 **67. a. and b.** Answers will vary.

EXERCISE SET 4.4 *page 218*

1. 332_{five} **3.** 6562_{seven} **5.** 1001000_{two} **7.** 1271_{twelve} **9.** $D036_{\text{sixteen}}$ **11.** 1124_{six} **13.** 241_{five} **15.** 6542_{eight} **17.** 1111_{two} **19.** 1111001_{two} **21.** $411A_{\text{twelve}}$ **23.** 384_{nine} **25.** 523_{six} **27.** 1201_{three} **29.** 45234_{eight} **31.** 1010100_{two} **33.** 14207_{eight} **35.** 321222_{four} **37.** $3A61_{\text{sixteen}}$ **39.** 33_{four} **41.** Quotient 33_{four}; remainder 0_{four} **43.** Quotient 1223_{six}; remainder 1_{six} **45.** Quotient 1110_{two}; remainder 0_{two} **47.** Quotient $A8_{\text{twelve}}$; remainder 3_{twelve} **49.** Quotient 14_{five}; remainder 11_{five} **51.** Eight **53. a.** 629 **b.** $384 = 110000000_{\text{two}}$; $245 = 11110101_{\text{two}}$ **c.** 1001110101_{two} **d.** 629 **e.** Same **55. a.** 6422 **b.** $247 = 11110111_{\text{two}}$; $26 = 11010_{\text{two}}$ **c.** $1100100010110_{\text{two}}$ **d.** 6422 **e.** Same **57.** Base seven **59.** In a base one numeration system, 0 would be the only numeral, and the place values would be $1^0, 1^1, 1^2, 1^3, \ldots$, each of which equals 1. Thus 0 is the only number you could write using a base one numeration system. **61.** M = 1, A = 4, S = 3, and O = 0

EXERCISE SET 4.5 *page 228*

1. 1, 2, 4, 5, 10, 20 **3.** 1, 5, 13, 65 **5.** 1, 41 **7.** 1, 2, 5, 10, 11, 22, 55, 110 **9.** 1, 5, 7, 11, 35, 55, 77, 385 **11.** Composite **13.** Prime **15.** Prime **17.** Prime **19.** Composite **21.** 2, 3, 5, 6, and 10 **23.** 3 **25.** 2, 3, 4, 6, and 8 **27.** 2, 5, and 10 **29.** $2 \cdot 3^2$ **31.** $2^3 \cdot 3 \cdot 5$ **33.** $5^2 \cdot 17$ **35.** 2^{10} **37.** $2^3 \cdot 3 \cdot 263$ **39.** $2 \cdot 3^2 \cdot 1013$ **41.** 2, 3, 5, 7, 11, 13, 17, 19, 23, 29, 31, 37, 41, 43, 47, 53, 59, 61, 67, 71, 73, 79, 83, 89, 97, 101, 103, 107, 109, 113, 127, 131, 137, 139, 149, 151, 157, 163, 167, 173, 179, 181, 191, 193, 197, 199 **43.** 3 and 5, 5 and 7, 11 and 13, 17 and 19, 29 and 31, 41 and 43, 59 and 61, 71 and 73, 101 and 103, 107 and 109, 137 and 139, 149 and 151, 179 and 181, 191 and 193, 197 and 199 **45.** 311 and 313, or 347 and 349 In Exercise 47, parts a–f, only one possible sum is given. **47. a.** $24 = 5 + 19$ **b.** $50 = 3 + 47$ **c.** $86 = 3 + 83$ **d.** $144 = 5 + 139$ **e.** $210 = 11 + 199$ **f.** $264 = 7 + 257$ **49.** Yes **51.** Yes **53.** No **55.** Yes **57.** Yes **59.** Yes **61.** No **63.** Yes **65. a.** $n = 3$ **b.** $n = 4$ **67.** To determine whether a given number is divisible by 17, multiply the ones digit of the given number by 5. Find the difference between this result and the number formed by omitting the ones digit from the given number. Keep repeating this procedure until you obtain a small final difference. If the final difference is divisible by 17, then the given number is divisible by 17. If the final difference is not divisible by 17, then the given number is not divisible by 17. **69.** 12 **71.** 8 **73.** 24

EXERCISE SET 4.6 *page 240*

1. Abundant **3.** Deficient **5.** Deficient **7.** Abundant **9.** Deficient **11.** Deficient **13.** Abundant **15.** Abundant **17.** Prime **19.** Prime **21.** $2^{126}(2^{127} - 1)$ **23.** 6 **25.** 420,921 **27.** 2,098,960 **29.** $9^5 + 15^5 = 818{,}424$. Because $15^5 < 818{,}424$ and $16^5 > 818{,}424$, we know there is no natural number z such that $z^5 = 9^5 + 15^5$. **31. a.** False. For instance, if $n = 11$, then $2^{11} - 1 = 2047 = 23 \cdot 89$. **b.** False. Fermat's Last Theorem was the last of Fermat's theories (conjectures) that other mathematicians were able to establish. **c.** True **d.** Conjecture. **33. a.** $12^7 - 12 = 35{,}831{,}796$, which is divisible by 7. **b.** $8^{11} - 8 = 8{,}589{,}934{,}584$, which is divisible by 11. **35.** $8128 = 1^3 + 3^3 + 5^3 + \cdots + 13^3 + 15^3$. See the *Student Solutions Manual* for the verifications in Exercise 37. **39.** The first five Fermat numbers formed using $m = 0, 1, 2, 3$, and 4 are all prime numbers. In 1732, Euler discovered that the sixth Fermat number 4,294,967,297, formed using $m = 5$, is not a prime number because it is divisible by 641.

CHAPTER 4 REVIEW EXERCISES *page 242*

1. 𓁨𓁨𓁨𓁨𓆐𓆐𓆐𓆼𓆼𓆼𓆼𓆼𓆼𓍢𓍢𓍢𓎆𓎆𓏺𓏺𓏺𓏺𓏺 [Sec. 4.1] **2.** 𓁨𓁨𓁨𓆐𓂭𓂭𓆼𓆼𓆼𓆼𓎆𓎆𓎆𓎆𓏺𓏺𓏺 [Sec. 4.1]
3. 223,013 [Sec. 4.1] **4.** 221,354 [Sec. 4.1] **5.** 349 [Sec. 4.1] **6.** 774 [Sec. 4.1] **7.** 9640 [Sec. 4.1]
8. 92,444 [Sec. 4.1] **9.** DLXVII [Sec. 4.1] **10.** DCCCXXIII [Sec. 4.1] **11.** MMCDLXXXIX. [Sec. 4.1]
12. MCCCXXXV [Sec. 4.1] **13.** $(4 \times 10^2) + (3 \times 10^1) + (2 \times 10^0)$ [Sec. 4.2]
14. $(4 \times 10^5) + (5 \times 10^4) + (6 \times 10^3) + (3 \times 10^2) + (2 \times 10^1) + (7 \times 10^0)$ [Sec. 4.2] **15.** 5,038,204 [Sec. 4.2]
16. 387,960 [Sec. 4.2] **17.** 801 [Sec. 4.2] **18.** 1603 [Sec. 4.2] **19.** 76,441 [Sec. 4.2]
20. 87,393 [Sec. 4.2] **21.** 𒌋𒁹𒁹 𒁹 [Sec. 4.2] **22.** 𒌋𒁹𒁹𒁹𒁹𒁹𒁹𒁹𒁹 ⁝ [Sec. 4.2]
23. 𒁹𒁹𒁹 𒌋𒌋𒁹𒁹𒁹𒁹𒁹𒁹𒁹𒁹𒁹 𒁹𒁹𒁹 [Sec. 4.2] **24.** 𒁹𒁹𒁹𒁹𒁹 𒌋𒌋𒁹 𒌋𒌋𒁹 [Sec. 4.2]
25. 194 [Sec. 4.2] **26.** 267 [Sec. 4.2] **27.** 2178 [Sec. 4.2] **28.** 6580 [Sec. 4.2]
29. 𝋡
𝋨
𝋢 [Sec. 4.2] **30.** 𝋱
𝋦 [Sec. 4.2] **31.** 𝋥
𝋣
𝋢 [Sec. 4.2] **32.** 𝋥
𝋩
𝋧 [Sec. 4.2] **33.** 29 [Sec. 4.3]
34. 146 [Sec. 4.3] **35.** 227 [Sec. 4.3] **36.** 286 [Sec. 4.3] **37.** 1153_{six} [Sec. 4.3]
38. 640_{eight} [Sec. 4.3] **39.** 458_{nine} [Sec. 4.3] **40.** $B62_{\text{twelve}}$ [Sec. 4.3] **41.** 34_{eight} [Sec. 4.3]
42. 124_{eight} [Sec. 4.3] **43.** $38D_{\text{sixteen}}$ [Sec. 4.3] **44.** 754_{sixteen} [Sec. 4.3] **45.** 10101_{two} [Sec. 4.3]
46. 1100111010_{two} [Sec. 4.3] **47.** 1001010_{two} [Sec. 4.3] **48.** $110001110010_{\text{two}}$ [Sec. 4.3]
49. 423_{six} [Sec. 4.4] **50.** 1240_{eight} [Sec. 4.4] **51.** 536_{nine} [Sec. 4.4] **52.** 1113_{four} [Sec. 4.4]
53. 16412_{eight} [Sec. 4.4] **54.** 324203_{five} [Sec. 4.4] **55.** Quotient 11100_{two}; remainder 1_{two} [Sec. 4.4]
56. Quotient 21_{four}; remainder 3_{four} [Sec. 4.4] **57.** $3^2 \cdot 5$ [Sec. 4.5] **58.** $2 \cdot 3^3$ [Sec. 4.5]
59. $3^2 \cdot 17$ [Sec. 4.5] **60.** $3 \cdot 5 \cdot 19$ [Sec. 4.5] **61.** Composite [Sec. 4.5] **62.** Composite [Sec. 4.5]
63. Composite [Sec. 4.5] **64.** Composite [Sec. 4.5] **65.** Perfect [Sec. 4.6] **66.** Deficient [Sec. 4.6]
67. Abundant [Sec. 4.6] **68.** Abundant [Sec. 4.6] **69.** $2^{60}(2^{61} - 1)$ [Sec. 4.6] **70.** $2^{1278}(2^{1279} - 1)$ [Sec. 4.6]
71. 368 [Sec. 4.1] **72.** 513 [Sec. 4.1] **73.** 1162 [Sec. 4.1] **74.** 3003 [Sec. 4.1] **75.** 410 [Sec. 4.3]
76. 277 [Sec. 4.3] **77.** 1041 [Sec. 4.3] **78.** 1616 [Sec. 4.3] **79.** A base ten number is divisible by 3 provided the sum of the digits of the number is divisible by 3. [Sec. 4.5] **80.** A number is divisible by 6 provided the number is divisible by 2 and by 3. [Sec. 4.5] **81.** Every composite number can be written as a unique product of prime numbers (disregarding the order of the factors). [Sec. 4.5] **82.** Zero [Sec. 4.6] **83.** 39,751 [Sec. 4.6] **84.** 895,932 [Sec. 4.6]

CHAPTER 4 TEST *page 244*

1. 𓆼𓆼𓆼𓍢𓎆𓎆𓏺𓏺𓏺𓏺 [Sec. 4.1] **2.** 4263 [Sec. 4.1] **3.** 1447 [Sec. 4.1] **4.** MMDCIX [Sec. 4.1]
5. $(6 \times 10^4) + (7 \times 10^3) + (4 \times 10^2) + (8 \times 10^1) + (5 \times 10^0)$ [Sec. 4.2] **6.** 530,284 [Sec. 4.2]
7. 37,274 [Sec. 4.2] **8.** 𒁹𒁹 𒌋𒌋𒌋𒌋𒁹 𒌋𒁹𒁹𒁹𒁹𒁹 [Sec. 4.2] **9.** 1305 [Sec. 4.2] **10.** 𝋡
𝋧
𝋢 [Sec. 4.2]
11. 854 [Sec. 4.3] **12. a.** 4144_{eight} **b.** $12B0_{\text{twelve}}$ [Sec. 4.3] **13.** $100101110111_{\text{two}}$ [Sec. 4.3]
14. $AB7_{\text{sixteen}}$ [Sec. 4.3] **15.** 112_{five} [Sec. 4.4] **16.** 313_{eight} [Sec. 4.4] **17.** 11100110_{two} [Sec. 4.4]
18. Quotient 61_{seven}; remainder 3_{seven} [Sec. 4.4] **19.** $2 \cdot 5 \cdot 23$ [Sec. 4.5] **20.** Composite [Sec. 4.5]
21. a. No **b.** Yes **c.** No [Sec. 4.5] **22. a.** Yes **b.** No **c.** No [Sec. 4.5]
23. Abundant [Sec. 4.6] **24.** $2^{16}(2^{17} - 1)$ [Sec. 4.6]

CHAPTER 5

EXERCISE SET 5.1 *page 258*

1. An equation expresses the equality of two mathematical expressions. An equation contains an equals sign. An expression does not.
3. Substitute the solution back into the original equation and confirm the equality. **5.** 12 **7.** 22 **9.** −8

11. -20 **13.** 8 **15.** -1 **17.** 1 **19.** $-\frac{1}{3}$ **21.** $\frac{2}{3}$ **23.** -2 **25.** 4 **27.** 2
29. 2 **31.** $\frac{1}{4}$ **33.** -2 **35.** 4 **37.** 8 **39.** 1 **41.** -32 **43.** $16,859.34
45. 1350 inches **47.** 60 feet **49.** 1952 **51.** 168 feet **53.** 18.6 degrees Celsius **55.** 175
57. $22,000 **59. a.** $2395 **b.** $4694 **61.** $117.75 **63. a.** 163,000 kilograms **b.** 1930 kilograms
65. $12.50 **67.** 1280 horizontal pixels **69.** more than 16 minutes but not over 17 minutes **71.** $b = P - a - c$
73. $R = \frac{E}{I}$ **75.** $r = \frac{I}{Pt}$ **77.** $C = \frac{5}{9}(F - 32)$ **79.** $t = \frac{A - P}{Pr}$ **81.** $f = \frac{T + gm}{m}$ **83.** $S = C - Rt$
85. $b_2 = \frac{2A}{h} - b_1$ **87.** $h = \frac{S}{2\pi r} - r$ **89.** $y = 2 - \frac{4}{3}x$ **91.** $x = \frac{y - y_1}{m} + x_1$ **93.** 0
95. Every real number is a solution. **97.** $x = \frac{d - b}{a - c}$. $a \neq c$ or the denominator equals zero and the expression is undefined.

EXERCISE SET 5.2 *page 276*

1. Examples will vary. **3.** The purpose is to allow currency from one country to be converted into the currency of another country. **5.** Explanations will vary. **7.** The cross-products method is a shortcut for multiplying each side of the proportion by the least common multiple of the denominators. **9.** 23 miles per gallon **11.** 12.5 meters per second **13.** 400 square feet per gallon **15.** 272,160 kilograms **17.** A 24-ounce jar of mayonnaise for $2.09 **19.** $16.50 per hour **21. a.** Australia **b.** 850 more people per square mile **23.** 60,374 krona **25.** 434,834 pesos **27.** For each state, the ratio is 3.125 to 1. **29.** 13:1 or 13 to 1. There is one faculty member for every 13 students at Syracuse University. **31.** University of Connecticut **33.** No **35.** 17.14 **37.** 25.6 **39.** 20.83 **41.** 2.22 **43.** 13.71 **45.** 39.6 **47.** 0.52 **49.** 6.74 **51.** $45,000 **53.** 5.5 milligrams **55.** 24 feet; 15 feet **57.** 160,000 people **59.** 63,000 miles **61.** 11.25 grams. Explanations will vary. **63. a.** True **b.** True **c.** True **d.** False **65. a.–c.** Explanations will vary.

EXERCISE SET 5.3 *page 297*

1. Answers will vary. **3.** 3 **5.** Employee B's salary is now the highest because Employee B had the highest initial salary, and the percent raises were the same for all employees. **7.** 0.5; 50% **9.** $\frac{2}{5}$; 0.4 **11.** $\frac{7}{10}$; 70% **13.** 0.55; 55% **15.** $\frac{5}{32}$; 0.15625 **17. a.** 73 fans **b.** More fans approved. **c.** 7%; 100% − 73% − 20% = 7% **19.** $26.6 billion **21.** 23.7% **23. a.** $1381 **b.** $1628 **25. a.** 48.2 hours **b.** 33.6 hours **c.** 3.4 hours **27. a.** 1998 to 2000 **b.** 2004 to 2006 **c.** More slowly **29. a.** Arizona: 118,153; California: 625,041; Colorado: 165,038; Maryland: 142,718; Massachusetts: 246,833; Minnesota: 229,543; New Hampshire: 51,714; New Jersey: 229,543; Vermont: 17,274; Virginia: 185,900 **b.** New Hampshire; California **c.** More than half **d.** Massachusetts: 7.5%; New Jersey: 5.5%; Virginia: 5.2% **e.** 5.3% **f.** The rate is 10 times the percent. **31. a.** 900% **b.** 60% **c.** 1500% **d.** Explanations will vary. **33. a.** home health aides **b.** software engineers (applications) **c.** 63,200 people **d.** 320,580 people **e.** computer support specialists **f.** The percent increases are based on different original employment figures. **g.** Answers will vary. **35.** Less than **37.** 39 months **39. a.** 10,700,000 TV households; 62,800,000 TV households **b.** 5,900,000 TV households; 53,600,000 TV households **c.** 2.5 people **d.** Answers will vary.

EXERCISE SET 5.4 *page 316*

1. The Principle of Zero Products states that if the product of two factors is zero, then one of the two factors equals zero. A second-degree equation must be written in standard form so that the variable expression is equal to zero. Then, when the variable expression is factored, the Principle of Zero Products can be used to set each factor equal to zero. **3.** Answers will vary. **5.** -2 and 5

7. $\frac{1+\sqrt{5}}{2}$ and $\frac{1-\sqrt{5}}{2}$; -0.618 and 1.618 **9.** $3+\sqrt{13}$ and $3-\sqrt{13}$; -0.606 and 6.606 **11.** 0 and 2 **13.** $\frac{1+\sqrt{17}}{2}$ and $\frac{1-\sqrt{17}}{2}$; -1.562 and 2.562 **15.** $\frac{2+\sqrt{14}}{2}$ and $\frac{2-\sqrt{14}}{2}$; -0.871 and 2.871 **17.** $2+\sqrt{11}$ and $2-\sqrt{11}$; -1.317 and 5.317 **19.** $-\frac{3}{2}$ and 6 **21.** No real number solutions **23.** -4 and $\frac{1}{4}$ **25.** $-\frac{1}{2}$ and $\frac{4}{3}$ **27.** $1+\sqrt{5}$ and $1-\sqrt{5}$; -1.236 and 3.236 **29.** 0.75 second and 3 seconds **31.** $T = 0.5(1)^2 + 0.5(1) = 1$; $T = 0.5(2)^2 + 0.5(2) = 3$; $T = 0.5(3)^2 + 0.5(3) = 6$; $T = 0.5(4)^2 + 0.5(4) = 10$; 10 rows **33. a.** 240 feet **b.** 30 miles per hour **35.** 5.51 seconds **37.** 244.10 feet **39.** No **41.** 1.74 seconds and 10.76 seconds **43.** 71 cents **45.** If the discriminant is not a perfect square, the radical expression in the quadratic formula will not simplify to a whole number. **47.** $-12a$ and $-4a$ **49.** $-\frac{b}{2}$ and $-b$ **51.** $-y$ and $\frac{3}{2}y$ **53.** $b^2 - 4ac = b^2 - 4(1)(-1) = b^2 + 4$. If $b^2 - 4ac \geq 0$, then the equation has real number solutions. Because $b^2 \geq 0$ for all real numbers b, $b^2 + 4 \geq 4$ for all real numbers b. Therefore, the equation has real number solutions regardless of the value of b.

CHAPTER 5 REVIEW EXERCISES *page 320*

1. 4 [Sec. 5.1] **2.** $\frac{1}{8}$ [Sec. 5.1] **3.** -2 [Sec. 5.1] **4.** $\frac{10}{3}$ [Sec. 5.2] **5.** No real number solutions [Sec. 5.4] **6.** -5 and 6 [Sec. 5.4] **7.** $2+\sqrt{3}$ and $2-\sqrt{3}$ [Sec. 5.4] **8.** $\frac{1+\sqrt{13}}{2}$ and $\frac{1-\sqrt{13}}{2}$ [Sec. 5.4] **9.** $y = -\frac{4}{3}x + 4$ [Sec. 5.1] **10.** $t = \frac{f-v}{a}$ [Sec. 5.1] **11.** 2450 feet [Sec. 5.1] **12.** 3 seconds [Sec. 5.1] **13.** 60°C [Sec. 5.1] **14.** 39 minutes [Sec. 5.1] **15.** 28.4 miles per gallon [Sec. 5.2] **16.** $\frac{1}{4}$ [Sec. 5.2] **17.** \$1260 [Sec. 5.1] **18. a.** New York, Chicago, Philadelphia, Los Angeles, Houston **b.** 21,596 more people per square mile [Sec. 5.2] **19. a.** 15:1, 15 to 1. There are 15 students for every one faculty member at the university. **b.** Grand Canyon University, Arizona State University **c.** Embry-Riddle Aeronautical University, Northern Arizona University, and University of Arizona [Sec. 5.2] **20.** Department A: \$105,000; Department B: \$245,000 [Sec. 5.2] **21.** 7.5 tablespoons [Sec. 5.2] **22. a.** No **b.** 41:9 **c.** \$1,828.6 billion **d.** \$490.6 billion [Sec. 5.2] **23. a.** 51.0% **b.** Less than [Sec. 5.3] **24.** 735,000 people [Sec. 5.3] **25.** 69.0% [Sec. 5.3] **26. a.** 83.3% **b.** 100% **c.** 50% [Sec. 5.3] **27.** 16.4% [Sec. 5.3] **28. a.** Ages 9–10 **b.** Ages 15–16 **c.** 46.5%; less than **d.** 7230 boys **e.** 549,400 young people [Sec. 5.3] **29.** 1 second and 5 seconds [Sec. 5.4] **30.** 0.5 second and 1.5 seconds [Sec. 5.4]

CHAPTER 5 TEST *page 324*

1. 14 [Sec. 5.1] **2.** 10 [Sec. 5.1] **3.** $\frac{21}{4}$ [Sec. 5.2] **4.** 3 and 9 [Sec. 5.4] **5.** $\frac{2+\sqrt{7}}{3}$ and $\frac{2-\sqrt{7}}{3}$ [Sec. 5.4] **6.** $y = \frac{1}{2}x - \frac{15}{2}$ [Sec. 5.1] **7.** $F = \frac{9}{5}C + 32$ [Sec. 5.1] **8.** 2.5 minutes [Sec. 5.1] **9.** 10 days [Sec. 5.1] **10.** 54.8 miles per hour [Sec. 5.2] **11.** 843 acres [Sec. 5.1] **12. a.** 2.727, 2.905, 2.777, 2.808, 2.901, 2.904 **b.** Ty Cobb, Rogers Hornsby, Joe Jackson, Tris Speaker, Ted Williams, Billy Hamilton [Sec. 5.2] **13.** $\frac{4}{7}$ [Sec. 5.2] **14.** \$112,500 and \$67,500 [Sec. 5.2] **15.** 2.75 pounds [Sec. 5.2] **16. a.** Miami-Dade **b.** 9420 violent crimes [Sec. 5.2] **17.** 14.4% [Sec. 5.3] **18. a.** 20.6% **b.** Between 2004 and 2005 **c.** 39.1% [Sec. 5.3] **19. a.** 20% **b.** 1.6 million working farms **c.** Answers will vary. [Sec. 5.3] **20.** 0.2 second and 1.6 seconds [Sec. 5.4]

CHAPTER 6

EXERCISE SET 6.1 *page 338*

1.
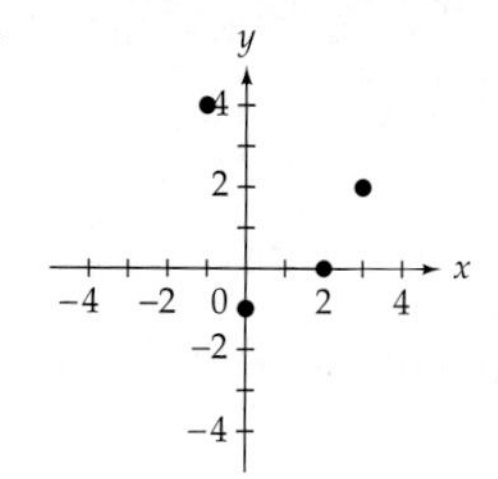

3.
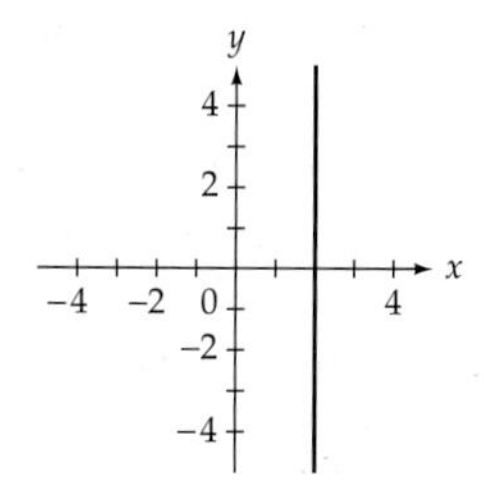

5.
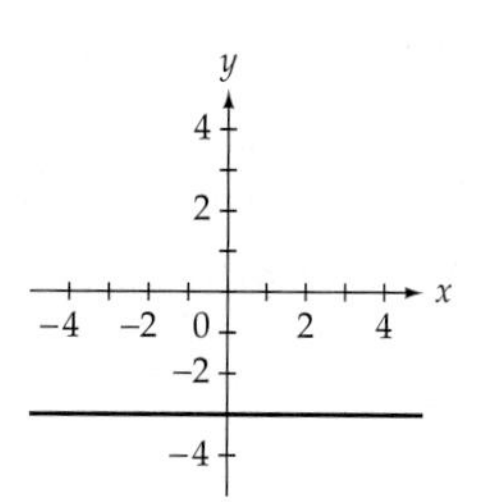

7.
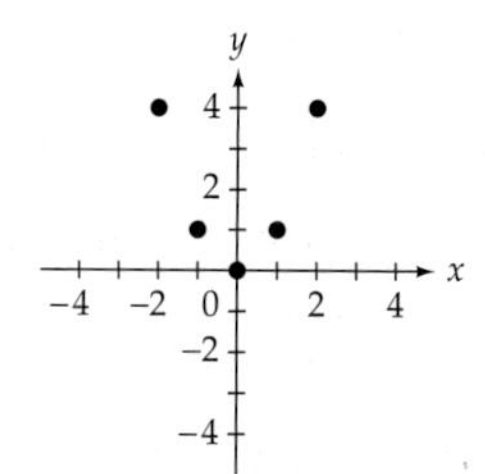

9.
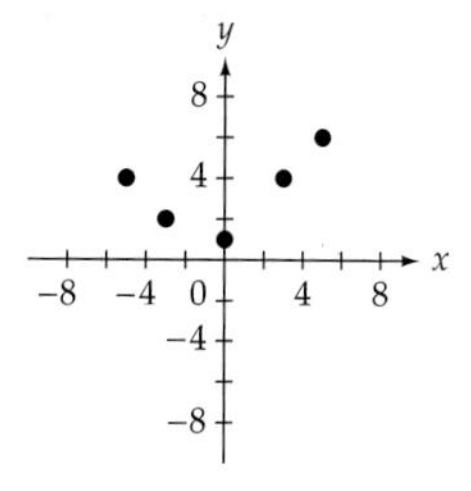

11.
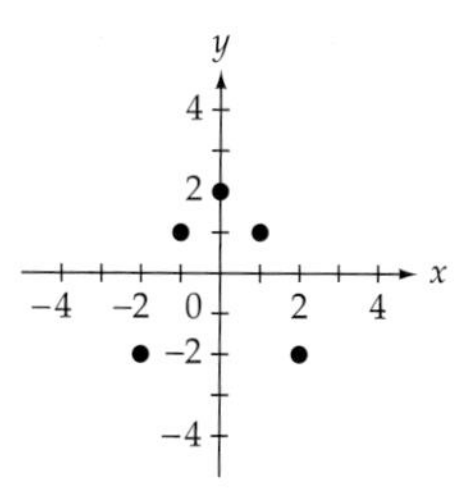

13.
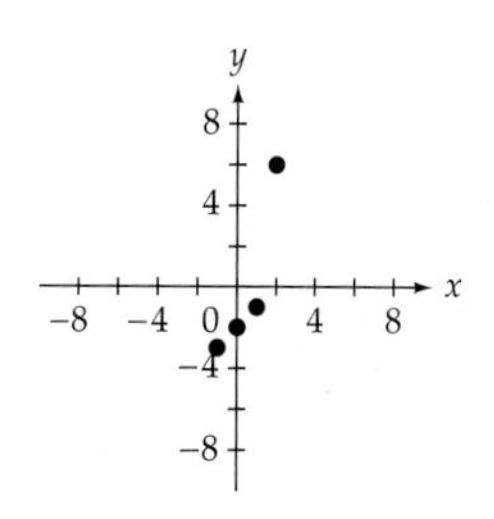

15.
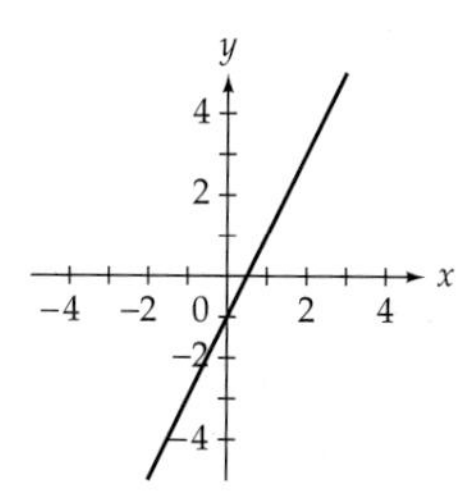

17.
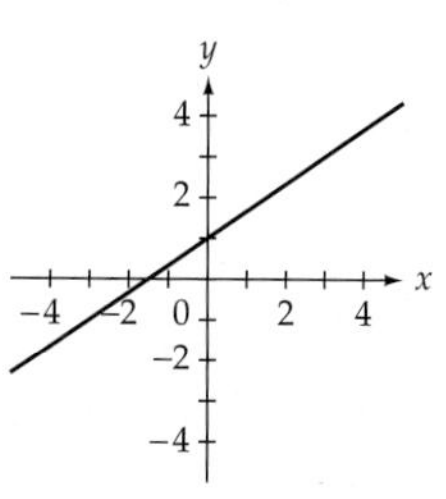

19.
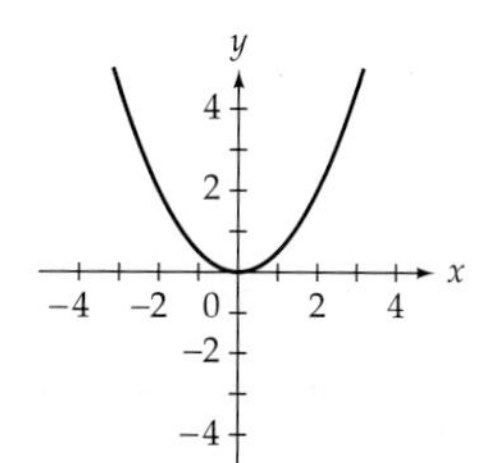

21.
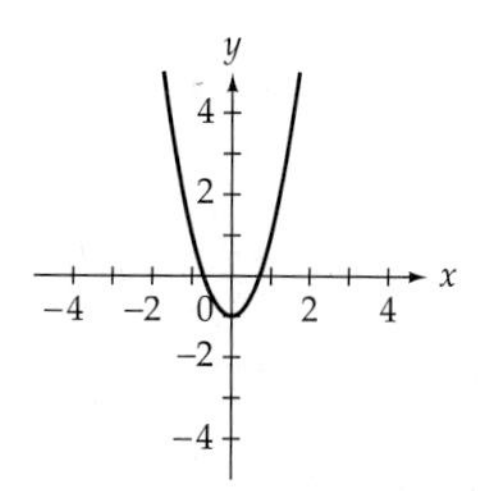

23.
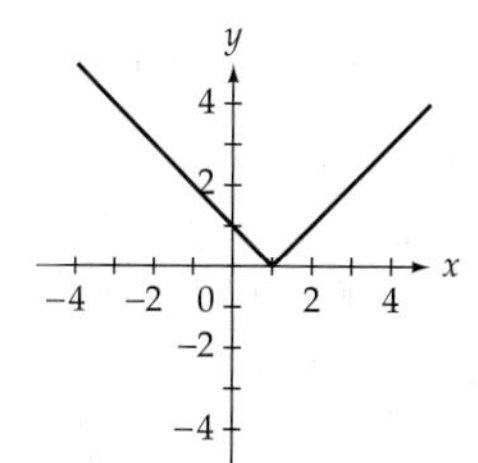

25. 3 **27.** 3 **29.** 10 **31.** 0 **33. a.** 16 meters **b.** 20 feet
35. a. 100 feet **b.** 68 feet **37. a.** 1087 feet per second **b.** 1136 feet per second
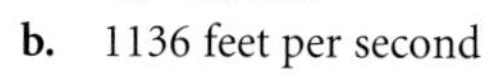
39. a. 10% **b.** 40%

41.
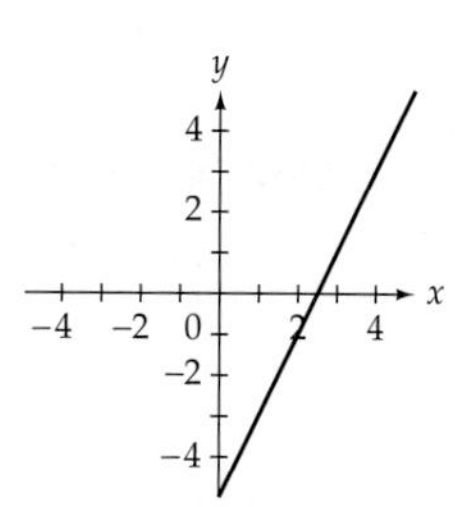

43.
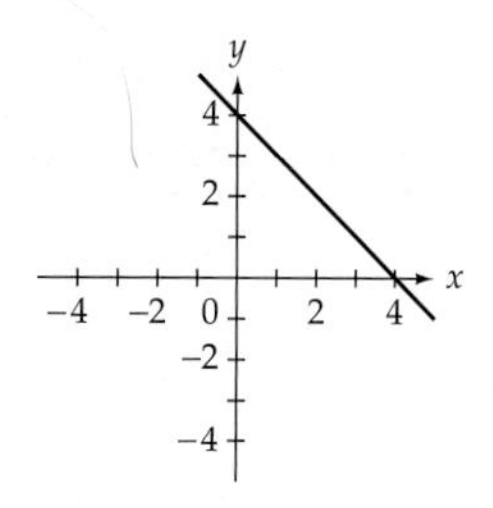

45.
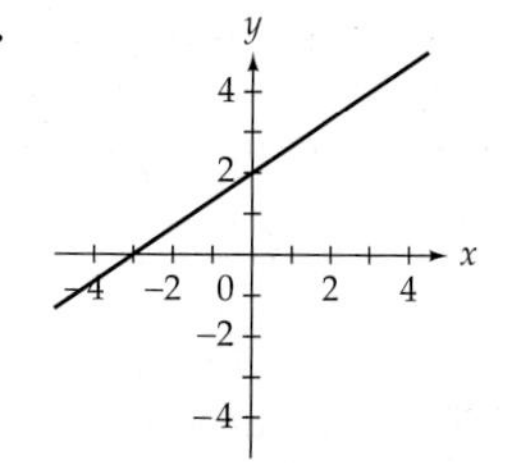

47.
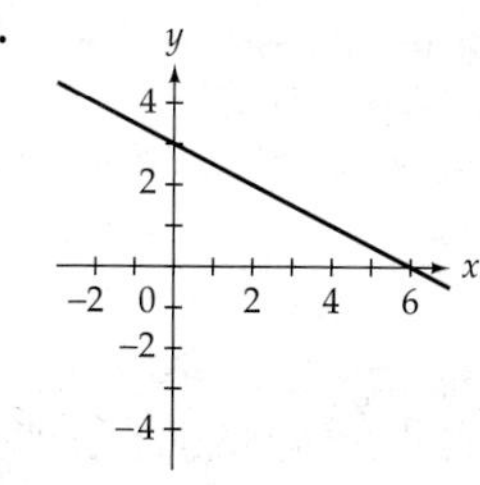

49.
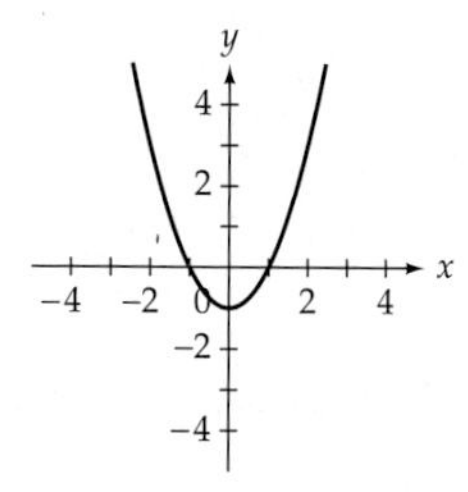

51.
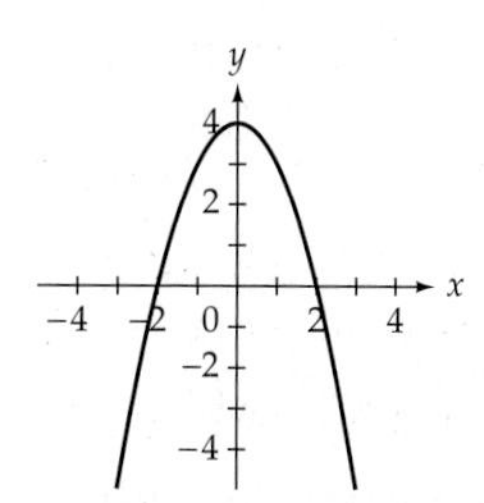

53.
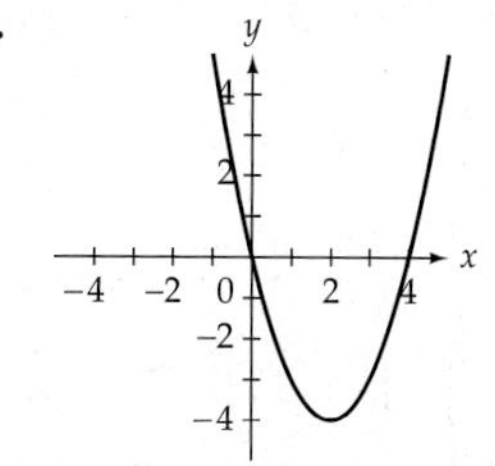

55.
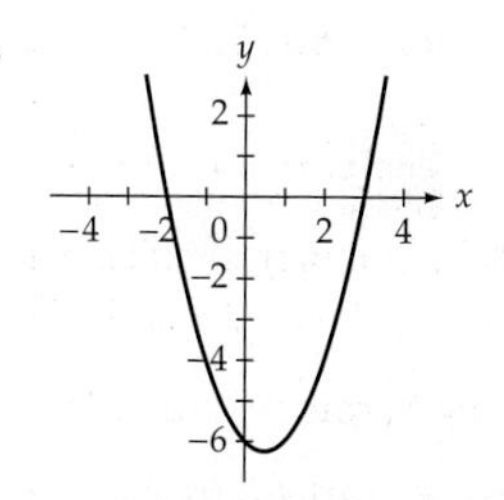

57. 48 square units **59.** No. A function cannot have different elements in the range corresponding to one element in the domain.
61. 2 **63.** 17 **65.** $x = 0$

EXERCISE SET 6.2 *page 350*

1. (2, 0), (0, −6) **3.** (6, 0), (0, −4) **5.** (−4, 0), (0, −4) **7.** (4, 0), (0, 3) **9.** $\left(\frac{9}{2}, 0\right)$, (0, −3) **11.** (2, 0), (0, 3) **13.** (1, 0), (0, −2) **15.** $\left(\frac{30}{7}, 0\right)$; At $\frac{30}{7}$°C the cricket stops chirping. **17.** The intercept on the vertical axis is (0, −15). This means that the temperature of the object is −15°F before it is removed from the freezer. The intercept on the horizontal axis is (5, 0). This means that it takes 5 minutes for the temperature of the object to reach 0°F. **19.** −1 **21.** $\frac{1}{3}$ **23.** $-\frac{2}{3}$ **25.** $-\frac{3}{4}$ **27.** Undefined **29.** $\frac{7}{5}$ **31.** 0 **33.** $-\frac{1}{2}$ **35.** Undefined **37.** The slope is 40, which means the motorist was traveling at 40 miles per hour. **39.** The slope is 0.25, which means the tax rate for an income range of \$29,050 to \$70,350 is 25%. **41.** The slope is approximately 343.9, which means the runner traveled at a rate of 343.9 meters per minute.

43.

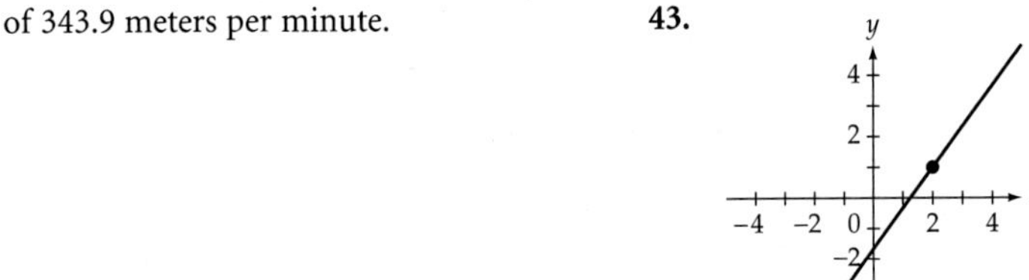

45.

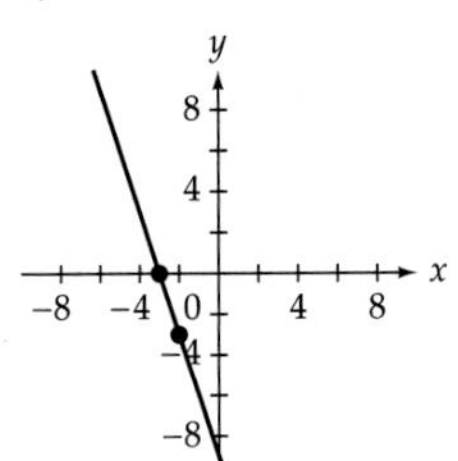

47.

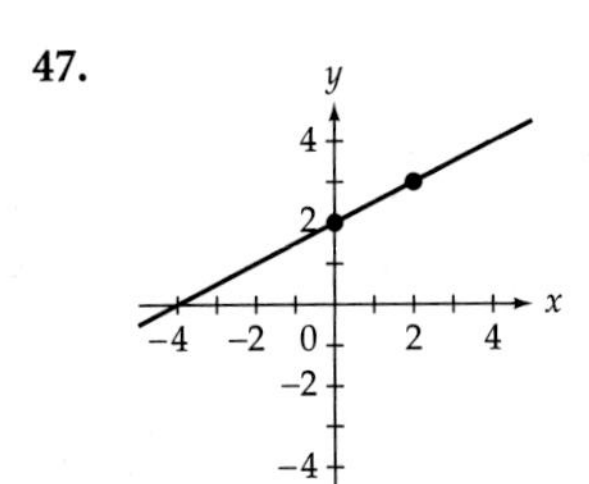

49.

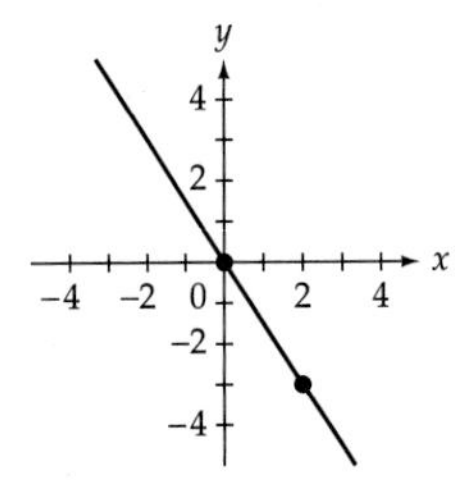

51.

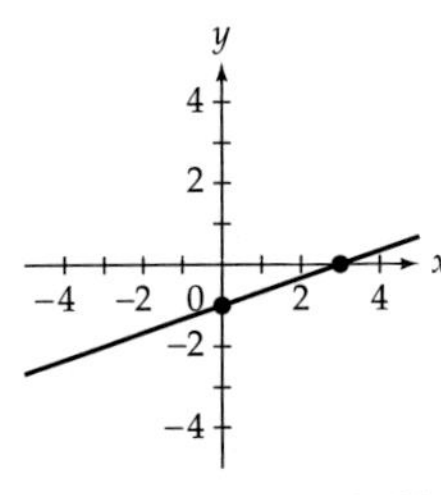

53. Line A represents the distance traveled by Lois in x hours, line B represents the distance traveled by Tanya in x hours, and line C represents the distance between them in x hours. **55.** No **57.** 7 **59.** 2 **61.** It rotates the line counterclockwise. **63.** It raises the line on the rectangular coordinate system. **65.** No. A vertical line through (4, 0) does not have a y-intercept.

EXERCISE SET 6.3 *page 360*

1. $y = 2x + 5$ **3.** $y = -3x + 4$ **5.** $y = -\frac{2}{3}x + 7$ **7.** $y = -3$ **9.** $y = x + 2$ **11.** $y = -\frac{3}{2}x + 3$ **13.** $y = \frac{1}{2}x - 1$ **15.** $y = -\frac{5}{2}x$ **17.** $R(x) = -\frac{3}{5}x + 545$; 485 rooms **19.** $D(t) = 415t$; 1867.5 miles **21.** $N(x) = -20x + 230{,}000$; 60,000 cars **23. a.** $y = 0.56x + 41.71$ **b.** 89 **25. a.** $y = 42.50x + 2613.76$ **b.** 3209 thousand students **27. a.** $y = -1.35x + 106.98$ **b.** 46°F **29.** Answers will vary. For example, (0, 3), (1, 2), and (3, 0). **31.** 7 **33.** No. The three points do not lie on a straight line. **35.** 3°. The car is climbing.

EXERCISE SET 6.4 *page 372*

1. (0, −2) **3.** (0, −1) **5.** (0, 2) **7.** (0, −1) **9.** $\left(\frac{1}{2}, -\frac{9}{4}\right)$ **11.** $\left(\frac{1}{4}, -\frac{41}{8}\right)$ **13.** (0, 0), (2, 0) **15.** (−2, 0), $\left(-\frac{3}{4}, 0\right)$ **17.** $(-1 + \sqrt{2}, 0), (-1 - \sqrt{2}, 0)$ **19.** None **21.** $(2 + \sqrt{5}, 0), (2 - \sqrt{5}, 0)$ **23.** $\left(-\frac{1}{2}, 0\right)$, (3, 0) **25.** Minimum: 2 **27.** Maximum: −3

29. Minimum: -3.25 **31.** Maximum: $\frac{9}{4}$ **33.** c **35.** 150 feet **37.** 100 lenses
39. 24.36 feet **41.** Yes **43.** 141.6 feet **45. a.** 41 miles per hour **b.** 33.658 miles per gallon
47. 7 **49.** 4

EXERCISE SET 6.5 *page 386*

1. a. 9 **b.** 1 **c.** $\frac{1}{9}$ **3. a.** 16 **b.** 4 **c.** $\frac{1}{4}$ **5. a.** 1 **b.** $\frac{1}{8}$ **c.** 16
7. a. 7.3891 **b.** 0.3679 **c.** 1.2840 **9. a.** 54.5982 **b.** 1 **c.** 0.1353
11. a. 16 **b.** 16 **c.** 1.4768 **13. a.** 0.1353 **b.** 0.1353 **c.** 0.0111
15. **17.** **19.**

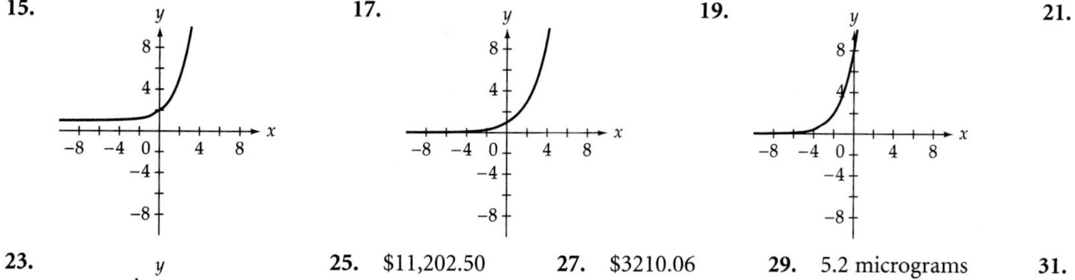

21.

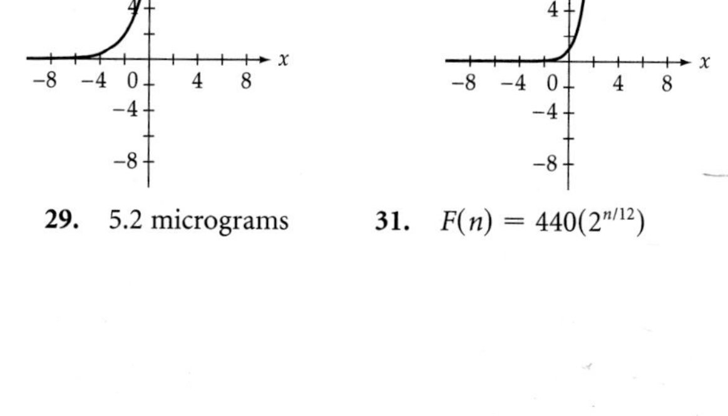

23.

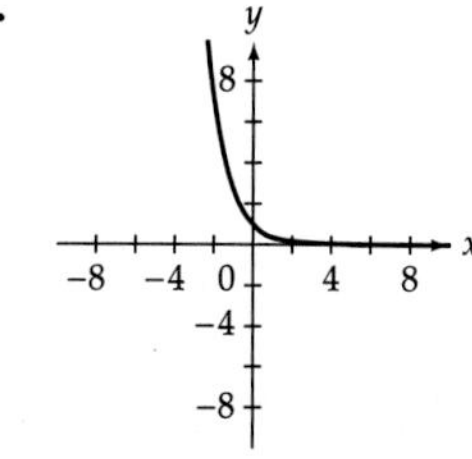

25. \$11,202.50 **27.** \$3210.06 **29.** 5.2 micrograms **31.** $F(n) = 440(2^{n/12})$

33. $y = 8000(1.1066^t)$; 552,200,000 automobiles **35.** $y = 100(0.99^t)$ **37. a.** $y = 4.959(1.063^x)$ **b.** 12.4 milliliters
39. a. 5.9 billion people **b.** 7.2 billion people **c.** The maximum population that Earth can support is 70.0 billion people.

EXERCISE SET 6.6 *page 400*

1. $\log_7 49 = 2$ **3.** $\log_5 625 = 4$ **5.** $\log 0.0001 = -4$ **7.** $\log x = y$ **9.** $3^4 = 81$
11. $5^3 = 125$ **13.** $4^{-2} = \frac{1}{16}$ **15.** $e^y = x$ **17.** 4 **19.** 2 **21.** -2 **23.** 6 **25.** 9
27. $\frac{1}{7}$ **29.** $\frac{1}{9}$ **31.** 1 **33.** 316.23 **35.** 7.39 **37.** 2.24 **39.** 14.39
41. **43.** **45.** **47.** 79%

49. 65 decibels **51.** 1.35 **53.** 6.8 **55.** $794{,}328{,}235I_0$ **57.** 100 times as strong **59.** 6.0 parsecs
61. 7805.5 billion barrels **63.** $x = \frac{10^{0.47712}}{10^{0.30103}} = 10^{0.17609} \approx 1.5$; $\log x = 0.17609$ **65.** Answers will vary.

CHAPTER 6 REVIEW EXERCISES *page 403*

1. 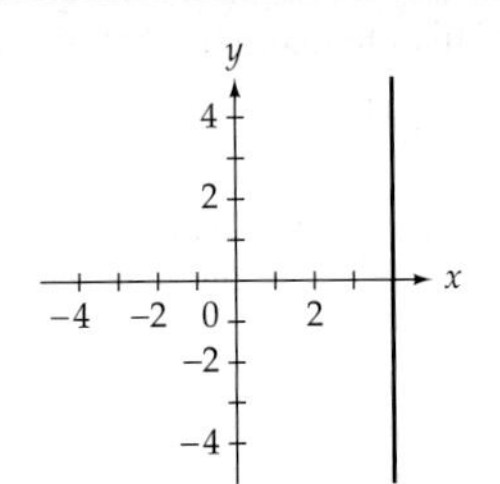 [Sec. 6.1]

2. 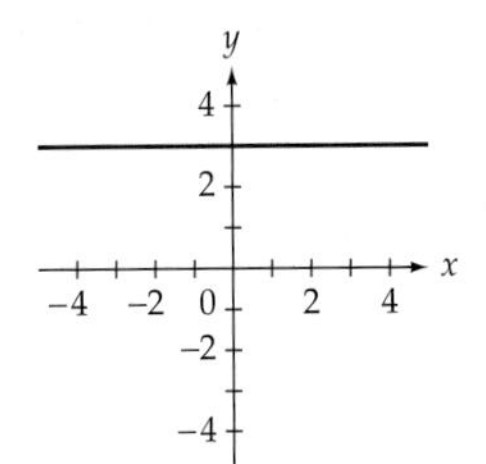 [Sec. 6.1]

3. 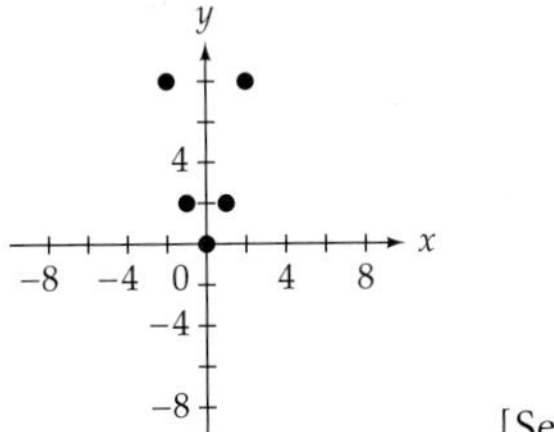 [Sec. 6.1]

4. 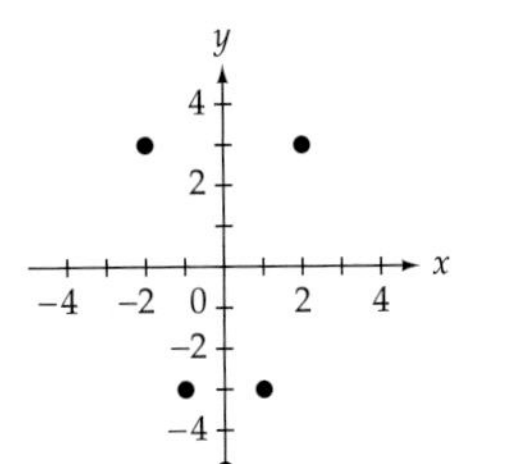 [Sec. 6.1]

5. 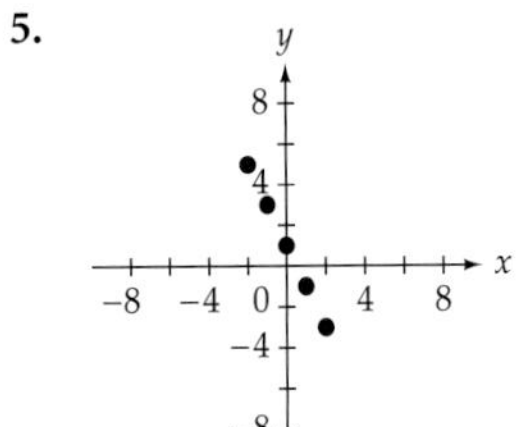 [Sec. 6.1]

6. [Sec. 6.1]

7. 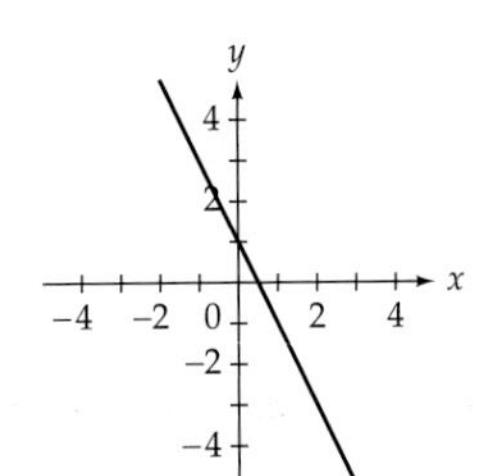 [Sec. 6.1]

8. 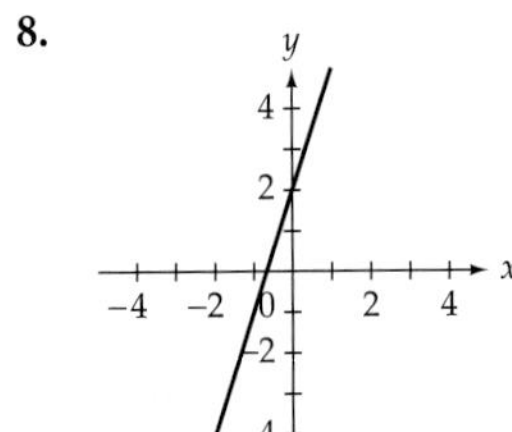 [Sec. 6.1]

9. [Sec. 6.1]

10. 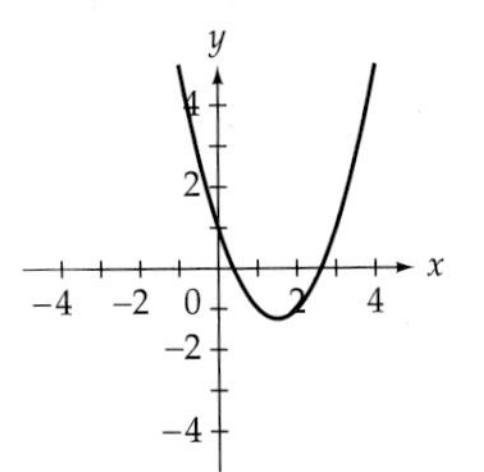 [Sec. 6.1]

11. 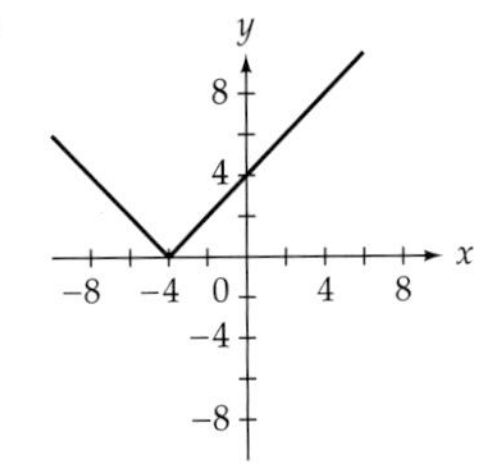 [Sec. 6.1]

12. 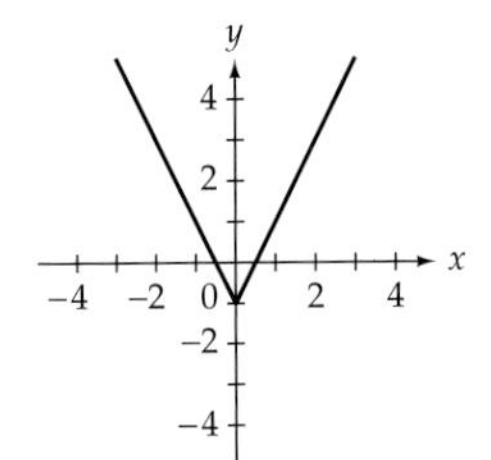 [Sec. 6.1]

13. 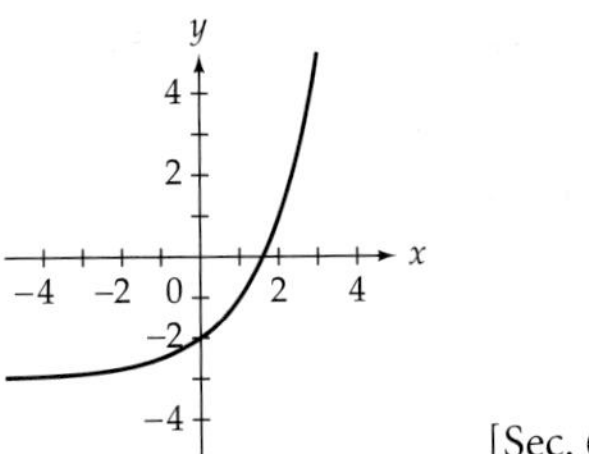 [Sec. 6.5]

14. 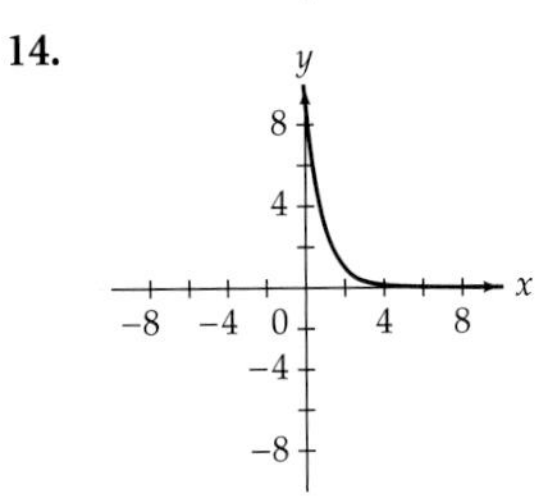 [Sec. 6.5]

15. 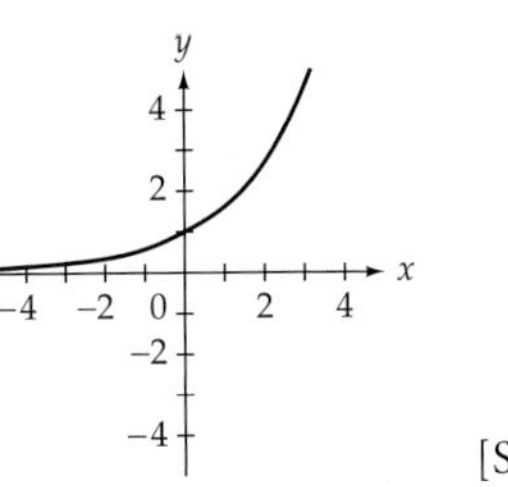 [Sec. 6.5]

16. 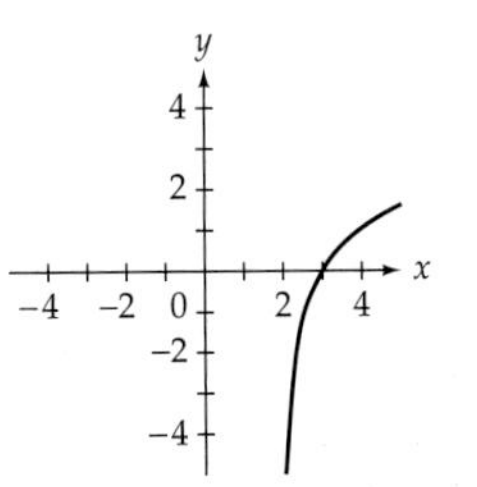 [Sec. 6.6]

17. -13 [Sec. 6.1] **18.** 13 [Sec. 6.1] **19.** $-\frac{1}{2}$ [Sec. 6.1]

20. -21 [Sec. 6.1] **21.** 4 [Sec. 6.5] **22.** $\frac{4}{9}$ [Sec. 6.5] **23.** 15.78 [Sec. 6.5] **24. a.** 113.1 cubic inches **b.** 7238.2 cubic centimeters [Sec. 6.1] **25. a.** 133 feet **b.** 133 feet [Sec. 6.1]

26. a. 5% **b.** 36.7% [Sec. 6.1] **27.** $(-5, 0), (0, 10)$ [Sec. 6.2] **28.** $(12, 0), (0, -9)$ [Sec. 6.2] **29.** $(5, 0), (0, -3)$ [Sec. 6.2] **30.** $(6, 0), (0, 8)$ [Sec. 6.2] **31.** The intercept on the vertical axis is (0, 25,000). This means that the value of the truck was $25,000 when it was new. The intercept on the horizontal axis is (5, 0). This means that after 5 years the truck will be worth $0. [Sec. 6.2] **32.** 5 [Sec. 6.2] **33.** $\frac{5}{2}$ [Sec. 6.2] **34.** 0 [Sec. 6.2] **35.** Undefined [Sec. 6.2] **36.** The slope is -0.6, which means that the revenue from home video rentals is decreasing $600,000,000 annually. [Sec. 6.2] **37.**

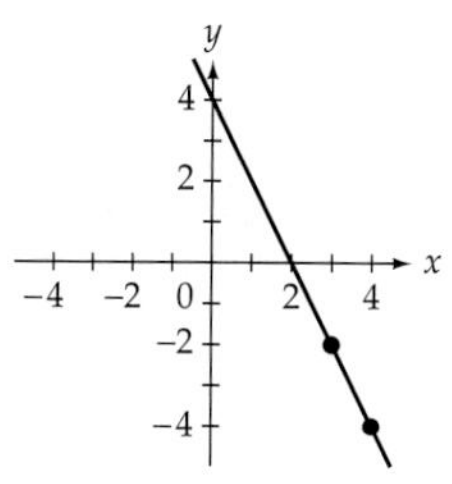

[Sec. 6.2] **38.**

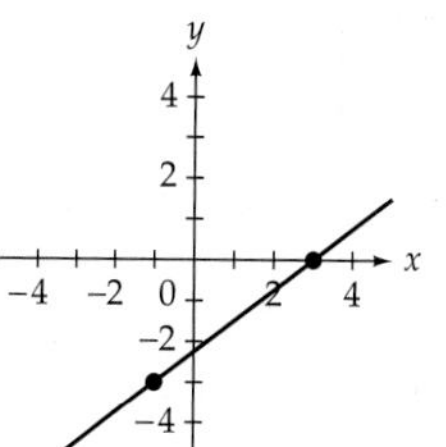

[Sec. 6.2]

39. $y = 2x + 7$ [Sec. 6.3] **40.** $y = x - 5$ [Sec. 6.3] **41.** $y = \frac{2}{3}x + 3$ [Sec. 6.3] **42.** $y = \frac{1}{4}x$ [Sec. 6.3]

43.

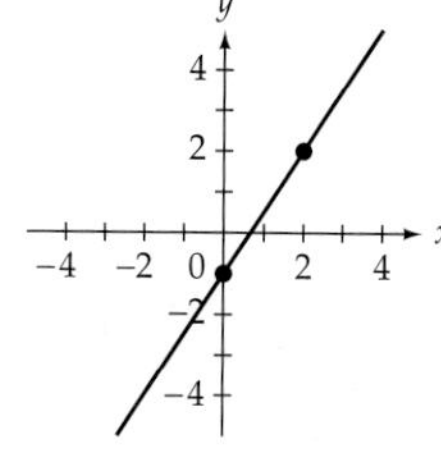

[Sec. 6.2]

44. a. $f(x) = 25x + 100$ **b.** $900 [Sec. 6.3]

45. a. $A(p) = -25{,}000x + 10{,}000$ **b.** 10,750 gallons [Sec. 6.3] **46. a.** $y = 0.16204x - 0.12674$ **b.** 25 [Sec. 6.3] **47.** $(-1, 3)$ [Sec. 6.4] **48.** $\left(-\frac{3}{2}, \frac{11}{2}\right)$ [Sec. 6.4] **49.** $(1, 2)$ [Sec. 6.4] **50.** $(-2.5, -7.25)$ [Sec. 6.4] **51.** $(4, 0), (-5, 0)$ [Sec. 6.4] **52.** $(-1 + \sqrt{2}, 0), (-1 - \sqrt{2}, 0)$ [Sec. 6.4] **53.** $\left(-\frac{1}{2}, 0\right), (-4, 0)$ [Sec. 6.4] **54.** None [Sec. 6.4] **55.** 5, maximum [Sec. 6.4] **56.** $-\frac{15}{2}$, minimum [Sec. 6.4] **57.** -5, minimum [Sec. 6.4] **58.** $\frac{1}{8}$, maximum [Sec. 6.4] **59.** 125 feet [Sec. 6.4] **60.** 2000 CD-RWs [Sec. 6.4] **61.** $12,297.11 [Sec. 6.5] **62.** 7.94 milligrams [Sec. 6.5] **63.** 3.36 micrograms [Sec. 6.5] **64. a.** $H(n) = 6\left(\frac{2}{3}\right)^n$ **b.** 0.79 foot [Sec. 6.5] **65. a.** $N = 250.2056(0.6506)^t$ **b.** 19 thousand people [Sec. 6.5] **66.** 5 [Sec. 6.6] **67.** -4 [Sec. 6.6] **68.** -1 [Sec. 6.6] **69.** 6 [Sec. 6.6] **70.** 64 [Sec. 6.6] **71.** 1.4422 [Sec. 6.6] **72.** 12.1825 [Sec. 6.6] **73.** 251.1886 [Sec. 6.6] **74.** 43.7 parsecs [Sec. 6.6] **75.** 140 decibels [Sec. 6.6]

CHAPTER 6 TEST *page 406*

1. -21 [Sec. 6.1] **2.** $\frac{1}{9}$ [Sec. 6.5] **3.** 3 [Sec. 6.6] **4.** 36 [Sec. 6.6]

5.

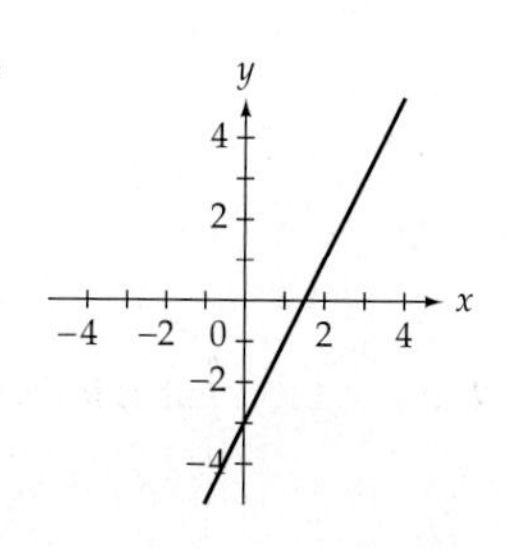

[Sec. 6.1]

6.

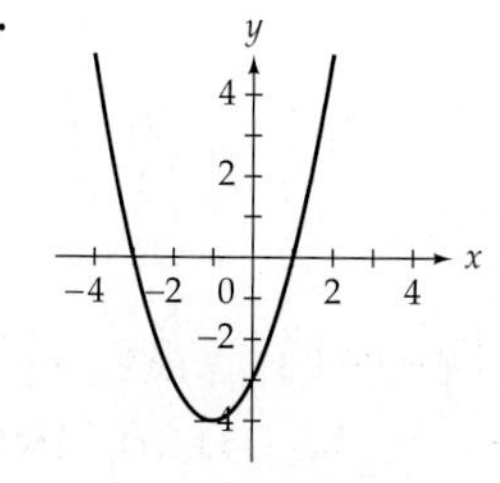

[Sec. 6.1]

7.

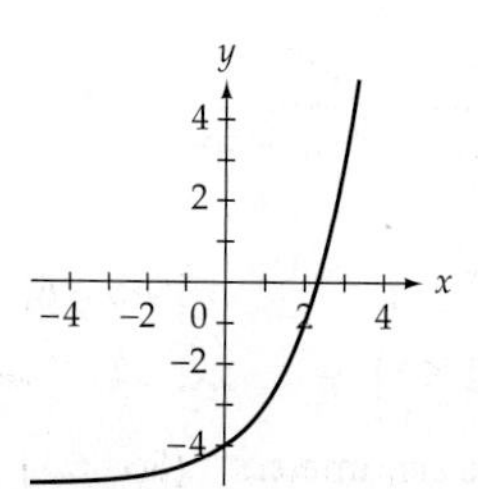

[Sec. 6.5]

8. 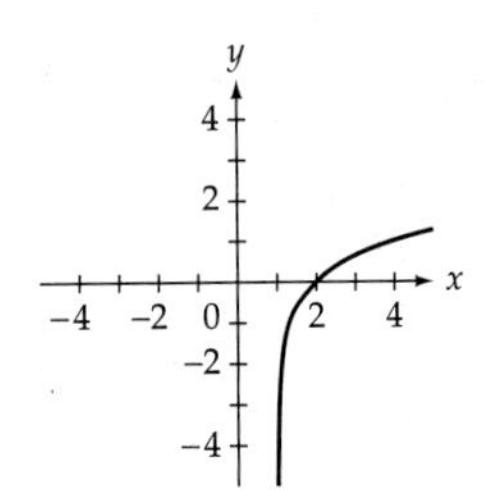

[Sec. 6.6]

9. $\frac{3}{5}$ [Sec. 6.2] **10.** $y = \frac{2}{3}x + 3$ [Sec. 6.3] **11.** $(-3, -10)$ [Sec. 6.4]

12. $(-4, 0), (2, 0)$ [Sec. 6.4] **13.** $\frac{49}{4}$, maximum [Sec. 6.4] **14.** The vertical intercept is (0, 250). This means that the plane starts 250 miles from its destination. The horizontal intercept is (2.5, 0). This means that it takes the plane 2.5 hours to reach its destination. [Sec. 6.2] **15.** 148 feet [Sec. 6.4] **16.** 3.30 grams [Sec. 6.5] **17.** 2.0 times as great [Sec. 6.6] **18.** $y = -1.5x + 740$ [Sec. 6.3] **19. a.** $y = 40.5x + 659$ **b.** 983 pounds [Sec. 6.3] **20.** 2.5 meters [Sec. 6.6]

CHAPTER 7

EXERCISE SET 7.1 *page 416*

1. 8 **3.** 12 **5.** 2 **7.** 4 **9.** 4 **11.** 7 **13.** 11 **15.** 7 **17.** 0300 **19.** 0400 **21.** 2000 **23.** 2100 **25.** 3 **27.** 6 **29.** True **31.** False **33.** True **35.** False **37.** True **39.** Possible answers are 2, 8, 14, 20, 26, 32, 38, ... **41.** 3 **43.** 3 **45.** 2 **47.** 10 **49.** 3 **51.** 3 **53.** 5 **55.** 4 **57.** 3 **59.** 7 **61.** 2 **63. a.** 6 o'clock **b.** 5 o'clock **65. a.** Tuesday **b.** Monday **67.** Saturday **69.** Friday **71.** 1, 4, 7, 10, 13, 16, . . . **73.** 1, 6, 11, 16, 21, 26, . . . **75.** 0, 2, 4, 6, 8, 10, 12, . . . **77.** No solutions **79.** 0, 2, 4, 6, 8, 10, 12, . . . **81.** No solutions **83.** 5, 7 **85.** 3, 3 **87.** 5, 3 **89.** 6 **91.** 6 **93.** 2 **97.** 11:00 **99.** 4

EXERCISE SET 7.2 *page 429*

1. No **3.** No **5.** Yes **7.** 9 **9.** 7 **11.** 3 **13.** 0 **15.** 6 **17.** 7 **19.** 5 **21.** 1 **23.** 7 **25.** 3 **27.** Yes **29.** Yes **31.** Yes **33.** No **35.** No **37.** Yes **39.** BPZMM UCASMBMMZA **41.** UF'E M SUDX **43.** VWLFNV DQG VWRQHV **45.** AGE OF ENLIGHTENMENT **47.** FRIEND IN NEED **49.** DANGER WILL ROBINSON **51.** FORTUNE COOKIE **53.** PHQ ZLOOLQJOB EHOLHYH ZKDW WKHB ZLVK **55.** JUSQD UT LURNUR **57.** PODONNQN NSBQK **59.** TURN BACK THE CLOCK **61.** BARREL OF MONKEYS **63.** Because the check digit is simply the sum of the first 10 digits mod 9, the same digits in a different order will give the same sum and hence the same check digit.

EXERCISE SET 7.3 *page 442*

1. a. Yes **b.** No **3. a.** Yes **b.** No **5.** Yes **7.** Yes **9.** No; property 4 fails. **11.** Yes **13.** Yes **15.** No; property 4 fails. **17.** Yes **19.** R_I **21.** R_{120} **23.** R_{120} **25.** R_r

27. R_l **29.** $I = \begin{pmatrix} 1 & 2 & 3 & 4 \\ 1 & 2 & 3 & 4 \end{pmatrix}$, $R_{90} = \begin{pmatrix} 1 & 2 & 3 & 4 \\ 2 & 3 & 4 & 1 \end{pmatrix}$, $R_{180} = \begin{pmatrix} 1 & 2 & 3 & 4 \\ 3 & 4 & 1 & 2 \end{pmatrix}$, $R_{270} = \begin{pmatrix} 1 & 2 & 3 & 4 \\ 4 & 1 & 2 & 3 \end{pmatrix}$, $R_v = \begin{pmatrix} 1 & 2 & 3 & 4 \\ 4 & 3 & 2 & 1 \end{pmatrix}$, $R_h = \begin{pmatrix} 1 & 2 & 3 & 4 \\ 2 & 1 & 4 & 3 \end{pmatrix}$, $R_r = \begin{pmatrix} 1 & 2 & 3 & 4 \\ 3 & 2 & 1 & 4 \end{pmatrix}$, $R_l = \begin{pmatrix} 1 & 2 & 3 & 4 \\ 1 & 4 & 3 & 2 \end{pmatrix}$ **31.** R_r **33.** R_{90} **35.** I **37.** D

39. E **41.** B **43.** $\begin{pmatrix}1&2&3&4\\1&2&3&4\end{pmatrix}, \begin{pmatrix}1&2&3&4\\1&2&4&3\end{pmatrix}, \begin{pmatrix}1&2&3&4\\1&3&2&4\end{pmatrix}, \begin{pmatrix}1&2&3&4\\1&3&4&2\end{pmatrix}, \begin{pmatrix}1&2&3&4\\1&4&2&3\end{pmatrix}, \begin{pmatrix}1&2&3&4\\1&4&3&2\end{pmatrix},$
$\begin{pmatrix}1&2&3&4\\2&1&3&4\end{pmatrix}, \begin{pmatrix}1&2&3&4\\2&1&4&3\end{pmatrix}, \begin{pmatrix}1&2&3&4\\2&3&1&4\end{pmatrix}, \begin{pmatrix}1&2&3&4\\2&3&4&1\end{pmatrix}, \begin{pmatrix}1&2&3&4\\2&4&1&3\end{pmatrix}, \begin{pmatrix}1&2&3&4\\2&4&3&1\end{pmatrix}, \begin{pmatrix}1&2&3&4\\3&1&2&4\end{pmatrix}, \begin{pmatrix}1&2&3&4\\3&1&4&2\end{pmatrix},$
$\begin{pmatrix}1&2&3&4\\3&2&1&4\end{pmatrix}, \begin{pmatrix}1&2&3&4\\3&2&4&1\end{pmatrix}, \begin{pmatrix}1&2&3&4\\3&4&1&2\end{pmatrix}, \begin{pmatrix}1&2&3&4\\3&4&2&1\end{pmatrix}, \begin{pmatrix}1&2&3&4\\4&1&2&3\end{pmatrix}, \begin{pmatrix}1&2&3&4\\4&1&3&2\end{pmatrix}, \begin{pmatrix}1&2&3&4\\4&2&1&3\end{pmatrix}, \begin{pmatrix}1&2&3&4\\4&2&3&1\end{pmatrix},$
$\begin{pmatrix}1&2&3&4\\4&3&1&2\end{pmatrix}, \begin{pmatrix}1&2&3&4\\4&3&2&1\end{pmatrix}$ **45.** $\begin{pmatrix}1&2&3&4\\2&1&3&4\end{pmatrix}$ **47.** $\begin{pmatrix}1&2&3&4\\3&1&4&2\end{pmatrix}$ **49.** $\begin{pmatrix}1&2&3&4\\4&3&2&1\end{pmatrix}$ **51.** d
53. c **55.** Answers will vary. **57.** Yes **59. a. and b.** Answers will vary. **c.** Values of n that are prime
61. a.

$\oplus$	1	2	3	4	5
1	1	2	3	4	5
2	2	3	4	5	1
3	3	4	5	1	2
4	4	5	1	2	3
5	5	1	2	3	4

b. Yes **c.** 1 **d.** 1 is its own inverse; 2 and 5 are inverses; 3 and 4 are inverses.

CHAPTER 7 REVIEW EXERCISES *page 446*

1. 2 [Sec. 7.1] **2.** 2 [Sec. 7.1] **3.** 5 [Sec. 7.1] **4.** 6 [Sec. 7.1] **5.** 9 [Sec. 7.1]
6. 4 [Sec. 7.1] **7.** 11 [Sec. 7.1] **8.** 7 [Sec. 7.1] **9.** 3 [Sec. 7.1] **10.** 4 [Sec. 7.1]
11. True [Sec. 7.1] **12.** False [Sec. 7.1] **13.** False [Sec. 7.1] **14.** True [Sec. 7.1] **15.** 2 [Sec. 7.1]
16. 4 [Sec. 7.1] **17.** 0 [Sec. 7.1] **18.** 3 [Sec. 7.1] **19.** 8 [Sec. 7.1] **20.** 3 [Sec. 7.1]
21. 7 [Sec. 7.1] **22.** 5 [Sec. 7.1] **23. a.** 2 o'clock **b.** 6 o'clock [Sec. 7.1] **24.** Monday [Sec. 7.1]
25. 3, 7, 11, 15, 19, 23, ... [Sec. 7.1] **26.** 7, 16, 25, 34, 43, 52, ... [Sec. 7.1] **27.** 0, 5, 10, 15, 20, 25, 30, ... [Sec. 7.1]
28. 4, 15, 26, 37, 48, 59, 70, ... [Sec. 7.1] **29.** 2, 3 [Sec. 7.1] **30.** 5, 7 [Sec. 7.1] **31.** 6 [Sec. 7.1]
32. 2 [Sec. 7.1] **33.** 6 [Sec. 7.2] **34.** 6 [Sec. 7.2] **35.** 2 [Sec. 7.2] **36.** 1 [Sec. 7.2]
37. No [Sec. 7.2] **38.** Yes [Sec. 7.2] **39.** No [Sec. 7.2] **40.** No [Sec. 7.2] **41.** THF AOL MVYJL IL DPAO FVB [Sec. 7.2] **42.** NLYNPW LWW AWLYD [Sec. 7.2] **43.** GOOD LUCK TOMORROW [Sec. 7.2]
44. THE DAY HAS ARRIVED [Sec. 7.2] **45.** UVR YX NDU PGVU [Sec. 7.2] **46.** YOU PASSED THE TEST [Sec. 7.2] **47.** Yes [Sec. 7.3] **48.** Yes [Sec. 7.3] **49.** No, properties 1, 3, and 4 fail. [Sec. 7.3]
50. Yes [Sec. 7.3] **51.** R_{240} [Sec. 7.3] **52.** R_l [Sec. 7.3] **53.** R_r [Sec. 7.3] **54.** R_{270} [Sec. 7.3]
55. R_{180} [Sec. 7.3] **56.** R_r [Sec. 7.3] **57.** There are only four distinct ways to place the rectangle in the reference rectangle. [Sec. 7.3] **58.** $\begin{pmatrix}1&2&3&4\\1&2&3&4\end{pmatrix}, \begin{pmatrix}1&2&3&4\\3&4&1&2\end{pmatrix}, \begin{pmatrix}1&2&3&4\\2&1&4&3\end{pmatrix}, \begin{pmatrix}1&2&3&4\\4&3&2&1\end{pmatrix}$ [Sec. 7.3]
59. Yes [Sec. 7.3] **60.** Answers will vary. [Sec. 7.3] **61.** Answers will vary. [Sec. 7.3] **62.** A [Sec. 7.3]
63. A [Sec. 7.3] **64.** $\begin{pmatrix}1&2&3&4&5\\3&4&1&5&2\end{pmatrix}$ [Sec. 7.3] **65.** $\begin{pmatrix}1&2&3&4&5\\3&5&1&4&2\end{pmatrix}$ [Sec. 7.3]

CHAPTER 7 TEST *page 448*

1. a. 3 **b.** 5 [Sec. 7.1] **2. a.** 0200 **b.** 1300 [Sec. 7.1] **3. a.** True **b.** False [Sec. 7.1]
4. 4 [Sec. 7.1] **5.** 6 [Sec. 7.1] **6.** 8 [Sec. 7.1] **7. a.** 6 o'clock **b.** 5 o'clock [Sec. 7.1]
8. 5, 14, 23, 32, 41, 50, ... [Sec. 7.1] **9.** 1, 3, 5, 7, 9, 11, ... [Sec. 7.1] **10.** 4, 2 [Sec. 7.1] **11.** 5 [Sec. 7.2]
12. 0 [Sec. 7.2] **13.** Yes [Sec. 7.2] **14.** BOZYBD LKMU [Sec. 7.2] **15.** NEVER QUIT [Sec. 7.2]
16. a. Yes **b.** No [Sec. 7.3] **17.** No, Property 4 fails; many elements do not have an inverse. [Sec. 7.3]
18. a. R_t **b.** R_r [Sec. 7.3] **19.** $\begin{pmatrix}1&2&3\\1&3&2\end{pmatrix}$ [Sec. 7.3] **20.** $\begin{pmatrix}1&2&3&4\\2&4&1&3\end{pmatrix}$ [Sec. 7.3]

CHAPTER 8

EXERCISE SET 8.1 *page 464*

1. $\angle O$, $\angle AOB$, and $\angle BOA$ 3. 40°, acute 5. 30°, acute 7. 120°, obtuse 9. Yes 11. No 13. A 28° angle 15. An 18° angle 17. 14 cm 19. 28 ft 21. 30 m 23. 86° 25. 71° 27. 30° 29. 36° 31. 127° 33. 116° 35. 20° 37. 20° 39. 20° 41. 141° 43. 106° 45. 11° 47. $m\angle a = 38°$, $m\angle b = 142°$ 49. $m\angle a = 47°$, $m\angle b = 133°$ 51. 20° 53. 47° 55. $m\angle x = 155°$, $m\angle y = 70°$ 57. $m\angle a = 45°$, $m\angle b = 135°$ 59. $90° - x$ 61. 60° 63. 35° 65. 102° 67. The three angles form a straight angle. The sum of the measures of the angles of a triangle is 180°. 69. Zero dimensions; one dimension; one dimension; one dimension; two dimensions 71. 360° 73. $\angle AOC$ and $\angle BOC$ are supplementary angles; therefore, $m\angle AOC + m\angle BOC = 180°$. Because $m\angle AOC = m\angle BOC$, by substitution, $m\angle AOC + m\angle AOC = 180°$. Therefore, $2(m\angle AOC) = 180°$, and $m\angle AOC = 90°$. Hence $\overline{AB} \perp \overline{CD}$.

EXERCISE SET 8.2 *page 486*

1.

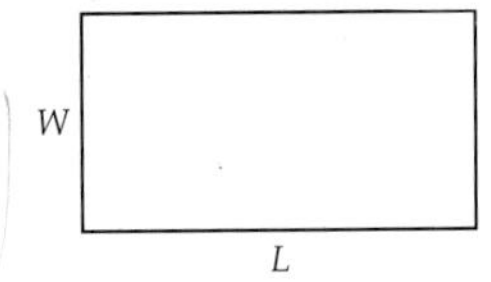

3. a. Perimeter is not measured in square units. b. Area is measured in square units.

5. Heptagon 7. Quadrilateral 9. Isosceles 11. Scalene 13. Right 15. Obtuse 17. a. 30 m b. 50 m² 19. a. 16 cm b. 16 cm² 21. a. 40 km b. 100 km² 23. a. 40 ft b. 72 ft² 25. a. 8π cm; 25.13 cm b. 16π cm²; 50.27 cm² 27. a. 11π mi; 34.56 mi b. 30.25π mi²; 95.03 mi² 29. a. 17π ft; 53.41 ft b. 72.25π ft²; 226.98 ft² 31. $29\frac{1}{2}$ ft 33. $10\frac{1}{2}$ mi 35. 68 ft 37. 20 in. 39. 214 yd 41. 10 mi 43. 2 packages 45. 144 m² 47. 9 in. 49. 10 in. 51. 8 m 53. 96 m² 55. 607.5 m² 57. 2 bags 59. 2 qt 61. 20 tiles 63. $40 65. $34 67. 120 ft² 69. 13.19 ft 71. 1256.6 ft² 73. 94.25 ft 75. 144π in² 77. 113.10 in² 79. 266,281 km 81. $8r^2 - 2\pi r^2$ 83. 4 times as large 85. $(a + 4)$ by $(a - 4)$ 87. a. 18 cm; 8 cm² b. 20 cm; 16 cm² c. 32 cm; 64 cm²

EXERCISE SET 8.3 *page 502*

1. $\frac{1}{2}$ 3. $\frac{3}{4}$ 5. 7.2 cm 7. 3.3 m 9. 12 m 11. 12 in. 13. 56.3 cm² 15. 18 ft 17. 16 m 19. $14\frac{3}{8}$ ft 21. 15 m 23. 8 ft 25. 13 cm 27. 35 m 29. Yes, SAS Theorem 31. Yes, SSS Theorem 33. Yes, ASA Theorem 35. No 37. Yes, SAS Theorem 39. No 41. No 43. 13 in. 45. 11.4 cm 47. 8.7 ft 49. 7.9 m 51. 7.4 m 53. 21.6 mi 55. 24 in. 57. Yes. Explanations will vary.

EXERCISE SET 8.4 *page 518*

1. 840 in³ 3. 15 ft³ 5. 4.5π cm³; 14.14 cm³ 7. 94 m² 9. 56 m² 11. 96π in²; 301.59 in² 13. 34 m³ 15. 15.625 in³ 17. 36π ft³ 19. 8143.01 cm³ 21. 75π in³ 23. 120 in³ 25. 7.80 ft³ 27. 6416 cm² 29. 13.5 in² 31. 50.27 in² 33. 2.88π m² 35. 874.15 in² 37. 832 m² 39. 8.5 in. 41. 3 ft 43. 3217 ft² 45. 881.22 cm² 47. 1.67 m³ 49. 115.43 cm³ 51. 4580.44 cm³ 53. 19 m² 55. 622.65 m² 57. 19,405.66 m² 59. 888.02 ft³ 61. 95,000 L 63. 79.17 g 65. $4860 67. $V = \frac{2}{3}\pi r^3$; $SA = 3\pi r^2$

69. Surface area of the sphere $= 4\pi r^2$. Surface area of the side of the cylinder $= 2\pi rh = 2\pi r(2r) = 4\pi r^2$.
71. a. Doubled $+ 4WH$ **b.** Quadrupled **c.** 8 times as large **d.** Quadrupled

EXERCISE SET 8.5 *page 530*

1. a. $\frac{a}{c}$ **b.** $\frac{b}{c}$ **c.** $\frac{b}{c}$ **d.** $\frac{a}{c}$ **e.** $\frac{a}{b}$ **f.** $\frac{b}{a}$

3. $\sin\theta = \frac{5}{13}, \cos\theta = \frac{12}{13}, \tan\theta = \frac{5}{12}$ **5.** $\sin\theta = \frac{24}{25}, \cos\theta = \frac{7}{25}, \tan\theta = \frac{24}{7}$

7. $\sin\theta = \frac{8}{\sqrt{113}}, \cos\theta = \frac{7}{\sqrt{113}}, \tan\theta = \frac{8}{7}$ **9.** $\sin\theta = \frac{1}{2}, \cos\theta = \frac{\sqrt{3}}{2}, \tan\theta = \frac{1}{\sqrt{3}}$

11. 0.6820 **13.** 1.4281 **15.** 0.9971 **17.** 1.9970 **19.** 0.8878 **21.** 0.8453
23. 0.8508 **25.** 0.6833 **27.** 38.6° **29.** 41.1° **31.** 21.3° **33.** 38.0°
35. 72.5° **37.** 0.6° **39.** 66.1° **41.** 29.5° **43.** 841.79 ft **45.** 13.6° **47.** 29.14 ft
49. 52.92 ft **51.** 13.59 ft **53.** 1056.63 ft **55.** 29.58 yd **57.** No. Explanations will vary.

59. $\frac{\sqrt{5}}{3}$ **61.** $\frac{\sqrt{7}}{3}$ **63.** $\sqrt{1 - a^2}$

EXERCISE SET 8.6 *page 544*

1. a. Through a given point not on a given line, exactly one line can be drawn parallel to the given line. **b.** Through a given point not on a given line, there are at least two lines parallel to the given line. **c.** Through a given point not on a given line, there exist no lines parallel to the given line. **3.** Carl Friedrich Gauss **5. a.** The sum equals 180°. **b.** The sum is less than 180°. **c.** The sum is greater than 180° but less than 540°. **7.** Imaginary geometry
9. A geodesic is a curve on a surface such that for any two points of the curve the portion of the curve between the points is the shortest path on the surface that joins these points. **11.** An infinite saddle surface **13.** π square units
15. 1,370,000 mi² **17.** $d_E(P, Q) = \sqrt{49} = 7$ blocks, $d_C(P, Q) = 7$ blocks
19. $d_E(P, Q) = \sqrt{89} \approx 9.4$ blocks, $d_C(P, Q) = 13$ blocks **21.** $d_E(P, Q) = \sqrt{72} \approx 8.5$ blocks, $d_C(P, Q) = 12$ blocks
23. $d_E(P, Q) = \sqrt{37} \approx 6.1$ blocks, $d_C(P, Q) = 7$ blocks **25.** $d_C(P, Q) = 7$ blocks **27.** $d_C(P, Q) = 5$ blocks
29. $d_C(P, Q) = 5$ blocks **31.** A city distance may be associated with more than one Euclidean distance. For example, if $P = (0, 0)$ and $Q = (2, 0)$, then the city distance between the points is 2 blocks and the Euclidean distance is also 2 blocks. However, if $P = (0, 0)$ and $Q = (1, 1)$, then the city distance between the points is still 2 blocks, but the Euclidean distance is $\sqrt{2}$ blocks.

33.

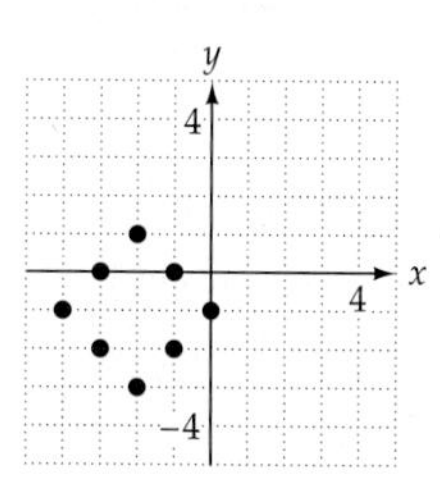

35.

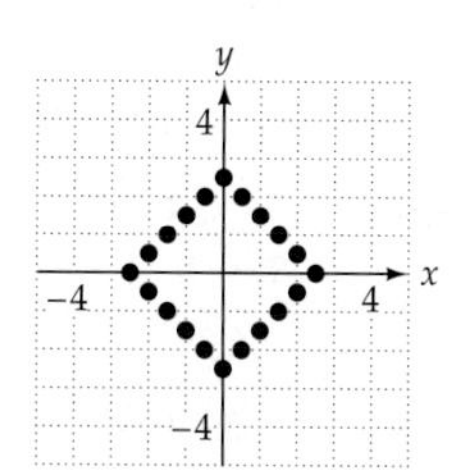

37. $4n$

39. a.

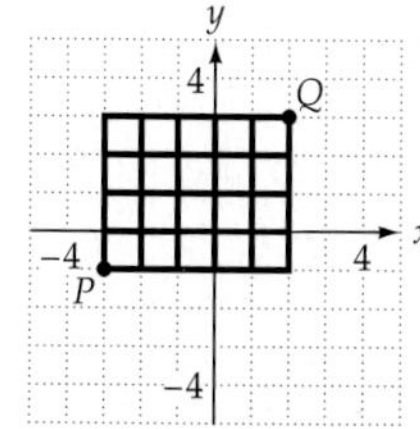

b.

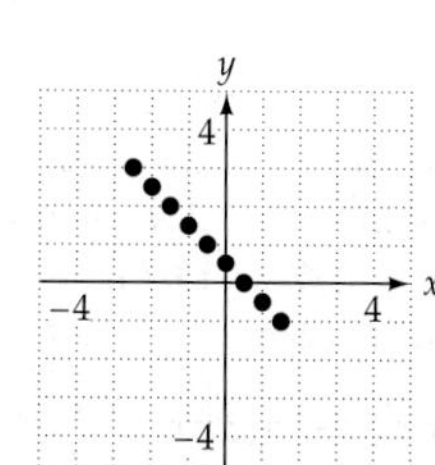

41. a. 10 **b.** 3

EXERCISE SET 8.7 *page 558*

1. Stage 2 — — — —

Stage 3 - - - - - - - -

3. Stage 2

5.

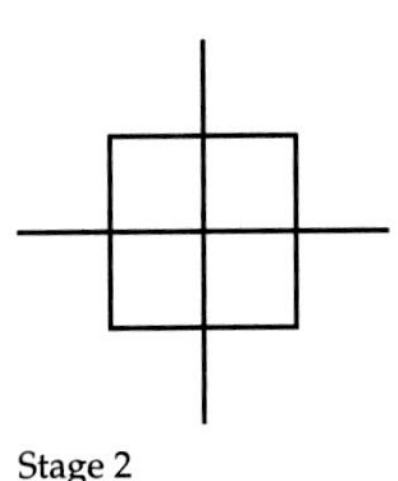

Stage 2

7.

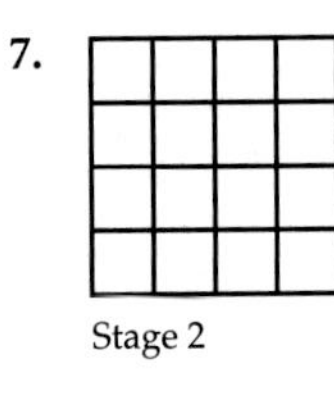

Stage 2

9.

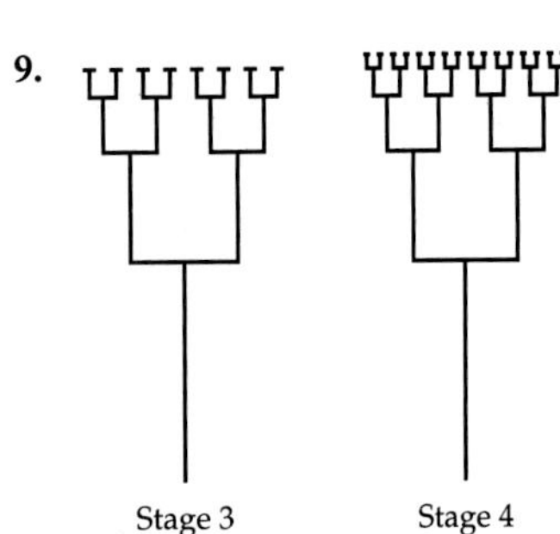

Stage 3 Stage 4

11. 0.631 **13.** 1.465 **15.** 2.000 **17.** 2.000

19. 1.613 **21. a.** Sierpinski carpet, 1.893; Variation 2, 1.771; Variation 1, 1.465 **b.** The Sierpinski carpet

23. The binary tree fractal is not a strictly self-similar fractal.

CHAPTER 8 REVIEW EXERCISES *page 563*

1. $m\angle x = 22°$; $m\angle y = 158°$ [Sec. 8.1] **2.** 24 in. [Sec. 8.3] **3.** 240 in^3 [Sec. 8.4] **4.** 68° [Sec. 8.1]
5. 220 ft^2 [Sec. 8.4] **6.** 40π m^2 [Sec. 8.4] **7.** 44 cm [Sec. 8.1] **8.** 19° [Sec. 8.1] **9.** 27 in^2 [Sec. 8.2]
10. 96 cm^3 [Sec. 8.4] **11.** 14.14 m [Sec. 8.2] **12.** $m\angle a = 138°$; $m\angle b = 42°$ [Sec. 8.1]
13. a 148° angle [Sec. 8.1] **14.** 39 ft^3 [Sec. 8.4] **15.** 95° [Sec. 8.1] **16.** 8 cm [Sec. 8.2]
17. 288π mm^3 [Sec. 8.4] **18.** 21.5 cm [Sec. 8.2] **19.** 4 cans [Sec. 8.4] **20.** 208 yd [Sec. 8.2]
21. 90.25 m^2 [Sec. 8.2] **22.** 276 m^2 [Sec. 8.2] **23.** Carl Friedrich Gauss [Sec. 8.6]
24. The triangles are congruent by the SAS theorem. [Sec. 8.3] **25.** 9.75 ft [Sec. 8.3]
26. $\sin\theta = \dfrac{5\sqrt{89}}{89}$, $\cos\theta = \dfrac{8\sqrt{89}}{89}$, $\tan\theta = \dfrac{5}{8}$ [Sec. 8.5] **27.** $\sin\theta = \dfrac{\sqrt{3}}{2}$, $\cos\theta = \dfrac{1}{2}$, $\tan\theta = \sqrt{3}$ [Sec. 8.5]
28. 25.7° [Sec. 8.5] **29.** 29.2° [Sec. 8.5] **30.** 53.8° [Sec. 8.5] **31.** 1.9° [Sec. 8.5] **32.** 100.1 ft [Sec. 8.5]
33. 153.2 mi [Sec. 8.5] **34.** 56.0 ft [Sec. 8.5] **35.** Nikolai Lobachevsky [Sec. 8.6]
36. Spherical geometry or elliptical geometry [Sec. 8.6] **37.** Hyperbolic geometry [Sec. 8.6]
38. Lobachevskian or hyperbolic geometry [Sec. 8.6] **39.** Riemannian or spherical geometry [Sec. 8.6]
40. 120π in^2 [Sec. 8.6] **41.** $\dfrac{25\pi}{3}$ ft^2 [Sec. 8.6] **42.** $d_E(P, Q) = 5$ blocks, $d_C(P, Q) = 7$ blocks [Sec. 8.6]
43. $d_E(P, Q) = \sqrt{113} \approx 10.6$ blocks, $d_C(P, Q) = 15$ blocks [Sec. 8.6]
44. $d_E(P, Q) = \sqrt{37} \approx 6.1$ blocks, $d_C(P, Q) = 7$ blocks [Sec. 8.6]
45. $d_E(P, Q) = \sqrt{89} \approx 9.4$ blocks, $d_C(P, Q) = 13$ blocks [Sec. 8.6] **46. a.** P and Q **b.** P and R [Sec. 8.6]

47.

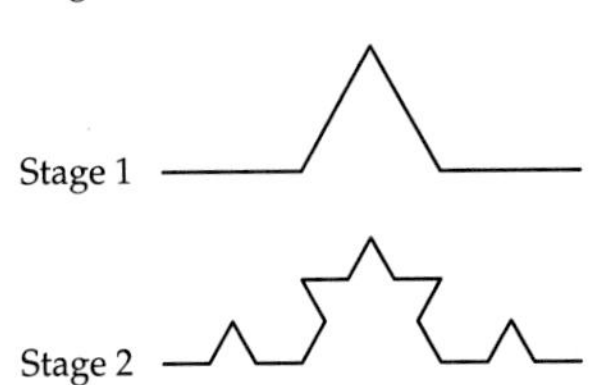

[Sec. 8.7]

48.

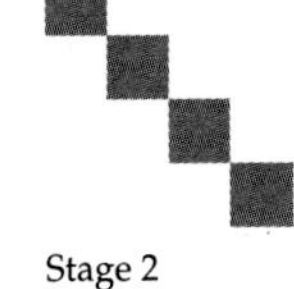

Stage 2 [Sec. 8.7]

49. $\dfrac{\log 5}{\log 4} \approx 1.161$ [Sec. 8.7]

50. 1 [Sec. 8.7]

CHAPTER 8 TEST *page 566*

1. 169.65 m^3 [Sec. 8.4] **2.** 6.8 m [Sec. 8.2] **3.** a 58° angle [Sec. 8.1] **4.** 3.14 m^2 [Sec. 8.2] **5.** 150° [Sec. 8.1] **6.** $m\angle a = 45°$; $m\angle b = 135°$ [Sec. 8.1] **7.** 5.0625 ft^2 [Sec. 8.2] **8.** $448\pi \text{ cm}^3$ [Sec. 8.4] **9.** $1\frac{1}{5}$ ft [Sec. 8.3] **10.** 90° and 50° [Sec. 8.1] **11.** 125° [Sec. 8.1] **12.** 32 m^2 [Sec. 8.2] **13.** 25 ft [Sec. 8.3] **14.** 113.10 in^2 [Sec. 8.2] **15.** The triangles are congruent by the SAS theorem. [Sec. 8.3] **16.** 7.55 cm [Sec. 8.3] **17.** $\sin\theta = \frac{4}{5}$, $\cos\theta = \frac{3}{5}$, $\tan\theta = \frac{4}{3}$ [Sec. 8.5] **18.** 127 ft [Sec. 8.5] **19.** 103.87 ft^2 [Sec. 8.2] **20.** 780 in^3 [Sec. 8.4] **21. a.** Through a given point not on a given line, exactly one line can be drawn parallel to the given line. **b.** Through a given point not on a given line, there exist no lines parallel to the given line. [Sec. 8.6] **22. a.** 1 **b.** 3 [Sec. 8.6] **23.** A great circle of a sphere is a circle on the surface of the sphere whose center is at the center of the sphere. [Sec. 8.6] **24.** $80\pi \text{ ft}^2 \approx 251.3 \text{ ft}^2$ [Sec. 8.6]

25. $d_E(P, Q) = \sqrt{82} \approx 9.1$ blocks, $d_C(P, Q) = 10$ blocks [Sec. 8.6] **26.** 16 [Sec. 8.6]

27.

Stage 2 [Sec. 8.7]

28.

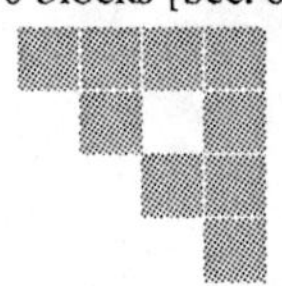

Stage 2 [Sec. 8.7]

29. Replacement ratio: 2; scale ratio: 2; similarity dimension: 1 [Sec. 8.7]

30. Replacement ratio: 3; scale ratio: 2; similarity dimension: $\frac{\log 3}{\log 2} \approx 1.585$ [Sec. 8.7]

APPENDIX

APPENDIX *page 936*

1. Meter, liter, gram **3.** Kilometer **5.** Centimeter **7.** Gram **9.** Meter **11.** Gram **13.** Milliliter **15.** Milligram **17.** Millimeter **19.** Milligram **21.** Gram **23.** Kiloliter **25.** Milliliter **27. a.** Column 2: k, c, m; column 3: 10^9, 10^3, 10^2, $\frac{1}{10^3}$, $\frac{1}{10^{12}}$; column 4: 1 000 000 000, 10, 0.1, 0.01, 0.000 001, 0.000 000 001 **b.** Answers will vary. **29.** 910 **31.** 1.856 **33.** 7.285 **35.** 8 000 **37.** 0.034 **39.** 29.7 **41.** 7.530 **43.** 9 200 **45.** 36 **47.** 2 350 **49.** 83 **51.** 0.716 **53.** 6.302 **55.** 458 **57.** 9.2 **59.** 2 grams **61.** 24 liters **63.** 16 servings **65.** 4 tablets **67.** The case containing 12 one-liter bottles **69.** 500 seconds **71.** $17,430 **73.** $66.50

PHOTO CREDITS

Chapter 1
p. 1, Comstock Images/PictureQuest; p. 1, Courtesy of *Parade* Magazine; p. 3, Hulton Archive/Getty Images; p. 7, Michael Newman/PhotoEdit, Inc.; p. 10, Photo copyright Robert Matthews/Office of Communications, Princeton University; p. 12, Courtesy of Istituto e Museo di Storia della Scienza; p. 19, Bettmann/CORBIS; p. 21, Dynamic Graphics/PictureQuest; p. 23, Hulton Archive/Getty Images; p. 27, Science Photo Library/Photo Researchers; p. 27, Science Photo Library/Photo Researchers; p. 29, Courtesy of Stanford University News Service Library; p. 35, Margaret Ross/Stock Boston, LLC; p. 37, Walter Sanders/Getty Images; p. 44, Michael Newman/PhotoEdit, Inc.

Chapter 2
p. 52, Dennis MacDonald/PhotoEdit, Inc.; p. 53, The Granger Collection; p. 53, The Granger Collection; p. 58, Science Photo Library/Photo Researchers; p. 59, Lotfi Zadeh; p. 69, CORBIS/PictureQuest; p. 72, Index Stock Images/PictureQuest; p. 73, *Jacksonville Journal Courier*/Steve Warnowski/The Image Works, Inc; p. 74, Nicholas Devore III/Network Aspen; p. 75, Bernard Wolff/PhotoEdit, Inc.; p. 79, Ted Spiegel/CORBIS; p. 80, Charles Gupton/Stock Boston, LLC/PictureQuest; p. 86, Brand X Pictures/PictureQuest; p. 88, Conrad Zobel/CORBIS; p. 89, Courtesy of Sylvia Wiegand; p. 92, David Young Wolff/PhotoEdit, Inc.; p. 92, Bob Daemmrich/Stock Boston, LLC/PictureQuest; p. 93, The Granger Collection; p. 94, Dale O'Dell/Stock Connection/PictureQuest; p. 95, Mark Bernett/Stock Boston, LLC; p. 103, Henry Kaiser/eStock Photography/PictureQuest; p. 108, H. Carol Moran/Index Stock Imagery/PictureQuest; p. 111, David Young Wolff/PhotoEdit, Inc.

Chapter 3
p. 113, Mark Richards/PhotoEdit, Inc.; p. 113, The Granger Collection; p. 114, Jean Marc Keystone/Hulton Archive/Getty Images; p. 116, The Granger Collection; p. 116, The Granger Collection; p. 118, Eric Meola/Getty Images; p. 121, Keystone/Getty Images; p. 125, Raymond Smullyan the logician; p. 127, Copyright Tribune Media Services, Inc. All Rights Reserved. Reproduced with permission; p. 137, The Everett Collection; p. 139, Reprinted with permission of Texas Instruments; p. 149, Bettmann/CORBIS; p. 149, Don Boroughs/The Image Works, Inc.; p. 152, The Granger Collection; p. 152, Bettmann/CORBIS; p. 157, The Granger Collection; p. 160, Reprinted with permission of Simon & Schuster from *The Unexpected Hanging and Other Mathematical Diversions* by Martin Gardner. Copyright © 1969 by Martin Gardner; p. 165, Kean Collection/Getty Images; p. 166, Pierre-Auguste Renoir, French, 1841–1919, *Dance at Bougival,* 1883. Oil on canvas. (71 5/8 × 38 5/8 in.) Museum of Fine Arts, Boston Picture Fund, 37.375/Photograph © 2005 Museum of Fine Arts, Boston.

Chapter 4
p. 177, Getty Images; p. 179, Art Resource, NY; p. 182, Image Source/PictureQuest; p. 183, The Granger Collection; p. 187, Doug Plummer/Stock Connection; p. 206, Michael Newman/PhotoEdit, Inc.; p. 215, The International Mathematical Union; p. 224, The Granger Collection; p. 224, AP/Wide World Photos; p. 226, Photo of Paul Erdos by George Csicsery from his film *N is a Number: A Portrait of Paul Erdos* (1993). All Rights Reserved; p. 226, © College de France; p. 232, The Granger Collection; p. 236, Bettmann/CORBIS; p. 237, Professor Peter Goddard/Science Photo Library/Photo Researchers.

Chapter 5
p. 245, Richard Hutchins/PhotoEdit, Inc.; p. 246, Lambert/Getty Images; p. 251, NASA/Getty Images; p. 261, AP/Wide World Photos; p. 265, Bob Daemmrich/Stock Boston, LLC; p. 266, Mike & Carol Weiner/Stock Boston, LLC; p. 269, Reprinted with permission of George E. Slye; p. 274, Olivier LeClerc/Gamma Presse; p. 276, Blank Archives/Getty Images; p. 312, AP/Wide World Photos; p. 317, Steve Allen/Brand X Pictures/Getty Images; p. 317, Hisham F. Ibrahim/Photodisc Blue/Getty Images.

Chapter 6
p. 327, Bob Stefko/The Image Bank/Getty Images; p. 328, Richard Norwitz/Photo Researchers; p. 330, The Granger Collection; p. 335, AP/Wide World Photos; p. 338, Russell Illig/Photodisc Green/Getty Images; p. 347, Hideo Kurihara/Getty Images; p. 364, Hulton Archive/Getty Images; p. 366, Joe Sohn/Photo Researchers; p. 366, Rafael Macia/Photo Researchers; p. 371, Bill W. Marsh/Photo Researchers; p. 373, Michael Medford/Getty Images; p. 381, Bettmann/CORBIS; p. 383, Courtesy of The History Factory. Image in the public domain; p. 384, Photo by Luc Norvitch/Reuters; p. 387, Keren Su/Getty Images; p. 395, AP/Wide World Photos; p. 400, Bill Aron/PhotoEdit, Inc.

Chapter 7
p. 407, Susan Van Etten/PhotoEdit, Inc.; p. 416, Robert Brenner/PhotoEdit, Inc.; p. 417, Jonathan Nourok/PhotoEdit, Inc.; p. 421, The Everett Collection; p. 424, The Granger Collection; p. 427, Hulton Archive/Getty Images; p. 428, The Granger Collection; p. 429, AP/Wide World Photos; p. 430, Bonnie Kamin/PhotoEdit, Inc.; p. 432, The Granger Collection; p. 433, The Granger Collection; p. 438, Bonnie Kamin/PhotoEdit, Inc.

Chapter 8

p. 449, Royalty-Free/CORBIS; p. 449, The Granger Collection; p. 450, Science Photo Library/Photo Researchers; p. 453, Hideo Kurihara/Getty Images; p. 463, The Everett Collection; p. 470, AP/Wide World Photos; p. 474, The Granger Collection; p. 475, Comstock Images/PictureQuest; p. 510, John Elk III/Getty Images; p. 516, Leonard Lee Rue III/Stock Boston, LLC; p. 517, Hulton Archive/Getty Images; p. 534, The Granger Collection; p. 535, The Granger Collection; p. 547, Phyllis Picardi/Stock Boston, LLC; p. 547, Phyllis Picardi/Stock Boston, LLC; p. 547, Phyllis Picardi/Stock Boston, LLC; p. 551, Hank Morgan/Science Photo Library/Photo Researchers; p. 555, Gregory Sams/Science Photo Library/Photo Researchers; p. 555, Dr. Fred Espanek/Science Photo Library/Photo Researchers; p. 555, Alfred Pasicks/Science Photo Library/Photo Researchers; p. 555, Mike & Carol Werner/Stock Boston, LLC; p. 556, Copyright Andy Ryan; p. 556, Reprinted with permission of Fractal Antenna Systems, Inc. © 1997; p. 558, Visuals Unlimited.

Chapter 9

p. 569, Stewart Cohen/Stone/Getty Images; p. 573, Courtesy of Intenix Software; p. 574, Bettmann/CORBIS; p. 595, William Cook; p. 612, Laima Druskis/Stock Boston, LLC; p. 625, Vic Bider/PhotoEdit, Inc.; p. 631, Creatas/PictureQuest.

Chapter 10

p. 641, Lori Adamski Peek/Stone/Getty Images; p. 655, Courtesy of Western Currency Facility; p. 657, Alex Wong/Getty Images; p. 661, Mark Wilson/Getty Images; p. 674, Tom Grill/CORBIS; p. 692, Topham/The Image Works, Inc.

Chapter 11

p. 719, AP/Wide World Photos; p. 720, Mark Burnett/Stock Boston, LLC; p. 733, Vaugh Youtz/Getty Images; p. 740, Image in the public domain; p. 741, Darren McCollester/Getty Images; p. 742, Superstock/PictureQuest; p. 744, Michael Simpson/Getty Images; p. 744, Richard Kaylin/Stone/Getty Images; p. 748, Topham/The Image Works, Inc.; p. 750, AP/Wide World Photos; p. 754, Visuals Unlimited; p. 755, "Wheel of Fortune," courtesy of Califon Productions, Inc.; p. 756, Rachel Epstein/The Image Works, Inc.; p. 756, Hank de Lespinasse; p. 766, Tim Boyle/Getty Images; p. 774, John E. Kelly/Getty Images; p. 775, Johnny Crawford/The Image Works, Inc.; p. 776, Michael Newman/PhotoEdit, Inc.; p. 779, Photodisc Green/Getty Images; p. 782, Visuals Unlimited; p. 783, Barbara Alper/Stock Boston, LLC; p. 784, Steve Smith/Getty Images; p. 785, Billy Husace/Getty Images.

Chapter 12

p. 793, Jeff Greenberg/PhotoEdit, Inc.; p. 794, Bettmann/CORBIS; p. 796, Spencer Grant/PhotoEdit, Inc.; p. 798, AP/Wide World Photos; p. 808, Bettmann/CORBIS; p. 816, AP/Wide World Photos; p. 817, Francis Miller/Getty Images; p. 824, Bettmann/CORBIS; p. 857, Negative # 324393/Courtesy Department of Library Services/American Museum of Natural History.

Chapter 13

p. 869, Paul Conklin/PhotoEdit, Inc.; p. 885, David Young Wolff/PhotoEdit, Inc.; p. 896, Tim Boyle/Getty Images; p. 898, The Granger Collection; p. 900, AP/Wide World Photos; p. 915, Stan Honda/AFP/Getty Images; p. 924, *Jacksonville Journal Courier*/Steve Warnowski/The Image Works, Inc.; p. 925, AP/Wide World Photos.

Appendix

p. 938, Timothy A. Clary/AFP/Getty Images.

Photo credits for back endpapers:

Rhind papyrus: Art Resource, NY; Aristotle: Bettmann/CORBIS; Euclid: Science Photo Library/Photo Researchers; Archimedes: Hulton Archive/Getty Images; Hypatia: Bettmann/CORBIS; Fibonacci: Bettmann/CORBIS; Solar System: Science Photo Library/Photo Researchers; Galileo: Hulton Archive/Getty Images; Einstein: Lambert/Getty Images.

INDEX

A Brief History of Mathematics

From the beginning to 4000 B.C.

There is very little evidence to create a definitive picture of mathematics prior to 4000 B.C. Early humans were more focused on survival than on counting possessions. Much of the suppositions about early mathematics are based on findings by anthropologists who study existing primitive cultures. For instance, there are clans in Australia that count only to two. Any quantity beyond that is many. Some South American clans count to higher numbers but do so by using only the words for one and two. For instance, four is two-two and five is one-two-two. Generally these clans use many for numbers beyond eight. Historians suspect that early human cultures had the same math abilities as these clans.

4000 to 300 B.C.

Counting days was an important function to early communities. As hunters and gatherers banded together to form tribes, the passage of time was important for planting crops and preparing for changes in the seasons. Around 4000 B.C. the Egyptians had created a calendar. This is significant in the history of mathematics because it is the first record of humans organizing number concepts.

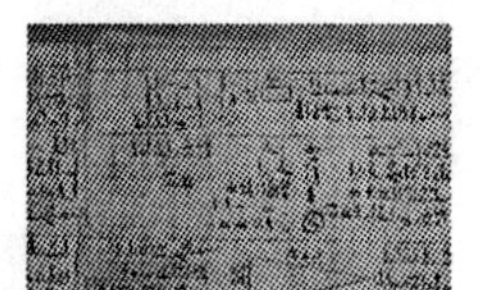
Rhind papyrus

By 3000 B.C. there is evidence arithmetic had evolved and was used by merchants. The Rhind papyrus, written around 1650 B.C., contained methods for working with fractions and solving equations. In some sense, it is an example of a mathematics textbook.

Prior to 600 B.C., mathematics was prescriptive. That is, there were procedures for doing certain calculations or solving some equations. The procedures were based on "this works for this type of problem" and not on underlying principles or logic that could be applied to a broad class of problems. Then, around 600 B.C., the merchant Thales of Miletus, who had studied Egyptian geometry and Babylonian astronomy, concluded that certain facts could be deduced from other facts. He is generally regarded as the first person to use logical reasoning as a way of doing mathematics. Thales was one of the Seven Sages of Greece and is credited with the saying "Know thyself."

Aristotle

The creation of Plato's Academy around 385 B.C. provided a place where scholars could meet and discuss problems from many fields and in particular mathematics. One of the scholars to study at Plato's Academy was Aristotle. Although Aristotle contributed to many areas of science and philosophy, his work on logic laid the foundation for modern mathematics.

300 B.C. to A.D. 200

Euclid used the principles of logic as stated by Aristotle to write the *Elements,* certainly one of the most important works of mathematics in all of history. This book established mathematics as a discipline of reasoning that required mathematical theorems to have irrefutable arguments to confirm their validity.

Euclid

Approximately a generation after Euclid, Archimedes applied some of the theorems of geometry to mathematics and to physics. He gave us the approximation $\pi \approx \frac{22}{7}$ and the Archimedean screw, a device still used in some parts of the world to lift water from a canal onto land to irrigate farms.

Claudius Ptolemy used the geometry of Euclid to write what is now called the *Almagest,* a complete account of Greek astronomy. This book established that a real-world situation (for instance, the orbits of planets) could be described by a mathematical model.

Archimedes

200 to 1000

Egyptian and Babylonian mathematics were presented in words. Then, around A.D. 250, Diophantes of Alexandria wrote a series of 13 books called *Arithmetica.* The major focus of these books was algebra and the solutions of equations. The importance of *Arithmetica* to modern mathematics is that he changed from a verbal description of mathematics to a symbolic one. For instance, he would write $\Delta^{Y}\theta$ which, in our modern notation, is $9x^2$. In his book, Diophantes stated some of rules of solving equations such as "add the same term to each side of an equation."

Hypatia

After the first publication of the major works we have discussed, other scholars would translate, annotate, or amplify the concepts contained in these works. Hypatia is the first known woman to undertake such a task. She lectured at the Museum of Alexandria, the most prestigious school of ancient Egypt, and wrote commentaries on the *Elements,* the *Almagest,* and *Arithmetica.*

There was not much innovation in mathematics from around 500 to A.D. 1000. Much of the work was refinements and restatements of earlier works. One of these restatements was a book by Boethius called *De Institutione Arthmetica.* It was really a translation of a work by Nicomachus from 350 years earlier. However, it became one of the most important early textbooks and was used in various forms over a period of 1000 years.

Another important event of this period was the publication of a text by al-Khwarizmi, a Persian mathematician. From the title of his book, we get the world *algebra* and from his name, the word *algorithm.*

1000 to 1400

Fibonacci

Omar Khayyam is best known in the Western world as the author of the *Rubaiyat* but he also made contributions to mathematics and astronomy. His text, *Demonstrations of Problems of Algebra and Almucabola,* was significant because it used algebraic methods rather than geometric methods to solve cubic equations.

One of the most significant events in mathematics during this period was a book by Fibonacci. He was the son of a merchant who, during travels for his father, studied the mathematics of the cultures he visited. After returning to Pisa, we wrote a number of treatises on mathematics. One in particular, *Liber Abbaci,* introduced the Western world to the Hindu-Arabic numerals 0, 1, 2, 3, 4, 5, 6, 7, 8, and 9. Besides introducing these symbols, Fibonacci went on to describe how these numbers can be used to simplify calculations. This was an improvement on the symbolic notation employed by Diophantes.

By the end of the fourteenth century, mathematics was being applied to physical situations. Notable among these applications was the analytical description of motion. These early studies eventually resulted in the formulation of calculus.

1400 to 1800

Solar system

Galileo

The Renaissance not only brought a new vitality to art and music but to mathematics and science as well. One of the most controversial ideas of the early Renaissance was stated by Nicholas Copernicus. His *De Revolutionibus,* published in the year of his death, proposed a heliocentric solar system as opposed to the Earth-centered system presented in the *Almagest.*

One proponent of Copernicus' heliocentric solar system was Galileo Galilei. Besides defending the heliocentric system, Galileo stated his own ideas on motion and acceleration that challenged long-standing Aristotelian principles. It was, however, left to Isaac Newton to bring together the apparent disparate principles of a heliocentric system and acceleration of objects into one coherent theory. The theory was called the Universal Law of Gravitation. The cornerstone of the theory rested on calculus, which Newton developed between 1664 and 1666. Around the same time, Wilhelm Gottfried Leibniz independently invented calculus. A bitter feud between Newton and Leibniz as to who invented calculus contined until the death of Leibniz.

Although the development of calculus was a tremendous feat, there were other extremely important developments in mathematics. The idea of decimal fractions (numbers such as 2.47) and how to calculate with them were popularized by Simon Stevin; Scot John Napier invented logarithms to assist astronomers with calculations; Francois Viete revolutionized algebraic symbolism. These are just some of the major mathematical ideas of this time.

1800 to present

Einstein

The Industrial Revolution brought with it a tremendous increase in the number of colleges and universities. The demands of industry for new technology and the growth of institutions of higher education provided a fertile ground for the expansion of mathematical knowledge. Group theory arose to solve problems in geometry and certain types of equations. It was not until later, however, that physicists used the concepts in group theory to advance their understanding of the atom. Mathematicians have used group theory to create secure communication systems to send information over the Internet.

The theory of matrices was created to solve certain types of equations. Today matrices are applied to problems in areas as diverse as physics, economics, and biology. A type of geometry that challenged the Euclidean view of parallel lines was studied. This non-Euclidean geometry provided Albert Einstein with the mathematics necessary to create the General Theory of Relativity.

This brief history of mathematics has omitted some important mathematicians and their contributions. All of them have contributed to the rich field of mathematics, and present-day mathematicians continue to expand mathematics beyond its current scope.